IPC®

2015

CODE AND COMMENTARY

The complete IPC with commentary after each section

2015 International Plumbing Code® Commentary

First Printing: October 2014
Second Printing: February 2016

ISBN: 978-1-60983-290-2

COPYRIGHT © 2014
by
INTERNATIONAL CODE COUNCIL, INC.

ALL RIGHTS RESERVED. This 2015 *International Plumbing Code® Commentary* is a copyrighted work owned by the International Code Council, Inc. ("ICC"). Without advance written permission from the ICC, no part of this book may be reproduced, distributed or transmitted in any form or by any means, including, without limitation, electronic, optical or mechanical means (by way of example, and not limitation, photocopying, or recording by or in an information storage retrieval system). For information on use rights and permissions, please contact: ICC Publications, 4051 Flossmoor Road, Country Club Hills, IL 60478. Phone 1-888-ICC-SAFE (422-7233).

Trademarks: "International Code Council," the "International Code Council" logo, "ICC," the "ICC" logo, "International Plumbing Code," "IPC" and other names and trademarks appearing in this book are registered trademarks of the International Code Council, Inc., and/or its licensors (as applicable), and may not be used without permission.

PRINTED IN THE USA

PREFACE

Significant changes in the plumbing industry, as well as in manufacturing technology, had become so commanding that a radically new approach to the design and installation of plumbing systems seemed an imperative. The reply to that imperative was the *International Plumbing Code*, a document emphasizing prescriptive and performance-related provisions.

As a follow-up to the *International Plumbing Code*, we offer a companion document, the *International Plumbing Code Commentary*. The basic appeal of the Commentary is thus: it provides in a small package and at reasonable cost thorough coverage of many issues likely to be dealt with when using the *International Plumbing Code* — and then supplements that coverage with historical and technical background. Reference lists, information sources and bibliographies are also included.

Throughout all of this, strenuous effort has been made to keep the vast quantity of material accessible and its method of presentation useful. With a comprehensive yet concise summary of each section, the Commentary provides a convenient reference for plumbing regulations. In the chapters that follow, discussions focus on the full meaning and implications of the code text. Guidelines suggest the most effective method of application, and the consequences of not adhering to the code text. Illustrations are provided to aid understanding; they do not necessarily illustrate the only methods of achieving code compliance.

The format of the Commentary includes the full text of each section, table and figure in the code, followed immediately by the commentary applicable to that text. Each section's narrative includes a statement of its objective and intent, and usually includes a discussion about why the requirement commands the conditions set forth. Code text and commentary text are easily distinguished from each other. All code text is shown as it appears in the *International Plumbing Code*, and all commentary is indented below the code text and begins with the symbol ❖.

Readers should note that the Commentary is to be used in conjunction with the *International Plumbing Code* and not as a substitute for the code. **The Commentary is advisory only;** the code official alone possesses the authority and responsibility for interpreting the code.

Comments and recommendations are encouraged, for through your input, we can improve future editions. Please direct your comments to the Codes and Standards Development Department at the Chicago District Office.

TABLE OF CONTENTS

CHAPTER 1	SCOPE AND ADMINISTRATION	1-1 – 1-20
CHAPTER 2	DEFINITIONS	2-1 – 2-32
CHAPTER 3	GENERAL REGULATIONS	3-1 – 3-48
CHAPTER 4	FIXTURES, FAUCETS AND FIXTURE FITTINGS	4-1 – 4-76
CHAPTER 5	WATER HEATERS	5-1 – 5-20
CHAPTER 6	WATER SUPPLY AND DISTRIBUTION	6-1 – 6-112
CHAPTER 7	SANITARY DRAINAGE	7-1 – 7-66
CHAPTER 8	INDIRECT/SPECIAL WASTE	8-1 – 8-14
CHAPTER 9	VENTS	9-1 – 9-72
CHAPTER 10	TRAPS, INTERCEPTORS AND SEPARATORS	10-1 – 10-26
CHAPTER 11	STORM DRAINAGE	11-1 – 11-24
CHAPTER 12	SPECIAL PIPING AND STORAGE SYSTEMS	12-1 – 12-4
CHAPTER 13	NONPOTABLE WATER SYSTEMS	13-1 – 13-18
CHAPTER 14	SUBSURFACE LANDSCAPE IRRIGATION SYSTEMS	14-1 – 14-6
CHAPTER 15	REFERENCED STANDARDS	15-1 – 15-14
APPENDIX A	PLUMBING PERMIT FEE SCHEDULE	A-1 – A-2
APPENDIX B	RATES OF RAINFALL FOR VARIOUS CITIES	B-1 – B-2
APPENDIX C	STRUCTURAL SAFETY	C-1 – C-2
APPENDIX D	DEGREE DAY AND DESIGN TEMPERATURES	D-1 – D-8
APPENDIX E	SIZING OF WATER PIPING SYSTEM	E-1 – E-24
INDEX		INDEX-1 – INDEX-10

Chapter 1: Scope and Administration

General Comments

The law of building regulation is grounded on the police power of the state. In terms of how it is used, this is the power of the state to legislate for the general welfare of its citizens. This power enables passage of such laws as a plumbing code. It is from the police power delegated by the state legislature that local governments are able to enact building regulations. If the state legislature has limited this power in any way, the municipality may not exceed these limitations. Although the municipality may not further delegate its police power (e.g., by delegating the burden of determining code compliance to the building owner, contractor or architect), it may turn over the administration of building regulations to a municipal official, such as a code official, provided that he or she is given sufficient criteria to clearly establish the basis for decisions as to whether or not a proposed building, including its plumbing systems, conforms to the code.

Chapter 1 is largely concerned with maintaining "due process of law" in enforcing the performance criteria contained in the code. Only through careful observation of the administrative provisions can the code official reasonably hope to demonstrate that "equal protection under the law" has been provided. Although it is generally assumed that the administrative and enforcement sections of a code are geared toward the code official, this is not entirely true. The provisions also establish the rights and privileges of the design professional, contractor and building owner. The position of the code official is merely to review the proposed and completed work and determine whether a plumbing installation conforms to the code requirements. The design professional is responsible for the design of a safe, sanitary plumbing system.

The contractor is responsible for installing the system in strict accordance with the plans.

During the course of the construction of a plumbing system, the code official reviews the activity to make sure that the spirit and intent of the law are being met and that the plumbing system provides adequate protection of public health. As a public servant, the code official enforces the code in an unbiased, proper manner. Every individual is guaranteed equal enforcement of the code. Furthermore, design professionals, contractors and building owners have the right of due process for any requirement in the code.

Purpose

A plumbing code, as with any other code, is intended for adoption as a legally enforceable document to safeguard health, safety, property and public welfare. A plumbing code cannot be effective without adequate provisions for its administration and enforcement. The official charged with the administration and enforcement of plumbing regulations has a great responsibility, and with this responsibility goes authority. No matter how detailed the plumbing code may be, the code official must, to some extent, exercise judgment in determining compliance. The code official has the responsibility for establishing that the homes in which the citizens of the community reside and the buildings in which they work are designed and constructed to be reasonably free from hazards associated with the presence and use of plumbing appliances, appurtenances, fixtures and systems. The code is intended to establish a minimum acceptable level of safety.

PART 1—SCOPE AND APPLICATION

SECTION 101
GENERAL

[A] 101.1 Title. These regulations shall be known as the *International Plumbing Code* of **[NAME OF JURISDICTION]** hereinafter referred to as "this code."

❖ This section sets forth the scope and intent of the code as it applies to new and existing structures. The adopted regulations are identified by inserting the name of the adopting jurisdiction into the code.

[A] 101.2 Scope. The provisions of this code shall apply to the erection, installation, alteration, repairs, relocation, replacement, addition to, use or maintenance of plumbing systems within this jurisdiction. This code shall regulate nonflammable medical gas, inhalation anesthetic, vacuum piping, nonmedical oxygen systems and sanitary and condensate vacuum collection systems. The installation of fuel gas distribution piping and equipment, fuel-gas-fired water heaters and water heater venting systems shall be regulated by the *International Fuel Gas Code*. Provisions in the appendices shall not apply unless specifically adopted.

> **Exception:** Detached one- and two-family dwellings and multiple single-family dwellings (townhouses) not more than three stories high with separate means of egress and their accessory structures shall comply with the *International Residential Code*.

❖ This section describes the types of plumbing system construction-related activities to which the code is intended to apply. The applicability of the code encompasses the initial design of plumbing systems, the installation and construction phases and the maintenance of operating systems. Section 101.2 excludes nothing plumbing related and does not limit applicability of the code to any device, fixture, system and associated equipment that could fall under, or is construed to fall under, the definition of "Plumbing" (see the definition of "Plumbing" in Chapter 2). The code is intended to govern plumbing systems provided for use by and for the general safety and well being of occupants of a building. The code intends to regulate any and all plumbing-related appliances, systems and associated equipment that can affect the health, safety and welfare of building occupants insofar as they are affected by the installation, operation and maintenance of such appliances and systems. Plumbing systems include the associated equipment by definition of "Plumbing system" in Chapter 2.

In addition, nonflammable medical gas, inhalation anesthetic, vacuum piping, nonmedical oxygen systems and sanitary and condensate vacuum collection systems are regulated by the code. Other than the actual connections to the potable water system, the code does not regulate hydronic piping for space heating or cooling, lawn sprinkler (irrigation) systems or automatic fire sprinkler and standpipe systems. These are not considered to be plumbing systems because they have nothing to do with supplying potable water or the transport of liquid wastes and water-borne solid wastes. Flammable/combustible liquid piping, hydronic piping, fire suppression system piping and nonplumbing-related systems are typically addressed in the *International Building Code*® (IBC®), the *International Fire Code*® (IFC®) and the *International Mechanical Code*® (IMC®). Most hydronic heating and cooling systems and fire suppression systems have one or more connections to the plumbing system. Such connections involve direct connections to the water supply system and indirect connections to the drainage system. For example, an automatic fire sprinkler system may be supplied by the public potable water supply and may have one or more drains and test apparatus that discharge water to the building drainage system. In the case of irrigation systems, for example, the plumbing system terminates at the backflow prevention device that joins the potable water supply to the irrigation piping system. All interfaces between plumbing systems and nonplumbing systems are subject to the requirements of the code.

This section references the *International Fuel Gas Code*® (IFGC®) for all fuel-gas-related regulations. This is the result of an agreement between the International Code Council® (ICC®) and the American Gas Association (AGA) to develop the IFGC.

The exception is actually a distinct requirement that sends the user to the *International Residential Code*® (IRC®) for one- and two-family dwellings that are less than four stories in height and townhouses that are less than four stories in height as these structures are within the scope of the IRC. It is the intent of the *International Codes*® that the *International Plumbing Code*® (IPC®) regulate plumbing in all structures that do not fall within the scope of the IRC. Structures falling within the scope of the IRC are to be regulated by the IRC.

[A] 101.3 Intent. The purpose of this code is to establish minimum standards to provide a reasonable level of safety, health, property protection and public welfare by regulating and controlling the design, construction, installation, quality of materials, location, operation and maintenance or use of plumbing equipment and systems.

❖ The intent of the code is to set forth regulations that establish minimum standards to provide a reasonable level of safety and to protect public health and general welfare of building occupants. The intent becomes important in the application of such sections as Sections 102, 104.2, 105.2 and 108, as well as any enforcement-oriented interpretive action or judgment. Like any code, the written text is subject to interpretation. Interpretations should not be affected by economics or the potential impact on any party. The only considerations should be occupants' safety, protection of occupants' health and welfare and emergency responder safety.

[A] 101.4 Severability. If any section, subsection, sentence, clause or phrase of this code is for any reason held to be unconstitutional, such decision shall not affect the validity of the remaining portions of this code.

❖ Only invalid sections of the code (as established by the court of jurisdiction) can be set aside. This is essential to safeguard the application of the code text to situations in which a provision of the code is declared illegal or unconstitutional. This section preserves the legislative action that put the legal provisions in place.

SECTION 102
APPLICABILITY

[A] 102.1 General. Where there is a conflict between a general requirement and a specific requirement, the specific requirement shall govern. Where, in any specific case, different sections of this code specify different materials, methods of construction or other requirements, the most restrictive shall govern.

❖ Specific requirements of the code override or take precedence over general requirements. For example, while the code specifies the types of piping materials that can be used for vent systems, the specific requirements for chemical waste systems require the vent piping be suitable for the service intended and be approved by the code official.

[A] 102.2 Existing installations. Plumbing systems lawfully in existence at the time of the adoption of this code shall be permitted to have their use and maintenance continued if the use, maintenance or repair is in accordance with the original design and hazard to life, health or property is not created by such plumbing system.

❖ An existing plumbing system is generally considered to be "grandfathered" with code adoption if the system meets a minimum level of safety. Frequently the criteria for this level are the regulations (or code) under which the existing building was originally constructed. If there are no previous code criteria to apply, the code official is to apply those provisions that are reasonably applicable to existing buildings. A specific level of safety is dictated by provisions dealing with hazard abatement in existing buildings and maintenance provisions, as contained in the code and the *International Property Maintenance Code*® (IPMC®), *the International Existing Building Code*® (IEBC®) and the IFC.

[A] 102.3 Maintenance. Plumbing systems, materials and appurtenances, both existing and new, and parts thereof, shall be maintained in proper operating condition in accordance with the original design in a safe and sanitary condition. Devices or safeguards required by this code shall be maintained in compliance with the edition of the code under which they were installed.

The owner or the owner's authorized agent shall be responsible for maintenance of plumbing systems. To determine compliance with this provision, the code official shall have the authority to require any plumbing system to be reinspected.

❖ All plumbing systems and equipment are subject to deterioration resulting from aging, wear, accumulation of dirt and debris, corrosion and other factors. Maintenance is necessary to keep plumbing systems and equipment in proper operating condition. Required safety devices and controls must be maintained to continue providing the protection that they afford. Existing equipment and systems could have safety devices or other measures that were necessary because of the nature of the equipment, and such safeguards may have been required by a code that predates the current code. Safeguards required by previous or present codes must be maintained for the life of the equipment or system.

The maintenance of plumbing systems as prescribed in this section is the responsibility of the owner of the property. The owner may authorize another party to be responsible for the property, in which case that party is responsible for the maintenance of the plumbing systems involved.

The reinspection authority of the code official is needed to assure compliance with the maintenance requirements in this section.

[A] 102.4 Additions, alterations or repairs. Additions, alterations, renovations or repairs to any plumbing system shall conform to that required for a new plumbing system without requiring the existing plumbing system to comply with all the requirements of this code. Additions, alterations or repairs shall not cause an existing system to become unsafe, insanitary or overloaded.

Minor additions, alterations, renovations and repairs to existing plumbing systems shall meet the provisions for new construction, unless such work is done in the same manner and arrangement as was in the existing system, is not hazardous and is *approved*.

❖ Simply stated, new work must comply with current code requirements. Any alteration or addition to an existing system involves some new work, and therefore is subject to the requirements of the code. Additions or alterations to an existing system can place additional loads or different demands on the system, which could necessitate changing all or part of the existing system. For example, the addition of plumbing fixtures to an existing system may necessitate an increase in drain piping size and water distribution piping size. Additions and alterations must not cause an existing system to be any less in compliance with the code than it was before the changes.

Repair of an existing nonconforming plumbing system is permitted without having to completely replace the nonconforming portion. This typically occurs when repairing a fixture or piping. Although some types of fixtures or piping arrangements are no longer permitted, existing fixtures or piping can be repaired and remain in service if a health hazard or insanitary condition is not maintained or created. This section distinguishes between alterations (subject to applicable provisions of the code) and ordinary repairs (maintenance activities not requiring a permit). The intent of this section is to allow the continued use of existing plumbing systems and equipment that may or may not be designed and constructed as required for new installations.

Existing plumbing systems and equipment will normally require repair and component replacement to remain operational. This section permits repair and component replacements without requiring the redesign, alteration or replacement of the entire system. In other words, the plumbing system is allowed to stay as it was if it is not hazardous. It is important to note that the word "minor" in this section is intended to modify "additions," "alterations," "renovations" and "repairs." It is not the intent of this section to waive code requirements for the replacement of all or major portions of systems under the guise of repair. Any work other than minor repairs or replacement of minor portions of a system must be considered as new work subject to all applicable provisions of the code. Repairs and minor component replacements are permitted in a manner that is consistent with the existing system if those repairs or replacements are approved by the code official; are not hazardous; do not cause the system or equipment to be any less in compliance with the code than before; and are, to the extent practicable, in compliance with the provisions of the code applicable to new work.

[A] 102.5 Change in occupancy. It shall be unlawful to make any change in the *occupancy* of any structure that will subject the structure to any special provision of this code applicable to the new *occupancy* without approval of the code official. The code official shall certify that such structure meets the intent of the provisions of law governing building construction for the proposed new *occupancy* and that such change of *occupancy* does not result in any hazard to the public health, safety or welfare.

❖ When a building undergoes a change of occupancy, the plumbing system must be evaluated to determine what effect the change has on the plumbing system. For example, if a mercantile building was converted to a restaurant, additional plumbing fixtures would be required for the public based on the increased occupant load. If an existing system serves an occupancy that is different from the one it served when the code went into effect, the plumbing system must comply with the applicable code requirements for the newer occupancy. Depending on the nature of the previous occupancy, changing a building's occupancy classification could result in the requirement for additional or different types of plumbing fixtures.

[A] 102.6 Historic buildings. The provisions of this code relating to the construction, alteration, repair, enlargement, restoration, relocation or moving of buildings or structures shall not be mandatory for existing buildings or structures identified and classified by the state or local jurisdiction as historic buildings where such buildings or structures are judged by the code official to be safe and in the public interest of health, safety and welfare regarding any proposed construction, alteration, repair, enlargement, restoration, relocation or moving of buildings.

❖ This section gives the code official the widest possible flexibility in enforcing the code when the building in question has historic value. This flexibility, however, does not come without conditions. The most important criterion for application of this section is that the building must be specifically classified as being of historic significance by a qualified party or agency. Usually this is done by a state or local authority after considerable scrutiny. Most, if not all, states have such authorities, as do many local jurisdictions. Agencies with such authority typically exist at the state or local government level.

[A] 102.7 Moved buildings. Except as determined by Section 102.2, plumbing systems that are a part of buildings or structures moved into or within the jurisdiction shall comply with the provisions of this code for new installations.

❖ Buildings that have been relocated are subject to the requirements of the code as if they were new construction. Placing a building where one did not previously exist is analogous to constructing a new building. It is the intent of this section to require alteration of the existing plumbing systems to the extent necessary to bring them into compliance with the provisions of the code applicable to new construction or make the existing plumbing system comply with Section 102.2.

[A] 102.8 Referenced codes and standards. The codes and standards referenced in this code shall be those that are listed in Chapter 15 and such codes and standards shall be considered as part of the requirements of this code to the prescribed extent of each such reference and as further regulated in Sections 102.8.1 and 102.8.2.

❖ The code references many standards and codes promulgated and published by other organizations. A complete list of referenced standards appears in Chapter 15. The wording in this provision, "shall be those that are listed in Chapter 15," was carefully chosen to establish the edition of the standard or code that is enforceable under the code.

[A] 102.8.1 Conflicts. Where conflicts occur between provisions of this code and the referenced standards, the provisions of this code shall apply.

❖ The code takes precedence when the requirements of the standard conflict with the requirements of the code. For example, Section 412.1 requires all floor drains to conform to ASME A112.3.1, ASME A112.6.3 or CSA B79. These standards regulate the design and construction of a floor drain and permit both fixed and removable strainers; however, Section 412.2 requires all floor drains to have removable strainers. Although a floor drain without a removable cover may conform to ASME A112.3.1, ASME A112.6.3 or CSA B79, removable strainers would be required because the code takes precedence over the standard. Although it is the intent of the code to be in harmony with referenced standards, the code text governs if conflicts occur.

[A] 102.8.2 Provisions in referenced codes and standards. Where the extent of the reference to a referenced code or standard includes subject matter that is within the scope of this code, the provisions of this code, as applicable, shall take precedence over the provisions in the referenced code or standard.

❖ Although a standard or code is referenced, its full scope and content is not necessarily applicable. The standard (or code) is applicable only to the extent indicated in the text in which the standard is specifically referenced. A referenced standard or a portion thereof is an enforceable extension of the code as if the content of the standard were included in the body of the code. The use and application of referenced standards are limited to those portions of the standards that are specifically identified.

[A] 102.9 Requirements not covered by code. Any requirements necessary for the strength, stability or proper operation of an existing or proposed plumbing system, or for the public safety, health and general welfare, not specifically covered by this code shall be determined by the code official.

❖ Evolving technology in our society will inevitably result in a situation in which the code is comparatively silent on an identified hazard. The reasonable application of the code to such hazardous, unforeseen conditions is addressed in this section. Clearly this section and the code official's judicious and reasonable application of it is necessary. The purpose of this section, however, is not to impose requirements that may be preferred when the code provides alternative methods or is not silent on the circumstances. Additionally, this section can be used to implement the general performance-oriented language of the code to specific enforcement situations.

[A] 102.10 Other laws. The provisions of this code shall not be deemed to nullify any provisions of local, state or federal law.

❖ Other laws enacted by the local, state or federal government may be applicable to a condition that is also governed by a requirement in the code. In such circumstances, the requirements of the code are in addition to those other laws, even though the code official may not be responsible for the enforcement of those laws. For example, the health department might require the handsinks in a restaurant to be indirectly connected whereas the code requires these sinks to be directly connected.

[A] 102.11 Application of references. Reference to chapter section numbers, or to provisions not specifically identified by number, shall be construed to refer to such chapter, section or provision of this code.

❖ In a situation where the code may make reference to a chapter or section number or to another code provision without specifically identifying its location in the code, assume that the referenced section, chapter or provision is in the code and not in a referenced code or standard.

PART 2—ADMINISTRATION AND ENFORCEMENT

SECTION 103
DEPARTMENT OF PLUMBING INSPECTION

[A] 103.1 General. The department of plumbing inspection is hereby created and the executive official in charge thereof shall be known as the code official.

❖ This section describes the plumbing inspection department and the duties of its employees. The executive official in charge of the plumbing department is named the "code official" by this section. In actuality, the person who is in charge of the department may hold a different title such as building commissioner, plumbing inspector, construction official, etc. For the purpose of the code, the person is referred to as the "code official" on being appointed.

[A] 103.2 Appointment. The code official shall be appointed by the chief appointing authority of the jurisdiction.

❖ This section establishes the code official as an appointed position from which he or she cannot be removed, except for cause subject to a due process review.

[A] 103.3 Deputies. In accordance with the prescribed procedures of this jurisdiction and with the concurrence of the appointing authority, the code official shall have the authority to appoint a deputy code official, other related technical officers, inspectors and other employees. Such employees shall have powers as delegated by the code official.

❖ This section gives the code official the authority to appoint other individuals to assist with the administration and enforcement of the code. These individuals have authority and responsibility as designated by the code official.

[A] 103.4 Liability. The code official, member of the board of appeals or employee charged with the enforcement of this code, while acting for the jurisdiction in good faith and without malice in the discharge of the duties required by this code or other pertinent law or ordinance, shall not thereby be rendered civilly or criminally liable personally, and is hereby relieved from all personal liability for any damage accruing to persons or property as a result of any act or by reason of an act or omission in the discharge of official duties.

❖ The code official, other department employees and members of the appeals board are not intended to be held liable, either civilly or criminally, for those actions performed in accordance with the code in a reasonable and lawful manner. The responsibility of the building official in this regard is subject to local, state and federal laws that may supersede this provision.

[A] 103.4.1 Legal defense. Any suit or criminal complaint instituted against any officer or employee because of an act performed by that officer or employee in the lawful discharge of duties and under the provisions of this code shall be defended by the legal representative of the jurisdiction until the final termination of the proceedings. The code official or any subordinate shall not be liable for costs in any action, suit or proceeding that is instituted in pursuance of the provisions of this code.

❖ This section establishes that code officials (or subordinates) must not be liable for costs in any legal action instituted in response to the performance of lawful duties. These costs are to be borne by the state, county or municipality. The best way to be certain that the code official's action is a "lawful duty" is always to cite the applicable code section on which the enforcement action is based.

SECTION 104
DUTIES AND POWERS OF THE CODE OFFICIAL

[A] 104.1 General. The code official is hereby authorized and directed to enforce the provisions of this code. The code official shall have the authority to render interpretations of this code and to adopt policies and procedures in order to clarify the application of its provisions. Such interpretations, policies and procedures shall be in compliance with the intent and purpose of this code. Such policies and procedures shall not have the effect of waiving requirements specifically provided for in this code.

❖ The duty of the code official is to enforce the code, and he or she is the "authority having jurisdiction" for all matters relating to the code and its enforcement. It is the duty of the code official to interpret the code and to determine compliance. Code compliance will not always be easy to determine and will require judgment and expertise. In exercising this authority, however, the code official cannot set aside or ignore any provision of the code.

[A] 104.2 Applications and permits. The code official shall receive applications, review construction documents and issue permits for the installation and alteration of plumbing systems, inspect the premises for which such permits have been issued, and enforce compliance with the provisions of this code.

❖ The code enforcement process is normally initiated with an application for a permit. The code official is responsible for processing the applications and issuing permits for the installation, replacement, addition to or modification of plumbing systems in accordance with the code.

[A] 104.3 Inspections. The code official shall make all the required inspections, or shall accept reports of inspection by *approved* agencies or individuals. Reports of such inspections shall be in writing and be certified by a responsible officer of such *approved agency* or by the responsible individual. The code official is authorized to engage such expert opinion as deemed necessary to report on unusual technical issues that arise, subject to the approval of the appointing authority.

❖ The code official is required to make inspections as necessary to determine compliance with the code or to accept written reports of inspections by an approved agency. The inspection of the work in progress or accomplished is another significant element in determining code compliance. Even though a department may not have the resources to inspect every aspect of all work, the required inspections are those that are dictated by administrative rules and procedures based on many parameters, including available inspection resources. To expand the available inspection resources, the code official may approve an inspection agency that, in his or her opinion, possesses the proper qualifications. When unusual, extraordinary or complex technical issues arise relative to plumbing installations or the safety of an existing plumbing system, the code official has the authority to seek the opinion and advice of experts. A technical report from an expert can be used to assist the code official in the approval process.

[A] 104.4 Right of entry. Where it is necessary to make an inspection to enforce the provisions of this code, or where the code official has reasonable cause to believe that there exists in any building or upon any premises any conditions or violations of this code that make the building or premises unsafe, insanitary, dangerous or hazardous, the code official shall have the authority to enter the building or premises at all reasonable times to inspect or to perform the duties imposed upon the code official by this code. If such building or premises is occupied, the code official shall present credentials to the occupant and request entry. If such building or premises is unoccupied, the code official shall first make a reasonable effort to locate the owner, the owner's authorized agent or other person having charge or control of the building or premises and request entry. If entry is refused, the code official shall have recourse to every remedy provided by law to secure entry.

Where the code official shall have first obtained a proper inspection warrant or other remedy provided by law to secure entry, no owner, owner's authorized agent, occupant or person having charge, care or control of any building or premises shall fail or neglect, after proper request is made as herein provided, to promptly permit entry therein by the code official for the purpose of inspection and examination pursuant to this code.

❖ The first part of this section establishes the right of the code official to enter the premises to make the permit inspections required by Section 107. Permit application forms typically include a statement signed by the applicant (who is the owner or owner's agent) granting the code official the authority to enter specific areas to enforce code provisions related to the permit. The right to enter other structures or premises is more limited. First, to protect the right of privacy, the owner or occupant must grant the code official permission before the interior of the property can be inspected. Permission is not required for inspections that can be accomplished from within the public right-of-way. Second, access may be denied by the owner or occupant. Unless the inspector has reasonable cause to believe that a violation exists, access may be unattainable. Third, code officials must present proper identification (see commentary, Section 104.5) and request admittance during reasonable hours—usually the normal business hours of the establishment. Fourth, inspections must be aimed at securing or determining compliance with the provisions and intent of the regulations that are specifically within the established scope of the code official's authority.

Searches to gather information for the purpose of enforcing the other codes, ordinances or regulations are considered unreasonable and are prohibited by the Fourth Amendment to the U.S. Constitution. "Reasonable cause" in the context of this section must be distinguished from "probable cause," which is required to gain access to property in criminal cases. The burden of proof establishing reasonable cause may vary among jurisdictions. Usually, an inspector must show that the property is subject to inspection under the provisions of the code; that the interests of the public health, safety and welfare outweigh the individual's right to maintain privacy; and that such an inspection is required solely to determine compliance with the provisions of the code.

Many jurisdictions do not recognize the concept of an administrative warrant and may require the code official to prove probable cause in order to gain access upon refusal. This burden of proof is usually more substantial, often requiring the code official to stipulate in advance why access is needed (usually access is restricted to gathering evidence for seeking an indictment or making an arrest); what specific items or information is sought; its relevance to the case against the individual subject; how knowledge of the relevance of the information or items sought was obtained; and how the evidence sought will be used. In all such cases, the right to privacy must always be weighed against the right of the code official to conduct an inspection to establish public health, safety and welfare. Such important and complex constitutional issues should be discussed with the jurisdiction's legal counsel. Jurisdictions should establish procedures for securing the necessary court orders when an inspection is deemed necessary following a refusal.

The last paragraph in this section requires the owner or occupant to permit entry for inspection if a proper warrant or other documentation required by law has been obtained.

[A] 104.5 Identification. The code official shall carry proper identification when inspecting structures or premises in the performance of duties under this code.

❖ This section requires the code official (including by definition all authorized designees) to carry identification in the course of conducting the duties of the position. This identification removes any question concerning the purpose and authority of the inspector.

[A] 104.6 Notices and orders. The code official shall issue all necessary notices or orders to ensure compliance with this code.

❖ An important element of code enforcement is the necessary advisement of deficiencies and correction, which is accomplished through notices and orders. The code official is required to issue orders to abate illegal or unsafe conditions. Sections 108.7, 108.7.1, 108.7.2 and 108.7.3 contain additional information for these notices.

[A] 104.7 Department records. The code official shall keep official records of applications received, permits and certificates issued, fees collected, reports of inspections, and notices and orders issued. Such records shall be retained in the official records for the period required for the retention of public records.

❖ In keeping with the need for an efficiently conducted business practice, the code official must keep records pertaining to permit applications, permits, fees collected, inspections, notices and orders issued. Such documentation provides a valuable resource if questions arise regarding the department's actions with respect to a building. It requires that other documents be kept for the length of time mandated by a jurisdiction's, or its state's, laws or administrative rules for retaining public records.

SECTION 105
APPROVAL

[A] 105.1 Modifications. Where there are practical difficulties involved in carrying out the provisions of this code, the code official shall have the authority to grant modifications for individual cases, upon application of the owner or owner's authorized agent, provided the code official shall first find that special individual reason makes the strict letter of this code impractical and the modification conforms to the intent and purpose of this code and that such modification does not lessen health, life and fire safety requirements. The details of action granting modifications shall be recorded and entered in the files of the plumbing inspection department.

❖ The code official may amend or make exceptions to the code as needed where strict compliance is impractical. Only the code official has authority to grant modifications. Consideration of a particular difficulty is to be based on the application of the owner and a demonstration that the intent of the code is accomplished. This section is not intended to permit setting aside or ignoring a code provision; rather, it is intended to provide acceptance of equivalent protection. Such modifications do not, however, extend to actions that are necessary to correct violations of the code. In other words, a code violation or the expense of correcting one cannot constitute a practical difficulty.

[A] 105.2 Alternative materials, methods and equipment. The provisions of this code are not intended to prevent the installation of any material or to prohibit any method of construction not specifically prescribed by this code, provided that any such alternative has been *approved*. An alternative material or method of construction shall be *approved* where the code official finds that the proposed alternative material, method or equipment complies with the intent of the provisions of this code and is not less than the equivalent of that prescribed in this code. Where the alternative material, design or method of construction is not *approved*, the code official shall respond in writing, stating the reasons why the alternative was not *approved*.

❖ The code is not intended to inhibit innovative ideas or technological advances. A comprehensive regulatory document such as a plumbing code cannot envision and then address all future innovations in the industry. As a result, a performance code must be applicable to and provide a basis for the approval of an increasing number of newly developed, innovative materials, systems and methods for which no code text or referenced standards yet exist. The fact that a material, product or method of construction is not addressed in the code is not an indication that the material, product or method is prohibited. The code official is expected to apply sound technical judgment in accepting materials, systems or methods that, although not anticipated by the drafters of the current code text, can be demonstrated to offer equivalent performance. By virtue of its text, the code regulates new and innovative construction practices while addressing the relative safety of building occupants. The code official is responsible for determining whether a requested alternative provides an equivalent level of protection of the public health, safety and welfare as required by the code. In order to ensure effective communication and due process of law, if an alternative is not approved, the code official must state, in writing, the reasons for the disapproval.

The most common application of an alternative approval occurs with the proposed use of new material. For example, if a new piping material is produced, the manufacturer may gain approval for use by submitting adequate technical data indicating it is equivalent in quality, strength, effectiveness, fire resistance, durability and safety to the piping material listed as acceptable in the code. At the same time, the manufacturer may submit a proposed code change to recognize the new piping material. If the code official rejects the request for an alternative approval, the applicant may appeal the decision, as regulated by Section 109.

[A] 105.2.1 Research reports. Supporting data, where necessary to assist in the approval of materials or assemblies not specifically provided for in this code, shall consist of valid research reports from *approved* sources.

❖ When an alternative material or method is proposed for construction, it is incumbent upon the code official to determine whether this alternative is, in fact, an equivalent to the methods prescribed by the code. Reports providing evidence of this equivalency are required to be supplied by an approved source, meaning a source that the code official finds to be reliable and accurate. The ICC Evaluation Service is one example of an agency that provides research reports for alternative materials and methods.

[A] 105.3 Required testing. Where there is insufficient evidence of compliance with the provisions of this code, or evidence that a material or method does not conform to the requirements of this code, or in order to substantiate claims for alternate materials or methods, the code official shall have the authority to require tests as evidence of compliance to be made at no expense to the jurisdiction.

❖ To provide the basis on which the code official can make a decision regarding an alternative material or type of equipment, sufficient technical data, test reports and documentation must be provided for evaluation. If evidence satisfactory to the code official proves that the alternative equipment, material or construction method is equivalent to that required by the code, the code official is obligated to approve it for use. Any such approval cannot have the effect of waiving any requirements of the code. The burden of proof of equivalence lies with the applicant who proposes the use of alternative equipment, materials or methods.

[A] 105.3.1 Test methods. Test methods shall be as specified in this code or by other recognized test standards. In the absence of recognized and accepted test methods, the code official shall approve the testing procedures.

❖ The code official must require the submission of any appropriate information and data to assist in the determination of equivalency before a permit can be issued. The type of information required includes test data in accordance with the referenced standards, evidence of compliance with the referenced standard specifications, and design calculations. An evaluation report issued by an authoritative agency, such as ICC Evaluation Service is particularly useful in providing the code official with the technical basis for evaluation and approval of new and innovative plumbing materials and components. The use of authoritative research reports can greatly assist the code official by reducing the time-consuming engineering analysis necessary to review materials and products. Failure to adequately substantiate a request for the use of an alternative is a valid reason for the code official to deny a request.

[A] 105.3.2 Testing agency. Tests shall be performed by an *approved agency*.

❖ The testing agency must be approved by the code official. The testing agency should have technical expertise, test equipment and quality assurance to properly conduct and report the necessary testing.

[A] 105.3.3 Test reports. Reports of tests shall be retained by the code official for the period required for retention of public records.

❖ Test reports substantiating the modification must be retained in accordance with public record laws. The attorney of the jurisdiction could be asked to verify the specific time period in applicable laws of the jurisdiction.

[A] 105.4 Approved materials and equipment. Materials, equipment and devices *approved* by the code official shall be constructed and installed in accordance with such approval.

❖ The code is a compilation of criteria with which materials, equipment, devices and systems must comply to be suitable for a particular application. The code official has a duty to evaluate such materials, equipment, devices and systems for code compliance and when compliance is determined, approve the same for use. The materials, equipment, devices and systems must be constructed and installed in compliance with, and all conditions and limitations considered as a basis for, that approval. For example, the manufacturer's instructions and recommendations are to be followed if the approval of the material was based, even in part, on those instructions and recommendations. The approval authority given to the code official is a significant responsibility and is a key to code compliance. The approval process is first technical and then administrative and must be approached as such. For example, if data to determine code compliance is required, such data should be in the form of test reports or engineering analysis and not simply taken from a sales brochure.

[A] 105.4.1 Material and equipment reuse. Materials, equipment and devices shall not be reused unless such elements have been reconditioned, tested, placed in good and proper working condition and *approved*.

❖ The code criteria for materials and equipment have changed over the years. Evaluation of testing and materials technology has permitted the development of new criteria, which the old materials may not satisfy. As a result, used materials must be evaluated in the same manner as new materials. Used (previously installed) equipment must be equivalent to that required by the code if it is to be used in a new installation.

SECTION 106
PERMITS

[A] 106.1 When required. Any owner, owner's authorized agent or contractor who desires to construct, enlarge, alter, repair, move, demolish or change the *occupancy* of a building or structure, or to erect, install, enlarge, alter, repair, remove, convert or replace any plumbing system, the installation of which is regulated by this code, or to cause any such work to be performed, shall first make application to the code official and obtain the required permit for the work.

❖ This section contains the administrative rules governing the issuance, suspension, revocation or modification of plumbing permits. It also establishes how and by whom the application for a plumbing permit is to be made, how it is to be processed and what information it must contain or have attached to it. In general, a permit is required for all activities that are regulated by the code, and these activities cannot begin until the permit is issued.

A plumbing permit is required for the installation, replacement, alteration or modification of all plumbing systems and components that are in the scope of applicability of the code. Replacement of an existing fixture, piece of equipment or related piping is treated no differently than a new installation in new building construction. The purpose of a permit is to cause the work to be inspected to determine compliance with the intent of the code.

[A] 106.1.1 Annual permit. Instead of an individual construction permit for each alteration to an already *approved* system or equipment or appliance installation, the code official is authorized to issue an annual permit upon application therefor to any person, firm or corporation regularly employing one or more qualified tradespersons in the building, structure or on the premises owned or operated by the applicant for the permit.

❖ Many large facilities need plumbing alteration work performed often and on short notice. (Note that there is not a definition in the code for what constitutes an "alteration.") The obtaining of permits could occur nearly every day and might take more time than performing the work. If a qualified individual will be performing work on an already approved plumbing system, the code official can issue an annual permit to cover all the alteration work that occurs. This is sometimes referred to as a "blanket" permit. The person in charge of getting the work accomplished needs to be a qualified tradesperson. The code is not specific as to what is meant by qualified. Often this is someone who has worked in the plumbing trades for several years and has attained a level of understanding that is commensurate of the knowledge possessed by a journeyman plumbing tradesperson. However, the code official will have to decide the level of qualifications needed before issuing an annual permit.

Note that the person in charge of performing the work needs to be someone who is regularly employed by the person, firm or organization that owns or operates the facility. The term "employed" usually means that the person is on the payroll of the facility's owner (or operator) and is not a contractor. The intent is that the annual permit not be used to cover the work performed at the facility by outside contractors. This employer/employee arrangement provides some assurance that the employee (the tradesperson) will "do the right thing" when performing the work in order to maintain future employment. This also makes the owner or operator of the facility directly responsible for the work performed. Where the work is performed by contractors, incorrect work is sometimes more difficult to get corrected as the owner/operator originally paid for correct work and might be reluctant to "pay again" to resolve violations found months later. Thus, contractors would be required to obtain a permit for each "job" except for the exempt work indicated in Section 106.2. The term "regularly" is somewhat ambiguous but generally, a facility's owner or operator will usually employ at least one tradesperson on a full-time basis because of the frequency of (unknown) events that occur in the facility.

[A] 106.1.2 Annual permit records. The person to whom an annual permit is issued shall keep a detailed record of alterations made under such annual permit. The code official shall have access to such records at all times or such records shall be filed with the code official as designated.

❖ The issuance of an annual permit makes the owner or operator responsible for detailed logging of all the work performed during that year. Ideally, when the code official has slack time periods in performing regular inspections, he or she can call on facilities with annual permits and inspect some of the work performed since the last visit. Where the code official cannot make periodic visits to the facility, he or she should periodically request a copy of the records to identify items that fall outside the scope of "alterations" (as defined by the code official) such that an inspection can be scheduled.

[A] 106.2 Exempt work. The following work shall be exempt from the requirement for a permit:

1. The stopping of leaks in drains, water, soil, waste or vent pipe provided, however, that if any concealed trap, drainpipe, water, soil, waste or vent pipe becomes defective and it becomes necessary to remove and replace the same with new material, such work shall be considered as new work and a permit shall be obtained and inspection made as provided in this code.

2. The clearing of stoppages or the repairing of leaks in pipes, valves or fixtures, and the removal and reinstallation of water closets, provided such repairs do not involve or require the replacement or rearrangement of valves, pipes or fixtures.

Exemption from the permit requirements of this code shall not be deemed to grant authorization for any work to be done in violation of the provisions of this code or any other laws or ordinances of this jurisdiction.

❖ Repair work, maintenance procedures and similar work are exempt from the permit requirement, provided that such work does not involve the replacement of any system components other than minor parts.

Literally, the only plumbing activities that are exempt from the requirement for a permit are drain cleaning and rodding and the repair of faucets, fill valves, flushometers and similar fixture trim and fittings. Any plumbing work that results in the alteration or disassembly of existing drain, waste, vent and water distribution piping would require a permit. Additionally, any plumbing work that involves the alteration of a plumbing fixture would require a permit.

[A] 106.3 Application for permit. Each application for a permit, with the required fee, shall be filed with the code official on a form furnished for that purpose and shall contain a general description of the proposed work and its location. The application shall be signed by the owner or owner's authorized agent. The permit application shall indicate the proposed *occupancy* of all parts of the building and of that portion of the site or lot, if any, not covered by the building or structure and shall contain such other information required by the code official.

❖ This section limits persons who may apply for a permit to the building owner or an authorized agent. An owner's authorized agent could be anyone who is given written permission to act in the owner's interest for the purpose of obtaining a permit, such as an architect, an engineer, a contractor, a tenant or other. Permit forms generally have sufficient space to write a very brief description of the work to be accomplished, which is acceptable for small jobs. For larger projects, the description will be augmented by construction documents.

[A] 106.3.1 Construction documents. Construction documents, engineering calculations, diagrams and other such data shall be submitted in two or more sets with each application for a permit. The code official shall require construction documents, computations and specifications to be prepared and designed by a registered design professional where required by state law. Construction documents shall be drawn to scale and shall be of sufficient clarity to indicate the location, nature and extent of the work proposed and show in detail that the work conforms to the provisions of this code. Construction documents for buildings more than two stories in height shall indicate where penetrations will be made for pipes, fittings and components and shall indicate the materials and methods for maintaining required structural safety, fire-resistance rating and fireblocking.

Exception: The code official shall have the authority to waive the submission of construction documents, calculations or other data if the nature of the work applied for is such that reviewing of construction documents is not necessary to determine compliance with this code.

❖ A detailed description of the work for which application is made must be submitted. When the work is of a "minor nature," either in scope or needed description, the code official may use judgement in determining the need for a detailed description of the work. An example of minor work that may not involve a detailed description is the replacement of an existing fixture in a plumbing system or the replacement or repair of a defective portion of a piping system.

These provisions are intended to reflect the minimum scope of information needed to determine code compliance. A statement on the construction documents such as, "All plumbing work must comply with the 2015 IPC," is not an acceptable substitute for showing the required information.

This section also requires the code official to determine compliance with any state professional registration laws as they apply to the preparation of construction documents.

[A] 106.3.2 Preliminary inspection. Before a permit is issued, the code official shall be authorized to inspect and evaluate the systems, equipment, buildings, devices, premises and spaces or areas to be used.

❖ Some projects might require a preliminary inspection by the code official prior to a permit being issued. This is especially useful for remodel and addition projects where the conditions of the existing building plumbing systems are unknown or are of questionable condition. This section authorizes the code official to make such inspections.

[A] 106.3.3 Time limitation of application. An application for a permit for any proposed work shall be deemed to have been abandoned 180 days after the date of filing, unless such application has been pursued in good faith or a permit has been issued; except that the code official shall have the authority to grant one or more extensions of time for additional periods not exceeding 180 days each. The extension shall be requested in writing and justifiable cause demonstrated.

❖ Once an application for a permit has been submitted for proposed work, a time limit of 180 days is established for issuance of the permit. This prevents the code official from having to hold onto incomplete or delayed applications for an indefinite amount of time. The code official can grant extensions for this time period if provided with a written request with justifiable reasons for the extension request.

[A] 106.4 By whom application is made. Application for a permit shall be made by the person or agent to install all or part of any plumbing system. The applicant shall meet all qualifications established by statute, or by rules promulgated by this code, or by ordinance or by resolution. The full name and address of the applicant shall be stated in the application.

❖ This section specifies that the permit applicant is to meet all of the qualifications that may be established by the governmental enforcing agency. For example, some governmental enforcing agencies may require the permit applicant to be a registered design professional or licensee.

[A] 106.5 Permit issuance. The application, construction documents and other data filed by an applicant for permit shall be reviewed by the code official. If the code official finds that the proposed work conforms to the requirements of this code and all laws and ordinances applicable thereto, and that the fees specified in Section 106.6 have been paid, a permit shall be issued to the applicant.

❖ This section requires the code official to review all submittals for a permit for compliance with the code and to verify that the project will be carried out in accordance with any other applicable laws. This may involve interagency communication and cooperation so that all laws are being obeyed. Once the code official verifies this, a permit may be issued upon payment of the required fees.

[A] 106.5.1 Approved construction documents. When the code official issues the permit where construction documents are required, the construction documents shall be endorsed in writing and stamped "APPROVED." Such *approved* construction documents shall not be changed, modified or altered without authorization from the code official. Work shall be done in accordance with the *approved* construction documents.

The code official shall have the authority to issue a permit for the construction of a part of a plumbing system before the entire construction documents for the whole system have been submitted or *approved*, provided adequate information and detailed statements have been filed complying with all pertinent requirements of this code. The holders of such permit shall proceed at their own risk without assurance that the permit for the entire plumbing system will be granted.

❖ Construction documents that reflect compliance with code requirements form an integral part of the permit process. Successful execution of the work depends on these documents. This section requires the code official to stamp the complying construction documents as being "approved." Once approved, no further revisions to the documents may be made without the express authorization of the code official in order to maintain code compliance.

In the interest of saving time and coordinating construction phases, it is common practice for contractors to seek permits solely applicable to the installation of site work, such as water services and sewers. This practice allows the project to proceed before the final construction documents are completed, thus minimizing delays in the construction process. This also allows the builder to perform site work while the weather permits.

The holder of a partial permit must realize that a permit for the remainder of the plumbing system or structure may not be granted for various reasons. Issuance of a partial permit in no way guarantees issuance of a permit for the entire scope of the project.

[A] 106.5.2 Validity. The issuance of a permit or approval of construction documents shall not be construed to be a permit for, or an approval of, any violation of any of the provisions of this code or any other ordinance of the jurisdiction. A permit presuming to give authority to violate or cancel the provisions of this code shall not be valid.

The issuance of a permit based upon construction documents and other data shall not prevent the code official from thereafter requiring the correction of errors in said construction documents and other data or from preventing building operations being carried on thereunder where in violation of this code or of other ordinances of this jurisdiction.

❖ An important code section, this section states the fundamental premise that the permit is only "a license to proceed with the work." It is not a license to "violate, cancel or set aside any provisions of the code." This is important because it means that, despite any errors in the approval process, the permit applicant is responsible for code compliance.

[A] 106.5.3 Expiration. Every permit issued by the code official under the provisions of this code shall expire by limitation and become null and void if the work authorized by such permit is not commenced within 180 days from the date of such permit, or if the work authorized by such permit is suspended or abandoned at any time after the work is commenced for a period of 180 days. Before such work can be recommenced, a new permit shall be first obtained and the fee therefor shall be one-half the amount required for a new permit for such work, provided changes have not been made and will not be made in the original construction documents for such work, and provided further that such suspension or abandonment has not exceeded 1 year.

❖ The permit becomes invalid under two distinct situations, both based on a six-month period. The first situation is when no work has started six months from issuance of the permit. The second situation is when there is no continuation of authorized work for six months. The person who was issued the permit should be notified in writing that it is invalid and what steps must be taken to restart the work.

This section also provides the administrative authority with a means of offsetting the costs associated with expired permits by charging a nominal fee for permit reissuance. If, however, the nature or scope of the work to be resumed is different from that covered by the original permit, the permit process essentially starts from "scratch" and full fees are charged. The same procedure would also apply if the work has not commenced within one year of the date of permit issuance or if work has to be suspended for a year or more.

[A] 106.5.4 Extensions. Any permittee holding an unexpired permit shall have the right to apply for an extension of the time within which the permittee will commence work under that permit when work is unable to be commenced within the time required by this section for good and satisfactory reasons. The code official shall extend the time for action by the permittee for a period not exceeding 180 days if there is reasonable cause. A permit shall not be extended more than once. The fee for an extension shall be one-half the amount required for a new permit for such work.

❖ Although it is typical for a project to begin immediately following issuance of a permit, there are occasions when an unforeseen delay may occur. This section intends to afford the permit holder the opportunity to apply for and receive a single, 180-day extension within which to begin a project under a still-valid permit

SCOPE AND ADMINISTRATION

(i.e., less than 180 days old). The applicant must, however, provide the code official with an adequate explanation for the delay in starting a project, which could include such things as the need to obtain approvals or permits from other agencies having jurisdiction. This section requires the code official to determine what constitutes "good and satisfactory" reasons for any delay and further allows the jurisdiction to offset its administrative costs for extending the permit by charging one-half the fee for a new permit for the extension.

[A] 106.5.5 Suspension or revocation of permit. The code official shall have the authority to suspend or revoke a permit issued under the provisions of this code wherever the permit is issued in error or on the basis of incorrect, inaccurate or incomplete information, or in violation of any ordinance or regulation or any of the provisions of this code.

❖ A permit is in reality a license to proceed with the work. The code official, however, must revoke all permits shown to be based, all or in part, on any false statement or misrepresentation of fact. An applicant may subsequently reapply for a permit with the appropriate corrections or modifications made to the application and construction documents.

[A] 106.5.6 Retention of construction documents. One set of *approved* construction documents shall be retained by the code official for a period of not less than 180 days from date of completion of the permitted work, or as required by state or local laws.

One set of *approved* construction documents shall be returned to the applicant, and said set shall be kept on the site of the building or work at all times during which the work authorized thereby is in progress.

❖ Once the code official has stamped or endorsed as approved the construction documents on which the permit is based (see commentary, Section 106.3.1), one set of approved construction documents must be kept on the construction site to serve as the basis for all subsequent inspections. To avoid confusion, the construction documents on the site must be precisely the documents that were approved and stamped. This is because inspections are based on the approved documents. Additionally, the contractor cannot determine compliance with the approved construction documents unless those documents are readily available. Unless the approved construction documents are available, the inspection should be postponed and work on the project halted.

[A] 106.5.7 Previous approvals. This code shall not require changes in the construction documents, construction or designated *occupancy* of a structure for which a lawful permit has been heretofore issued or otherwise lawfully authorized, and the construction of which has been pursued in good faith within 180 days after the effective date of this code and has not been abandoned.

❖ This section provides the code official with a useful tool to protect the continuity of permits issued under previous codes or code editions, as long as such permits are being actively executed subsequent to the effective date of the ordinance adopting the newer code.

[A] 106.5.8 Posting of permit. The permit or a copy shall be kept on the site of the work until the completion of the project.

❖ This section requires the permit (or a copy of the permit) to be on the work site until the project is completed. Having the permit at the jobsite provides project information and evidence to anyone needing to know if the project has been duly authorized.

[A] 106.6 Fees. A permit shall not be issued until the fees prescribed in Section 106.6.2 have been paid, and an amendment to a permit shall not be released until the additional fee, if any, due to an increase of the plumbing systems, has been paid.

❖ All fees are to be paid prior to permit issuance. This requirement establishes that the permit applicant intends to proceed with the work, as well as facilitates payment.

[A] 106.6.1 Work commencing before permit issuance. Any person who commences any work on a plumbing system before obtaining the necessary permits shall be subject to 100 percent of the usual permit fee in addition to the required permit fees.

❖ This section is intended to serve as a deterrent to proceeding with work on a plumbing system without a permit (except as provided in Section 106.2). As a punitive measure, it doubles the cost of the permit fee. This section does not, however, intend to penalize a contractor called upon to do emergency work after hours, provided that he or she makes prompt notification to the code official the next business day and obtains the requisite permit for the work done and has the required inspections performed.

[A] 106.6.2 Fee schedule. The fees for all plumbing work shall be as indicated in the following schedule:

[JURISDICTION TO INSERT APPROPRIATE SCHEDULE]

❖ A published fee schedule must be established for plans examination, permits and inspections. Ideally, the department should generate revenues that cover operating costs and expenses. The permit fee schedule is an integral part of this process.

[A] 106.6.3 Fee refunds. The code official shall authorize the refunding of fees as follows:

1. The full amount of any fee paid hereunder that was erroneously paid or collected.

2. Not more than **[SPECIFY PERCENTAGE]** percent of the permit fee paid where work has been done under a permit issued in accordance with this code.

3. Not more than [SPECIFY PERCENTAGE] percent of the plan review fee paid where an application for a permit for which a plan review fee has been paid is withdrawn or canceled before any plan review effort has been expended.

The code official shall not authorize the refunding of any fee paid except upon written application filed by the original permittee not later than 180 days after the date of fee payment.

❖ This section allows for a partial refund of fees resulting from the revocation, abandonment or discontinuance of a plumbing project for which a permit has been issued and fees have been collected. The incomplete work for which the excess fees are to be refunded refers to the work that would have been required by the department had the permit not been terminated. The refund of fees should be related to the cost of enforcement services not provided because of termination of the project.

SECTION 107
INSPECTIONS AND TESTING

[A] 107.1 General. The code official is authorized to conduct such inspections as are deemed necessary to determine compliance with the provisions of this code. Construction or work for which a permit is required shall be subject to inspection by the code official, and such construction or work shall remain accessible and exposed for inspection purposes until *approved*. Approval as a result of an inspection shall not be construed to be an approval of a violation of the provisions of this code or of other ordinances of the jurisdiction. Inspections presuming to give authority to violate or cancel the provisions of this code or of other ordinances of the jurisdiction shall not be valid. It shall be the duty of the permit applicant to cause the work to remain accessible and exposed for inspection purposes. Neither the code official nor the jurisdiction shall be liable for expense entailed in the removal or replacement of any material required to allow inspection.

❖ The inspection function is one of the more important aspects of building department operations. This section authorizes the code official to inspect the work for which a permit has been issued and requires that the work to be inspected remain accessible to the code official until inspected and approved. Any expense incurred in removing or replacing material that conceals an item to be inspected is not the responsibility of the code official or the jurisdiction. As with the issuance of permits (see Section 106.5.2), an approval as a result of an inspection is not a license to violate the code. Any work approved that might contain a violation of the code does not relieve the applicant from complying with the code.

[A] 107.2 Required inspections and testing. The code official, upon notification from the permit holder or the permit holder's agent, shall make the following inspections and such other inspections as necessary, and shall either release that portion of the construction or shall notify the permit holder or an agent of any violations that must be corrected. The holder of the permit shall be responsible for the scheduling of such inspections.

1. Underground inspection shall be made after trenches or ditches are excavated and bedded, piping installed, and before any backfill is put in place.

2. Rough-in inspection shall be made after the roof, framing, fireblocking, firestopping, draftstopping and bracing is in place and all sanitary, storm and water distribution piping is roughed-in, and prior to the installation of wall or ceiling membranes.

3. Final inspection shall be made after the building is complete, all plumbing fixtures are in place and properly connected, and the structure is ready for occupancy.

❖ This section requires that all portions of the plumbing system be inspected before and after fixtures are installed. The code official has the authority to require noncomplying plumbing to be brought into compliance and reinspected.

Inspections are necessary to determine that an installation conforms to all code requirements. Because the majority of a plumbing system is hidden within the building enclosure or is underground, periodic inspections are necessary before the systems are concealed. The code official is required to determine that plumbing systems and equipment are installed in accordance with the applicable code requirements and any special arrangements that fall outside of the code requirements, but have been approved during the jurisdiction's plan review (see Section 106.3). All inspections that are necessary to provide such verifications must be conducted. The code official is not required to inspect the work against the construction drawings (except the drawing for any special arrangements that were approved during plan review). The inspecting code official is not a quality control inspector insofar as making sure that the plumbing system is installed exactly like the construction drawings that the jurisdiction approved during plan review. As long as the installed work complies with the code, the work should be acceptable to the inspecting code official. Note that some jurisdictions could have different policies regarding site inspections. Contractors should always check with the local code official to determine how inspections are handled for projects.

Generally, the administrative rules of a department indicate the required interim inspections. Construction projects that occur in steps or phases may necessitate multiple inspections; therefore, an exact number of required inspections cannot be specified. Where violations are noted and corrections are required, reinspections may be necessary. As time permits, frequent inspections of some job sites, especially where the work is complex, can be beneficial in detecting code violations or other potential problems before they develop or become more difficult to resolve.

It is the responsibility of the contractor, the builder, the owner or other authorized party to arrange for the required inspections and to coordinate them to prevent work from being concealed before it is inspected.

1. Inspection of underground plumbing is especially important because once it is covered, it is the most challenging part of a plumbing system in which to detect a leak. If repairs are necessary, underground repairs are proportionately more expensive because of the need for heavy equipment and the more labor-intensive nature of working below grade level (see Section 306 for trenching, excavation and backfill requirements).

2. A rough-in inspection is a visual observation of all parts of the plumbing system that will eventually be concealed in the building structure. Rough-in inspections also include verification that the applicable test pressures are applied to the system and that leaks do not exist. The inspection must be made before any of the system is covered by building finish materials or hidden by future work.

    A rough-in inspection may be completed in one visit or as a series of inspections. This is administratively determined by the local inspections department and is typically dependent on the size of the job.

3. A final inspection may be done as a series of inspections or in one visit, similar to a rough-in inspection. A final inspection is required prior to the approval of plumbing work and installations. For the construction of a new building, final approval is required prior to the issuance of the certificate of occupancy as specified in the IBC. To verify that all previously issued correction orders have been complied with and to determine whether subsequent violations exist, a final inspection must be made. All violations observed during the final inspection must be noted and the permit holder must be advised.

    The final inspection follows the completion of the work or installation. Typically, the final inspection is an inspection of all that was installed after the rough-in inspection and not concealed in the building construction. Subsequent reinspections are necessary if the final inspection generates a notice of violation.

[A] 107.2.1 Other inspections. In addition to the inspections specified above, the code official shall be authorized to make or require other inspections of any construction work to ascertain compliance with the provisions of this code and other laws that are enforced.

❖ Any item regulated by the code is subject to inspection by the code official to determine compliance with the applicable code provision, and no list can include all types of work in a given building. Also, other inspections before, during or after the rough-in could be necessary. This section gives the code official the authority to inspect any regulated work.

[A] 107.2.2 Inspection requests. It shall be the duty of the holder of the permit or their duly authorized agent to notify the code official when work is ready for inspection. It shall be the duty of the permit holder to provide *access* to and means for inspections of such work that are required by this code.

❖ This section clarifies that it is the responsibility of the permit holder to arrange for the required inspections when the completed work is ready. It also establishes his or her responsibility for keeping the work open for inspection and providing all means needed to accomplish the inspections.

[A] 107.2.3 Approval required. Work shall not be done beyond the point indicated in each successive inspection without first obtaining the approval of the code official. The code official, upon notification, shall make the requested inspections and shall either indicate the portion of the construction that is satisfactory as completed, or notify the permit holder or his or her agent wherein the same fails to comply with this code. Any portions that do not comply shall be corrected and such portion shall not be covered or concealed until authorized by the code official.

❖ This section establishes that work cannot progress beyond the point of a required inspection without the code official's approval. Upon making the inspection, the code official must either approve the completed work or notify the permit holder or other responsible party of that which does not comply with the code. Approvals and notices of noncompliance must be in writing, as required by Section 104.3, to avoid any misunderstanding as to what is required. Any work not approved cannot be concealed until it has been corrected and approved by the code official.

[A] 107.2.4 Approved agencies. The code official is authorized to accept reports of *approved* inspection agencies, provided that such agencies satisfy the requirements as to qualifications and reliability.

❖ The determination as to whether to accept an agency test report rests with the code official and the reporting agency must be acceptable to the code official.

[A] 107.2.5 Evaluation and follow-up inspection services. Prior to the approval of a closed, prefabricated plumbing system and the issuance of a plumbing permit, the code official shall require the submittal of an evaluation report on each prefabricated plumbing system indicating the complete details of the plumbing system, including a description of the system and its components, the basis upon which the plumbing system is being evaluated, test results and similar information, and other data as necessary for the code official to determine conformance to this code.

❖ As an alternative to a physical inspection in the plant or location where prefabricated components are fabricated (such as modular homes, prefabricated structures, etc.), the code official has the option of accepting an evaluation report from an approved agency detailing such inspections. These evaluation reports can serve as the basis for code compliance.

[A] 107.2.5.1 Evaluation service. The code official shall designate the evaluation service of an *approved agency* as the evaluation agency, and review such agency's evaluation report for adequacy and conformance to this code.

❖ The code official is required to review all submitted reports for conformity to the applicable code requirements. If, in the judgment of the code official, the submitted reports are acceptable, he or she should document the basis for the approval.

[A] 107.2.5.2 Follow-up inspection. Except where *ready access* is provided to all plumbing systems, service equipment and accessories for complete inspection at the site without disassembly or dismantling, the code official shall conduct the frequency of in-plant inspections necessary to ensure conformance to the *approved* evaluation report or shall designate an independent, *approved* inspection agency to conduct such inspections. The inspection agency shall furnish the code official with the follow-up inspection manual and a report of inspections upon request, and the plumbing system shall have an identifying label permanently affixed to the system indicating that factory inspections have been performed.

❖ The owner is required to provide special inspections of fabricated assemblies at the fabrication plant. The code official or an approved inspection agency must conduct periodic in-plant inspections to ensure conformance to the approved evaluation report described in Section 107.2.5.1. Such inspections would not be required where the plumbing systems can be inspected completely at the job site.

[A] 107.2.5.3 Test and inspection records. Required test and inspection records shall be available to the code official at all times during the fabrication of the plumbing system and the erection of the building, or such records as the code official designates shall be filed.

❖ All testing and inspection records related to a fabricated assembly must be filed with the code official so he or she can maintain a complete and legal record of the assembly and erection of the building.

[A] 107.3 Special inspections. Special inspections of *alternative engineered design* plumbing systems shall be conducted in accordance with Sections 107.3.1 and 107.3.2.

❖ This section establishes that the design professional has to periodically inspect the alternative engineered design, keep records of such inspections and submit a final report to the code official certifying that all work conforms to the construction documents. Because of the unusual nature and possible complexity of alternative engineered plumbing systems, it is necessary for the system designer to be involved in the inspection process.

[A] 107.3.1 Periodic inspection. The registered design professional or designated inspector shall periodically inspect and observe the *alternative engineered design* to determine that the installation is in accordance with the *approved* construction documents. Discrepancies shall be brought to the immediate attention of the plumbing contractor for correction. Records shall be kept of all inspections.

❖ The registered design professional must periodically inspect the engineered plumbing system during installation to determine that the system conforms to the approved construction documents. This is an important step because the design professional can identify any deviations from the approved plans in the early stages of the plumbing work. The design professional must then advise the plumbing contractor of any problems so that corrective measures can be taken before needless costs are incurred, and labor and materials are wasted.

The design professional must compile a complete legal record of the project, which must include all inspections made, discrepancies found and resolutions of discrepancies. It is the responsibility of the design professional to document and submit inspection records and written certification in accordance with Section 107.3.2.

[A] 107.3.2 Written report. The registered design professional shall submit a final report in writing to the code official upon completion of the installation, certifying that the *alternative engineered design* conforms to the *approved* construction documents. A notice of approval for the plumbing system shall not be issued until a written certification has been submitted.

❖ After all work is completed, the design professional is required to inspect the entire alternative engineered plumbing system. The details of that inspection, including verification of compliance with the approved construction documents, must be submitted in writing to the code official before final approval can be granted.

[A] 107.4 Testing. Plumbing work and systems shall be tested as required in Section 312 and in accordance with Sections 107.4.1 through 107.4.3. Tests shall be made by the permit holder and observed by the code official.

❖ Visual inspection is not all that is required in the determination of a plumbing system's compliance with the code. This section establishes where and how testing is to be performed to disclose leaks and defects.

[A] 107.4.1 New, altered, extended or repaired systems. New plumbing systems and parts of existing systems that have been altered, extended or repaired shall be tested as prescribed herein to disclose leaks and defects, except that testing is not required in the following cases:

1. In any case that does not include addition to, replacement, alteration or relocation of any water supply, drainage or vent piping.

2. In any case where plumbing equipment is set up temporarily for exhibition purposes.

❖ Every plumbing system must be tested before it is placed into service. Testing is necessary to make sure that the system is free from leaks or other defects. Testing is also required, to the extent practicable, for portions of existing systems that have been repaired, altered or extended.

[A] 107.4.2 Equipment, material and labor for tests. Equipment, material and labor required for testing a plumbing system or part thereof shall be furnished by the permit holder.

❖ The permit holder is responsible for performing tests, as well as for supplying all of the labor, equipment and apparatus necessary to conduct such tests. The code official observes but never performs the test.

[A] 107.4.3 Reinspection and testing. Where any work or installation does not pass any initial test or inspection, the necessary corrections shall be made to comply with this code. The work or installation shall then be resubmitted to the code official for inspection and testing.

❖ If a system or portion thereof does not pass the initial test or inspection, all violations must be corrected and the system must be reinspected.

To encourage code compliance and cover the expense of the code official's time, many jurisdictions charge fees for inspections that are required subsequent to the first reinspection.

[A] 107.5 Approval. After the prescribed tests and inspections indicate that the work complies in all respects with this code, a notice of approval shall be issued by the code official.

❖ After the code official has performed the required inspections and observed the required equipment and system tests (or has received written reports of the results of such tests), he or she must determine whether the installation or work is in compliance with all applicable sections of the code. The code official must issue a written notice of approval if it has been determined that the subject plumbing work or installation is in apparent compliance with the code. The notice of approval is given to the permit holder, and a copy of the notice is retained on file by the code official.

[A] 107.5.1 Revocation. The code official is authorized to, in writing, suspend or revoke a notice of approval issued under the provisions of this code wherever the notice is issued in error, or on the basis of incorrect information supplied, or where it is determined that the building or structure, premise or portion thereof is in violation of any ordinance or regulation or any of the provisions of this code.

❖ This section is needed to give the code official the authority to revoke a notice of approval for the reasons indicated in the code text. The code official can suspend the notice until all of the code violations are corrected.

[A] 107.6 Temporary connection. The code official shall have the authority to authorize the temporary connection of the building or system to the utility source for the purpose of testing plumbing systems or for use under a temporary certificate of occupancy.

❖ The typical procedure for a local jurisdiction is to withhold the issuance of the certificate of occupancy until approvals have been received from each code official responsible for inspection of the structure. The code official is permitted to issue a temporary authorization to make connections to the public utility system prior to the completion of all work. The certification is intended to acknowledge that, because of seasonal limitations, time constraints, the need for testing or partial operation of equipment, some building systems may be connected even though the building is not suitable for final occupancy. The intent of this section is that a request for temporary occupancy or the connection and use of plumbing equipment or systems should not be denied when the requesting permit holder has demonstrated to the code official's satisfaction that the public health, safety and welfare will not be endangered.

The code official should view the issuance of a "temporary authorization or certificate of occupancy" as substantial an act as the issuance of the final certificate. Indeed, the issuance of a temporary certificate of occupancy offers a greater potential for conflict because once the building or structure is occupied, it is very difficult to remove the occupants through legal means.

[A] 107.7 Connection of service utilities. A person shall not make connections from a utility, source of energy, fuel, power, water system or *sewer* system to any building or system that is regulated by this code for which a permit is required until authorized by the code official.

❖ This section establishes the authority of the code official to approve utility connections to a building, such as water, sewer, electricity, gas and steam, and to require their disconnection when such approval has not been granted. For the protection of building occupants, including workers, such systems must have had final inspection approvals, except as allowed by Section 110.3 for temporary connections.

SECTION 108
VIOLATIONS

[A] 108.1 Unlawful acts. It shall be unlawful for any person, firm or corporation to erect, construct, alter, repair, remove, demolish or utilize any plumbing system, or cause same to be done, in conflict with or in violation of any of the provisions of this code.

❖ This section describes the citing, recording and subsequent actions pursuant to observed code violations. Violations of the code are prohibited; this is the basis for all citations and correction notices.

[A] 108.2 Notice of violation. The code official shall serve a notice of violation or order to the person responsible for the erection, installation, alteration, extension, repair, removal or demolition of plumbing work in violation of the provisions of this code, or in violation of a detail statement or the *approved* construction documents thereunder, or in violation of a permit or certificate issued under the provisions of this code. Such order shall direct the discontinuance of the illegal action or condition and the abatement of the violation.

❖ The code official is required to notify the person responsible for the erection or use of a building found to be in violation of the code. The section that is allegedly being violated must be cited so that the responsible party can respond to the notice.

[A] 108.3 Prosecution of violation. If the notice of violation is not complied with promptly, the code official shall request the legal counsel of the jurisdiction to institute the appropriate proceeding at law or in equity to restrain, correct or abate such violation, or to require the removal or termination of the unlawful occupancy of the structure in violation of the provisions of this code or of the order or direction made pursuant thereto.

❖ The code official must pursue, through the use of legal counsel of the jurisdiction, legal means to correct the violation. This is not optional.

Any extensions of time for voluntary correction of the violations must be for a reasonable, bona fide cause, or the code official may be subject to criticism for "arbitrary and capricious" actions. In general, it is better to have a standard time limitation for correction of violations. Departures from this standard must be for a clear and reasonable purpose, usually stated in writing by the violator.

[A] 108.4 Violation penalties. Any person who shall violate a provision of this code or shall fail to comply with any of the requirements thereof or who shall erect, install, alter or repair plumbing work in violation of the *approved* construction documents or directive of the code official, or of a permit or certificate issued under the provisions of this code, shall be guilty of a **[SPECIFY OFFENSE]**, punishable by a fine of not more than **[AMOUNT]** dollars or by imprisonment not exceeding **[NUMBER OF DAYS]**, or both such fine and imprisonment. Each day that a violation continues after due notice has been served shall be deemed a separate offense.

❖ A standard fine or other penalty as deemed appropriate by the jurisdiction is prescribed in this section. Additionally, this section identifies a principle that "each day that a violation continues shall be deemed a separate offense" for the purpose of applying the prescribed penalty in order to facilitate prompt resolution.

[A] 108.5 Stop work orders. Upon notice from the code official, work on any plumbing system that is being performed contrary to the provisions of this code or in a dangerous or unsafe manner shall immediately cease. Such notice shall be in writing and shall be given to the owner of the property, or to the owner's authorized agent, or to the person performing the work. The notice shall state the conditions under which work is authorized to resume. Where an emergency exists, the code official shall not be required to give a written notice prior to stopping the work. Any person who shall continue any work in or about the structure after having been served with a stop work order, except such work as that person is directed to perform to remove a violation or unsafe condition, shall be liable to a fine of not less than **[AMOUNT]** dollars or more than **[AMOUNT]** dollars.

❖ Upon receipt of a violation notice from the code official, the owner of the property, the owner's agent or the person doing the work must immediately cease all construction activities identified in the notice, except as expressly permitted to correct the violation. A stop work order can prevent a violation from becoming worse and more difficult or expensive to correct. However, it can result in inconvenience and monetary loss to the affected parties; therefore, justification must be evident and judgment must be exercised before such an order is issued.

A stop work order may be issued where work is proceeding without a permit. Hazardous conditions could develop where the code official is unaware of the nature of the work and a permit has not been issued. The issuance of a stop work order on a plumbing system often results from work done by the plumbing contractor that affects a nonplumbing component. For example, if a plumbing contractor cuts a structural element to install piping or fixtures, the structure may be weakened enough to cause a partial or complete structural failure. As determined by the adopting jurisdiction, a penalty may be assessed for failure to comply with this section, and it is to be inserted in the blanks provided.

[A] 108.6 Abatement of violation. The imposition of the penalties herein prescribed shall not preclude the legal officer of the jurisdiction from instituting appropriate action to prevent unlawful construction or to restrain, correct or abate a violation, or to prevent illegal occupancy of a building, structure or premises, or to stop an illegal act, conduct, business or utilization of the plumbing on or about any premises.

❖ Despite the assessment of a penalty in the form of a fine or imprisonment against a violator, the violation itself must still be corrected. Failure to make the necessary corrections will result in the violator being subject to additional penalties as described in the preceding section.

[A] 108.7 Unsafe plumbing. Any plumbing regulated by this code that is unsafe or that constitutes a fire or health hazard, insanitary condition, or is otherwise dangerous to human life is hereby declared unsafe. Any use of plumbing regulated by this code constituting a hazard to safety, health or public welfare by reason of inadequate maintenance, dilapidation, obsolescence, fire hazard, disaster, damage or abandonment is hereby declared an unsafe use. Any such unsafe equipment is hereby declared to be a public nuisance and shall be abated by repair, rehabilitation, demolition or removal.

❖ Unsafe conditions include those that constitute a health hazard, fire hazard, explosion hazard, shock hazard, asphyxiation hazard, physical injury hazard or are otherwise dangerous to human life and property.

In the course of performing duties, the code official may identify a hazardous condition. Such condition must be declared in violation of the code and, therefore, must be abated. For example, while inspecting a plumbing installation, the code official may notice a leaking soil pipe that is causing structural decay, a missing cleanout plug or a defective water-heater relief valve. Even though the defective plumbing may be unrelated to the installation that the code official was called to inspect, he or she is obligated to address the hazard and have it corrected.

[A] 108.7.1 Authority to condemn equipment. Whenever the code official determines that any plumbing, or portion thereof, regulated by this code has become hazardous to life, health or property or has become insanitary, the code official shall order in writing that such plumbing either be removed or restored to a safe or sanitary condition. A time limit for compliance with such order shall be specified in the written notice. A person shall not use or maintain defective plumbing after receiving such notice.

Where such plumbing is to be disconnected, written notice as prescribed in Section 108.2 shall be given. In cases of immediate danger to life or property, such disconnection shall be made immediately without such notice.

❖ When a plumbing system or plumbing equipment is determined to be unsafe, the code official is required to notify the owner or agent of the building as the first step in correcting the difficulty. Such notice is to describe the repairs and improvements necessary to correct the deficiency or require removal or replacement of the unsafe equipment or system. All such notices must specify a time frame in which the corrective actions must occur. Additionally, such notice should require the immediate response of the owner or agent. If he or she is not available, public notice of such declaration should suffice for the purposes of complying with this section. The code official may also determine that disconnection of the system is necessary to correct an unsafe condition and must give written notice to that effect (see commentary, Section 108.2) unless immediate disconnection is essential for public health and safety reasons (see commentary, Section 108.7.2).

[A] 108.7.2 Authority to disconnect service utilities. The code official shall have the authority to authorize disconnection of utility service to the building, structure or system regulated by the technical codes in case of an emergency, where necessary, to eliminate an immediate danger to life or property. Where possible, the owner or the owner's authorized agent and occupant of the building, structure or service system shall be notified of the decision to disconnect utility service prior to taking such action. If not notified prior to disconnecting, the owner, the owner's authorized agent or occupant of the building, structure or service systems shall be notified in writing, as soon as practical thereafter.

❖ The code official should have the authority to order disconnection of any plumbing supplied to a building, structure or equipment regulated by the code when it is determined that the equipment or any portion thereof has become an immediate danger. Written notice of an order to disconnect service and the causes therefore, should be given to the owner and the occupant of the building, structure or premises. However, disconnection should be done without such notice in cases of immediate danger to life or property.

[A] 108.7.3 Connection after order to disconnect. A person shall not make connections from any energy, fuel, power supply or water distribution system or supply energy, fuel or water to any equipment regulated by this code that has been disconnected or ordered to be disconnected by the code official or the use of which has been ordered to be discontinued by the code official until the code official authorizes the reconnection and use of such equipment.

Where any plumbing is maintained in violation of this code, and in violation of any notice issued pursuant to the provisions of this section, the code official shall institute any appropriate action to prevent, restrain, correct or abate the violation.

❖ Once the reason for discontinuation of use or disconnection of the plumbing system no longer exists, only the code official can authorize resumption of use or reconnection of the system after it is demonstrated to his or her satisfaction that all repairs or other work are in compliance with applicable sections of the code. This section also requires him or her to take action to abate code violations (see commentary, Section 108.2).

SECTION 109
MEANS OF APPEAL

[A] 109.1 Application for appeal. Any person shall have the right to appeal a decision of the code official to the board of appeals. An application for appeal shall be based on a claim that the true intent of this code or the rules legally adopted thereunder have been incorrectly interpreted, the provisions of this code do not fully apply, or an equally good or better form of construction is proposed. The application shall be filed on a form obtained from the code official within 20 days after the notice was served.

❖ This section holds that any aggrieved party with a material interest in the decision of the code official may challenge such a decision before a board of appeals. This provides a forum, other than the court of jurisdiction, in which to review the code official's actions.

This section literally allows any person to appeal a decision of the code official. In practice, this section has been interpreted to permit appeals only by aggrieved parties with a material or definitive interest in the decision of the code official. An aggrieved party may not appeal a code requirement per se. The intent of the appeal process is not to waive or set aside a code requirement; rather, it is to provide a means of reviewing a code official's decision on an interpretation or application of the code or to review the equivalency of protection to the code requirements.

[A] 109.2 Membership of board. The board of appeals shall consist of five members appointed by the chief appointing authority as follows: one for 5 years, one for 4 years, one for 3 years, one for 2 years and one for 1 year. Thereafter, each new member shall serve for 5 years or until a successor has been appointed.

❖ The board of appeals is to consist of five members appointed by the "chief appointing authority"—typically, the mayor or city manager. One member is to be appointed for five years, one for four, one for three, one for two and one for one year. This method of appointment allows for a smooth transition of board of appeals members, allowing continuity of action over the years.

[A] 109.2.1 Qualifications. The board of appeals shall consist of five individuals, one from each of the following professions or disciplines:

1. Registered design professional who is a registered architect; or a builder or superintendent of building construction with not less than 10 years' experience, 5 years of which shall have been in responsible charge of work.

2. Registered design professional with structural engineering or architectural experience.

3. Registered design professional with mechanical and plumbing engineering experience; or a mechanical and plumbing contractor with not less than 10 years' experience, 5 years of which shall have been in responsible charge of work.

4. Registered design professional with electrical engineering experience; or an electrical contractor with not less than 10 years' experience, 5 years of which shall have been in responsible charge of work.

5. Registered design professional with fire protection engineering experience; or a fire protection contractor with not less than 10 years' experience, 5 years of which shall have been in responsible charge of work.

❖ The board of appeals consists of five persons with the qualifications and experience indicated in this section. One must be a registered design professional (see Item 2) with structural or architectural experience. The others must be registered design professionals, construction superintendents or contractors with experience in various areas of building construction. These requirements are important in that technical people rule on technical matters. The board of appeals is not the place for policy or political deliberations. It is intended that these matters be decided purely on their technical merits, with due regard for state-of-the-art construction technology.

[A] 109.2.2 Alternate members. The chief appointing authority shall appoint two alternate members who shall be called by the board chairman to hear appeals during the absence or disqualification of a member. Alternate members shall possess the qualifications required for board membership, and shall be appointed for 5 years or until a successor has been appointed.

❖ This section authorizes the chief appointing authority to appoint two alternate members who are to be available if the principal members of the board are absent or disqualified. Alternate members must possess the same qualifications as the principal members and are appointed for a term of five years or until such time that a successor is appointed.

[A] 109.2.3 Chairman. The board shall annually select one of its members to serve as chairman.

❖ It is customary to determine chairmanship annually so that a regular opportunity is available to evaluate and either reappoint the current chairman or appoint a new one.

[A] 109.2.4 Disqualification of member. A member shall not hear an appeal in which that member has any personal, professional or financial interest.

❖ All members must disqualify themselves regarding any appeal in which they have a personal, professional or financial interest.

[A] 109.2.5 Secretary. The chief administrative officer shall designate a qualified clerk to serve as secretary to the board. The secretary shall file a detailed record of all proceedings in the office of the chief administrative officer.

❖ The chief administrative officer is to designate a qualified clerk to serve as secretary to the board. The secretary is required to record the proceedings using detailed records.

[A] 109.2.6 Compensation of members. Compensation of members shall be determined by law.

❖ Members of the board of appeals need not be compensated unless required by the local municipality or jurisdiction.

[A] 109.3 Notice of meeting. The board shall meet upon notice from the chairman, within 10 days of the filing of an appeal or at stated periodic meetings.

❖ The board must meet within 10 days of the filing of an appeal or at regularly scheduled meetings.

[A] 109.4 Open hearing. Hearings before the board shall be open to the public. The appellant, the appellant's representative, the code official and any person whose interests are affected shall be given an opportunity to be heard.

❖ All hearings before the board must be open to the public. The appellant, the appellant's representative, the code official and any person whose interests are affected must be heard.

[A] 109.4.1 Procedure. The board shall adopt and make available to the public through the secretary procedures under which a hearing will be conducted. The procedures shall not require compliance with strict rules of evidence, but shall mandate that only relevant information be received.

❖ The board is required to establish and make available to the public written procedures detailing how hearings are to be conducted. Additionally, this section provides that although strict rules of evidence are not applicable, the information presented must be deemed relevant.

[A] 109.5 Postponed hearing. When five members are not present to hear an appeal, either the appellant or the appellant's representative shall have the right to request a postponement of the hearing.

❖ When all five members of the board are not present, either the appellant or the appellant's representative may request a postponement of the hearing.

[A] 109.6 Board decision. The board shall modify or reverse the decision of the code official by a concurring vote of three members.

❖ A concurring vote of three members of the board is needed to modify or reverse the decision of the code official.

[A] 109.6.1 Resolution. The decision of the board shall be by resolution. Certified copies shall be furnished to the appellant and to the code official.

❖ A formal decision in the form of a resolution is required to provide an official record. Copies of this resolution are to be furnished to both the appellant and the code official. The code official is bound by the action of the board of appeals unless he or she thinks that the board of appeals has acted improperly. In such cases, relief through the court having jurisdiction may be sought by corporate council.

[A] 109.6.2 Administration. The code official shall take immediate action in accordance with the decision of the board.

❖ To avoid any undue hindrance in the progress of construction, the code official is required to act without delay based on the board's decision. This action may be to enforce the decision or to seek legislative relief if the board's action can be demonstrated to be inappropriate.

[A] 109.7 Court review. Any person, whether or not a previous party of the appeal, shall have the right to apply to the appropriate court for a writ of certiorari to correct errors of law. Application for review shall be made in the manner and time required by law following the filing of the decision in the office of the chief administrative officer.

❖ This section allows any person to request a review by the court of jurisdiction of perceived errors of law. Application for such review must be made after the decision of the board is filed with the chief administrative officer. This helps to establish the observance of due process for all concerned.

SECTION 110
TEMPORARY EQUIPMENT, SYSTEMS AND USES

[A] 110.1 General. The code official is authorized to issue a permit for temporary equipment, systems and uses. Such permits shall be limited as to time of service, but shall not be permitted for more than 180 days. The code official is authorized to grant extensions for demonstrated cause.

❖ The code official is permitted to issue temporary authorization to make connections to a public utility system prior to completion of all work. This acknowledges that, because of seasonal limitations, time constraints or the need for testing or partial operations of equipment, some building systems may be safely connected even though the building is not suitable for final occupancy. The temporary connection and utilization of connected equipment should be approved when the requesting permit holder has demonstrated to the code official's satisfaction that public health, safety and welfare will not be endangered.

[A] 110.2 Conformance. Temporary equipment, systems and uses shall conform to the structural strength, fire safety, means of egress, accessibility, light, ventilation and sanitary requirements of this code as necessary to ensure the public health, safety and general welfare.

❖ Even though a utility connection may be temporary, the only way to make sure that the public health, safety and general welfare are protected is for those temporary connections to comply with the code.

[A] 110.3 Temporary utilities. The code official is authorized to give permission to temporarily supply utilities before an installation has been fully completed and the final certificate of completion has been issued. The part covered by the temporary certificate shall comply with the requirements specified for temporary lighting, heat or power in the code.

❖ Commonly, the utilities on many construction sites are installed and energized long before all aspects of the system are completed. This section would allow such temporary systems to continue provided that they comply with the applicable safety provisions of the code.

[A] 110.4 Termination of approval. The code official is authorized to terminate such permit for temporary equipment, systems or uses and to order the temporary equipment, systems or uses to be discontinued.

❖ This section provides the code official with the necessary authority to terminate the permit for temporary equipment, systems and uses if conditions of the permit have been violated or if temporary equipment or systems pose an imminent hazard to the public. This enables the code official to act quickly when time is of the essence in order to protect public health, safety and welfare.

Bibliography

The following resource materials are referenced in this chapter or are relevant to the subject matter addressed in this chapter.

ASME A112.6.3-01, *Floor and Trench Drains*. New York: American Society of Mechanical Engineers, 2001.

CSA B79-94(2000), *Floor, Area and Shower Drains, and Cleanouts for Residential Construction*. Canadian Standards Association, 2000.

Chapter 2: Definitions

General Comments

The words or terms defined in this chapter are deemed to be of prime importance in both specifying the subject matter of code provisions and giving meaning to certain terms used throughout the code for administrative or enforcement purposes.

Purpose

Codes, by their very nature, are technical documents. As such, literally every word, term and punctuation mark can add to or change the meaning of the intended result. This is even more so with a performance code where the desired result often takes on more importance than the specific words.

Furthermore, the code, with its broad scope of applicability, includes terms inherent in a variety of construction disciplines. These terms can often have multiple meanings depending on the context or discipline being used at the time.

For these reasons, it is necessary to maintain a consensus on the specific meaning of terms contained in the code. Chapter 2 performs this function by stating clearly what specific terms mean for the purpose of the code.

SECTION 201 GENERAL

201.1 Scope. Unless otherwise expressly stated, the following words and terms shall, for the purposes of this code, have the meanings shown in this chapter.

❖ This section contains language and provisions that are supplemental regarding the use of Chapter 2. The subsections give guidance to the use of the defined words relevant to tense, gender, etc. Finally, this chapter provides the means to resolve those terms not defined.

201.2 Interchangeability. Words stated in the present tense include the future; words stated in the masculine gender include the feminine and neuter; the singular number includes the plural and the plural the singular.

❖ Although the definitions contained in Chapter 2 are to be taken literally, gender and tense are considered to be interchangeable. This is so that any grammatical inconsistencies within the code do not hinder the understanding or enforcement of the requirements.

201.3 Terms defined in other codes. Where terms are not defined in this code and are defined in the *International Building Code, International Fire Code, International Fuel Gas Code* or the *International Mechanical Code*, such terms shall have the meanings ascribed to them as in those codes.

❖ When a word or term appears in the code that is not defined in this chapter, other references may be used to find its definition, such as the other *International Codes®* in the International Code Council® (ICC®) family of codes, which are coordinated to prevent conflict among documents.

201.4 Terms not defined. Where terms are not defined through the methods authorized by this section, such terms shall have ordinarily accepted meanings such as the context implies.

❖ Another resource for defining words or terms not defined herein or in other codes is their "ordinarily accepted meanings." The intent of this statement is that a dictionary definition could suffice, provided that such definition refers to the context.

Some of the construction terms used throughout the code may not be defined in Chapter 2 or in a dictionary. In such a case, one would first turn to the definitions contained in the referenced standards (see Chapter 15) and then to published textbooks on the subject in question.

SECTION 202 GENERAL DEFINITIONS

ACCEPTED ENGINEERING PRACTICE. That which conforms to accepted principles, tests or standards of nationally recognized technical or scientific authorities.

❖ The code makes frequent reference to a variety of consensus standards. Where a requirement is not based on a specific standard but on a body of knowledge of a particular engineering or construction discipline, this term is used.

It is also understood that where the code is silent on a subject, any rules that could be applicable in the interest of public safety must be in accordance with that body of knowledge called "accepted engineering practice." The code official can enforce this concept through the use of Section 102.9, titled "Requirements not covered by code." Care and caution need to be used by the code official when enforcing the "accepted engineering practice" concept to be sure that the concept is sound and safe.

[M] ACCESS (TO). That which enables a fixture, appliance or equipment to be reached by ready *access* or by a means that first requires the removal or movement of a panel, door or similar obstruction (see "Ready *access*").

❖ Access to plumbing fixtures, appliances and appurtenances is necessary to facilitate inspection, observation, maintenance, adjustment, repair or replacement. Access to something means that it can be easily physically reached without having to remove a permanent portion of the structure. It is acceptable, for example, to install a device in a space that would require removal of lay-in suspended ceiling panels to gain access. A fixture, appliance or device would not be considered accessible if it were necessary to remove or open any portion of a structure other than panels, doors, covers or similar obstructions intended to be removed or opened normally, without special tools or knowledge. See the definition of "Ready access."

Access can be described as the capability of being reached or approached for inspection, observation, maintenance, adjustment, repair or replacement. Achieving access could require the removal or opening of a panel, door or similar obstruction and might require overcoming an obstacle such as elevation. When access is being provided to an emergency shut-off valve, the designer should be aware that more damage can occur during the time it takes to actually "access" the valve.

ACCESS COVER. A removable plate, usually secured by bolts or screws, to permit *access* to a pipe or pipe fitting for the purposes of inspection, repair or cleaning.

❖ A removable access cover is normally installed to conceal a plumbing fitting that must be inspected, serviced, repaired or cleaned through a finished material or surface. Where the floor or paved surface is subject to pedestrian or vehicular traffic, the access cover must be structurally capable of supporting the load applied to it. Most access covers are secured with either fasteners, such as bolts or screws, or a simple latching mechanism. In special installations where security or safety is a concern, the method for securing the access cover could become more extensive and complicated. These special installations should be limited in number wherever possible because of the concern over the time it takes to access the fitting behind the cover.

Access covers placed in finished walls are often set flush with the face of the wall. Some access covers have provisions that allow for insertion of a portion of the actual wall finish material to help conceal the access cover. Where the access cover is in a floor surface, it should be installed flush so that it does not create a tripping hazard.

ADAPTER FITTING. An *approved* connecting device that suitably and properly joins or adjusts pipes and fittings that do not otherwise fit together.

❖ Where approved by the code official, these fittings are used to join pipes that are of different diameters, different materials or both. Adapter fittings are used to convert from one joining method to another. Refer to Sections 605 and 705 for specific information related to joints.

AIR ADMITTANCE VALVE. One-way valve designed to allow air to enter the plumbing drainage system when negative pressures develop in the piping system. The device shall close by gravity and seal the vent terminal at zero differential pressure (no flow conditions) and under positive internal pressures. The purpose of an air admittance valve is to provide a method of allowing air to enter the plumbing drainage system without the use of a vent extended to open air and to prevent *sewer* gases from escaping into a building.

❖ An air admittance valve (AAV) is a device used as an alternative method for the venting of traps for fixtures and waste receptors. AAVs reduce the amount of vent piping required to terminate outdoors. An AAV is designed to open and admit air into the drainage system when negative drain system pressures occur. An AAV closes by gravity and seals the vent terminal when the drain system pressure is equal to or exceeds atmospheric pressure. The seal prevents sewer gas from entering the building [see Commentary Figures 202(1) and 202(2)].

It is important to note that a mechanical vent is not considered an air admittance valve [see Commentary Figure 202(3)]. A mechanical vent is a device that operates in a manner similar to that of an air admittance valve. The mechanical vent opens to admit air under negative pressure conditions and closes under zero and positive pressure conditions. It cannot, however, be classified as an air admittance valve because it closes against the force of gravity by a lightweight wire spring. The use of a spring to overcome the forces of gravity to return the seal to the closed position introduces a higher probability of device failure than that of an air admittance valve.

AIR BREAK (Drainage System). A piping arrangement in which a drain from a fixture, appliance or device discharges indirectly into another fixture, receptacle or interceptor at a point below the *flood level rim* and above the trap seal.

❖ An air break is an indirect drainage method in which waste discharges to the drainage system through piping that terminates below the flood level rim and above the trap seal of an approved receptor. An air break is commonly used to protect mechanical equipment from sewage backup in the event that stoppage occurs. It also protects the drainage system from adverse pressure conditions caused by pumped discharge [see Commentary Figure 202(4)].

AIR GAP (Drainage System). The unobstructed vertical distance through the free atmosphere between the outlet of the waste pipe and the *flood level rim* of the receptacle into which the waste pipe is discharging.

❖ An air gap (drainage system) is a type of indirect waste arrangement where the waste is discharged to the drainage system through piping that terminates at a specified distance above the flood level rim of an approved receptor.

An air gap is commonly installed in drain lines serving equipment that is used in the preparation, storage and service of food; the conveyance of potable water; and the sterilization of medical equipment. The air gap serves as an impossible barrier for sewage to overcome in the event that stoppage occurs in the receptor drain, because sewage backup would overflow the receptor drain flood level before it came in contact with the drain line above [see Commentary Figure 202(4)].

DEFINITIONS

Figure 202(1)
AIR ADMITTANCE VALVE OPENS

Figure 202(2)
AIR ADMITTANCE VALVE CLOSES

DEFINITIONS

Figure 202(4)
AIR GAP, AIR BREAK

AIR GAP (Water Distribution System). The unobstructed vertical distance through the free atmosphere between the lowest opening from any pipe or faucet supplying water to a tank, plumbing fixture or other device and the *flood level rim* of the receptacle.

❖ An air gap is the most reliable and effective means of backflow protection. It is simply the vertical airspace between the potable water supply outlet and the possible source of contamination. This air gap prevents possible contamination of the potable water supply by preventing the supply outlet, such as a faucet, from backsiphoning waste water, such as from a basin. Many manufacturers have developed air gap fittings that provide a rigid connection to the drainage system while maintaining the minimum level of protection from contamination.

ALTERNATE ON-SITE NONPOTABLE WATER. Nonpotable water from other than public utilities, on-site surface sources and subsurface natural freshwater sources. Examples of such water are gray water, on-site reclaimed water, col-

Figure 202(3)
MECHANICAL VENT
NOT CODE COMPLIANT

lected rainwater, captured condensate and rejected water from reverse osmosis systems.

❖ A common problem in potable water-saving designs that use nonpotable water in buildings and on building sites is the understanding and naming of the different classes of nonpotable water. The term "reclaimed water" (also known as "recycled water") has been reserved for nonpotable water that is delivered to the building site by municipal water delivery systems (see the definition of "Reclaimed water"). The term "on-site nonpotable water" is too encompassing, as this means well water that has not been treated to be potable, and gray water from showers and lavatories. The handling and treatment protocols and concerns for these two sources are quite different. Thus, municipally reclaimed water, on-site surface sources and subsurface natural freshwater sources are excluded from this definition. All other on-site nonpotable water is considered to be alternate on-site nonpotable water.

ALTERNATIVE ENGINEERED DESIGN. A plumbing system that performs in accordance with the intent of Chapters 3 through 12 and provides an equivalent level of performance for the protection of public health, safety and welfare. The system design is not specifically regulated by Chapters 3 through 12.

❖ Plumbing systems designed by accepted engineering practice (as previously defined in this section) to provide the same performance as achieved by conventional plumbing design methods are considered as alternative engineered designs. An alternative engineered design could include water supply and distribution, sanitary and storm drainage, venting, traps, fixtures and connections, devices and appurtenances. Alternative engineered designs must be at least equivalent to that prescribed by Chapters 3 through 14 (see commentary, Section 105.2).

ANCHORS. See "Supports."

❖ Anchors are the principal components of a support device or system that affix the support device or system to the building structure. Perhaps the most common use of the term "anchor" in the building industry is for indicating a "concrete anchor," which is an imbedded (either integrally cast-in-place or drilled-and-inserted) fastener. However, the term can also include nails, screws, bolts or adhesives used in conjunction with fastening the support device or system to the structure.

ANTISIPHON. A term applied to valves or mechanical devices that eliminate siphonage.

❖ This term refers to the function of certain valves and devices used to break or prevent the siphon effect that can be created in plumbing systems.

[A] APPROVED. Acceptable to the code official.

❖ As related to the process of acceptance of building installations, including materials, equipment and construction systems, this definition identifies where the ultimate authority rests. Wherever this term is used, it intends that only the enforcing authority can accept a specific installation or component as complying with the code.

[A] APPROVED AGENCY. An established and recognized agency that is regularly engaged in conducting tests or furnishing inspection services, where such agency has been *approved* by the code official.

❖ Third-party testing or inspections may be needed for elements within the built environment. The basis for the code official's approval of any agency for a particular activity may include, but is not necessarily limited to, the capacity and capability of the agency to perform the work in accordance with Section 303 and other applicable sections. This is typically accomplished through a review of the résumés and references of the agency and its personnel.

AREA DRAIN. A receptacle designed to collect surface or storm water from an open area.

❖ An area drain is a storm water and surface drainage collection point installed in outdoor open areas and paved areas, such as driveways, patios, outside stairwells, window wells and parking lots. These devices are placed at points of lower elevation to allow water to flow toward them. Additionally, this location causes debris, ice, etc., to flow toward them. Therefore, attention must be given to their design and installation to prevent them from becoming inoperative, often at the time when they are most needed.

ASPIRATOR. A fitting or device supplied with water or other fluid under positive pressure that passes through an integral orifice or constriction, causing a vacuum. Aspirators are also referred to as suction apparatus, and are similar in operation to an ejector.

❖ Within the context of this code, an aspirator is a device that uses a flowing liquid to create a reduced pressure condition. A simple example of an aspirator is a laboratory filter pump. Where water is forced through a smooth tube having a zone with an abrupt restricted area, the flow velocity will necessarily increase in the restricted area, thereby creating a reduced pressure condition at the restriction. Highly effective aspirators can produce vacuums approaching zero atmospheres (perfect vacuum).

Since an aspirator has no moving parts and only requires a flow of liquid to function, they are an economic method for drawing liquids (or gases) into the flow stream. Common examples of aspirators can be found in many food service facilities to "inject" cleaning chemicals for the dishwasher or for mop bucket filling. The flow of water through the aspirator draws the chemical into the flow stream without the need for pumps. Where an aspirator device is connected to a potable water supply, a backflow protection device must be installed between the water supply and the aspirator due to the connection of a contaminant to the aspirator. Most manufacturers of chemical dispensing equipment that are designed to use internal aspirators will also provide integral backflow protection within the equipment.

BACKFLOW. Pressure created by any means in the water distribution system, which by being in excess of the pressure in the water supply mains causes a potential backflow condition.

❖ Backflow is the reversal of flow from its intended direction. In the context of this code, backflow refers to a reversed flow condition that occurs in a closed water piping system caused by an event or device within the system. Since movement of liquids in a pipe is caused by a difference in pressure, in order for flow to reverse, the event or device must cause a change in pressure that is either greater than or less than the differential pressure conditions that cause flow in a normal direction (see "Backflow preventer").

Backpressure, low-head. A pressure less than or equal to 4.33 psi (29.88 kPa) or the pressure exerted by a 10-foot (3048 mm) column of water.

❖ Low-head backpressure is created where the source of backpressure comes from a hose elevated to a level that produces 10 feet (3048 mm) or less of water column at the outlet. A hose connection, such as a wall hydrant or faucet, is where "low-head backpressure" is typically experienced (also see the commentary to Section 608).

Backsiphonage. The backflow of potentially contaminated water into the potable water system as a result of the pressure in the potable water system falling below atmospheric pressure of the plumbing fixtures, pools, tanks or vats connected to the potable water distribution piping.

❖ Backsiphonage occurs where the pressure within the potable water distribution system drops below atmospheric (subatmospheric) pressure or negative gauge pressure. It, like backpressure, allows the normal direction of flow to be reversed. A siphon can result in an unprotected cross connection, causing contamination or pollution of the potable water supply. The main difference between backsiphonage and backpressure is the pressure in the system. Water distribution systems having a pressure less than zero gauge have the potential to create a siphon, resulting in backsiphonage. Where backpressure typically occurs in a closed system, backsiphonage can occur in an open system or a system that is open to the atmosphere.

Drainage. A reversal of flow in the drainage system.

❖ Backflow drainage is the flow of waste in the drainage system opposite the direction intended, regardless of the cause. Backflow is generally the result of drainage system stoppages, overloads or a higher pressure on the drainage pipe side of the trap. Drainage backflow can result in interior flooding, contamination of fixtures and contamination of the potable water system.

Water supply system. The flow of water or other liquids, mixtures or substances into the distribution pipes of a potable water supply from any source except the intended source.

❖ The flow of water or other liquids, mixtures or substances into the distribution pipes of a potable water supply from any source except the intended source.

BACKFLOW CONNECTION. Any arrangement whereby backflow is possible.

❖ Backflow is a possibility in most, if not all, piping systems. The theory of backflow prevention is based on the possibility, not on the probability or likelihood, of an event.

BACKFLOW PREVENTER. A backflow prevention assembly, a backflow prevention device or other means or method to prevent backflow into the potable water supply.

❖ There are three distinct classes of backflow equipment: assemblies, devices and other means or methods. This class distinction clarifies that an assembly is field testable, a device is not testable in the field and a mean or method is observable for their adequacy to protect the potable water supply from backflow. Table 608.1 was reorganized for the 2015 edition to group backflow preventers into these three classes. This change (along with changes to Table 608.1) should make the code user more aware of the suitability of particular backflow equipment after the initial installation. In other words, if a verification of the backflow protection is needed year after year, either an assembly would be chosen for its field testability or other means or methods would be chosen so that installation conditions can be observed.

BACKWATER VALVE. A device or valve installed in the *building drain* or *sewer* pipe where a *sewer* is subject to backflow, and that prevents drainage or waste from backing up into a lower level or fixtures and causing a flooding condition.

❖ Backwater valves are a type of check valve designed for installation in drainage piping. The valves have a lower invert downstream of the flapper to help prevent solids from interfering with valve closure.

[BS] BASE FLOOD ELEVATION. A reference point, determined in accordance with the building code, based on the depth or peak elevation of flooding, including wave height, which has a 1 percent (100-year flood) or greater chance of occurring in any given year.

❖ The base flood elevation is the height to which floodwaters are predicted to rise during passage or occurrence of a base flood. The Federal Emergency Management Agency (FEMA) uses commonly accepted computer models that estimate hydrologic and hydraulic conditions to determine the 1-percent annual chance (base) flood. Along rivers and streams, statistical methods and computer models may have been used to estimate runoff and develop flood elevations. The models take into consideration watershed characteristics and the shape and nature of the flood plain, including natural ground contours and the presence of buildings, bridges and culverts. Along coastal areas, base flood elevations can be developed using models that take into account offshore bathymetry, historical storms and typical wind patterns. In many coastal areas, the base flood elevation includes wave heights.

BATHROOM GROUP. A group of fixtures consisting of a water closet, lavatory, bathtub or shower, including or excluding a bidet, an *emergency floor drain* or both. Such fixtures are located together on the same floor level.

❖ This term is used in Table 709.1 and Section 912.1. Special consideration is given to a bathroom group when sizing drain, waste and vent piping because of the probability of fixture usage. A bathroom group has historically referred to a single water closet, a lavatory and a bathtub or shower that are all located within a single room. Such an arrangement would normally allow only one occupant in the room; therefore, the likelihood of simultaneous fixture discharge is remote. For the fixture usage theory of the bathroom group to be valid, the fixtures in the group must be located within the same bathroom. A bidet has been added to the group of fixtures considered as a bathroom group. A bathroom group may or may not contain a bidet, an emergency floor drain or both, and in any case, the drainage fixture unit (dfu) value is the same (see commentary, Table 709.1).

BEDPAN STEAMER OR BOILER. A fixture utilized for scalding bedpans or urinals by direct application of steam or boiling water.

❖ A bedpan steamer or boiler is a special health care fixture designed to sterilize bedpans by the direct application of steam or boiling water. Bedpan washers and sterilizers combine the functions of a human waste elimination fixture and a sterilizing fixture.

As with all hose connections to the potable water supply, the potential for backflow is present, and the flushing devices and hand-held hose apparatus must be protected against backflow in accordance with Section 608.

BEDPAN WASHER AND STERILIZER. A fixture designed to wash bedpans and to flush the contents into the sanitary drainage system. Included are fixtures of this type that provide for disinfecting utensils by scalding with steam or *hot water*.

❖ See the definition of "Bedpan steamer or boiler."

BEDPAN WASHER HOSE. A device supplied with hot and cold water and located adjacent to a water closet or clinical sink to be utilized for cleansing bedpans.

❖ A bedpan washer hose consists of a flexible hose connected on one end to a source of water and having a spray head with a shutoff valve on the other end. The flexibility of the hose allows the user to flush all inside areas of a bedpan without having to change the position of the bedpan being held above a water closet or clinical sink.

BRANCH. Any part of the piping system except a riser, main or *stack*.

❖ Risers, stacks and mains (including a building drain) are the principal arteries of water and waste plumbing systems. Pipes connecting to these arteries are called branches. Pipes connecting to those branches are also called branches except a fixture drain pipe does not become a fixture branch until it merges with other fixture drain piping (see commentary to "Fixture drain"). A branch could be the same size or smaller than the pipe to which it connects. Branches in plumbing systems are analogous to branches of trees.

BRANCH INTERVAL. A vertical measurement of distance, 8 feet (2438 mm) or more in *developed length*, between the connections of horizontal *branches* to a drainage *stack*. Measurements are taken down the *stack* from the highest horizontal *branch* connection.

❖ One branch interval corresponds to a length of a drainage stack that has two horizontal branch connections at least 8 feet (2438 mm) apart. In typical multistory construction practice with 8-foot (2438 mm) ceilings, horizontal branch drains connect just below each floor level to a vertical stack; therefore, a branch interval will typically correspond to each floor in the building, below the top floor. If the lowest floor fixtures drain into a horizontal branch drain (and not the building drain), the number of branch intervals is the number of floors minus one, assuming that each floor has fixtures that discharge to the stack. The numbering of branch intervals starts at the uppermost horizontal branch connection to the stack and proceeds down the stack.

Note that the number of branch intervals does not correspond to the number of stories in a building. Because stories are not limited in height, multiple horizontal branches could enter the stack within a story and some floors might not have any horizontal branch connections. See Commentary Figures 202(5) and 202(6) illustrating where branch intervals do and do not exist.

Branch intervals are a design factor for drainage pipe sizing and venting design. Drain, waste and vent (DWV) system design must consider the nature of waste and airflow in a stack and the effects that branch connections have on such flow. [see Commentary Table 710.1(2) and Sections 904.2 and 908].

BRANCH VENT. A vent connecting one or more individual vents with a vent *stack* or *stack* vent.

❖ Commentary Figure 202(7) illustrates a branch vent.

[A] BUILDING. Any structure occupied or intended for supporting or sheltering any *occupancy*.

❖ Where this term is used in the code, it means a structure used to provide shelter or support for some activity or occupancy.

DEFINITIONS

For SI: 1 inch = 25.4 mm, 1 foot = 304.8 mm.

**Figure 202(5)
DEFINITION OF A BRANCH INTERVAL**

BUILDING DRAIN. That part of the lowest piping of a drainage system that receives the discharge from soil, waste and other drainage pipes inside and that extends 30 inches (762 mm) in *developed length* of pipe beyond the exterior walls of the building and conveys the drainage to the *building sewer*.

❖ Building drains are usually considered to be the main drain of a drainage system within a structure. Building drains are horizontal (including vertical offsets) and are the portion of the drainage system that is at the lowest elevation in the structure. All horizontal drains above the elevation of the building drain are horizontal branches. The building drain terminates at the point where it exits the building [see commentary, Table 710.1(1)].

Combined. A *building drain* that conveys both sewage and storm water or other drainage.

❖ See the definitions of "Building drain" and "Building sewer, combined."

Sanitary. A *building drain* that conveys sewage only.

❖ The sanitary building drain is the main drain of the sanitary drainage system and does not conduct storm drainage [see Commentary Figure 202(8)].

Storm. A *building drain* that conveys storm water or other drainage, but not sewage.

❖ The storm building drain is the main drain of the storm drainage system and does not conduct sewage (see the definition of "Building drain").

BUILDING SEWER. That part of the drainage system that extends from the end of the *building drain* and conveys the discharge to a *public sewer*, *private sewer*, individual sewage disposal system or other point of disposal.

❖ The building sewer is the extension of the building drain and is located entirely outside of the building exterior walls [see commentary, Table 710.1(1) and Commentary Figure 202(8)].

Combined. A *building sewer* that conveys both sewage and storm water or other drainage.

❖ Combination storm and sanitary sewers are no longer constructed because of the burden placed on sewage treatment plants. More than 1,100 cities housing 42 million people nationwide, however, are served by existing combination storm and sanitary sewers. Storm water does not require treatment; therefore, combining it with sanitary drainage unnecessarily increases the volume of influent to treatment facilities (see commentary, Section 1101.3). Controlled-flow storm drainage systems should be considered in all combined systems to minimize the impact on municipal sewage disposal plants.

Sanitary. A *building sewer* that conveys sewage only.

❖ The sanitary sewer extends from the point of connection to the sanitary building drain to the point of connection to the public sewer system or the private sewage disposal system. Most jurisdictions arbitrarily set a distance such as 3 or 5 feet (914 or 1524 mm) outside of the building foundation as the point at which the building drain becomes the building sewer [see Commentary Figure 202(8)].

DEFINITIONS

Figure 202(6)
BRANCH INTERVALS

DEFINITIONS

Figure 202(7)
BRANCH VENT

Figure 202(8)
DRAINAGE SYSTEM COMPONENTS

Storm. A *building sewer* that conveys storm water or other drainage, but not sewage.

❖ The storm sewer extends from the point of connection to the storm building drain to the point of connection to the public storm sewer or other point of disposal (see the definition of "Sewer, sanitary"). Storm sewers are prohibited from being combined with a sanitary sewer.

BUILDING SUBDRAIN. That portion of a drainage system that does not drain by gravity into the *building sewer*.

❖ Building subdrains must discharge to sumps or receivers equipped with pumps or ejectors to lift the influent to the elevation of the gravity sewer.

BUILDING TRAP. A device, fitting or assembly of fittings installed in the *building drain* to prevent circulation of air between the drainage system of the building and the *building sewer*.

❖ A building trap (house trap) is a "running" trap that prevents the circulation of sewer gases between the building drain and the building sewer. The installation of building traps is prohibited in Chapter 10 as they could be a source of drainage blockages.

CIRCUIT VENT. A vent that connects to a horizontal drainage *branch* and vents two traps to a maximum of eight traps or trapped fixtures connected into a battery.

❖ A circuit vent is a means of venting multiple fixtures using only one or two vents. The horizontal drainage branch actually serves as a wet vent for the fixtures located downstream of the circuit vent connection (see Commentary Figure 914.1.1).

CIRCULATING HOT WATER SYSTEM. A specifically designed water distribution system where one or more pumps are operated in the service hot water piping to circulate heated water from the water-heating equipment to fixture supply and back to the water-heating equipment.

❖ This is a piping arrangement to provide hot (or tempered) water at plumbing fixtures in a manner that does not require the user to wait any appreciable amount of time for the heated water to begin discharging from the water outlet. The hot (or tempered) water is circulated (by a pump) in a "loop" of piping that is routed in close proximity to each fixture and then back to the heated water source. Ideally, the fixture supply piping from this "loop" piping to the fixture is relatively short such that it only takes a few seconds for the heated water to arrive at the outlet when the faucet or mixing valve is opened. Large buildings such as hotels with centralized water heaters and hot water storage tanks are often piped in this manner so that the occupants don't have to wait for hot (or tempered) water. Wasting cool potable water down the drain (to get hot water to arrive) is costly for the building owner and time consuming for the occupants. Generally, the hot (or tempered) water is circulated in the piping "loop" for 24 hours a day for buildings that are occupied on a 24-hour basis. For buildings that are only occupied during "work hours," the circulation pumps can be shut off when building occupancy is at a minimum. Time switches typically control these type of intermittently operated pumps (see the definition of "Demand recirculation water system").

CISTERN. A small covered tank for storing water for a home or farm. Generally, this tank stores rainwater to be utilized for purposes other than in the potable water supply, and such tank is placed underground in most cases.

❖ A cistern is a reservoir, usually placed underground, used for the storage of rainwater. Once common, cisterns are rarely used today because of the availability of potable water supplies from wells and public water distribution systems. Cisterns were used as the primary or supplemental source of water for drinking, cooking, bathing and washing. When cisterns are used, care needs to be taken to prevent contamination during construction, maintenance and use.

CLEANOUT. An access opening in the drainage system utilized for the removal of obstructions. Types of cleanouts include a removable plug or cap, and a removable fixture or fixture trap.

❖ Cleanouts are broadly defined as any access opening point into the drainage system for the purpose of removing a clog or other obstruction in the drain line. A cleanout is intended to provide convenient access to the piping interior without significant disassembly of the plumbing installation. A "P" trap with a removable U-bend is typically acceptable as a cleanout for the fixture drain or branch to which the trap discharges. Although a water closet connection (flange) is permitted to serve as a cleanout, such use will require removal and reinstallation of the water closet. The installation of a water closet is an activity that requires a permit, whereas the clearance of drain stoppages does not (see commentary, Sections 106.1 and 106.2 and Commentary Figure 708.1). Removal of "fixed" objects should occur only as a last resort when no other cleanouts are available.

[A] CODE. These regulations, subsequent amendments thereto, or any emergency rule or regulation that the administrative authority having jurisdiction has lawfully adopted.

❖ The adopted regulations are generally referred to as "the code" and include not only the *International Plumbing Code*® (IPC®), but any adopted modifications to the code and all other associated building codes, related rules and regulations promulgated and enacted by the jurisdiction.

[A] CODE OFFICIAL. The officer or other designated authority charged with the administration and enforcement of this code, or a duly authorized representative.

❖ The statutory power to enforce the code is normally vested in a building department (or the like) of a state, county or municipality whose designated enforcement officer is termed the "code official" (see commentary, Section 103).

COLLECTION PIPE. Unpressurized pipe used within the collection system that drains on-site nonpotable water or rainwater to a storage tank by gravity.

❖ A different term, other than drainage pipe, is used to describe piping that conveys, by gravity, on-site nonpotable water and rainwater to a storage tank to distinguish such piping from drainage piping that carries waste water.

COMBINATION FIXTURE. A fixture combining one sink and laundry tray or a two- or three-compartment sink or laundry tray in one unit.

❖ These fixtures have more than one waste outlet and, depending on their design, are allowed to be served by a single trap (see commentary, Section 1002.1, Exceptions 2 and 3). Combination fixtures include two- and three-well sinks, such as pots-and-pans sinks, laundry sinks, bar sinks and culinary sinks.

COMBINATION WASTE AND VENT SYSTEM. A specially designed system of waste piping embodying the horizontal wet venting of one or more sinks, lavatories, drinking fountains or floor drains by means of a common waste and vent pipe adequately sized to provide free movement of air above the flow line of the drain.

❖ This term refers to a system of waste piping in which the horizontal wet venting of floor drains, drinking fountains, sinks and lavatories is accomplished by means of a common waste and vent pipe that is oversized to provide free movement of air above the waste flow line of the drain.

COMBINED BUILDING DRAIN. See "*Building drain, combined.*"

COMBINED BUILDING SEWER. See "*Building sewer, combined.*"

COMMON VENT. A vent connecting at the junction of two *fixture drains* or to a fixture *branch* and serving as a vent for both fixtures.

❖ Where two fixtures are connected to the same drainage pipe, a common vent could be used. This common vent is sized and classified as an individual vent. Any two fixtures may be common vented to either a vertical or horizontal drainage pipe.

The most typical form of common venting is where two fixtures are connected at the same level. Two fixtures connecting at different levels but within the same story are still considered to be common vented. When one of the fixtures connected is at a different level, the vertical drain between the fixtures is oversized because it functions as both a drain for the upper fixture and a vent for the lower fixture.

Additional information on common venting is found in the commentary to Section 911 and Commentary Figure 911.1.

CONCEALED FOULING SURFACE. Any surface of a plumbing fixture that is not readily visible and is not scoured or cleansed with each fixture operation.

❖ The concealed fouling surface is any part of the plumbing fixture that is not readily visible and cleanable by the occupants. This is of concern because of the possible contamination of these surfaces by waste and the resulting growth of bacteria. The amount of concealed fouling surface is to be kept to a minimum. This is controlled in the standards for each type of fixture. As the amount of water used with each operation of the fixture is reduced for conservation purposes, the possibility for contamination of these concealed fouling surfaces increases. This concern needs to be addressed in the related fixture standards.

CONDUCTOR. A pipe inside the building that conveys storm water from the roof to a storm or combined *building drain*.

❖ A conductor is an interior vertical drain pipe (including horizontal offsets) that generally serves roof drains (also see the definition of "Leader").

[A] CONSTRUCTION DOCUMENT. All of the written, graphic and pictorial documents prepared or assembled for describing the design, location and physical characteristics of the elements of the project necessary for obtaining a building permit. The construction drawings shall be drawn to an appropriate scale.

❖ For the code official to determine that proposed construction is in compliance with code requirements, sufficient information must be submitted for review. This typically consists of the drawings (floor plans, elevations, sections, details, etc.), specifications and product information describing the proposed work. This term replaces the less descriptive "plans and specifications" used in earlier editions of the codes.

CONTAMINATION. An impairment of the quality of the potable water that creates an actual hazard to the public health through poisoning or the spread of disease by sewage, industrial fluids or waste.

❖ The Environmental Protection Agency (EPA) sets maximum levels for various chemicals and bacteria in drinking water. When an unacceptable level of one or more contaminants is present, the water is considered to be nonpotable. One of the primary purposes of the code is to protect potable water supplies from contamination. Contamination consists of either hazardous chemicals or raw sewage and is considered by the code to represent a high hazard when present in the water supply. Refer to Table 608.1 for the required potable protection requirements related to high-hazard situations.

The purpose of the code and other related regulations, such as those from the EPA, is to reduce, if not totally eliminate, the risk to the general public of poisoning or health impairment through sewage, industrial fluids or waste entering the potable water supply. It should be noted that these contaminants would have a harmful effect on the occupants of a building should such water be consumed. Contamination is quite different from pollution in that polluted water is still considered to be nonpotable, but would not kill or harm if ingested. Refer to the definition of "Pollution."

CRITICAL LEVEL (C-L). An elevation (height) reference point that determines the minimum height at which a backflow preventer or vacuum breaker is installed above the *flood level rim* of the fixture or receptor served by the device. The critical level is the elevation level below which there is a potential for backflow to occur. If the critical level marking is not indicated on the device, the bottom of the device shall constitute the critical level.

❖ The critical level determines the installation height of vacuum breaker backflow-preventer devices. This is referred to as critical because whether or not the vacuum breaker will function as intended is dependent on that level (see commentary, Section 608.13.6). The critical level (C-L) will be either marked on the device or stated in the manufacturer's installation instructions.

CROSS CONNECTION. Any physical connection or arrangement between two otherwise separate piping systems, one of which contains potable water and the other either water of unknown or questionable safety or steam, gas or chemical, whereby there exists the possibility for flow from one system to the other, with the direction of flow depending on the pressure differential between the two systems (see "Backflow").

❖ Cross connections are the links through which it is possible for contaminating materials to enter a potable water supply. The contaminant enters the potable water supply when the pressure of the polluted source exceeds the pressure of the potable source. The action may be called "backsiphonage" or "backpressure." Many serious outbreaks of illness and disease have been traced to cross connections. The intent of the code is to eliminate cross connections or prevent backflow where cross connections cannot be eliminated. Also refer to the definitions and commentary for the terms "Backflow," "Backsiphonage" and "Backpressure."

DEMAND RECIRCULATION WATER SYSTEM. A water distribution system where one or more pumps prime the service hot water piping with heated water upon a demand for hot water.

❖ This definition mirrors the same definition in the *International Energy Conservation Code®* (IECC®). This type of circulating hot (or tempered) water system has substantially less energy consumption than a circulating hot (or tempered) water system that has a constant (pumped) flow of water whenever the building is occupied. The demand type of system operates the circulation pump only when the presence of a user is detected. The detection could be a motion sensor, a door switch or a button that the user presses when entering the room where the fixture, requiring hot (or tempered) water, is located. A pump is activated to cause the hot (or tempered) water to flow towards the fixture until a temperature sensor in (or on) the piping detects a small increase in the temperature of the water. The pump then shuts off and the pump control locks out pump operation for a certain amount of time. After the "lockout" period, the pump can again be activated by the detection method to increase the water temperature by a small amount and so forth. If the hot (or tempered) water is actually used at the fixture, the temperature of the water for the very first user might not be very warm at first, but warmer water is fairly close to the fixture(s). With additional activations, or with multiple fixture uses, hot (or tempered) water is readily available at the fixtures.

A significant amount of energy is consumed by running pumps constantly and there is a large amount of heat loss from piping (even though the piping is insulated). By keeping the pump run time low and only bringing the hot (or tempered) water just a little closer to the fixture upon each "user detection," energy is conserved. Demand recirculation water systems can utilize cold water supply pipe or a dedicated pipe for water return back to the hot (or tempered) water source. See the definition of "Circulating hot water system."

DEPTH OF TRAP SEAL. The depth of liquid that would have to be removed from a full trap before air could pass through the trap.

❖ This is the distance measured from the bottom of the "dip" to the top of the "crown of the trap" (crown weir) [see Commentary Figure 202(9)]. The definition prior to this edition of the code referred to water. But with the advent of liquid seal traps for nonwater urinal designs, this definition changed "water" to "liquid."

**Figure 202(9)
DEPTH OF TRAP SEAL**

[BS] DESIGN FLOOD ELEVATION. The elevation of the "design flood," including wave height, relative to the datum specified on the community's legally designated flood hazard map. In areas designated as Zone AO, the *design flood elevation* shall be the elevation of the highest existing grade of the building's perimeter plus the depth number (in feet) specified on the flood hazard map. In areas designated as Zone AO where a depth number is not specified on the map, the depth number shall be taken as being equal to 2 feet (610 mm).

❖ The design flood elevation is the height to which floodwaters are predicted to rise during passage or occurrence of the design flood. Unless a community has adopted another flood hazard map or otherwise designated the design flood, the design flood elevation is the base flood elevation shown on the Flood Insurance Rate Maps (FIRM). The datum specified on the flood hazard map is important because it may differ from the datum used locally for other purposes. Also, the datum shown on the FIRM may change over time, as FEMA updates and revises the maps.

Some flood hazard maps have areas denoted as AO Zones, identifying areas subject to sheet flow flooding, shallow flooding or ponding. Specific instruction for these zones is provided in the code because the maps specify a flood depth above grade rather than a height above datum.

FEMA refers to some areas subject to shallow flooding and/or unpredictable flow paths as AH Zones. AH Zones have specified base flood elevations and they are treated the same as the other A Zones with base flood elevations.

DEVELOPED LENGTH. The length of a pipeline measured along the centerline of the pipe and fittings.

❖ This term identifies a concept necessary for computing the actual length of piping in a plumbing system. The developed length is measured along the actual flow path of piping and includes the piping lengths in all offsets and changes in direction. Several code requirements are dependent on the actual length of plumbing piping (see commentary, Sections 708, 802.2, 906 and 917).

DISCHARGE PIPE. A pipe that conveys the discharge from plumbing fixtures or appliances.

❖ This type of pipe typically refers to waste piping that either empties into a sump or conveys the discharge of a pump (see commentary, Sections 712 and 1111).

DRAIN. Any pipe that carries waste water or water-borne wastes in a building drainage system.

❖ A drain is any pipe in a plumbing system that carries sanitary waste, clear-water waste or storm water. In the case of wet vents, common vents, waste stack vents, circuit vents and combination drain and vents, the drain may also be conducting airflow.

DRAINAGE FITTING. The type of fitting or fittings utilized in the drainage system. Drainage fittings are similar to cast-iron fittings, except that instead of having a bell and spigot, drainage fittings are recessed and tapped to eliminate ridges on the inside of the installed pipe.

❖ The second sentence of the definition above is outdated. Fittings for galvanized steel drain piping were called drainage fittings. Drainage fittings were made substantially different from standard cast pipe fittings so that restrictions to drainage flow were minimized. These special fittings were made of cast iron with recessed internal threads to enable connection with Schedule 40 galvanized pipe. The threads on the fittings were cut in a manner that a horizontal pipe would be sloped $^1/_4$ inch per foot (20.8 mm/m) to cause gravity flow. A drain piping system using galvanized pipe and these special threaded cast fittings was called a Durham system, named after prominent manufacturer's product line of threaded drainage fittings. Thus, all threaded drainage pattern fittings were coined "Durham" fittings. These special drainage fittings are no longer used in modern construction. Today, the term "drainage fitting" applies to any fitting that has been made to a recognized (ASTM) drainage pattern for use in a gravity flow drain system.

DRAINAGE FIXTURE UNIT.

Drainage (dfu). A measure of the probable discharge into the drainage system by various types of plumbing fixtures. The drainage fixture-unit value for a particular fixture depends on its volume rate of drainage discharge, on the time duration of a single drainage operation and on the average time between successive operations.

❖ The conventional method of designing a sanitary drainage system is based on drainage fixture unit (dfu) load values. The fixture unit approach takes into consideration the probability of load on a drainage system. The dfu is an arbitrary loading factor assigned to each fixture relative to its impact on the drainage system. The dfu values are determined based on:

- Average rate of water discharge by a fixture;
- Duration of a single operation; and
- Frequency of use or interval between each operation.

Because dfu values have a built-in probability factor, they cannot be directly translated into flow rates or discharge rates.

A dfu is not the same as a water supply fixture unit (wsfu) in Table E103.3(2), Appendix E.

DRAINAGE SYSTEM. Piping within a *public* or *private* premise that conveys sewage, rainwater or other liquid waste to a point of disposal. A drainage system does not include the mains of a *public sewer* system or a private or public sewage treatment or disposal plant.

❖ The drainage system is the network of piping starting from the waste outlets of fixtures or receptors and ending at the public sewer system or other point of disposal [see Commentary Figure 202(8)].

Building gravity. A drainage system that drains by gravity into the *building sewer*.

❖ Plumbing systems are designed to drain by gravity wherever possible. Where gravity drainage is not possible, pumps or ejectors must be used to lift the waste to higher elevations.

Sanitary. A drainage system that carries sewage and excludes storm, surface and ground water.

❖ See the commentary to Section 701.

Storm. A drainage system that carries rainwater, surface water, subsurface water and similar liquid waste.

❖ The storm drainage system conducts water from roof drains, area drains, subsurface drains and clear-water waste sources to an approved location or to the public storm sewer (see Section 1101). The storm drainage system is independent of the sanitary drainage system.

DRINKING FOUNTAIN. A plumbing fixture that is connected to the potable water distribution system and the drainage system. The fixture allows the user to obtain a drink directly from a stream of flowing water without the use of any accessories.

❖ Although such a definition might seem unnecessary to veterans in the plumbing industry, there are so many different types of water dispensers on the market today that the distinction between those dispensers and what the code intended as a drinking fountain becomes blurred. Simply, a drinking fountain provides a stream of drinking water quality (potable) water from which to consume water. Excess water is captured and directed to the drainage system.

EFFECTIVE OPENING. The minimum cross-sectional area at the point of water supply discharge, measured or expressed in terms of the diameter of a circle or, if the opening is not circular, the diameter of a circle of equivalent cross-sectional area. For faucets and similar fittings, the *effective opening* shall be measured at the smallest orifice in the fitting body or in the supply piping to the fitting.

❖ The effective opening is used to determine the minimum air gap required between a potable water supply outlet or opening and the flood level rim of a receptacle, fixture or other potential source of contamination (see Table 608.15.1).

EMERGENCY FLOOR DRAIN. A floor drain that does not receive the discharge of any drain or indirect waste pipe, and that protects against damage from accidental spills, fixture overflows and leakage.

❖ A floor drain installed in a toilet facility is an example of an emergency drain.

ESSENTIALLY NONTOXIC TRANSFER FLUID. Fluids having a Gosselin rating of 1, including propylene glycol; mineral oil; polydimethylsiloxane; hydrochlorofluorocarbon, chlorofluorocarbon and carbon refrigerants; and FDA-approved boiler water additives for steam boilers.

❖ See the commentary to Section 608.16.3.

ESSENTIALLY TOXIC TRANSFER FLUID. Soil, waste or gray water and fluids having a Gosselin rating of 2 or more, including ethylene glycol, hydrocarbon oils, ammonia refrigerants and hydrazine.

❖ Transfer fluids are liquids or gases that transfer heat to or remove heat from another fluid, such as water. The exchange of heat energy takes place through a heat exchanger material that separates the transfer fluid from the fluid being heated or cooled (see commentary, Section 608.16.3).

EXISTING INSTALLATION. Any plumbing system regulated by this code that was legally installed prior to the effective date of this code, or for which a permit to install has been issued.

❖ Existing installations is a term that applies to all plumbing work that has been legally installed, and for which a permit to install has been issued prior to the effective date of the code. Plumbing that has been illegally installed prior to the effective date of the code is not considered existing and is subject to all code requirements for new installations. Sections that address existing plumbing systems and installations include Sections 102 and 108.

FAUCET. A valve end of a water pipe through which water is drawn from or held within the pipe.

❖ A faucet is a type of valve or terminal fitting that controls the volume and directional flow of water. In the context of the code, faucets are installed in the potable water distribution system. Faucets terminate to the atmosphere by means of a spout or similar opening and are considered as outlets for the potable water distribution system.

FILL VALVE. A water supply valve, opened or closed by means of a float or similar device, utilized to supply water to a tank. An antisiphon fill valve contains an antisiphon device in the form of an *approved air gap* or vacuum breaker that is an integral part of the fill valve unit and that is positioned on the discharge side of the water supply control valve.

❖ A fill valve is a float-actuated valve used in flush tanks. The fill valve serves to control the water supply for refilling the flush tank and restoring the trap seal in the water closet bowl. All flush tanks must be equipped with an antisiphon fill valve installed with the critical level 1 inch (25 mm) above the overflow pipe.

FIXTURE. See "Plumbing fixture."

❖ See the commentary to "Plumbing fixture."

FIXTURE BRANCH. A drain serving two or more fixtures that discharges to another drain or to a *stack*.

❖ See Commentary Figure 202(10).

FIXTURE DRAIN. The drain from the trap of a fixture to a junction with any other drain pipe.

❖ A fixture drain is the section of pipe that connects the outlet weir of a trap to a stack, fixture branch or any other drain [see Commentary Figures 202(10) and 202(11)]. A fixture drain conveys the discharge of only a single trap and can consist of both horizontal and vertical sections of pipe. A fixture drain ends at the point where the discharge from any other fixture is introduced. Although commonly referred to as a "trap arm," a fixture drain can extend beyond the connection of the vent for the fixture drain.

DEFINITIONS

**Figure 202(11)
FIXTURE DRAIN**

FIXTURE FITTING.

❖ Fixture fittings include faucets, spouts, terminal fittings, diverters, supply stops, wastes, traps and other "trim" items that attach to, or in the case of supply fittings are accessible from, a plumbing fixture. These items are necessary to allow the fixture to function as intended.

Supply fitting. A fitting that controls the volume, direction of flow or both, of water and is either attached to or accessible from a fixture, or is used with an open or atmospheric discharge.

❖ These fittings are associated with the supply and control of flow of water to a fixture. Supply fittings can also include fittings that serve the same purpose but are not mounted directly to the fixture.

Waste fitting. A combination of components that conveys the sanitary waste from the outlet of a fixture to the connection to the sanitary drainage system.

❖ These fittings include the fixture traps that are not an integral part of the fixture (see commentary, Section 1002) and the parts used to connect piping to the sanitary drainage system (see Section 424.1.2).

FIXTURE SUPPLY. The water supply pipe connecting a fixture to a *branch* water supply pipe or directly to a main water supply pipe.

❖ A fixture supply is a water supply pipe that serves a single fixture (see commentary, Section 604.5).

[BS] FLOOD HAZARD AREA. The greater of the following two areas:

1. The area within a flood plain subject to a 1-percent or greater chance of flooding in any given year.

2. The area designated as a *flood hazard area* on a community's flood hazard map or as otherwise legally designated.

❖ The FIRMs that are prepared by FEMA show areas that are predicted to be flooded under conditions of the 1-percent annual chance (base) flood. Some states

**Figure 202(10)
FIXTURE DRAINS AND FIXTURE BRANCHES**

and local jurisdictions develop and adopt maps of flood hazard areas that are more extensive than the areas shown on FEMA's maps. For the purposes of the code, the flood hazard area within which the requirements are to be applied is the greater of the two delineated areas.

FLOOD LEVEL RIM. The edge of the receptacle from which water overflows.

❖ The flood level rim corresponds to the highest elevation that liquid can be contained in a receptacle without spilling over the side. The flood level rim is the point above which liquid will flow over and out of a fixture or receptor. For example, the flood level rim of a sink is the top edge of the basin, and the flood level rim of a floor drain is the floor surface at the floor drain [see Commentary Figure 608.15.1(1)].

FLOW CONTROL (Vented). A device installed upstream from the interceptor having an orifice that controls the rate of flow through the interceptor and an air intake (vent) downstream from the orifice that allows air to be drawn into the flow stream.

❖ Referenced standard ASME A112.14.3 and Section 1003.4.2 use this term in connection with grease interceptors. This device is necessary to control the flow rate of waste through the interceptor and to entrain air to assist in separation of grease and oily waste from the water. Grease interceptors depend on having the necessary waste retention time to allow the fats, oils and grease to separate. If the influent flow rate is too high, the waste will pass through the interceptor too quickly, not allowing sufficient time for separation to occur. Entraining air in the waste influent will cause air bubbles to attach to grease globules, making them more buoyant and thereby enhancing the separation process (see Commentary Figure 1003.3.4.2).

FLOW PRESSURE. The pressure in the water supply pipe near the faucet or water outlet while the faucet or water outlet is wide open and flowing.

❖ Each fixture requiring water has a minimum flow pressure that is necessary for the fixture to operate properly and to help protect the potable water supply from contamination. Flow pressure is a factor used in the sizing of the water supply system. The flow pressure of a plumbing system is less than the static pressure because energy is lost in putting the fluid in motion (see Table 604.3).

FLUSH TANK. A tank designed with a fill valve and flush valve to flush the contents of the bowl or usable portion of the fixture.

❖ The flush tank for a water closet or urinal must be able to supply water in sufficient quantity to flush the contents of the fixture, cleanse the fixture bowl and refill the fixture trap. The definition states clearly that the flush valve is considered to be part of the flush tank.

FLUSHOMETER TANK. A device integrated within an air accumulator vessel that is designed to discharge a predetermined quantity of water to fixtures for flushing purposes.

❖ These devices are a hybrid between flushometer valves and gravity flush tanks. A flushometer tank uses a compression tank that holds water under a pressure equivalent to the water supply pressure. The flushing action created by the flushometer tank is similar to that of the flushometer valve. The amount of water used is limited, as it is with a gravity flush tank.

FLUSHOMETER VALVE. A valve attached to a pressurized water supply pipe and so designed that when activated it opens the line for direct flow into the fixture at a rate and quantity to operate the fixture properly, and then gradually closes to reseal fixture traps and avoid water hammer.

❖ These flushing devices are a type of metering valve and are used primarily in public occupancy applications because of the more powerful flushing action.

GRAY WATER. Waste discharged from lavatories, bathtubs, showers, clothes washers and laundry trays.

❖ The collection of water only from the fixtures indicated deserves a special and distinct name so designers and installers know to keep this water separate from blackwater (waste water from water closets, urinals and kitchen sinks), where necessary to do so. In fact, graywater usually does have a grayish color. Although the code does not require the marking of gray water collection piping, it may be prudent to mark such piping systems to avoid future connection of blackwater (containing fecal matter) piping to gray water piping.

GREASE INTERCEPTOR.

Fats, oils and greases (FOG) disposal system. A plumbing appurtenance that reduces nonpetroleum fats, oils and greases in effluent by separation or mass and volume reduction.

Gravity. Plumbing appurtenances of not less than 500 gallons (1893 L) capacity that are installed in the sanitary drainage system to intercept free-floating fats, oils and grease from waste water discharge. Separation is accomplished by gravity during a retention time of not less than 30 minutes.

Hydromechanical. Plumbing appurtenances that are installed in the sanitary drainage system to intercept free-floating fats, oils and grease from waste water discharge. Continuous separation is accomplished by air entrainment, buoyancy and interior baffling.

❖ Prior to the 2012 edition, the code did not distinguish between the two of these types of grease interceptors, gravity and hydromechanical. Many in the plumbing industry still use the old term "grease trap" without giving a preference to meaning either type. Now more than ever, correct terminology for these appurtenances must be used because with the 2015 edition, a new type of grease interceptor, the FOG disposal system, is introduced.

Gravity grease interceptors are typically large tanks, usually buried below a parking lot or driveway, that provide for quiescent flow of grease-laden waste through the tank. Internal baffles and changes in flow direction cause solids to settle out and grease to float.

All are retained in the tank until pumped out on a periodic basis by a grease waste removal/disposal contractor.

Hydromechanical grease interceptors are much smaller in size than the typical gravity grease interceptor. These units are more often located indoors and can be somewhat automated to draw off grease, making maintenance easier.

FOG disposal systems are relatively new and are converters of fats, oils and greases. Rather than simply capturing FOG like the other type of grease interceptors, this type of system converts a majority of the FOG into grey water through a bioremediation process. In other words, these systems are an on-site, in-line treatment process specifically designed for grease-laden waste. FOG disposal systems can attain lower FOG concentrations in the effluent than the other types of grease interceptors.

These definitions, along with their proper application in the code text of Chapter 10, helps to clarify which type of grease interceptor the code is regulating.

GREASE-LADEN WASTE. Effluent discharge that is produced from food processing, food preparation or other sources where grease, fats and oils enter automatic dishwater prerinse stations, sinks or other appurtenances.

❖ The effluent discharge from food preparation and processing of animal products will usually produce grease-laden waste. Additionally, many industrial processes that use animal fats or petroleum products will produce grease-laden waste. This waste has two main deleterious effects on sewer systems: frequent stoppages resulting from congealed grease and overloading of the sewage treatment system.

GREASE REMOVAL DEVICE, AUTOMATIC (GRD). A plumbing appurtenance that is installed in the sanitary drainage system to intercept free-floating fats, oils and grease from waste water discharge. Such a device operates on a time- or event-controlled basis and has the ability to remove free-floating fats, oils and grease automatically without intervention from the user except for maintenance.

❖ Referenced standard ASME A112.14.4 and Section 1003.3.4 use this term in connection with grease interceptors. An automatic grease removal device serves the same function as a grease interceptor with the distinction that the unpleasant aspect of grease removal is performed automatically. There are two main types of automatic GRD: pump-out and skimmer type. The pump-out type has an internal sensor that senses a buildup of collected grease, liquefies it and pumps it out to an external container for proper environmental disposal. The skimmer type has a timer-activated skimmer mechanism to mechanically collect and transport the collected grease to an external container. Both types usually have internal heaters to keep the grease in liquid form for easy pumping or skimming.

Although the solids interceptor section of the GRD still requires manual maintenance to remove waste stream particulates such as food scraps, broken glass and plastics, the task is relatively easy, thus having the likelihood of being performed regularly [see Commentary Figures 202(12) through 202(14)].

Figure 202(12)
SENSOR-CONTROLLED GREASE RECOVERY DEVICE

DEFINITIONS

Figure 202(13)
TIMER-CONTROLLED GREASE RECOVERY DEVICE

Figure 202(14)
GREASE-REMOVAL DEVICE, AUTOMATIC

DEFINITIONS

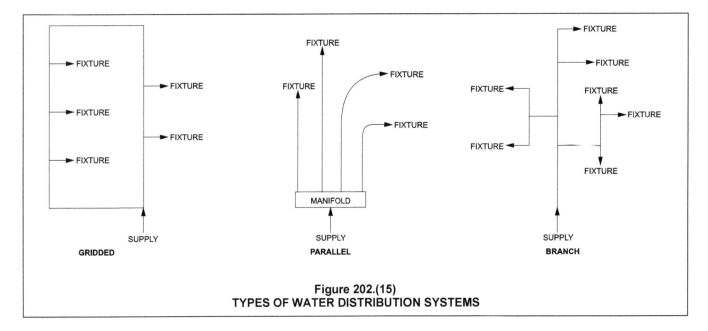

Figure 202.(15)
TYPES OF WATER DISTRIBUTION SYSTEMS

GRIDDED WATER DISTRIBUTION SYSTEM. A water distribution system where every water distribution pipe is interconnected so as to provide two or more paths to each fixture supply pipe.

❖ A gridded water distribution system is more hydraulically efficient than a parallel or branch line layout but not necessarily more cost efficient. The design balances pressure and equalizes flow so that fixtures farthest from the water service entry point will operate under the same pressure and flow conditions as those closest to the water service entry point. Commentary Figure 202(15) illustrates various distribution layouts.

HANGERS. See "Supports."

HORIZONTAL BRANCH DRAIN. A drainage *branch* pipe extending laterally from a soil or waste *stack* or *building drain*, with or without vertical sections or *branches*, that receives the discharge from two or more *fixture drains* or *branches* and conducts the discharge to the soil or waste *stack* or to the *building drain*.

❖ See Commentary Figure 202(10) and Sections 912 and 914.

HORIZONTAL PIPE. Any pipe or fitting that makes an angle of less than 45 degrees (0.79 rad) with a horizontal plane.

❖ This definition is needed for the application of many code provisions that apply only to horizontal piping, such as those contained in Chapters 3, 7, 9 and 11.

HOT WATER. Water at a temperature greater than or equal to 110°F (43°C).

❖ The code does not specify a maximum temperature for hot water; however, it should be limited to minimize the risk of burn injury and thermal stress to plumbing system components. The designer, the installer or the owner has the choice of determining the temperature. Underwriters Laboratories (UL) standard UL 174 for electric water heaters requires that the temperature-regulating thermostat or control be set at the factory at not greater than 125°F (51.7°C).

Water temperature below the minimum temperature of 110°F (43°C) is considered to be either tempered or cold. Care must be used when installing a system that has temperature capability of scalding.

HOUSE TRAP. See "Building trap."

INDIRECT WASTE PIPE. A waste pipe that does not connect directly with the drainage system, but that discharges into the drainage system through an *air break* or *air gap* into a trap, fixture, receptor or interceptor.

❖ Often referred to as "open site" or "safe waste," indirect waste pipe installations are intended to protect fixtures, equipment and systems from the backflow or backsiphonage of waste (see commentary, Sections 801.1 and 802.1).

INDIVIDUAL SEWAGE DISPOSAL SYSTEM. A system for disposal of domestic sewage by means of a septic tank, cesspool or mechanical treatment, designed for utilization apart from a public *sewer* to serve a single establishment or building.

❖ When installed in accordance with the *International Private Sewage Disposal Code®* (IPSDC®) and approved by the administrative authority, these systems are acceptable as a point of disposal for sewage when a public sewer system is not available.

INDIVIDUAL VENT. A pipe installed to vent a fixture trap and that connects with the vent system above the fixture served or terminates in the open air.

❖ An individual vent is the simplest and most common method of venting a single plumbing fixture (see commentary, Section 907.1).

INDIVIDUAL WATER SUPPLY. A water supply that serves one or more families, and that is not an *approved* public water supply.

❖ This type of water supply includes wells, springs, streams and cisterns. Surface bodies of water and land cisterns can be used when properly treated and approved (see commentary, Sections 602.3 and 608.17).

INTERCEPTOR. A device designed and installed to separate and retain for removal, by automatic or manual means, deleterious, hazardous or undesirable matter from normal wastes, while permitting normal sewage or wastes to discharge into the drainage system by gravity.

❖ These devices are often necessary for the initial treatment of sewage before it is discharged to an approved point of disposal. Interceptors are intended to eliminate or reduce the amounts of substances in sewage that can be detrimental to drainage and sewage treatment systems. When the interceptor is designed for a specific use, it often bears that name, such as a grease interceptor.

JOINT.

❖ For a piping system, either supply or drainage, to be installed in a structure, there must be a series of pipe sections and pipe fittings. The connection of these two types of components is a joint in its most general definition. The four types of joints listed below are special types used for a particular installation. These joints are very specific and should be made using the proper components and procedures.

Expansion. A loop, return bend or return offset that provides for the expansion and contraction in a piping system and is utilized in tall buildings or where there is a rapid change of temperature, as in power plants, steam rooms and similar occupancies.

❖ See the commentary to Sections 305.2 and 308.8.

Flexible. Any joint between two pipes that permits one pipe to be deflected or moved without movement or deflection of the other pipe.

❖ See the commentary to Section 305.2.

Mechanical. See "Mechanical joint."

Slip. A type of joint made by means of a washer or a special type of packing compound in which one pipe is slipped into the end of an adjacent pipe.

❖ A slip joint is a joint commonly found in tubular-type fixture tailpieces and "P" traps. Even though the term "slip" is used, there is not the intent that such joints be suitable for accommodating expansion and contraction of piping or differential movement. In the early use of slip joints, sealing rings were made of various fibrous materials and then later, rubber. Joints using these materials would "slip" if the tubes were pushed around. The rings adequately "sealed" the connections when new but when aged, they would leak if the tubes were bumped. Newer designs of tubular-type tailpieces and "P" traps allow for cone-shaped plastic sealing rings to be used for the connections. When the joint is assembled and the connection nut is still loose, the tubes can be "slipped" for final adjustment into position. However, once the nut is tightened, the "wedge action" of the sealing ring into the tube opening offers a tight, superior, long-lasting seal for the connection. Even though the joint does not "slip," these joints are still called slip joints (see commentary, Section 405.8 and Figure 1002.2).

LEAD-FREE SOLDER AND FLUX. Containing not more than 0.2-percent lead.

❖ A soldered joint is one method of joining copper pipe and tubing. ASTM B32 covers many grades of solder, including tin-silver solders. Many decades ago the grades used in the plumbing industry were typically 40-60 (40-percent tin and 60-percent lead) and 50-50 (50-percent tin and 50-percent lead). However, solders for joining copper tubing used for the distribution of potable water and having more than 0.2 percent have not been allowed for several decades. Most solders for potable water service are 95-5 (95-percent tin and 5-percent antimony). There are other "lead-free" solders available that sometimes offer better characteristics for certain installer's techniques.

Soldering requires a flux be applied to the joint surfaces prior to heating and applying the solder. The term "flux" in Latin is *fluxus*, meaning "flow." The flux removes oxidation on the surfaces to be joined even though the procedure for preparing a joint requires brush cleaning of the surfaces to a bright finish. The flux encourages capillary action of the melted solder throughout the joint.

Lead-bearing solders are still acceptable for joining copper and copper alloy drain, waste and vent piping.

Installers using both "lead free" and "leaded" solder and flux in their work need to carefully control truck and shop stock of the leaded products so they are not inadvertently used for a potable water piping application (see commentary, Section 705.7.3).

LEADER. An exterior drainage pipe for conveying storm water from roof or gutter drains to an *approved* means of disposal.

❖ Otherwise known as a downspout, this type of drain serves the storm drainage system and typically connects to hanging or built-in gutters and scuppers (see the definition of "Conductor").

LOCAL VENT STACK. A vertical pipe to which connections are made from the fixture side of traps and through which vapor or foul air is removed from the fixture or device utilized on bedpan washers.

❖ Local vents are designed to conduct moisture vapor and foul air to the exterior of the building. Local vents connect to a fixture on the house side (fixture side) of the fixture trap; therefore, such vents must be entirely independent of the sanitary drainage system vents.

DEFINITIONS

MACERATING TOILET SYSTEM. An assembly consisting of a water closet and sump with a macerating pump that is designed to collect, grind and pump wastes from the water closet and up to two other fixtures connected to the sump.

❖ This assembly includes a water closet and a sump facilitating the connection of two other fixtures and is normally used where fixtures are required below the building sewer invert. Section 712.4.1 specifically requires these assemblies to be installed according to the manufacturers' instructions and ASME A112.3.4 or CSA B45.9. The assembly includes a grinder pump that reduces the size of solids, allowing the use of discharge piping that is smaller than conventional size.

MAIN. The principal pipe artery to which branches are connected.

❖ This is a generic item that applies to the primary (and typically the largest) water supply system piping or drainage system piping to which all other piping is connected. The main is the trunk line of branch system piping designs.

MANIFOLD. See "Plumbing appurtenance."

❖ A manifold is a multiple-opening header to which one or more branch lines connect (see the definition of "Plumbing appurtenance" and Section 604.10).

[M] MECHANICAL JOINT. A connection between pipes, fittings, or pipes and fittings that is not screwed, caulked, threaded, soldered, solvent cemented, brazed, welded or heat fused. A joint in which compression is applied along the centerline of the pieces being joined. In some applications, the joint is part of a coupling, fitting or adapter.

❖ Mechanical joints form a seal by compressing a ferrule or seal around the outside circumference of the pipe or tube or by compressing the tubing around a fitting inserted into the tubing. Mechanical joints are often proprietary and must be installed in accordance with the manufacturers' instructions. Heat fused joints are not considered to be a mechanical joint.

MEDICAL GAS SYSTEM. The complete system to convey medical gases for direct patient application from central supply systems (bulk tanks, manifolds and medical air compressors), with pressure and operating controls, alarm warning systems, related components and piping networks extending to station outlet valves at patient use points.

❖ Nonflammable medical gases are distributed through piping systems from central supply systems, such as cylinders or medical air compressors, to station outlets throughout the medical facility (i.e., patient rooms, operating/delivery rooms or emergency treatment facilities). Medical gases include, but are not limited to, oxygen, nitrogen, nitrous oxide, carbon dioxide and compressed air.

Because more than one medical gas is usually delivered to an outlet station, care must be exercised so that the correct gas is being used for a given procedure. Accidental cross connection can result in injury or death of a patient. All piping, supply systems and outlet stations must be clearly labeled with the name of the gas being supplied. In addition, color-coded or noninterchangeable fittings, keyed to the specific gas, are usually incorporated to further safeguard against cross connections.

The delivery of medical gases is critical to the operation of a health care facility, and provisions must be made to ensure continuous operation of the system. Secondary (backup) cylinders and air compressors are required, and the compressors should be connected to the emergency electrical power grid.

Detailed guidance for the design, installation and maintenance of a medical gas system is provided in NFPA 99.

MEDICAL VACUUM SYSTEM. A system consisting of central-vacuum-producing equipment with pressure and operating controls, shutoff valves, alarm-warning systems, gauges and a network of piping extending to and terminating with suitable station inlets at locations where patient suction may be required.

❖ Central medical vacuum systems, including surgical and bedside vacuum systems and waste gas scavenging systems, have become a standard feature in most health care facilities. Inlet stations are typically installed in patient rooms, operating/delivery rooms, nurseries, emergency rooms and other patient care facilities. A vacuum system may also be provided in the laboratory, but is not usually connected to the medical vacuum system.

Each vacuum inlet station requires a collection bottle for collecting fluids and a trap bottle to prevent any overflow fluids from entering and clogging the suction line. This dual bottle system helps prevent the spread of contaminants within the medical facility.

Because the medical vacuum system has become so critical to patient care, continuous service from the system must be assured. At least two vacuum pumps should be installed, each sized to meet all of the facility's needs if the other fails. The pumps should also be connected to the facility's emergency electrical power grid.

Detailed guidance for the design, installation and maintenance of a medical vacuum system is contained in NFPA 99.

METER. A measuring device used to collect data and indicate water usage.

❖ Just about everyone understands what a water (utility) meter is—it is the device that a public water purveyor installs on the water service line to measure water use for a building in order to charge the user for the water. Most meters are a totalizer type, that is, they simply indicate the number of gallons that have passed through the device. Periodic readings (such as once per month) allow the water supplier to subtract the lower reading from the higher reading to know how much water was used in a month. Over the years, totalizer-type meters have been fitted with electronic readout devices to make the reading process quicker and sometimes by remote means.

Newer meters are now available that can store data on water usage patterns for reporting to a remote location when activated to do so. Such data has been found to be very useful for alerting users of wasteful water usage, helping them curtail their use of water. The data is also useful to building owners for spotting patterns indicating leaks, malfunctioning fixtures and defective equipment. Measuring is the first step in the process of water conservation.

NONPOTABLE WATER. Water not safe for drinking, personal or culinary utilization.

❖ Myriad substances can make water unfit for consumption, washing or bathing. Water not meeting the parameters for quality as defined in the EPA Clean Drinking Water Act is considered to be nonpotable. The code has numerous requirements for preventing the cross connection of nonpotable water to the potable water supply (see commentary, Section 608).

NUISANCE. Public nuisance as known in common law or in equity jurisprudence; whatever is dangerous to human life or detrimental to health; whatever structure or premises is not sufficiently ventilated, sewered, drained, cleaned or lighted, with respect to its intended *occupancy*; and whatever renders the air, or human food, drink or water supply unwholesome.

❖ The meaning of the term "nuisance" can be interpretive. Although there are certainly situations wherein a health hazard is clearly present, other conditions that represent an unreasonable interference with "the enjoyment of life and property" require a subjective decision on the part of the administrative authority. An indirect waste discharge onto a floor adjacent to a floor receptor could be a slipping hazard and therefore judged to be a nuisance. Additionally, any insanitary condition caused by a plumbing installation or lack of a plumbing installation could be considered a nuisance.

[A] OCCUPANCY. The purpose for which a building or portion thereof is utilized or occupied.

❖ The occupancy classification of a building is an indication of the level of hazard to which occupants are exposed as a function of the building's use. Occupancy in terms of an occupancy group classification is one of the primary considerations in the development and application of many code requirements that are designed to offset the hazards specific to each occupancy group designation.

OFFSET. A combination of *approved* bends that makes two changes in direction bringing one section of the pipe out of line but into a line parallel with the other section.

❖ Offsets are necessary to route piping around or through structural elements such as beams, joists, trusses and columns. An offset always involves at least two changes in direction to keep the piping heading in the same direction. Any other arrangement would be considered a bend or a simple change in direction.

ON-SITE NONPOTABLE WATER REUSE SYSTEM. A water system for the collection, treatment, storage, distribution and reuse of nonpotable water generated on site, including but not limited to a gray water system. This definition does not include a rainwater harvesting system.

❖ There are a number of water sources that can be generated by the operation or occupancy of a building. The first source that comes to mind is usually gray water (see the definition for "Gray water"). But there could be many other sources such as heat exchanger cooling water, air-conditioning condensate, cooling tower discharge water, steam condensate and other sources that are specific to unique processes. In other words, if it is water-based liquid and is not blackwater (that which contains fecal matter or urine), it could possibly be used for another purpose. Whether or not all sources can be used depends on how much the water needs to be treated to make it suitable for an end use, the feasibility of collecting the water and the jurisdiction's allowances for using such water for end uses.

OPEN AIR. Outside the structure.

❖ This term clarifies the intent of the code with regard to the location of vent terminals. "Open air" refers to the atmosphere that is exterior to the envelope or outside of a building.

PLUMBING. The practice, materials and fixtures utilized in the installation, maintenance, extension and alteration of all piping, fixtures, plumbing appliances and plumbing appurtenances, within or adjacent to any structure, in connection with sanitary drainage or storm drainage facilities; venting systems; and public or private water supply systems.

❖ The term refers to the collective system of piping, fixtures, fittings, components or devices and appurtenances that transport potable water and liquid and solid wastes associated with cleaning, washing, bathing, food preparation, drinking and the elimination of bodily wastes.

The word "plumbing" comes from the Latin word for the element lead (plumbum), because lead was used extensively in the construction of piping systems. The practice of installing piping systems and the materials used in such systems became known as plumbing.

Piping, fittings, devices, faucets, vessels, containers and receptacles that are used to supply, distribute, receive or transport potable water, liquid wastes or solid wastes suspended in water are considered as plumbing. Plumbing also includes appliances, equipment and systems such as water conditioners, water heaters, storm drainage systems, water coolers, water filters and waste treatment systems.

PLUMBING APPLIANCE. Water or drain-connected devices intended to perform a special function. These devices have their operation or control dependent on one or

DEFINITIONS

more energized components, such as motors, controls or heating elements. Such devices are manually adjusted or controlled by the owner or operator, or are operated automatically through one or more of the following actions: a time cycle, a temperature range, a pressure range, a measured volume or weight.

❖ Examples of plumbing appliances include water heaters, hot water dispensers, garbage disposals, dishwashers, clothes washers, water purifiers and water softeners.

PLUMBING APPURTENANCE. A manufactured device, prefabricated assembly or on-the-job assembly of component parts that is an adjunct to the basic piping system and plumbing fixtures. An appurtenance demands no additional water supply and does not add any discharge load to a fixture or to the drainage system.

❖ Examples of plumbing appurtenances include water closet seats, hand-held showers, manifolds, backflow preventers, water-hammer arrestors, strainers and filters.

PLUMBING FIXTURE. A receptacle or device that is connected to a water supply system or discharges to a drainage system or both. Such receptacles or devices require a supply of water; or discharge liquid waste or liquid-borne solid waste; or require a supply of water and discharge waste to a drainage system.

❖ This definition was changed for the 2012 edition of the code in order to fully encompass the variety of plumbing fixtures that could be installed in a building. The most common type of plumbing fixture has a supply of water connected to it and is connected to the drain system. Examples are water closets, lavatories and bath-

Figure 202(16)
PARTS OF A PLUMBING SYSTEM

tubs. Nonwater-supplied (waterless) urinals, floor sinks and floor drains are fixtures that only require connection to the drain system. An outdoor shower discharging to the ground is an example of a plumbing fixture that is connected only to a supply of water.

PLUMBING SYSTEM. A system that includes the water distribution pipes; plumbing fixtures and traps; water-treating or water-using equipment; soil, waste and vent pipes; and *building drains*; in addition to their respective connections, devices and appurtenances within a structure or premises; and the water service, *building sewer* and building storm *sewer* serving such structure or premises.

❖ The plumbing system includes all piping, fixtures and components that transport potable water and convey liquid and liquid-borne solid wastes [see Commentary Figure 202(16) and the definition of "Plumbing"]. Also included would be devices that treat the water prior to its use.

POLLUTION. An impairment of the quality of the potable water to a degree that does not create a hazard to public health but that does adversely and unreasonably affect the aesthetic qualities of such potable water for domestic use.

❖ Many books have been written trying to define the term "pollution." For use in the code, pollution is anything that reduces the quality of the potable water supply so that it is undesirable for consumption or use. A pollutant in the water supply causes the water to look, smell or taste bad, but does not make drinking the water harmful. This does not change the fact that polluted water is still considered to be nonpotable. Pollutants are thus not considered potentially harmful, whereas a contaminant would be considered potentially harmful. Refer to the definition of the term "Contamination" to find additional information and the differences between pollution and contamination.

The methods for determining potable and nonpotable water supplies are very closely regulated by the EPA. For the purposes of protecting a water supply, the code considers a polluted source to represent a low hazard when selecting a backflow preventer. Refer to Table 608.1 for potable protection requirements related to low-hazard situations.

It should be noted that most forms of pollution are the direct result of some action by humans on their environment and thus, humans are responsible for preventing this pollution.

POTABLE WATER. Water free from impurities present in amounts sufficient to cause disease or harmful physiological effects and conforming to the bacteriological and chemical quality requirements of the Public Health Service Drinking Water Standards or the regulations of the public health authority having jurisdiction.

❖ The EPA Clean Drinking Water Act defines the quality requirements for water to be classified as potable. The word "potable" means "fit to drink."

PRIVATE. In the classification of plumbing fixtures, "*private*" applies to fixtures in residences and apartments, and to fixtures in nonpublic toilet rooms of hotels and motels and similar installations in buildings where the plumbing fixtures are intended for utilization by a family or an individual.

❖ Because there are code requirements specific to public plumbing fixtures and facilities, the distinction must be made between public and private (see commentary, Section 403.3).

PUBLIC OR PUBLIC UTILIZATION. In the classification of plumbing fixtures, "*public*" applies to fixtures in general toilet rooms of schools, gymnasiums, hotels, airports, bus and railroad stations, public buildings, bars, public comfort stations, office buildings, stadiums, stores, restaurants and other installations where a number of fixtures are installed so that their utilization is similarly unrestricted.

❖ Public or public utilization is a descriptor for toilet facilities that are available to the general public. Employee plumbing fixtures are considered to be public fixtures unless their use is restricted to an individual (see the definition of "Private").

PUBLIC WATER MAIN. A water supply pipe for public utilization controlled by public authority.

❖ See the definition of and commentary to "Water main."

QUICK-CLOSING VALVE. A valve or faucet that closes automatically when released manually or that is controlled by a mechanical means for fast-action closing.

❖ A quick-closing valve would be any type of solenoid-actuated valve; spring-loaded, self-closing faucets or any other device capable of instantaneously reducing water flow from full flow to no flow. This rapid change from a full-flow to no-flow condition can cause water-hammer-related damage. Although most faucets can be manually closed fast enough to produce water hammer, it is not the intent of the code to consider manually closed valves or faucets as "quick closing."

Typical quick-closing valves include spring-loaded faucets and drinking fountain control valves that abruptly close when manually released; electrically actuated valves such as those found in dishwashing machines, clothes washing machines, boiler makeup water feeders, self-washing air cleaners or grease extractors; and process equipment and electronically controlled "touchless" faucets. Some self-closing valves and faucets are not considered as quick closing. For example, typical metering faucets and typical flushometer valves are slowly, automatically closed by a diaphragm/bleed orifice or pilot valve assembly. Additionally, the typical float-actuated valve is automatic, yet slow in closing (see commentary, Section 604.9).

RAINWATER. Water from natural precipitation.

❖ Water falling from the sky is an abundant source of water in some areas. However, some jurisdictions located in "dry" climate areas could have legal "rights" to rainwater such that rainwater cannot be collected for a specific use on the building site. As water reservoirs and rivers dry up in these areas, any use that takes away from precipitation finding its way into those water impoundments would make a limited water problem even worse. Note that once rainwater contacts a surface at, or below grade, it is then called storm water (see the definition of "Storm water").

[M] READY ACCESS. That which enables a fixture, appliance or equipment to be directly reached without requiring the removal or movement of any panel, door or similar obstruction and without the use of a portable ladder, step stool or similar device.

❖ Ready access can be described as the capability of being quickly reached or approached for the purpose of operation, inspection, observation or, typically, an emergency action. Plumbing installations requiring ready access must not be covered or concealed by a door, panel or similar obstruction, nor are they allowed to be obstructed by physical obstacles, including elevation above a floor such that something must be stood upon to reach the item.

The term "door" sometimes causes questions about whether a room, that has something that requires ready access, is allowed to have a passage door on the room. For example, a door to a mechanical room where there is a floor drain. Section 412.2 requires that floor drains be provided with ready access. Does the code intend to allow for the room to have a passageway door? Probably so. Conversely, if a valve is required to have ready access, can the valve be located behind the door panel of a cabinet? Probably not.

RECLAIMED WATER. Nonpotable water that has been derived from the treatment of waste water by a facility or system licensed or permitted to produce water meeting the jurisdiction's water requirements for its intended uses. Also known as "recycled water."

❖ For decades, reclaimed water has been used for golf course irrigation in many jurisdictions. In the early days, those jurisdictions prohibited the use of reclaimed water for any purpose inside of buildings because of concerns about disinfection, the potential for the spread of what was thought to be "harmful" components in the water and the uncertainty about how to prevent such water from being inadvertently used for a potable water application. Many studies have been performed to allay the initial concerns and the color purple has been chosen to identify the piping and components that convey nonpotable water.

REDUCED PRESSURE PRINCIPLE BACKFLOW PREVENTION ASSEMBLY. A backflow prevention device consisting of two independently acting check valves, internally force-loaded to a normally closed position and separated by an intermediate chamber (or zone) in which there is an automatic relief means of venting to the atmosphere, internally loaded to a normally open position between two tightly closing shutoff valves and with a means for testing for tightness of the checks and opening of the relief means.

❖ This device is considered to be one of the most reliable mechanical devices (second only to an air gap) for the prevention of backflow. These devices use two spring-loaded check valves with a relief valve in between that monitors the system pressure upstream and downstream of the device. The zone between the check valves is maintained at a pressure that is less than the water supply pressure (see commentary, Section 608.13.2).

[A] REGISTERED DESIGN PROFESSIONAL. An individual who is registered or licensed to practice professional architecture or engineering as defined by the statutory requirements of the professional registration laws of the state or jurisdiction in which the project is to be constructed.

❖ Legal qualifications for engineers and architects are established by the state having jurisdiction. Licensing and registration of engineers and architects is accomplished by written or oral examinations offered by states or by reciprocity (licensing in other states).

RELIEF VALVE.

❖ In general, a relief valve is a safety device designed to prevent the development of a potentially dangerous condition in a closed piping system. This is normally associated with a closed system that has a heat source (see commentary, Section 504.4).

Pressure relief valve. A pressure-actuated valve held closed by a spring or other means and designed to relieve pressure automatically at the pressure at which such valve is set.

❖ This valve is designed to be actuated by excessive pressures to prevent rupture or explosion of tanks, vessels or piping. The degree of valve opening is directly proportional to the pressure acting on the valve disk.

Temperature and pressure relief (T&P) valve. A combination relief valve designed to function as both a temperature relief and a pressure relief valve.

❖ These devices combine the components of pressure and temperature relief valves and are designed to be actuated by excessive pressures, excessive temperatures or both.

Temperature relief valve. A temperature-actuated valve designed to discharge automatically at the temperature at which such valve is set.

❖ This valve is designed to be actuated by excessive temperatures to prevent dangerously high water temperatures that can cause tank, vessel or pipe explosions.

RELIEF VENT. A vent whose primary function is to provide circulation of air between drainage and vent systems.

❖ A relief vent is required for some circuit venting configurations and for stacks of more than 10 branch intervals. The relief vent acts to relieve pressures that develop in stacks as a result of the flow of waste (see commentary, Section 914).

RIM. An unobstructed open edge of a fixture.

❖ The rim of a fixture or receptor is the reference point for the measurement of installation dimensions, such as for critical levels, air gaps, air breaks and horizontal vent piping (see the definition of "Flood level rim").

RISER. See "Water pipe, riser."

ROOF DRAIN. A drain installed to receive water collecting on the surface of a roof and to discharge such water into a leader or a conductor.

❖ Roof drains act as a receptor for storm water flowing across the surface of a building roof. Roof drains are equipped with inlet strainers that are designed to prevent the entry of leaves, debris and roof ballast materials.

ROUGH-IN. Parts of the plumbing system that are installed prior to the installation of fixtures. This includes drainage, water supply, vent piping and the necessary fixture supports and any fixtures that are built into the structure.

❖ Rough-in refers to the stage of construction prior to the installation of any materials that would conceal plumbing installations in the building structure. This term generally refers to the installation of drain, waste, vent and water distribution piping that will be concealed in floor, ceiling and wall cavities or underground or under slab. At this stage of construction, the plumbing systems are tested for leaks and the entire plumbing "rough-in" is inspected for code compliance.

SELF-CLOSING FAUCET. A faucet containing a valve that automatically closes upon deactivation of the opening means.

❖ Use of self-closing faucets is increasing, primarily in public restroom facilities, as public awareness of water conservation efforts increases. The flow of water automatically shuts off when the faucet handle is released, resulting in less water consumption per use.

SEPARATOR. See "Interceptor."

SEWAGE. Any liquid waste containing animal or vegetable matter in suspension or solution, including liquids containing chemicals in solution.

❖ Sewage is the general term referring to the combined discharge from all plumbing fixtures and primarily includes human bodily wastes and the wastes associated with cleaning, washing, bathing and food preparation.

SEWAGE EJECTOR. A device for lifting sewage by entraining the sewage in a high-velocity jet of steam, air or water.

❖ Ejectors are used to lift sewage from sumps serving subdrains. Ejectors are distinct from sewage pumps that use motor-driven impellers to lift the sewage [see Figure 712.1(1)].

SEWER.

❖ This is the general term used to describe piping that conducts waste from structures to the point of disposal. It also refers to piping within a public system that transports sewage to treatment facilities.

Building sewer. See "Building sewer."

❖ See the commentary to "Building sewer."

Public sewer. That part of the drainage system of pipes, installed and maintained by a city, township, county, public utility company or other public entity, and located on public property, in the street or in an approved dedicated easement of public or community use.

❖ A public sewer is the piping system used to collect sewage from building sewers and to transport it to treatment facilities.

Sanitary sewer. A *sewer* that carries sewage and excludes storm, surface and ground water.

❖ See the definition of "Building sewer, sanitary."

Storm sewer. A *sewer* that conveys rainwater, surface water, subsurface water and similar liquid wastes.

❖ This is the general term used to describe the piping that conducts waste other than sewage from structures to the point of disposal.

SLOPE. The fall (pitch) of a line of pipe in reference to a horizontal plane. In drainage, the slope is expressed as the fall in units vertical per units horizontal (percent) for a length of pipe.

❖ Commonly referred to as pitch, fall or grade, slope is the gradual change in elevation of horizontal piping. Slope is necessary to cause waste and sewage to flow by gravity (see commentary, Section 704.1).

SOIL PIPE. A pipe that conveys sewage containing fecal matter to the *building drain* or *building sewer*.

❖ In the plumbing trade, "soil pipe" is a common name for a cast-iron drainage pipe. The term, however, is not specific to cast-iron pipe and refers to any drain pipe that conveys the discharge of water closets or urinals or any other fixture that receives human waste.

SPILLPROOF VACUUM BREAKER. An assembly consisting of one check valve force-loaded closed and an air-inlet vent valve force-loaded open to atmosphere, positioned downstream of the check valve, and located between and including two tightly closing shutoff valves and a test cock.

❖ Spillproof vacuum breakers are designed primarily for indoor use where leakage or spillage caused by a loss of system pressure could cause problems such as water damage or health risks, but the performance requirements of a pressure-type vacuum breaker are still necessary.

STACK. A general term for any vertical line of soil, waste, vent or inside conductor piping that extends through at least one story with or without offsets.

❖ Traditionally, the term "stack" has been used to describe vertical piping [90 degrees (1.57 rad) from the horizontal] that extends through one or more stories of a building and constitutes a main to which branch piping connects. Vertical vent branch piping that extends to the open air or through two or more stories is considered a stack.

STACK VENT. The extension of a soil or waste *stack* above the highest horizontal drain connected to the *stack*.

❖ A stack vent is the dry extension of a soil or waste stack. Generally, a stack vent extends to the open air and can serve as a vent to which branch vents connect.

STACK VENTING. A method of venting a fixture or fixtures through the soil or waste *stack*.

❖ This is a method of wet venting where a uniformly oversized vertical drain serves as the vent for one or more fixtures. The fixtures being vented are independently connected to the stack. This venting method is limited to use with fixtures discharging waste only (see the definition of "Waste").

STERILIZER.

❖ A sterilizer is a plumbing appliance that uses heat to disinfect instruments, utensils and other equipment used in health care facilities.

Boiling type. A boiling-type sterilizer is a fixture of a nonpressure type utilized for boiling instruments, utensils or other equipment for disinfection. These devices are portable or are connected to the plumbing system.

❖ Boiling water is the oldest and most widely used form of microbial destruction. However, temperatures of water boiling at atmospheric pressure might not be high enough [212°F (100°C) or less] to kill all forms of bacteria. Thus, these types of sterilizers are not used widely where serious bacteria elimination is required.

Instrument. A device for the sterilization of various instruments.

❖ Usually this refers to a small bench top pressure-type, self-contained sterilizer used to disinfect hand instruments used in dental or minor surgery offices.

Pressure (autoclave). A pressure vessel fixture designed to utilize steam under pressure for sterilizing.

❖ Pressure sterilizers (autoclave) achieve temperatures well above 212°F (100°C) for complete bacteria elimination. The size of these devices can range from bench top size to room size units. The basic principle of operation is similar to that of a household pressure cooker. Water in a closed vessel having an airspace above the water is heated. Steam generated from the water rises into the airspace creating a pressure in the vessel. The increased pressure in the vessel raises the boiling point of the water. Thus, a steam temperature of greater than 212°F (100°C) is generated. For example, water heated in a closed vessel under a pressure of 10 psi (69.9 kPa) above atmospheric pressure will generate steam at a temperature of 240°F (115.6°C).

Pressure instrument washer sterilizer. A pressure vessel fixture designed to both wash and sterilize instruments during the operating cycle of the fixture.

❖ This device is a combination washer and a pressure-type sterilizer, all in one cabinet or unit (similar to a dishwasher). It removes debris from instruments and specialized surgical equipment (e.g, endoscopes).

Utensil. A device for the sterilization of utensils as utilized in health care services.

❖ This type of sterilizer is used for devices that are larger than small hand-held instruments and includes trays, pans, buckets, saws and drills.

Water. A device for sterilizing water and storing water.

❖ Sterilized water is required for many dental, medical and scientific procedures to prevent bacterial contamination. Water sterilizers are generally of the distillation type that condenses wet steam to produce sterile water.

STERILIZER VENT. A separate pipe or *stack*, indirectly connected to the building drainage system at the lower terminal, that receives the vapors from nonpressure sterilizers, or the exhaust vapors from pressure sterilizers, and conducts the vapors directly to the open air. Also called vapor, steam, atmospheric or exhaust vent.

❖ Sterilizer vents are independent of the sanitary venting system and serve only sterilizer appliances. Furthermore, sterilizer vents cannot serve more than one type of sterilizer (see commentary, Section 713.8).

STORM DRAIN. See "Drainage system, storm."

STORM WATER. Natural precipitation, including snowmelt, that has contacted a surface at or below grade.

❖ Although rainwater and snowmelt water make up storm water, for the purposes of Chapter 13, this water is not called storm water until it contacts grade areas such as sidewalks, earth, parking lots, paved areas, etc. Similarly rainwater directed into areas below grade such as underground piping or culverts leading to a river, stream, lake or pond, is considered storm water. Where there is a rainwater harvesting system having underground piping leading to a rainwater storage tank, the water in that piping and storage tank is not called storm water but rainwater (see the definition of "Rainwater").

[A] STRUCTURE. That which is built or constructed or a portion thereof.

❖ This definition is intentionally broad to include within its scope, and therefore the scope of the code (see commentary, Section 101.2), everything that is built as an improvement to real property. The phrase "or a portion thereof" is included so that those words do not have to be inserted at each location in the code where a provision applies to a portion of a structure.

SUBSOIL DRAIN. A drain that collects subsurface water or seepage water and conveys such water to a place of disposal.

❖ These drains are generally installed adjacent to the foundation of buildings to alleviate problems caused by subsurface (ground) water. Additionally, any other piping that collects either subsurface water or seepage would be termed a "subsoil drain."

SUMP. A tank or pit that receives sewage or liquid waste, located below the normal grade of the gravity system and that must be emptied by mechanical means.

❖ This term refers to the receiving tank or vessel used to collect and store waste from drainage systems that are incapable of draining by gravity. A sump generally houses an ejector or pump used to evacuate the contents of the sump. A sump can refer to a receiver of either waste water or storm water.

SUMP PUMP. An automatic water pump powered by an electric motor for the removal of drainage, except raw sewage, from a sump, pit or low point.

❖ These pumps are used for removing waste from sumps other than sewage sumps. Sump pumps handle storm water only.

SUMP VENT. A vent from pneumatic sewage ejectors, or similar equipment, that terminates separately to the open air.

❖ A sump vent is needed to neutralize pressures that develop in closed sumps as a result of either the rise and fall of waste levels or the operation of pneumatic ejectors.

SUPPORTS. Devices for supporting and securing pipe, fixtures and equipment.

❖ Types of supports include hangers, anchors, braces, brackets, strapping and any other material or device used to support plumbing system piping and components (see commentary, Section 308).

SWIMMING POOL. Any structure, basin, chamber or tank containing an artificial body of water for swimming, diving or recreational bathing having a depth of 2 feet (610 mm) or more at any point.

❖ Swimming pools, for the application of the code, are man-made structures designed to hold water. Bodies of water such as ponds, streams, lakes or retention ponds are not considered to be swimming pools.

Swimming pools are generally filled with potable water. The water in swimming pools is filtered to remove contaminants and debris and is chemically treated to control bacterial growth, plant life growth and acidity/alkalinity levels. Swimming pools are used for recreational purposes and include private and public swimming pools, indoor and outdoor swimming pools, hot tubs and spas. The water discharged from a swimming pool having water that meets the quality requirements for recreational swimming is considered to be clear-water waste.

Note that the IBC and the IRC reference the ISPSC for all pool and spa requirements. The ISPSC does not establish a minimum depth for classification as a swimming pool. If the pool has a filtration system and is intended for recreational use, then the pool qualifies as a swimming pool according to the ISPSC.

Bathtubs and therapeutic whirlpools used for hygienic or medical purposes are not considered to be swimming pools.

TEMPERED WATER. Water having a temperature range between 85°F (29°C) and 110°F (43°C).

❖ Tempered water (warm water) is at a temperature suitable for bathing, handwashing and similar purposes and is defined as a range between 85°F (29°C) and 110°F (43°C) (see commentary, Section 607.1).

THIRD-PARTY CERTIFICATION AGENCY. An *approved* agency operating a product or material certification system that incorporates initial product testing, assessment and surveillance of a manufacturer's quality control system.

❖ These agencies evaluate products, verify compliance with specific standards or criteria and conduct periodic inspections of the manufacturer's facilities to determine ongoing compliance with manufacturing processes, standards or criteria (see Section 303.4).

THIRD-PARTY CERTIFIED. Certification obtained by the manufacturer indicating that the function and performance characteristics of a product or material have been determined by testing and ongoing surveillance by an *approved third-party certification agency*. Assertion of certification is in the form of identification in accordance with the requirements of the *third-party certification agency*.

❖ A product that is certified by a third-party certification agency (see commentary, "Third-party certification agency") to comply with requisite standards or criteria, usually identified by a certification mark attached to or permanently marked on it. Additionally, the agency usually provides a list of the products it certifies, which provides information relative to the certification (see Section 303.4).

TOILET FACILITY. A room or space that contains not less than one water closet and one lavatory.

❖ The common household name for a water closet is "toilet." Most manufacturers of water closets call the product a "toilet." Thus, the tendency by many readers of the code is to assume that a toilet facility is a room or space that has a "toilet" (water closet). This misunderstanding can create a problem where the code requires separate (male/female) "toilet" facilities. One might (incorrectly) interpret this to mean that only separate (male/female) water closet compartments are required. The intent of the code is that a toilet room is a room or space that has not less than one water closet and one lavatory.

TRAP. A fitting or device that provides a liquid seal to prevent the emission of *sewer* gases without materially affecting the flow of sewage or waste water through the trap.

❖ The sole purpose of a trap is to isolate the interior of occupiable spaces from the sanitary drainage and vent system. A trap "traps" liquid waste and retains it to form a seal or barrier through which gases and vapors in the drainage and vent system cannot pass under normal operating conditions. Traps are either separate devices or integral with fixtures. Other than a grease trap or trap that is integral with a fixture, the only type of trap permitted by the code is the "P" type (see commentary, Sections 1002.1, 1002.3, 1002.4 and 1002.5).

TRAP SEAL. The vertical distance between the weir and the top of the dip of the trap.

❖ The trap seal is a measurement of the quantity of liquid held in the dip of a trap. Trap seals are required to have a minimum depth (see commentary, Section 1002.4).

UNSTABLE GROUND. Earth that does not provide a uniform bearing for the barrel of the *sewer* pipe between the joints at the bottom of the pipe trench.

❖ Any earth that will be used to support underground sewer piping that is not capable of providing adequate, uniform bearing so that the sewer piping and its related fittings are not damaged is defined as "unstable ground." This term can relate to either earth that exists as is or any earth disturbed by construction that is not properly replaced and compacted.

Typically, when unstable ground is encountered during the course of construction, the portion of the ground that is unstable is either removed and replaced, or is made stable through various engineering methods, such as dewatering or chemical injections.

VACUUM. Any pressure less than that exerted by the atmosphere.

❖ Commonly referred to as "negative pressure" and "suction," vacuum is measured in inches of mercury or inches of water column. Any pressure below atmospheric pressure is a partial vacuum. A perfect vacuum is a pressure of zero pounds per square inch (psi) (0 kPa) absolute. Atmospheric pressure at sea level is approximately 14.7 psi (101 kPa), which converts to approximately 29.92 inches (760 mm) of mercury or approximately 34 feet (10 363 mm) of water. The phenomenon of siphonage depends on vacuum (negative pressures).

VACUUM BREAKER. A type of backflow preventer installed on openings subject to normal atmospheric pressure that prevents backflow by admitting atmospheric pressure through ports to the discharge side of the device.

❖ A vacuum breaker is a device used as a backflow preventer. Whenever this device senses a system pressure that is less than atmospheric pressure (e.g, a vacuum), a port opens to allow atmospheric pressure to enter the device to relieve (break) the vacuum condition in the piping.

VENT PIPE. See "Vent system."

❖ The term "vent pipe" refers to any pipe in the vent system that conducts air to the drainage system, from the drainage system or both.

VENT STACK. A vertical vent pipe installed primarily for the purpose of providing circulation of air to and from any part of the drainage system.

❖ A dry vent that extends from the base of a waste or soil stack and typically runs parallel and adjacent to the waste or soil stack it serves. The vent stack can connect either to the horizontal drain downstream of the drain stack or directly to the drainage stack (see commentary, Section 903.2).

VENT SYSTEM. A pipe or pipes installed to provide a flow of air to or from a drainage system, or to provide a circulation of air within such system to protect trap seals from siphonage and backpressure.

❖ A piping arrangement that is designed to maintain pressure fluctuation at fixture traps to within plus or minus 1 inch of water column (249 Pa) (see commentary, Section 901.2).

VERTICAL PIPE. Any pipe or fitting that makes an angle of 45 degrees (0.79 rad) or more with the horizontal.

❖ A vertical pipe is any pipe that is not defined as a horizontal pipe (see the definition of "Horizontal pipe"). Vertical pipes are pipes that are angled 45 degrees (0.79 rad) or more above the horizontal.

WALL-HUNG WATER CLOSET. A wall-mounted water closet installed in such a way that the fixture does not touch the floor.

❖ A wall-hung water closet is designed for installation above the floor. The water closet discharges horizontally through the wall (see Commentary Figure 405.4.3).

WASTE. The discharge from any fixture, appliance, area or appurtenance that does not contain fecal matter.

❖ Any plumbing fixture discharge other than that from a water closet, urinal, clinical sink or bedpan washer is defined as waste. This material does not contain human fecal matter.

WASTE PIPE. A pipe that conveys only waste.

❖ A waste pipe is any waste-conducting pipe not falling under the definition of "Soil pipe." Waste pipes do not convey human fecal matter.

WASTE RECEPTOR. A floor sink, standpipe, hub drain or floor drain that receives the discharge of one or more indirect waste pipes.

❖ A common assumption is that only floor sinks and hub drains are waste receptors. Standpipes and floor drains are rarely thought of as being waste receptors but they are commonly installed for capturing the flow from indirect waste pipes. A floor drain is only considered to be a waste receptor where it receives the discharge of one or more indirect waste pipes.

WATER COOLER. A drinking fountain that incorporates a means of reducing the temperature of the water supplied to it from the potable water distribution system.

❖ There has been much confusion over terminology regarding drinking fountains, water coolers and bottled water dispensers. A water cooler is simply a drinking fountain that reduces the temperature of the water from the potable water supply system before sending the stream of drinking water to the user (see the definition of "Drinking fountain").

WATER DISPENSER. A plumbing fixture that is manually controlled by the user for the purpose of dispensing potable drinking water into a receptacle such as a cup, glass or bottle. Such fixture is connected to the potable water distribution system of the premises. This definition also includes a freestanding apparatus for the same purpose that is not connected to the potable water distribution system and that is supplied with potable water from a container, bottle or reservoir.

❖ This term is intended to encompass fixtures, other than drinking fountains and water coolers, that provide drinking water into a drinking water container such as a cup, glass or bottle. The potable water dispensed is from a connection to the potable water distribution system of the building. The term also encompasses devices that are not connected to the potable water system where the water is from a container, bottle or reservoir on the device.

WATER-HAMMER ARRESTOR. A device utilized to absorb the pressure surge (water hammer) that occurs when water flow is suddenly stopped in a water supply system.

❖ Water-hammer arrestors are factory-built mechanical devices that use expandable elements or piston-type compression chambers to dissipate the energy of rapidly decelerating water (see commentary, Section 604.9). Site-built air chambers do not qualify as acceptable water-hammer arresters.

[M] WATER HEATER. Any heating appliance or equipment that heats potable water and supplies such water to the potable *hot water* distribution system.

❖ The types of water heaters include storage (tank) type, circulating type and instantaneous type. Point-of-use water heaters typically supply hot water to a single fixture or outlet and are located in close proximity to it.

WATER MAIN. A water supply pipe or system of pipes, installed and maintained by a city, township, county, public utility company or other public entity, on public property, in the street or in an approved dedicated easement of public or community use.

❖ A water main is owned and operated by municipalities, rural water districts, privately owned water purveyors and other such entities.

WATER OUTLET. A discharge opening through which water is supplied to a fixture, into the atmosphere (except into an open tank that is part of the water supply system), to a boiler or heating system, or to any devices or equipment requiring water to operate but which are not part of the plumbing system.

❖ Literally, a water outlet is any point where water is withdrawn from the potable water distribution system. An outlet, such as a faucet or spigot, discharges water through the atmosphere into a fixture or receptacle. Outlets are not direct connections to other systems except into an open tank that is part of the water supply system (see commentary, Section 608.15).

WATER PIPE.

Riser. A water supply pipe that extends one full story or more to convey water to *branches* or to a group of fixtures.

❖ A riser extends vertically to supply water to upper stories.

Water distribution pipe. A pipe within the structure or on the premises that conveys water from the water service pipe, or from the meter when the meter is at the structure, to the points of utilization.

❖ Water distribution pipe refers to all potable water piping that is downstream of the water meter or termination of the water service.

Water service pipe. The pipe from the water main or other source of potable water supply, or from the meter when the meter is at the public right of way, to the water distribution system of the building served.

❖ Water pipe is any code-accepted pipe material used to convey potable water.

WATER SUPPLY SYSTEM. The water service pipe, water distribution pipes, and the necessary connecting pipes, fittings, control valves and all appurtenances in or adjacent to the structure or premises.

❖ The water supply system includes all piping and components that convey potable water from the public main or other source to the points of water usage.

WELL.

❖ A well is a vertical shaft that extends down into the earth for the purpose of extracting water from underground formations or strata. Wells can be constructed by drilling, boring or digging in the earth or created by driving a pipe (casing) into the earth. Wells are the source of water for many public water supply systems and are also used to serve private (individual) water supply systems. Depending on the location, some water from wells may be potable without requiring any treatment, whereas some must be treated to make it potable. The following four methods are the most common or typical forms of well construction and are self-defining.

Bored. A well constructed by boring a hole in the ground with an auger and installing a casing.

❖ See the commentary to "Well."

Drilled. A well constructed by making a hole in the ground with a drilling machine of any type and installing a casing and screen.

❖ See the commentary to "Well."

DEFINITIONS

Driven. A well constructed by driving a pipe in the ground. The drive pipe is usually fitted with a well point and screen.

❖ See the commentary to "Well."

Dug. A well constructed by excavating a large-diameter shaft and installing a casing.

❖ See the commentary to "Well."

WHIRLPOOL BATHTUB. A plumbing appliance consisting of a bathtub fixture that is equipped and fitted with a circulating piping system designed to accept, circulate and discharge bathtub water upon each use.

❖ Whirlpool bathtubs are equipped with circulator pumps, suction fittings, piping and related appurtenances designed to agitate water in the tub for relaxation or therapeutic purposes (see commentary, Section 421). In recent years, consumers have embraced multihead luxury showers, which can discharge large volumes of water during a showering event. Unfortunately, there is the direct cost of the water and sewer charges and the indirect cost of high-capacity water heating systems and larger piping for high flows. Those facts coupled with not being a conservative water practice, have led some manufacturers to also offer "whirlpool tub/showers." These multipiece units utilize a bathtub with a vertical panel having multiple spray heads and jets. The bathtub serves as the water reservoir for a whirlpool-type pump to convey water to the shower heads and jets in a closed-system arrangement. An in-line heater maintains the desired water temperature. Once the tub is filled with a sufficient volume of water, the shower can be used as long as desired without further introduction of water, resulting in much less water use than a shower, where all the water discharged goes to the drain system.

YOKE VENT. A pipe connecting upward from a soil or waste *stack* to a vent *stack* for the purpose of preventing pressure changes in the *stacks*.

❖ This vent derives its name from the manner in which it connects a drainage stack to the attendant vent stack. Its purpose is to prevent pressure changes in the stack.

Chapter 3:
General Regulations

General Comments

The content of Chapter 3 is often referred to as "miscellaneous," rather than general regulations. Chapter 3 received that label because it is the only chapter in the code whose requirements do not interrelate. If a requirement cannot be located in another chapter, it should be located in this chapter.

Some nonplumbing regulations merely reference other codes that have the specific requirements. The requirements provide a cross reference to the appropriate document, recognizing that it affects the plumbing system but the details are not specifically contained in the code (Sections 307, 309, 310 and 313 reference other *International Codes*®).

The jurisdictional requirements specify that the water and sewer must connect to the public system when a public system is provided (Sections 602.1 and 701.2 are more specific on this issue).

Nonplumbing requirements include surface requirements for walls and floors in a toilet room, light and ventilation, floodproofing and rodentproofing.

Purpose

Chapter 3 contains safety requirements for the installation of plumbing and nonplumbing requirements for toilet rooms, including requirements for the identification of pipe, pipe fittings, traps, fixtures, materials and devices used in plumbing systems.

The safety requirements provide protection for the building's structural members, stress and strain of pipe and sleeving. The building's structural stability is protected by the regulations for cutting and notching of structural members. Additional protection for the building occupants includes requirements to maintain the plumbing in a safe and sanitary condition, as well as privacy for those occupants.

SECTION 301
GENERAL

301.1 Scope. The provisions of this chapter shall govern the general regulations regarding the installation of plumbing not specific to other chapters.

❖ The requirements included in Chapter 3 are not interrelated, as is typical with other chapters. Many regulations are not specific plumbing requirements, but relate to the overall plumbing system.

301.2 System installation. Plumbing shall be installed with due regard to preservation of the strength of structural members and prevention of damage to walls and other surfaces through fixture usage.

❖ Plumbing components and materials are to be installed in accordance with the requirements of the applicable standard listed in the code.

Where a standard is not provided, the manufacturer's instructions must be followed. For example, because there are very few standards available that regulate the installation of valves, the manufacturer's instructions must be used to install these components.

301.3 Connections to drainage system. Plumbing fixtures, drains, appurtenances and appliances used to receive or discharge liquid waste or sewage shall be directly connected to the sanitary drainage system of the building or premises, in accordance with the requirements of this code. This section shall not be construed to prevent indirect waste systems required by Chapter 8.

Exception: Bathtubs, showers, lavatories, clothes washers and laundry trays shall not be required to discharge to the sanitary drainage system where such fixtures discharge to an *approved* system in accordance with Chapters 13 and 14.

❖ All waste water captured or generated within a building is required to be directed into the sanitary drainage system. In most cases, the connection to the drainage system is a direct connection, meaning that each drain pipe is routed in a continuous manner to its connecting point to a branch of the sanitary drainage system. However, there are some situations where it is not desirable to have a direct connection. For example, a direct connection of the drainage pipe of a commercial kitchen food preparation sink is not desirable because a waste-water backup could contaminate the contents of the sink, perhaps without being noticed by kitchen staff. In these special cases, the drainage pipe is required to discharge through an air break or an air gap into a waste receptor (usually a floor sink), which is indirectly connected to the sanitary drainage system. The outlet of the waste receptor is directly connected to the sanitary drainage system. Chapter 8 covers which fixtures and appliance drains are required to be indirect connected.

The exception to this section recognizes that certain waste-water flows can be a source of water that can be reused for the purposes of water-closet flushing, urinal flushing or subsurface irrigation. These waste-water flows must be collected by a drainage system that is wholly separate from the sanitary drainage system so that extensive treatment of such water is not required before reuse.

301.4 Connections to water supply. Every plumbing fixture, device or appliance requiring or using water for its proper operation shall be directly or indirectly connected to the water supply system in accordance with the provisions of this code.

❖ Fixtures, devices and appliances that require potable water or nonpotable water must be connected, directly or indirectly, to the water supply. In other words, a sink (other than a floor sink) cannot be installed without a faucet installed on the sink.

Indirect connections include faucets or fixture fittings discharging into fixtures such as tubs and lavatories. Direct connections occur at water closets and urinals. Water closets and urinals can be supplied with treated gray water through a direct connection to the fixture if the system meets the requirements of Chapter 13.

301.5 Pipe, tube and fitting sizes. Unless otherwise indicated, the pipe, tube and fitting sizes specified in this code are expressed in nominal or standard sizes as designated in the referenced material standards.

❖ Pipe, tube and fitting sizes referenced in the code refer to the inside diameter (ID) of the pipe, tube or fitting. The ID measurement in the text is expressed in both English and metric units (inches and millimeters). Systeme International d'Unites (SI) metric unit conversions are indicated at the bottom of each table.

301.6 Prohibited locations. Plumbing systems shall not be located in an elevator shaft or in an elevator equipment room.

Exception: Floor drains, sumps and sump pumps shall be permitted at the base of the shaft, provided that they are indirectly connected to the plumbing system and comply with Section 1003.4.

❖ Plumbing systems are prohibited in elevator shafts and elevator equipment rooms because of inaccessibility for repairs and the possible water damage that could be caused to the elevator equipment if a leak developed in the plumbing piping or components. The exception allows floor drains, sumps and sump pumps at the bottom of elevator shafts as long as they are indirectly connected to the plumbing system. An indirect connection is required to prevent waste from a plumbing system from backing up into the elevator shaft. Note that a back-water valve cannot be used as a substitute for the indirect connection.

Because the defined term "plumbing system" includes both sanitary drains and storm sewers, the designer has to make a decision of whether to put the discharge from the floor drain, sump or sump pump into the sanitary system or storm sewer. Two reasons for water to be in the base of an elevator shaft are: 1) the base of many elevator shafts is below grade where ground water (from rain events or a seasonally high water table) might enter through cracks and seams in the walls and floors of the shaft and 2) water from an activated fire sprinkler system could enter through elevator doors. If the water is considered to be storm water, Section 1101.3 would prohibit connection to the sanitary system. If the water is considered no different than what would enter a floor drain, then Section 301.3 would require its discharge to the sanitary system. Other considerations could be whether or not local storm water regulators or waste-water plant operators have authority to specify where such water should be discharged.

The exception references Section 1003.4 to alert the code user that if the elevator is a hydraulic type, an oil separator is required to be installed before the discharge of the floor drain, sump or sump pump enters the plumbing system.

301.7 Conflicts. In instances where conflicts occur between this code and the manufacturer's installation instructions, the more restrictive provisions shall apply.

❖ A conflict refers to instances where the code and manufacturer's instructions differ. The code official must evaluate each circumstance of perceived conflict and identify the requirements that provide the greatest level of protection for life and property.

SECTION 302
EXCLUSION OF MATERIALS DETRIMENTAL TO THE SEWER SYSTEM

302.1 Detrimental or dangerous materials. Ashes, cinders or rags; flammable, poisonous or explosive liquids or gases; oil, grease or any other insoluble material capable of obstructing, damaging or overloading the building drainage or *sewer* system, or capable of interfering with the normal operation of the sewage treatment processes, shall not be deposited, by any means, into such systems.

❖ This section prohibits the disposal of detrimental or dangerous materials into the sewer system. Such materials can cause the pipes to clog or accelerate the clogging of pipes, which prevents the proper disposal of sewage waste. Section 1003 contains design and installation details for the use of interceptors, grease traps and separators to remove oil, grease, sand and other detrimental substances.

Discharge of materials that are flammable or combustible into the public sewer system is prohibited because an accumulation of these types of materials poses a fire and explosion hazard. Insoluble chemicals that are not processed before disposal could react with other discharged chemicals to cause damage to the piping and components of the drainage, sewer and waste treatment systems. Section 803.2 provides details for using approved dilution or neutralizing devices to process harmful chemicals prior to disposal.

302.2 Industrial wastes. Waste products from manufacturing or industrial operations shall not be introduced into the *public sewer* until it has been determined by the code official or other authority having jurisdiction that the introduction thereof will not damage the public sewer system or interfere with the functioning of the sewage treatment plant.

❖ Harmful or hazardous industrial waste must be treated before it is discharged to the sewer. This can require the complete removal or neutralization of certain chemicals or substances.

SECTION 303
MATERIALS

303.1 Identification. Each length of pipe and each pipe fitting, trap, fixture, material and device utilized in a plumbing system shall bear the identification of the manufacturer and any markings required by the applicable referenced standards.

❖ The manufacturer is given the option of determining the type of marking for the material. If there is no applicable standard or the applicable standard does not require that a material be identified, identification of the manufacturer is still required by the code. Where the code indicates compliance with an approved standard, the manufacturer must comply with the requirements for marking in accordance with the applicable standard [see Commentary Figures 303.1(1) and 303.1(2)].

303.2 Installation of materials. All materials used shall be installed in strict accordance with the standards under which the materials are accepted and *approved*. In the absence of such installation procedures, the manufacturer's instructions shall be followed. Where the requirements of referenced standards or manufacturer's installation instructions do not conform to minimum provisions of this code, the provisions of this code shall apply.

❖ Plumbing components and materials are to be installed in accordance with the requirements of the applicable standard listed in the code. Where a standard is not provided, the manufacturer's instructions must be followed. For example, because there are very few standards available that regulate the installation of valves, the manufacturer's instructions must be used to install these components.

303.3 Plastic pipe, fittings and components. All plastic pipe, fittings and components shall be third-party certified as conforming to NSF 14.

❖ Plastic piping, fittings and plastic pipe-related components, including solvent cements, primers, tapes, lubricants and seals used in plumbing systems, must be tested and certified as conforming to NSF 14. This includes all water service, water distribution, drainage piping and fittings and plastic piping system components, including but not limited to pipes, fittings, valves, joining materials, gaskets and appurtenances. This section does not apply to components that only include plastic parts such as brass valves with a plastic stem, or to fixture fittings such as fixture stop valves. Plastic piping systems, fittings and related components intended for use in the potable water supply system must comply with NSF 61 in addition to NSF 14.

303.4 Third-party certification. All plumbing products and materials shall be listed by a third-party certification agency as complying with the referenced standards. Products and materials shall be identified in accordance with Section 303.1.

❖ The term "third party" refers to an outside organization with no financial or other interest in the outcome.
The code requirements for testing and certification have frequently confused code officials and manufacturers over the years. Securing and submitting the necessary documentation for certain products and materials is often a challenge for contractors and engineers. The code official was also burdened with trying to keep up with the myriad of products he or she sees in the field and the documentation that has (or has not) been submitted. To simplify inspections and approvals, the code requires that all products and materials

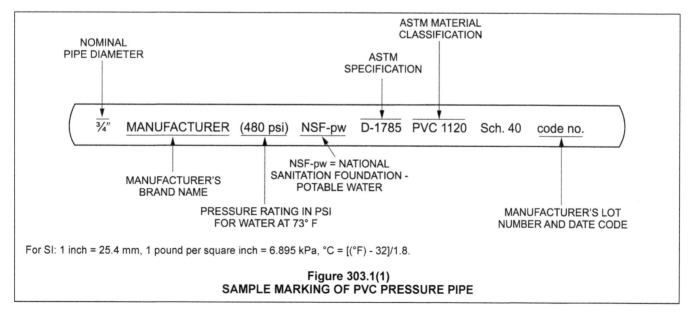

For SI: 1 inch = 25.4 mm, 1 pound per square inch = 6.895 kPa, °C = [(°F) - 32]/1.8.

**Figure 303.1(1)
SAMPLE MARKING OF PVC PRESSURE PIPE**

GENERAL REGULATIONS

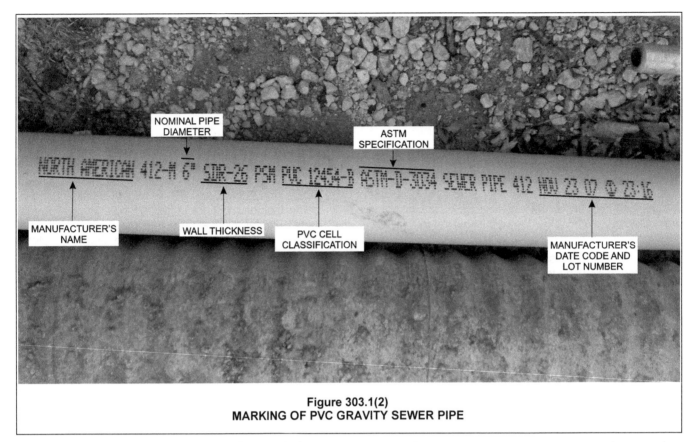

Figure 303.1(2)
MARKING OF PVC GRAVITY SEWER PIPE

be third-party certified. The code official only has to confirm that the product or material has the mark of the third-party certifying agency.

SECTION 304
RODENTPROOFING

304.1 General. Plumbing systems shall be designed and installed in accordance with Sections 304.2 through 304.4 to prevent rodents from entering structures.

❖ Rodents are known to be carriers of diseases and present serious health risks to humans. To prevent the spread of disease, Sections 304.2 through 304.4 require plumbing systems to be installed in a manner that will reduce the potential for rodent entry into structures.

304.2 Strainer plates. All strainer plates on drain inlets shall be designed and installed so that all openings are not greater than $^1/_2$ inch (12.7 mm) in least dimension.

❖ Rodents often travel and live within sanitary sewer systems. The limitation for opening size in strainer plates for floor and shower drains as well as receptor strainers provides two forms of protection. If rodents are in the sewer system, strainer plates prevent them from entering the building through the floor or shower drain. If rodents are within the structure itself, the strainer plate prevents rodent access to the drainage system.

304.3 Meter boxes. Meter boxes shall be constructed in such a manner that rodents are prevented from entering a structure by way of the water service pipes connecting the meter box and the structure.

❖ The water service pipe may be tunneled into the building from the meter box. When such an installation occurs, the annular space around the pipe must be protected to prevent rodents from getting into the building. This can be accomplished by a barrier to block their entry, such as a corrosion-resistant heavy wire screen or metal plate that is securely fastened in place.

304.4 Openings for pipes. In or on structures where openings have been made in walls, floors or ceilings for the passage of pipes, the annular space between the pipe and the sides of the opening shall be sealed with caulking materials or closed with gasketing systems compatible with the piping materials and locations.

❖ Rodents can enter a building through the annular spaces around piping that penetrates a wall, ceiling or floor. These spaces must be blocked using caulking or gasketing (see also commentary, Section 315).

SECTION 305
PROTECTION OF PIPES AND
PLUMBING SYSTEM COMPONENTS

305.1 Corrosion. Pipes passing through concrete or cinder walls and floors or other corrosive material shall be protected against external corrosion by a protective sheathing or wrapping or other means that will withstand any reaction from the lime and acid of concrete, cinder or other corrosive material.

Sheathing or wrapping shall allow for movement including expansion and contraction of piping. The wall thickness of the material shall be not less than 0.025 inch (0.64 mm).

❖ Metallic piping must be protected from external corrosion in locations where the piping is in direct contact with materials known to be corrosive.

This section specifically identifies walls and floors built of cinder(s) or concrete as being corrosive. Cinders are a byproduct created from the burning of coal. Cinders contain sulfur compounds and, where exposed to moisture, these compounds can form acids that can be corrosive to metals. Cinders are sometimes used as an aggregate in lightweight (nonstructural) concrete for slab-on-grade floors and lightweight (nonstructural) "cinder blocks."

Concrete, in a wet slurry state, is caustic (pH level of 12 to 13) due to the limestone-based portland cement in the concrete reacting with water. Because concrete (and masonry units made from concrete) could contain acid-producing cinders and cured concrete is still somewhat caustic, metallic piping must be protected from direct contact with concrete or concrete masonry units.

Naturally corrosive soils, such as found in swamps, peat bogs or tidal marshes, can corrode metallic piping. Fill materials containing furnace slag, cinders, ash or spent industrial byproducts can also be corrosive to metallic piping. The designer or installer should consult with soils engineers, piping manufacturers as well as local code officials concerning the need for corrosion protection of metallic piping in contact with soil or fill materials of potentially corrosive nature.

Direct exposure of metallic piping to corrosive materials can be prevented by sheathing or wrapping of the piping. Sheathing involves inserting the piping into a flexible plastic sleeve of material. Wrapping of piping is where a flexible product is "wrapped" around the pipe circumference. Sheets of flexible plastic material can be wrapped around the circumference of the piping with sufficient overlap at the ends to ensure complete protection of the pipe.

Where sheathing or wrapping protects piping, it must allow for pipe movement. In other words, the sheathing or wrapping should not be adhered to the pipe so that the pipe cannot expand and contract within the protective material. This section requires the thickness of the sheathing or wrapping to be at least 0.025 inch (0.64 mm). Note that this thickness value is much greater that the commonly available flexible plastic sleeve materials (usually colored red or blue for pipe identification) of either 0.006 or 0.010 inch (0.15 mm or 0.25 mm) that are found to be stocked in the majority of plumbing supply houses. Although use of these thinner sheathing materials appears to be a widely accepted practice, these materials do not meet the minimum thickness requirement stated in this section. Therefore, the code official must approve the use of sheathing materials having a material thickness of less than 0.025 inch (0.64 mm) thick.

305.2 Stress and strain. Piping in a plumbing system shall be installed so as to prevent strains and stresses that exceed the structural strength of the pipe. Where necessary, provisions shall be made to protect piping from damage resulting from expansion, contraction and structural settlement.

❖ A plumbing system must not be damaged by stresses, strains or movement of the building components. In piping systems, provisions must be made for the expansion and contraction of the pipes themselves. Each piping material has a different rate of expansion and contraction that must be considered when designing the restraint system for the piping and the structure that surrounds the piping system.

Changes in temperature can cause distortion of the pipe material; for example, heat causes the material to expand. The greatest amounts of expansion and contraction in piping will occur along the length of the pipe. Hot water piping can experience significant movements in short runs of piping. Though the amount of expansion per unit length is low, large movements can occur in long lengths of pipe. Commentary Figure 305.2 contains the expansion rate for various pipe materials.

The most common method to absorb thermal expansion in plumbing piping systems is through the installation of one or more offsets in the piping. The typical offset piping arrangements used are the "L" bend (1-elbow change in direction), the "Z" bend (2-elbow offset) and the "U" bend (4-elbow offset) [see Commentary Figures 305.2(1) through 305.2(3)]. The

PIPING MATERIAL	RATE OF THERMAL EXPANSION in/in/°F
Brass-red	0.000009
Copper	0.00001
Cast iron	0.0000056
Carbon steel	0.000005
Ductile iron	0.0000067
Stainless steel	0.0000115
Borosilicate (glass)	0.0000018
ABS	0.00005
CPVC	0.000035
HDPE	0.00011
PE	0.00008
PEX	0.00093
PP	0.000065
PVC	0.00004
PVDF	0.000096

Source: *Facility Piping Systems Handbook*, 2nd Edition, Frankel, McGraw Hill, 2002.

For SI: 1 inch = 25.4 mm, °F = 68°C + 32.

Figure 305.2
THERMAL EXPANSION RATES FOR PIPING MATERIALS

length, L, in each of these figures is determined from the following equation:

$$L = [1.5 \times (E/S) \times D_o \times (G)]^{1/2}$$

where:

E = Modulus of elasticity at pipe temperature (psi)

S = Maximum allowable stress for the pipe material at the highest in-service pipe temperature (psi)

D_o = Outside diameter of pipe (inches)

G = Change in length of piping due to temperature change (inches). Thermal expansion rate of the pipe material from Commentary Table 305.2 × Change in temperature of the pipe (installation day to highest in-service temperature expected) × Length of pipe in feet between points where the pipe is restricted from expanding in length × 12 inches per foot.

The values for E and S are obtained from piping manufacturers, engineering publications or the material standards for the pipe products.

Sample Problem: A 160-foot-long straight run of 1-inch copper tube size CPVC pipe is to be installed to convey hot water at 140°F. The ends of the piping will be attached to equipment. The temperature at the time of installation will be 50°F. Determine the required offset length, L, to accommodate the thermal expansion that will occur when the system is operating.

Problem Approach

Obtain the thermal expansion rate from Commentary Table 305.2, calculate the amount of thermal expansion of the piping run, obtain the modulus of elasticity and allowable stress values from the piping manufacturer, and calculate the required offset length, L.

Solution

For CPVC piping material, the thermal expansion rate listed in Commentary Table 305.2 is 0.000035 inches/inch/°F. The change in temperature of the pipe from installation to service condition is 140°F minus 50°F, which equals 90°F. The change in length of the pipe, G, is calculated as follows:

G = 0.000035 inches/inch/°F × 90°F × 160 feet × 12 inches per foot

G = 6.1 inches

Figure 305.2(1)
"L" BEND

OFFSET PIPING MUST BE SUPPORTED WITH HANGERS, SLIDE PLATES OR SPRING HANGERS.

Figure 305.2(2)
"Z" BEND

Figure 305.2(3)
"U" BEND

The pipe manufacturer provides a modulus of elasticity of 3.23×10^5 psi and an allowable stress of 1000 psi, both at the service temperature of 140°F. The outside diameter of the pipe, D_o, is 1.125 inches. Therefore,

$L = [1.5 \times (E/S) \times D_o \times G]^{1/2}$

$L = [1.5 \times (3.23 \times 10^5/1000) \times 1.125 \times 12.5]^{1/2}$

$L = 57.7$ inches

Expansion joints can also be used to absorb thermal expansion. These products are preengineered or can be custom engineered for any application. The code does not regulate the design or installation of expansion joints.

The change in dimension due to temperature change is not limited to the length of the pipe but affects the diameter as well. As the temperature of a pipe increases, the pipe will "swell" to a larger diameter. If the thermal expansion is restrained such as by casting of the pipe in a concrete wall, significant stresses can develop in the pipe wall.

305.3 Pipes through foundation walls. Any pipe that passes through a foundation wall shall be provided with a relieving arch, or a pipe sleeve pipe shall be built into the foundation wall. The sleeve shall be two pipe sizes greater than the pipe passing through the wall.

❖ Piping installed through a foundation wall must be structurally protected from any transferred loading from the foundation wall. This protection must be provided by either a relieving arch or a pipe sleeve.

Where a pipe sleeve is used, it must be sized to be two pipe sizes larger than the pipe that passes through the pipe sleeve. For example, a 4-inch (102 mm) drainage pipe would require a 6-inch (152 mm) pipe sleeve. The annular space between the pipe sleeve and the pipe that passes through the sleeve will allow for any differential movement between the pipe and the wall. The pipe sleeve or relieving arch protects the pipe so that it will not be subjected to undue stresses that could cause it to rupture and leak [see Commentary Figures 305.3(1) and 305.3(2)].

In rare circumstances, it is sometimes necessary for a pipe to pass though a footing. For example, the slope of a building drain might not be able to be changed, resulting in the drainage pipe needing to pass through the footing. Although the code does not specifically address pipes passing through footings, the need for protecting pipes passing though footings is no different than for pipes passing through foundation walls. In these situations, protection of the pipe is not the only concern as the strength of footing could be compromised by the installation of a pipe sleeve or relieving arch. Any footing to be altered, either before or after placement of the footing, should be reviewed by a design professional to determine what footing design changes might be necessary to maintain the required footing strength.

Prior editions of the code required a relieving arch or a pipe sleeve for a pipe that passes under a footing. This was determined to be unnecessary as the footing already acts as a relieving arch to "span over" the area where the pipe passes under the footing. This is not to say that routing any size of or number of pipes under a footing is without consequences. Footings are designed to be uniformly supported by undisturbed or highly compacted soil. A trench for a pipe passing under the footing disrupts the uniform support for the footing. If the trench can be backfilled and compacted to the same level of compaction as the surrounding undisturbed or highly compacted soil, then the footing will be uniformly supported. But if the trench cannot be compacted sufficiently, such as where the trench is dug after the footing is in place, then the footing will not be uniformly supported by the soil; the footing will experience more stress than it was designed for. In actual practice, in a low-rise building where one or two "small" diameter pipes [≤ 8 inches (≤ 203 mm)] pass

GENERAL REGULATIONS

Figure 305.3(1)
PIPE PROTECTION

Figure 305.3(2)
BUILDING DRAIN PENETRATION THROUGH FOUNDATION WALL

under a footing in the same area, the footing will most likely have sufficient reserve capacity to span the trench. Larger buildings typically requiring larger pipes will have footings that are already designed for spanning over the trenches of these larger pipes. There are no "rules of thumb" to know when a footing design should be analyzed for adequacy. Installers and code officials need to be aware of the possible consequences to the footings and, where necessary, consult with a design professional concerning the footing design.

305.4 Freezing. Water, soil and waste pipes shall not be installed outside of a building, in attics or crawl spaces, concealed in outside walls, or in any other place subjected to freezing temperatures unless adequate provision is made to protect such pipes from freezing by insulation or heat or both. Exterior water supply system piping shall be installed not less than 6 inches (152 mm) below the frost line and not less than 12 inches (305 mm) below grade.

❖ Water, soil and waste pipes must be protected from freezing. Although vent piping is not required to be protected from freezing, it may be prudent to insulate vent pipes in extreme freezing climates to prevent condensation freezing and blocking of vents. Where a water or drain pipe is installed in an exterior wall or ceiling of a space intended for occupancy (in other words, a heated space), some degree of freeze protection can be achieved by making sure that the thermal insulation for the wall (or ceiling) is installed between the outdoor surface and the piping [see Commentary Figures 305.4(1) and 305.4(2)]. Whether or not this arrangement will prevent freezing temperatures at the piping location depends on the climatic conditions, the thickness of insulation (between the outdoor surface and the piping) and the room temperature.

Note that there must always be a heat source along with an appropriate insulation thickness in order to protect pipes from freezing conditions. Insulation by itself (without a heat source) cannot protect a pipe from freezing; insulation only slows the rate of heat loss. The closer a pipe is to the heat source and the less insulation there is between the pipe and the heat source, the warmer the pipe will be. Commentary Figure 305.4(3) shows piping in an exterior wall of a building. The water distribution piping has been covered

Figure 305.4(1)
PROTECTION AGAINST FREEZING

Figure 305.4(2)
WATER DISTRIBUTION PIPING IN EXTERIOR WALL

GENERAL REGULATIONS

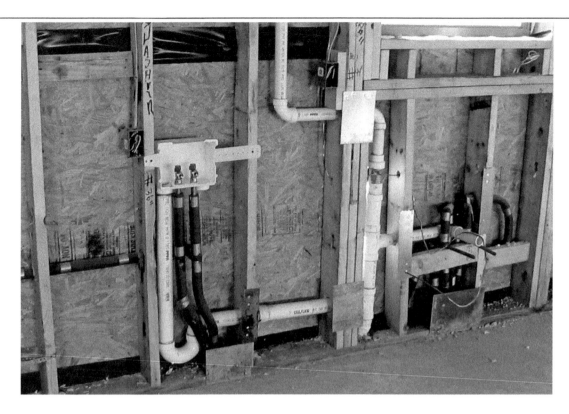

Figure 305.4(3)
PIPING IN EXTERIOR WALL

with foam tube insulation. On the left of the photo, note the close proximity of the horizontal water pipe to the exterior sheathing. While the insulation on the pipe between the pipe and the sheathing does help protect the pipe from outdoor temperatures, the insulation on the pipe between the pipe and the room will insulate the pipe from the heat coming from the room. In all likelihood, the wall insulation will be installed over the insulated pipe resulting in further insulation of the pipe from the room heat. On the right side of this photo, note the multitude of water distribution pipes behind the fixture location (probably a laundry tub). Again, while the pipe insulation between the pipes and the sheathing is beneficial for freeze protection of the pipes, the insulation on the room side of the pipes is detrimental because it insulates the pipes from the room heat. The congestion in this area will undoubtedly cause difficulty for installing wall insulation, especially between the piping and the sheathing. In the center of this photo, note that the trap for the clothes washer stand pipe is in close proximity to the sheathing so much so that it will be difficult to install any insulation between the trap and the sheathing. The arrangement and insulation of pipes shown in this figure would most likely only be suitable for a building located where freezing climatic conditions are unlikely or are of limited duration (such as hours).

Where piping is not directly adjacent to heated spaces in a building, electric resistance heat tapes or cables can be used to supply heat to the piping. Generally, heat tapes should not be used on piping in concealed spaces as the tapes can burn out and require replacement. However, as the code is silent with regard to the use of heat tape in a concealed area, the only recourse the code official has to prohibit such installations is to determine if the manufacturer's installation instructions limit the product use to unconcealed areas. There are heavy duty electric heat-tracing products that are suitable for concealed and earth burial applications. Note that plastic piping requires self-limiting-type heat tape to prevent overheating of the pipe.

The code does not provide any guidelines for insulating methods to ensure that piping will not be exposed to freezing temperatures. The code official is responsible for determining whether water and drain piping is adequately protected for the climatic conditions expected in the geographic area.

Freezeproof Hose Faucets

Hose bibbs and wall hydrants located on the exterior wall must be protected when installed in areas subject to freezing temperature. This can be accomplished by installing devices, such as freezeproof hose bibbs, that locate the valve seat within the heated space and allow residual water within the hydrant to drain after the valve is closed. A freezeproof hose bibb cannot be installed with the valve seat located in an unheated garage or storage room. The valve seat must extend through to the heated side of the exterior wall. Valve casings on these devices are available in various lengths to accommodate a variety of wall thicknesses [see Commentary Figure 305.4(4)].

Freezer Floor Drains

Where floor drains are located in areas (e.g., walk-in freezers or frozen food warehouses) maintained at temperatures at or below 32°F (0°C), the traps for the floor drains could freeze and damage the piping.

Figure 305.4(4)
TYPICAL FREEZEPROOF HOSE FAUCET INSTALLATION

NOTE: FROST PENETRATION IN INCHES
SOURCE: U.S. WEATHER BUREAU

For SI: 1 inch = 25.4 mm.

Figure 305.4(5)
SEVERE FROST PENETRATION LEVELS

Therefore, floor drains in these areas must not have traps and must be indirect connected in accordance with Section 802.1.2.

Underground Water Service Piping

Water service piping that is installed exterior to a building and underground must be buried at least 12 inches (305 mm) below grade or 6 inches (152 mm) below the established frost penetration depth for the geographic area [see Commentary Figure 305.4(5)], whichever is the greater depth. For example, if the frost depth for the region is 24 inches (610 mm) below grade, the water service pipe must be installed so that the top surface of the pipe is at least 30 inches (762 mm) below grade. If the frost depth is 5 inches (127 mm) below grade, the water service pipe must be installed such that the top surface of the pipe is at least 12 inches (305 mm) below grade. Where frost penetration depth controls the burial depth, the 6-inch buffer protects the pipe from forces caused by the freezing and expansion of moisture in the above soil (i.e., "frost heave") [see Commentary Figure 305.4(6)]. The minimum burial depth of 12 inches (305 mm) protects the pipe from the most common accidental damage: shallow hand digging for landscape plantings.

For SI: 1 inch = 25.4 mm.

**Figure 305.4(6)
REQUIRED BURIAL DEPTH**

Plumbing in Seasonal-use Buildings and Outdoor-use Areas

The potable water plumbing systems for poolside bathhouses, summer-use-only cabins and other unheated areas not used during freezing temperature conditions are typically designed in some way to be "winterized" so that water in the piping does not freeze and break the piping. Two common ways of accomplishing winterization of plumbing systems are: 1) gravity draining the potable water system to an outlet at the system's lowest point or 2) pumping an antifreeze solution into the potable water system to displace the water. Although the code is silent about winterization, there are a few code sections to consider: 1) stop and waste valves below grade (Section 608.7) and 2) introduction of chemicals into a potable water system (Sections 608.1 and 608.5).

Consider a poolside bathhouse built on a slab-on-grade foundation. Gravity draining of the water from the potable water system would require a drain outlet located below grade so that water in the system can drain down to an elevation below the frost line. But Section 608.7 prohibits stop and waste valves below grade because there is a high probability for the drain outlet valve to be left open during the off season such that ground water could enter and contaminate the potable water system. Therefore, gravity draining of the bathhouse potable water system through the below-grade outlet is not an acceptable method for winterizing. One possible solution for evacuating the water from the potable water system could be to design the system so that the water could be vacuumed out from one or more specifically designed above-grade openings that are connected directly to the low points in the system. Other parts of the potable water system would require careful piping installation to ensure that the water in the system gravity drains to the low points of the system.

A potable water system could also be winterized by pumping antifreeze into the system to displace the water throughout all of the piping. However, many commonly available antifreeze solutions (e.g., automobile cooling system coolants) are toxic and must not be ingested by humans. Even if toxic solutions are used and then flushed out of the system with potable water, the potential for contamination of the water is still too great to consider the use of toxic antifreezes. In general, plastic piping manufacturers recommend using only glycerin-based antifreeze solution as it is generally regarded as nontoxic and is chemically compatible with most types of plastic pipe materials listed in the code. The use of glycerin-based antifreeze for nonplastic piping should be equally acceptable. Waste glycerin antifreeze could be captured and stored for reuse or if disposed of, it should be done so in an environmentally safe manner.

305.4.1 Sewer depth. *Building sewers* that connect to private sewage disposal systems shall be installed not less than **[NUMBER]** inches (mm) below finished grade at the point of septic tank connection. *Building sewers* shall be installed not less than **[NUMBER]** inches (mm) below grade.

❖ A building sewer is not subject to the same constraints as a water pipe because a sewer carries heat to the surrounding soil as the drainage flows through it. A building sewer also has an intermittent flow. As a result, the frost depth is higher in the immediate area of the building sewer; therefore, the burial depth of a building sewer does not have to be below the recorded frost depth. A typical burial depth for a building sewer ranges from 6 to 18 inches (152 to 457 mm).

The installation depth of the sewer pipe connecting to a septic tank could be different to facilitate the proper operation of the private sewage disposal system. The connection at the septic tank is typically 18 inches (457 mm) below grade; however, the code official is to determine the depths necessary for the locality.

305.5 Waterproofing of openings. Joints at the roof and around vent pipes shall be made water tight by the use of lead, copper, galvanized steel, aluminum, plastic or other *approved* flashings or flashing material. Exterior wall openings shall be made water tight.

❖ Where a pipe penetrates a roof or an exterior wall, the annular space around the pipe must be sealed, or the pipe must be provided with flashing to prevent the entry of moisture. If this annular space is improperly sealed or not sealed at all, moisture from precipitation can enter the structure and damage the surrounding structure and finishes.

Because of direct exposure to rain and, in colder climates, exposure to snow accumulation, piping that penetrates a roof must be provided with flashing. Many types of flashings have been developed to prevent leakage around penetrations. In a typical installation, the flashing surrounds the pipe and rises above the roof surface, and is made water tight by wrapping over the top of the pipe or applying an elastomeric seal around the pipe. The installation of materials, such as lead, that wrap over the top of the pipe must not reduce the required cross-sectional area of the vent terminal (see Commentary Figure 305.5).

Several of the flashing materials listed in Section 305.5 have been used for many years and are recognized as effective moisture penetration protection. This section also provides for the use of new and innovative materials that are not specifically listed when approved by the code official. Whenever an alternative material is proposed for use as a flashing, the material must be proven to be corrosion resistant and durable against the effects of exposure to the elements, including but not limited to resistance to impact, freeze-thaw and ultraviolet exposure and aging. Another consideration in the selection of the flashing material is that it must be compatible with the roof-covering material, the assembly penetrated and the pipe that it will be in contact with.

Because exterior walls are typically not subjected to the same severity of exposure as a roof, the space around pipes that penetrate exterior walls need not be sealed with flashing; rather, it must only be made water tight. Openings in exterior walls, therefore, can be sealed with an approved sealant, mechanical device, flashing or other approved method that will prevent penetration into the structure.

305.6 Protection against physical damage. In concealed locations where piping, other than cast-iron or galvanized steel, is installed through holes or notches in studs, joists, rafters or similar members less than $1^1/_2$ inches (38 mm) from the nearest edge of the member, the pipe shall be protected by steel shield plates. Such shield plates shall have a thickness of not less than 0.0575 inch (1.463 mm) (No. 16 gage). Such plates shall cover the area of the pipe where the member is notched or bored, and shall extend not less than 2 inches (51 mm) above sole plates and below top plates.

❖ This section is intended to minimize the possibility of drainage and water pipe damage caused by nails, screws or other fasteners. Because nails and screws are driven "blindly" after wall covering is placed, the shield must protect the pipe through the full width of the member on each side and must extend not less than 2 inches (51 mm) above or below the sole or top wall plates. Commentary Figure 305.6(1) shows typical shield plates. Commentary Figure 305.6(2) shows shield plates improperly installed because the plate(s) do not extend 2 inches (51 mm) below the bottom member of the double top plate. Cast-iron and galvanized steel pipe each have wall thicknesses greater than the required thicknesses for the shield plate, which makes them inherently resistant to nail and screw penetrations.

The shield plate minimum thickness specified represents the "low end" of the thickness tolerance range for No. 16 gage galvanized sheet metal in accordance with Sheet Metal and Air Conditioning Contractors National Association (SMACNA) heating, ventilation and air-conditioning (HVAC) duct construction standards. The "low end" thickness dimension is stated so

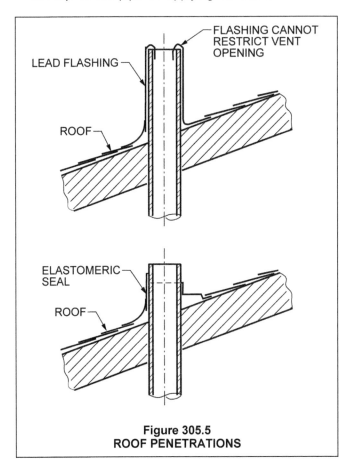

Figure 305.5
ROOF PENETRATIONS

GENERAL REGULATIONS

For SI: 1 inch = 25.4 mm.

Figure 305.6(1)
REQUIRED SHIELD PLATES

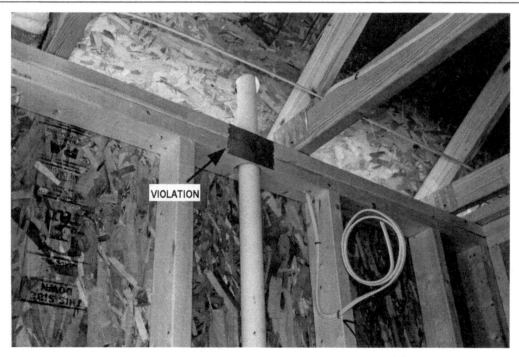

Figure 305.6(2)
PIPE PROTECTION VIOLATION

GENERAL REGULATIONS

Figure 305.7
PROTECTION OF PLUMBING SYSTEM COMPONENTS

that if a field measurement of shield plate thickness is necessary, the absolute minimum thickness will allow all possible thicknesses of No. 16 gage shield plates to be acceptable. No. 16 gage has been proven to be the minimum thickness to prevent penetration by commonly used fasteners without causing noticeable "bulge" appearance problems in the covering materials.

305.7 Protection of components of plumbing system. Components of a plumbing system installed along alleyways, driveways, parking garages or other locations exposed to damage shall be recessed into the wall or otherwise protected in an *approved* manner.

❖ Plumbing system components such as water supply pipes, vents, stacks, leaders or downspouts that are subject to possible motor vehicle impact must be adequately protected (see Commentary Figure 305.7). The method of protection must be approved by the code official.

SECTION 306
TRENCHING, EXCAVATION AND BACKFILL

306.1 Support of piping. Buried piping shall be supported throughout its entire length.

❖ Piping that is installed below grade must be continuously supported to prevent a number of possible problems including, but not limited to, reverse slope (of gravity flow pipelines), flexible connector misalignment/failure and pipe breakage.

306.2 Trenching and bedding. Where trenches are excavated such that the bottom of the trench forms the bed for the pipe, solid and continuous load-bearing support shall be provided between joints. Bell holes, hub holes and coupling holes shall be provided at points where the pipe is joined. Such pipe shall not be supported on blocks to grade. In instances where the materials manufacturer's installation instructions are more restrictive than those prescribed by the code, the material shall be installed in accordance with the more restrictive requirement.

❖ A trench must be wide enough to allow for proper alignment of the piping system. Piping is best supported when it rests directly on the bottom of a solid, continuous trench for its entire length, as required by Section 306.1 (see Commentary Figure 306.2). Sections 306.2.1 through 306.2.3 contain specific requirements for excavating and rebedding the trench below the installation level of the pipe.

If bell or hub pipe is installed, the bottom of the trench must be dug out around the hub to maintain continuous support of the pipe (see Commentary Figure 306.2). For additional information on burial techniques, consult installation standards such as ASTM D2321 (plastic sewer) and ASTM D2774 (plastic pressure pipe). The standards include criteria such as the size of backfill particles, classification and compaction density requirements of various embedment materials and the deflection requirements of different pipe materials. They also address the minimum backfill cover required before heavy equipment can traverse the pipe trench. When the manufacturer's instructions are more restrictive than the requirements of this section, the manufacturer's instructions govern. The rationale for this is that the manufacturer has tested the assembly for various characteristics, including proper support of the piping system for various areas where the piping may be installed.

306.2.1 Overexcavation. Where trenches are excavated below the installation level of the pipe such that the bottom of the trench does not form the bed for the pipe, the trench shall be backfilled to the installation level of the bottom of the pipe

GENERAL REGULATIONS

Figure 306.2
PIPE ON BOTTOM OF TRENCH

For SI: 1 inch = 25.4 mm.

Figure 306.2.1
OVEREXCAVATED TRENCH

with sand or fine gravel placed in layers not greater than 6 inches (152 mm) in depth and such backfill shall be compacted after each placement.

❖ Trench overexcavation occurs frequently because it is sometimes difficult to maintain an even grade line. The removal of large rocks, construction debris or abandoned piping can create voids below the desired grade line. Where overexcavation occurs, the trench bottom must be built up with compacted layers of sand or fine gravel (see Commentary Figure 306.2.1). The layers of fill material must not exceed 6 inches (152 mm) in thickness so that compaction is effective throughout the fill layer. Compaction is necessary to prevent future settling and to decrease the permeability of the fill material so that any ground-water flow does not washout the fill material.

306.2.2 Rock removal. Where rock is encountered in trenching, the rock shall be removed to not less than 3 inches (76 mm) below the installation level of the bottom of the pipe, and the trench shall be backfilled to the installation level of the bottom of the pipe with sand tamped in place so as to provide uniform load-bearing support for the pipe between joints. The pipe, including the joints, shall not rest on rock at any point.

❖ Piping must not be supported intermittently by hard surfaces, such as rock or concrete as this creates "point loads" on the pipe wall. Bedding provides a continuous "smooth" cradled support to ensure that the pipe is not exposed to concentrated "point" loads, which could ultimately cause pipe failure.

306.2.3 Soft load-bearing materials. If soft materials of poor load-bearing quality are found at the bottom of the trench, stabilization shall be achieved by overexcavating not less than two pipe diameters and backfilling to the installation level of the bottom of the pipe with fine gravel, crushed stone or a concrete foundation. The concrete foundation shall be bedded with sand tamped into place so as to provide uniform load-bearing support for the pipe between joints.

❖ If soil at the bottom of an excavated trench is of poor load-bearing quality, such as peat or muck, the trench must be overexcavated by two pipe diameters and the area backfilled with material that will properly support the piping (see commentary, Section 306.2.1). For example, where a 4-inch (102 mm) pipe is to be installed in a trench where there is soggy soil, the trench must be overexcavated at least 8 inches (203 mm) below the level where the pipe is to be installed. The overexcavated area must be backfilled with either crushed stone or fine gravel.

In some cases, the top of abandoned concrete foundations might be found at the bottom of the trench or new concrete might have to be placed to span "bottomless" soft earth locations. In either case, the pipe must be isolated from the concrete with a layer of tamped or compacted sand (two pipe diameters deep) to protect the pipe from the hard and often uneven concrete surface.

306.3 Backfilling. Backfill shall be free from discarded construction material and debris. Loose earth free from rocks, broken concrete and frozen chunks shall be placed in the trench in 6-inch (152 mm) layers and tamped in place until the crown of the pipe is covered by 12 inches (305 mm) of tamped earth. The backfill under and beside the pipe shall be compacted for pipe support. Backfill shall be brought up evenly on both sides of the pipe so that the pipe remains aligned. In instances where the manufacturer's instructions for materials are more restrictive than those prescribed by the code, the material shall be installed in accordance with the more restrictive requirement.

❖ This section provides a general approach for backfilling piping that has been properly bedded in a trench. Proper backfilling technique is important to ensure that the pipe remains aligned in its bedded position especially where there will be loads from additional backfill above the 12 inches (305 mm) of tamped backfill cover over the pipe. Note that the last line of text in this section requires that the installer consider "manufacturer's installation instructions" to determine if more restrictive requirements exist. Different types of soil conditions, deep trenches or additional loads anticipated at grade level (such as rolling trucks) could require different backfilling methods.

It is not the intent of this section to require that all piping be buried with at least 12 inches (305 mm) of cover. Sections 305.4 and 305.4.1 indicate the required burial depths for some types of piping systems. Piping in trenches under a slab-on-ground building need not be buried any more than what is necessary to provide for a full thickness of the concrete slab above. Piping in trenches in a crawl space need not have any bedding or backfill installed as long as the piping is supported at the same required intervals as if the piping was suspended from the structure above.

Discarded construction materials such as broken concrete block, brick and lumber scraps must be cleared out of the backfill material to prevent the pipe from being misaligned or the pipe becoming "point loaded" from above when the backfill is compacted. Large rocks, frozen clumps of soil and broken concrete must also be kept out of the initial backfill (haunching) because these materials do not easily flow into the voids next to the pipe. The backfill for the sides of the pipe must readily flow and be easily placed to completely fill the open trench space next to the pipe. An even distribution of backfill along both sides of the pipe is required to prevent the pipe from being pushed sideways. Backfilling above the pipe is to be accomplished by placing 6 inches (152 mm) of backfill material in the trench, tamping that material firm and then placing another 6 inches (152 mm) of material in the trench and tamping that layer of material firm. The process is repeated until the tamped material above the top of the pipe is at least 12 inches thick (305 mm) (see Commentary Figure 306.3) or the trench is filled to grade. Any remaining backfill can then be placed into the trench in any layer thickness that can be adequately compacted with the available equipment or placed in the trench without compacting and left to naturally settle over time.

306.4 Tunneling. Where pipe is to be installed by tunneling, jacking or a combination of both, the pipe shall be protected from damage during installation and from subsequent uneven loading. Where earth tunnels are used, adequate supporting structures shall be provided to prevent future settling or caving.

❖ When installing pipe under a paved area or sidewalk, tunneling, jacking or a combination of both is required. The pipe must be protected by proper backfilling in the tunneled area to prevent caving or settlement around the piping system. The length of tunneling should be limited to only that required to clear the obstacle above. Tunnels of any length are difficult to refill with proper backfill and the longer the tunnel, the more difficult it is to prevent settlement or caving. If an uneven loading condition exists, the piping system could eventually fail.

SECTION 307
STRUCTURAL SAFETY

307.1 General. In the process of installing or repairing any part of a plumbing and drainage installation, the finished floors, walls, ceilings, tile work or any other part of the building or premises that must be changed or replaced shall be left in a safe structural condition in accordance with the requirements of the *International Building Code*.

❖ Quite often, the installation of a plumbing system affects other systems in the building. Structural components, for example, may have to be modified to allow for a plumbing system to be installed in accordance with code requirements. Any change to a building system or component that is necessary for the proper installation of the plumbing system must result in the modified building system or component still complying with the code requirements regulating that particular system. Structural components must meet the *International Building Code*® (IBC®) minimums.

For SI: 1 inch = 25.4 mm.

Figure 306.3
TAMPING LOOSE FILL IN 6-INCH LAYERS

307.2 Cutting, notching or bored holes. A framing member shall not be cut, notched or bored in excess of limitations specified in the *International Building Code*.

❖ The code recognizes that it is occasionally necessary to modify framing members by boring holes, notching or cutting in the course of installing plumbing piping or plumbing fixture supports. While the original intent of this section was to address modifications to solid sawn wood members (commonly called "dimensional lumber"), modern building construction includes a mix of "engineered" wood members and assemblies, cold-formed steel light-framed members, structural steel members and log members. To maintain a building's structural integrity, the load-carrying capacity of these members must not be reduced. If a structural member is modified beyond what is specifically permitted by the IBC, the member must be either replaced with an unmodified member or a structural analysis of that member must be performed by a registered design professional (and approved by the code official) to verify that the modified member will have the required structural capacity for the loading conditions. As either solution could be costly, the plumbing installer should carefully consider the specified limitations for modification of all framing members before such alterations are performed.

Bearing Walls Versus Nonbearing Walls

The terms "bearing" and "nonbearing" are used in the prescriptive allowances for the alteration of studs. In order to apply the allowances in the least restrictive manner, the plumbing installer must understand how the roof and floor loads of a building are supported by the wall framing. Exterior walls are always considered to be (load) bearing walls. Where floor joists, trusses or I-joists are supported by an interior wall, the wall is a (load) bearing wall. Roof trusses might also require support by interior walls and as such, those walls would be (load) bearing. However, not all walls located under the bottom chord of a roof truss are intended to support roof trusses. Where a wall only supports itself and no other framing member, it is non (load) bearing. If the plumbing installer has doubt about whether a wall is bearing or nonbearing, he should either consult with the party responsible for the framing design or assume that the wall is (load) bearing.

Sawn Wood Members

The IBC offers prescriptive allowances for cutting, notching and boring of sawn wood studs, joists and rafters. These allowances apply to wood-constructed buildings that have been designed in accordance with the allowable stress design method, the load-resistance design method or conventional light-frame construction design method. However, in rare cases, some portions of or entire sawn wood constructed structures might be specially designed such that standard limitations for cutting, notching or boring of members are restricted further or prohibited altogether. The plumbing contractor should always ask the general contractor or framing contractor where, if any, such special designs are located in the building.

Commentary Figure 307.2(1) shows the IBC limitations for modifications to wood studs. Commentary Figure 307.2(2) shows the IBC limitations for modifications to wood joists and rafters.

In Commentary Figure 307.2(3), the bearing wall studs have $2^1/_8$-inch (54 mm) holes drilled in 2 by 4 studs for passage of the $1^1/_2$-inch (38 mm) drain and vent pipes. The holes exceed 40 percent of the $3^1/_2$-inch (89 mm) actual depth of the stud and therefore, according to the IBC, the modified studs should have been evaluated by a design professional for structural adequacy. However, in this situation, the code official allowed the use of metal stud brackets (often called "stud shoes") to reinforce the studs at the oversized hole locations. Note that the IBC is silent on the use of metal stud shoes to reinforce wood studs where holes are larger than the IBC limitations. Although it may be common practice for stud shoes to be installed in these situations, the code official would have to approve their use. Commentary Figure 307.2(3) also shows notching on the face of the studs for $^3/_4$-inch (19 mm) copper water pipe. The allowable notch depth for a bearing stud is 25 percent of the stud depth or $^7/_8$ inch (22 mm). The notches in these studs are deeper than $^7/_8$ inch (22 mm) and thus are in violation of the notching rules.

Commentary Figures 307.2(4) through 307.2(9) show various violations of the cutting, notching and boring limitations for sawn wood members.

Sawn Wood Wall Top Plate Reinforcement

It is common for piping to be routed through holes or notches in the top plate of a (load) bearing wall (exterior or interior). Where this occurs, the IBC requires that the top wall plate be reinforced with a metal tie (reinforcing strap). The IBC is silent as to what size of hole can exist without a metal tie being installed. Where the plate is completely severed, the IBC is also silent as to whether a metal tie is required on both faces of the wall. The IBC is also silent as to how wide a notch or missing section of top plate can be. The ICC Plan Review Department provided the following guidance based upon common sense and general knowledge of the types of loads carried by the top plate of a prescriptively designed (load) bearing wall:

1. Where a hole in the top plate leaves less than 1 inch (25 mm) between the edge of the hole and a face of the wall, that face of the top plate requires a metal tie to be installed. If the hole leaves less than 1 inch (25 mm) between the edge of the hole and both faces of the wall, both faces of the top plate require a metal tie to be installed.

GENERAL REGULATIONS

2. Where the top plate is notched 50 percent or less of the width of the top plate, a metal tie is only required to be installed on the notched face of the top plate. If the notch exceeds 50 percent of the width of the top plate (including complete severing of the top plate), both faces of the top plate require a metal tie.

3. If a metal tie is required to be installed on the face of a wall but that face has structural sheathing installed, the metal tie is not required to be installed on that wall face.

4. The length of a notch (or the length of the section of plate completely removed) should not exceed the length allowed by the metal tie manufacturer's installation instructions. Typically, metal tie manufacturers offer ties for a maximum notch length of 12 inches (305 mm).

5. The final determination on the extent of the reinforcement required for a top wall plate that has been notched or bored is the responsibility of the code official.

The metal tie must be at least $1^{1}/_{2}$ inch (38 mm) wide by 0.058 inch (1.47 mm) No. 16 galvanized gage thick.

Each end of the tie must be fastened to the top plate with six 16d nails [see Commentary Figure 307.2(10)].

Note that the installation of reinforcing ties on both faces of a wall presents a complication. As each tie requires six 16d [diameter of 0.162 inch (4 mm)] nails [$3^{1}/_{2}$ inches (89 mm) long] at each end of the tie, nailing a tie on the same plate member on both faces of the wall will split the wood member. From this, one could conclude that where a single top plate is used, it would be prudent to limit the depth of a notch or size of a bored hole in the plate such that only one metal tie is necessary. Where the top plate is a double (member) plate, one face of the wall could have the tie installed on the top member of the top plate and the other face of the wall have the tie installed on the lower member of the top plate. Some tie manufacturers might allow for the use of shorter (and smaller diameter) nails so as to permit nailing from both sides of the same top plate member without splitting the wood. Refer to the tie manufacturer's installation instructions for alternative fastener sizes. While shorter, smaller diameter nails might slightly decrease the load-carrying capacity of the tie, the consensus is that it is better to have the ties securely nailed to the plate member (with the

For SI: 1 inch = 25.4 mm.

Figure 307.2(1)
CUTTING, NOTCHING AND BORING OF WALL STUDS

For SI: 1 inch = 25.4 mm.

Figure 307.2(2)
CUTTING, NOTCHING AND BORING OF FLOOR JOISTS, CEILING JOISTS AND RAFTERS

GENERAL REGULATIONS

tie manufacturer's prescribed alternate fasteners) rather than the ties having little to no load-carrying capacity due to the plate member being split by 16d nailing on both faces of the plate member. Screws must not be used to fasten the ties to the plates unless specified by the tie manufacturer.

Where copper or plastic piping passes through wall frame members (including plates) within $1^1/_2$ inches (38 mm) of the edge of the plate (face of the wall), steel shield plates are required to be installed to protect the pipes from fastener penetration (see Section 305.6). The installation of the steel shield plates and the top plate reinforcing ties must be coordinated. If the top plate is a single member top plate (or a double mem-

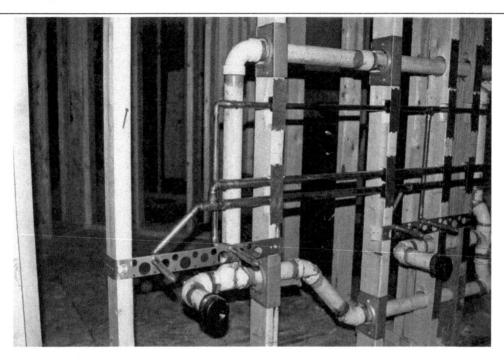

Figure 307.2(3)
NOTCHING AND BORING IN A WOOD-BEARING WALL

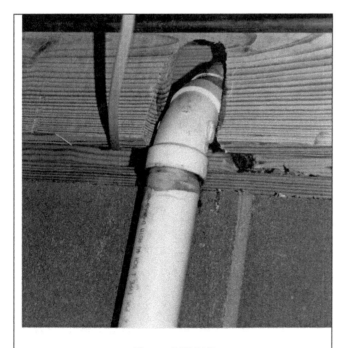

Figure 307.2(4)
VIOLATION: NOTCH IN FLOOR JOIST TOO DEEP

Figure 307.2(5)
VIOLATION: HOLES TOO CLOSE TO BOTTOM OF JOIST

ber top plate requiring a tie on the lower member of the top plate) then, in order to prevent a bulge in the wall covering material, the edges of the wall plate should be notched for the combined thicknesses of the shield plate, the tie, and the heads of the tie fastening nails.

The shield plate would then be installed first with the tie installed over the shield plate. Although some tie/shield plate manufacturers offer wide shield plates that have the required height dimension to cover a double member top plate plus 2 inches (50 mm) below the

Figure 307.2(6)
VIOLATION: NOTCHES TOO DEEP IN WALL STUDS

Figure 307.2(7)
VIOLATION: EXTERIOR WALL STUD NOTCHED TOO DEEP

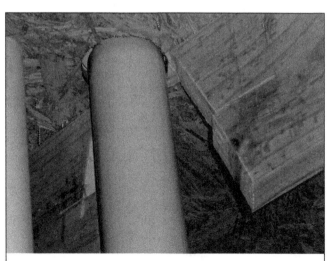

Figure 307.2(8)
VIOLATION: FLOOR JOIST COMPLETELY SEVERED

GENERAL REGULATIONS

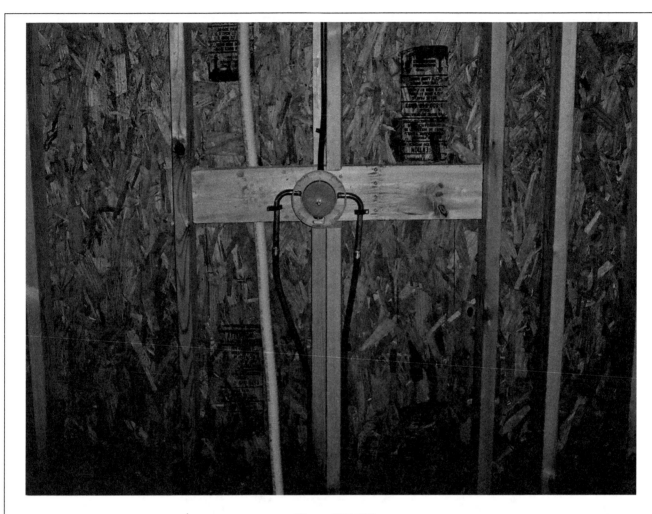

Figure 307.2(9)
VIOLATION: EXTERIOR WALL STUD COMPLETELY SEVERED

plate, these plates might not be suitable for use as top plate reinforcing ties. Use of a single shield plate to serve as the top plate reinforcing tie would require alternative approval by the code official in accordance with Section 105.2.

In summary, the plumbing designer or installer must carefully consider the impact of routing piping through the top plates of walls with respect to the requirements for metal ties (reinforcing straps). Commentary Figure 307.2(11) shows the severed top plates of an interior (load) bearing wall. Note that the close proximity of other pipes penetrating the plates would prevent the installation of the nails required for installation of the required reinforcing ties.

Engineered Wood Members

Engineered wood members are manufactured in many ways. The most common products used are laminated veneer lumber (LVL) beams [see Commentary Figure 307.2(12)], trusses of dimensional lumber, and I-joists of oriented strand board (OSB) webs with dimensional lumber top and bottom flanges (chords). LVL members must not be cut or notched. Because most LVL manufacturers either prohibit the installation of holes or severely limit the size, location and number of holes in LVL members, as a general rule, holes should never be bored in LVL members. Truss members must not be altered in any manner.

The top and bottom chords of I-joist members must never be cut or notched. Although most I-joist manufacturers allow holes to be cut in the web of the member, the maximum size and location of holes must be in accordance with the specific manufacturer's product instructions. The cutting, notching and boring limitations indicated for sawn lumber [Commentary Figure 307.2(1)] do not apply to I-joists. The installer and inspector must have access to the product instructions in order to determine if hole sizes and locations are acceptable. A review of several I-joist manufacturers' installation instructions does provide

GENERAL REGULATIONS

Figure 307.2(10)
REQUIRED TOP PLATE REINFORCEMENT

Figure 307.2(11)
VIOLATION: WALL TOP PLATES NOT REINFORCED

some generalizations that can help the plumbing installer generally plan piping routes as well as alert the inspector that a violation might exist. The generalizations are:

1. The closer the hole is located to the bearing locations (supports) of the I-joist, the smaller the allowable size of the hole becomes. Typically, holes near the bearing ends are limited to $1^1/_2$ inches (38 mm) in diameter with a spacing (hole edge to hole edge) of not less than 12 inches (305 mm).

2. As the hole location is placed farther from the bearing locations of the I-joist, the allowable size of the hole becomes larger. Large holes should be spaced apart (hole edge to hole edge) not less than the largest diameter or longest dimension of the largest hole.

3. Holes should be cut round, square or rectangular. The edge of any hole should not be any

GENERAL REGULATIONS

closer than $^1/_4$ inch (6.4 mm) to a top or bottom chord of the I-joists.

Always refer to the I-joist manufacturer's installation instructions for specific hole limitations. Commentary Figures 307.2(13) through 307.2(16) show prohibited boring, cutting and notching of trusses and I-joists.

Cold-formed Steel Light-framed Members

Cold-formed steel light-framed members of trusses must not be altered in any manner. Flanges and lips of stud, floor joist, ceiling joist and rafter members must not be cut, notched, bored or punched [see Commentary Figure 307.2(18) for identification of flanges, lips and webs]. Holes in the webs of studs, floor joists, ceiling joists and rafter members are prohibited except where:

1. The holes are punched at the factory,
2. The design drawings for the building specifically indicate the location and size of holes or
3. The prescriptive design method was used for the building framing design.

Because the IBC limits the use of the prescriptive design method to only one- and two-family detached dwellings or townhouses up to three stories in height, a building outside that scope cannot be designed using the prescriptive method. [In most jurisdictions, the *International Residential Code®* (IRC®) is adopted as the controlling code for one- and two-family detached dwellings or townhouses up to three stories in height. Where the IRC is the adopted code for these types of structures, refer to the IRC for all structural and plumbing requirements]. The IBC references the

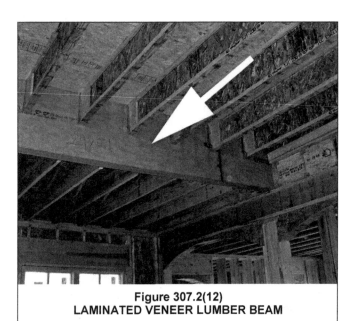

Figure 307.2(12)
LAMINATED VENEER LUMBER BEAM

Figure 307.2(13)
VIOLATION: TOP CHORD OF TRUSS NOTCHED

Figure 307.2(14)
VIOLATION: HOLE IN BOTTOM CHORD OF ROOF TRUSS

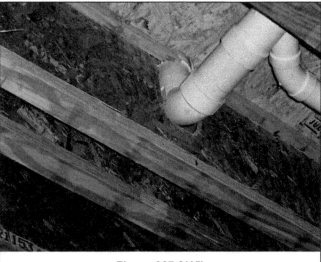

Figure 307.2(15)
VIOLATION: TOP CHORD OF
I-JOIST COMPLETELY SEVERED

standard AISI A230 for the prescriptive design method for cold-formed steel light-framed construction. This standard has prescriptive allowances for holes and their location in the webs of studs, floor joists, ceiling joists and rafters [see Commentary Figures 307.2(19) and 307.2(20)]. The reader should bear in mind that one- and two-family detached dwellings or townhouses up to three stories in height could be designed using methods other than the prescriptive design and, as such, the prescriptive allowances for holes cannot be used for those buildings.

Structural Steel Members and Structural Steel Trusses

Any modifications to structural steel members [see Commentary Figure 307.2(17)] or structural steel trusses must be approved by a registered design professional.

Log Members

The IBC references the standard ICC-400 for the construction of log structures. ICC-400 Section 302.2.4 covers the limitations for notching and boring of holes in logs used in structural applications. Commentary

Figure 307.2(16)
VIOLATION: I-JOIST WEB HOLES IRREGULAR SHAPE, TOO LARGE, TOO CLOSE TOGETHER

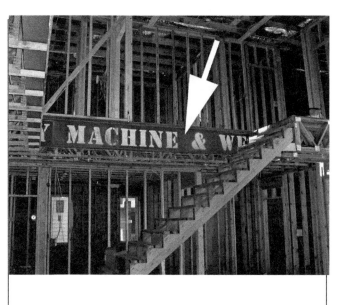

Figure 307.2(17)
STRUCTURAL STEEL BEAM

Figure 307.2(18)
FLANGE, LIP AND WEB OF COLD-FORMED STEEL LIGHT-FRAME MEMBER

For SI: 1 inch = 25.4 mm.

Figure 307.2(19)
ALLOWABLE WEB HOLES IN STUDS
(PRESCRIPTIVE DESIGN ONLY)

GENERAL REGULATIONS

Figure 307.2(21) illustrates the notching and boring limitations.

307.3 Penetrations of floor/ceiling assemblies and fire-resistance-rated assemblies. Penetrations of floor/ceiling assemblies and assemblies required to have a fire-resistance rating shall be protected in accordance with the *International Building Code*.

❖ This section states that openings for plumbing pipe in fire-resistance-rated building assemblies must be fire-resistance rated. In addition to maintaining the structural integrity of walls, floors, ceilings and roofs when these elements are penetrated by the plumbing pipe, the penetrations must also be protected to prevent the passage of fire and other products of combustion through the fire-resistance-rated assembly because of the pipe penetrating the assembly. These penetrations

For SI: 1 inch = 25.4 mm.

Figure 307.2(20)
ALLOWABLE WEB HOLES IN MEMBERS OTHER THAN STUDS (PRESCRIPTIVE DESIGN ONLY)

For SI: 1 inch = 25.4 mm.

Figure 307.2(21)
ALLOWABLE NOTCHING AND BORING OF STRUCTURAL LOG MEMBERS

GENERAL REGULATIONS

Figure 307.3(1)
COMBUSTIBLE PIPE PENETRATION OF FIRE-RATED WALL

Figure 307.3(2)
ANNULAR SPACE PROTECTION

Figure 307.3(3)
PIPE PENETRATION OF FIRE-RESISTANCE-RATED ROOF/CEILING ASSEMBLY

Figure 307.3(4)
PIPE PENETRATION OF FIRE-RESISTANCE-RATED FLOOR/CEILING ASSEMBLY

GENERAL REGULATIONS

must be protected in accordance with the IBC [see Section 714 of the IBC and Commentary Figures 307.3(1) through 307.3(4)].

[BS] 307.4 Alterations to trusses. Truss members and components shall not be cut, drilled, notched, spliced or otherwise altered in any way without written concurrence and approval of a registered design professional. Alterations resulting in the addition of loads to any member (e.g., HVAC equipment, water heater) shall not be permitted without verification that the truss is capable of supporting such additional loading.

❖ Trusses are engineered products, and any alteration to specific elements may prevent the truss from per-

Figure 307.4
VIOLATION: TOP CHORD OF FLOOR TRUSS REMOVED

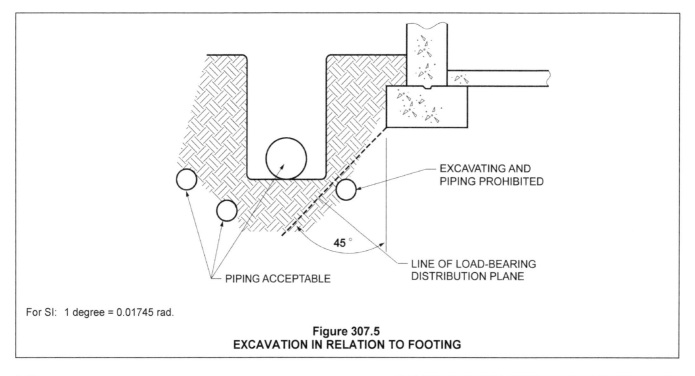

For SI: 1 degree = 0.01745 rad.

Figure 307.5
EXCAVATION IN RELATION TO FOOTING

forming as intended. The code does not permit drilling and notching of truss members and components without documentation that the design has taken this possibility into account. Additionally, trusses are designed to resist specific loads; the loading on the truss or on any of its members should not be changed without approval of a registered design professional (see Commentary Figure 307.4).

307.5 Protection of footings. Trenching installed parallel to footings and walls shall not extend into the bearing plane of a footing or wall. The upper boundary of the bearing plane is a line that extends downward, at an angle of 45 degrees (0.79 rad) from horizontal, from the outside bottom edge of the footing or wall.

❖ A footing requires a minimum load-bearing area to distribute the weight of the building. This load-bearing distribution plane extends downward at approximately a 45-degree (0.79 rad) angle from the base of the footing. Water, building drainage and building sewer piping must not be installed below this load-bearing plane. Excavation for the installation of pipe below the plane could affect the load capacity of the footing or cause the excavation to collapse (see Commentary Figure 307.5).

307.6 Piping materials exposed within plenums. Piping materials exposed within plenums shall comply with the provisions of the *International Mechanical Code*.

❖ Materials exposed within a plenum must be noncombustible or have a flame spread index of 25 or less and a smoke-developed index of 50 or less when tested in accordance with ASTM E84. Typically, the plenum space is often used to accommodate components of the plumbing system and other building systems such as electrical, fire protection and mechanical. The code permits exposure of limited types of combustibles used for these systems within the plenum. A flame spread index of 25 or less and a smoke-developed index of 50 or less are viewed as acceptable values for combustibles in a plenum because of the minimal risk associated with those materials.

SECTION 308
PIPING SUPPORT

308.1 General. Plumbing piping shall be supported in accordance with this section.

❖ There are many variables to consider during the design and installation of hangers and supports. Consideration must be given to the type and size of the pipe material; the weight of the pipe, fittings and contents; the structural element available for attachment; earthquake motions; vibration-sensitive applications and thermal expansion and contraction. Other load conditions to consider include the weight of valves and appurtenances; insulating materials; hanger and support components and wind, snow and ice loads.

308.2 Piping seismic supports. Where earthquake loads are applicable in accordance with the building code, plumbing piping supports shall be designed and installed for the seismic forces in accordance with the *International Building Code*.

❖ While the IBC prescribes the level of seismic forces to be considered in the design of buildings in seismic zones, it does not provide prescriptive methods for designing, sizing, or installing piping supports. A registered design professional should be consulted to provide the required seismic support designs and installation instructions.

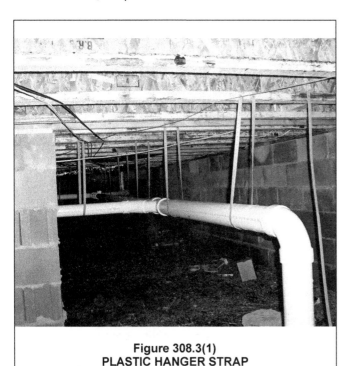

Figure 308.3(1)
PLASTIC HANGER STRAP

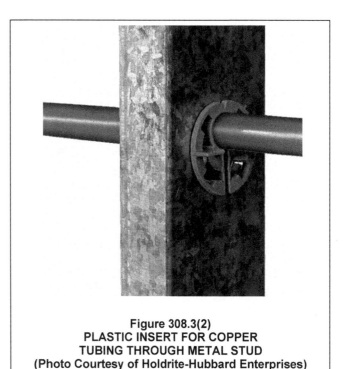

Figure 308.3(2)
PLASTIC INSERT FOR COPPER
TUBING THROUGH METAL STUD
(Photo Courtesy of Holdrite-Hubbard Enterprises)

308.3 Materials. Hangers, anchors and supports shall support the piping and the contents of the piping. Hangers and strapping material shall be of *approved* material that will not promote galvanic action.

❖ The hangers and supports for plumbing system piping must be capable of supporting the load imposed by the piping system and not be detrimental to the pipe they support. When the hanger or strapping material is not compatible with the piping material it supports, corrosion caused by galvanic action can occur. This happens when dissimilar metals are in contact with one another and sufficient moisture is present to carry an electrical current. Such corrosion can deteriorate the hanger, anchor or piping to the point of failure.

Where metallic pipe is installed, the hangers or supports must be of similar material to prevent any corrosive galvanic action. For example, if the water distribution system is copper tubing or copper pipe, copper hangers, straps and clamps are required or such supports must be of a material or clad by a material (such as plastic) that will not react with copper. Some support devices have special plastic coatings or are made entirely of plastic to prevent contact of dissimilar metals and are acceptable alternatives [see Commentary Figures 308.3(1) and 308.3(2)].

308.4 Structural attachment. Hangers and anchors shall be attached to the building construction in an *approved* manner.

❖ The selection of an appropriate hanger or support depends on the structural element available for hanger attachment. The integrity of all hangers and anchors depends on the method of attachment to the building construction. The means of attachment must provide the required strength and must not adversely affect or overstress the construction to which it is attached. Ultimately, a support is only as strong as its means of attachment and the member to which it is attached (see Commentary Figure 308.4)

308.5 Interval of support. Pipe shall be supported in accordance with Table 308.5.

Exception: The interval of support for piping systems designed to provide for expansion/contraction shall conform to the engineered design in accordance with Section 316.1.

❖ Piping must be supported at distances not to exceed those indicated in Table 308.5. The exception allows for design of special piping support systems in order to accommodate the movements due to thermal expansion of piping.

The horizontal support intervals listed in the table are based on the limiting mid-span deflection (sag) of pipes. The maximum amount of sag occurs at a point equidistant from the supports when the pipe is full of water. The table is designed to limit sag (and the resulting pipe stress) as well as provide for the maintaining of slopes required for gravity drainage.

Because any given length of pipe becomes more resistant to mid-span deflection as the pipe diameter becomes larger, the spacing in the table is calculated

Figure 308.4
POOR HANGER ATTACHMENT

for the pipe diameters at the low end of the pipe size ranges stated [see Commentary Figures 308.5(1) and 308.5(2)]. As this is a generalized table intended to cover a wide range of manufacturers, the installation instructions from the specific manufacturer should also be consulted to determine if support spacing is required to more be restrictive.

The spacing intervals indicated will result in the minimum number of supports allowed. Extra supports should be used where necessary to accommodate items such as fittings, connections, changes in direction or the connection to a vertical pipe.

Note a allows the hangers for cast-iron pipe to be increased to 10 feet (3048 mm) where 10-foot-long (3048 mm) sections of pipe are installed.

Note b specifies that smaller diameter vertical flexible piping cannot "bow" out of the initially installed straight line, in between required vertical supports, especially when filled with water (water distribution piping) or when subjected to hot water flow (drainage piping). The bowed piping can come in contact with adjacent piping, fittings or other objects (such as wall coverings) that might cause noise or damage to the piping. If the piping was allowed to bow unrestricted, high bending stresses could occur at the required vertical pipe supports. The purpose of the mid-story guide is to restrict the movement of the pipe in a direction perpendicular to the axis of the pipe. As a guide allows movement of the pipe in a direction parallel to the pipe axis, the guide is not required to firmly grip the pipe or absolutely prevent any movement whatsoever. For example, in wood frame construction, a mid-story pipe guide could be a block of wood (between studs) that has a clearance hole for the pipe to pass through. Other arrangements that substantially limit pipe movement in any direction perpendicular to the axis of the pipe will serve the intended purpose such

TABLE 308.5
HANGER SPACING

PIPING MATERIAL	MAXIMUM HORIZONTAL SPACING (feet)	MAXIMUM VERTICAL SPACING (feet)
Acrylonitrile butadiene styrene (ABS) pipe	4	10[b]
Aluminum tubing	10	15
Brass pipe	10	10
Cast-iron pipe	5[a]	15
Chlorinated polyvinyl chloride (CPVC) pipe and tubing, 1 inch and smaller	3	10[b]
Chlorinated polyvinyl chloride (CPVC) pipe and tubing, 1¼ inches and larger	4	10[b]
Copper or copper-alloy pipe	12	10
Copper or copper-alloy tubing, 1¼-inch diameter and smaller	6	10
Copper or copper-alloy tubing, 1½-inch diameter and larger	10	10
Cross-linked polyethylene (PEX) pipe	2.67 (32 inches)	10[b]
Cross-linked polyethylene/aluminum/cross-linked polyethylene (PEX-AL-PEX) pipe	2.67 (32 inches)	4
Lead pipe	Continuous	4
Polyethylene/aluminum/polyethylene (PE-AL-PE) pipe	2.67 (32 inches)	4
Polyethylene of raised temperature (PE-RT) pipe	2.67 (32 inches)	10[b]
Polypropylene (PP) pipe or tubing 1 inch and smaller	2.67 (32 inches)	10[b]
Polypropylene (PP) pipe or tubing, 1¼ inches and larger	4	10[b]
Polyvinyl chloride (PVC) pipe	4	10[b]
Stainless steel drainage systems	10	10[b]
Steel pipe	12	15

For SI: 1 inch = 25.4 mm, 1 foot = 304.8 mm.

a. The maximum horizontal spacing of cast-iron pipe hangers shall be increased to 10 feet where 10-foot lengths of pipe are installed.

b. For sizes 2 inches and smaller, a guide shall be installed midway between required vertical supports. Such guides shall prevent pipe movement in a direction perpendicular to the axis of the pipe.

as a pipe strap or a pipe clamp. Using flexible strapping around a pipe and fastening the ends of the strapping to both adjacent studs only limits pipe movement in one direction (i.e., side to side) but not movement in a direction 90 degrees (1.57 rad.) from the strapped direction. Thus, this strapping arrangement would not be acceptable to meet the requirement of this note.

308.6 Sway bracing. Rigid support sway bracing shall be provided at changes in direction greater than 45 degrees (0.79 rad) for pipe sizes 4 inches (102 mm) and larger.

❖ This section contains lateral support requirements for pipe 4 inches (102 mm) and larger in diameter. The flow of waste in drainage piping subjects the piping to the forces that result from the momentum of the waste.

These forces are a product of the mass and velocity of the flow together with the change in direction of the pipe and can be quite large, especially in piping that is 4 inches (102 mm) and larger. Hangers alone might not be sufficient to resist such forces; therefore, rigid bracing is required at all changes of direction greater than 45 degrees (0.79 rad) for piping that is 4 inches (102 mm) or larger in size. Without adequate bracing, the piping could be damaged or joint failure could occur. Bracing is intended to restrict or eliminate lateral movement of both horizontal and vertical piping (see commentary, Section 308.7.1 and Commentary Figure 308.6).

308.7 Anchorage. Anchorage shall be provided to restrain drainage piping from axial movement.

❖ This section requires a method of resisting axial movement of piping systems to prevent joint separation, regardless of the type of fittings or connections used. In particular, mechanical couplings using an elastomeric seal (typically hubless piping systems) have a limited ability to resist axial movement (pulling apart); therefore, pipe restraints must be installed to prevent joint separation. The required anchorage locations of axial restraints are specified in Section 308.7.1. Such joints also have a limited ability to resist shear forces. The hanger and support system must, therefore, prevent the couplings and connections from being subjected to shear forces that can damage the joint (see Commentary Figure 308.7). Although mechanical couplings using elastomeric seals are most often the type of joint to experience damage caused by axial movement, this requirement pertains to all portions of drainage piping systems.

308.7.1 Location. For pipe sizes greater than 4 inches (102 mm), restraints shall be provided for drain pipes at all changes in direction and at all changes in diameter greater than two pipe sizes. Braces, blocks, rodding and other suitable methods as specified by the coupling manufacturer shall be utilized.

❖ In drain pipes that are large in diameter, particularly those having a diameter greater than 4 inches (102 mm), the flow of drainage can create sizable forces in the system resulting from the mass and velocity of the waste. The force could have the effect of pushing apart mechanical and other types of couplings when there is a change in direction or diameter. To resist such forces in the piping, additional anchorage and supports would be necessary to hold the pipe and couplings in place and prevent joint separation caused by this dynamic force (see Commentary Figure 308.7.1).

308.8 Expansion joint fittings. Expansion joint fittings shall be used only where necessary to provide for expansion and contraction of the pipes. Expansion joint fittings shall be of the typical material suitable for use with the type of piping in which such fittings are installed.

❖ Plumbing and drainage pipe materials experience changes in length caused by the changing temperature of the pipe material, typically as a result of a change in water temperature or ambient air temperature. Repeated expansion and contraction of the pipe material can induce stress into the pipe and its attached fittings. To relieve this stress and thus prevent damage to the pipe, its fittings and the surrounding building components, properly designed expansion joints are installed in the piping runs. The designer of the piping system must determine whether an expansion joint in the pipe run is required based on the length of the run, the pipe material and diameter, the expected temperature range of the material inside the pipe and the exact configuration of the installed pipe and fittings.

Where installed, expansion joints must be of a material compatible with the pipe. The installer of the expansion joint must follow the manufacturer's instructions so that the finished joint is leak free while still providing the required expansion. Special attention needs to be given to the number and type of pipe hangers used when a pipe run is equipped with an expansion joint, so that the hangers do not restrict the movement of either the pipe or the expansion joint while providing the required support to both.

Figure 308.5(1)
POINT OF MAXIMUM SAG BETWEEN HANGERS

For SI: 1 inch = 25.4 mm.

Figure 308.5(2)
LOCATION OF HANGERS

308.9 Parallel water distribution systems. Piping bundles for manifold systems shall be supported in accordance with Table 308.5. Support at changes in direction shall be in accordance with the manufacturer's instructions. Where hot water piping is bundled with cold or hot water piping, each hot water pipe shall be insulated.

❖ A parallel water distribution system consists of a central distribution manifold having independent water distribution tubes supplying each plumbing fixture. This is analogous to an electrical circuit breaker panel feeding independent circuits. A typical central manifold is shown in Commentary Figure 604.10(2). As individual tubes lead from the central manifold, the tubing is often bundled for ease of routing. Bundles are typically loosely bound with plastic ties at intervals. The bundle is supported by hangers in accordance with the spacing requirements for the type of tubing as indicated in Table 308.5.

Where the tubing makes a small radius bend (turn), the bend must be supported according to the manufacturer's instructions. Special formed plastic brackets are often used to protect tight radius pipe bends. Otherwise, the tubing could flatten or kink over time and restrict flow to the fixtures [see Commentary Figures 308.9(1) through 308.9(4)].

For SI: 1 inch = 25.4 mm, 1 degree = 0.0175 rad.

For SI: 1 inch = 25.4 mm, 1 degree = 0.01745 rad.

**Figure 308.6
SWAY BRACING**

**Figure 308.7
RESTRAINED COUPLING**

GENERAL REGULATIONS

Because tubes in contact with one another will transfer heat from one to the other, special attention must be given to bundling hot (or tempered) water tubes with other hot (or tempered) water tubes and especially cold water tubes. For the 2012 edition of the code, it is required that each individual hot water tube be insulated to protect against heat transfer to other tubes carrying water of a different temperature. Although this section does not require that tempered water tubing be individually insulated, it would seem to be prudent to also individually insulate such tubing to prevent heat loss to adjacent cooler temperature piping or heat gain from adjacent hotter temperature piping.

For SI: 1 inch = 25.4 mm.

**Figure 308.7.1
ANCHORAGE LOCATION**

SECTION 309
FLOOD HAZARD RESISTANCE

309.1 General. Plumbing systems and equipment in structures erected in *flood hazard areas* shall be constructed in accordance with the requirements of this section and the *International Building Code*.

❖ Buildings and structures erected in flood hazard areas must be designed to resist the forces caused by flooding, including flotation, collapse and lateral movement from flowing water, standing water or wave action. This section sets a broad requirement that plumbing systems and equipment are to be constructed according to the provisions of the IBC. Section 1612 of the IBC references ASCE 24, *Flood Resistant Design and Construction*. Where possible, plumbing fixtures and equipment are to be elevated above the elevation required by the IBC or they must be specifically designed and intended for placement in locations that may be inundated. For additional guidance, refer to FEMA 348, *Protecting Building Utilities from Flood Damage*.

[BS] 309.2 Flood hazard. For structures located in *flood hazard areas*, the following systems and equipment shall be located and installed as required by Section 1612 of the *International Building Code*.

1. Water service pipes.
2. Pump seals in individual water supply systems where the pump is located below the *design flood elevation*.
3. Covers on potable water wells shall be sealed, except where the top of the casing well or pipe sleeve is elevated to not less than 1 foot (305 mm) above the *design flood elevation*.
4. Sanitary drainage piping.
5. Storm drainage piping.
6. Manhole covers shall be sealed, except where elevated to or above the *design flood elevation*.
7. Other plumbing fixtures, faucets, fixture fittings, piping systems and equipment.
8. Water heaters.
9. Vents and vent systems.

Exception: The systems listed in this section are permitted to be located below the elevation required by Section 1612 of the *International Building Code* for utilities and attendant equipment, provided that the systems are designed and installed to prevent water from entering or accumulating within their components and the systems are constructed to resist hydrostatic and hydrodynamic loads and stresses, including the effects of buoyancy, during the occurrence of flooding up to such elevation.

❖ Although most plumbing facilities and system elements within a building or structure must be located at or above the elevation required by the code, certain plumbing system elements and equipment are expected to extend below that elevation. In such instances, the installation must be designed and installed to withstand anticipated flood forces. The flood forces may include buoyancy as well as dynamic loads imposed by moving water and/or waves and the impact of floating debris. This section specifically addresses those plumbing systems and equipment that are to withstand flood forces. It also identifies those systems and equipment that are to be sealed to prevent infiltration of floodwaters into the systems or loss of contents out of the systems during the conditions of flooding. For additional guidance, refer to FEMA 348.

1. Water service pipes located below elevated buildings and those buried below ground are to be water tight to prevent floodwater from entering the pipes. Water service pipes could develop leaks when subjected to the additional stress from floodwater. This means that the entire piping system subject to floodwater con-

GENERAL REGULATIONS

ALLOW 1/8" TO 3/16" OF SLACK PER FOOT OF PEX PIPE
(4'-0" × 3/16" = 3/4" OF SLACK).

For SI: 1 inch = 25.4 mm, 1 foot = 304.8 mm.

**Figure 308.9(1)
PIPE SUPPORT**

**Figure 308.9(2)
PIPE BEND SUPPORT MAINTAINING BEND RADIUS**
(Photo courtesy of Holdrite-Hubbard Enterprises)

**Figure 308.9(3)
PIPE BEND SUPPORT MAINTAINING BEND RADIUS**
(Photo courtesy of Holdrite-Hubbard Enterprises)

Figure 308.9(4)
PIPE BEND SUPPORT MAINTAINING BEND RADIUS
(Photo courtesy of Holdrite-Hubbard Enterprises)

tact must be designed accordingly; for example, using hangers that are capable of resisting not only the normal downward load of the pipe and its contents, but also lateral or upward forces on the pipe caused by floodwaters.

2. Pump seals in individual water supply systems are not normally considered water tight from the "outside." This item requires that these be capable of keeping out the floodwater.

3. To prevent contamination of the below-ground water supply, private well heads that are subject to submersion below the floodwater must be water tight. This should also include any attachment to the well head, such as the electrical connection.

4. Sanitary drainage piping should be designed so that the additional stress of the floodwater does not cause it to fail and thus contaminate the floodwater.

5. Storm drainage systems are not normally connected to any other system, such as the sanitary drainage system. The potential exists that the storm drainage system could carry water from the flood to an area currently not flooded. This item is trying to contain the potential damage from the floodwater to the areas immediately subject to that water.

6. As noted in Item 5 above, the concern is potential for floodwater to get into another area or system, which may cause backup into buildings. Therefore, there is a requirement for sealing these potential avenues for floodwater to travel.

7. This item is intended to cover any other item not specifically addressed in the six items above. It requires both the code official and the designer to verify that all plumbing systems subject to contact with the floodwater have been properly designed to withstand the forces of that water.

8. Water heaters and mechanical equipment must be protected from water or elevated above the elevation required by the code. Exposure to water can cause deterioration of most system components and serious malfunctions.

9. The venting system is no different from any other plumbing system component that must be protected from flood levels. Floodwaters can cause damage if the venting system is not properly anchored, supported, sealed, installed and elevated above the elevation required by the code or designed to prevent water from entering or accumulating within such system.

[BS] 309.3 Coastal high-hazard areas and coastal A zones. Structures located in coastal high-hazard areas and coastal A zones shall meet the requirements of Section 309.2. The plumbing systems, pipes and fixtures shall not be mounted on or penetrate through walls intended to break away under flood loads.

❖ Breakaway walls are required if elevated buildings in certain flood hazard areas (V zones and coastal A zones) have enclosed areas beneath the elevated floors. Breakaway walls are designed and constructed to fail under flood loads in order to avoid transferring the loads and damaging the primary structural support of the building. When plumbing systems, pipes and fixtures are mounted on or penetrate through the breakaway walls, they can prevent the walls from breaking away, thus transferring additional load to the building and increasing flood damage. Also, damage to the mechanical systems is certain to occur when the walls fail as intended.

SECTION 310
WASHROOM AND TOILET ROOM REQUIREMENTS

310.1 Light and ventilation. Washrooms and toilet rooms shall be illuminated and ventilated in accordance with the *International Building Code* and *International Mechanical Code*.

❖ In the early part of the 20th century, it was discovered that poorly ventilated bathing and toilet rooms in tenement housing were a leading cause of the spread of disease to inhabitants of those structures. Poor illumination in those rooms compounded the problem by making it difficult to see spills and provide for proper housekeeping. Even though the requirement for illumination and ventilation is an *International Plumbing Code*® (IPC®) requirement, these features are specifically controlled by the IBC and the *International Mechanical Code*® (IMC®), respectively.

310.2 Location of fixtures and compartments. The location of plumbing fixtures and the requirements for compartments and partitions shall be in accordance with Section 405.3.

❖ See the commentary for Section 405.3.

310.3 Interior finish. Interior finish surfaces of toilet rooms shall comply with the *International Building Code.*

❖ Specifically, Section 1210 of the IBC has detailed requirements for the interior finishes of toilet rooms. The IBC sections require that, in other than dwelling units, the floor finish of toilet and bathing rooms must have a smooth, hard, nonabsorbent surface with a vertical base extending upward onto the walls at least 4 inches (102 mm). For all other toilet and bathing rooms, the IBC requires that the walls within 2 feet (610 mm) of urinals and water closets must have a smooth, hard, nonabsorbent surface to a height of 4 feet (1219 mm) above the floor. Two exceptions are provided for the wall requirement: 1) dwelling units and sleeping units and 2) toilet rooms not accessible to the public and having only one water closet.

A question that often arises is, what constitutes a "hard" wall surface? As the hard wall surface is required for public toilet facilities, it is generally expected that the walls will be subject to some abuse and vandalism. Gypsum board covering wall framing is not sufficiently strong (hard) enough to withstand the all too common abuse that is observed in many public toilet rooms. Another question that arises is, what constitutes a nonabsorbent surface? As gypsum board with any type of paint coating is easily scratched or the paper layer broken by a jab or punch, it is unlikely that such a surface will remain nonabsorbent over the life of the building. Concrete block, while hard, contains pores and crevices where bacteria is harbored and not easily removed by normal maintenance cleaning. Thus, painted concrete block that has not had the pores and crevices filled and smoothed by suitable block filler is not a nonabsorbent surface. Typically, grouted tile, cultured marble sheet and fiberglass beaded sheet have been used in these applications with good success. These types of finishes will allow thorough cleaning of the surfaces (see commentary, Section 419.3).

SECTION 311
TOILET FACILITIES FOR WORKERS

311.1 General. Toilet facilities shall be provided for construction workers and such facilities shall be maintained in a sanitary condition. Construction worker toilet facilities of the nonsewer type shall conform to ANSI Z4.3.

❖ Construction employees must have plumbing facilities available to them during the construction of a building. These facilities may be either permanent or portable. Portable facilities are regulated by ANSI Z4.3, which specifies various construction requirements. Commentary Figure 311.1 has been provided to show the minimum number of fixtures required by ANSI Z4.3.

NUMBER OF EMPLOYEES	MINIMUM NUMBER OF TOILET FACILITIES
	If serviced once per week[a]
1 - 10	1
11 - 20	2
21 - 30	3
31 - 40	4
Over 40	1 additional facility for each 10 additional employees
	If serviced more than once per week[a]
1 - 15	1
16 - 35	2
36 - 55	3
56 - 75	4
76 - 95	5
Over 95	1 additional facility for each 20 additional employees

a. "Servicing" refers to the emptying of waste and the cleaning of the toilet facility. A camp equipped with flush toilets shall meet the standard for "minimum number of toilet facilities if serviced more than once per week."

(Source: ANSI Z4.3-95)

Figure 311.1
MINIMUM NUMBER OF TOILET FACILITIES

SECTION 312
TESTS AND INSPECTIONS

312.1 Required tests. The permit holder shall make the applicable tests prescribed in Sections 312.2 through 312.10 to determine compliance with the provisions of this code. The permit holder shall give reasonable advance notice to the code official when the plumbing work is ready for tests. The equipment, material, power and labor necessary for the inspection and test shall be furnished by the permit holder and he or she shall be responsible for determining that the work will withstand the test pressure prescribed in the following tests. All plumbing system piping shall be tested with either water or, for piping systems other than plastic, by air. After the plumbing fixtures have been set and their traps filled with water, the entire drainage system shall be submitted to final tests. The code official shall require the removal of any cleanouts if necessary to ascertain whether the pressure has reached all parts of the system.

❖ This is a general section requiring the testing of plumbing systems in accordance with subsequent sections, as applicable. For example, if only a sewer line is installed, then only the sewer line test must be performed. If there is no storm drainage system, then obviously no storm drainage system test can be performed.

Sections 312.2 through 312.8 describe various types of pressure tests to piping systems. Section 312.9 describes leak testing of field-installed shower receptor linings. Section 312.10 describes inspection and testing of backflow preventers.

Water or waste-water leaks from installed piping can have serious health consequences for the inhabitants of a building. Leaks can cause bacteria and mold to grow in concealed areas. If a leak continues long enough, structural members of the building can be compromised, rendering the building unsafe to occupy. Thus, it is very important that piping systems be tested at stages where leaks can be readily identified and repaired before piping is concealed

Backflow prevention assemblies, devices and methods are critical components in making sure that the potable water distribution system remains safe from contamination and pollution. As backflow prevention assemblies are field testable, Section 312.10 requires testing at the time of installation and every year thereafter. Although backflow devices and methods can only be field inspected, not field tested, their use is typically limited to low-hazard applications.

The code official's duty is only to witness the testing as the permit holder is responsible for conducting the tests. As code officials often have many project sites to visit in a day, the permit holder must provide reasonable advance notice that the project is ready for testing. All test equipment, power, water, materials and labor to conduct the test(s) are the permit holder's responsibility.

Because the minimum length of time for the pressure test is 15 minutes, the code official will probably take the opportunity to perform a visual inspection of the plumbing system under test to verify that the system has been installed in accordance with the requirements of the code. As this will likely be his or her only opportunity to spot violations before concealment, it is imperative that all the work that needs to be performed is done so before concealment. Nail/screw shield plates, sealing caulking, all required hangers/supports and piping identification must be done prior to the code official arriving at the project site for the test.

The medium for pressure testing of piping systems is ideally water because it is not compressible and a failure of a plumbing system under test will not be dangerous to nearby personnel; however, this code section allows for pipe system testing by forcing air into the piping as long as the piping is not of plastic material. There is an inherent danger in testing plastic piping with compressed air. Where compressed gas (air) is used, the energy stored in the compressed gas is significant. A pipe/fitting failure could result in pieces and shards of plastic being propelled at high velocity toward personnel.

The last two lines of this code section are somewhat mislocated. The requirement for final tests are covered in Section 312.4 and the requirement for having the plumbing fixtures set and traps filled with water should be part of that section. The removal of a cleanout plug to determine if test pressure has reached certain parts of the drainage system is only applicable to drainage and sewer system (rough-in) testing. Note, however, it would not be wise to remove any cleanout plugs while a system is pressurized.

312.1.1 Test gauges. Gauges used for testing shall be as follows:

1. Tests requiring a pressure of 10 pounds per square inch (psi) (69 kPa) or less shall utilize a testing gauge having increments of 0.10 psi (0.69 kPa) or less.

2. Tests requiring a pressure of greater than 10 psi (69 kPa) but less than or equal to 100 psi (689 kPa) shall utilize a testing gauge having increments of 1 psi (6.9 kPa) or less.

3. Tests requiring a pressure of greater than 100 psi (689 kPa) shall utilize a testing gauge having increments of 2 psi (14 kPa) or less.

❖ It is common to encounter test gauges that read in 1 or 2 pounds per square inch (psi) increments while attempting to perform a drain, waste and vent (DWV) inspection that only requires a 5 psi (34.5 kPa) test. A small drop in pressure will be difficult to notice on a gauge with 1 or 2 psig (6.895 or 13.79 kPa) increments and, therefore they are not appropriate for low test pressures.

In each case, the user needs to ensure that the proper pressure gauge was selected with respect to indicating range and design. Most standard dial-type pressure gauges use a bourdon tube sensing element generally made of a copper alloy (brass) or stainless steel. The construction is simple and operation does not require any additional power source. The C-shaped or spirally wound bourdon tube flexes when pressure is applied, producing a rotational movement, which in turn causes the pointer to indicate the measured pressure. Such gauges are the most accurate when reading in the midrange of their scale; therefore, the test gauge should be chosen to reflect the test pressure in the midrange of the scale. A shutoff valve should be installed between the measuring point and the pressure gauge with another valve to depressurize the gage only. This will allow an exchange of the pressure gauge and checks on the zero setting while the system remains under test pressure or operating pressure.

312.2 Drainage and vent water test. A water test shall be applied to the drainage system either in its entirety or in sections. If applied to the entire system, all openings in the piping shall be tightly closed, except the highest opening, and the system shall be filled with water to the point of overflow. If the system is tested in sections, each opening shall be tightly plugged except the highest openings of the section under test, and each section shall be filled with water, but no section shall be tested with less than a 10-foot (3048 mm) head of water. In testing successive sections, at least the upper 10 feet (3048 mm) of the next preceding section shall be tested so that no joint or pipe in the building, except the uppermost 10 feet (3048 mm) of the system, shall have been submitted to a test of less than a 10-foot (3048 mm) head of water. This pressure shall be held for not less than 15 minutes. The system shall then be tight at all points.

❖ This section discusses the procedure for performing a water test on portions of or the entire DWV system. A rough-in test is commonly performed on the entire DWV system. In multistory buildings, this could be a dangerous practice because the lower floors are exposed to much higher head pressure than that for which the system was designed. This excess pressure could be detrimental to the plumbing system. Testing should be conducted in stages so that no one section of piping is exposed to excessive pressures.

This section requires that, except for the uppermost 10 feet (3048 mm) of the system, no section be tested with less than a 10-foot (3048 mm) head of water. This 10-foot (3048 mm) head of water is equivalent to 4.33 psi (30 kPa) and must be held tight for at least 15 minutes. The water test allows system defects to be found more readily than in air tests but introduces the risks of water damage to the structure, freeze damage to traps and piping system damage caused by excessive head pressures and excessive weight. This portion of the testing could be performed with air in compliance with Section 312.3 if the piping is not of plastic material.

312.3 Drainage and vent air test. Plastic piping shall not be tested using air. An air test shall be made by forcing air into the system until there is a uniform gauge pressure of 5 psi (34.5 kPa) or sufficient to balance a 10-inch (254 mm) column of mercury. This pressure shall be held for a test period of not less than 15 minutes. Any adjustments to the test pressure required because of changes in ambient temperatures or the seating of gaskets shall be made prior to the beginning of the test period.

❖ For other than systems made of plastic piping materials, this section allows pressure testing of DWV piping using compressed air. The closed piping system is filled with air until there is either a steady gauge pressure of 5 psi (34.5 kPa) or sufficient pressure to balance a 10-inch (254 mm) column of mercury. This pressure must be held for a period of not less than 15 minutes without the introduction of additional air.

Air tests are easy to apply; however, there are disadvantages to air testing. Defects in the system are more difficult to locate using an air test. System temperature changes can also cause misleading changes in pressure, which might be interpreted as a leak. In some cases, gasket or o-ring materials within valves will properly seal only after they are allowed a brief period to seal under pressure. Adjustments for changes in pressure resulting from ambient temperature fluctuations or the sealing of gaskets must be made prior to the start of the 15-minute test period.

Testing with a compressed gas (air) is inherently more hazardous to personnel than water testing because of the energy release that can occur from a system rupture or failure. Note that Section 312.1 prohibits the testing of all plastic piping with compressed air. The 5 psi (34.5 kPa) testing pressure stated in this section is often exceeded in the field by the persons performing the test to make the location of leaks easier. Increasing the test pressure beyond the safe limits of the material can result in a dangerous situation and must not be allowed.

312.4 Drainage and vent final test. The final test of the completed drainage and vent systems shall be visual and in sufficient detail to determine compliance with the provisions of this code. Where a smoke test is utilized, it shall be made by filling all traps with water and then introducing into the entire system a pungent, thick smoke produced by one or more smoke machines. When the smoke appears at *stack* openings on the roof, the *stack* openings shall be closed and a pressure equivalent to a 1-inch water column (248.8 Pa) shall be held for a test period of not less than 15 minutes.

❖ In preparation for the final drainage and vent test, the plumbing fixtures must be fully assembled, set and sealed to walls and floors, connected to the water supply system (if required), connected to the drain system (if required) and the fixture traps filled with liquid. Venting applications requiring an air admittance valve must have the valve installed. The items checked in the final visual test of the drainage and vent system is up to the discretion of the code official.

A smoke test of the completed drainage and vent system could be required by the code official where there is justification for such testing. Smoke testing requires that the outlet of the DWV system be plugged and a thick smoke be pumped into the DWV system so that smoke circulates throughout the entire system. After all vent openings are capped off, the system is then continuously pressurized to 1-inch water column (248.8 Pa) for a period of 15 minutes. During the test period, smoke must not be visually detected inside of the building in order for the system to pass the test [see Commentary Figures 312.4(1) and 312.4(2)].

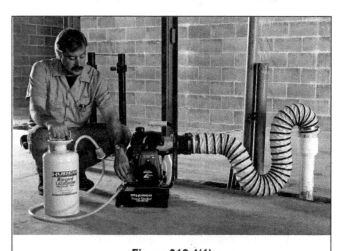

**Figure 312.4(1)
SMOKE-GENERATING EQUIPMENT
(Photo courtesy of Hurco Technologies, Inc.)**

**Figure 312.4(2)
SMOKE INDICATING PRESENCE OF LEAK
(Photo courtesy of Hurco Technologies, Inc.)**

312.5 Water supply system test. Upon completion of a section of or the entire water supply system, the system, or portion completed, shall be tested and proved tight under a water pressure not less than the working pressure of the system; or, for piping systems other than plastic, by an air test of not less than 50 psi (344 kPa). This pressure shall be held for not less than 15 minutes. The water utilized for tests shall be obtained from a potable source of supply. The required tests shall be performed in accordance with this section and Section 107.

❖ The water supply system consists of piping from the water meter at the curb, the stop valve at the curb or the well system to the ends of the water distribution piping in the building. Where only portions of the water supply system have been installed or worked on, only those portions are required to be tested.

The term "working pressure" is not defined in the code. It it understood to be the maximum pressure in the water supply system under normal operating conditions. The working pressure can be different depending on where the pressure is measured in a system. For example, the working pressure in a water service line from the curb stop valve to the entry point into the building might be 120 psi (827 kPa). As this pressure is greater than 80 psi (551 kPa), in accordance with Section 604.8, a pressure-reducing valve must be installed to limit the pressure in the building's water distribution system to no greater than 80 psi (551 kPa). Therefore, the working pressure in the water distribution system portion of the water supply system is 80 psi (551 KPa) or less.

The phrase "the system shall be proved tight" means that by visual inspection, no evidence of leakage from the piping system is observed. Evidence of leakage is typically determined by attaching a pressure gage to the system, pressurizing the system to the test pressure and, without further addition of test water or air to the system, verifying after 15 minutes that the pressure gage indication has not changed from the reading taken at the beginning of the test. Note the critical test gage requirements of Section 312.1.1. Where minor repairs or modifications are made to a water supply system and the system is tested with water, the pressure gage method is not necessary because evidence of leakage could be determined simply by observing each piping connection for the presence of leaking water.

If air is used to test a water supply system of other than plastic material, the test pressure need only be 50 psi (344 kPa). Where testing with compressed air, it is advisable to allow the system under test pressure to thermally stabilize before the start of the observation period. Warm air from a compressor introduced into cold piping will cool and result in a decrease in the test pressure, falsely indicating a leak. Gaskets and o-rings in shower mixing valves are intended to seal against water pressure and typically do not initially seat or seal well when first pressurized with air. In most cases, leaks from these locations will eventually cease after a "seating in" period of time. Adjustments for changes in pressure resulting from ambient temperature fluctuations or the seating of gaskets must be made prior to the start of the 15-minute test period.

Air testing of water supply systems of plastic material is prohibited by the code and many plastic piping manufacturers due to the risk of personal injury caused by flying shards of plastic should the piping rupture. Compressed air stores potential energy not unlike a compressed coil spring stores energy. If the piping ruptures, the stored potential energy becomes kinetic energy that can propel plastic pieces with great force and velocity.

Section 107 specifies three general construction stages that require inspection: 1) underground rough-in; 2) above-ground rough-in and 3) final (completion of plumbing work). Depending on the size and required sequence of a project, multiple inspections may be necessary for each stage. The intent of this code section is to ensure that all underground and above-ground water supply system piping is tested prior to concealment so that any leaks can be readily located and repaired. The final inspection and "test" (Section 312.4) provides for visual inspection to find leaks in exposed water supply system connections, such as water heater piping, stop valves, supply tube connections and fixture/faucet assemblies.

312.6 Gravity sewer test. Gravity *sewer* tests shall consist of plugging the end of the *building sewer* at the point of connection with the public sewer, filling the *building sewer* with water, testing with not less than a 10-foot (3048 mm) head of water and maintaining such pressure for 15 minutes.

❖ The building sewer must be subjected to the same testing as required for the DWV system, except that air testing is not an option for building sewers. This section states specific points of procedure for conducting this test, including plugging the sewer at its point of connection with the public sewer, the minimum test water head and the time duration of the test.

312.7 Forced sewer test. Forced *sewer* tests shall consist of plugging the end of the *building sewer* at the point of connection with the public sewer and applying a pressure of 5 psi (34.5 kPa) greater than the pump rating, and maintaining such pressure for 15 minutes.

❖ Although not specifically defined in the code, the term "forced sewer" refers to any sewer system that cannot drain by gravity. The forced sewer has its contents placed under pressure for some length of travel. This length of travel is often a short vertical rise to allow the sewage to again travel using the forces of gravity. This type of arrangement normally involves the use of a sewage pump or sewage ejector (refer to Section 712 for requirements related to sumps and ejectors).

This section requires that the forced sewer be tested in a manner similar to that required for the gravity sewer, except that this test is to be conducted with an applied test pressure of 5 psi (34.5 kPa) greater than the highest pressure that the pump can generate. Many times, this pressure is called the "shutoff" pressure because for many pumps, the highest pressure is when the flow is zero. The pump pressure curve or data table must be reviewed to determine the maximum pump pressure.

Although this section does not specify the test medium (air or water), the intent is that forced sewers be tested using the same methods and precautions as gravity sewer testing, except that the pressure will be according to the maximum pump pressure.

312.8 Storm drainage system test. *Storm drain* systems within a building shall be tested by water or air in accordance with Section 312.2 or 312.3.

❖ Piping used for drainage of roof water that is located inside a building has the same potential for leaks and defects as building sanitary drainage system. Because of this potential, this section requires that piping used for storm drains, conductors or downspouts and related branch lines that are located inside a building be tested. This testing is to be performed in a manner similar to the methods that would be used to test building DWV systems.

312.9 Shower liner test. Where shower floors and receptors are made water tight by the application of materials required by Section 417.5.2, the completed liner installation shall be tested. The pipe from the shower drain shall be plugged water tight for the test. The floor and receptor area shall be filled with potable water to a depth of not less than 2 inches (51 mm) measured at the threshold. Where a threshold of at least 2 inches (51 mm) high does not exist, a temporary threshold shall be constructed to retain the test water in the lined floor or receptor area to a level not less than 2 inches (51 mm) deep measured at the threshold. The water shall be retained for a test period of not less than 15 minutes, and there shall not be evidence of leakage.

❖ Although Section 417.5.2 always required field-fabricated shower floors and receptors to be made "water tight" (i.e., free of any leaks when exposed to water) by the application of liner materials. In previous editions the code did not actually require testing to prove that the installation was leak free. Many jurisdictions have required testing of field-fabricated shower pans for many years due to the potential for water damage if the installation of liner material is not performed properly. The resulting water damage, increased difficulty in making repairs and additional expense involved with the discovery of a leak after a building has been completed prompted the addition of this testing requirement.

This section clearly spells out how to perform the test. The evidence of leakage could be either the lowering of the water level from the full threshold level or water drips/seepage outside of the receptor area (see Commentary Figure 312.9).

312.10 Inspection and testing of backflow prevention assemblies. Inspection and testing shall comply with Sections 312.10.1 and 312.10.2.

❖ Backflow prevention assemblies are critical components that protect the potable water system from pollution or contamination sources. As such, these assemblies must be inspected and tested after installation at the project site to ensure that the devices are providing the level of protection intended.

312.10.1 Inspections. Annual inspections shall be made of all backflow prevention assemblies and *air gaps* to determine whether they are operable.

❖ Backflow prevention assemblies, including air gaps, must be inspected annually. Air gaps must be inspected to determine that proper clearances are maintained in accordance with Table 608.15.1. Additionally, backflow prevention assemblies are to be inspected for damage or evidence of tampering that could cause the device to malfunction.

Because these devices must be periodically inspected, they cannot be located in concealed spaces. Depending on the type of device and the installed location, they might be located behind access panels. Backflow devices must not be allowed to be installed in a pit or other below grade locations as flooding could prevent the device from properly operating.

312.10.2 Testing. Reduced pressure principle, double check, pressure vacuum breaker, reduced pressure detector fire protection, double check detector fire protection, and spill-resistant vacuum breaker backflow preventer assemblies and hose connection backflow preventers shall be tested at the time of installation, immediately after repairs or relocation and at least annually. The testing procedure shall be performed in accordance with one of the following standards: ASSE 5013, ASSE 5015, ASSE 5020, ASSE 5047, ASSE 5048, ASSE 5052, ASSE 5056, CSA B64.10 or CSA B64.10.1.

❖ Most sections in the code that concern backflow prevention devices have been aligned to reflect the

Figure 312.9
TESTING OF FIELD-FABRICATED SHOWER LINER

names of the devices in accordance with the American Society of Sanitary Engineers (ASSE) standards names. There is a method to the naming convention:

1. *Backflow prevention assemblies* are those devices that have test ports and test valves to enable precise field testing with highly sensitive testing instruments to determine if the assembly will perform backflow prevention under various test conditions.

2. *Backflow preventers* are devices that do not have test ports to enable field testing, although some devices such as hose connection backflow preventers can be subjected to field conditions to cause a visible indication that the device is functional.

3. *Other methods of backflow prevention* include air gaps and barometric loops. Testing is not required as field dimensional checking is used to verify that the method will perform the intended function.

Each of the ASSE standards listed, with the exception of ASSE 5052, is the test procedure for a specific type of backflow prevention assembly. ASSE 5052 is the procedure for testing hose connection backflow preventers. The code requires that backflow prevention assemblies and hose connection backflow preventers be tested when the assembly is installed, after it is relocated, after repairs and at intervals not greater than one year apart. In most jurisdictions, backflow prevention assembly testers are required to be licensed by the state. This is to ensure that test personnel are qualified by initial training, initial skill testing and periodic training and testing for as long as they wish to maintain their certification. Although the code does not require assembly testing by certified testers, the code official will be obligated to follow any state laws concerning who can legally test backflow prevention assemblies. Because testing of hose connection backflow preventers does not involve specialized training, certified testers may not be required for testing of these simple devices.

How is the code official to enforce the annual and repair testing requirement of the code? Testing after the installation and initial testing of the assembly is the responsibility of the building owner as he or she is responsible for the safety of the potable water supply within his or her building. In most cases, the code official assumes that the building owner has the required tests performed just as the building owner would perform periodic maintenance and tests on other types of building equipment, such as HVAC units, filter replacements, etc. However, if the code official has: 1) doubts that testing is being performed; 2) heard of complaints about the condition of the building potable water supply or 3) observed during the course of other inspections that backflow prevention assemblies do not have up-to-date test tags, this section of the code gives the code official the authority to request to see the latest test records and require testing of the assemblies if inspection reveals testing has not been performed within the past year.

SECTION 313
EQUIPMENT EFFICIENCIES

313.1 General. Equipment efficiencies shall be in accordance with the *International Energy Conservation Code*.

❖ The *International Energy Construction Code*® (IECC®) requires minimum efficiencies for appliances and hot water storage tanks.

SECTION 314
CONDENSATE DISPOSAL

[M] 314.1 Fuel-burning appliances. Liquid combustion byproducts of condensing appliances shall be collected and discharged to an *approved* plumbing fixture or disposal area in accordance with the manufacturer's instructions. Condensate piping shall be of *approved* corrosion-resistant material and shall not be smaller than the drain connection on the appliance. Such piping shall maintain a horizontal slope in the direction of discharge of not less than one-eighth unit vertical in 12 units horizontal (1-percent slope).

❖ Condensation must be collected from appliances in accordance with the manufacturers' installation instructions and this section. High-efficiency condensing-type appliances (Category IV) and some midefficiency appliances produce liquid as a combustion byproduct. This condensate is collected at various points in the appliance's heat exchangers and venting system and must be disposed of. Because of impurities in the fuel and combustion air, the condensate can be acidic, thus corrosive, to many materials. For example, such condensate can contain hydrochloric and also sulfuric acids. Such drain piping is part of the mechanical system and, as such, is also covered in the IMC. Although condensate piping is not plumbing, it is covered in the code because of the possibility of plumbers installing such piping.

[M] 314.2 Evaporators and cooling coils. Condensate drain systems shall be provided for equipment and appliances containing evaporators or cooling coils. Condensate drain systems shall be designed, constructed and installed in accordance with Sections 314.2.1 through 314.2.5.

❖ Appliances and equipment containing evaporators or cooling coils, including refrigeration, dehumidification and comfort cooling equipment, can produce condensate from the water vapor in the atmosphere. A drainage system is necessary to dispose of the condensate and prevent damage to the structure.

[M] 314.2.1 Condensate disposal. Condensate from all cooling coils and evaporators shall be conveyed from the drain pan outlet to an *approved* place of disposal. Such piping shall maintain a horizontal slope in the direction of discharge of not less than one-eighth unit vertical in 12 units horizontal (1-percent slope). Condensate shall not discharge into a street, alley or other areas so as to cause a nuisance.

❖ The "approved place of disposal" specified in this section will vary from jurisdiction to jurisdiction. While condensate is typically "clear" water, it can gather dust, bacteria and mold spores, which can result in a solids buildup in the waste stream. A slope of at least 1 percent creates a minimum flow velocity to aid in transport of these solids towards the point of disposal.

Some jurisdictions strictly forbid condensate from being drained into the sanitary sewer system primarily because of overloading the system, but could allow it, depending on the quantity of effluent produced. Common points of disposal are to grade in areas where a nuisance will not be created, onto rooftops where the condensate will evaporate or drain to grade, to seepage pits where soil condition allows or to a storm drain. Section 802.1.5 prohibits condensate drains from being directly connected to a sanitary sewer because of the possibility of sewer gases or sewage entering the equipment or system.

Additional consideration must be given to the disposal of condensate from fuel-burning appliances (see commentary, Section 314.1).

[M] 314.2.2 Drain pipe materials and sizes. Components of the condensate disposal system shall be cast iron, galvanized steel, copper, cross-linked polyethylene, polyethylene, ABS, CPVC or PVC or polypropylene pipe or tubing. All components shall be selected for the pressure and temperature rating of the installation. Joints and connections shall be made in accordance with the applicable provisions of Chapter 7 relative to the material type. Condensate waste and drain line size shall be not less than $^3/_4$-inch (19.1 mm) internal diameter and shall not decrease in size from the drain pan connection to the place of condensate disposal. Where the drain pipes from more than one unit are manifolded together for condensate drainage, the pipe or tubing shall be sized in accordance with Table 314.2.2.

❖ Condensate drains must be constructed of a material listed in this section. The previous editions of this section required that horizontal sections of the condensate piping be installed in uniform alignment at a uniform slope. This was to prevent dips and sags that could impede flow and also prevented the use of flexible piping that could not maintain a uniform slope. Section 314.2.1 requires a slope of at least 1 percent [$^1/_8$ inch per foot (10 mm per meter)]. Condensate drains must have the minimum prescribed slope to promote gravity drainage and lessen the possibility of clogging.

GENERAL REGULATIONS

Greater slopes are allowed, and in all cases, the slope must be maintained in the direction flow such that no dips or reverse slopes occur.

While the code is clear on a minimum condensate pipe size, designers previously had no approved method for sizing condensate lines connecting multiple units. This table provides minimum sizing for those condensate piping system connecting multiple units.

**[M] TABLE 314.2.2
CONDENSATE DRAIN SIZING**

EQUIPMENT CAPACITY	MINIMUM CONDENSATE PIPE DIAMETER (inch)
Up to 20 tons of refrigeration	$^3/_4$ inch
Over 20 tons to 40 tons of refrigeration	1 inch
Over 40 tons to 90 tons of refrigeration	$1^1/_4$ inch
Over 90 tons to 125 tons of refrigeration	$1^1/_2$ inch
Over 125 tons to 250 tons of refrigeration	2 inch

For SI: 1 inch = 25.4 mm, 1 ton of capacity = 3.517 kW.

[M] 314.2.3 Auxiliary and secondary drain systems. In addition to the requirements of Section 314.2.1, where damage to any building components could occur as a result of overflow from the equipment primary condensate removal system, one of the following auxiliary protection methods shall be provided for each cooling coil or fuel-fired appliance that produces condensate:

1. An auxiliary drain pan with a separate drain shall be provided under the coils on which condensation will occur. The auxiliary pan drain shall discharge to a conspicuous point of disposal to alert occupants in the event of a stoppage of the primary drain. The pan shall have a depth of not less than $1^1/_2$ inches (38 mm), shall be not less than 3 inches (76 mm) larger than the unit or the coil dimensions in width and length and shall be constructed of corrosion-resistant material. Galvanized sheet metal pans shall have a thickness of not less than 0.0236-inch (0.6010 mm) (No. 24 gage) galvanized sheet metal. Nonmetallic pans shall have a thickness of not less than 0.0625 inch (1.6 mm).

2. A separate overflow drain line shall be connected to the drain pan provided with the equipment. Such overflow drain shall discharge to a conspicuous point of disposal to alert occupants in the event of a stoppage of the primary drain. The overflow drain line shall connect to the drain pan at a higher level than the primary drain connection.

3. An auxiliary drain pan without a separate drain line shall be provided under the coils on which condensate will occur. Such pan shall be equipped with a water-level detection device conforming to UL 508 that will shut off the equipment served prior to overflow of the pan. The auxiliary drain pan shall be constructed in accordance with Item 1 of this section.

4. A water-level detection device conforming to UL 508 shall be provided that will shut off the equipment served in the event that the primary drain is blocked.

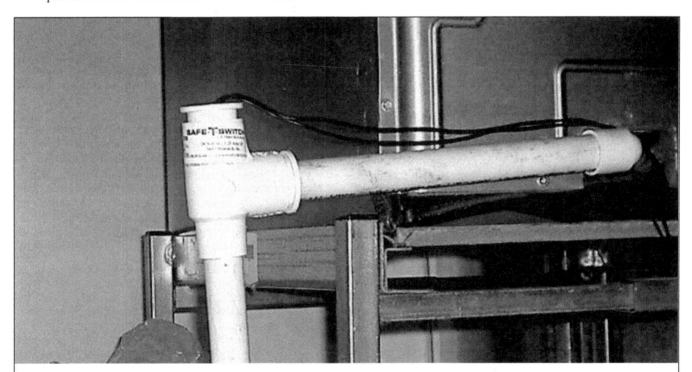

Figure 314.2.3(1)
CONDENSATE OVERFLOW SWITCH
(Photo courtesy of SMD Research, Inc.)

GENERAL REGULATIONS

The device shall be installed in the primary drain line, the overflow drain line or in the equipment-supplied drain pan, located at a point higher than the primary drain line connection and below the overflow rim of such pan.

Exception: Fuel-fired appliances that automatically shut down operation in the event of a stoppage in the condensate drainage system.

❖ An auxiliary or secondary drain system is required for equipment locations where condensate overflow would cause damage to a building or its contents. These systems are intended to catch condensate spilling from the primary drain or drain pan if the primary drain becomes clogged. These methods protect the building from structural damage. This section applies to cooling and evaporator coils and condensing-type fuel-fired appliances.

Condensate drains are notorious for clogging because of debris (lint and dust) from air-handling systems and the natural affinity to produce slime growth in drain pans and pipes.

Note that four distinct options are provided for the auxiliary/secondary system. Option 1 is a redundant drain pan and drain piping system that is completely independent of the primary system. Option 2 is the simplest method because it involves only an overflow drain connected to the overflow opening on the primary pan. Such an opening is typically found on factory-built primary pans. The overflow drain must be completely independent of the primary drain. Option 3 involves an auxiliary pan without a drain and depends on a float switch or sensor to detect water in the pan and shut off the equipment producing the condensate [see Commentary Figure 314.2.3(3)]. Option 4 involves a float-switch assembly designed to be threaded into the overflow tapping provided on the primary pan or installed at some point in a drain line. Like Option 3, the float-switch assembly will detect overflow from the primary pan and shut off the equipment [See Commentary Figures 314.2.3(1) and (2)].

The exception is for fuel-fired appliances having integral controls to prevent operation of the appliance

Figure 314.2.3(3)
PRIMARY OR AUXILIARY DRAIN PAN FLOAT SWITCH
(Photo courtesy of SMD Research, Inc.)

Figure 314.2.3(2)
CONDENSATE OVERFLOW SWITCH
(Photo courtesy of SMD Research, Inc.)

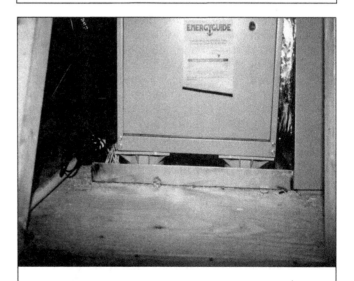

Figure 314.2.3.2
FURNACE ELEVATED ABOVE TOP OF CONDENSATE PAN

if condensate is not draining. The standards for some appliances, such as gas-fired furnaces, will permit either of two performance modes for an appliance with a failed condensate drainage system: 1) the appliance will shut down or 2) the appliance will continue to operate, provided that such operation does not in any way present a hazard.

[M] 314.2.3.1 Water-level monitoring devices. On downflow units and all other coils that do not have a secondary drain or provisions to install a secondary or auxiliary drain pan, a water-level monitoring device shall be installed inside the primary drain pan. This device shall shut off the equipment served in the event that the primary drain becomes restricted. Devices installed in the drain line shall not be permitted.

❖ Some appliances (downflow type) and coil arrangements cannot be fitted with a secondary drain or installed with an auxiliary drain pan. This is particularly true for the typical rooftop HVAC unit where drain pan overflow could damage ducts and duct insulation. An internal water monitoring device must be used in these situations so that the device shuts off the unit prior to overflow of the primary drain pan. If the device was installed external to the appliance, a drain blockage in or near the appliance might occur upstream of the device, thereby rendering it useless.

[M] 314.2.3.2 Appliance, equipment and insulation in pans. Where appliances, equipment or insulation are subject to water damage when auxiliary drain pans fill such portions of the appliances, equipment and insulation shall be installed above the *flood level rim* of the pan. Supports located inside of the pan to support the appliance or equipment shall be water resistant and *approved*.

❖ If a condensate overflow situation occurs, appliances sitting directly on the bottom of the pan will be exposed to the condensate water, even though the pan may have a working drain. Most appliances are not suitable for constantly damp or intermittent wet environments. Sheet metal enclosures will quickly rust under these conditions. Exterior and interior insulation will be degraded by the water, providing ideal conditions for mold growth in concealed locations. This section requires that any such appliance be elevated above the flood rim level of the pan below so that the appliance is not directly exposed to wet conditions. The code is silent on what type of material is to be used for support other than requiring it to be water resistant and approved by the code official (see Commentary Figure 314.2.3.2).

[M] 314.2.4 Traps. Condensate drains shall be trapped as required by the equipment or appliance manufacturer.

❖ The appliance or equipment manufacturer determines the need for a trap and often specifies its depth and configuration. The traps addressed in this section are unrelated to plumbing traps and serve a different purpose. Condensate drain traps do not directly connect in any way to the plumbing DWV system of a building. Condensate drain traps are installed to prevent air from being pushed or pulled through the drain piping. Such airflow can impede condensate flow, causing overflow or abnormal water depth in drain pans. Some drain pans, such as those under pull-through cooling coils, might not drain without a trap to block airflow in the drain piping.

[M] 314.2.4.1 Ductless mini-split system traps. Ductless mini-split equipment that produces condensation shall be provided with an in-line check valve located in the drain line or a trap.

❖ See the definition of "Ductless Mini-split system" in the IMC. Such equipment may use a liquid seal trap or a check valve, depending on the design of the indoor unit, as both prevent airflow from interfering with condensate flow and prevent the escape of air through the drain.

[M] 314.2.5 Drain line maintenance. Condensate drain lines shall be configured to permit the clearing of blockages and performance of maintenance without requiring the drain line to be cut.

❖ Drains that convey condensate water from cooling coils and evaporators are known to develop blockages as a result of debris and biological growth in the system. These drains are commonly cleared of blockages by a compressed gas such as air or nitrogen being forced through the drain. It is inherently hazardous to pressurize plastic piping such as PVC and CPVC with a compressed gas because of the potential for violent rupture and propelled shards of plastic. The drains are seldom large enough to accommodate mechanical drain-cleaning (rodding) equipment. The code permits any arrangement that provides access to the drain interior without the drain being severed or cut. This includes capped or plugged tees and cross fittings, unions, removable mechanical couplings and specialty devices made specifically for the attachment of compressed-gas hoses. The intent is to prevent the spillage of condensate that would cause damage.

SECTION 315
PENETRATIONS

315.1 Sealing of annular spaces. The annular space between the outside of a pipe and the inside of a pipe sleeve or between the outside of a pipe and an opening in a building envelope wall, floor, or ceiling assembly penetrated by a pipe shall be sealed in an *approved* manner with caulking material, foam sealant or closed with a gasketing system. The caulking material, foam sealant or gasketing system shall be designed for the conditions at the penetration location and shall be compatible with the pipe, sleeve and building materials in contact with the sealing materials. Annular spaces created by pipes penetrating fire-resistance-rated assemblies or membranes of such assemblies shall be sealed or closed in accordance with Section 714 of the *International Building Code*.

❖ The annular space created between a pipe sleeve and a pipe or an opening in the building envelope and a pipe must be sealed to prevent airflow through the space and to block the space from insects. The material used must be somewhat elastic to prevent structural loading of the pipe (see commentary, Sections 304.4 and 305.3) should there be some movement between the pipe and the opening. Caulking, expanding foam sealant or a gasketing system can be used to seal the ends of the annular space. It is not required to fill the entire annular space with material. The materials used to seal the annular space must be compatible with the pipe as well as the pipe sleeve material or building envelope materials [see Commentary Figure 307.3(2)].

Annular spaces created by penetrations in fire-resistance-rated assemblies must only be sealed in accordance with materials and methods specific to the firestop assembly chosen for the through penetration. All fire-resistance-rated through-penetration assemblies must be qualified by testing in accordance with ASTM E814 as required by the IBC [see Commentary Figures 307.3(3) and 307.3(4)].

SECTION 316
ALTERNATIVE ENGINEERED DESIGN

316.1 Alternative engineered design. The design, documentation, inspection, testing and approval of an *alternative engineered design* plumbing system shall comply with Sections 316.1.1 through 316.1.6.

❖ This section permits an engineer or architect to design a plumbing system that may not comply with all of the provisions found in Chapters 3 through 14. The design must be approved by the code official and must conform to accepted engineering principles. The engineered plumbing system must provide the level of protection of the public health, safety and welfare intended by the code.

316.1.1 Design criteria. An *alternative engineered design* shall conform to the intent of the provisions of this code and shall provide an equivalent level of quality, strength, effectiveness, fire resistance, durability and safety. Material, equipment or components shall be designed and installed in accordance with the manufacturer's instructions.

❖ Although an engineered plumbing system may not comply with all of the minimum requirements set forth in Chapters 3 through 14, it must comply with the intent of these provisions. This section permits use of standard engineering principles in the design of an innovative system as long as there is no sacrifice of quality, strength, effectiveness, fire resistance, durability and safety. This section further reinforces the intent of Section 105.2 for the acceptance of alternative materials and equipment. The requirement for compliance with the manufacturer's installation instructions is generally intended to address entire engineered systems, such as the siphonic roof drain system. The manufacturer or appropriate industry association provides criteria contained in design and installation handbooks. The manufacturer's instructions must be followed for all innovative fittings or products regulated by this section.

316.1.2 Submittal. The registered design professional shall indicate on the permit application that the plumbing system is an *alternative engineered design*. The permit and permanent permit records shall indicate that an *alternative engineered design* was part of the *approved* installation.

❖ The permit and permanent permit records must indicate that an alternative engineered design is part of the proposed plumbing design. This is essential information to have on file to maintain a complete legal record of the plumbing system. When future permits are applied for regarding alterations or modifications, appropriate measures can then be taken to determine that the future work will not adversely affect the system design.

316.1.3 Technical data. The registered design professional shall submit sufficient technical data to substantiate the proposed *alternative engineered design* and to prove that the performance meets the intent of this code.

❖ The appropriate information and data must be submitted to the code official to assist in the approval of the alternative engineered design. This is not an option. Acceptable data to substantiate the performance of the proposed plumbing system or components include results of tests performed by an approved third-party testing agency, design calculations or an evaluation report issued by an authoritative agency, such as ICC Evaluation Service.

316.1.4 Construction documents. The registered design professional shall submit to the code official two complete sets of signed and sealed construction documents for the *alternative engineering design*. The construction documents shall include floor plans and a riser diagram of the work. Where appropriate, the construction documents shall indicate the direction of flow, all pipe sizes, grade of horizontal piping, loading and location of fixtures and appliances.

❖ This section is used in conjunction with Section 106.3.1. The option for the code official to waive the requirements for filing construction documents, however, is not applicable. The required detailing of such documents is needed to provide the code official with the necessary information to review and approve the plans.

316.1.5 Design approval. Where the code official determines that the *alternative engineered design* conforms to the intent of this code, the plumbing system shall be *approved*. If the *alternative engineered design* is not *approved*, the code official shall notify the registered design professional in writing, stating the reasons thereof.

❖ The code official is responsible for determining whether the requested alternative engineered design provides the equivalent level of protection of public health, safety and welfare as required in the code. The code official's response to the design professional must be in writing, stating the reason for either accepting or rejecting the request. If the code official rejects the request for the alternative engineered system, the registered design professional may appeal the decision, as regulated by Section 109.

GENERAL REGULATIONS

316.1.6 Inspection and testing. The *alternative engineered design* shall be tested and inspected in accordance with the requirements of Sections 107 and 312.

❖ As is the case for all plumbing installations, the code official must inspect the alternative engineered plumbing to verify that the work is in compliance with the construction documents. Section 107 requires the code official to witness the testing of the plumbing system before it is placed in service to verify that it is free from leaks or other defects, in accordance with Section 312.

Bibliography

The following resource materials are referenced in this chapter or are relevant to the subject matter addressed in this chapter.

ANSI Z4.3-95, *Minimum Requirements for Nonsewered Waste-Disposal Systems.* New York: American National Standards Institute, Inc., 1995.

Facility Piping Systems Handbook, 2nd Edition, Frankel, McGraw Hill, 2002.

Chapter 4:
Fixtures, Faucets and Fixture Fittings

General Comments

Determining the exact number of plumbing fixtures needed by the maximum number of possible users in any given building continues to be a challenging problem for the plumbing industry. Various methods have been used to evaluate the number and type of plumbing fixtures needed for an occupancy. Studies of office buildings have produced design guidelines based on occupancy times, arrival rates and patterns of fixture use that offer insight into the number of required plumbing fixtures for a desired level of service. The Building Technology Research Laboratory at the Stevens Institute of Technology conducted a study based on the Queueing Theory, which involves probabilities for waiting times and fixture use times based upon a preferred level of service. The study resulted in design guidelines for the quantities of fixtures needed in public toilet facilities. For residential-type buildings and health care facilities, plumbing fixture numbers are based on the minimum need, resulting in at least one water closet and one lavatory for each dwelling unit, guestroom or hospital room.

Studies completed by the U.S. military have been used for dormitories and prisons to determine the number of fixtures required based on a simultaneous need in a regimented society. This assumes that everyone rises at approximately the same time and has a limited amount of time to shower and use the water closet and lavatory.

The National Restaurant Association conducted a study based on the difference in use between a restaurant and a nightclub. It should be noted that the study did not take into account today's fast-food-style restaurant, nor did it allow for restaurants located along heavily traveled routes, such as those at highway rest stops or oases.

Fixture requirements for factory and industrial uses are based on the same requirements as for storage facilities. This method establishes a realistic minimum requirement for factory occupancies. The reasonableness in the number of plumbing fixtures was established through a limited study of factory projects in Henrico County, Virginia.

The fixture needs for the remaining occupancies were determined based on empirical data, experience and tradition. There are no exact studies providing values that are definitively supportable; the values have been modified periodically based on general observations. Various studies will most likely continue to be performed in order to develop a completely rational method for determining the minimum number of plumbing fixtures for any occupancy.

Purpose

The purpose of Chapter 4 is to provide a building with the necessary number of plumbing fixtures of a specific type and quality. The fixtures must be properly installed to be usable by the individuals occupying the building. The quality and design of every fixture must conform to the applicable referenced standard.

SECTION 401
GENERAL

401.1 Scope. This chapter shall govern the materials, design and installation of plumbing fixtures, faucets and fixture fittings in accordance with the type of *occupancy*, and shall provide for the minimum number of fixtures for various types of occupancies.

❖ This section contains the scoping requirements for Chapter 4. Compliance with this chapter will result in a building or structure having adequate plumbing fixtures for the sanitary, hygienic, cleaning, washing and food preparation needs of the occupants. Many of the requirements of this chapter have been duplicated in the *International Building Code®* (IBC®), Chapter 29, for the convenience of the architect, who is often responsible for the layout and number of plumbing fixtures when creating the floor plan of a building.

401.2 Prohibited fixtures and connections. Water closets having a concealed trap seal or an unventilated space or having walls that are not thoroughly washed at each discharge in accordance with ASME A112.19.2/CSA B45.1 shall be prohibited. Any water closet that permits siphonage of the contents of the bowl back into the tank shall be prohibited. Trough urinals shall be prohibited.

❖ Water closet fixture designs that do not adequately clean the inside bowl surfaces with flushing water can harbor bacteria, resulting in an insanitary condition. The inspection and testing requirements of ASME A112.19.2/CSA B45.1 ensure that water closet designs have the necessary scouring action as well as a visible trap seal. Water closets must not be located in unventilated spaces as stagnant air conditions in toilet rooms are known to have caused widespread public disease resulting in high mortality for building occupants. Tank-type water closets must be designed so that the bowl contents cannot be siphoned into the tank as this can expose the occupants to water-borne diseases. Trough urinals are not allowed to be installed as they do not offer the necessary privacy to the users and urine stream splashing creates insanitary conditions for adjacent users.

401.3 Water conservation. The maximum water flow rates and flush volume for plumbing fixtures and fixture fittings shall comply with Section 604.4.

❖ The impetus for the plumbing code to begin regulating the amount of water used in the operation of certain plumbing fixtures was the Federal Energy Policy Act of 1992 (EPAct). The process of transporting raw water to treatment plants, the treatment of water, the conveyance of treated water to the users, the transportation of waste water to treatment plants, and the treatment of waste water consume tremendous amounts of energy. As the amount of water processed through this never-ending cycle becomes greater, the energy consumed increases proportionally. The production of energy for this purpose (primarily electricity) depletes the planet's nonrenewable resources of coal, gas, oil and fissionable materials. The purpose of the Energy Act of 1992 was to put controls in place to reduce the consumption rate of nonrenewable resources used for power generation. Among the many restrictions required by EPAct, certain plumbing fixtures were targeted for a reduction in the rate of water use. These flow and volume restrictions are reflected in Table 604.4.

Water is a resource that all life forms on this planet must have to survive. While there is an abundance of water on the Earth, the total amount of water is finite and is not necessarily located where water is needed. As the number of people on this planet continues to increase and as more countries become industrialized, the demand for potable water continues to rise. Many regions of the world are quickly depleting the once plentiful fresh water resources. Water conversation practices are now a major factor in building designs as there is already a real crisis of exhausting fresh water resources in some regions of the world.

SECTION 402
FIXTURE MATERIALS

402.1 Quality of fixtures. Plumbing fixtures shall be constructed of *approved* materials, with smooth, impervious surfaces, free from defects and concealed fouling surfaces, and shall conform to standards cited in this code. All porcelain enameled surfaces on plumbing fixtures shall be acid resistant.

❖ This section is provided to direct plumbing fixture standard creators as to the core requirements that all plumbing fixtures must meet. Most mass-produced plumbing fixtures for installation in accordance with the code have the design and quality regulated by the various standards as specified in the code. Where a plumbing fixture that doesn't have a standard is desired to be used, the information in this section assists the code official in evaluating the product in accordance with Section 105.2.

402.2 Materials for specialty fixtures. Materials for specialty fixtures not otherwise covered in this code shall be of stainless steel, soapstone, chemical stoneware or plastic, or shall be lined with lead, copper-base alloy, nickel-copper alloy, corrosion-resistant steel or other material especially suited to the application for which the fixture is intended.

❖ Most commercially available fixtures are designed and made in compliance with nationally recognized standards that are referenced in the code. Where a fixture is produced to a fixture standard not referenced by the code or no fixture standard exists, the code official must evaluate and approve the fixture in accordance with Section 105.2, regardless of whether the fixture is made of the materials listed in this section. For example, consider a stainless steel kitchen sink that is made from one of the materials listed in this section but is in compliance only with an unrecognized fixture standard. The faucet hole(s) and drain outlet dimensions could be such that fittings complying with the referenced standards in the code will not fit properly. The stainless steel material used to make the sink might be of a grade that will quickly corrode or be stained in normal service. Therefore, the code official might deem the fixture not suitable since he or she is charged with the responsibility to ensure that the sink will provide for reasonable durability, sanitary conditions and ease of maintenance.

Shower bases and group wash-up sinks made of terrazzo are fixtures that are not covered by standards referenced in the code and, as such, must be specifically approved by the code official.

Other specialty fixtures including lavatory bowls made of exotic metals, glass, and antique wash basins; bathtubs of hollowed-out blocks of granite or other natural materials such as wood and custom-fabricated copper kitchen sinks. If these products are not made to a referenced standard listed in the code, these products must be approved by the code official.

402.3 Sheet copper. Sheet copper for general applications shall conform to ASTM B152 and shall not weigh less than 12 ounces per square foot (3.7 kg/m^2).

❖ In the past, general-use sheet copper was predominantly used as a lining in plumbing fixtures. Today, the material is rarely used. The ASTM International (ASTM) standard regulates cold-rolled tempered sheet copper, hot-rolled tempered sheet copper and annealed copper sheet. Note that Section 425.3.3 permits the use of 10-ounce-per square-foot (0.026 kg/m^2) sheet copper to line flush tanks, and Section 902.2 permits the use of 8-ounce-per-square-foot (0.021 kg/m^2) sheet copper for vent flashings.

402.4 Sheet lead. Sheet lead for pans shall not weigh less than 4 pounds per square foot (19.5 kg/m²) and shall be coated with an asphalt paint or other *approved* coating.

❖ In years past, general-use sheet lead was used to field fabricate shower pans. It was previously used as a liner on a more extensive basis in many other areas of plumbing systems. The approximate thickness of 4-pound-per-square-foot (20 kg/m²) sheet lead is $^1/_{16}$ inch (1.6 mm).

SECTION 403
MINIMUM PLUMBING FACILITIES

403.1 Minimum number of fixtures. Plumbing fixtures shall be provided in the minimum number as shown in Table 403.1, based on the actual use of the building or space. Uses not shown in Table 403.1 shall be considered individually by the code official. The number of occupants shall be determined by the *International Building Code*.

❖ This section requires that the type and number of plumbing fixtures be based on the actual use of an occupancy and the occupant load as determined for the means of egress for a building or space. Occupancy loads are from Section 1004 of the IBC. Table 403.1 includes ratios indicating the maximum number of occupants that can be served by one fixture of each type. For example, "1 per 75" means one fixture will serve up to 75 occupants. These "ratios" are applied to the occupant load of the building or space in order to determine the total number of each required fixture type.

The "description" of plumbing fixture use determines which set of fixture ratios is used for calculating plumbing fixture requirements. In most cases, the description of plumbing fixture use will match the IBC occupancy classification. However, there are situations where a description of fixture use is different from the IBC occupancy classification.

Educational Facilities

Consider an educational facility (Grades 1-12) with a gymnasium with a stage. The IBC states that assembly areas that are accessory to Group E occupancies are not to be considered as a separate occupancy. Therefore, the number of water closets should be determined from Section 3 of Table 403.1, which requires one water closet per 50 occupants. Because this gymnasium has a stage, a greater occupant density than for a gymnasium-only space must be chosen as nonfixed chairs will most likely be set up for viewing on-stage activities. Therefore, for egress purposes, the gymnasium is considered to be an assembly area with an occupant density of 7 square feet (0.65 m²) per person. The resulting occupant load number applied with a Group E occupancy plumbing fixture ratio would result in an excessive number of plumbing fixtures for the building. The correct approach is to consider the plumbing fixture ratios in the fourth row of Table 403.1 for gymnasium space, since these ratios reflect how the space is intended to be used with respect to the occupancy density chosen for the space. In other words, fixture ratios must be chosen to "agree" with the actual use of the space.

TABLE 403.1
MINIMUM NUMBER OF REQUIRED PLUMBING FIXTURES[a]
(See Sections 403.1.1 and 403.2)

NO.	CLASSIFICATION	OCCUPANCY	DESCRIPTION	WATER CLOSETS (URINALS: SEE SECTION 419.2)		LAVATORIES		BATHTUBS/ SHOWERS	DRINKING FOUNTAIN (SEE SECTION 410)	OTHER
				MALE	FEMALE	MALE	FEMALE			
1	Assembly	A-1[d]	Theaters and other buildings for the performing arts and motion pictures	1 per 125	1 per 65	1 per 200		—	1 per 500	1 service sink
		A-2[d]	Nightclubs, bars, taverns, dance halls and buildings for similar purposes	1 per 40	1 per 40	1 per 75		—	1 per 500	1 service sink
			Restaurants, banquet halls and food courts	1 per 75	1 per 75	1 per 200		—	1 per 500	1 service sink
		A-3[d]	Auditoriums without permanent seating, art galleries, exhibition halls, museums, lecture halls, libraries, arcades and gymnasiums	1 per 125	1 per 65	1 per 200		—	1 per 500	1 service sink
			Passenger terminals and transportation facilities	1 per 500	1 per 500	1 per 750		—	1 per 1,000	1 service sink
			Places of worship and other religious services	1 per 150	1 per 75	1 per 200		—	1 per 1,000	1 service sink

(continued)

TABLE 403.1 —continued
MINIMUM NUMBER OF REQUIRED PLUMBING FIXTURES[a]
(See Sections 403.1.1 and 403.2)

NO.	CLASSIFICATION	OCCUPANCY	DESCRIPTION	WATER CLOSETS (URINALS: SEE SECTION 419.2)		LAVATORIES		BATHTUBS/ SHOWERS	DRINKING FOUNTAIN (SEE SECTION 410)	OTHER
				MALE	FEMALE	MALE	FEMALE			
1 (cont.)	Assembly	A-4	Coliseums, arenas, skating rinks, pools and tennis courts for indoor sporting events and activities	1 per 75 for the first 1,500 and 1 per 120 for the remainder exceeding 1,500	1 per 40 for the first 1,520 and 1 per 60 for the remainder exceeding 1,520	1 per 200	1 per 150	—	1 per 1,000	1 service sink
		A-5	Stadiums, amusement parks, bleachers and grandstands for outdoor sporting events and activities	1 per 75 for the first 1,500 and 1 per 120 for the remainder exceeding 1,500	1 per 40 for the first 1,520 and 1 per 60 for the remainder exceeding 1,520	1 per 200	1 per 150	—	1 per 1,000	1 service sink
2	Business	B	Buildings for the transaction of business, professional services, other services involving merchandise, office buildings, banks, light industrial and similar uses	1 per 25 for the first 50 and 1 per 50 for the remainder exceeding 50		1 per 40 for the first 80 and 1 per 80 for the remainder exceeding 80		—	1 per 100	1 service sink[e]
3	Educational	E	Educational facilities	1 per 50		1 per 50		—	1 per 100	1 service sink
4	Factory and industrial	F-1 and F-2	Structures in which occupants are engaged in work fabricating, assembly or processing of products or materials	1 per 100		1 per 100		(see Section 411)	1 per 400	1 service sink
5	Institutional	I-1	Residential care	1 per 10		1 per 10		1 per 8	1 per 100	1 service sink
		I-2	Hospitals, ambulatory nursing home care recipient	1 per room[c]		1 per room[c]		1 per 15	1 per 100	1 service sink per floor
			Employees, other than residential care[b]	1 per 25		1 per 35		—	1 per 100	—
			Visitors, other than residential care	1 per 75		1 per 100		—	1 per 500	—
		I-3	Prisons[b]	1 per cell		1 per cell		1 per 15	1 per 100	1 service sink
			Reformatories, detention centers, and correctional centers[b]	1 per 15		1 per 15		1 per 15	1 per 100	1 service sink
			Employees[b]	1 per 25		1 per 35		—	1 per 100	—
		I-4	Adult day care and child day care	1 per 15		1 per 15		1	1 per 100	1 service sink

(continued)

FIXTURES, FAUCETS AND FIXTURE FITTINGS

TABLE 403.1 —continued
MINIMUM NUMBER OF REQUIRED PLUMBING FIXTURES[a]
(See Sections 403.1.1 and 403.2)

NO.	CLASSIFICATION	OCCUPANCY	DESCRIPTION	WATER CLOSETS (URINALS: SEE SECTION 419.2)		LAVATORIES		BATHTUBS/ SHOWERS	DRINKING FOUNTAIN (SEE SECTION 410)	OTHER
				MALE	FEMALE	MALE	FEMALE			
6	Mercantile	M	Retail stores, service stations, shops, salesrooms, markets and shopping centers	1 per 500		1 per 750		—	1 per 1,000	1 service sink[e]
7	Residential	R-1	Hotels, motels, boarding houses (transient)	1 per sleeping unit		1 per sleeping unit		1 per sleeping unit	—	1 service sink
		R-2	Dormitories, fraternities, sororities and boarding houses (not transient)	1 per 10		1 per 10		1 per 8	1 per 100	1 service sink
		R-2	Apartment house	1 per dwelling unit		1 per dwelling unit		1 per dwelling unit	—	1 kitchen sink per dwelling unit; 1 automatic clothes washer connection per 20 dwelling units
		R-3	Congregate living facilities with 16 or fewer persons	1 per 10		1 per 10		1 per 8	1 per 100	1 service sink
		R-3	One- and two-family dwellings and lodging houses with five or fewer guestrooms	1 per dwelling unit		1 per dwelling unit		1 per dwelling unit	—	1 kitchen sink per dwelling unit; 1 automatic clothes washer connection per dwelling unit
		R-4	Congregate living facilities with 16 or fewer persons	1 per 10		1 per 10		1 per 8	1 per 100	1 service sink
8	Storage	S-1 S-2	Structures for the storage of goods, warehouses, storehouse and freight depots. Low and Moderate Hazard.	1 per 100		1 per 100		See Section 411	1 per 1,000	1 service sink

a. The fixtures shown are based on one fixture being the minimum required for the number of persons indicated or any fraction of the number of persons indicated. The number of occupants shall be determined by the *International Building Code*.
b. Toilet facilities for employees shall be separate from facilities for inmates or care recipients.
c. A single-occupant toilet room with one water closet and one lavatory serving not more than two adjacent patient sleeping units shall be permitted provided that each patient sleeping unit has direct access to the toilet room and provision for privacy for the toilet room user is provided.
d. The occupant load for seasonal outdoor seating and entertainment areas shall be included when determining the minimum number of facilities required.
e. For business and mercantile occupancies with an occupant load of 15 or fewer, service sinks shall not be required.

A question that is often asked about Group E occupancies is whether the concept of nonsimultaneous occupancy could be considered in order to further reduce the number of required plumbing fixtures. For example, where a school auditorium is occupied by only the students within the school, the students are in either the classroom or the auditorium but not in both places at the same time. Therefore, why should each area require a number of fixtures as if both areas were occupied simultaneously? The code does not recognize a nonsimultaneous use concept for any building as simultaneous use could easily occur. For example, while students are in the classrooms, a school auditorium could be used temporarily for blood drives, regional science fairs, and town meetings. However, this is not to say that in some circumstances, a nonsimultaneous use could, in some way, be completely guaranteed such that the local code official might entertain a proposed reduction in the required number of plumbing fixtures for the building.

Business Facilities

Consider a barber college where the IBC classifies the entire building as a Group B occupancy. The building has several large assembly rooms where the intent is to have training sessions for large groups of students. Clearly, these areas are used for assembly and therefore, the use of the fixture ratios in Section 1, Row 4 of Table 403.1 should be used.

The choice of an occupancy use for the purposes of determining plumbing fixtures does not affect the occupancy group chosen for IBC purposes, that being for egress. In other words, using a previous example, the school gymnasium with stage space should be chosen to be "assembly use" for plumbing fixture requirements, but the entire building is still a Group E occupancy for the purposes of egress as far as the IBC is concerned.

Note that Section 1004.1.2 of the IBC has an exception that allows for the designer to present an "actual" occupancy load number to the code official for approval instead of the calculated load for the occupancy square footage. While this exception would allow a smaller occupancy load to be chosen for the purposes of determining the required number of plumbing fixtures, code officials must carefully consider and appropriately justify the reduction. The code official should consider the potential for the space to be loaded with more persons than stated by the designer, future use by different tenants of the same occupancy classification and the difficulty of enforcement of the maximum occupancy load based upon the limited number of fixtures provided

In Table 403.1, the brief wording in the "Description" column is not intended to be complete or all inclusive of all uses for a particular building or space. However, the descriptions do identify a majority of the types of uses encountered in the design of most buildings. Because the descriptions "narrow in" on what row of ratios to use, users of this table will benefit by using the description column as the entry point into the table. The number, classification and occupancy columns are intended only for reference purposes. The requirements for type and number of plumbing fixtures are driven by the actual use of a building or space, which is usually consistent with the occupancy group classification but not always.

Some use description rows have a different ratio for male and female water closets, and in some cases, lavatories. The smaller ratio for female fixtures provides an "equality of fixture availability" in those particular occupancies. The occupancies described by these use descriptions have historically had long lines of females waiting to use toilet facilities while male facilities had no lines. The reasons for this include the following:

1. For a variety of social and physical reasons, women generally take a longer period of time to use the facilities;

2. Women outnumber men in the general population and this becomes especially evident in large groups of people and

3. Women, in general, tend to use the facilities more frequently.

The term "potty parity" was coined in general sense to mean that a sufficient number of female plumbing fixtures were available such that women did not have to wait any longer than men (in a queue or line) to use an equivalent type of fixture as men. In a specific sense, "potty parity" also indicates the ratio of the number of male to the number of female fixtures of the same type. For example, if 24 male and 48 female water closets are installed in a building, then the potty parity is stated to be "1 to 2."

The code is silent concerning a requirement for child-size water closets. In a child care facility having children age six years old and under, the provision of child-size water closets (and sinks mounted at child height) could be beneficial to a facility's operation. Generally, for children over six years old, standard-size fixtures are suitable because most children have learned to use the same size fixtures in a home setting. Note that the accessibility standard, ICC/ANSI A117.1, as referenced by Section 404, has specific dimensional requirements for accessible child-size water closets.

Some school facilities are designed to have a single occupant toilet room that can be accessed only from within the classroom. The code is silent on whether the fixtures in this limited access toilet room can be counted toward the required number of plumbing fixtures for the building.

The intent of table Note d concerning outdoor seating and entertainment areas is to require that outdoor patios, decks, balconies, beer gardens and similar areas, whether those areas have seating or not, be included for calculating the number of plumbing fixtures. While seating does provide a definitive occupant load for these types of areas, there are many situations where these areas are not provided with seating in order to accommodate as many customers as possible. Therefore, the occupant load should be based upon standing space density. This occupant load number must be applied to the ratios for appropriate use description row in Table 403.1. For example, a restaurant has a fixed seating area with fixed booths and an outdoor entertainment area that has no seating or tables. The fixed seating area load is simply a count of the number of seats according to Section 1004.7 of the IBC. The restaurant fixture ratios (Section 1, Row 3 of Table 403.1) are then applied to this load to determine the number of plumbing fixtures for the fixed seating area. The outdoor area requires a different approach as it is a standing space area, not fixed seating. The occupant load is calculated from the area using Table 1004.4 of the IBC. Even though the building is a restaurant, this outdoor entertainment area might be used, not as a restaurant, but as a nightclub/dance floor. Therefore, the fixture ratios in Row 2 of Table 403.1 (for nightclub/dance halls) are applied to the standing occupant load. Using Row 3 fixture ratios for this area would be inappropriate. The use of proper occupant loads and use ratios will ensure that an adequate number of plumbing fixtures are provided where the occupant load is substantially increased by the addition of outdoor seating or standing areas.

Choosing the proper use description for a restaurant having a bar or for a nightclub serving food can be challenging. Some restaurant operations evolve into a nightclub/dance hall in the late evening. Bars with dance floors often have kitchens for preparing food to serve to patrons at the bar. The code is silent on how to determine which use description (restaurant or night club) is the best "fit" for the demand on the toilet facilities. The answer to this question lies in determining what use is primary. Is the bar's main purpose to support the restaurant operation? Or is the serving of food in support of the bar operation? In situations where there is no clear answer, perhaps considering the occupancy as a mixed use arrangement is a practical answer.

The drinking fountain ratio indicated for a use description does not necessarily imply that every tenant space must have a drinking fountain within that building. Just as toilet facilities are only required to be provided when a building or space is in use, the provision of a drinking fountain follows similar logic. For example, consider a strip center with multiple tenant spaces, each tenant space having the required number of plumbing fixtures for and within each space. The drinking fountain for that space (and all the adjacent spaces) could be located exterior to all the tenant spaces in a common area. The calculation for the required number of drinking fountains to be used by multiple occupancies is performed in the same manner as for other plumbing fixtures. See the Sample Problems in this commentary section.

The "service sinks" required by Table 403.1 are intended to be of a type suitable for janitorial and building maintenance purposes. Service sinks include mop sinks/basins, utility tub/sinks, janitor sinks, slop sinks and laundry trays. In all occupancy classifications, except Groups I-2, R-2 and R-3, only one service sink is required for the entire building, except that hospitals are required to have a service sink on each floor. The one service sink per building must be available from all portions of the building. The intent of the code is that employees and tenants of the building be able to access the required sink. This does not mean that the service sink is required to be in an open area of the building for use by any occupant (e.g., visitors). The service sink can be located in a locked janitor's closet for which all employees have access by key or door lock code to gain entry. However, the required service sink cannot be located in areas that are access controlled by a single tenant. For example, if the service sink is located on floor one within a tenant's space, it would be possible for the tenant of floor one to lock out (exclude) other tenants from accessing the required service sink. Therefore, that service sink within one tenant controlled space cannot be the required service sink for the building. Note e to the table allows business and mercantile occupancies of 15 or fewer persons to not require a service sink. Service sinks in these small occupancies are rarely used.

Although the IBC classifies parking garages as a Group S occupancy, there is no entry in Table 403.1 for parking structures. Furthermore, the exception in Section 403.3 exempts parking garages from having public toilet facilities. A parking garage is similar to a parking lot in that it is not normally the final destination of a person parking a car. The only reason a person is in a parking garage is to park their car and then leave. A parking lot, having no enclosing structure, is not required to be provided with toilet facilities for the occupants who are parking their vehicles, so it stands to reason that a parking garage would not be required to be provided with toilet facilities. When people drive into a parking garage, their normal expectation is that the toilet facilities will be located in the building or buildings served by the parking garage. If public toilet facilities were provided in a parking garage, significant maintenance, vandalism and security issues would probably result in the eventual closure of such facilities. However, not requiring public toilet facilities in a parking garage does not provide justification for not providing toilet facilities for parking garage employees (such as ticket kiosk attendants). These employees must be provided with toilet facilities in accordance with Section 403.3. However, these facilities are not required to be located in the parking garage but only within the travel limitations of Section 403.3.3.

403.1.1 Fixture calculations. To determine the occupant load of each sex, the total occupant load shall be divided in half. To determine the required number of fixtures, the fixture ratio or ratios for each fixture type shall be applied to the occupant load of each sex in accordance with Table 403.1. Fractional numbers resulting from applying the fixture ratios of Table 403.1 shall be rounded up to the next whole number. For calculations involving multiple *occupancies*, such frac-

FIXTURES, FAUCETS AND FIXTURE FITTINGS

tional numbers for each *occupancy* shall first be summed and then rounded up to the next whole number.

Exception: The total occupant load shall not be required to be divided in half where *approved* statistical data indicates a distribution of the sexes of other than 50 percent of each sex.

❖ This section requires that the occupant load be divided by two in order to determine the quantities of males and females. This is known as a 50-50 gender distribution meaning that 50 percent of the occupants are male and 50 percent of occupants are female. While a 50-50 distribution is used most of the time, there might be situations where this assumption is inaccurate. The exception to this section allows for a different gender distribution if statistical data is submitted and approved by the code official for a building or space. Examples that might require a modified distribution would be an all-women's health club, a males-only boarding school, a convent, or a monastery.

Once the appropriate use description row in Table 403.1 is chosen, the ratios for each fixture type and for each gender are used to determine the minimum number of plumbing fixtures for a building or space.

Although not specifically indicated in this code section, prior to performing any plumbing fixture calculations for a building or space, the exceptions of Section 403.2 must be reviewed for applicability. For example, a small restaurant could have two employees and 12 seats. If the fixture calculations were performed without consideration of Section 403.2, this restaurant would be required to have a toilet facility for each gender. However, because the occupancy load is less than 15, a single toilet facility is all that is required.

The following sample problems illustrate calculation methods for various situations.

Sample Problem 1: A mixed-use building has a business-use occupant load of 538, a library-use occupant load of 115 and storage-use occupant load of 82. The occupancies are not separate tenant spaces. Determine the total number of required plumbing fixtures for the building.

Problem Approach

Check Section 403.2 for applicability and, if not applicable, proceed with calculations.

Apply the gender distribution for each space use, calculate the number of fixtures required by each gender, sum the gender totals for each type of fixture from all spaces and round up the number of each type of fixture for each gender to a whole number. Note that drinking fountains are not gender specific. Therefore, the total number of occupants for the building is applied to the drinking fountain fixture ratio.

Solution—Part I

For business-use spaces, find the required ratios in Table 403.1, Section 2. Because a gender distribution was not specified, the number of occupants must be divided by two to obtain the number of occupants for each gender (refer to Section 403.1.1). Therefore, in this business-use space, there will be:

538/2 = 269 occupants of each gender

1. Calculate the number of water closets for 269 males:

    For the first 50 males, the ratio of 1 per 25 is applied:

    $50 \times 1/25 = 2$ water closets

    For the remaining number of males, the ratio of 1 per 50 is applied:

    $(269 - 50) \times 1/50 = 219 \times 1/50 = 4.38$ water closets

    The required number of water closets for males is:

    $2 + 4.38 = 6.38$ water closets for males

    Because the gender distribution is equal and the water closet ratio for females is the same as for males, the number of water closets for females is also 6.38.

2. Calculate the number of lavatories for 269 males:

    For the first 80 males, the ratio of 1 per 40 is applied:

    $80 \times 1/40 = 2$ lavatories

    For the remaining males, the ratio of 1 per 80 is applied:

    $(269 - 80) \times 1/80 = 189 \times 1/80 = 2.36$ lavatories

    The required number of lavatories for males is:

    $2 + 2.36 = 4.36$ lavatories for males

    Because the gender distribution is equal and the lavatory ratio for females is the same as for males, the number of lavatories for females is also 4.36.

3. Calculate the number of drinking fountains for 538 occupants:

    The ratio of 1 per 100 is applied:

    $538 \times 1/100 = 5.38$ drinking fountains for male and female use

Solution—Part II

For the library-use spaces, find the required ratios in Table 403.1, Section 1. Because a gender distribution was not specified, the number of occupants must be divided by two to obtain the number of occupants for

FIXTURES, FAUCETS AND FIXTURE FITTINGS

USE	SPACE LOAD	WATER CLOSETS RATIO	MALE	RATIO	FEMALE	LAVATORIES RATIO	MALE	FEMALE	DF RATIO	DRINKING FOUNTAINS	SERVICE SINK
Business	538	1 per 25 for the first 50 and 1 per 50 for the remainder exceeding 50	6.38	1 per 25 for the first 50 and 1 per 50 for the remainder exceeding 50	6.38	1 per 40 for the first 80 and 1 per 80 for the remainder exceeding 80	4.36	4.36	1 per 100	5.38	Note 1
Libraries, Halls, Museums, etc.	115	1/125	0.46	1/65	0.88	1/200	0.29	0.29	1 per 500	0.23	
Storage	82	1/100	0.41	1/100	0.41	1/100	0.41	0.41	1/1000	0.082	
Subtotals			7.25		7.67		5.06	5.06		5.69	
Required totals			8		8		6	6		6	

Notes:

1. The code requires only one service sink per building if all occupancies have access to the service sink at all times.

Figure 403.1(1)
SOLUTION SUMMARY FOR SAMPLE PROBLEM 1

each gender (refer to Section 403.1.1). Therefore, in this library-use space, there will be:

115/2 = 57.5 occupants of each gender

1. Calculate the number of water closets for males:

    The ratio of 1 per 125 is applied:

    57.5 × 1/125 = 0.46 water closets for males

2. Calculate the number of water closets for females:

    The ratio of 1 per 65 is applied:

    57.5 × 1/65 = 0.88 water closets for females

3. Calculate the number of lavatories for 57.5 males:

    The ratio of 1 per 200 is applied:

    57.5 × 1/200 = 0.29 lavatories for males

    Because the gender distribution is equal and the lavatory ratio for females is the same as for males, the number of lavatories for females is also 0.29.

4. Calculate the number of drinking fountains for 115 occupants:

    The ratio of 1 per 500 is applied:

    115 × 1/500 = 0.23 drinking fountains of both males and females.

Solution—Part III

For the storage-use space, find the required ratios in Table 403.1, Section 8. Because a gender distribution was not specified, the number of occupants must be divided by two to obtain the number of occupants for each gender (refer to Section 403.1.1). Therefore, in this storage-use space, there will be:

82/2 = 41 occupants of each gender

1. Calculate the number of water closets for 41 males:

    The ratio of 1 per 100 is applied:

    41 × 1/100 = 0.41 water closets for males

    Because the gender distribution is equal and the water closet ratio for females is the same as for males, the number of water closets for females is also 0.41.

2. Calculate the number of lavatories for 41 males:

    The ratio of 1 per 100 is applied:

    41 × 1/100 = 0.41 lavatories for males

    Because the gender distribution is equal and the lavatory ratio for females is the same as for males, the number of lavatories for females is also 0.41.

3. Calculate the number of drinking fountains for 82 occupants:

    The ratio of 1 per 1000 is applied:

    82 × 1/1000 = 0.082 drinking fountains for males and females.

Solution—Summary

See Commentary Figure 403.1(1) for the solution summary for Sample Problem 1. Note that if the raw fixture numbers were rounded prior to summation for the building, the calculated minimum fixture requirements would have been increased by one water closet

FIXTURES, FAUCETS AND FIXTURE FITTINGS

for each gender, one lavatory for each gender and two drinking fountains.

Sample Problem 2: A mixed-use building has a business-use occupant load of 940 and a restaurant-use occupant load of 360. The occupancies are separate tenant spaces. The toilet facilities are to be located in each tenant space. Determine the total number of required plumbing fixtures for the building.

Problem Approach

Check Section 403.2 for applicability and if not applicable, proceed with calculations.

Note that the plumbing fixtures required by each occupancy will be located in each tenant space. Thus, the required number of plumbing fixtures for each tenant space is calculated as if the space stood alone, in other words, in its own building.

Solution—Part I

For the business-use space, find the required ratios in Table 403.1, Section 2. Because a gender distribution was not specified, the number of occupants must be divided by two to obtain the number of occupants for each gender (refer to Section 403.1.1). Therefore, in this business-use space, there will be:

940/2 = 470 occupants of each gender

1. Calculate the number of water closets for 470 males:

For the first 50 males, the ratio of 1 per 25 is applied:

$50 \times 1/25 = 2$ water closets

For the remaining number of males, the ratio of 1 per 50 is applied:

$(470 - 50) \times 1/50 = 420 \times 1/50 =$ 8.4 water closets

The required number of water closets for males is:

2 + 8.4 = 10.4 water closets for males

Because the gender distribution is equal and the water closet ratio for females is the same as for males, the number of water closets for females is also 10.4.

2. Calculate the number of lavatories for 470 males:

For the first 80 males, the ratio of 1 per 40 is applied:

$80 \times 1/40 = 2$ lavatories

For the remaining males, the ratio of 1 per 80 is applied:

$(470 - 80) \times 1/80 = 390 \times 1/80$ = 4.88 lavatories.

The required number of lavatories for males is:

2 + 4.88 = 6.88 lavatories for males

SPACE USE	LOAD	WATER CLOSETS RATIO	MALE	RATIO	FEMALE	LAVATORIES RATIO	MALE	FEMALE	DF RATIO	DRINKING FOUNTAINS	SERVICE SINK
Business	940	1 per 25 for the first 50 and 1 per 50 for the remainder exceeding 50	10.4	1 per 25 for the first 50 and 1 per 50 for the remainder exceeding 50	10.4	1 per 40 for the first 80 and 1 per 80 for the remainder exceeding 80	6.88	6.88	1 per 100	9.4	Note 1
Business Required Totals			11		11		7	7		10	
Restaurants, Banquet Halls, and Food Courts	360	1 per 75	2.4	1 per 75	2.4	1 per 200	0.9	0.9	1 per 500	0.72	
Restaurant Required Total			3		3		1	1		1	
Building Required Totals			14		14		8	8		11	

Notes:

1. The code only requires one service sink per building if all occupancies have access at all times. If this building has a central area (such as janitorial closet for the building) that both tenants can access at all times, only one service sink is required for the building. But if the service sink is located in either one of the tenant spaces, one tenant would not have access; therefore, the building would require two service sinks: one for each tenant.

Figure 403.1(2)
SOLUTION SUMMARY FOR SAMPLE PROBLEM 2

FIXTURES, FAUCETS AND FIXTURE FITTINGS

Because the gender distribution is equal and the lavatory ratio for females is the same as for males, the number of lavatories for females is also 6.88.

3. Calculate the number of drinking fountains for 940 occupants:

   The ratio of 1 per 100 is applied:

   940 × 1/100 = 9.4 drinking fountains for males and females

Solution—Part II

For the restaurant-use space, find the required ratios in Table 403.1, Section 1. Because a gender distribution was not specified, the number of occupants must be divided by two to obtain the number of occupants for each gender (refer to Section 403.1.1). Therefore, in this restaurant-use space, there will be:

360/2 = 180 occupants of each gender

1. Calculate the number of water closets for males:

   The ratio of 1 per 75 is applied:

   180 × 1/75 = 2.4 water closets for males

   Therefore, 2.4 male water closets are needed.

   Because the gender distribution is equal and the water closet ratio for females is the same as for males, the number of water closets for females is also 2.4.

2. Calculate the number of lavatories for 180 males:

   180 × 1/200 = 0.9 lavatories for males are required.

   Therefore, 1 male lavatory is needed.

Because the gender distribution is equal and the lavatory ratio for females is the same as for males, the number of lavatories for females is also 1.

3. Calculate the number of drinking fountains for 360 occupants:

   The ratio of 1 per 500 is applied:

   360 × 1/500 = 0.72 drinking fountains of both males and females.

Solution—Summary

See Commentary Figure 403.1(2) for the solution summary for Sample Problem 2. Note that if the raw fixture numbers were rounded prior to summation for the building, the calculated minimum fixture requirements would have increased by one water closet for each gender, one lavatory for each gender and two drinking fountains.

Sample Problem 3: A stadium has an occupant load of 5,200. Urinals are to be substituted for male water closets to the maximum extent allowed. Determine the total number of required plumbing fixtures for the stadium.

Problem Approach

Check Section 403.2 for applicability and if not applicable, proceed with calculations.

Apply the gender distribution, calculate the number of fixtures required by each gender, sum the gender totals for each type of fixture and round up the number of each type of fixture for each gender to a whole number. Apply the urinal substitution (Section 419.2) to reduce the number of male water closets.

OCCUPANCY		WATER CLOSETS				LAVATORIES				DRINKING FOUNTAINS	
USE	LOAD	MALE RATIO	MALE	FEMALE RATIO	FEMALE	MALE RATIO	MALE	FEMALE RATIO	FEMALE	RATIO	ALL GENDERS
Stadiums	5200	1 per 75 for the first 1500 and 1 per 120 for the remainder exceeding 1500	29.17	1 per 40 for the first 1520 and 1 per 60 for the remainder exceeding 1520	56	1 per 200	13	1 per 150	17.3	1 per 1000	5.3
Required Totals			30		56		13		18		6

Required Totals with Urinal Substitution	URINALS	MALE W/C	FEMALE W/C	MALE LAV	FEMALE LAV	DRINKING FOUNTAINS	SERVICE SINKS
	20	10	56	13	18	6	1

Figure 403.1(3)
SOLUTION SUMMARY FOR SAMPLE PROBLEM 3

FIXTURES, FAUCETS AND FIXTURE FITTINGS

OCCUPANCY		WATER CLOSETS				LAVATORIES				DRINKING FOUNTAINS	
USE	LOAD	MALE RATIO	MALE	FEMALE RATIO	FEMALE	MALE RATIO	MALE	FEMALE RATIO	FEMALE	RATIO	ALL GENDERS
Theater	520	1 per 125	208	1 per 65	4	1 per 200	1.3	1 per 200	1.3	1 per 500	1.04
Required Totals			3		4		2		2		2

Figure 403.1(4)
SOLUTION SUMMARY FOR SAMPLE PROBLEM 4

Theater Building Totals with Urinal Substitution	MALE TOILET FACILITY			FEMALE TOILET FACILITY		DRINKING FOUNTAINS	SERVICE SINK	FAMILY OR ASSISTED-USE	
	URINAL	WATER CLOSETS	LAVATORIES	WATER CLOSETS	LAVATORIES			WATER CLOSETS	LAVATORIES
	1	1	1	4	2	2	1	1	1

Figure 403.1(5)
BUILDING PLUMBING FIXTURE TOTALS FOR SAMPLE PROBLEM 4

Solution

For the stadium-use occupancy, find the required ratios in Table 403.1 in row 8. Because a gender distribution was not specified, the number of occupants must be divided by two to obtain the number of occupants for each gender (refer to Section 403.1.1). Therefore, in this stadium-use building, there will be:

5,200/ 2 = 2,600 occupants of each gender

1. Calculate the number of water closets for 2,600 males:

   For the first 1,500 males, the ratio of 1 per 75 is applied:

   1,500 × 1/75 = 20 water closets

   For the remaining number of males, the ratio of 1 per 120 is applied:

   (2,600 - 1,500) × 1/120 = 110 x 1/120 = 9.17 water closets

   The required number of water closets for males is:

   20 + 9.17 = 29.17 → Round to 30 water closets for males

2. Calculate the number of water closets for 2,600 females:

   For the first 1,520 females, the ratio of 1 per 40 is applied:

   1,520 × 1/ 40 = 38 water closets

   For the remaining number of females, the ratio of 1 per 60 is applied:

   (2,600 - 1,520) × 1/60 = 1,080 × 1/60 = 18 water closets

   The required number of water closets for females is:

   38 + 18 = 56 water closets for females.

3. Calculate the number of lavatories for 2,600 males:

   The ratio of 1 per 200 is applied:

   2,600 × 1/200 = 13 lavatories for males

4. Calculate the number of lavatories for 2,600 females:

   The ratio of 1 per 150 is applied:

   2,600 × 1/150 = 17.3 lavatories → Round to 18 lavatories for females

5. Calculate the number of drinking fountains for 5,200 occupants

   The ratio of 1 per 1,000 is applied:

   5,200 × 1/1,000 = 5.2 → Round to 6 drinking fountains for males and females

6. Calculate the number of urinals to be substituted for male water closets. Because a stadium is in the assembly classification, 67 percent of the water closets can be replaced with urinals.

   The ratio of 67/100 is applied:

   30 × 67/100 = 20.1 urinals → Round *down* to 20 urinals because rounding up would result in the number of urinals being in excess of 67 percent of the required number of water closets.

Solution—Summary

See Commentary Figure 403.1(3) for the solution summary for Sample Problem 3.

403.1.2 Family or assisted-use toilet and bath fixtures. Fixtures located within family or assisted-use toilet and bathing rooms required by Section 1109.2.1 of the *International Building Code* are permitted to be included in the number of required fixtures for either the male or female occupants in assembly and mercantile *occupancies*.

❖ A family or assisted-use toilet room is required by the IBC in assembly or mercantile occupancies having an aggregate total of six or more required male and female water closets for the assembly or mercantile occupancy. The term "aggregate total" means the combined number of required male and female water closets before any urinal substitutions. A family or assisted-use bathing room is required by the IBC in recreational facilities having separate-sex bathing rooms except where the separate-sex bathing rooms only have one shower or fixture. Family or assisted-use toilet (and bathing room) facilities are intended for use by individuals who need the assistance of a family member or care provider to properly use plumbing fixtures in a private and dignified manner. Fixtures located in family or assisted-use toilet or bathing rooms in assembly and mercantile occupancies are allowed to reduce the number of required fixtures for the building or space in either the male or female standard toilet rooms, but not both.

Sample Problem 4: A theater has an occupant load of 520. Urinals are to be substituted for male water closets to the maximum extent allowed. Determine the total number of plumbing fixtures that will be in building.

Problem Approach

Apply the gender distribution, calculate the number of fixtures required by each gender, sum the gender totals for each type of fixture and round up the number of each type of fixture for each gender to a whole number. Consider the IBC requirement for a family or assisted-use toilet room. Apply the urinal substitution (Section 419.2) to reduce the number of male water closets to the maximum extent possible.

Solution

Calculation method is similar to Sample Problem 1—Part II

Solution—Summary

See Commentary Figures 403.1(4) and 403.1(5) for the solution summary for Sample Problem 4.

403.2 Separate facilities. Where plumbing fixtures are required, separate facilities shall be provided for each sex.

Exceptions:

1. Separate facilities shall not be required for dwelling units and sleeping units.

2. Separate facilities shall not be required in structures or tenant spaces with a total occupant load, including both employees and customers, of 15 or fewer.

3. Separate facilities shall not be required in mercantile *occupancies* in which the maximum occupant load is 100 or fewer.

❖ Separate facilities are required for males and females. Exception 1 is redundant as Table 403.1 already indicates the types of occupancies where one single-occupancy toilet room is sufficient. Exception 2 allows one, single-occupant toilet room to be provided in small establishments because the space requirement and installation cost is considered to be a significant hardship for the minimal floor area that results in a 15-person occupant load. Exception 3 allows one single-occupant toilet room to be provided in what are considered to be small floor area mercantile establishments so as to not create a significant hardship in both space and installation cost. There is no specific documentation as to why a 15-person threshold was chosen for Exception 2 other than it was perceived to be reasonable for such small spaces. The 100-person threshold for Exception 3 was justified because small mercantile occupancies [1,500 to 3,000 square feet (139 to 279 m^2)] were unduly burdened by the requirement for two accessible toilet rooms. For a 1,500-square-foot (139 m^2) tenant space, two accessible toilet rooms occupy more than 5 percent of the total square footage. Since it is rare that these small establishments will ever experience the full design occupant load, there is no need to burden the tenant with two toilet rooms.

The separate facilities requirement for males and females addresses two main concerns: privacy and safety. Users of toilet facilities often experience embarrassment if members of the opposite sex are in the same room. This increased embarrassment can lead to difficulty or prevention of waste elimination for many users. While some of these inhibitions are often temporarily "given up" under special conditions such as in co-ed college dormitories or where one gender's facilities are inadequate under demand situations, these inhibitions quickly return for most people as it is innate to the human species to desire privacy during waste elimination. In a public environment, safety for female users of toilet facilities is of paramount concern. While a female restroom placard is no barrier to those intent on harming female occupants, a female user confronted by (or even hearing) a male in the female-only toilet facility will immediately recognize the potential threat and take action. A toilet facility intended to be used by both sexes simultaneously does not offer the same level of immediate situational awareness for females and therefore, it is not perceived as to be as safe as a facility intended for one gender only. Also, some individuals may be inhibited to use a facility that is used by the opposite sex because of notions of cleanliness and perceived appropriateness.

403.2.1 Family or assisted-use toilet facilities serving as separate facilities. Where a building or tenant space requires a separate toilet facility for each sex and each toilet facility is required to have only one water closet, two family or assisted-use toilet facilities shall be permitted to serve as the required separate facilities. Family or assisted-use toilet facilities shall not be required to be identified for exclusive use by either sex as required by Section 403.4.

❖ For a building or space having only one male water closet and one female water closet, two-family/assisted-use toilet rooms increase the overall availability of a toilet room when needed. A single-gender-based toilet can be unavailable for periods of up to 15 minutes when, for example, the current occupant is using it for companion care, to change diapers or to change a colostomy bag. Also, there is less impact to availability when one toilet room is being cleaned or

FIXTURES, FAUCETS AND FIXTURE FITTINGS

serviced. For decades, males and females have used the same toilet facility on airliners.

403.3 Required public toilet facilities. Customers, patrons and visitors shall be provided with *public* toilet facilities in structures and tenant spaces intended for public utilization. The number of plumbing fixtures located within the required toilet facilities shall be provided in accordance with Section 403 for all users. Employees shall be provided with toilet facilities in all *occupancies*. Employee toilet facilities shall be either separate or combined employee and *public* toilet facilities.

Exception: *Public* toilet facilities shall not be required in:

1. Open or enclosed parking garages where there are no parking attendants.
2. Structures and tenant spaces intended for quick transactions, including takeout, pickup and drop-off, having a public access area less than or equal to 300 square feet (28 m^2).

❖ Toilet facilities must be available for all public establishments that are used by persons engaged in activities involved with the purpose of the establishment. Public establishments include but are not limited to restaurants, nightclubs, theaters, offices, retail shops, stadiums, libraries and churches. Persons engaged in the activities of the establishment include, but are not limited to, buyers of merchandise, recipients of services, viewers of displays, receivers of information materials, employees and those persons in attendance with those engaging in the activities. The code is silent about whether the toilet facilities are for use by persons not engaged in the activities involved with the purpose of the establishment, such as passersby, street maintenance workers or vagrants. However, because the number of plumbing fixtures for any establishment is based only upon either the square footage or number of seats of the establishment's space, the intent of the code is to serve only the people involved with the activities of the establishment. The photo of the front door of the restaurant shown in Commentary Figure 403.3(1) might first appear to be in conflict with the requirements of this section; however, this sign actually captures the intent of this code section: "If you are engaged with the activities involved with the purpose of this establishment, toilet facilities are available for you."

The quantities of plumbing fixtures required for an establishment are specified in Section 403 of the code. Employees of the establishment must also have access to the toilet facilities for the establishment whether those facilities are separate from or combined with the facilities for the other users. Except for the specific requirements of I-2 occupancies, I-3 occupancies and Note b of Table 403.1, employee toilet facilities are not required to be separate from other toilet facilities.

Item 1 of the exception clarifies that even though a parking structure is classified as a Group S occupancy (storage structure) and Table 403.1 requires storage occupancies to have toilet facilities, parking structures are exempt from this requirement. Note that if a parking structure has attendants, they must be provided with toilet facilities within the travel limitations of the code.

Item 2 of the exception allows for small public-access areas in buildings and tenant spaces used for quick transactions, such as automated teller machine lobbies, shoe repair drop-off/pickup, dry cleaners, take-out only counters at restaurants (such as delicatessens, pizza and other takeout food) and entertainment ticket purchase/pickup lobbies, to not have public toilet facilities.

The intent is that the exception is for establishments where customers walk in, make a relatively fast transaction and walk out. In other words, the customer does not spend any time waiting, such as for orders to be prepared (food ordered upon walking in), filling out extensive forms, waiting on an order to be pulled or loaded (such as for parts and materials supply houses), waiting for a fuel tank to be filled, waiting for things to be repaired ("while-u-wait" services), waiting on estimates for repairs to be determined or shopping for merchandise on racks or shelves within the public access area of 300 square feet (28 m^2) or less. These small public-access areas typically would not have chairs, stools, benches, bars, merchandise shelving, catalog viewing/ordering desks or racks for products. The presence of these items might cause the customer to be inclined to spend more time than needed to make a simple and quick transaction.

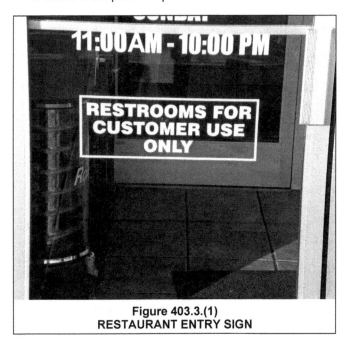

**Figure 403.3.(1)
RESTAURANT ENTRY SIGN**

403.3.1 Access. The route to the *public* toilet facilities required by Section 403.3 shall not pass through kitchens, storage rooms or closets. Access to the required facilities shall be from within the building or from the exterior of the building. Routes shall comply with the accessibility requirements of the *International Building Code*. The public shall have access to the required toilet facilities at all times that the building is occupied.

❖ It is inappropriate and dangerous to locate toilet facilities intended for use by persons who are not employees of the establishment, such that the path of travel is through kitchens, storage rooms or closets. This provision avoids placing occupants from the general public in an unfamiliar area of the building where they could be confused about the path of egress. Especially during an emergency situation, public users of the toilet facility will instinctively want to try to exit the building along the same path of travel used to access the toilet facilities, which could be through an area where the emergency condition exists such as a kitchen [see Commentary Figure 403.3.1(2)]. Note that Section 403.3.2 prohibits toilet rooms from directly opening into a room used for the preparation of food to be served to the public.

It has been common practice in many areas that the required toilet facilities were located on the outside of a building such as sometimes found in automotive service stations and convenience store buildings. Outlet mall complexes sometimes elect to have a central toilet facility building separate from the tenant spaces. While the code is silent as to whether the required toilet facilities are allowed to be in another building not under the direct control of the owner or tenant of the building from where the toilet user travels to the toilet facilities, the intent of the code is that public toilet facilities will be available (and open for use) when anyone is using a building or tenant space. However, it is unreasonable for toilet facilities in a building under different ownership or in a tenant space under different control to serve as the required toilet facilities for a building or tenant space.

Walk-up and drive-through service windows are unique applications with regard to the requirement for toilet facilities. Where these establishments do not allow customers to enter the building, the code is silent on whether toilet facilities for the public are required. However, if facilities were required, how would one calculate the occupant loading to determine the required number of fixtures? Examples are ice cream shops with only a walk-up window, coffee shops with only a drive-up window and 24-hour fast food drive-throughs that close the building entrances to walk-in customers during late night/early morning hours. Where outdoor seating for customers is made available by these types of businesses, it is the author's opinion that public toilet facilities for these walk-up/drive-through establishments should be available [see Commentary Figure 403.3.1(1)]. If such outdoor seating areas can be "closed for public use," such as by a perimeter "fence" and locked entry gates, the seating area is not available for use and no requirement would exist to provide public toilet facilities.

Variations of the above service window applications are dry cleaners, pizza pickups, and banks that provide service or products to customers from indoor counters or kiosks. Because the public must walk into these buildings for service, public toilet facilities are required. Obviously, these establishments are required to have toilet facilities for the employees; however, those facilities might not be properly located for public access. For example, if the location of the toilet facilities in a dry-cleaning establishment requires that the public walk a convoluted path around machinery and hanging clothes racks, the user could become confused about how to safely exit the toilet facilities and return to where he or she started. Public toilet facilities must be provided along an obvious and safe path of travel. Note that this path of travel must also meet the accessible route requirements for disabled persons. In the pizza pickup situation, because the public is not allowed to pass through a kitchen, the toilet facilities would have to be located on the customer side of the pickup counter. Banks (and check cashing establishments) do not have code restrictions for the areas that the public cannot pass through; however, the operators of these establishments might wish to restrict public access to "backroom" areas for security reasons. These self-imposed restrictions do not justify denying the public access to the toilet facilities located in restricted areas. If the toilet facilities are located in restricted areas, public toilet facilities must be provided outside of those self-imposed restricted areas.

[BG] 403.3.2 Prohibited toilet room location. Toilet rooms shall not open directly into a room used for the preparation of food for service to the public.

❖ This prohibition has been in the IBC for some time. The language is repeated here for convenience to the code user. This section provides a safeguard against a toilet room being used for food preparation functions, such as dumping of waste liquids and food scraps in a water closet.

403.3.3 Location of toilet facilities in occupancies other than malls. In occupancies other than covered and open mall buildings, the required *public* and employee toilet facilities shall be located not more than one story above or below the space required to be provided with toilet facilities, and the path of travel to such facilities shall not exceed a distance of 500 feet (152 m).

Exception: The location and maximum distances of travel to required employee facilities in factory and industrial *occupancies* are permitted to exceed that required by this section, provided that the location and maximum distance of travel are *approved*.

❖ To prevent health problems for building occupants related to difficulty in accessing a needed toilet facility in a timely fashion, the code clearly provides two conditions that must be met: 1) the required toilet facilities must be located within a certain travel distance and 2) the building occupant must not be required to travel beyond the next adjacent story above or below his or her location.

The distribution of plumbing fixtures within a building must satisfy the maximum travel limitations for occupants in the building. Sections 403.3.3 and 403.3.4

limit the horizontal and vertical distances that an occupant has to travel to use a toilet facility. A 20-story high-rise office building cannot have all of the required plumbing fixtures located on the 10th floor, as the path of travel must neither exceed a travel distance of 500 feet (152 m) nor require travel beyond the adjacent story above or below. While this travel path limitation can result in toilet facilities not being located on every floor of a multistory building, the intent of the code is for a proportional number of fixtures to be available for the intended occupant use. For example, occupants on the 20th, 19th, and 18th floors could reasonably be expected to utilize toilet facilities located only on the 19th floor [as long as the 500-foot (152 m) travel distance limitation is not exceeded]. However, that toilet facility cannot have just one fixture of each type to serve three floors of occupants. The distribution of fixtures must be proportioned according to the number of occupants that will be using the toilet facility. To illustrate, assume that each of the 20 floors in this high-rise business occupancy building has an occupant load of 100. Applying the business occupancy water closet fixture ratio to the occupant load for the entire building results in 21 male water closets required for the building. Three floors of occupants will require (150 floor occupants/1,000 total occupants) × 21 w/c = 3.15 or 4 male water closets on the 19th floor.

The exception in this section provides for the unique nature of the workplace factory and industrial occupancies where the designer is better equipped to determine the location of the employee toilet facilities.

403.3.4 Location of toilet facilities in malls. In covered and open mall buildings, the required *public* and employee toilet facilities shall be located not more than one story above or below the space required to be provided with toilet facilities, and the path of travel to such facilities shall not exceed a distance of 300 feet (91 m). In mall buildings, the required facilities shall be based on total square footage within a covered mall building or within the perimeter line of an open mall building, and facilities shall be installed in each individual store or in a central toilet area located in accordance with this section. The maximum distance of travel to central toilet facilities in mall buildings shall be measured from the main entrance of any store or tenant space. In mall buildings, where employees' toilet facilities are not provided in the individual store, the maximum distance of travel shall be measured from the employees' work area of the store or tenant space.

❖ In covered and open malls, the path of travel to required toilet facilities must not exceed a distance of 300 feet (91 mm) and is measured from the main

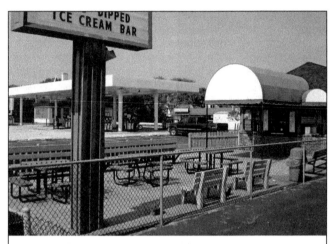

Figure 403.3.1(1)
WALK-UP ICE CREAM SHOP WITH OUTDOOR SEATING

Figure 403.3.1(2)
CUSTOMER TOILET ACCESS

entrance of any store or tenant space. Note, however, that if the tenant space does not have toilets for the employees, the starting point for measuring the travel path is the employee's work area. As for other occupancies, the facilities must be located no more than one story above or below the occupant's location. This shorter travel distance addresses the fact that malls are frequently very congested and occupants are unfamiliar with their surroundings. It is also logical to reduce the travel time for employees for security reasons. The minimum number of required plumbing facilities is based on total square footage of the mall, including tenant spaces. This section does not prohibit the installation of separate employee toilet facilities in individual tenant spaces; however, such facilities are not deductible from the total number of common facilities required (see commentary, Section 403.4.1).

403.3.5 Pay facilities. Where pay facilities are installed, such facilities shall be in excess of the required minimum facilities. Required facilities shall be free of charge.

❖ Pay facilities have been included in some public areas to prevent vagrants from loitering in the public bathrooms. They may also be installed to offer customers greater amenities where the customer desires such upgrades. This section does not prevent pay facilities from being installed; rather, it requires that such facilities be in excess of those required by Section 403.

403.3.6 Door locking. Where a toilet room is provided for the use of multiple occupants, the egress door for the room shall not be lockable from the inside of the room. This section does not apply to family or assisted-use toilet rooms.

❖ A multiple-occupant toilet room could have a locking door to facilitate closure of the toilet facility for maintenance or for security when the toilet room is closed after a building's occupancy hours. This section prohibits that lock or any other lock from being engaged on the inside of the door so that users cannot lock themselves in the toilet room for whatever reason. This requirement is to prevent multiple-occupant toilet rooms from becoming safe havens for illicit activities. Because family/assisted-use toilet rooms are provided to offer privacy to a user and a family member or caretaker, the lock prohibition does not apply to those types of toilet rooms.

403.4 Signage. Required *public* facilities shall be provided with signs that designate the sex, as required by Section 403.2. Signs shall be readily visible and located near the entrance to each toilet facility. Signs for accessible toilet facilities shall comply with Section 1111 of the *International Building Code*.

❖ Public facilities must be designated by a legible sign for each sex and located at the entrance to such facilities. The code is silent on the specifics of how gender designation is displayed or stated. Where a toilet or bathing room is accessible, the signage must comply with Section 1111.1 of the IBC, which requires the International Symbol of Accessibility to be displayed at any accessible toilet or bathing room, including a family/assisted-use toilet or bathing room. In addition, Section 1111.2 requires directional signage (with the International Symbol of Accessibility) at inaccessible toilet and bathing rooms indicating the direction of accessible toilet or bathing facilities and at separate-sex toilet and bathing facilities indicating the direction of family or assisted-use toilet and bathing facilities, if provided.

403.4.1 Directional signage. Directional signage indicating the route to the required *public* toilet facilities shall be posted in a lobby, corridor, aisle or similar space, such that the sign can be readily seen from the main entrance to the building or tenant space.

❖ This section reinforces the intent of Section 403.3, which is to make sure that public toilet facilities are provided for spaces intended for public use. Where the public facilities are not clearly visible from within the public areas or the facilities are locked, patrons and visitors are sometimes told that public toilet facilities are not available or are for employee use only. Mandated directional signs will inform the public that public restrooms do exist and where they are located. It is difficult to tell a customer or patron that there are no public facilities when there is an obvious sign advertising their existence. This code requirement provides the code official with the means for issuing a citation to the building owner if the signs are found to be removed in an attempt by the building owner or tenant to limit public access to toilet facilities. The location of the sign must be near the main entrance to the building or tenant space. The code is silent on specific details of the signage such as wording, size, color or mounting height (see Commentary Figure 403.4.1).

Note that IBC Section 1111 covering signage for accessibility requires that directional signage be provided at the location of separate-sex toilet facilities indicating the location of family or assisted-use toilet facilities, if such separate-sex facilities are required.

403.5 Drinking fountain location. Drinking fountains shall not be required to be located in individual tenant spaces provided that *public* drinking fountains are located within a distance of travel of 500 feet (152 m) of the most remote location in the tenant space and not more than one story above or below the tenant space. Where the tenant space is in a covered or open mall, such distance shall not exceed 300 feet (91 m). Drinking fountains shall be located on an accessible route.

❖ This section clarifies the required location for drinking fountains with respect to the spaces that require access to drinking fountains. Drinking fountains are not required to be near toilet facilities, but because of the similar travel distance limitations, they might be in the same area as the toilet facility. Because all installed drinking fountains are required to be of Accessible design (either for standing or wheelchair-seated persons), they must be on an accessible route (see Chapter 11 of the IBC).

Previous to the addition of this section to the code, the designer of the building had the choice as to whether drinking fountains were to be installed in individual tenant spaces or in a common public area outside of the tenant space. Because the code did not specify a travel distance limitation, drinking fountains located in the common area for serving the tenant spaces could be located at any distance away from a tenant space. It was, therefore, acceptable to put all of the common area drinking fountains in a single location instead of spread around in the common area to better serve the demand for drinking water. Now, if common area drinking fountains are to be installed, they must be apportioned such that they serve the demands of the tenants within the travel distance limitations.

SECTION 404
ACCESSIBLE PLUMBING FACILITIES

404.1 Where required. Accessible plumbing facilities and fixtures shall be provided in accordance with the *International Building Code*.

❖ In almost all occupancies, toilet rooms, bathing rooms and kitchens as well as the plumbing fixtures contained within are required to be constructed in such a manner that persons with physical disabilities will have "access" to those facilities in a manner similar to persons without disabilities. A portion of the plumbing fixtures in each facility must be designed and installed so that persons with physical disabilities have the "ability" to use fixtures of a similar type to those used by persons not having disabilities. One- and two-family detached dwellings are exempt from the requirements to provide for accessibility. The requirement for "accessibility" is mandated by several Federal and state laws as well by the IBC.

In nearly all jurisdictions in the United States using the *International Codes*®, the federal requirements for accessibility are generally deemed to be accomplished if a building, site, or facility complies with the applicable provisions of Chapter 11 of IBC and ICC A117.1. Chapter 11 of the IBC regulates the extent (scoping) of required accessible features to be provided in various occupancies and ICC A117.1 regulates the (technical) characteristics of those accessible features.

While the adoption of the IBC is nearly 100 percent for all International Code Council® (ICC®) member jurisdictions, the reader must be made aware that there are jurisdictions that choose to delete, replace, or modify Chapter 11 of the IBC in favor of using other (or slightly modified) accessibility requirements or standards. Users of the code are cautioned to thoroughly review each jurisdiction's (and project's) specific requirements for accessibility to determine if Chapter 11 of the IBC and ICC A117.1 can be used for design and installation of accessible building features, or whether additional or alternative standards apply.

404.2 Accessible fixture requirements. Accessible plumbing fixtures shall be installed with the clearances, heights, spacings and arrangements in accordance with ICC A117.1.

❖ Prior to the 2015 edition of the code, the required clearances, heights, spacings and arrangements for accessible plumbing fixtures were covered by the IBC referencing the standard ICC A117.1. This code did not have a direct connection to ICC A117.1, which made responsibilities for compliance and inspection unclear. This new section for 2015 directly makes the connection to the referenced standard so responsibilities are clear.

Although ICC A117.1 is clear about many aspects of accessible plumbing fixture installations, the designer, the contractor and the inspector need to have an understanding of the impact that the standard's requirements have on piping layout and fixture installation. The following information is advisory. The requirements of the A117.1 standard and its interpretation thereof are the controlling requirements.

Dimension Reference Points and Dimension Ranges

The dimensions to floor and wall surfaces in all of the commentary figures for Section 404 (which are dimensionally identical to those in the ICC A117.1 standard) are to *finished* wall and floor surfaces and not to the face of a framed wall or top of a subfloor, respectively. Although this seems like a minor detail, this is often the cause of many plumbing fixture and plumbing trim location errors made by designers as well as plumbing contractors. Many of the dimensions shown in the standard (and in the commentary figures) are given as a dimension range [e.g., 16 to 18 inches (405 to 455 mm) or 34 inches (432 mm) maximum]. Improper understanding of the purpose and use of these dimension ranges could result in violations at final inspection.

The ranges on dimensions are not intended to give the designer or installer the initial "latitude" on where to locate the item. The ranges are also not intended to absorb errors due to misinterpreting specific measurement reference points or to permit a general carelessness in making measurements. The ranges are intended to allow for normal variances associated with field construction practices so that upon final inspection, a measurement of the location of the item has a very high probability of being within the required dimension range. The following examples illustrate some of the most commonly occurring problems.

Example A

Consider the required 16- to 18-inch (405 to 455 mm) dimension range for the centerline of a water closet for Accessible design to the finished surface of the adjacent sidewall. Suppose the designer chooses to show 18 inches (457 mm) on the drawing or the contractor chooses 18 inches (457 mm) (from the dimension range shown in the standard) because "that's the dimension that we have used for years." The 18-inch (457 mm) dimension complies with the accessibility standard. The contractor necessarily makes adjustments to this dimension to be able to measure to reference points that he or she has available to work from such as a string line indicating where the face of a framed wall will be located or an existing block wall that will eventually be "furred out," insulated and covered with finish materials (e.g., gypsum board, cement board with tile, etc). Even though the contractor uses "correct" nominal thicknesses of materials to adjust the 18-inch (457 mm) dimension in order to locate the center of the drain piping for the water closet, small variations in measurement accuracy, actual material thicknesses, straightness of materials, placement of the future wall, the "plumbness" of the wall, the final mounting position of the fixture, and the perceived center of the water closet bowl could cause the final inspection measurement to exceed 18 inches (457 mm), some percentage of the time.

The lesson to be learned from this example is that the "target" dimension to be used for the initial layout should be the value that is midway between the extremes of the range. Therefore, in this example, the target dimension is 17 inches (432 mm).

Example B

Consider the 34-inch (864 mm) maximum dimension required for the height of a carrier-mounted lavatory above a finished floor. Suppose the designer shows 34 inches (864 mm) maximum on the drawing. If the contractor uses the 34-inch (864 mm) dimension as the basis for figuring out a dimension to an unfinished floor surface that he or she must use as a reference point, for similar reasons as given in Example A, the final inspection measurement could exceed 34 inches (864 mm) some percentage of the time.

The lesson to be learned from this example is that the "target" dimension for maximum (or minimum) dimensions to be used for the initial layout should be a value that slightly less than the maximum (or slightly more than the minimum). How much less (or more) must be based on an understanding of the possible variations that could be encountered in each situation.

Why is compliance to the dimensions in the ICC A117.1 standard so critical? Although the standard indicates that the dimensions "are subject to conventional industry tolerances," there has never been a consensus on what those tolerances should be. The standard also states "dimensions not stated as maximum or minimum are absolute." This seems to indicate that there is "no tolerance" for installations that have measurements that fall outside of the indicated dimension limits. Because of past litigation that has occurred in many jurisdictions, it has become commonplace for some code officials to expect plumbing fixtures for all levels of accessibility to be installed exactly to the dimensions shown on the project drawings or in the ICC A117.1 standard, if the project drawings lack the correct critical dimensions. The accessibility details on the project drawings should show the same dimensional limits and ranges as indicated in ICC A117.1. The plumbing contractor should interpret the dimensional limits and ranges appropriately to provide for the highest probability that the fixtures, upon final inspection, will be located within the dimensional limits or ranges.

Clear Floor Space and Approach Direction

The operation of a wheelchair requires ample floor area for turning and positioning. This required floor area is termed "clear floor space." The size and orientation of clear floor space around plumbing fixtures can impact the location and features required for the fixtures.

The direction of approach by a user of a wheelchair can also impact the location and features required for the fixtures. Fixtures are intended to be approached by either the forward or parallel methods. A forward approach is where the user approaches in a manner that is perpendicular to the wall that the back of the fixture is against. A parallel approach is where the user approaches in a manner that is parallel to the wall that the back of the fixture is against.

Required Clearances Around Grab Bars

The information on grab bar locations is provided in this commentary to illustrate potential interferences with plumbing components. The first potential for interference is between the required blocking in the walls for the grab bars and piping installed (or to be installed) in the walls. This is especially critical where framed walls are only $3^1/_2$ inches (89 mm) thick, because 2-inch (51 mm) nominal thickness blocking is typically installed for grab bar attachments. The second potential for interference is between the grab bars and plumbing components that are located within the finished area. The following provides the clearance requirements and examples of typical plumbing components located in finished areas.

For water closet grab bar installations, the minimum clearance between the top of a grab bar and any element protruding from the wall above is 12 inches (305 mm). A minimum clearance of $1^1/_2$ inches (38 mm) is required below, and at the ends of grab bars. For example, if a flushometer valve location is required to be below the rear-mounted grab bar of a water closet, the top of the flushometer valve is required to be located at least $1^1/_2$ inches (38 mm) below the bottom surface of the grab bar. A greater clearance might be necessary to enable proper access for repairing the valve without removal of the grab bar.

For grab bars in bathtubs and showers, the minimum clearance between the grab bar top and bottom surfaces and ends and other objects such as shower controls/shower fittings is $1^1/_2$ inches (38 mm). For example, the outermost point of a shower control handle must not pass any closer than $1^1/_2$ inches (38 mm) to a grab bar surface.

FIXTURES, FAUCETS AND FIXTURE FITTINGS

Four Levels of Accessibility

There are four designs for accessibility that can be required: Accessible, Type A, Type B and Type C. The Accessible design provides the greatest level of accessibility. Accessible design is used for public facilities as well as for some sleeping units and dwelling units. The standards in ICC A117.1 for the Accessible design are intended to be equivalent to the accessibility requirements of the Americans with Disabilities Act (ADA). The Type B design provides the most design flexibility for accessibility. The standards in ICC A117.1 for the Type B design are intended to comply with the requirements of the Federal Fair Housing Act. The Type A design provides a level of accessibility that is between the Accessible and Type B designs. Type A and B designs are used only for sleeping units and dwelling units. The Type C design is intended to be for single-family homes and townhouses. Where "aging in place" is a desired design feature, Type C design will not be discussed in this commentary. The architect or designer of a building is responsible for determining which of the designs for accessibility are required in a building as well as the locations of accessible toilet and bathing facilities and the quantities of plumbing fixtures required in those facilities.

Installation of Plumbing Fixtures in the Accessible Design

Water Closets for Wheelchair Access (Accessible Design)

Toilet and bathing rooms in the Accessible design must have accessible plumbing features installed in accordance with Sections 603 through Section 610 of ICC A117.1. Sections 603 and 604 of ICC A117.1 cover water closets (and grab bar locations). Commentary Figures 404(1) through 404(5) indicate the required dimensions for locating a water closet as well as the minimum clear floor space and grab bar locations for a water closet. The required dimension from the centerline of the water closet to the side wall having the grab bar must be from 16 to 18 inches (405 to 455 mm) to the finished wall surface. For reasons stated previously in this commentary, the recommended "target" dimension for installation is 17 inches (432 mm).

The heights of the water closets shown in the commentary figures are to the top of the seat and not to the top of the bowl rim or to the top of a lid (cover) for the seat. (Note that Section 420 of the code requires that any public or employee water closet must be of the elongated type and have an open-front seat.) A water closet in a single-occupant toilet room accessed only through a private office and not for public use is not required to have a seat height of 17 to 19 inches (430 to 480 mm) (see Section 1109.2 of the IBC, Exception 1, and Section 604.4 of ICC A117.1).

Even though the commentary figures depict a flush tank-type water closet, floor or wall-mount water closet bowls with flushometer valves could be used in these applications. As previously indicated, the ICC A117.1 standard specifies the minimum allowable clearances above, below and at the ends of grab bars. The standard intends that the area bound by the top of the clearance zone above the bar and the bottom of the clearance zone below the bar, extending in a direction that is horizontal and perpendicular away from the wall surface, is not to be blocked. Some operators of health care facilities require flushometer valves be located at or near the maximum allowable flush handle height elevation of 48 inches (1219 mm). This is to protect a seated user from abrasions that could be caused by the flushometer valve body located at lower elevations. This higher mounting height location will create an obstruction in front of the rear grab bar. In these situa-

Figure 404(1)
WATER CLOSET LOCATIONS

Figure 404(2)
WATER CLOSET HEIGHT

FIXTURES, FAUCETS AND FIXTURE FITTINGS

Figure 404(3)
SIZE OF CLEARANCE FOR WATER CLOSET

Figure 404(5)
SIDEWALL GRAB BAR FOR WATER CLOSET

Figure 404(4)
REAR WALL GRAB BAR FOR WATER CLOSET

tions, the ICC A117.1 standard allows for the rear grab bar to be supplied in two sections or the one-piece bar shifted towards the open side of the water closet to eliminate interference with the flush valve.

The ICC A117.1 standard requires that a hand-actuated flush handle (for either a flush tank or flushometer) be on the "open side" of the water closet location. For example, where facing the water closet, if the wall is on the right, then the flush lever must be on the left. If the wall is on the left, the flush lever must be on the right. For flushometer valves, attention to this right/left detail at the rough-in stage will eliminate last minute corrective actions in order to pass final inspection. The required flush handle height is 15 to 48 inches (381 to 1219 mm) above the finished floor. This height range eliminates the possibility of a foot-operated flush device. Where flush tank water closets are utilized, right-hand flush lever configurations might be a special order, long delivery time item.

Flush valves and flush tank controls are required to meet accessible (operational) requirements. Flushometer and water closet tank manufacturers are responsible for determining which flushometers and water closet tank flush controls in their product lines comply with the requirements. Packaging, product labeling or specification sheets can be used to verify compliance. The indication for compliance might be the words "ADA-Compliant," "Compliant with (ICC) ANSI A117.1" or the International Symbol for Accessibility [see Commentary Figure 404(6)].

Commentary Figures 404(7) and 404(8) indicate the required dimensions for an accessible child-size water closet installation, should one be specified by the designer. The centerline-to-sidewall dimension range for an accessible child water closet is 12 to 18 inches (305 to 457 mm.) The required floor clearance area for the accessible child water closet is the same as for the standard accessible water closet as indicated in Commentary Figure 404(3). However, the sidewall grab bar height is lower than for a standard accessible water closet [see Commentary Figure 404(8)]. The flush handle height for a child water closet is required to be not more than 36 inches (914 mm) above the finished floor.

Water Closets for Ambulatory Access (Accessible Design)

Water closets for ambulatory access have a wall or partition on both sides of the water closet such that the compartment width is 36 inches (914 mm). This enables an ambulatory disabled user to use both side grab bars to assist in lowering onto and rising from the water closet seat. The ICC A117.1 standard allows for the water closet centerline to finished sidewall (or partition) dimension to range from 17 to 19 inches (430 to 485 mm). For the reasons stated previously in this commentary, the recommended "target" dimension for installation is 18 inches (457 mm). The required flush handle height is 15 to 48 inches (381 mm to 1219 mm) above the finished floor. This height range eliminates the possibility of a foot-operated flush device.

Note that the IBC only requires a water closet for ambulatory access where there are six or more water closets in the same toilet room (or bathing room). This requirement does not increase the number of water closets but only requires one of the water closets to have ambulatory access.

Urinals for Wheelchair Access (Accessible Design)

Section 605 of ICC A117.1 specifies that urinals either be of the stall type (where the drain is very near or at floor level) or of a wall-hung type having the "rim" located at a maximum dimension of 17 inches (430 mm) above the finished floor [see Commentary Figures 404(9) and 404(10)]. The 17-inch (430 mm) dimension is a maximum height; a lesser dimension is acceptable. The "rim" of the urinal is understood to be that part of the fixture that is the highest point of the "bowl" that is the farthest point away from the wall mounting surface.

Flush handles can be located on either side of the centerline of a urinal. The required flush handle height is 15 to 48 inches (381 to 1219 mm) above the finished floor. This height range eliminates the possibility of a foot-operated flush device.

A critical point that is often overlooked by designers and contractors is the requirement for the distance between urinal partitions for the wheelchair access urinal to be at least 36 inches (914 mm). The centerline of the urinal must be not less than 18 inches (457 mm) from the face of each partition.

Note that where there is only one urinal in a toilet or bathing room, the urinal is not required to be accessible (see Section 1109.2 of the IBC, Exception 4). However, it might be prudent to install the urinal at the accessible height to allow use by wheelchair seated users as well as children.

Figure 404(6)
INTERNATIONAL SYMBOL FOR ACCESSIBILITY

Figure 404(8)
CHILDREN'S WATER CLOSET HEIGHT
AND SIDEWALL GRAB BARS

Figure 404(7)
CHILDREN'S WATER CLOSET LOCATION

Figure 404(9)
WALL HUNG-TYPE URINAL

FIXTURES, FAUCETS AND FIXTURE FITTINGS

Wall-hung urinals installed as an accessible fixture must have a rear-wall-to-outermost-rim dimension of not less than 13$^1/_2$ inches (343 mm). This provides for ease of use for wheelchair users.

Bathtubs (Accessible Design)

Section 607 of ICC A117.1 covers the requirements for bathtubs and grab bar locations. Commentary Figures 404(11) and 404(12) illustrate the two possible types of bathtub arrangements, one without a permanent seat and one with a permanent seat. The bathtub without a permanent seat requires a clear floor space in front of the tub of 30 inches (760 mm) by the length of the tub. For bathtubs with a permanent seat, the seat end of the tub must be located on the end where the 12-inch (305 mm) extended area of the required clear floor space is located [see Commentary Figure 404(12)]. The control end of the bathtub unit must be located on the end that is opposite of the 12-inch (305 mm) extended area of required clear floor space.

Commentary Figures 404(13) and (14) show the grab bar locations for bathtubs with and without permanent seats, respectively. This information is especially critical where the building walls are only 3$^1/_2$ inches (89 mm) (rough dimension) deep, because 2-inch (50 mm) nominal thickness blocking is typically installed for grab bar attachments. Where a prefabricated Accessible design bathtub module already having factory-installed grab bars is being installed, blocking for the grab bars is typically not required to be installed in the building wall framing.

Bathtubs must have the water control located in the hatched area as shown in Commentary Figure 404(15). The depicted zone should not be interpreted as requiring that the entire control face plate (escutcheon) be located within the zone. For a single handle control, locating the centerline of the control handle (as well as a diverter spout) on the centerline of the tub meets the

Figure 404(10)
ACCESSIBLE HEIGHT URINAL ON LEFT

Figure 404(12)
BATHTUB WITH PERMANENT SEAT

Figure 404(13)
GRAB BARS FOR BATHTUBS WITH PERMANENT SEATS

Figure 404(11)
BATHTUB WITHOUT PERMANENT SEAT

FIXTURES, FAUCETS AND FIXTURE FITTINGS

intent of ICC A117.1. Where multiple handle controls are installed, the centerline of the control handle farthest from the open side of the tub must be located no farther away than the centerline of the tub. Note that any control centerline should probably not be located very near to or on the top edge of the indicated hatched area due to the potential for interference of the control handle with the required grab bar location. The minimum allowable clearance of the path of the control handle to the grab bar is 1$^1/_2$ inches (38 mm). The location of the control should also provide ample space for service access to the control without requiring removal of the grab bar. For the location of hand showers in bathtubs, see the section "Hand Showers (Accessible design)" within this commentary section.

Commentary Figures 404(11) through (15) and the text of ICC A117.1 do not specify a minimum or maximum width for the bathtub [e.g., 30 inches (762 mm), 36 inches (914 mm) and 42 inches (1067 mm)]. The ICC A117.1 standard originally based the control location requirements on a 30-inch-wide (762 mm) bathtub so that floor plans requiring accessible bathtubs would not require a floor space larger than that required for the minimum required bathtub size for a nonaccessible unit. This does not mean that a bathtub of a width greater than 30 inches (762 mm) cannot be used. However, it is advisable to consider limiting the most remote vertical edge of the designated control zone at a dimension that does not exceed 18 inches (457 mm) from the open side of the bathtub. This limitation is suggested in order to enable the control to still be within an easy side reach range for wheelchair-seated users who are parked outside the bathtub.

The maximum height of the bathtub wall on the open side of the tub (i.e., the "apron" side) is controlled by the type of seat used for the installation. Where a removable seat is used, the apron side tub wall cannot be higher than the underside of the removable seat located with its top at 19 inches (483 mm) above the finished floor. Where a permanent seat is used, the wall can be no higher than 19 inches (483 mm) above the finished floor. While most standard bathtubs have apron sidewall heights less than 17 inches (432 mm), some models of soaker tubs, whirlpool tubs or freestanding tubs might have apron heights that exceed the allowable height dimension.

Figure 404(15)
LOCATION OF BATHTUB CONTROLS

Showers (Accessible Design)

Section 608 of ICC A117.1 covers the requirements for showers. There are three types of shower designs: 1) transfer type, 2) standard roll-in type and 3) alternate roll-in type.

Commentary Figure 404(16) illustrates the dimensional and clear floor space requirements for a transfer-type shower. The seat of the shower compartment must be arranged such that there is a clear floor space to position the wheelchair and align the wheelchair seat with the shower seat for a side transfer. This point is made to draw attention to the necessity for locating the control on the wall opposite the seat wall. Reversing the arrangement in some circumstances might not provide adequate wheelchair "rear parking area" [the 12-inch (305 mm) area past the opening of the shower] to enable alignment of the wheelchair seat to the shower seat for a proper and safe transfer of the user.

Commentary Figures 404(17) and (18) are provided to show the required grab bar installation so that the plumbing contractor can avoid installing piping in locations where blocking is required to be installed in the building framing.

Commentary Figure 404(18) also indicates the required zone to locate the control for a transfer shower. Refer to the previous commentary for bathtubs in this section for general information on control placement in the designated control zone.

Figure 404(14)
GRAB BARS FOR BATHTUBS WITHOUT PERMANENT SEATS

FIXTURES, FAUCETS AND FIXTURE FITTINGS

Commentary Figure 404(19) shows the minimum dimensions for a standard roll-in shower. An accessible lavatory can be located in the 30-inch (762 mm) by 60-inch (1524 mm) clear floor space as shown in the figure. Note that the lavatory must be located on the wall opposite the seat wall.

Commentary Figure 404(20) is provided to show the required grab bar installation so that the plumbing contractor can avoid installing piping in locations where blocking might be required to be installed in the building framing.

Commentary Figure 404(21) indicates the allowable designated control location zone for a standard roll-in shower. The installer must determine which end of the shower will have the seat in order to properly locate the control. Refer to the previous commentary for bathtubs in this section for general information on control placement in the designated control zone.

Commentary Figure 404(22) shows the minimum dimensions for an alternate roll-in shower. The ICC A117.1 standard requires that this type of shower arrangement have a seat or at least have the blocking in the wall so that a seat can be installed at some future point in time.

Commentary Figure 404(23) is provided to show the required grab bar installation so that the plumbing contractor can avoid installing piping in locations where blocking might be required to be installed in the building framing.

The diagrams in Commentary Figure 404(24) illustrate where controls and hand-shower locations are required to be located, depending on the depth of the shower. If the shower depth is the minimum dimension of 36 inches (915 mm), the controls and hand shower can be located in either of the positions shown in Diagrams "a" and "b." The shower-seated user is able to reach the controls in either position. If the

Figure 404(16)
TRANSFER-TYPE SHOWER COMPARTMENT SIZE AND CLEARANCE

Figure 404(18)
TRANSFER-TYPE SHOWER CONTROL AND HAND SHOWER LOCATION

Figure 404(17)
GRAB BARS IN TRANSFER-TYPE SHOWERS

Figure 404(19)
STANDARD ROLL-IN-TYPE SHOWER COMPARTMENT SIZE AND CLEARANCE

FIXTURES, FAUCETS AND FIXTURE FITTINGS

depth of the shower is greater than 36 inches (915 mm), the controls and hand shower must be located according to Diagram "a" because the shower-seated user will not able to reach controls on a wall that is farther away than 36 inches (915 mm). The control and hand-shower locations are dictated by the seat location. Refer to the previous commentary under bathtubs in this section for general information on control placement in the designated control zone.

Hand Showers (Accessible Design)

All bathtubs and showers for the Accessible design must be provided with a hand shower. (It is optional for the hand shower to be placed in a vertically adjustable sliding mount.) The hand-shower mount (the holder) must be located in the designated control zone.

The hand-shower hose length must be at least 59 inches (1500 mm) long. ICC A117.1 does not specify a required location for the hose connect supply elbow. Because this item is not a "control," it is not required to be located in the designated control zone. However, one must consider where a hand-shower user will be seated in order to provide the user with ample usable hose length for showering of all body parts. For example, for a bathtub arrangement, locating the supply elbow near the traditional shower arm location might result in limited usability of the hand shower for the seated user. Another example would be locating the supply elbow too close to the grab bar surfaces such that the fittings or the hose, being relatively stiff and immovable, at the connection to the fitting presents an interference for use of the grab bars. Refer to previous commentary concerning required clearances around grab bars. The designer or installer has the responsibility to intelligently locate the supply fitting for the hand shower hose connection.

Hand-shower heads must have a control that provides a "leaky" shut off of the water flow. In other words, the "pause" control on the hand-shower head must not completely shut off the water flow so as to alert the user that the main water control is still open, to prevent the water in the hose from cooling off and to reduce the potential for cross flow between the hot and cold supply to the control valve. Note that Section

Figure 404(20)
GRAB BARS IN ALTERNATE ROLL-IN-TYPE SHOWERS

Figure 404(21)
STANDARD ROLL-IN-TYPE SHOWER CONTROL AND HAND SHOWER LOCATION

Note: Inside finished dimensions measured at the center points of opposing sides

Figure 404(22)
ALTERNATE ROLL-IN-TYPE SHOWER COMPARTMENT SIZE AND CLEARANCE

Figure 404(23)
GRAB BARS IN ALTERNATE ROLL-IN SHOWERS

FIXTURES, FAUCETS AND FIXTURE FITTINGS

404 of the code has requirements for hand showers as well as requirements for backflow protection for hand showers.

The requirements of ICC A117.1 do not directly address other possible arrangements for providing hand-shower water control. For example, some designs of shower controls use either an integral or separate transfer valve for controlling the water flow from the main water control to a fixed shower head, a body spray array or a hand shower. Should the transfer valve be located in the designated control zone indicated in Commentary Figure 404(15)? The answer is "yes" because it is a control.

Hand-shower manufacturers are responsible for determining which hand showers in their product lines comply with the Accessible design (operational) requirements. Packaging, product labeling or specification sheets can be used to verify compliance. The indication for compliance might be the words "ADA-Compliant," "Compliant with (ICC) ANSI A117.1" or the International Symbol for Accessibility [see Commentary Figure 404(6)].

A hand shower connected to a "push-pull"-type diverter valve assembly located on a shower head arm does not comply with the requirements of ICC A117.1 for the following reasons: 1) these types of diverter valves require significant grasping or twisting to operate and 2) the mounting height would be well outside of the designated control zone (e.g., they are usually intended to be attached to a shower head arm located for a standing person's height).

Lavatories and Sinks (Accessible Design)

Section 606 of ICC A117.1 covers the requirements for lavatories and sinks for the Accessible design. In the context of the ICC A117.1 standard, the term "sink" includes a sink in a kitchen, a wet bar or a break/lunch room. All lavatories and most sinks for the Accessible design must be capable of being used by a person in a wheelchair using a "forward approach" to the fixture. In other words, the feet and legs of the user seated in a wheelchair must be provided a clearance of specific dimensions under the plumbing fixture. Where a double bowl (kitchen) sink is installed, knee and toe clearances are only required under one of the bowls. There are a few limited exceptions for sink applications that allow for a parallel approach by a person in a wheelchair. Kitchen sinks (not in a space having a cooktop or range) and wet bar sinks are permitted to have a parallel approach. Although a parallel approach eliminates the need for providing knee/toe clearances and pipe protection under those fixtures,

Figure 404(24)
ALTERNATE ROLL-IN-TYPE SHOWER CONTROL AND HAND SHOWER LOCATION

proper location of the sink with respect to the minimum clear floor space requirements is essential. The center of the sink bowl or the bowl of a double bowl sink for parallel approach can be at any location that is not less than 24 inches (610 mm) from at least one end of the clear floor space length dimension.

Commentary Figure 404(25) shows the maximum allowable height for a lavatory or a sink as measured from the finished floor to the rim of the lavatory or sink; or the counter surface that the sink is mounted to, whichever is higher. This requirement applies to sinks with forward or parallel approach. The "rim" is understood to be that part of the fixture that is the highest point of the "bowl" that is the closest to the user. The dimension is a maximum; the rim height could be located slightly lower to provide some reasonable installation tolerance for practicality. However, the installed rim height cannot be too low because of the 27-inch (685 mm) minimum knee clearance and the 8 inches (203 mm) minimum from the front edge to the beginning of the knee clearance under the sink as shown in Diagram "a" of Commentary Figure 404(26). Most wall-mount lavatories marketed for the Accessible design have "underside" bowl dimensions that easily accommodate a slightly lower rim mounting height [e.g., $33^1/_2$ inches (851 mm)] without violating the minimum required knee clearance under the lavatory [see Commentary Figure 404(27)]. However, where lavatories and sinks for the Accessible design are constructed using countertops with sink bowls attached, the designer/installer must carefully review the bowl projection below the countertop to ascertain whether adequate knee clearance will be provided at the chosen installation height of the bowl rim or countertop.

Commentary Figures 404(26) and 404(28) show the required knee and toe clearances for forward approach. This information is provided so that the plumbing installer realizes that plumbing components such as stops, traps and supply piping, as well as any exterior surfaces of chase coverings for piping, cannot be located in zones required for knee and toe clearances. One possible situation where piping might have to be located near to knee or toe clearance areas is where drain or supply piping cannot be installed in the wall that the lavatory is mounted because of firewall restrictions, freeze protection or retrofit considerations. Supply and drain piping can extend up from the floor in these applications, but only where placement does not violate the required clearance zones. Any piping under lavatories or sinks must be configured so that user contact is prevented (e.g., by the installation of a removable panel) or if piping (as well as stops, traps and other plumbing devices) is exposed, those items must have a smooth covering installed so as to protect the user from those surfaces (see Section 404.3).

Figure 404(25)
HEIGHT OF LAVATORIES AND SINKS

Figure 404(26)
KNEE CLEARANCE

Commentary Figure 404(29) shows an arrangement for a forward approach kitchen sink for Accessible design in an employee break room. A standard countertop height for kitchens and employee break rooms is approximately 36 inches (914 mm) above the floor. However, Accessible design requires that the portion of the countertop having the sink must have a height no greater than 34 inches (864 mm) from the top of the countertop (or bowl rim, whichever is higher) to the finished floor.

For a toilet or bathing room that can only be accessed through a private office, not for common or public use and intended for use by a single occupant, the lavatory height, knee and toe clearances do not apply (see Section 1109.2 of the IBC, Exception 1). However, the ICC A117.1 standard requires that the toilet and bathing room space and door width be sized so that the room can be easily converted to an Accessible design. Although not required by the standard, it would be prudent to locate the lavatory supply and drain piping such that if a conversion would occur, the required plumbing rework would be minimal.

Where a lavatory or sink will be used primarily by children of ages 6 through 12, Section 606.2 of ICC A117.1, Exception 3, allows the 27-inch (685 mm) minimum knee clearance limitation to be reduced to 24 inches (610 mm) and limits the bowl rim or countertop height to a maximum of 31 inches (785 mm) above the finished floor.

Drinking Fountains (Accessible Design)

Most occupancies require the installation of drinking fountains in common or public areas of the building (see Table 403.1 as well as Section 410.2). Where drinking fountains are provided, Chapter 11 of the IBC requires those drinking fountains to serve wheelchair-seated persons and standing persons.

For wheelchair-seated users, drinking fountains must be located with respect to the approach method to

Figure 404(27)
WALL-MOUNTED ACCESSIBLE LAVATORY

Figure 404(29)
ACCESSIBLE KITCHEN SINK

Figure 404(28)
TOE CLEARANCE

the drinking fountain. Drinking fountains are required to have a forward approach direction and require a clear floor space of 30 inches wide by 48 inches deep (762 mm by 1219 mm). The drinking fountain must be centered on the 30-inch (762 mm) width of the clear floor space. The required knee and toe clearances for a forward approach drinking fountain are the same as for forward approach lavatories and sinks [see Commentary Figures 404(26) and (28)]. The spout outlet of a wheelchair-accessible drinking fountain can be no greater than 36 inches (915 mm) above the finished floor. The spout height can be slightly lower (to provide a reasonable installation tolerance) as long as the minimum knee and toe clearances below the unit can still be achieved. A review of most drinking fountain manufacturer's drawings shows plumbing rough-in and mounting dimensions based upon the minimum knee clearance height of 27 inches (685 mm) above finished floor. Although this guarantees that the spout outlet height will be well below the 36-inch (915 mm) maximum height required by ICC A117.1, careful consideration must be given to the thickness of floor coverings that might cause the knee clearance height to be violated.

Where a drinking fountain having only parallel approach is being replaced, the replacement must allow for a forward approach.

Drinking fountains to be used by standing persons must have the spout outlet installed at a height not less than 38 inches (965 mm) and no greater than 43 inches (1090 mm) above the finished floor.

Where drinking fountains are to be used primarily by wheelchair-seated children, a spout height no greater than 30 inches (760 mm) above the finished floor is allowed to be used but only where the drinking fountain is provided with the clear floor space for the parallel approach method. The lower spout height for this application does not allow for the required knee and toe clearance for a forward approach to the drinking fountain. Note that a parallel-approach drinking fountain is required to have the spout located 3 inches (76 mm) or less from the front of the drinking fountain.

Drinking fountains must comply with the accessible (operational) requirements of ICC A117.1 as well as meet the requirements of Section 410 of the code. Drinking fountain manufacturers are responsible for determining which drinking fountains in their product lines comply with the accessible (operational) requirements. Packaging, product labeling or specification sheets can be used to verify compliance. The indication for compliance might be the words "ADA-Compliant," "Compliant with (ICC) ANSI A117.1" or the International Symbol for Accessibility.

A typical arrangement of drinking fountains for both standing persons and wheelchair-seated persons is shown in Commentary Figure 404(30).

Enhanced Reach Range Lavatories (Accessible Design)

The IBC requires an enhanced reach range lavatory in toilet and bathing rooms having six or more lavatories. This requirement does not increase the number of lavatories; it only requires one of the lavatories to have enhanced reach range characteristics. It is a lavatory for use by those of short stature having short arm lengths. The requirements of Section 606.5 of ICC A117.1 are that the faucet controls and soap dispenser controls be located not greater than 11 inches (279 mm) from the front edge of the lavatory. This also means that water or soap streams from their respective spouts must also be positioned at a dimension not greater than 11 inches (279 mm) from the front edge of the lavatory. As this is a relatively new requirement, wall-mount lavatory manufacturers have not yet created special designs that meet this requirement. However, using certain currently available electronic sensor-controlled faucets and soap dispensers with certain Accessible design-compliant wall-mount lavatories, the 11-inch (279 mm) requirement could be achieved. Where a lavatory is constructed using a countertop with a separately mounted lavatory bowl, Accessible design-compliant faucets and soap dispensers can be located along the sides of the bowl to provide for the 11-inch (279 mm) reach range. The enhanced reach range lavatory is not required to be mounted at the Accessible design height [34 inches (864 mm) maximum]. Although not addressed in ICC A117.1, it is suggested by proponents of enhanced reach range lavatories that these types of lavatories have a rim height of not greater than 28 inches (711 mm) above the finished floor.

**Figure 404(30)
ACCESSIBLE DRINKING FOUNTAIN**

FIXTURES, FAUCETS AND FIXTURE FITTINGS

Faucets (Accessible Design)

Faucet manufacturers are responsible for determining which faucets in their product lines comply with the accessible (operational) requirements. Packaging, product labeling or specification sheets can be used to verify compliance. The indication for compliance might be the words "ADA-Compliant," "Compliant with (ICC) ANSI A117.1" or the International Symbol for Accessibility [see Commentary Figure 404(6)]. Note that metering faucets are required to stay open for at least 10 seconds.

Installation of Plumbing Fixtures in the Type A Design

The plumbing fixture installation requirements for the Type A design are covered by Section 1003.11 of ICC A117.1. The Type A design could be called the "adaptable design" as all of the "prep work" must be done to allow future changes to be easily accomplished in order for the access to plumbing fixtures to be similar to that provided in the Accessible design. Type A-designed facilities have door widths, clear floor spaces and other built-in features that are identical to what is required for the Accessible design, but the original finish construction has a residential look and feel. This residential look and feel is important in order to market Type A design units to persons who do not require or immediately want some of the obvious Accessible design features in their living environment. Because all buildings covered by the IBC are required to be designed for accessibility, the Type A designed unit satisfies the intent for accommodating someone who requires an Accessible design unit as well as satisfies the builder's, contractor's or owner's need to be able sell, rent or lease the unit to someone who does not want to live in an Accessible designed unit.

Water Closets for Wheelchair Access (Type A Design)

The only difference between a water closet for the Type A design and a water closet for the Accessible design is that the low end for the dimension range for the top of the water closet seat to the floor is 15 inches (381 mm) as opposed to 17 inches (432 mm). This allows for a more "standard height" water closet to be installed providing for a more aesthetically pleasing "residential-like" appearance. Grab bars are also not required to be initially installed but the blocking for the bars must be in place. A Type A design allows for a lavatory to be located on the rear wall alongside the water closet if the following conditions exist: 1) the side of the lavatory is located not less than 18 inches (455 mm) to the centerline of the water closet and 2) the clear floor space dimension for the water closet is not less than 66 inches (1676 mm), measured perpendicular from the rear wall, by a width of not less than 15 inches (381 mm), measured to either side of the centerline of the water closet.

Urinals for Wheelchair Access (Type A Design)

There is no requirement for urinals for the Type A design. However, there is nothing to prohibit the installation of a urinal in a Type A design unit other than the urinal must not be a substitute for an accessible water closet. The urinal does not have to be installed in accordance with the Accessible design dimensional requirements because the water closet serves as the required accessible fixture. However, it would be prudent to mount the urinal at or below the Accessible design height so that wheelchair users (as well as children) could benefit from the urinal installation.

Bathtubs (Type A Design)

Bathtubs are required to comply with the Accessible design except that countertops and cabinetry (with floor covering/finish under such cabinetry) are allowed to be installed at the control end of the bathtub's clear floor space. The countertops and cabinetry must be removable. The 12-inch (305 mm) extended area beyond the head end of the bathtub's clear floor space remains a requirement.

Showers (Type A Design)

The requirements for showers are the same as for the Accessible design for showers. Countertops and cabinetry (with floor covering/finish under such cabinetry) are allowed to be installed at the control end of the shower's clear floor space, provided that countertops and cabinetry are removable. Although not specifically stated in ICC A117.1, it should be intuitively obvious that this countertop/cabinet allowance must not be used where a transfer shower is installed. It is intended to be for standard roll-in shower designs only. The 12-inch (305 mm) extended area of the shower's clear floor space beyond the seat end of the shower (whether the seat is actually installed or not) remains a requirement.

Hand Showers (Type A Design)

The hand-shower requirements for Type A design are the same as for the Accessible design.

Lavatories (Type A Design)

The requirements for lavatories for Type A design are the same as for lavatories for Accessible design. However, the Type A design allows cabinetry to be installed under the lavatory provided that: 1) the cabinetry can be removed without removal of the lavatory (i.e., the lavatory or lavatory countertop is wall mounted or wall supported); 2) the current floor finish material extends up under such cabinetry and 3) the wall behind and surrounding the cabinet is finished. These provisions allow for the installation to initially have a more "residential-like" appearance but have the capability to be easily converted allowing forward approach to the lavatory. Refer to the previous discussion of water closets for the Type A design for an allowance for locating a lavatory adjacent to a water closet.

Sinks (Type A Design)

For a multiple-bowl sink, the knee and toe clearances according to Commentary Figures 404(27) and 404(29) are only required under one bowl. This allows ample area for the required drain piping connections under an adjacent bowl that ordinarily might not be easily accommodated if the knee and toe clearance were required under all bowls of a multiple-bowl sink. If the kitchen sink is of a single-bowl design, the knee and toe clearances must be provided under the entire sink (bowl).

Cabinetry is also allowed to be installed under the kitchen sink provided that: 1) the cabinetry can be removed without removal of the sink or countertop; 2) the current floor finish must extend under such cabinetry and 3) the wall behind and the sides of the cabinetry (adjacent to the cabinet to be removed) be finished before the cabinetry is initially installed under the kitchen sink.

Although not required by ICC A117.1, plumbing piping should not interfere with the removal of cabinetry, as the intent of the removable cabinet feature is that the cabinetry can be easily and economically removed to convert a Type A design sink installation to an Accessible design sink installation. Providing this ease of cabinet removability requirement might be problematic if plumbing piping is required to extend up from the floor, outside the wall cavity. Therefore, prior planning and coordination with the cabinet installer are recommended. There are no requirements for the removable cabinetry to be removed "intact"; it could be destructively dismantled. However, prudent design of the removable cabinetry would allow for nondestructive removal so that reinstallation is possible to return the room environment back to a more standard-looking "residential-like" environment.

Plumbing components such as stops and supply piping must not be in required knee and toe clearance areas and where located near to knee and toe clearances, those components must have a smooth covering installed to protect the user. It is the intent of ICC A117.1 that these coverings be installed upon initial installation of the plumbing installation so as to make the installation compliant without the cabinet in place. In other words, no components would be required to be added to the installation to comply with forward approach requirements.

Drinking Fountains (Type A Design)

There is no requirement for drinking fountains in a Type A design.

Enhanced Reach Range Lavatories (Type A Design)

There are no requirements for this type of lavatory in a Type A design.

Faucets (Type A Design)

Faucets must be located and comply with the requirements for Accessible design.

Installation of Plumbing Fixtures in the Type B Design

The plumbing fixture installation requirements for a Type B design are covered by Section 1004.11 of ICC A117.1. Although the Type B design could be considered to provide a lesser degree of accessibility than the designs previously discussed, the Type B design should be viewed as a design that provides for basic accessibility requirements rather than a downgraded Type A design or Accessible design. The Type B design can be either of an Option A or an Option B configuration.

Water Closets for Wheelchair Access (Type B, Option A Design)

The water closet centerline is required to be located at 16 (406 mm) to 18 inches (457 mm) from the finished surface of an adjacent wall or edge of adjacent fixture. However, where a swing up grab bar is to be installed adjacent to the water closet, the centerline of the water closet must be not less than 18 inches (457 mm) from the finished surface of an adjacent wall or edge of adjacent fixture [see Commentary Figure 404(31)]. The requirement for swing up grab bar is dependent on the approach direction to the water closet and the presence (or absence) of an adjacent sidewall. Refer to Sections 1004.11.1 and 1004.11.1.1 of ICC A117.1 for further details about grab bar installations around water closets.

Type B, Option A design requires a water closet seat height of 15 to 19 inches (380 to 483 mm) above the finished floor. Flush handle height must comply with that for Accessible design and the flush handle must be located on the open side of the water closet.

Figure 404(31)
SWING UP GRAB BAR FOR WATER CLOSET

Urinals for Wheelchair Access (Type B, Option A Design)

There are no requirements for urinals in the Type B design. A urinal could be installed in a Type B design but it must not be a substitute for the accessible water closet. The urinal does not have to be installed in accordance with the Accessible design dimensional requirements because the water closet serves as the required accessible fixture. However, it would be prudent to install the urinal at or below the Accessible design height so that wheelchair users (as well as children) could benefit from the urinal installation.

Bathtubs (Type B, Option A Design)

For Type B, Option A design, Commentary Figures 404(32) and 404(33) show the required clear floor space for the two possible approaches to the bathtub. Diagram "a" of Commentary Figure 404(32) shows the required clear floor space where there is either no fixture located in the clear floor space or a lavatory in compliance with the Accessible design is located in the clear floor space. Diagram "b" of Commentary Figure 404(32) shows the required clear floor space where there is a countertop (with or without a vanity below) having a lavatory bowl. The lavatory in Diagram "b" must have the required clear space for a parallel approach to the lavatory [see commentary for "Lavatories (Type B, Option A Design)"]. The "jog" in the control end wall is the result of providing at least the required minimum length of clear floor space of 48 inches (1219 mm) plus the depth of the lavatory countertop.

Commentary Figure 404(33) shows the required clear floor space where there is a forward approach to the bathtub and a water closet is adjacent to the bathtub [see the previous commentary for "Water Closets for Wheelchair Access (Type B, Option A Design)"].

Blocking for bathtub grab bars is required to be installed in walls in accordance with the dimensional requirements for the Accessible design.

Showers (Type B, Option A Design)

If the only bathing fixture in the Type B, Option A designed unit is a shower, it must be at least the size of an Accessible design transfer-type shower as shown in Commentary Figure 404(16) with the following modifications:

1. The clear floor space is only required to be 30 inches (760 mm) in width.
2. The blocking for the seat is not required if the shower is larger in size than both of the minimum shower dimensions.

Blocking for grab bars is required to be installed in accordance with the dimensional requirements for the Accessible design transfer shower.

Figure 404(33)
FORWARD APPROACH BATHTUB IN TYPE B UNITS

Figure 404(32)
PARALLEL APPROACH BATHTUB IN TYPE B UNITS

FIXTURES, FAUCETS AND FIXTURE FITTINGS

Hand Showers (Type B, Option A Design)

Hand showers are not required in either showers or bathtubs. Standard shower heads are acceptable. If a hand shower is installed, there are no requirements for supplying or installing the hand shower in accordance with previously discussed Accessible designs or that the hand shower be compliant with the operational requirements for accessible controls.

Lavatories (Type B, Option A Design)

There are three possible arrangements for lavatories in Type B, Option A units.

1. Where there is at least a 30-inch (762 mm) by 48-inch (1219 mm) clear floor space for parallel approach in front of the lavatory, with or without permanent cabinetry below, a lavatory can be installed [see Commentary Figure 404(32), Diagram "b"]. The lavatory bowl center must be not less than 24 inches (610 mm) from one end of the clear floor space length dimension. The height of the lavatory must be not greater than 34 inches (864 mm) above the finished floor.

2. Where there is at least a 30-inch (762 mm) by 60-inch (1219 mm) clear floor space for forward approach to the lavatory, a wall-mounted lavatory that complies with all of the requirements for an Accessible design lavatory can be installed [see Commentary Figure 404(32), Diagram "a"]. The lavatory bowl center must be not less than 15 inches (381 mm) from one end of the clear floor space width dimension.

3. Where removable cabinetry is planned for below a lavatory, a clear floor space of at least 30 inches (762 mm) by 60 inches (1219 mm) for forward approach (after the cabinet is removed) must be provided. The removable cabinet must not support the lavatory/countertop and the wall/floor must be finished under/behind the cabinet. The lavatory bowl center must be not less than 15 inches (381 mm) from either edge of the clear floor space width dimension. The lavatory/countertop must comply with the height, knee/toe clearances and pipe protection requirements required for an Accessible design lavatory. The intent of the ICC A117.1 standard is for the pipe protection to be installed at the time of the initial installation such that only cabinet removal is required to comply with forward approach lavatory requirements.

Drinking Fountains (Type B, Option A Design)

There is no requirement for drinking fountains in Type B, Option A units.

Sinks (Type B, Option A Design)

There are two possible arrangements for kitchen sinks:

1. Where the approach is parallel to the sink and knee/toe clearances are not provided under the sink, the sink can be any height. A clear floor space of at least 30 inches (762 mm) by 48 inches (1219 mm) for parallel approach must be provided in front of the sink. The center of the bowl (or one of the bowls of a two-bowl sink) must be not less than 24 inches (610 mm) from one end of the clear floor space length dimension.

2. A sink that complies with all of the requirements for Accessible design can be installed. A clear floor space of at least 30 inches (762 mm) by 48 inches (1219 mm) for forward approach must be provided in front of the bowl (or one of the bowls of a two-bowl sink). The center of the bowl (or one of the bowls of a two-bowl sink) must be not less than 15 inches (381 mm) from either edge of the clear floor space width dimension. The sink is also permitted to have parallel approach configuration having a clear floor space of at least 30 inches (762 mm) by 48 inches (1219 mm). The center of the bowl (or one of the bowls of a two-bowl sink) must be not less than 24 inches (610 mm) from one end of the clear floor space length dimension.

Enhanced Reach Range Lavatories (Type B, Option A Design)

There are no requirements for this type of lavatory in Type B, Option A design.

Faucets (Type B, Option A Design)

Faucets are required be compliant with the operational limitations required for accessible controls.

Water Closets for Wheelchair Access (Type B, Option B Design)

The water closet requirements are the same as for the Type B, Option A design.

Urinals for Wheelchair Access (Type B, Option B Design)

There is no discussion of urinals for the Type B design. However, there is nothing to prohibit the installation of a urinal in a Type B design unit other than the urinal must not be a substitute for an accessible water closet. The urinal does not have to be installed in accordance with the Accessible design dimensional requirements because the water closet serves as the required accessible fixture. However, it would be prudent to mount the urinal at the Accessible design height so that male wheelchair users (as well as children) could benefit from the urinal installation.

Bathtubs (Type B, Option B Design)

The bathtub requirements are the same as for the Type B, Option A design.

FIXTURES, FAUCETS AND FIXTURE FITTINGS

Showers (Type B, Option B Design)

The shower requirements are the same as for the Type B, Option A design.

Hand Showers (Type B, Option B Design)

Hand showers are required in compliance with Accessible design requirements.

Lavatories (Type B, Option B Design)

The lavatory requirements are the same as for the Type B, Option A design.

Sinks (Type B, Option B Design)

The sink requirements are the same as for the Type B, Option A design.

Drinking Fountains (Type B, Option B Design)

There is no requirement for drinking fountains in the Type B, Option B design.

Enhanced Reach Range Lavatories (Type B, Option B Design)

There are no requirements for this type of lavatory in the Type B, Option B design.

Faucets (Type B, Option B Design)

❖ Faucets are required be compliant with the operational limitations required for accessible controls.

404.3 Exposed pipes and surfaces. Water supply and drain pipes under accessible lavatories and sinks shall be covered or otherwise configured to protect against contact. Pipe coverings shall comply with ASME A112.18.9.

❖ Persons seated in a wheelchair and using a sink or lavatory designed for accessibility could have their legs come in contact with the drain trap and water supply tubing and valves under those plumbing fixtures. The traps, tubing and valves could have sharp edges that could abrade skin. One way to protect the wheelchair-seated person from coming in contact with those components is to build cabinetry to conceal the components. The cabinetry would need to be designed to allow the necessary leg and toe clearance for a wheelchair-seated person. Figures 306.2 and 306.3 in ICC A117.1 indicate the required clearance. Another way to protect the wheelchair-seated user is to install "covers" on the components. The covers must comply with ASME A112.18.9.

SECTION 405
INSTALLATION OF FIXTURES

405.1 Water supply protection. The supply lines and fittings for every plumbing fixture shall be installed so as to prevent backflow.

❖ This section identifies minimal installation requirements for water supply protection. Backflow prevention is perhaps the most important aspect of a plumbing code (see commentary, Section 608). All plumbing fixtures, plumbing appliances and water distribution system openings, outlets and connections are potentially capable of contaminating the potable water supply.

405.2 Access for cleaning. Plumbing fixtures shall be installed so as to afford easy access for cleaning both the fixture and the area around the fixture.

❖ Poorly designed facilities lead to challenges for housekeeping professions. For proper sanitation, every plumbing fixture must be capable of being cleaned. The installation of any fixture must not result in concealed spaces that do not facilitate proper cleaning.

405.3 Setting. Fixtures shall be set level and in proper alignment with reference to adjacent walls.

❖ The operational design of plumbing fixtures and the connections to those fixtures requires that the fixture be installed with certain surfaces either level or plumb. This level fixture setting is required regardless of any out-of-level or out-of-plumb condition of the building structure that the fixture attaches to or rests upon. For example, a floor surface around a water closet connection may be out of level, but when the water closet is installed, the base of the water closet must be either shimmed or grouted such that the fixture is in level condition. Where a water closet is in an out-of-level condition, the integral trap may not provide the required water seal depth and the flushing action within the fixture may be compromised. Where a wall-mounted lavatory is mounted to an out-of-plumb wall, water in the basin may not completely drain and the trap below may not have the required water seal depth due to also being in an out-of-plumb condition. The lavatory must be installed so as to be level, regardless of the wall condition.

Proper alignment to adjacent walls is a subjective requirement, but the intent is that the installed fixture appear to be "square" and "in alignment" with the walls that are in close proximity to the fixture. An example would be the clearance from the back of a water closet flush tank to the wall behind it. The gap should appear to be even. Another example is one-piece lavatory countertop with bowl on a vanity. The backsplash of the countertop should appear to be even against the wall so as to prevent large gaps that are difficult to caulk and maintain a seal against possible splashing. While the code is not specific on the precision of the alignments, the intent is to ensure that fixtures are installed with attention to details that could affect the operation and future sanitary condition of the fixture.

405.3.1 Water closets, urinals, lavatories and bidets. A water closet, urinal, lavatory or bidet shall not be set closer than 15 inches (381 mm) from its center to any side wall, partition, vanity or other obstruction, or closer than 30 inches (762 mm) center to center between adjacent fixtures. There shall be not less than a 21-inch (533 mm) clearance in front of the water closet, urinal, lavatory or bidet to any wall, fixture or door. Water closet compartments shall be not less than 30 inches (762 mm) in width and not less than 60 inches (1524 mm) in depth for floor-mounted water closets and not less than 30 inches (762 mm) in width and 56 inches (1422 mm) in depth for wall-hung water closets.

❖ For the 2012 edition, Code Figure 405.3.1 was removed because the figure fell short of providing a clear understanding of the requirements and, in some cases, made the requirements more confusing. In the context of this section only, bathtubs and showers are not considered to be "fixtures" but "obstructions" for which the minimum clearance dimensions from the indicated fixtures must be observed. Required minimum clearances in front of bathtubs and showers are not addressed in the code.

The requirements for placement and installation of plumbing fixtures for accessibility in toilet and bathing facilities "override" this section's dimensional requirements for "nonaccessible" plumbing fixture installation (see commentary, Section 404). A critical point that is often overlooked by designers and contractors is the accessibility requirement that the distance between urinal partitions for the wheelchair-accessible urinal must be at least 36 inches (914 mm). The centerline of the wheelchair-accessible urinal must be at least 18 inches (457 mm) from the face of each partition and not the minimum of 15 inches (381 mm) as required for a nonaccessible urinal.

Even though the code section is silent on exactly how the dimensions indicated in this section are to be measured, the intent is that the dimensions are to be measured to finished surfaces of walls (including tile or other coverings) and to imaginary vertical planes located against the most exterior feature of a fixture or obstruction. The code is also silent on whether minor protrusions from walls such as paper product dispensers, hand dryers, cove moldings, diaper-changing units, display cases, vending machines and grab bars impact the measurements discussed in this section.

The centers of the fixtures discussed in this section are not specific points on the fixtures, but a point that is "derived." For example, a centerline of a lavatory bowl is typically ascertained by calculating the "halfway" dimension between the widest outside dimensions of the bowl. Given the difficulty in locating an exact center of a fixture and a lack of coordination between trades, establishing an inspection tolerance for the spacing and clearance dimensions required by this code section is next to impossible. However, code officials will necessarily impart their own acceptance criteria (i.e., tolerances) during an inspection in order to accept or reject a required clearance dimension. Therefore, designers of toilet facilities are cautioned against creating layouts, which require building to the exact minimum clearance dimensions. A fixture layout having greater-than-the-minimum-clearance dimensions is inherently easier to build and results in far less inspection difficulties.

Note that all of the minimum dimensions stated in this code section might be overridden by the accessibility requirements of Chapter 11 of the IBC. See the commentary for Section 404 for more information.

Good design practices for toilet facilities and bathrooms usually provide clearances in excess of the minimum requirements in order to provide for appealing aesthetics and user friendliness.

The 21-inch (533 mm) minimum clearance from the front edge of a fixture to any wall, fixture or door applies only to water closets, urinals, lavatories and bidets. This dimension has been determined to be the necessary space required by a user to properly and safely use these fixtures. A question that is frequently asked is: "Does the intent of the requirement for providing at least 21 inches clearance in front of a water closet, urinal, lavatory or bidet to any wall, fixture or door include the door in any position of its swing path?" The answer is no. The purpose of the 21-inch (533 mm) clearance is to provide the user adequate space in front of the fixture to use it comfortably. As the door in any open position is an object that can be moved out of the way to provide the user space, there is no need to provide additional clearance from the door in any position of its opening swing path. Toilet room doors are normally closed when the fixtures are being used, which is another reason why it is assumed that the user will have adequate space to be able to use the fixture after moving the door out of his or her way.

There are many real-world examples of where doors swing into this 21-inch (533 mm) clearance zone. One example is a typical three-fixture, minimum size, hotel/motel bathroom where the door swings through the clearance area in front of all three fixtures in the bathroom. Typical powder rooms in dwellings and single-occupant toilet rooms in commercial occupancies are commonly designed with doors that swing into the space in front of water closets and lavatories. Another example is the typical (nonaccessible) minimum size- [30-inch by 60-inch (762 mm by 1524 mm)] water closet compartment, where the door swings into the compartment. If the many users of the code enforced this section to require the clearance for the door swing path, there would have been many attempts over numerous past code cycles to change the code text to specifically state that requirement; however, there have been no such proposals brought forth. And, because it has been common practice to allow door swing paths into the fixture user area (as evidenced by the examples presented), the commentary authors have concluded that this code section does not intend to prohibit doors from swinging into the 21-inch (553 mm) clearance. In other words, if the code intended to measure from the door swing, there would be evidence that the code is being enforced that way, but it is apparently not being enforced that way.

Could there be some multiple-user toilet room arrangements where allowing the door swing path to encroach into the 21-inch (553 mm) user clearance area would, in some cases, be aggravating to the users of the toilet facility? There are bound to be some; however, in most cases, this potential problem is rarely encountered in real-world scenarios because accessibility travel path and clear floor space requirements automatically require larger spaces in toilet rooms. If the code intended to apply to the door swing path, it would be stated outright, similar to the way the IBC addresses doors that swing into corridors, for example.

Note that the egress requirements of the IBC require that egress doors not be prevented from fully opening (see Commentary Figure 405.3.2).

If there are two fixtures facing each other, the 21-inch (533 mm) clearance in front of each fixture can overlap; however, the minimum egress width requirements in the IBC would most likely prevent a complete overlap of the frontal minimum clearances. Accessibility requirements for fixtures that are required to be accessible will require greater frontal approach dimensions. Door swing paths, in some accessible situations, are not allowed to overlap the accessible frontal approach dimensions.

The 15-inch (381 mm) minimum clearance dimension from the center of a fixture to any sidewall, partition, vanity or other obstruction only applies to water closets, urinals, lavatories and bidets. Note that the requirement does not include "to any fixture" as the 30-inch (762 mm) minimum center to center for fixtures, as discussed below, controls the spacing between fixtures. Fixtures that are required to be accessible require greater clearances from centerline to sidewall, partition, vanity or other obstructions (see commentary, Section 404).

The 30-inch (762 mm) minimum center-to-center dimension is only between the centers of water closets, urinals, lavatories and bidets. Public restrooms, having the requirement for partitions and compartments separating fixtures, cannot have fixtures located at the minimum of 30 inches (762 mm) center to center. The controlling aspect is the 15-inch (381 mm) minimum dimension from the center of the fixture to the required partition or compartment wall, not the center of the adjacent fixture. In other words, the thicknesses of partitions must be considered, thereby necessitating dimensions greater than 30 inches (762 mm) center to center.

Lavatories, both private and public, that are part of vanities require special attention. As indicated in the previous paragraph, a vanity, an "obstruction" in the context of this section, can be no closer than 15 inches (381 mm) to the centerline of an adjacent fixture. The center of the lavatory bowl, which is in the lavatory top that rests on the vanity, must be no closer than 30 inches (762 mm) to the centerline of an adjacent fixture. If the center of the lavatory is less than 15 inches (381 mm) from the side edge of the lavatory top, then the 30-inch (762 mm) center-to-center minimum dimension controls. If the center of the lavatory is greater than 15 inches (381 mm), then the 15 inches (381 mm) from the centerline of the adjacent fixture to the nearest obstruction (i.e., the edge of the lavatory top, not necessarily the vanity cabinet) controls (see Commentary Figure 405.3.1(1). Note that locating the vanity cabinet at 15 inches (381 mm) from the centerline of the adjacent fixture would not necessarily provide the required minimum clearance, as most lavatory tops overhang the edge of the vanity cabinet to some extent [e.g., $^3/_4$ inch (19.1 mm)]. For example, a 48-inch-wide (1219 mm) vanity cabinet is provided with a 49-inch-wide (1245 mm) cultured marble integral lavatory bowl countertop. Theoretically, this countertop could overhang along one side as much as 1 inch (25 mm).

Improper location of lavatories and lavatory vanities narrower than 30 inches (762 mm), offset-to-one-side bowls in lavatory countertops, and corner lavatory arrangements can unknowingly cause code violations. Where there is a strong emphasis on minimizing bathroom spaces, designers might be tempted to reduce the space for the lavatory by choosing a narrow vanity width or providing a wall mount or pedestal lavatory of a narrow dimension with the intent of locating this fixture as close as possible to a side wall. This can be a violation. Bowls in the lavatory countertops are often located offset to one side in order to provide clearance for vanity cabinet drawers beneath. Again, a violation can be created as there is not sufficient clearance from the center of the lavatory (bowl) to the adjacent sidewall or to an adjacent fixture to provide for proper use of the lavatory.

The code does not offer any specific insight concerning corner sinks. One possible method for determining the required clearances would be to establish a vertical plane perpendicular to a horizontal centerline extending from the corner though the center of the lavatory bowl and positioning this plane at the front edge of the fixture. This "user width" could be determined by measuring horizontally within the established plane from the centerline of the lavatory to a point where the plane insects the adjacent wall. The intent of this section appears to be to provide a "user width" of at least 30 inches (762 mm). If the measured dimensions left and right of the lavatory centerline are 15 inches (381 mm) or more, then the "user width" requirement of this section is satisfied. The front clearance measurement for this type of fixture could also utilize the imaginary front edge plane as a reference point. The intent of this section appears to be to provide a "user depth" of at least 21 inches (533 mm). Therefore, for a corner lavatory, the "user space" of 30 inches (762 mm) wide by 21 inches (533 mm) deep must not be encroached upon by other fixtures, closed doors or walls (see Commentary Figure 405.3.1(2).

For the 2012 edition of the code, the language of this section was amended to make a distinction between the compartment depths for wall-hung water closets [56 inches (1422 mm)] versus floor-mounted water closets [60 inches (1524 mm)]. The reason for this change was to align the code with the minimum requirement of the accessibility standard ANSI A117.1, which is referenced by the IBC. Note, however, that where accessible water closet compartments for children's use are concerned, the required minimum compartment depth according to ANSI A117.1 is 59 inches (1499 mm).

FIXTURES, FAUCETS AND FIXTURE FITTINGS

For SI: 1 inch = 25.4 mm.

**Figure 405.3.1(1)
MINIMUM CLEARANCE BETWEEN FIXTURES**

405.3.2 Public lavatories. In employee and *public* toilet rooms, the required lavatory shall be located in the same room as the required water closet.

❖ In employee and public toilet rooms, the required lavatory must be located in the same room as the required water closet or in an adjacent room connected by openings without doors. This provides the user with necessary sanitary facilities to promote proper hygiene and to prevent the spread of infectious diseases. In some public toilet arrangements, the water closets and urinals are located in a separate room from the lavatories, thereby requiring the user to engage doors to access the lavatories after using a urinal or water closet. Such an arrangement would be a violation of this section as the intent is that the lavatories be available to allow the user to wash his/her hands prior contacting any door handle or push plate for exiting the toilet facility.

405.3.3 Location of fixtures and piping. Piping, fixtures or equipment shall not be located in such a manner as to interfere with the normal operation of windows, doors or other means of egress openings.

❖ Fixtures and piping located in washrooms or toilet rooms cannot adversely affect the operation of building components such as windows and doors. The location of the fixtures and piping should not interfere with or block an occupant's egress to or from that space (see Commentary Figure 405.3.2). Obstructions caused by fixtures or piping can seriously hamper egress in an emergency condition or cause injury to the occupants.

405.3.4 Water closet compartment. Each water closet utilized by the *public* or employees shall occupy a separate compartment with walls or partitions and a door enclosing the fixtures to ensure privacy.

Exceptions:

1. Water closet compartments shall not be required in a single-occupant toilet room with a lockable door.

2. Toilet rooms located in child day care facilities and containing two or more water closets shall be permitted to have one water closet without an enclosing compartment.

3. This provision is not applicable to toilet areas located within Group I-3 housing areas.

❖ Psychological studies have proven that lack of privacy places a burden on an individual's physical ability to use bathroom facilities. This is caused by uneasiness, inhibition or indignation. A partitioned compartment can

FIXTURES, FAUCETS AND FIXTURE FITTINGS

For SI: 1 inch = 25.4 mm.

Figure 405.3.1(2)
CORNER SINK USER CLEARANCE

For SI: 1 degree = 0.0174.5 rad.

Figure 405.3.2
FIXTURE AND PIPING OBSTRUCTION

provide the necessary privacy. Also, a single-occupant toilet room with a lockable door can provide the required privacy as stated in Exception 1. Exception 2, allowing an unenclosed water closet in a child care facility, recognizes the need for people to be able to assist children. Exception 3 is to allow for security personal to better monitor for illicit activities of inmates in Group I-3 occupancies.

405.3.5 Urinal partitions. Each urinal utilized by the *public* or employees shall occupy a separate area with walls or partitions to provide privacy. The walls or partitions shall begin at a height not greater than 12 inches (305 mm) from and extend not less than 60 inches (1524 mm) above the finished floor surface. The walls or partitions shall extend from the wall surface at each side of the urinal not less than 18 inches (457 mm) or to a point not less than 6 inches (152 mm) beyond the outermost front lip of the urinal measured from the finished backwall surface, whichever is greater.

Exceptions:

1. Urinal partitions shall not be required in a single-occupant or family/assisted-use toilet room with a lockable door.

2. Toilet rooms located in child day care facilities and containing two or more urinals shall be permitted to have one urinal without partitions.

❖ Users of urinals in a public toilet room need to be afforded a level of privacy for the reasons stated in the commentary to Section 405.3.4. In addition, partitions aid in protecting adjacent users from splashing. Because of the likelihood of splashing, the partition surfaces must be designed so that they can be easily cleaned and sanitized during normal restroom maintenance. Exception 1 assumes that in a single-occupant toilet room or a family/assisted-use toilet room, privacy is afforded by locking the door. Exception 2 allows day

FIXTURES, FAUCETS AND FIXTURE FITTINGS

care and child care centers to have one urinal out of two or more urinals in a toilet room to be without partitions. This affords care providers the ability to assist or supervise certain children with the use of the urinal (e.g., those just learning to use a urinal or those with difficult clothing arrangements). Where urinal partitions are provided, occupants are more inclined to use a urinal rather than using a water closet for urination. This will result in less soiling of water closet seats and surrounding surfaces [see Commentary Figures 405.3.5(1) and 405.3.5(2)]. Note that where partitions are used, the distance from the center of the urinal to the face of the partition must be at least 15 inches (381 mm) to be in compliance with Section 405.3.1. The thickness of partitions cannot encroach into the required spaces for fixtures.

405.4 Floor and wall drainage connections. Connections between the drain and floor outlet plumbing fixtures shall be made with a floor flange or a waste connector and sealing gasket. The waste connector and sealing gasket joint shall comply with the joint tightness test of ASME A112.4.3 and shall be installed in accordance with the manufacturer's instructions. The flange shall be attached to the drain and anchored to the structure. Connections between the drain and wall-hung water closets shall be made with an *approved* extension nipple or horn adaptor. The water closet shall be bolted to the hanger with corrosion-resistant bolts or screws. Joints shall be sealed with an *approved* elastomeric gasket, flange-to-fixture connection complying with ASME A112.4.3 or an *approved* setting compound.

❖ For the 2012 edition of the code, another type of connection arrangement for floor outlet fixtures was added to this section. The water closet manufacturing industry refers to this "new" method as the "gasketed waste tube connection method" [see Commentary Figure 405.4(1)]. While this method is new for the North American plumbing industry, the gasketed waste tube connection method has been used in European countries (and countries using European standards) for many decades. This new connection method will allow for a wider selection of water closet designs that are already being sold in markets outside of North America. Note that this method does not require a floor flange. In some cases, the presence of a floor flange could interfere with the water closet installation. Note that without a floor flange to support the pipe at the floor, it will be necessary for the piping installer to adequately support/provide blocking for the pipe to remain stationary when the sealing gasket/waste tube assembly is pressed into the pipe opening (or pulled out from the opening). Also note that the rough-in dimension for gasketed waste tube connected water closets may not necessarily be standardized in the familiar 14-, 12- and 10-inch (356, 305 and 254 mm) center-of-outlet to face-of-finished-wall dimensions. Piping installers will need to know the exact model of water closet to be

Figure 405.3.5(1)
URINAL PARTITIONS
(Photo courtesy of Falcon Waterfree Technologies)

FIXTURES, FAUCETS AND FIXTURE FITTINGS

Figure 405.3.5(2)
URINAL PARTITIONS

installed prior to installing the outlet pipe to ensure that the gasketed waste tube connection arrangement will work.

The flange connection method is the most common connection method used in North America. The bottom outlet of the water closet forms one half of the flange and the floor flange forms the other half of the connection [see Commentary Figure 405.4(2)]. The setting compound most commonly used is a ring made of a synthetic wax. In the past, putty was commonly used but this practice is rare as putty dries out and cracks. Elastomeric gaskets are also used between fixture outlets (horns) and flanges. Corrosion-resistant bolts or screws are required in this application so that there is not a reduction of strength of the fastener over time due to corrosion. Failure of the bolting of a wall-hung water closet could result in serious personal injury and fixture damage. The flange-to-fixture connection complying with ASME A112.4.3 is a special plastic fitting that is only suitable for floor-mounted, bottom discharge water closets. The device seals to the inside of the drain pipe with an elastomeric o-ring and seals to the outlet of the water closet bowl with an elastomeric gasket. The gasket does not adhere to the bowl but seals to the bowl when backpressure in the drain pipe occurs. Where this device is used, the floor flange only serves as an attachment point for the closet bowl bolts.

405.4.1 Floor flanges. Floor flanges for water closets or similar fixtures shall be not less than 0.125 inch (3.2 mm) thick for brass, 0.25 inch (6.4 mm) thick for plastic and 0.25 inch (6.4 mm) thick and not less than a 2-inch (51 mm) caulking depth for cast iron or galvanized malleable iron.

Floor flanges of hard lead shall weigh not less than 1 pound, 9 ounces (0.7 kg) and shall be composed of lead alloy with not less than 7.75-percent antimony by weight. Closet screws and bolts shall be of brass. Flanges shall be secured to the building structure with corrosion-resistant screws or bolts.

❖ Closet bolts must be made of brass because of its strength and corrosion resistance, because they alone

Figure 405.4(1)
GASKETED WASTE TUBE CONNECTION
FOR WATER CLOSET

Figure 405.4(2)
WATER CLOSET FLOOR FLANGE

secure the water closet in place. Closet bolts are commonly available in other materials such as nylon and brass-plated steel; however, the code is specific on the use of brass for closet bolts.

Closet flanges must be firmly secured to the building structure because they receive the closet bolts that hold down the water closet. This ensures that the water closet is adequately supported and no load is transferred to any piping material attached to the flange. The fasteners securing the closet flange to the floor must also be corrosion resistant so that there is not a reduction of strength of the fasteners over time. Failure of these fasteners can allow the water closet to shift around and break down the wax ring seal resulting in undetectable water leakage under the water closet. Since the secure mounting of the water closet depends on the secure connection to the flange with brass closet bolts, it also depends on the secure mounting of the closet flange to the structure. The use of nails or drywall screws for installing closet flanges on wood subfloors is not code compliant and can result in structural damage soon after the initial installation.

With respect to securing closet flanges to concrete slabs, self-tapping, hex head, alloy carbon steel concrete screws are frequently used. The code does not define corrosion-resistant materials; therefore, it is unclear if alloy carbon steel has sufficient corrosion resistance to satisfy the intent of the code section or provide for a long service life. Other concrete attachment methods include drop-in-type lead or plastic anchors with flat head brass or stainless steel sheet metal screws. Solid brass and stainless steel screw materials have historically performed well in this application.

Occasionally, in light-frame construction, the required location for a water closet flange interferes with a structural member below the subfloor. Typically, if the structural member is "dead center" of where the water closet flange is to be located, the only two options are to either move the location of the water closet flange or move/modify the structural member. In some of these interference situations, only a small portion of the required hole in the subfloor is blocked by the structural member. Since the code does not prohibit the use of an "offset closet flange" in these situations, they can be used just as long as the structural member does not require notching in excess of the limitations. See the entire commentary to Section 307 for further information.

405.4.2 Securing floor outlet fixtures. Floor outlet fixtures shall be secured to the floor or floor flanges by screws or bolts of corrosion-resistant material.

❖ Floor outlet fixtures, such as water closets and bidets, are held in place by bolts or screws that attach to the floor structure or to a drain pipe floor flange, which is attached to the floor structure. These bolts and screws could be exposed to moisture and, therefore, must be corrosion resistant. The floor outlet fixture screws and bolts are usually constructed of brass (see commentary, Section 405.4.1).

405.4.3 Securing wall-hung water closet bowls. Wall-hung water closet bowls shall be supported by a concealed metal carrier that is attached to the building structural members so that strain is not transmitted to the closet connector or any other part of the plumbing system. The carrier shall conform to ASME A112.6.1M or ASME A112.6.2.

❖ Contrary to what the name implies, a wall-hung water closet is not actually supported by the wall that the water closet appears to be bolted to. The weight of the water closet and of the person sitting on the water closet is transferred through bolts or studs that are screwed into a robust metal carrier assembly. The carrier assembly is bolted to the floor behind the wall. The referenced standard ASME A112.6.1M or ASME A112.6.2 contains design and test requirements for the loading of the hanger (see Commentary Figure 405.4.3).

**Figure 405.4.3
WALL-HUNG WATER CLOSET**

405.5 Water-tight joints. Joints formed where fixtures come in contact with walls or floors shall be sealed.

❖ This section addresses the surface connection of the fixture to the floor or wall. The point of contact does not form a joint with any piping material; however, it is still an important area to seal where a fixture is installed to prevent a concealed fouling surface.

The contact joint is sealed with a substance such as plaster-of-paris, grout or caulking. The exact material selected to seal the joint should be capable of withstanding any anticipated movement of the fixture during its normal use. The seal prevents the accumulation of moisture, dirt or vermin that could lead to the growth of bacteria and structural deterioration in the concealed space [see Commentary Figures 405.5(1), 405.5(2) and 405.5(3)].

405.6 Plumbing in mental health centers. In mental health centers, pipes or traps shall not be exposed, and fixtures shall be bolted through walls.

❖ In certain mental health centers, care recipients are considered a danger to themselves; thus, special care is required to secure the plumbing fixtures and piping in these facilities.

FIXTURES, FAUCETS AND FIXTURE FITTINGS

Figure 405.5(1)
JOINTS BETWEEN FIXTURES AND WALLS OR FLOORS

Figure 405.5(2)
VIOLATION: LAVATORY NOT ATTACHED OR SEALED TO WALL

Figure 405.5(3)
LAVATORY PROPERLY SEALED TO WALL

405.7 Design of overflows. Where any fixture is provided with an overflow, the waste shall be designed and installed so that standing water in the fixture will not rise in the overflow when the stopper is closed, and no water will remain in the overflow when the fixture is empty.

❖ An overflow outlet is sometimes provided for certain fixtures to limit the potential for the water level in the fixture to rise above the flood level rim when objects are immersed in the water. For example, if a bathtub (with the drain outlet closed) is filled to an elevation just below the overflow outlet and an average-sized person sits in the tub, the displaced water will cause the water level to rise in the tub, but most likely it will not overflow the top (flood level rim) of the tub. If the overflow did not exist, the user might be tempted to fill the tub to the flood level rim and, upon entering the tub, would cause spillage of water onto the floor outside of the tub. While the overflow location (elevation) provides some ensurance to protect against spillage of water outside of the fixture, the overflow outlet cannot protect against flooding due to large objects being immersed into the fixture or the fixture being supplied with water at a faster rate than what the overflow is capable of accepting. In other words, an overflow might not be capable of accommodating the rate of water discharged by a faucet that was opened and forgotten about.

This section also concerns the configuration of the overflow whether it is integral or external to the fixture. The overflow passageway must always discharge downstream of the closure device for the fixture (i.e., the pop-up or stopper) so that dirty tub water is never allowed to stand or rise in the overflow (see Commentary Figure 405.7).

405.7.1 Connection of overflows. The overflow from any fixture shall discharge into the drainage system on the inlet or fixture side of the trap.

Exception: The overflow from a flush tank serving a water closet or urinal shall discharge into the fixture served.

❖ Section 405.7 requires that the discharge of the overflow outlet be downstream of the closure device for the fixture. This section requires that the discharge of the overflow be directed to the inlet of a plumbing trap.

This is necessary to prevent sewer gases from the drainage system from entering the building through the overflow outlet.

The exception recognizes that flush tank overflows, because of their location within the flush tank, must discharge to the fixture served by the flush tank. Flush tanks normally have an open standpipe that allows any overflow condition in the tank to discharge directly into the water closet or urinal.

Figure 405.7
FIXTURE OVERFLOW

405.8 Slip joint connections. Slip joints shall be made with an *approved* elastomeric gasket and shall only be installed on the trap outlet, trap inlet and within the trap seal. Fixtures with concealed slip-joint connections shall be provided with an *access* panel or utility space not less than 12 inches (305 mm) in its smallest dimension or other *approved* arrangement so as to provide *access* to the slip joint connections for inspection and repair.

❖ An access panel is required only for slip-joint connections using an elastomeric gasket to seal the joint. Fixtures such as bathtubs and bidets commonly use slip joints in concealed spaces that would require access in accordance with this section. Access is not required should all the joints be soldered, solvent cemented or screwed joints. For example, Schedule 40 ABS and PVC tub waste and overflow assemblies are now widely available and allow all joints to be solvent cemented so that an access panel is not required.

405.9 Design and installation of plumbing fixtures. Integral fixture fitting mounting surfaces on manufactured plumbing fixtures or plumbing fixtures constructed on site shall meet the design requirements of ASME A112.19.2/CSA B45.1 or ASME A112.19.3/CSA B45.4.

❖ The design of fixture fitting mounting surfaces of manufactured and site-built plumbing fixtures must comply with ASME A112.19.2/CSA B45.1 or ASME A112.19.3/CSA B45.4. These standards cover the required hole size, hole spacing and mounting area size to enable standard (and code compliant) fixture fittings (faucets and waste assemblies) to be installed onto plumbing fixtures.

SECTION 406
AUTOMATIC CLOTHES WASHERS

406.1 Water connection. The water supply to an automatic clothes washer shall be protected against backflow by an *air gap* that is integral with the machine or a backflow preventer shall be installed in accordance with Section 608. *Air gaps* shall comply with ASME A112.1.2 or A112.1.3.

❖ Any appliance or piece of equipment with a potable water supply connection is a potential source of backflow contamination to the potable water supply. An automatic clothes washer is no exception. Many editions of the code leading up to the 2012 edition required that automatic clothes washing machines have an internal air gap that conformed to ASSE 1007. However, a review of automatic clothes washing machines for residential (domestic) use revealed that machine manufacturers have ceased indicating that their machines comply with ASSE 1007. Because machines no longer bear ASSE 1007 compliance information, the code requirement was removed for the 2012 edition to prevent a compliance issue during inspection.

The lack of external indication on automatic clothes washing machines or the mention in equipment manuals concerning the presence of an internal air gap offers little comfort to the code official that the potable water supply is properly protected against backflow. While there is a high probability that the majority of clothes washer manufacturers are providing appropriately designed air gaps internal to their machines, the lack of confirming evidence might cause code officials to require the machine water connections to have backflow preventers in accordance with Section 608. In addition to the cost implications of adding external backflow protection, there are the issues of supplying backflow preventers suitable for hot water, capturing water from the preventer caused by an upset condition/failure and required annual inspections.

406.2 Waste connection. The waste from an automatic clothes washer shall discharge through an *air break* into a standpipe in accordance with Section 802.4 or into a laundry sink. The trap and *fixture drain* for an automatic clothes washer standpipe shall be not less than 2 inches (51 mm) in diameter. The *fixture drain* for the standpipe serving an automatic clothes washer shall connect to a 3-inch (76 mm) or larger diameter fixture *branch* or *stack*. Automatic clothes washers that discharge by gravity shall be permitted to drain to a waste receptor or an *approved* trench drain.

❖ The minimum required size of a trap and fixture drain for an automatic clothes washer is 2 inches (51 mm). An automatic clothes washer fixture drain of 2 inches (51 mm) has no length limit but at the point where the fixture drain becomes a branch drain (that is, combines with another fixture drain), an increase to a 3-inch (76 mm) pipe size is required. Most modern automatic clothes washers can pump waste at flows approaching 21 gallons (79.5 liters) per minute. These higher flows have caused trap siphonage and sudsing in other fixture drains downstream of the washing machine fixture drain connection where the drain piping remained a 2-inch (51 mm) size after the connection. Increasing the fixture branch drain size or stack size to 3-inch (76 mm) piping is necessary to eliminate these problems.

Where a clothes washer discharges to a laundry sink, the high rate of discharge does not affect the drain size for the laundry sink. The capacity of the laundry sink allows a slower discharge flow rate to the fixture drain. A clothes washer must discharge to the drainage system through an air break. A direct connection would allow sewage to back up into the appliance in the event of a blockage in the drain pipe. A clothes washer is usually discharged to an individually trapped standpipe or a laundry sink. The use of a laundry sink as a waste receptor is common practice. For commercial washers that discharge by gravity, trench drains are a suitable discharge point. Some large commercial machines drain by gravity by opening a solenoid-actuated valve located at the low point of the drum. Such machines could not use a standpipe (see Section 412 and Commentary Figure 406.2).

SECTION 407
BATHTUBS

407.1 Approval. Bathtubs shall conform to ASME A112.19.1/CSA B45.2, ASME A112.19.2/CSA B45.1, ASME A112.19.3/CSA B45.4 or CSA B45.5/IAPMO Z124.

❖ The referenced standards are for factory-made bathtubs of various materials. Bathtubs are available in many materials, including enameled cast iron, porcelain-enameled formed steel or plastic. Plastic bathtubs can be constructed of a variety of synthetic materials, including acrylonitrile butadiene styrene (ABS), polyvinyl chloride (PVC), gel-coated fiberglass-reinforced plastic, acrylic, cultured marble cast-filled fiberglass, polyester and cultured marble acrylic.

Plastic bathtub designs are subjected to an ignition test. Plastic bathtubs must be impregnated with fire-retardant chemicals to reduce their potential fuel contribution during a fire and to protect against accidental fire that might be caused by a plumber's torch or misplaced cigarette. However, it should be noted that plastic bathtub designs need not be subjected to flame spread and smoke-density tests such as ASTM E84.

Bathtubs are supplied in a variety of shapes and sizes designed for greater comfort in accommodating the human body. Some bathtubs are large enough to allow more than one user at a time. Large bathtubs

**Figure 406.2
CLOTHES WASHER TRENCH DRAIN**

could require special considerations for structural support. For example, consider a 67-inch (1702 mm) and 49-inch (1219 mm) rectangular two-person bathtub filled with water holding two people with a total weight in excess of 1,000 pounds (454 kg). With a support footprint area of only 37 inches by 44 inches (940 mm by 1118 mm), the resultant loading to the building floor would be approximately 90 pounds per square foot (439 kg/m²). Because normal design floor loading (for a residential structure) is only 40 pounds per square foot (195 kg/m²), the building structure might be required to be strengthened for large tubs.

For the 2012 code edition, all the standards in this section, with the exception of CSA B45.5, were updated because of a harmonization of standards that occurred since the last code edition. For more information on the harmonization of standards in the code, see the commentary to Section 405.9.

407.2 Bathtub waste outlets and overflows. Bathtubs shall be equipped with a waste outlet and an overflow outlet. The outlets shall be connected to waste tubing or piping not less than 1$\frac{1}{2}$ inches (38 mm) in diameter. The waste outlet shall be equipped with a water-tight stopper.

❖ Previous code editions did not require bathtubs to have an overflow outlet. Because the *International Residential Code*® (IRC®) requires overflow outlets on bathtubs and the IPC committee concurred that an overflow outlet would provide some level of protection against tub overfill, an overflow outlet is now required. However, the code user should realize that most, if not all, of the overflow waste assemblies on the market are not designed to prevent overfilling of a bathtub. The primary purpose of an overflow is to drain excess water from the tub when a bather enters a "full" tub (that is, a tub filled with water to the elevation of the overflow outlet). The water level at the overflow point allows the bather to calmly move about without too much concern for sloshing water outside the bathtub. If the overflow was not provided, bathers might try to use the bathtub with it filled to the flood level rim, thus flooding the bathroom with water when they entered the tub. In this case, the presence of an overflow provides a reasonable level of protection against flooding. The other potential flooding situation arises when a user turns on the tub faucet, leaves the bathroom and is delayed in getting back to the room in enough time to turn the water off. Overflow waste assemblies are not designed to accommodate the water-flow rate that could possibly be discharged from a tub faucet. Although the presence of overflow might slightly delay overfilling of a bathtub from unattended filling, it will not prevent flooding if the filling of the tub is completely forgotten.

Waste overflows are sometimes used to access the drain piping for the purpose of rodding the trap and the drain piping to clear blockages in the drain system. A device for blocking the flow of tub water from the drain must be provided. The device could be as simple as a "pull out"-type rubber plug that tightly fits in the opening of the waste assembly or other approved methods, such as trip lever waste units, cable-actuated closure devices or metal push-pull or pop-up devices.

It should be noted that a tub waste (and overflow) assembly using slip joints must be located in areas that can be accessed in accordance with Section 405.8.

407.3 Glazing. Windows and doors within a bathtub enclosure shall conform to the safety glazing requirements of the *International Building Code*.

❖ Glass (clear or otherwise), light-transmitting ceramic or light-transmitting plastic used in doors, windows and enclosures of hot tubs, whirlpools, saunas, steam rooms, bathtubs and showers where the bottom edge is less than 60 inches (1524 mm) above the floor surface of the fixture or floor of the enclosed fixture must comply with the safety glazing requirements of Chapter 24 of the IBC. Due to the likely presence of water on the standing or walking surfaces within these types of enclosed areas, there is high probability of user slips and falls such that the user might thrust his or her arm or leg through the panel. Materials that shatter into large sharp shards would be extremely dangerous if installed immediately adjacent to these areas. A glazing material is considered "safe" if it complies with the impact testing requirements of the specific standards for the type of product. Glass block masonry walls are covered by Chapter 21 of the IBC and not the safety glazing requirements of Chapter 24 of the IBC.

Partitions and windows outside the enclosure need not be of safety glazing, except as might be required by other sections of the IBC. Exterior windows in exterior walls that serve as part of the tub enclosure must be of safety glazing unless the bottom exposed edge of the glass is at least 60 inches (1524 mm) above the tub floor surface. Note that the height dimension is taken from the tub floor and not from the finished floor outside the tub area or from a deck that might surround a tub. A sitting ledge or other horizontal surface provided around a tub next to large glass windows does not decrease the hazard of someone falling into the windows; therefore, safety glazing is required. If the width of the sitting ledge is increased such that it becomes more of a walking surface, it might not be considered part of the tub enclosure; however, the glazing would then have to be evaluated in accordance with the other safety glazing requirements of the IBC.

Note that with regard to bathtub and shower enclosures, the requirement for safety glazing is not dependent on the area (size) of the glazing panels or panes. The intent is to require safety glazing wherever it is possible for the user to fall into or thrust a limb into a glass panel.

407.4 Bathtub enclosure. Doors in a bathtub enclosure shall conform to ASME A112.19.15.

❖ This section concerns bathtubs and whirlpool bathtubs that have pressure-sealed doors. Such bathing units allow for the user to simply walk in to the tub without having to step over the wall of the tub. The easy entry allows for these tubs to be much deeper than standard tubs. Once the bather is inside the fixture, the door is closed and a latch is actuated to positively lock the door water tight. Upon completion of bathing, the door can be opened only when the bathwater is drained from the vessel, thereby releasing the pressure on the door to enable the door latching/locking mechanism to be released and the door opened.

SECTION 408
BIDETS

408.1 Approval. Bidets shall conform to ASME A112.19.2/ CSA B45.1.

❖ The bidet is a small bathing fixture used by both sexes. It is not designed for the elimination of human waste. Although used primarily to clean the perineal area, it can also be used for bathing other parts of the body, including feet. A bidet is normally equipped with a water spray, designed for external body cleaning, that directs a jet of water upward to the body. The standard referenced by this code section requires that the bidet meet material, dimensional and performance requirements.

408.2 Water connection. The water supply to a bidet shall be protected against backflow by an *air gap* or backflow preventer in accordance with Section 608.13.1, 608.13.2, 608.13.3, 608.13.5, 608.13.6 or 608.13.8.

❖ Most bidets present the hazard of a submerged outlet because the typical design uses a water supply nozzle located below the flood level rim of the fixture. This outlet is considered to be subject to high-hazard backsiphonage. Therefore, the water supply to a bidet must be protected from backflow in accordance with the provisions of Section 608 (see commentary, Section 608).

408.3 Bidet water temperature. The discharge water temperature from a bidet fitting shall be limited to a maximum temperature of 110°F (43°C) by a water temperature limiting device conforming to ASSE 1070 or CSA B125.3.

❖ Users of bidets would be at significant risk for scalding of sensitive areas of the body, if not for the code section's requirement for a water temperature limiting device. Temperature-limiting devices meeting ASSE 1070 or CSA B125.3 provide for temperature limiting control for flow rates as small as 0.5 gallons per minute (gpm) (1.89 L/m).

SECTION 409
DISHWASHING MACHINES

409.1 Approval. Commercial dishwashing machines shall conform to ASSE 1004 and NSF 3.

❖ The ASSE 1004 standard requires that the water supply to a commercial dishwashing machine be protected against backflow by an air gap in accordance with ASME A112.2 or a backflow preventer complying with ASSE 1001. The installation of the air gap and backflow preventer on a commercial dishwashing machine is regulated by ASSE 1004.

Commercial dishwashing machines must also conform to NSF 3, which establishes minimum public health and sanitation requirements for the materials, design, construction and performance of spray-type commercial dishwashing machines and their related components. The NSF 3 standard also references ASSE 1004 for water supply protection and gives provisions for the location of water supply piping and valves in relation to the machine tank or hood.

409.2 Water connection. The water supply to a dishwashing machine shall be protected against backflow by an *air gap* that is integral with the machine or a backflow preventer shall be installed in accordance with Section 608. *Air gaps* shall comply with ASME A112.1.2 or A112.1.3.

❖ The potable water supply to a dishwasher must be protected by an air gap integral to the machine or by an externally installed backflow preventer. Most, if not all, domestic dishwasher machine designs include an air gap in the water supply system of the unit. However, one way to verify that a dishwasher has an air gap device is to actually see the device on the dishwasher unit.

409.3 Waste connection. The waste connection of a dishwashing machine shall comply with Section 802.1.6 or 802.1.7, as applicable.

❖ Requirements for waste connections for dishwashers are described in Sections 802.1.6 and 802.1.7.

SECTION 410
DRINKING FOUNTAINS

410.1 Approval. Drinking fountains shall conform to ASME A112.19.1/CSA B45.2 or ASME A112.19.2/CSA B45.1 and water coolers shall conform to AHRI 1010. Drinking fountains and *water coolers* shall conform to NSF 61, Section 9. Electrically operated, refrigerated drinking *water coolers* shall be listed and labeled in accordance with UL 399.

❖ Each of the ASME/CSA standards referenced in this section are primarily concerned about basic dimensional requirements and quality tests for many types of plumbing fixtures made of vitreous china, nonvitreous ceramic, and porcelain enameled cast iron. Only two of these standards have some reference to drinking fountains. Those standards simply require that the bubbler head mounting base be above the rim (flood level rim) of the bowl and provide a few dimensions for the basic wall mount (with rear boss), semirecessed, and fully recessed drinking fountains that were commonly installed in buildings many years ago.

Drinking fountains are not required to provide chilled water. Whether designers, contractors and building officials realize this or not, most modern drinking fountains that supply just "cold water" as supplied from a potable water system do not comply with any standards listed in this section (other than Section 9 of NSF 61). Where a "drinking fountain" supplies chilled water, this unit is called a "bubbler-type pressure water cooler" (as defined by the ARI 1010 stan-

FIXTURES, FAUCETS AND FIXTURE FITTINGS

dard). It is assumed that even though the code requirement is for drinking fountains, bubble-type pressure water coolers are equivalent.

The commentary to Section 405 discusses drinking fountain height and clear floor area for wheelchair-seated persons. Because the IBC references ICC A117.1 and that standard requires a specific height for drinking fountains for standing persons, that requirement is repeated here: For a drinking fountain to be used by standing persons, the spout outlet of a drinking fountain must be not less than 38 inches (965 mm) but not greater than 43 inches (1090 mm) above the finished floor.

See the commentary to Section 403.5 concerning the location of drinking fountains.

410.2 Small occupancies. Drinking fountains shall not be required for an occupant load of 15 or fewer.

❖ This section was previously a footnote to Table 403.1. Buildings or tenant spaces with 15 or fewer occupants are not required to have a drinking fountain. The requirement for drinking fountains in all occupancies puts an undue hardship on small establishments where the floor space is usually limited. Adding to this problem is the requirement in Section 410.3 stating that where one drinking fountain is provided, two drinking fountains must be installed, one high (for standing persons) and one low (for wheelchair-seated persons). One required drinking fountain in a small establishment turns into an expensive space-consuming fixture that, according to the proponents of the original code change, rarely, if ever, is used.

[BE] 410.3 Provide high and low drinking fountains. Where drinking fountains are required, not fewer than two drinking fountains shall be provided. One drinking fountain shall comply with the requirements for people who use a wheelchair and one drinking fountain shall comply with the requirements for standing persons.

Exceptions:

1. A single drinking fountain with two separate spouts that complies with the requirements for people who use a wheelchair and standing persons shall be permitted to be substituted for two separate drinking fountains.

2. Where drinking fountains are primarily for children's use, the drinking fountains for people using wheelchairs shall be permitted to comply with the children's provisions in ICC A117.1 and drinking fountains for standing children shall be permitted to provide the spout at 30 inches (762 mm) minimum above floor.

❖ This section duplicates a similar section in Chapter 11 of the IBC. Where the required number of drinking fountains (from Code Table 403.1) to be installed is equal to one, this section requires two drinking fountains: one installed at a height for standing persons and one installed for use by wheelchair-seated persons. An exception in the IBC allows a single "combination unit" that has two spouts at different heights, each with a separate bowl. Chapter 11 of the IBC is specific in requiring that whatever the number of drinking fountains to be installed, they are to be of Accessible design for standing persons and wheelchair-seated persons (see Commentary Figure 410.2).

410.4 Substitution. Where restaurants provide drinking water in a container free of charge, drinking fountains shall not be required in those restaurants. In other occupancies where drinking fountains are required, *water dispensers* shall be permitted to be substituted for not more than 50 percent of the required number of drinking fountains.

❖ Restaurants are not required to have drinking fountains as long as water is served in a container, free of

Figure 410.2
REQUIRED DRINKING FOUNTAINS AND CONFIGURATIONS

charge. Where a restaurant provides purchased beverages in disposable cups, drinking water must be still be free of charge even though the restaurant incurs a cost for the disposable cups. This substitution is only for restaurants and is not intended for other locations such as "quick" marts where a cup is used to obtain drinks from a fountain arrangement.

410.5 Prohibited location. Drinking fountains, *water coolers* and *water dispensers* shall not be installed in public restrooms.

❖ Installation of drinking fountains in public toilet rooms is prohibited. This prohibition is necessary to prevent the contamination of a drinking water unit. Biological aerosols (air-borne germs) present in toilet rooms can contaminate exposed surfaces.

SECTION 411
EMERGENCY SHOWERS AND EYEWASH STATIONS

411.1 Approval. Emergency showers and eyewash stations shall conform to ISEA Z358.1.

❖ Emergency showers and eyewash stations are sometimes installed in areas where an individual may come in contact with substances that are immediately harmful to the body. These emergency devices provide an instantaneous deluge of water to help neutralize any adverse reactions. Emergency showers and eyewash stations are typically located where people are exposed to chemicals, acids, nuclear products and fire. The water supply requirements for an emergency eyewash station depend on the manufacturer's requirements for the specific equipment installed (see Commentary Figure 411.1).

Although cold-flushing fluid temperatures provide immediate cooling after chemical contact, prolonged exposure to cold fluids affects the ability to maintain adequate body temperature and can result in the premature cessation of first aid treatment. The International Safety Equipment Association (ISEA) standard referenced (which is also ANSI Z358.1) covers minimum performance and use requirements for emergency shower, eyewash equipment, eye/face wash equipment, handheld drench hoses and combination shower and eyewash or eye/face equipment. The mandatory portion of the standard requires that the delivered flushing fluid temperature be tepid and defines tepid as being lukewarm; no temperature range is given. However, the voluntary (i.e., not enforceable) section of this standard contains a suggested temperature range. Ultimately, each installer of such equipment should consult a medical advisor for specific temperature settings based on the types of chemical exposure anticipated.

It is unclear whether emergency showers under this section are considered to be individual showers according to the code. Individual showers would require a mixing valve conforming to ASSE 1016 or CSA B125 (see Section 424.3); however, not just any ASSE 1016 valve may be suitable for the application because of the large flows and instantaneous temperature control needed. With respect to eyewash stations, the code and the ISEA/ANSI standard are silent on what type of temperature control is necessary.

Note that the code does not require the installation of emergency eyewash stations and showers. This section has been provided to address where they are installed. Usually, governmental agencies regulating worker and building occupant safety establish the circumstances under which these fixtures must be installed.

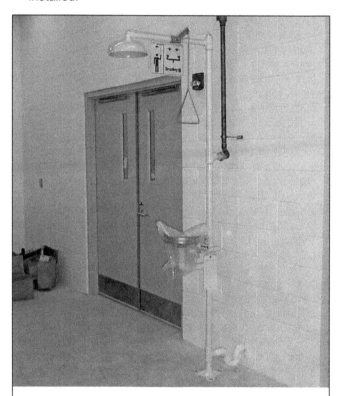

Figure 411.1
EMERGENCY SHOWER AND EYEWASH STATION
(Photo courtesy of B&I Contractors, Inc.)

411.2 Waste connection. Waste connections shall not be required for emergency showers and eyewash stations.

❖ The designer should take into account, where possible, the potential discharge volume from an emergency shower or eyewash with respect to cleanup procedures and property damage. Consideration should be given to the proper disposal of possible hazardous waste flushed from the user of emergency showers and eyewash stations.

SECTION 412
FLOOR AND TRENCH DRAINS

412.1 Approval. Floor drains shall conform to ASME A112.3.1, ASME A112.6.3 or CSA B79. Trench drains shall comply with ASME A112.6.3.

❖ Except as required by Section 412.4, floor drains and trench drains are typically installed at the discretion of the design professional where an accumulation of water on a floor or other horizontal area could create a personnel hazard or could cause damage to building finishes or structures.

ASME A112.6.3 specifies the design and quality criteria for floor drains and trench drains for buildings that are typically nonresidential. Materials covered by this standard are cast iron, copper alloy, leaded nickel bronze, ABS and PVC. Note that stainless steel is not included in the list of materials. ASME A112.3.1 specifies performance and installation requirements for floor drains of stainless steel material. Even though this standard also covers trench drains of stainless steel, the code only allows this standard to be used for floor drains. Trench drains must comply with ASME A112.6.3, which covers the design and quality criteria for drains made of cast iron, copper alloy, leaded nickel bronze, ABS and PVC. Because stainless steel is not in this list of materials, trench drains of stainless steel must be approved by the code official under Section 105.2 of the code.

CSA B79 specifies the requirements for residential floor drains constructed of cast iron, ductile iron, bronze, nickel bronze, white bronze, aluminum alloy, plastic and stainless steel.

Field-constructed floor drains and trench drains made of concrete are not covered by the above standards and therefore must be approved by the code official under Section 105.2, "Alternate materials, methods and equipment."

A floor drain is for draining liquids from indoor floor areas within a building. Floor drains are prohibited from connecting to a storm drain (see Section 1104.3) because building occupants typically assume that floor drains can be used for the disposal of any liquid. Therefore, a floor drain must be connected through a trap to a sanitary sewer system or a private sewage disposal system. Because Section 1101.3 prohibits storm water from being drained into a sanitary sewer, a floor drain must not be used for storm water disposal. For example, a rain water conductor must not discharge to a floor drain.

Trench drains are used in a variety of indoor and outdoor applications (see Commentary Figure 412.1). They are frequently used to intercept storm water at the toe of wide outdoor sloped areas such as driveways and parking lots. Where used to collect and convey only storm water, trench drain outlets are not required to be connected to a trap prior to connection to a storm water system. Trench drains are also used indoors where equipment and large areas are frequently washed down for maintenance and sanitization. A trench drain in this application must be connected through a trap to a sanitary sewer system or a private sewage disposal system.

Trench drains are also often located just inside the bay entrances of covered or indoor vehicle repair facilities. The primary intent of a trench drain in this application is to collect rainwater and ice/snow melt runoff from vehicles that are moved into a repair garage area. However, trench drains in these applications must be connected through a trap that is connected to an oil separator. The reason for this is the significant potential for the trench drain to receive grease-laden floor wash water and accidental or intentional spills of vehicle fluids. See Section 1003.4 concerning the requirement for an oil separator.

In vehicle repair facilities, multiple trench drain assemblies, each with an outlet, must not be connected together prior to connecting to a trap. Each outlet from a trench drain is considered to be a separate fixture and must have a trap installed and located in compliance with Section 1002.1. Because all traps must be vented, and trench drains are often used in large open areas without walls, a combination drain and vent system is typically used to connect the traps and carry the flow to the desired point of discharge.

The code does not provide a specific drainage fixture unit value for trench drains. Therefore, Table 709.2 must be used to determine the drainage fixture unit (dfu) value.

Commentary Figure 406.2 shows a trench drain serving as the waste discharge point for a commercial laundry washer.

412.2 Floor drains. Floor drains shall have removable strainers. The floor drain shall be constructed so that the drain is capable of being cleaned. *Access* shall be provided to the drain inlet. Ready *access* shall be provided to floor drains.

Exception: Floor drains serving refrigerated display cases shall be provided with *access*.

❖ The strainer of the floor drain is usually flush with the floor surface to alleviate tripping hazards; however, if the floor drain is located in a remote, typically untraveled area such as the corner or edge of a room, a dome-type strainer might be more suitable. All strainers must be removable to enable cleaning of the drain body and rodding of drain. Because floor drains are purposely located in areas where drainage is a necessity, either due to an emergency or normal maintenance, it is imperative that the drain be able to be serviced quickly and easily. Therefore, floor drains require ready access and must not be located under appliances, equipment or shelving, within wall cavities or covered by objects that might obscure their location.

The exception allows for floor drains located under refrigerated display cases because locating the drain exterior to the case footprint would most likely create a trip or fall hazard.

Some floor drain designs have an integral trap [see Commentary Figures 412.2(1) and 412.2(2)]. If a trap is not integral to the design, the drain must be concreate flow blockage problems. These appliances effectively "dispose of" the food by turning the food into a liquid flow stream.

Discharging fibrous or stringy food such as corn husks, celery, banana skins and onions in a food waste disposer is usually not recommended. Rather than reduce in size, the fibers tend to bypass the grinding hammers and slip down the drain, creating stoppages in the drainage piping.

Domestic food waste grinder disposers are supplied with water from a sink supply faucet and do not increase the dfu load of the kitchen sink. Domestic food waste disposers complying with ASSE 1008 are not intended for installation in food-handling establishments.

The requirement for listing and labeling to UL 430 is for electrical safety, as motors are subject to water leaks from the above grinding mechanism. Any electrical appliance connected to a system that conveys water or liquid waste has the potential to misdirect electrical current to appliance housings and onto other conductive paths that could come in contact with humans. The requirement for listing and labeling to a standard ensures that all necessary precautions have been taken by the manufacturer of food waste disposers to protect users of these appliances from hazards.

413.2 Domestic food waste disposer waste outlets. Domestic food waste disposers shall be connected to a drain of not less than $1^1/_2$ inches (38 mm) in diameter.

❖ A domestic food waste disposer is installed in conjunction with a kitchen sink. The drain pipe size required is consistent with that for a kitchen sink.

413.3 Commercial food waste disposer waste outlets. Commercial food waste disposers shall be connected to a drain not less than $1^1/_2$ inches (38 mm) in diameter. Commercial food waste disposers shall be connected and trapped separately from any other fixtures or sink compartments.

❖ Historically, the minimum size of a commercial food waste disposer drain has been 2 inches (51 mm). To help prevent contamination of other fixtures, commercial food waste disposers must be trapped and independently connected to the drainage system (see Commentary Figure 413.3). Although this section does not specifically state that an indirect connection is not allowed, the intent of the code is that the discharge be directly connected to the sanitary drain system (see Section 802.1). Food waste disposer discharge is often of a high-velocity nature such that an indirect connection would most likely result in significant splatter, clogging of a waste receptor, offensive odors, a potential health hazard and a maintenance problem. This would be analogous to providing an indirect connection for the discharge of a water closet. Obviously, water closets should not be indirect connected (see Section 802.1).

413.4 Water supply required. Food waste disposers shall be provided with a supply of cold water. The water supply shall be protected against backflow by an *air gap* or backflow preventer in accordance with Section 608.

❖ A supply of cold water is necessary to act as the transport vehicle and lubricant during the grinding operation. The water flushes the grinding chamber and carries the waste into the drain. If a waste disposer is used without running water, a drain blockage will result. While hot or warm water could be considered instead of cold water, most disposer manufacturers suggest cold water in order to keep greases more solid for best grinding action. The water supply serving a commercial food waste disposer must be protected from backflow by an air gap or backflow preventer in accordance with Section 608 (see Commentary Figure 413.3 and commentary to Section 608). In a residential kitchen sink application, the kitchen faucet spout mounted above the sink provides the necessary air gap for protection of the potable water supply.

SECTION 414
GARBAGE CAN WASHERS

414.1 Water connection. The water supply to a garbage can washer shall be protected against backflow by an *air gap* or a backflow preventer in accordance with Section 608.13.1, 608.13.2, 608.13.3, 608.13.5, 608.13.6 or 608.13.8.

❖ Water and waste connection requirements for garbage can washers are stated in this section. There are no

For SI: 1 inch = 25.4 mm.

**Figure 413.3
COMMERCIAL FOOD WASTE GRINDER**

referenced material or performance standards for these types of plumbing fixtures. These fixtures are typically floor sinks or receptors with a grated cover and a single-orifice spray nozzle protruding through the center of the grate. The spray nozzle directs a stream of water upward to clean the interior of an overturned garbage can.

A garbage can washer is a potential source of water supply contamination. The potable water supply to a garbage can washer is considered to be a submerged inlet subject to high-hazard backsiphonage and must be protected in accordance with the provisions of Section 608 (see commentary, Sections 608.13.1, 608.13.2, 608.13.3, 608.13.5, 608.13.6 and 608.13.8).

414.2 Waste connection. Garbage can washers shall be trapped separately. The receptacle receiving the waste from the washer shall have a removable basket or strainer to prevent the discharge of large particles into the drainage system.

❖ A removable basket on the drainage receptor is equivalent to a small waste interceptor. The basket collects the large garbage particles, preventing them from passing into the drainage system and creating a possible stoppage in the downstream drainage piping (see Commentary Figure 414.2).

SECTION 415
LAUNDRY TRAYS

415.1 Approval. Laundry trays shall conform to ASME A112.19.1/CSA B45.2, ASME A112.19.2/CSA B45.1, ASME A112.19.3/CSA B45.4 or CSA B45.5/IAPMO Z124.

❖ Material standards, along with waste outlet requirements for laundry trays, are provided in this section. A laundry tray is a sink typically used in a laundry room of a residential occupancy. The referenced standards regulate enameled cast-iron, plastic, stainless steel and nonvitreous ceramic laundry trays. Another common material used to construct laundry trays is soapstone. This type of laundry tray requires an alternative approval in accordance with Section 105.2 (see commentary, Section 402.2). It should be noted that concrete laundry trays are not permitted by this section because they do not provide a smooth, impervious sanitary surface. Besides serving as a sink in a laundry room, a laundry tray is commonly used to receive the discharge of an automatic clothes washer.

415.2 Waste outlet. Each compartment of a laundry tray shall be provided with a waste outlet not less than $1^1/_2$ inches (38 mm) in diameter and a strainer or crossbar to restrict the clear opening of the waste outlet.

❖ The outlet must be equipped with either a strainer or crossbar to catch debris or materials that could be discharged into the sink, causing drain clogging.

SECTION 416
LAVATORIES

416.1 Approval. Lavatories shall conform to ASME A112.19.1/CSA B45.2, ASME A112.19.2/CSA B45.1, ASME A112.19.3/CSA B45.4 or CSA B45.5/IAPMO Z124. Group wash-up equipment shall conform to the requirements of Section 402. Every 20 inches (508 mm) of rim space shall be considered as one lavatory.

❖ This section cites standards for lavatories, including cultured marble types, and their waste outlet requirements. This section also regulates group wash-up equipment.

Lavatory means wash basin. It is derived from the Latin word "lavatorium," which means "washing vessel," and the French word "laver," meaning "to wash." Lavatories are available in enameled cast iron, vitreous china, stainless steel, porcelain-enameled formed steel, plastic and nonvitreous ceramic. They also come in a variety of shapes and sizes such as wall-mounted, hanger-mounted, pedestal, rimmed, above-counter basins and under mounted types.

Standards previously required lavatories to have an overflow; however, that is not currently the case. The provision of an overflow and its location is an option of the manufacturer. The reason for eliminating the overflow requirement was the lack of use, which resulted in the growth of bacteria and microorganisms.

**Figure 414.2
GARBAGE CAN WASHER**

A group wash-up fixture, shown in Commentary Figure 416.1, is a single large basin accommodating more than one person. A group wash-up fixture has to comply with Sections 402.1 and 402.2 and be approved by the code official because it is generally not addressed by any of the referenced material or fixture standards. These fixtures are installed in areas subject to large volumes of traffic, such as facilities in sports arenas and schools.

The rim space calculation is used to determine the minimum number of required fixtures in accordance with Table 403.1 and the maximum permitted water consumption listed in Table 604.4. This calculation is not intended for use in determining the dfu value of the fixture, which is based on the trap size in accordance with Table 709.2.

416.2 Cultured marble lavatories. Cultured marble vanity tops with an integral lavatory shall conform to CSA B45.5/IAPMO Z124.

❖ Vanity tops with an integral lavatory that is all plastic must conform to the standard indicated. Lavatory units can be constructed of a variety of synthetic materials, including ABS, PVC, gel-coated fiberglass reinforced plastic, acrylic, cultured marble cast-filled fiberglass, polyester and cultured marble acrylic.

Cultured marble vanity tops are impregnated with fire-retardant chemicals, reducing the fuel contribution of the unit during a fire. They are also designed to resist the effect of a burning cigarette left unattended on the vanity top.

416.3 Lavatory waste outlets. Lavatories shall have waste outlets not less than $1^1/_4$ inches (32 mm) in diameter. A strainer, pop-up stopper, crossbar or other device shall be provided to restrict the clear opening of the waste outlet.

❖ The sizing requirements of Table 709.1 for a lavatory are emphasized. Each lavatory opening must have a restriction, preventing items such as rings, toothbrushes and cosmetic devices from inadvertently entering the drain.

416.4 Moveable lavatory systems. Moveable lavatory systems shall comply with ASME A112.19.12.

❖ Moveable and pivoting lavatories allow the user to adjust the location of a lavatory upward and downward, from side to side and front to back to maximize the ease of use of the fixture. Some products also pivot to allow greater ease of use. Commentary Figure 416.4 shows an adjustable shampoo bowl that complies with ASME A112.19.12.

416.5 Tempered water for public hand-washing facilities. *Tempered water* shall be delivered from lavatories and group wash fixtures located in public toilet facilities provided for customers, patrons and visitors. *Tempered water* shall be delivered through an *approved* water-temperature limiting device that conforms to ASSE 1070 or CSA B125.3.

❖ Prior to the 2012 code edition, it was not exceptionally clear what constituted a public hand-washing facility. Therefore, a conservative interpretation of this resulted in the inclusion of hand-washing sinks in commercial kitchens, restaurant bars, medical examination rooms, dental offices and other similar locations. The change for the 2012 edition clarifies that the requirement is only for lavatories and group wash fixtures in public toilet facilities for customers, patrons and visitors. Employee hand wash sinks, bar sinks and similar nonpublic, nontoilet room hand-washing fixtures do not require tempered water.

Hot water for public hand-washing facilities is rarely provided by a dedicated water heater set at a temperature that prevents scalding. In most cases, the hot water is provided from a central water heater that provides hot water for all purposes within the building. In occupancies that require high-temperature water for dishwashing, cleaning and other commercial applications, public users might be abruptly exposed to this excessively hot water without warning. Requiring the installation of a temperature-limiting device conforming to ASSE 1070 or CSA B125.3 will prevent scalding.

Note that an ASSE 1016 tempering valve cannot be used in this application because the testing protocol for

For SI: 1 inch = 25.4 mm.

**Figure 416.1
GROUP WASH-UP FIXTURE**

ASSE 1016 does not require the valve to perform the temperature control function at a flow less than 2.25 gpm (8.5 L/m). The ASSE 1070 and CSA B125.3 compliant tempering valves require temperature regulation at the manufacturer's stated minimum flow rate [usually 0.5 gpm (1.9 L/m)]. Because public hand-washing facilities are required to have flow regulated to no more than 0.5 gpm (1.9 L/m), an ASSE 1070 or CSA B125.3 compliant device is required.

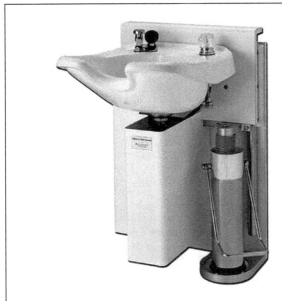

Figure 416.4
ADJUSTABLE HEIGHT SHAMPOO BOWL
(Photo courtesy of Accessible Systems LLC)

SECTION 417
SHOWERS

417.1 Approval. Prefabricated showers and shower compartments shall conform to ASME A112.19.2/CSA B45.1 or CSA B45.5/IAPMO Z124. Shower valves for individual showers shall conform to the requirements of Section 424.3.

❖ This section contains requirements governing prefabricated showers and field-constructed shower compartments. Material standards are cited, as are water supply riser, waste outlet and sizing limitations for such compartments. Provisions are also established for the shower floor or receptor wall area, along with specific support and pan requirements. As with bathtubs, a cross reference to the IBC is provided for safety glazing requirements.

The referenced standards for prefabricated shower compartments regulate plastic and nonvitreous ceramic units. Prefabricated units could include the entire shower compartment or the shower floor or base. The plastic units or floors can be constructed of a variety of synthetic materials, including ABS, PVC, gel-coated fiberglass-reinforced plastic, acrylic, cultured marble cast-filled fiberglass, polyester and cultured marble acrylic. The units are impregnated with fire-retardant chemicals to reduce the fuel contribution during a fire and to protect against an accidental fire from a plumber's torch.

417.2 Water supply riser. Water supply risers from the shower valve to the shower head outlet, whether exposed or concealed, shall be attached to the structure. The attachment to the structure shall be made by the use of support devices designed for use with the specific piping material or by fittings anchored with screws.

❖ The shower riser pipe must be firmly attached to a building structural element to protect the piping from leaks caused by stress fractures or joint failures. If the riser is not firmly attached, the movement of the shower head may result in movement of the riser piping or possible failure of the piping.

Previous editions of the code called for attachment of the water supply riser to the structure in an "approved manner," which was much too ambiguous. As a result, steel nails bent over piping, roofing nails, copper-coated J-nails or hanger strap were often used to restrain the riser resulting in undesirable long term effects. This section now requires fittings (such as a drop ear elbow) to be screwed to the structure or use of a support system specifically designed for riser support [see Commentary Figures 417.2(1) and (2)].

Where a shower riser pipe is of the exposed type, the manufacturer typically provides a clamp or other means of support that matches the finish of the exposed riser. The clamp or other support must be firmly secured to a structural member and attached to the riser.

Figure 417.2(2)
SPECIAL BRACKET FOR SHOWER RISER
(Photo courtesy of Holdrite-Hubbard Enterprises)

FIXTURES, FAUCETS AND FIXTURE FITTINGS

Figure 417.2(1)
SECURING SHOWER RISER

417.3 Shower waste outlet. Waste outlets serving showers shall be not less than $1^1/_2$ inches (38 mm) in diameter and, for other than waste outlets in bathtubs, shall have removable strainers not less than 3 inches (76 mm) in diameter with strainer openings not less than $^1/_4$ inch (6.4 mm) in least dimension. Where each shower space is not provided with an individual waste outlet, the waste outlet shall be located and the floor pitched so that waste from one shower does not flow over the floor area serving another shower. Waste outlets shall be fastened to the waste pipe in an *approved* manner.

❖ In gang or multiple showers, the shower room floor must slope toward the shower drains to prevent waste water from flowing from one shower area through another shower area. It would be undesirable and unhygienic for a bather to stand in waste water from someone else's shower (see Commentary Figure 417.3).

417.4 Shower compartments. Shower compartments shall be not less than 900 square inches (0.58 m²) in interior cross-sectional area. Shower compartments shall be not less than 30 inches (762 mm) in least dimension as measured from the finished interior dimension of the compartment, exclusive of fixture valves, showerheads, soap dishes and safety grab bars or rails. Except as required in Section 404, the minimum required area and dimension shall be measured from the finished interior dimension at a height equal to the top of the threshold and at a point tangent to its centerline and shall be continued to a height not less than 70 inches (1778 mm) above the shower drain outlet.

Exception: Shower compartments having not less than 25 inches (635 mm) in minimum dimension measured from the finished interior dimension of the compartment, provided that the shower compartment has not less than 1,300 square inches (0.838 m²) of cross-sectional area.

❖ Not less than 900 square inches (0.58 m²) of interior cross-sectional area is required to facilitate an average-size standing adult cleaning the lower body extremities while bending over. A smaller-size shower would not provide adequate space. The 30-inch (762 mm) minimum dimension is based on this movement of the body while cleaning oneself (see Commentary Figure 417.4).

The exception allows for shower compartments with one dimension being as small as 25 inches (635 mm) so long as the compartment has at least a 1,300-square-inch (0.838 m²) cross-sectional area. This is especially useful in remodeling situations where old bath tubs are removed and the space turned into a shower-only enclosure.

Figure 417.3
SHOWER FLOOR DESIGN

FIXTURES, FAUCETS AND FIXTURE FITTINGS

Figure 417.4
SHOWER COMPARTMENT

417.4.1 Floor and wall area. Bathtub floors, shower floors, wall areas above built-in tubs that have installed shower heads and walls in shower compartments shall be constructed of smooth, corrosion-resistant and nonabsorbent waterproof materials. Wall materials shall extend to a height of not less than 6 feet (1829 mm) above the room floor level, and not less than 70 inches (1778 mm) above the drain of the tub or shower. Such walls shall form a water-tight joint with each other and with either the tub or shower floor.

❖ The enclosure walls for tub/showers or showers must extend to a height of not less than 6 feet (1829 mm) above the floor level of the room or 70 inches (1778 mm) above the drain, whichever is greater (see Commentary Figure 417.4). The walls and bathtub or shower floors must be constructed of smooth, corrosion-resistant and nonabsorbent materials to protect the underlying building materials from water damage and to provide for a surface that can be easily cleaned. These requirements apply to both factory-fabricated enclosures and field-constructed enclosures.

Where field-constructed tub/shower or shower enclosures are concerned, the IBC (Sections 2509.2 and 1210.3) regulates the materials of construction. Where the finished surface of the enclosure wall will be tile, the backing for the tile must be made of glass mat water-resistant backing panels, discrete nonasbestos fiber-cement interior substrate or nonasbestos fiber-mat-reinforced cement substrate sheets. The walls above the required height of the tub/shower or shower enclosure are only required to be of water-resistant gypsum board (commonly called "green board" or "blue board"). Where water-resistant gypsum board is used within tub or shower compartment areas (including the ceiling), the IBC prohibits installation of a vapor barrier behind the board. Where water-resistant gypsum board is used on ceilings within tub or shower compartment areas, $^1/_2$-inch-thick (12.7 mm) board must be installed on a framing spacing of not greater than 12 inches (305 mm) center to center and $^5/_8$-inch-thick (15.9 mm) board must be installed on a framing spacing of not greater than 16 inches (406 mm) center to center. Water-resistant gypsum board may be used as a backing board for tile installed beyond the required tub/shower or shower enclosure walls.

After the tile has been installed and grouted, the corners of the enclosure walls and the walls to the tub or shower receptor must be caulked to make those joints water tight.

The height of the shower head above the floor is not regulated by the code. The common practice is to locate the riser fitting for the shower arm such that the shower head will be approximately 70 to 80 inches (1778 to 2032 mm) above the drain in the tub or shower. This allows for the shower arm to penetrate the wall above the required height of the enclosure wall and provides for ample adult standing height under the shower head.

417.4.2 Access. The shower compartment access and egress opening shall have a clear and unobstructed finished width of not less than 22 inches (559 mm). Shower compartments required to be designed in conformance to accessibility provisions shall comply with Section 404.1.

❖ Many injuries in the home are related to accidents caused by slips and falls occurring in the tub or shower. The minimum opening requirements for access to showers allow the user to have the room necessary to safely step in and out of the shower area without having to twist or turn through a narrow open-

FIXTURES, FAUCETS AND FIXTURE FITTINGS

ing. The 22 inches (559 mm) is based on the approximate shoulder width of an average adult. The minimum opening dimension also takes into account: 1) better access for servicing valves, shower head and drain; 2) emergency egress and 3) emergency response and rescue access.

417.5 Shower floors or receptors. Floor surfaces shall be constructed of impervious, noncorrosive, nonabsorbent and waterproof materials.

❖ The floor surfaces of the shower compartment must be made water tight with smooth, corrosion-resistant, nonabsorbent, waterproof materials. Joints created between the floor and walls of the shower must be properly flashed or caulked with sealants to avoid moisture penetration. The code does not currently require specialized slip-resistant floor surface.

417.5.1 Support. Floors or receptors under shower compartments shall be laid on, and supported by, a smooth and structurally sound base.

❖ An important requirement for shower floors is proper structural support. The support base for the shower floor must be designed for applicable dead and live loads.

417.5.2 Shower lining. Floors under shower compartments, except where prefabricated receptors have been provided, shall be lined and made water tight utilizing material complying with Sections 417.5.2.1 through 417.5.2.6. Such liners shall turn up on all sides not less than 2 inches (51 mm) above the finished threshold level. Liners shall be recessed and fastened to an *approved* backing so as not to occupy the space required for wall covering, and shall not be nailed or perforated at any point less than 1 inch (25 mm) above the finished threshold. Liners shall be pitched one-fourth unit vertical in 12 units horizontal (2-percent slope) and shall be sloped toward the fixture drains and be securely fastened to the waste outlet at the seepage entrance, making a water-tight joint between the liner and the outlet. The completed liner shall be tested in accordance with Section 312.9.

Exceptions:

1. Floor surfaces under shower heads provided for rinsing laid directly on the ground are not required to comply with this section.

2. Where a sheet-applied, load-bearing, bonded, waterproof membrane is installed as the shower lining, the membrane shall not be required to be recessed.

❖ Where a shower is built in place and not prefabricated, a shower pan is required under the shower's finished floor. Various materials are used for shower pans, including sheet lead, sheet copper, polyethylene sheet, chlorinated polyethylene sheet, preformed ABS, and bonded waterproof membranes. The shower pan is designed to collect water that leaks through the shower floor. The shower pan must connect to the shower drain with a flashing flange or clamping device and weep holes or seepage openings to allow the leaking water to enter the drainage system. The flashing flange or clamping device is used to clamp down the waterproof membrane and make a water-tight joint at the drain (see Commentary Figure 417.5.2).

A standard practice with lead and copper shower pans is to coat the finished shower pan with tar. If a plastic material or bonded waterproof membrane is installed, the shower pan must not be tarred because it may adversely affect or react with the plastic or membrane material. Care must be taken during the installation of a shower pan to prevent damage from occurring to the pan itself, which could result in a shower pan leak.

Exception 1 permits installing rinsing showers such as those used at beach facilities for sand removal without a shower pan liner. Note that this does not negate the requirements for shower floors in Section 417.5.

Because sheet-applied, load-bearing, bonded waterproof membranes are very thin, Exception 2 allows the installation of this material without recessing the material into the wall studs.

417.5.2.1 PVC sheets. Plasticized polyvinyl chloride (PVC) sheets shall meet the requirements of ASTM D4551. Sheets shall be joined by solvent welding in accordance with the manufacturer's installation instructions.

❖ PVC sheet material is a flexible plastic membrane that is fitted into a shower compartment on top of the subfloor to serve as the liner required by Section 417.5.2 to protect the building from water damage. The sheet is

For SI: 1 inch = 25.4 mm.

**Figure 417.5.2
SHOWER LINING CONNECTION**

cut at the job site to accommodate the shower drain and line the subfloor, threshold and rough sidewalls of the shower compartment to a point at least 2 inches (51 mm) above the finished threshold level. The PVC sheet should be fastened to solid sidewalls and should not be nailed or perforated at any point below the flood level rim of the shower compartment. For the 2012 code edition, the thickness requirement for the material was removed because the standard already has puncture-resistance tests. Testing of materials has shown that thickness is not necessarily any indicator of puncture resistance.

417.5.2.2 Chlorinated polyethylene (CPE) sheets. Nonplasticized chlorinated polyethylene sheet shall meet the requirements of ASTM D4068. The liner shall be joined in accordance with the manufacturer's installation instructions.

❖ Chlorinated polyethylene (CPE) sheet material is a flexible plastic membrane that is fitted into a shower compartment on top of the subfloor to serve as the liner required by Section 417.5.2 to protect the building from water damage. The sheet is cut at the job site to accommodate the shower drain and line the subfloor, threshold and rough sidewalls of the shower compartment to a point at least 2 inches (51 mm) above the finished threshold level. The CPE sheet should be fastened to solid sidewalls and should not be nailed or perforated at any point below the flood level rim of the shower compartment. For the 2012 code edition, the thickness requirement for the material was removed because the standard already has puncture-resistance tests. Testing of materials has shown that thickness is not necessarily any indicator of puncture resistance.

417.5.2.3 Sheet lead. Sheet lead shall weigh not less than 4 pounds per square foot (19.5 kg/m^2) and shall be coated with an asphalt paint or other *approved* coating. The lead sheet shall be insulated from conducting substances other than the connecting drain by 15-pound (6.80 kg) asphalt felt or an equivalent. Sheet lead shall be joined by burning.

❖ Various materials are used for shower pans, including sheet lead. Shower pans are designed to collect the water that leaks through the shower floor. The bottom layer is fitted to the formed subbase and each succeeding layer thoroughly (hot mopped) to the one below.

417.5.2.4 Sheet copper. Sheet copper shall conform to ASTM B152 and shall weigh not less than 12 ounces per square foot (3.7 kg/m^2). The copper sheet shall be insulated from conducting substances other than the connecting drain by 15-pound (6.80 kg) asphalt felt or an equivalent. Sheet copper shall be joined by brazing or soldering.

❖ The industry standard addressing this product is ASTM B152. There are 13 different types of coppers listed in this standard; each has a unique property making it suitable for the specific application. Annealed (bendable) copper sheets are typically used for pan liners.

417.5.2.5 Sheet-applied, load-bearing, bonded, waterproof membranes. Sheet-applied, load-bearing, bonded, waterproof membranes shall meet requirements of ANSI A118.10 and shall be applied in accordance with the manufacturer's installation instructions.

❖ Sheet applied, load-bearing, bonded, waterproof membranes are used as part of a system of waterproofing for a shower floor when installing tile. The installation of this material must be in compliance with the manufacturer's instructions. The membrane is a finely woven cloth of synthetic fibers that is impregnated with a proprietary waterproofing system. The membrane has a stiffness similar to that of heavy brown paper and a roughness similar to that of burlap cloth. It can be cut with scissors as well as folded with a minimum radius to fit well into corners. Installation involves spreading a thin layer of special mortar base adhesive, laying the membrane on the mastic layer and then troweling (embedding) the membrane into the mastic. Seams in the membrane are made by overlapping the material and using mastic to adhere the overlapping materials.

417.5.2.6 Liquid-type, trowel-applied, load-bearing, bonded waterproof materials. Liquid-type, trowel-applied, load-bearing, bonded waterproof materials shall meet the requirements of ANSI A118.10 and shall be applied in accordance with the manufacturer's instructions.

❖ Liquid-type, trowel-applied, load-bearing, bonded waterproof materials are used as a system of waterproofing for a shower floor when installing tile. The installation of this material must be in compliance with the manufacturer's instructions. The material is best described as a stiff liquid that can be troweled or rolled on. The cured material forms a crack-resistant, watertight membrane.

417.6 Glazing. Windows and doors within a shower enclosure shall conform to the safety glazing requirements of the *International Building Code*.

❖ If a glass or plastic window or door is installed in a tub/shower or shower compartment, the IBC (Section 2406) requires the glazing (the glass or plastic) to be of safety glazing materials. This is required so that if a bather slips or falls and breaks the glazing, he or she will not be exposed to sharp, broken glazing material. Safety glazing is made to break into small rounded pieces that have minimal capacity to severely cut human skin (see commentary, Section 407.3).

SECTION 418
SINKS

418.1 Approval. Sinks shall conform to ASME A112.19.1/CSA B45.2, ASME A112.19.2/CSA B45.1, ASME A112.19.3/CSA B45.4 or CSA B45.5/IAPMO Z124.

❖ Standards containing material and performance requirements are provided in this section of the code for plumbing fixtures known as sinks. This section also contains waste outlet requirements for sinks. Sinks are considered a different item than lavatories by the code, even though the terms can be used interchangeably.

Sinks are available in enameled cast iron, vitreous china, stainless steel, porcelain-enameled formed steel, plastic and nonvitreous ceramic materials. Soapstone sinks are not regulated by any standard; therefore, those sinks must be approved as an alternative in accordance with Section 105.2.

Although the most common classification is kitchen sinks, this section also regulates service sinks, bar sinks, mop sinks and wash sinks.

418.2 Sink waste outlets. Sinks shall be provided with waste outlets having a diameter not less than $1^1/_2$ inches (38 mm). A strainer or crossbar shall be provided to restrict the clear opening of the waste outlet.

❖ The waste discharge pipe from a sink must be not less that $1^1/_2$ inches (38 mm) in diameter. While some sinks are made with an integral waste discharge outlet, most have an opening that is suitable for the fitting of a waste outlet assembly that has the required waste outlet connection. For example, the typical stainless steel residential kitchen sink has an opening in the bottom of the sink that is approximately $3^1/_2$ inches (89 mm) in diameter. A waste outlet assembly (commonly called a "basket strainer assembly") or a food waste grinder attachment assembly is installed into this opening and the outlet of the assembly (or the disposer) is $1^1/_2$ inches (38 mm) in diameter. The requirement for a strainer or crossbar in the outlet is to catch dishcloths and other large pieces from entering and clogging the waste outlet piping.

418.3 Moveable sink systems. Moveable sink systems shall comply with ASME A112.19.12.

❖ See the commentary to Section 416.4.

SECTION 419
URINALS

419.1 Approval. Urinals shall conform to ASME A112.19.2/CSA B45.1, ASME A112.19.19 or CSA B45.5/IAPMO Z124. Urinals shall conform to the water consumption requirements of Section 604.4. Water-supplied urinals shall conform to the hydraulic performance requirements of ASME A112.19.2/CSA B45.1 or CSA B45.5/IAPMO Z124.

❖ A urinal is a fixture that has been designed specifically for urination; a water closet is primarily designed for defecation. Although urinals are normally associated with men, there have been urinals designed for women. Female urinals have not been popular because of clothing styles, the need for privacy and uncertainty regarding the proper use of the fixture.

Although it appears that the main feature in the design of the urinal was to have the user avoid contact with the fixture surface that can be contaminated, it was actually designed for practicality and speed. The installation of this fixture is the reason why there is frequently no line for the men's room, while a long line gathers for the women's room.

The design and construction of the urinal is regulated by the referenced standards. Trough urinals are prohibited because they do not conform to the referenced standards and are considered unhygienic.

There are four types of water-supplied urinals: stall, blowout, siphon jet and wash down. A stall urinal is floor mounted, with other urinals being predominantly wall hung. Both blowout and siphon-jet urinals flush completely by siphonic action; the contents of the bowl (and integral trap) are completely evacuated during the flushing cycle, and the integral trap of the fixture is refilled by water from the water flush tank or flushometer valve. Stall and wash-down urinals operate by having water calmly cascade down the inside back wall of the fixture to flood the urine in the drain area so that the mixed waste flow is pushed into the trap of the fixture. The contents remaining in the trap for the fixture is a mixture of urine and water. The water consumption limitation for water supplied urinals is 1.0 gallon (3.785 L) per flushing cycle (see commentary, Section 604.4).

Stall urinals require a separate trap to be installed below the floor line. In the past, many wall-hung washout urinals did not typically have an integral trap, but the required trap was a tubular type, exposed under the fixture. The exposed trap was often the source of leaks caused by the normal abuse often occurring in public toilet rooms. Almost all urinal manufacturers now design wall-hung wash-out urinals with an integral trap. While the new installation of a wall-hung, exposed trap wash-out urinal is not prohibited by the code, it is rarely done.

As water conservation has become more important in recent years, manufacturers have responded with waterless urinal designs. Waterless urinals are a popular consideration for sustainable "green" building projects. One design uses a special liquid that is less dense than the urine and the liquid creates the trap seal. Urine passes through the liquid seal and into the drain without affecting the liquid seal. The liquid is held in a special cartridge that is easily replaced for maintenance. Table 709.1 assigns a dfu value of 0.5 to "nonwater-supplied" urinals. Because of the obvious conservation of water, they are popular in areas having water shortages [see Commentary Figures 419.1(1) and (2)].

The absence of flushing water for nonwater-supplied urinals can present some unforeseen problems with respect to the drain piping for these fixtures. Concentrated urine has been known to deposit significant amounts of urea salts in the drain piping, resulting in restrictions of flow. The salts are especially difficult to remove and, in some cases, replacement of the piping is the only solution to the problem. Designers of toilet rooms that are planned to have nonwater-supplied urinals would be wise to arrange such fixtures so that the waste discharge from a water-supplied fixture is located and connected to the drainage piping upstream of the nonwater-supplied fixtures.

Where a wall-hung urinal is to be installed as an accessible fixture, the ICC A117.1-2009 standard requires the urinal be not less than 13 1/2 inches (343 mm) from the front outer rim to the rear wall. This is to allow ease of use by persons sitting in a wheelchair.

Figure 419.1(1)
WATERLESS URINAL
(Photo courtesy of Falcon Waterfree Technologies)

Figure 419.1(2)
WATERLESS URINAL CARTRIDGE
(Photo courtesy of Falcon Waterfree Technologies)

419.2 Substitution for water closets. In each bathroom or toilet room, urinals shall not be substituted for more than 67 percent of the required water closets in assembly and educational *occupancies*. Urinals shall not be substituted for more than 50 percent of the required water closets in all other *occupancies*.

❖ This section allows urinals to be substituted for a maximum of 67 percent of the required number of water closets in assembly and educational occupancies and a maximum of 50 percent in all other occupancies. For example, if five water closets are required in an educational occupancy men's room, the design professional may choose to install two water closets and three urinals, or three water closets and two urinals. If four urinals are installed, however, two water closets are also required. Experience has shown that water closet shortages result with a 67-percent substitution allowance in occupancies other than Groups E and A; thus, a 50-percent maximum allowance was introduced for those occupancies.

[BG] 419.3 Surrounding material. Wall and floor space to a point 2 feet (610 mm) in front of a urinal lip and 4 feet (1219 mm) above the floor and at least 2 feet (610 mm) to each side of the urinal shall be waterproofed with a smooth, readily cleanable, nonabsorbent material.

❖ Wall and floor space to a point 2 feet (610 mm) in front of a urinal lip and 4 feet (1219 mm) above the floor and at least 2 feet (610 mm) to each side of the urinal must be waterproofed with a smooth, readily cleanable, nonabsorbent material.

One of the major problems with the design of a urinal is the splash factor. When a forceful stream of urine hits the wall of the urinal or the water seal, there is a splash that could bounce out of the fixture. To prevent the splash from contaminating the building material surrounding the urinal, smooth, waterproof, nonabsorbent material is required on the surfaces most likely to be splashed. The finished material will allow for easy cleaning of the area, thus preventing any bacterial growth (see Commentary Figure 419.3). Note that there are requirements for urinal partitions in Section 310.5.

Section 1210.2 of the IBC also stipulates that the material be "hard." The intent of this section and the IBC section is that the surrounding material should not be of a type that is easily damaged such that the material could become absorbent and thus, not readily cleanable (and sanitized). Painted drywall, even where coated with highly durable paints, does not meet the intent of these sections. The typical materials used for this application are ceramic tile, fiberglass reinforced plastic (FRP) panels, phenolic panels, laminated plastic panels, painted steel and stainless steel. Painted concrete block and poured concrete walls that have had holes and cracks filled and smoothed with block filler material have also been used.

FIXTURES, FAUCETS AND FIXTURE FITTINGS

Figure 419.3
URINAL SURROUNDING MATERIAL

SECTION 420
WATER CLOSETS

420.1 Approval. Water closets shall conform to the water consumption requirements of Section 604.4 and shall conform to ASME A112.19.2/CSA B45.1, ASME A112.19.3/CSA B45.4 or CSA B45.5/IAPMO Z124. Water closets shall conform to the hydraulic performance requirements of ASME A112.19.2/CSA B45.1. Water closet tanks shall conform to ASME A112.19.2/CSA B45.1, ASME A112.19.3/CSA B45.4 or CSA B45.5/IAPMO Z124. Electro-hydraulic water closets shall comply with ASME A112.19.2/CSA B45.1. Water closets equipped with a dual flushing device shall comply with ASME A112.19.14.

❖ The most recognizable and most used plumbing fixture is the water closet. Material and performance requirements are included in the referenced standards for water closets.

A water closet is referred to by many other names, including toilet and commode. The term "water closet" remains the technical expression. The term "water closet" originates from the first attempt to bring plumbing indoors. The human elimination process was done in a closet that contained a cistern. The closet was shaped like a chair to accommodate a person in the squatting or sitting position. The expression was soon associated with the room because it was the size of a closet. The term, however, applies today as the name of the fixture itself.

The most common word used in the United States to identify a water closet is "toilet." The word "toilet" is derived from the French word "*toilette*," meaning "dressing." The word "toilet" was considered more socially acceptable than the term "water closet." The word "toilet" more accurately describes the room in which water closets are installed.

The water closet is the primary fixture responsible for the modernization of plumbing. The fixture effectively disposes of human waste in a sanitary manner.

Water closets must be of the water-conservation type. The applicable fixture standards require a maximum average of not greater than 1.6 gallons (6.1 L) of water per flushing cycle. No cycle during the testing can exceed 2.0 gallons (7.6 L) per flush (see Section 604.4). The exceptions to Section 604.4 give the cases where use of a water closet with an increased water usage per flush, up to a maximum of 3.5 gallons (13 L) per flushing cycle, is permitted. To determine conformance, a representative model of the water closet must be tested for its flushing efficiency. The flushing test uses polypropylene balls and polyethylene granule discs. There is also a test for the ability to remove an ink stain on the inside surface of the water closet bowl, and a dye test measuring the exchange rate of water in the bowl. Another test measures the capacity of the drain line attached to the water closet to carry away the solids from the bowl following flushing.

There are three styles of water closets: close coupled, one piece and those using a flushometer valve. The close-coupled water closet has a bowl and separate gravity-type tank or flushometer tank that is supported by the bowl. A one-piece water closet is constructed with the gravity-type tank or flushometer tank and the bowl as one integral unit. A flushometer valve water closet is a bowl with a flushometer valve.

Water closet bowls come in six styles: blowout, siphon jet, reverse trap, washdown, siphon vortex and siphon wash. The following provides a brief description of each style:

Blowout: A blowout water closet must have a flushometer valve because the flushing relies on the higher pressure of the entering water. The water pressure forces the bowl contents through the trapway with little reliance on siphonic action. A blowout is the noisiest of the water closets.

Siphon jet: The siphon-jet bowl has a jet of water entering the bottom of the bowl directed into the trapway. Water enters through the rim of the bowl creating the siphon action started by the trapway opening. The water rises in the bowl until the siphon action draws the contents out of the bowl.

Reverse trap: The reverse trap is similar to the siphon-jet bowl. The difference between the two fixtures is the water surface area and the depth of the trap seal. The reverse-trap bowl is smaller in area with less of a trap seal.

Washdown: The washdown bowl has the trapway in the front of the bowl with a small water surface area. The siphon action is started by the water entering the rim.

Siphon vortex: The siphon vortex is a one-piece water closet. The water entering the rim at an angle creates a swirling vortex in the center of the bowl. In some designs, there is a flush-water opening in the bowl near the trap opening that assists in the forming of the vortex. There is usually no flush-water opening in the trapway. The siphon action is created only by the water entering the rim of the bowl.

Siphon wash: The siphon wash is designed for maximum water conservation. The trap seal depth and water surface area are the smallest of all the designs. The flushing action is similar to the washdown principle.

Commentary Figure 420.1(1) shows a cutaway of four of the styles of water closet bowls. The top left is a siphon jet and the top right is a reverse trap. The bottom left is a blowout and the bottom right is a washdown [see Commentary Figure 420.1(2)].

The code does not specify the rim height of water closet bowls unless required to accommodate persons with disabilities (see commentary, Section 404). Although the referenced standards specify a bowl rim height between 14 and 15 inches (356 mm to 381 mm) above the floor, these standards also permit other rim heights for water closets intended for use by children, juveniles, the elderly and persons with disabilities [see Commentary Figure 420.1(3)].

Dual flushing devices for water closets have become a popular water closet design feature to achieve low-rated flush volumes. Dual-flush water closets offer a flushing mechanism that allows the user to choose between a standard flush volume (equal to or less than the federal and code-mandated maximum flush volume of 1.6 gallons (6 L) per flush) and a lower flush volume. Lower flush volumes can be used for flushing bowl contents that do not have solids (except for toilet paper), and the standard flush volume is used for flushing bowl contents having solids. Overall, dual-flush water closets can reduce water usage in a building. Some green building codes such as the *International Green Construction Code*® (IgCC®) require the installation of dual-flush water closets rated for 1.28 gallons (4.9 L) per flush. The rating of dual-flush water closets is the average of two low-flush volumes plus one standard-flush volume. The standard covering dual-flush water closets has been added to the IPC so that designers, installers and code officials will be ensured that these products will work as intended.

420.2 Water closets for public or employee toilet facilities. Water closet bowls for *public* or employee toilet facilities shall be of the elongated type.

❖ An elongated water closet bowl is required for public or employee use because it gives the user an extended area in which to eliminate human waste. The concept is to reduce the possibility of the user missing the bowl and soiling the water closet seat and surroundings.

The elongated water closet bowl also has a larger water surface area, which helps keep the inside surface of the bowl clean. An elongated water closet bowl is approximately 2 inches (51 mm) longer than a regu-

Figure 420.1(1)
BOWL DESIGN

Figure 420.1(2)
FLUSHING ACTION FOR CERTAIN WATER CLOSETS

Figure 420.1(3)
DETERMINING WATER CLOSET RIM HEIGHT

FIXTURES, FAUCETS AND FIXTURE FITTINGS

lar round bowl. The bowl extension is toward the front of the fixture. Elongated bowls are not required for various nonpublic or nonemployee uses, including hotel guestrooms and private facilities.

420.3 Water closet seats. Water closets shall be equipped with seats of smooth, nonabsorbent material. All seats of water closets provided for *public* or employee toilet facilities shall be of the hinged open-front type. Integral water closet seats shall be of the same material as the fixture. Water closet seats shall be sized for the water closet bowl type.

❖ A water closet seat must be designed to facilitate cleaning and prevent the accumulation of dirt or bacterial growth. The seat must also be designed for the bowl on which it is installed.

The requirement for a hinged open-front seat was originally to aid women. The open seat, however, is equally beneficial to men. The open seat allows women to wipe the perineal area after using the water closet, while it also eliminates an area that could be contaminated with urine when used by men. The open-front design also eliminates the user's genital contact with the seat.

420.4 Water closet connections. A 4-inch by 3-inch (102 mm by 76 mm) closet bend shall be acceptable. Where a 3-inch (76 mm) bend is utilized on water closets, a 4-inch by 3-inch (102 mm by 76 mm) flange shall be installed to receive the fixture horn.

❖ A 4-inch by 3-inch (102 mm by 76 mm) closet bend is not considered a reduction in drainage pipe size (see Commentary Figure 420.4 and commentary, Section 704.2).

For SI: 1 inch = 25.4 mm.

**Figure 420.4
ACCEPTABLE 4 × 3 CLOSET BENDS**

SECTION 421
WHIRLPOOL BATHTUBS

421.1 Approval. Whirlpool bathtubs shall comply with ASME A112.19.7/CSA B45.10 and shall be listed and labeled in accordance with UL 1795.

❖ This section of the code provides material and performance requirements for bathtubs in which there is continuous movement of water. Installation requirements along with drain and suction fitting requirements are also provided.

ASME A112.19.7/CSA B45.10 was developed because of an increase in the use of whirlpool bathtubs. This standard regulates fixture design and construction, installation instructions and field-testing requirements. Whirlpool bathtubs must conform to this standard.

Listing and labeling of whirlpool tubs are necessary for electrical safety. Any electrical appliance connected to a system that conveys water or liquid waste has the potential to misdirect electrical current to appliance housings and onto other conductive paths that could come in contact with humans. The requirement for listing and labeling whirlpool tubs to a standard ensures that all necessary precautions have been taken by the manufacturer to protect users of these appliances from hazards.

421.2 Installation. Whirlpool bathtubs shall be installed and tested in accordance with the manufacturer's instructions. The pump shall be located above the weir of the fixture trap.

❖ The referenced standard requires the manufacturer to indicate that each whirlpool bathtub must be filled with water and tested during the rough-in inspection. This is necessary to determine whether there are any leaks in the factory-piped whirlpool lines that might have been caused by shipping, handling or installation.

The pump, being a mechanical device, must be accessible to facilitate any repairs, maintenance or replacement. The pump (circulation) must be located above the tub trap weir to verify that the pump body drains each time the tub is drained.

421.3 Drain. The pump drain and circulation piping shall be sloped to drain the water in the volute and the circulation piping when the whirlpool bathtub is empty.

❖ The whirlpool tub, suction lines and pump must be designed to drain completely after each use. This improves the sanitation of a whirlpool bathtub. Wastewater left standing in the pump body or other waterways will promote the growth of bacteria.

421.4 Suction fittings. Suction fittings for whirlpool bathtubs shall comply with ASME A112.19.7/CSA B45.10.

❖ ASME A112.19.7/CSA B45.10 was developed to regulate suction fittings on whirlpool bathtubs. The standard contains performance criteria for suction fittings. The purpose of the standard is to prevent the use of a fitting that may create a potential hazard for hair or body entrapment.

For the 2012 code edition, the standard in this section was updated because of a harmonization of standards that occurred since the last code edition. For more information on the harmonization of standards in the code, see the commentary to Section 405.9.

421.5 Access to pump. *Access* shall be provided to circulation pumps in accordance with the fixture or pump manufacturer's installation instructions. Where the manufacturer's instructions do not specify the location and minimum size of

FIXTURES, FAUCETS AND FIXTURE FITTINGS

field-fabricated access openings, an opening not less than 12 inches by 12 inches (305 mm by 305 mm) shall be installed to provide *access* to the circulation pump. Where pumps are located more than 2 feet (609 mm) from the access opening, an opening not less than 18 inches by 18 inches (457 mm by 457 mm) shall be installed. A door or panel shall be permitted to close the opening. In all cases, the access opening shall be unobstructed and of the size necessary to permit the removal and replacement of the circulation pump.

❖ A whirlpool pump must be accessible for removal for repairs/replacement. The opening is needed for access to the mounting bolts, pipe unions and electrical connections to physically remove the pump without damaging it or the surrounding finished walls or ceiling. The larger opening is required when the pump is located deeper in the enclosure because the service personnel will have to position more of their body in the opening to reach and remove the pump. Note that the whirlpool tub manufacturer may dictate specific access requirements and such requirements prevail over what is specified in this section [see Commentary Figures 421.5(1) and (2)].

421.6 Whirlpool enclosure. Doors within a whirlpool enclosure shall conform to ASME A112.19.15.

❖ See the commentary to Section 407.4.

SECTION 422
HEALTH CARE FIXTURES AND EQUIPMENT

422.1 Scope. This section shall govern those aspects of health care plumbing systems that differ from plumbing systems in other structures. Health care plumbing systems shall conform to the requirements of this section in addition to the other requirements of this code. The provisions of this section shall apply to the special devices and equipment installed and maintained in the following *occupancies*: nursing homes, homes for the aged, orphanages, infirmaries, first aid stations, psychiatric facilities, clinics, professional offices of dentists and doctors, mortuaries, educational facilities, surgery, dentistry, research and testing laboratories, establishments manufacturing pharmaceutical drugs and medicines and other structures with similar apparatus and equipment classified as plumbing.

❖ In health care facilities, there are plumbing fixtures that have special requirements because of their unique application or use. This section outlines occupancies where special health care plumbing devices and equipment may be installed. Although these fixtures are typically found in hospitals, they might be located in other health care facilities.

Note that the list of occupancies is not all inclusive. It is intended only to be representative of the occupan-

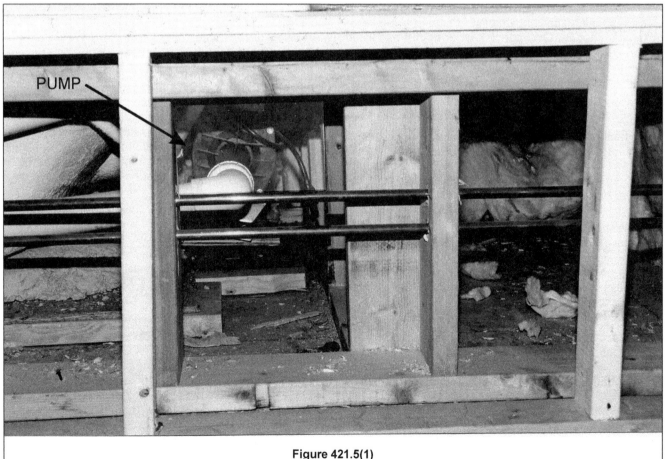

Figure 421.5(1)
NONCOMPLIANT WHIRLPOOL BATHTUB PUMP ACCESS

FIXTURES, FAUCETS AND FIXTURE FITTINGS

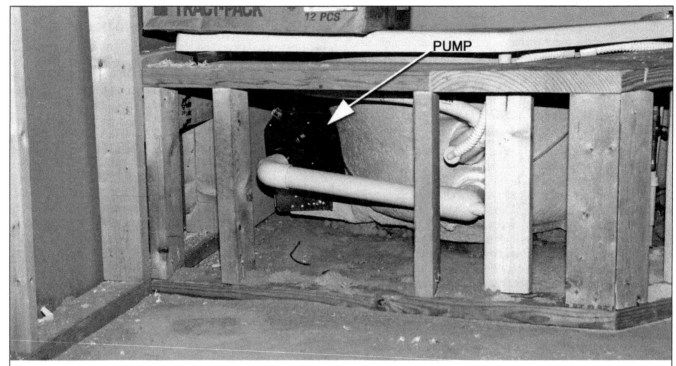

Figure 421.5(2)
NONCOMPLIANT WHIRLPOOL BATHTUB PUMP ACCESS

cies in which special plumbing fixtures are regulated by this chapter. Occupancies other than those listed can be equipped with some of these special fixtures. When this occurs, the requirement shown here would apply to those occupancies.

422.2 Approval. All special plumbing fixtures, equipment, devices and apparatus shall be of an *approved* type.

❖ Because many health care plumbing fixtures are not regulated by any fixture standards, each must be evaluated and approved by the code official. The code official should consider backflow prevention and the ability to maintain sanitary conditions of the special fixtures.

422.3 Protection. All devices, appurtenances, appliances and apparatus intended to serve some special function, such as sterilization, distillation, processing, cooling, or storage of ice or foods, and that connect to either the water supply or drainage system, shall be provided with protection against backflow, flooding, fouling, contamination of the water supply system and stoppage of the drain.

❖ This section requires protection of fixtures, appliances and equipment against contamination resulting from backflow. The water supply system must also be protected against contamination resulting from backflow from such fixtures, appliances and equipment.

This section combines the intent of Sections 608, 609 and 802.

422.4 Materials. Fixtures designed for therapy, special cleansing or disposal of waste materials, combinations of such purposes, or any other special purpose, shall be of smooth, impervious, corrosion-resistant materials and, where subjected to temperatures in excess of 180°F (82°C), shall be capable of withstanding, without damage, higher temperatures.

❖ This section requires the materials used for special plumbing fixtures to be cleanable and capable of resisting staining, corrosion and the accumulation of waste in any pores or other minute openings in the finish of the material. The material used for the fixture must also be capable of withstanding the temperatures to which it will be exposed during normal service conditions.

422.5 Access. *Access* shall be provided to concealed piping in connection with special fixtures where such piping contains steam traps, valves, relief valves, check valves, vacuum breakers or other similar items that require periodic inspection, servicing, maintenance or repair. *Access* shall be provided to concealed piping that requires periodic inspection, maintenance or repair.

❖ Any plumbing fixture, appliance or device having mechanical components must be accessible to facilitate any needed repair or maintenance. A mechanical component is always prone to failure and, therefore, access is required.

422.6 Clinical sink. A clinical sink shall have an integral trap in which the upper portion of a visible trap seal provides a water surface. The fixture shall be designed so as to permit complete removal of the contents by siphonic or blowout action and to reseal the trap. A flushing rim shall provide water to cleanse the interior surface. The fixture shall have the flushing and cleansing characteristics of a water closet.

❖ A clinical sink is the main special fixture used in health care facilities. It is designed to both wash a bedpan and remove its contents; therefore, the other name for a clinical sink is a bedpan washer.

A clinical sink is like a water closet in that it has a flushing rim, which washes the interior walls of the fixture and discharges urine and fecal matter by siphonic action.

A hose or washing arm is provided with a clinical sink to wash the contents of the bedpan into the fixture. In some installations, a water closet functions as both a clinical sink and a water closet. The common use of the dual-purpose fixture reduces the number of plumbing fixtures required in a health care facility.

422.7 Prohibited usage of clinical sinks and service sinks. A clinical sink serving a soiled utility room shall not be considered as a substitute for, or be utilized as, a service sink. A service sink shall not be utilized for the disposal of urine, fecal matter or other human waste.

❖ A service sink is not designed for the elimination of human waste, nor does it have a flushing rim to clean the fixture with each use. By not completely eliminating the waste material via a flushing method, a service sink could allow bacteria to grow and infect the surrounding areas. A service sink and a clinical sink, therefore, must not be interchanged.

422.8 Ice prohibited in soiled utility room. Machines for manufacturing ice, or any device for the handling or storage of ice, shall not be located in a soiled utility room.

❖ In rooms where fecal matter is eliminated, biological aerosols are released into the air, making contact with exposed surfaces. The feces of an unhealthy person could also contain pathogens that may release into the air.

To avoid the possible contamination of fluid that is to be ingested, locating ice machines in the same room as a clinical sink is prohibited. For the same reason, installation of drinking fountains in toilet rooms is prohibited.

422.9 Sterilizer equipment requirements. The approval and installation of all sterilizers shall conform to the requirements of the *International Mechanical Code*.

❖ The *International Mechanical Code*® (IMC®) regulates the installation of appliances, including sterilizers. The appliance must bear the label of an approved agency and be installed in accordance with the manufacturer's instructions. If a sterilizer is gas fired, the *International Fuel Gas Code*® (IFGC®) specifies the requirements for gas piping, combustion air and venting.

422.9.1 Sterilizer piping. *Access* for the purposes of inspection and maintenance shall be provided to all sterilizer piping and devices necessary for the operation of sterilizers.

❖ The term "access" is defined in Chapter 2. Access to piping and devices associated with sterilizers is essential to allow for periodic inspection, maintenance and repair.

422.9.2 Steam supply. Steam supplies to sterilizers, including those connected by pipes from overhead mains or branches, shall be drained to prevent any moisture from reaching the sterilizer. The condensate drainage from the steam supply shall be discharged by gravity.

❖ Because condensate could contain contaminants that interfere with the operation of the sterilizer, it must not be allowed to enter the sterilizer. The condensate must be drained by gravity, which is the most dependable method. Drainage of the condensate also reduces the likelihood of it entering the sterilizer.

422.9.3 Steam condensate return. Steam condensate returns from sterilizers shall be a gravity return system.

❖ Steam condensate returns from sterilizers are required to drain by gravity to reduce the likelihood of condensate backing up into a sterilizer, which could violate the sterile conditions.

422.9.4 Condensers. Pressure sterilizers shall be equipped with a means of condensing and cooling the exhaust steam vapors. Nonpressure sterilizers shall be equipped with a device that will automatically control the vapor, confining the vapors within the vessel.

❖ To avoid the release of steam into the space that houses the sterilizer, units that operate above atmospheric pressures must condense the exhaust steam vapors. For the same reason, sterilizers that operate at approximately atmospheric pressure must confine the steam vapors within the vessel.

422.10 Special elevations. Control valves, vacuum outlets and devices protruding from a wall of an operating, emergency, recovery, examining or delivery room, or in a corridor or other location where patients are transported on a wheeled stretcher, shall be located at an elevation that prevents bumping the patient or stretcher against the device.

❖ Many of the control devices used in health care plumbing systems protrude from the wall. They must be located in a way to prevent any accidental contact by a patient in a wheelchair or stretcher. Besides protecting an individual from hitting any control device, it also protects the device from any possible damage.

SECTION 423
SPECIALTY PLUMBING FIXTURES

423.1 Water connections. Baptisteries, ornamental and lily pools, aquariums, ornamental fountain basins, swimming pools, and similar constructions, where provided with water supplies, shall be protected against backflow in accordance with Section 608.

❖ This section addresses the myriad of specialty and decorative items that are not commonly thought of as plumbing fixtures. The most important requirement for these types of construction is to prevent water supply contamination. Note that Section 423.2 requires that specialty fixtures be approved by the code official. Requirements for specialty fixture materials are located in Section 402.2. Baptismal pools, ornamental ponds and fountains, swimming pools, specialty food service equipment, dispensers, pedicure chairs and ice makers are examples of items and equipment that, although not classified as plumbing fixtures, require a water supply connection or waste connection.

The majority of these "special fixtures" are not governed by standards; therefore, extraordinary precautions must be taken to verify that the potable water supply is not endangered. The code official must evaluate each installation independently to secure compliance with Section 608 (see commentary, Section 608).

A pedicure chair (having integral foot bathtub) that is permanently connected to the water supply and drain system is good example of a specialty plumbing fixture. While these factory-assembled units have water and waste connections and are conceptually similar to some appliances, such as a whirlpool tub, they might not be designed or tested to any plumbing code referenced standard. Therefore, the code official must review the equipment information prior to approving such fixtures for installation so he or she can advise the installer on special requirements for protecting the potable water supply from contamination.

423.2 Approval. Specialties requiring water and waste connections shall be submitted for approval.

❖ Fixtures or devices not conforming to referenced standards must be approved by the code official as an alternative material in accordance with Section 105.2. Water supplies serving such fixtures or devices must be protected against backflow to prevent contamination of the potable water supply (see Section 423.1).

423.3 Footbaths, pedicure baths and head shampoo sinks. The water supplied to specialty plumbing fixtures, such as pedicure chairs having an integral foot bathtub, footbaths, and head shampoo sinks, shall be limited to a maximum temperature of 120°F (49°C) by a water temperature limiting device that conforms to ASSE 1070 or CSA B125.3.

❖ Water heaters in establishments that include pedicure chairs with integral foot bathtubs, footbaths and head shampoo sinks are often overtaxed during periods of high usage. Because client satisfaction requires the use of very warm water, owners/operators of these establishments sometimes adjust water heater thermostats to high temperature settings to use less hot water volume for the desired mixed water temperatures. Clients' feet or heads could be inadvertently exposed to high water temperatures (greater than 120°F) that could cause skin burns. Temperature-limiting devices on the water supply to these specialty plumbing fixtures will prevent such exposures.

SECTION 424
FAUCETS AND OTHER FIXTURE FITTINGS

424.1 Approval. Faucets and fixture fittings shall conform to ASME A112.18.1/CSA B125.1. Faucets and fixture fittings that supply drinking water for human ingestion shall conform to the requirements of NSF 61, Section 9. Flexible water connectors exposed to continuous pressure shall conform to the requirements of Section 605.6.

❖ The indicated standard regulates plumbing supply fittings located between the supply line stop and the terminal fitting, inclusive. The terminal fitting is the fitting that has an open or atmospheric discharge (e.g., a faucet or hose bibb). This standard covers automatic compensating valves for individual wall-mounted showering systems, bath and shower supply fittings, bidet supply fittings, clothes washer supply fittings, drinking fountain supply fittings, humidifier supply stops, kitchen sink and lavatory supply fittings, laundry tub supply fittings, lawn and sediment faucets, metering and self-closing supply fittings and supply stops. The standard does not apply to pipes, tubes or the fittings for pipes and tubes.

The main focus of this standard is to ensure that mounting dimensions and connection interchangeability exist among manufacturers of these types of plumbing products. The standard also covers all of the quality requirements and uniform test methods including but not limited to, strength (pressure and structural), finish durability, operational life, back flow protection, force/torque to operate, and maximum leakage amounts for the fittings.

The second line of this code section requires that faucets and fittings that supply drinking water for human ingestion conform to the requirements of Section 9 of NSF 61. This is also a requirement of the ASME A112.18.1/CSA B125.1 standard. It is often assumed that because a potable water distribution system conveys potable water, water discharged from any point in the system is of drinking water quality. However, this is not the case as the following explanation of the Safe Drinking Water Act (SDWA) and NSF 61 will clarify.

In 1974, the SDWA was enacted by the United States Federal government to minimize chemical and bacterial contamination of drinking water. The Act authorized the U.S. Environmental Protection Agency (EPA) to set national health-based standards for drinking water to protect the public from naturally occurring and man-made contaminants. In 1986, an amendment to the SDWA addressed concerns for lead contamination of drinking water by requiring that all pipes, solders, pipe fittings and plumbing fixtures, intended to provide water for human consumption and sold after July 1986, be lead-free. Lead-free in this context is defined to mean that solder and flux cannot contain more than 0.2-percent lead and that pipes, fittings and fixtures cannot contain more than 8.0-percent lead (see commentary Sections 605.2 and 605.14.3).

Because studies indicated that plumbing (supply) fittings and fixtures within the last liter volume of the water discharge point were significant contributors of contaminants (especially lead) in dispensed water, the 1986 amendment to the SDWA also required that by August 1997, a set of voluntary consensus standards and testing protocols be developed for these products. In preparation for meeting the August 1997 deadline, the initial standard, NSF 61, *Drinking Water System Components—Health Effects*, was voluntarily developed in 1988 by a consortium of National Sanitation Foundation (NSF) International, the American Water Works Association and the Association of State Drinking Water Administrators with support from the U.S. EPA. In 1994, Section 9 of NSF 61 was completed to in order to comply with the August 1997 deadline.

The reader might question why plumbing (supply) fittings and fixtures within the last liter volume of the discharge point are required to comply with more stringent standards beyond that of a maximum lead content of 8 percent. Because the majority of plumbing (supply) fittings and fixtures are made from copper and brass alloys, the casting, forging, machining and finishing processes to produce those products can cause the lead in the alloys to be much more "leachable" than the lead in pipe and pipe fitting products. If water remains at rest in the plumbing (supply) fittings and fixtures for a long period of time such as might happen overnight, the next draw of water from the plumbing (supply) fittings or fixtures often contains a lead concentration much higher than the maximum allowable lead concentration established by the U.S. EPA for safe drinking water.

Section 9 of NSF 61 only covers plumbing (supply) fittings and fixtures intended by the manufacturer to supply water for human consumption. Examples are kitchen faucets, lavatory faucets, bar faucets, drinking fountains, residential icemakers, hot/cold water dispensers and commercial kitchen pot fillers. Section 9 of NSF 61 specifically excludes certain "end point" devices such as bath/shower valves, utility faucets, laboratory faucets, bidet controls, any faucets with hose thread fittings at the discharge point, nonlavatory hand wash stations, and faucets that are self-closing, metering or electronically activated. In summary, any end point device where the water is specifically intended to be drawn for consumption by humans must be certified to Section 9 of NSF 61.

The last line of this code section refers the reader to another section that specifically covers flexible water connectors as the requirements of this section do not apply to that type of product.

424.1.1 Faucets and supply fittings. Faucets and supply fittings shall conform to the water consumption requirements of Section 604.4.

❖ Faucets are the end point devices that allow the user to turn the water flow on and off, regulate the flow and in some cases, regulate the temperature at the point of use. A standard lavatory faucet is an example of an end point device that must have a water discharge rate of no greater than the maximum stated in Section 604.4. Faucets that are required to have the flow rate limited, utilize a removable flow restricting device at the outlet of the faucet, commonly called an aerator. Not all of these devices "aerate" (entrain air in) the flow stream.

The original use of aerators was to reduce the degree of flow stream splashing as it impacted the basin or objects being washed. Entrained air "softens" the impact of the flow stream. However, as water conservation was mandated (see Section 604.4), aerators took on the more important role of limiting flow from faucets and remain today as the key water flow limiting device for most faucets.

As stated previously, not all "aerators" entrain air into the flow stream. A common example is a laminar flow device found in many medical and dental faucet applications. In these environments, bacteria-laden room air entrained into the flow stream by an aerator is detrimental to infection control efforts. Since laminar flow devices do not entrain air into the flow stream, bacterial infection is reduced. Further towards the effort of infection control, some laminar flow devices now include minute amounts of silver (Ag), which has been proven to prevent bacteria growth within the laminar flow device during periods of nonuse.

In applications where water flow from faucets is significantly reduced such as in public lavatories [limited to 0.5 gallon per minute (1.89 L/m)], an aerated or laminar flow stream does not provide enough force to allow the user to effectively remove soap and dirt from the hands. A nonaerated spray pattern device is used that produces numerous small but powerful streams for better cleaning [see Commentary Figures 424.1.1(1) and (2)].

Supply fittings, as referenced in this section, are those devices downstream of faucets or control valves. Examples are shower heads and hand-held showers. Water conservation in these supply fittings is typically accomplished with internal flow restricting devices.

424.1.2 Waste fittings. Waste fittings shall conform to ASME A112.18.2/CSA B125.2, ASTM F409 or to one of the standards listed in Tables 702.1 and 702.4 for above-ground drainage and vent pipe and fittings.

❖ The indicated standards cover plumbing waste fittings of sizes 2 inches (51 mm) nominal and smaller and

Figure 424.1.1(1)
AERATED STREAM AERATOR
(Photo courtesy of Neoperl, Inc.)

Figure 424.1.1(2)
SPRAY AERATOR
(Photo courtesy of Neoperl, Inc.)

include traps, tailpieces, trap wall adapters, diversion tees, twin waste elbows, overflow elbows, drain elbows and strainer bodies of various materials, including plastic.

ASTM F409 covers waste fittings made of plastic materials only.

The main focus of these standards is to assure that mounting dimensions and connection interchangeability exist among manufacturers of these types of plumbing products. The standard also covers all of the quality requirements and uniform test methods including but not limited to, strength (pressure and structural), finish, durability and maximum leakage amounts for the fittings.

424.2 Hand showers. Hand-held showers shall conform to ASME A112.18.1/CSA B125.1. Hand-held showers shall provide backflow protection in accordance with ASME A112.18.1/CSA B125.1 or shall be protected against backflow by a device complying with ASME A112.18.3.

❖ Because it is possible for a hand-held shower to be submerged in the bathtub or shower compartment base, a cross connection to waste water could be created. The hand-held shower must have adequate protection against backflow. The backflow protection may be integral to the hand-held shower or may be provided by an external backflow device that meets ASME A112.18.3. These types of showers are commonly installed in accessible shower enclosures. Their use, however, has increased in popularity in showers within a dwelling unit. The referenced standards specify the requirements for backflow protection of the hand-held shower.

A hand-held shower, used in a hand-held mode, is not considered to be the type of shower that requires a means to protect the user from scalding. The reason for this comes from the justifications that eventually led to the code requirement for anti-scald shower valves (see commentary, Section 424.3). Therefore, if the hand-held shower is mounted for use in a hands-free mode with the user in a standing position, a hand-held shower is no different than any other fixed-type shower head and must be supplied through an anti-scald shower valve in accordance with Section 424.3.

However, where a hand-held shower is used in a hand-held mode, most users, when subjected to a high temperature surge of water spray, will either drop or simply redirect the hand-held shower. If the user is standing, the potential for a fall injury is minimal. If the user is sitting, fall injuries cannot occur. Therefore, where a hand-held shower is used only in a hand-held mode, the water discharged from it is not required to be supplied through an antiscald shower valve.

For the 2012 code edition, all the standards in this section, with the exception of ASME A112.18.3, were updated because of a harmonization of standards that occurred since the last code edition. For more information on the harmonization of standards in the code, see the commentary to Section 405.9.

424.3 Individual shower valves. Individual shower and tub-shower combination valves shall be balanced-pressure, thermostatic or combination balanced-pressure/thermostatic valves that conform to the requirements of ASSE 1016/ASME A112.1016/CSA B125.16 or ASME A112.18.1/CSA B125.1 and shall be installed at the point of use. Shower and tub-shower combination valves required by this section shall be equipped with a means to limit the maximum setting of the valve to 120°F (49°C), which shall be field adjusted in accordance with the manufacturer's instructions. In-line thermostatic valves shall not be utilized for compliance with this section.

❖ Commentary Figure 424.3 shows exposure times at various water temperatures to cause third-degree burns to human skin. Industry safety experts have chosen 120°F (49°C) to be the maximum safe water for bathing purposes. In a showering application where the shower head is fixed, a user subjected to a surge of high temperature water usually reacts by abruptly moving away from the hot spray. This abrupt movement often causes falls resulting in the stunned user not being able to escape the scalding water stream. Children and some elderly persons often have delayed or no physical reaction to the scalding water such that they continue to stand in the stream. Therefore, showers and combination tub/showers must have a shower control valve that is capable of protecting an individual from being exposed to water temperatures in excess of 120°F (49°C). The control valve must be installed at the point of use. In other words, the person in the shower must have access to the control handle(s) of the valve.

There are two types of events that the control valve must be designed to protect against: 1) extreme temperature fluctuations from the user set temperature caused by changes in hot or cold water distribution line pressures and 2) extreme temperature conditions caused by the user either purposely or accidentally adjusting the control valve to deliver the hottest water available from the hot water distribution system. Where water inlet pressures or temperatures fluctuate during shower use, control valves complying with the standard must automatically and rapidly adjust to maintain the water discharge temperature to ±3.6°F (±2°C) of the user selected temperature.

The standards also require that shower control valves have a maximum temperature limit device that is not adjustable by the user at the point of use. The high-limit stop is typically an adjustable set screw or cam that is manually set to limit the travel of the control valve handle. The high-limit stop must be field adjusted at the time of installation to limit the delivered water temperature to a maximum of 120°F (49°C). Most control valve manufacturer's installation instructions require that a thermometer be used to verify the maximum discharge temperature from the valve.

The three types of shower control valves available are pressure balancing, thermostatic and combination thermostatic/pressure balancing. A pressure-balancing valve senses changes in pressure of the hot and cold water supplies (up to 50-percent pressure change). If the pressure on one side changes, the valve reacts so that the flow from each side of the valve is adjusted to maintain the user's temperature selection. The thermostatic valve senses the discharge water temperature and adjusts flows of supply water to the valve to maintain the user set temperature. A thermostatic valve provides some limited protection against hot and cold supply pressure fluctuations (up to 20-percent pressure change). A combination thermostatic/pressure-balancing valve adjusts to changes in supply pressures (up to 50-percent pressure change) and discharge temperature.

Pressure-balancing control valves are used in a majority of shower applications. One slight disadvantage of the pressure balanced-type control valve is that any change in the temperature setting of the water heating system will affect the maximum discharge temperature available from the control valve. If the system temperature is lowered, some users might complain that they are not receiving hot enough water. If the system temperature is raised, the maximum water temperature available from the valve might be in excess of 120°F (49°C). Seasonal fluctuations in the cold water supply temperature can also affect the maximum discharge temperature. Thus, pressure balancing-type control valves might require periodic readjustment of the high-limit setting.

The maximum discharge temperature from a thermostatic-type control valve is not affected by changes in the temperatures of the hot and cold water supplies to the valve except for when the hot water supply temperature falls below the high-limit setting. Although this type of valve might significantly reduce the need for readjustments to the high-limit setting, due to its inherent limited protection against supply side pressure changes, a thermostatic valve might not be suitable for applications where large pressure fluctuations (greater than a 20-percent change in pressure) are expected.

A combination pressure balanced/thermostatic control valve offers the greatest degree of protection for the shower user. As the name implies, this type of valve offers the best features of both the pressure balanced-type valve and the thermostatic-type valve designs.

This section does not apply to control valves that supply water only for filling tubs (see Section 424.5) or emergency showers (see Section 411). Inline-type thermostatic and pressure balancing valves do not satisfy the requirements of this section because they are not the final control valve that the user adjusts at the point of use and they do not have a high-limit setting feature.

FIXTURES, FAUCETS AND FIXTURE FITTINGS

TIME AND TEMPERATURE RELATIONSHIP TO SERIOUS BURNS		
	Time required for a third-degree burn to occur	
Water temperature	Adults (skin thickness of 2.5 mm)	Children (skin thickness of 56 mm)
155°F (68°C)	1 second	0.5 second
148°F (64°C)	2 seconds	1 second
140°F (60°C)	5 seconds	1 second
133°F (56°C)	15 seconds	4 seconds
127°F (53°C)	1 minute	10 seconds
124°F (51°C)	3 minutes	1.5 minutes
120°F (49°C)	5 minutes	2.5 minutes
100°F (38°C)	Safe temperature for bathing	Safe temperature for bathing

**Figure 424.3
TEMPERATURE BURN CHART**

424.4 Multiple (gang) showers. Multiple (gang) showers supplied with a single-tempered water supply pipe shall have the water supply for such showers controlled by an *approved* automatic temperature control mixing valve that conforms to ASSE 1069 or CSA B125.3, or each shower head shall be individually controlled by a balanced-pressure, thermostatic or combination balanced-pressure/thermostatic valve that conforms to ASSE 1016/ASME A112.1016/CSA B125.16 or ASME A112.18.1/CSA B125.1 and is installed at the point of use. Such valves shall be equipped with a means to limit the maximum setting of the valve to 120°F (49°C), which shall be field adjusted in accordance with the manufacturers' instructions.

❖ In a shower facility that has multiple shower areas where all shower heads are supplied with tempered water from a single source (a master valve), the tempered water supply must be controlled by an automatic temperature control mixing valve conforming to either of the referenced standards. Master thermostatic mixing valves must be sized to meet the peak demand of showers located downstream of the valve so that there is sufficient water volume during peak usage, and more importantly, so the master thermostatic mixing valve works properly. Regardless of shower control valve operation or supply pressure fluctuations, the user will be protected from scalding injury because the temperature of the water supplied to the shower will not exceed 120°F (49°C) [see Commentary Figures 424.4(1) and 424.4(2)].

In lieu of using a master thermostatic valve, each shower head must have a control valve at the point of use that is capable of protecting an individual from being scalded when taking a shower.

The water heater thermostat cannot be used instead of a master thermostatic valve or individually protected showers to limit water temperature to 120°F (49°C).

424.5 Bathtub and whirlpool bathtub valves. The *hot water* supplied to bathtubs and whirlpool bathtubs shall be limited to a maximum temperature of 120°F (49°C) by a water-temperature limiting device that conforms to ASSE 1070 or CSA B125.3, except where such protection is otherwise provided by a combination tub/shower valve in accordance with Section 424.3.

❖ This section addresses faucets used only for the filling of whirlpool tubs or bathtubs. The water discharged from these "tub filler" faucets must be limited to a maximum temperature of 120°F (49°C) to prevent skin burns (see commentary, Section 424.3). It is not

PLUMBING FIXTURE	MAXIMUM DISCHARGE TEMPERATURE
1. Bidet	≤ 110°F
2. Public hand-washing facilities	> 85°F and < 110°F
3. Individual shower valves	≤ 120°F
4. Multiple (gang) showers	≤ 120°F
5. Bathtub/whirlpool bathtubs	≤ 120°F

For SI: °C = [(°F) - 32]/1.8.

**Figure 424.4(1)
MAXIMUM TEMPERATURES ALLOWED AT USE LOCATIONS**

FIXTURES, FAUCETS AND FIXTURE FITTINGS

For SI: °C = [(°F) - 32]/1.8.

Figure 424.4(2)
TEMPERATURE CONTROL STANDARDS RELATED TO FIXTURE APPLICATIONS

uncommon for a user to fill a bathtub in advance, with water hotter than he or she could stand, in an attempt to compensate for the cooling effect caused by the tub walls or a planned delay in entering the tub. These users typically avoid burns by testing the temperature of the water in the tub and add cold water to temper the hot water before getting into the tub. However, there have been cases in which the user was sitting on the side of the tub, reached to adjust the control for cold water, accidentally slipped into the tub and received third-degree burns. In other cases, children placed in a tub being filled have been scalded because the water became hotter some time after an adult tested the flowing water temperature at the beginning of tub filling. In some cases, an unsupervised child has adjusted a faucet and caused scalding. While it would seem that hot water supply temperatures might be simply controlled by setting the water heater thermostat to 120°F (49°C), experience has indicated that this is an unreliable means for limiting the temperature of hot water in any type of building because thermostats are inaccurate and are too easy for laypersons to adjust. The adjustment to a higher temperature setting is usually in response to users' complaints that there is not enough hot water during periods of high demand. Adjusting a water heater to a higher temperature effectively creates more hot water volume. This is because less, higher temperature, hot water flow is required for any particular mixed hot water use and thus, there is effectively, more hot water available to the users. Because the code does not regulate the temperature to which

water heater thermostats must be set and the thermostats are not tamperproof or have the necessary repeatable accuracy, water temperature control devices exterior to the water heater unit must be used to positively limit the discharge water temperature to 120°F (49°C).

The temperature-limiting device required by this section is a control valve that might require periodic adjustment, maintenance or replacement. Obviously, these valves should be located so that they can be accessed. Access panels in finished walls, floors or ceilings might be required in order to provide access.

This code section allows for a combination tub/shower valve (complying with ASME A112.18.1/CSA B125.1 or ASSE 1016/ASME A112.18.1016/CSA B125.16), with or without a fixed shower head, to be used for filling a tub, including a large whirlpool tub. However, in actual practice, the installation of a combination tub/shower valve as a tub filler for a large tub (such as a garden or whirlpool tub) is rarely done, for a few reasons: 1) faucet manufacturers generally do not provide or offer "extra long" wall-mount tub spouts to enable the discharge of the spout to be located over the tub well so as to not splash onto the tub deck and 2) the placement of the water control on the wall makes it difficult for a user sitting in a large tub to add more water if needed.

This "allowance" for a combination tub/shower valve for a tub application might lead the reader to assume that any device meeting ASME A112.18.1/CSA B125.1 is acceptable for meeting the requirements of this section but this is not the case. Combination tub/shower valves are the only ASME A112.18.1/CSA B125.1 device that can be used to meet the requirements of this section.

Why can't ASSE 1016 valves (other than combination tub/shower valves) be used for the ASSE 1070 or CSA B125.3 device required by this section? ASSE 1016 or CSA B125.3 devices, other than combination tub/shower valves, have a limited useful "maximum flow" rating. In other words, for flows in excess of 2.5 gpm (9.5 L/m), the pressure drop through these devices severely limits the user's ability to fill a large tub in short order. ASSE 1070 and CSA B125.3 devices have a much higher usable flow rate, typical of what is needed to rapidly fill a large bathtub or whirlpool tub. The code "allowance" to use a combination tub/shower valve complying with Section 424.3 for a bathtub filler faucet is a modest compromise directed at the most commonly installed, minimum size [30 inches (762 mm) by 60 inches (1524 mm) by 16 inches (406 mm) deep], standard tub found in nearly every private bathroom. The filling time of this standard tub size, using a combination tub/shower valve through a standard tub spout, is deemed "reasonable" based upon years of user experience with filling tubs through antiscald-type combination tub/shower valves. However, where tub sizes are larger, higher faucet flow rates are required to fill those tubs within an acceptable length of time. Requiring the use of an ASSE 1070 device with the higher flow through rating ensures that larger tubs can be filled quickly so that the tub water temperature has less time to "cool off" [since these devices are limiting the discharge temperature to 120°F (49°C)].

Where a hand-held shower is connected to a tub filler faucet (that is supplied by an ASSE 1070 or CSA B125.3 device), the hand shower is not required to be provided with water supplied through a combination tub/shower valve as is required for a standup-type "fixed" shower head application (see commentary, Section 424.2). The hand-held shower also does not "reclassify" a tub filler faucet as needing to be a "combination tub/shower valve" such that it would be required to comply with a combination tub/shower valve meeting ASSE 1016/ASME A112.18.1016/CSA B125.16 or a combination tub/shower valve meeting A112.18.1/CSA B125.1.

424.6 Hose-connected outlets. Faucets and fixture fittings with hose-connected outlets shall conform to ASME A112.18.3 or ASME A112.18.1/CSA B125.1.

❖ Typically, this is a deck-mounted fitting with a moveable spout or outlet connected to the body of the faucet through a flexible hose. Applications could include kitchen sinks, bar and lavatory sinks or shampoo bowls. The standard governing requirements for backflow protection devices in plumbing fixture fittings are found in ASME A112.18.3M or ASME A112.18.1/CSA B125.1. These standards establish the performance requirements for backflow prevention devices that provide protection consistent with the level of risk associated with the application.

424.7 Temperature-actuated, flow reduction valves for individual fixture fittings. Temperature-actuated, flow reduction devices, where installed for individual fixture fittings, shall conform to ASSE 1062. Such valves shall not be used alone as a substitute for the balanced-pressure, thermostatic or combination shower valves required in Section 424.3.

❖ Temperature-actuated flow reduction valves for individual fixture fittings must comply with the provisions set forth in ASSE 1062. These valves reduce flow to 0.25 gpm (0.95 L/m) or less automatically in response to outlet temperatures above a preset temperature not to exceed 120°F (49°C). This will limit exposure to high-temperature water discharge from an individual fixture fitting, such as bath and utility faucets, lavatory or sink faucets. These devices may not be used as a substitute for shower or tub/shower valves required in Section 424.3.

424.8 Transfer valves. Deck-mounted bath/shower transfer valves containing an integral atmospheric vacuum breaker shall conform to the requirements of ASME A112.18.1/CSA B125.1.

❖ Deck-mounted bath/shower transfer valves with integral backflow protection must comply with ASME A112.18.1/CSA B125.1. With the introduction of hand-held showers and other bathing appliances mounted on the deck of bathtubs and other fixtures, the need for backflow and backsiphonage protection became evident. This is because the accessories that are being furnished typically incorporate a hose or pull out features that could be submerged. The standard combines safety features such as a vent to the atmosphere and check members to protect against backflow or backsiphonage.

424.9 Water closet personal hygiene devices. Personal hygiene devices integral to water closets or water closet seats shall conform to the requirements of ASME A112.4.2.

❖ These devices can be integral to some water closet designs or part of a water closet seat. Typical functions of personal hygiene devices are bidet spray and blow drying. The standard ensures that products have proper potable water backflow prevention, strength and dimensional conformity to fit a variety of water closets. Some models have wireless programmable wall controls for ease of use.

SECTION 425
FLUSHING DEVICES FOR
WATER CLOSETS AND URINALS

425.1 Flushing devices required. Each water closet, urinal, clinical sink and any plumbing fixture that depends on trap siphonage to discharge the fixture contents to the drainage system shall be provided with a flushometer valve, flushometer tank or a flush tank designed and installed to supply water in quantity and rate of flow to flush the contents of the fixture, cleanse the fixture and refill the fixture trap.

❖ This section contains the applicability requirements for flushing devices recognized by the code. Material and performance requirements are specified in referenced standards, as are provisions for flush tanks.

There are three classifications of devices that are required to flush a water closet, urinal or other self-siphoning type of plumbing fixture: a flush tank, a flushometer valve and a flushometer tank. The flush tank discharges to the fixture by gravity, using only the head pressure created by the height of the tank. A flushometer valve is a controlling device discharging to the fixture at the full line pressure of the water supply. A flushometer tank is a hydropneumatic tank that creates a flushing action like a flushometer valve and fills like a gravity tank.

425.1.1 Separate for each fixture. A flushing device shall not serve more than one fixture.

❖ Each fixture must have its own flushing device. Individual control valves are necessary in an effort to conserve water.

425.2 Flushometer valves and tanks. Flushometer valves and tanks shall comply with ASSE 1037 or CSA B125.3. Vacuum breakers on flushometer valves shall conform to the performance requirements of ASSE 1001 or CSA B64.1.1. *Access* shall be provided to vacuum breakers. Flushometer valves shall be of the water conservation type and shall not be used where the water pressure is lower than the minimum required for normal operation. When operated, the valve shall automatically complete the cycle of operation, opening fully and closing positively under the water supply pressure. Each flushometer valve shall be provided with a means for regulating the flow through the valve. The trap seal to the fixture shall be automatically refilled after each flushing cycle.

❖ There is a variety of styles of flushometer valves and tanks, all of which must be equipped with a vacuum breaker to prevent backflow. The flushometer valve operation is either a diaphragm type or a piston type. The majority of flushometer valves are diaphragm type because of its reliability. The working components of a flushometer valve are either exposed or concealed. When concealed, the valve must be accessible through an access opening, cover or plate.

Flushometers are either manually or automatically operated. Common manual valves are activated by a hand lever, push button or foot valve. Automatic valves are activated electrically by infrared sensors or other devices that sense the presence of an individual using the fixture. A segment diaphragm separates the valve into an upper and lower chamber with the pressure being the same on both sides of the diaphragm, equalized through the bypass in the diaphragm. The greater surface area on the top of the diaphragm holds the valve closed. The slightest touch of the handle grip in any direction pushes the plunger inward, which tilts the relief valve, releasing the pressure in the upper chamber. Then, the pressure below raises the diaphragm and working parts as a unit, allowing the water that flushes the bowl to go down through the barrel of the valve. While this is occurring, a small amount travels up through the bypass, gradually filling the upper chamber to close the valve.

The newest category of flushing device is a flushometer tank. The flushing device incorporates a hydropneumatic tank to flush the fixture at full line pressure. The flushing of the fixture is similar to a flushometer valve; however, the water demand is the same as a flush tank. Rather than using the water supply pressure from the piping system, a flushometer tank stores the necessary water for flushing in the hydropneumatic accumulator tank. The water in the tank is stored under a pressure equivalent to the water supply pressure connected to the tank.

425.3 Flush tanks. Flush tanks equipped for manual flushing shall be controlled by a device designed to refill the tank after each discharge and to shut off completely the water flow to the tank when the tank is filled to operational capacity. The trap seal to the fixture shall be automatically refilled after each flushing. The water supply to flush tanks equipped for automatic flushing shall be controlled with a timing device or sensor control devices.

❖ The flush tank is designed to permit a large volume of water to discharge at once, creating the flushing action in the fixture. The tank then refills while simultaneously refilling the trap seal of the fixture. Most flush tanks have a float device (fill valve) to open and close the water supply mechanism. When the tank empties, the float lowers, opening the fill valve. As the float rises, the water supply closes at the predetermined height in the tank.

425.3.1 Fill valves. Flush tanks shall be equipped with an antisiphon fill valve conforming to ASSE 1002 or CSA B125.3. The fill valve backflow preventer shall be located not less than 1 inch (25 mm) above the full opening of the overflow pipe.

❖ The fill valve is the refilling component that controls the flow of water into the flush tank. Because the fill valve connects the potable water to the flush tank, it must have adequate protection against backflow. The referenced standards require an antisiphon device in the fill valve. An antisiphon device is a type of vacuum breaker that must be located with its critical level not less than 1 inch (25 mm) above the overflow pipe to prevent its submersion.

425.3.2 Overflows in flush tanks. Flush tanks shall be provided with overflows discharging to the water closet or urinal connected thereto and shall be sized to prevent flooding the tank at the maximum rate at which the tanks are supplied with water according to the manufacturer's design conditions. The opening of the overflow pipe shall be located above the *flood level rim* of the water closet or urinal or above a secondary overflow in the flush tank.

❖ The overflow in the flush tank prevents flooding of the tank and submersion of the fill valve. The fill valve antisiphon device must be located above water at all times to maintain the protection against backflow. The overflow discharges into the water closet or urinal, ultimately discharging to the drainage system.

Where the flush tank is a separate unit (two-piece water closets), the overflow is always above the flood level rim of the water closet or urinal. If the water closet or urinal drain is obstructed, the water discharges onto the floor before submerging the fill valve antisiphon device. In certain styles of one-piece water closets, the overflow opening is below the water closet flood level rim; therefore, openings are provided in the back of the tank to serve as a secondary overflow to prevent the submersion of the fill valve.

425.3.3 Sheet copper. Sheet copper utilized for flush tank linings shall conform to ASTM B152 and shall not weigh less than 10 ounces per square foot (0.03 kg/m^2).

❖ Years ago, flush tanks for urinals and water closets were made of wood and hung on the wall high above the fixture. Since wood was not impervious to water, sheet copper, formed and soldered, was used to line the wood tank. Today, perhaps only historical building renovations would require application of this code section.

425.3.4 Access required. All parts in a flush tank shall be accessible for repair and replacement.

❖ Flush pipes and fittings connect a water closet tank to the bowl or a flush tank to a urinal. ASME A112.19.5 specifies only the dimensional requirements for the flush pipe and fittings. For the material requirement, this section stipulates only that it must be nonferrous material, such as brass, copper alloy or plastic. These materials are intended to extend the life of the pipe and fittings.

425.4 Flush pipes and fittings. Flush pipes and fittings shall be of nonferrous material and shall conform to ASME A112.19.5/CSA B45.15.

❖ Flush pipes and fittings connect a water closet tank to the bowl or a flush tank to a urinal. The standard specifies only the dimensional requirements for the flush pipe and fittings. For the material requirement, this section stipulates only that it must be nonferrous material, such as brass, copper alloy or plastic. Use of these materials ensures long life of the pipe and fittings.

SECTION 426
MANUAL FOOD AND BEVERAGE DISPENSING EQUIPMENT

426.1 Approval. Manual food and beverage dispensing equipment shall conform to the requirements of NSF 18.

❖ NSF 18 establishes minimum food protection and sanitation requirements for the materials, design, construction and performance requirements for manual food and beverage dispensing equipment. This standard does not apply to vending machines, dispensing freezers and bulk milk-dispensing equipment covered by the scope of other NSF standards.

SECTION 427
FLOOR SINKS

427.1 Approval. Sanitary floor sinks shall conform to the requirements of ASME A112.6.7.

❖ Floor sinks must comply with ASME A112.6.7, which covers the material requirements for cast-iron and PVC sanitary floor sinks. All floor sinks must have a removable strainer or accessible grate.

Chapter 5:
Water Heaters

General Comments

A water heater is any appliance that heats potable water and supplies it to the hot water distribution system of the building. Some water heaters might also provide hot water for space heating purposes in addition to supplying hot water to the water distribution system. The code does not apply to appliances, equipment and devices that provide hot water for the sole purpose of space heating (hydronic heating systems) because: 1) by definition in this code, a water heater supplies water to the potable water distribution system and 2) Section 607.1 of the code requires a supply of hot or tempered water to all plumbing fixtures and equipment used for bathing, washing, culinary purposes, cleansing, laundry or building maintenance within occupied structures. Hydronic heating systems are regulated by the *International Mechanical Code*® (IMC®).

Historically, the main focus of this chapter has been to regulate the installation and safety features for automatic storage tank water heaters and unfired hot water storage tanks. Other water heating methods and equipment such as solar, electric heat pump, ground source thermal transfer, boiler water recirculation, steam heat transfer, solid-fuel-fired technology and the necessary controls/safety devices for those methods are not specifically addressed in the code but many are covered in the IMC.

Questions often arise as to how the regulations in this chapter are intended to apply to these alternative water heating methods. Where the code does not specifically address those questions, the manufacturer's instructions must serve as the basis for installation of these products (see Section 502.1).

Wherever hot water is stored in a closed tank, there is a potential for explosion. There are documented cases where improperly installed (or poorly maintained) water heaters have exploded with such force that they have been propelled through floor and roof assemblies and over 100 feet (30 480 mm) in the air. Thus, the intent of this chapter is to regulate the materials, design and installation of water heaters and related safety devices in order to protect property and life.

Purpose

Chapter 5 regulates the design, approval and installation of water heaters and related safety devices. The intent is to minimize the hazards associated with the installation and operation of water heaters.

SECTION 501
GENERAL

501.1 Scope. The provisions of this chapter shall govern the materials, design and installation of water heaters and the related safety devices and appurtenances.

❖ This section defines the scope of the chapter and contains requirements for water heaters used for space heating, the installation of drain valves, the location of water heaters, third-party certification of water heaters, tankless water heaters, marking of storage tanks and temperature controls for hot water supply systems.

The materials, design and installation of water heaters and related safety equipment and appurtenances are regulated by this chapter. The requirements in this chapter are intended to establish that water heaters and the necessary controls and safety devices are designed and installed so that they perform in an acceptable manner.

501.2 Water heater as space heater. Where a combination potable water heating and space heating system requires water for space heating at temperatures greater than 140°F (60°C), a master thermostatic mixing valve complying with ASSE 1017 shall be provided to limit the water supplied to the potable *hot water* distribution system to a temperature of 140°F (60°C) or less. The potability of the water shall be maintained throughout the system.

❖ Where a water heater has a dual purpose of supplying hot water to both a potable hot water distribution system and a hot water space heating system, the maximum water temperature allowed for the potable hot water distribution system is 140°F (60°C). A master thermostatic mixing valve conforming to ASSE 1017 must be installed at the source of the hot water supply to limit the potable hot water distribution system water temperature to 140°F (60°C). Master thermostatic valves must not be used for limiting the final hot or tempered water temperatures at fixtures where they are required by the code to have a discharge temperature limit. For example, public handwashing lavatory hot water must be controlled (tempered) to be not greater that 110°F (43°C). This application requires an additional thermostatic mixing valve of a different type.

A water heater used to supply hot water to a space heating system must be protected from any conditions that can cause contamination of the potable water supply system. A typical installation might be an underfloor radiant heating system. Because the water heater is part of the potable water system, materials used in the heating system must be approved for use in a potable water system, and all connections must be protected against contamination. In the summer months when the heating system is inactive, a method to prevent stagnation of the water should be employed. Small orifices in the isolation valves are often used to

WATER HEATERS

allow a small amount of water to circulate through the system to prevent stagnation. Chemicals of any type must not be added to the heating system, because this would directly contaminate the potable water supply. Protection of the potable water supply must be in accordance with Section 608 (see Commentary Figure 501.2).

501.3 Drain valves. Drain valves for emptying shall be installed at the bottom of each tank-type water heater and hot water storage tank. The drain valve inlet shall be not less than $^3/_4$-inch (19.1 mm) nominal iron pipe size and the outlet shall be provided with male garden hose threads.

❖ A storage tank-type water heater and a hot water storage tank must be drainable to facilitate service, sediment removal, repair or replacement. The ASSE 1005 standard was discontinued some time ago so the code language was changed to require the basic requirements for drain valves.

The reason the promulgator chose to discontinue publishing the ASSE 1005 standard is unknown. The standard did cover a number of details/requirements for water heater drain valves, including cycle testing and elevated temperature pressure testing. However, from a code-enforcement perspective, the code should not reference a standard that is no longer published. The two requirements that replace the compliance with ASSE 1005 are the minimum to ensure that the supplied/ installed valve provides for servicing of a water heater or hot water storage tank. The revised provision will allow other types of drain valves to be installed that, up to now, could not be installed because they did not comply with ASSE 1005. For example, some contractors install a $^3/_4$-inch (19.1 mm) full-port ball valve in this application so that at a future date, lime scale build-up at the drain outlet of the tank can be rodded through the valve to promote faster tank drain down.

501.4 Location. Water heaters and storage tanks shall be located and connected so as to provide *access* for observation, maintenance, servicing and replacement.

❖ Because water heaters and hot water storage tanks require routine inspections, maintenance, repairs and possible replacement, access is required. Further recommendations and requirements are stated in the manufacturer's instructions.

The code defines "Access" as being able to be reached but which first may require the removal of a panel, door or similar obstruction. Access to a water heater is not achieved if it is necessary to remove any portion of the structure's permanent finish materials, such as drywall, plaster, paneling, built-in furniture or cabinets or any other similar permanently affixed building component.

There is not always sufficient room for installation of a water heater in spaces such as basements, alcoves, utility rooms and furnace rooms. In an effort to save space or simplify an installation, water heaters are often installed in crawl spaces, attics or similar remote locations. Access to and servicing of water heaters could be difficult; therefore, a walkway, work space or

For SI: °C = [(°F) - 32]/1.8.

Figure 501.2
WATER-TO-AIR HEAT EXCHANGER

platform is necessary to provide a large enough path of travel to, and adequate space for, such activities in accordance with Section 502.3.

501.5 Water heater labeling. All water heaters shall be third-party certified.

❖ See the commentary to Section 303.4 for a general description of third-party certification of products. The code does not specify the standards that water heaters must comply with; it simply requires that all water heaters be third-party certified. Other *International Codes®* require compliance with the following standards as applicable to the type of water heating unit.

The *International Fuel Gas Code®* (IFGC®) requires that fuel gas-fired water heaters comply with one of the following:

ANSI Z21.10.1, *Gas Water Heaters–Volume I*. This standard covers gas-fired storage tank water heaters with a British thermal unit (Btu) input of 75,000 or less.

ANSI Z21.10.3, *Gas Water Heaters—Volume III*. This standard covers gas-fired storage tank water heaters with a Btu input greater than 75,000 as well as gas-fired circulating and (gas-fired) instantaneous water heaters.

The IMC requires that oil-fired water heaters comply with the following:

UL 732, *Oil-fired Storage Tank Water Heaters*. This standard covers oil-fired storage tank water heaters with a Btu input of no greater than 200,000, having a storage capacity no more than 120 gallons (454 L), and producing water temperatures no higher than 200°F (93°C).

The IMC requires that electric water heaters comply with one of the following as applicable:

UL 174, *Household Electric Storage Tank Water Heaters*. This standard covers household electric storage water heaters from 1 to 120 gallons (3.785 to 454 L) having an input of 12kW or less.

UL 1453, *Electric Booster and Commercial Storage Tank Water Heaters*. This standard covers commercial electric booster heaters and storage water heaters that meet at least one of the following criteria:

- Have a capacity over 120 gallons (454 L).
- Have an input greater than 12 kW.
- Have the capability of being set to discharge water temperature higher than 185°F (275°C).

Note that the IMC does not indicate a standard for hot water dispensers or electric instantaneous water heaters. This does not mean that these devices cannot be installed just because the codes do not specify a standard that is appropriate for a particular type of water heating unit. This code simply requires that water heaters be third-party certified so that any type of water heating unit that is not covered by the standards indicated by the IMC or the IFGC will be built and third party certified to a nationally recognized standard. A few examples are:

UL 499, *Electric Heating Appliances*. This standard covers hot water dispensers of less than 1-gallon (3.785 L) capacity and all electric instantaneous water heaters.

UL 1995, *Heating and Cooling Equipment*. This standard covers electric heat pump water heaters.

The certification mark is the primary, if not the only, indicator to the installer, the inspector and the end user that the water heater has been evaluated and tested by an approved agency and has been determined to perform in an acceptable manner where installed and operated in accordance with the manufacturer's instructions.

The approved agency can require the manufacturer to alter, add or delete information in the installation instructions as necessary to achieve compliance with the applicable standard. Because water heaters must be installed in accordance with the manufacturer's instructions, these instructions must be available to the code official at the time of inspection to verify that the water heater has been properly installed.

501.6 Water temperature control in piping from tankless heaters. The temperature of water from tankless water heaters shall be not greater than 140°F (60°C) where intended for domestic uses. This provision shall not supersede the requirement for protective shower valves in accordance with Section 424.3.

❖ The intent of this section is to prevent excessively high water temperatures from reaching plumbing fixtures. Tankless water heaters do not have a storage capacity and are often called "instantaneous" because the water is heated at the same rate it is used. Instantaneous heaters can discharge an uncertain range of temperatures at any given time, depending on the flow demanded from the unit. Therefore, some form of temperature control is necessary to protect the user against exposure to excessively hot water discharged from domestic fixtures, such as lavatories, kitchen sinks, tubs and laundry trays. This could be accomplished by installation of a tempering valve [set at 140°F (60°C)] on the outlet of the water heater or by equipping the tankless water heater with a temperature-limiting device or thermostat that has a maximum setting of 140°F (60°C).

Where a tankless water heater is used to supply hot water to a shower or a tub/shower combination, the maximum shower or tub/shower control valve outlet temperature must be limited to 120°F (49°C) in accordance with Section 424.3.

In recent years, tankless water heaters have become increasingly popular because they occupy less floor space than a storage-type tank water heater, use zero energy in the standby mode and if properly maintained, can last considerably longer than a storage tank water heater. But perhaps their most advertised feature is that a tankless water heater can provide an "unlimited" flow of hot water. This is opposed to a storage tank water heater that will eventually be depleted of its stored hot water, given enough time of hot water flowing from a hot water outlet. A tankless water heater instantly and continuously heats water as long as hot water is demanded at a fixture.

But if a tankless water heater can provide an "unlimited" flow of hot water, could installing a tankless water heater instead of a storage tank water heater be a solution to not running out of hot water? The answer depends on the flow rate demanded of the tankless water heater.

Tankless water heaters are rated to provide for a specific temperature increase (rise) of the incoming cold water at specific flow rate. For example, one rating might be: 50°F (27.8°C) temperature rise at 2.6 gallons per minute (gpm). If the cold water temperature flowing into the tankless heater is 60°F (15.6°C) and the flow through the heater is 2.6 gpm (15.6°C), the outlet water temperature will be 50°F (27.8°C) + 60°F (15.6°C) = 110°F (43.3°C). Therefore, if the hot water demand at the fixture is 2.6 gpm, hot water at a temperature of 110°F can be expected to be discharged at the hot water outlet (neglecting heat losses from the hot water piping system). This flow of hot water can continue indefinitely so, indeed, the user "never runs out of hot water."

However, one significant characteristic of tankless water heaters is that, as the flow through the water heater becomes greater, the temperature rise given to the water becomes less. Conversely, as the flow becomes less, the temperature rise given to the water becomes greater. Commentary Figure 501.6(1) shows typical capacity graphs for Model A and Model B electric tankless water heaters. Although some manufacturers might choose to show the same information in table form as shown in Commentary Figure 501.6(3), the concept is the same as that shown in a capacity graph. Commentary Figure 501.6(2) illustrates the operating point of 50°F (27.8°C) temperature rise at 2.6 gpm (9.8 L/m) for the Model A unit. If the flow through the unit is 3 gpm (11.4 L/m), the temperature rise would only be approximately 40°F (22.2°C). If the flow through the unit is only 1.5 gpm (5.7 L/m), the temperature rise would be 100°F (55.6°C) (ignoring any high-temperature limit control that might be present). What does this information mean in terms of the operation of plumbing fixtures in a building?

Consider a shower with one shower head that discharges at 2.5 gpm (9.8 L/m). Then consider a shower user desiring a shower water temperature of 110°F (43.3°C). If a Model A tankless water heater unit (as described previously) supplies hot water to the water distribution system, the shower user could turn the shower mixing valve almost all the way to the full hot position and enjoy a 110°F (43.3°C) shower, indefinitely. Now consider that, while the first user is showering, another user desires to wash his or her hands at a lavatory [flowing at 1.5 gpm (5.7 L/m)]. Assume that the lavatory user turns on only the hot water side of the

For SI: °C Rise = 1.8°F Rise.

Figure 501.6(1)
TYPICAL CAPACITY GRAPH FOR TANKLESS WATER HEATERS

Figure 501.6(2)
OPERATING POINT FOR TANKLESS WATER HEATER

FLOW	TEMPERATURE RISE, °F		
	1.0 gpm	2.0 gpm	4.0 gpm
Model A (9.5 kW)	65	32	16
Model B (19 kW)	130	65	32

Figure 501.6(3)
TYPICAL CAPACITY TABLE FOR TANKLESS WATER HEATERS

faucet. What is the result of this action? Now the flow rate through the tankless water heater is 2.5 + 1.5 = 4 gpm (15.1 L/m). According to the graph shown in Commentary Figure 501.6(2), the Model A unit, at 4 gpm (15.1 L/m), has a temperature rise of only 30°F (16.7°C)—a drop of 20°F (15.1°C) from the 50°F (27.8°C) temperature rise at 2.5 gpm (9.8 L/m). The result is that both the shower user and the lavatory user will only receive 110°F - 20°F = 90°F (11.1°C) hot water. The shower user will most likely notice the change in temperature and will wonder what happened to his or her "hot water that never runs out."

The previous example, although simplified, illustrates the fact that for any tankless water heater (electric or fuel-gas type), beyond the rating of a particular unit, the temperature of the hot water becomes flow dependent. Note that this is substantially different from a storage tank water heater that provides the same temperature of hot water at any flow rate but only for a limited amount of time. Thus, users of tankless water heater systems might have to adjust their expectations concerning when and to what extent simultaneous hot water demands can be expected before temperature of the hot water received is affected.

There are many different models of electric and fuel gas-powered tankless water heaters, each having a specific capacity graph or table. If the largest tankless water heater cannot accommodate the total demand, additional tankless units can be added to meet the total demand. For electric units, each unit can be installed to serve a specific fixture or group of fixtures that can be accommodated by that unit. Where additional fuel gas units are required to meet the total demand, each unit could be connected to serve only specific fixture groups within the building. However, many fuel gas-powered unit controls are quite sophisticated so that multiple units can be connected to a common header and the controls interconnected so as to cause additional units to "come on line" when hot water demand increases beyond the capacity of the first unit.

501.7 Pressure marking of storage tanks. Storage tanks and water heaters installed for domestic hot water shall have the maximum allowable working pressure clearly and indelibly stamped in the metal or marked on a plate welded thereto or otherwise permanently attached. Such markings shall be in an accessible position outside of the tank so as to make inspection or reinspection readily possible.

❖ Storage tank water heaters, like all pressure vessels, must be able to withstand the working pressures to which they will be subjected. The maximum working pressure of the water heater must be known so that a properly rated relief valve not exceeding the manufacturer's rated working pressure can be installed. The working pressure must be marked in such a way that it is permanent, not susceptible to damage and legible for the life of the water heater. Therefore, if the relief valve requires replacement, a properly sized valve can be reinstalled. The marking on the water heater must be located so that it is readily accessible during inspections and servicing.

501.8 Temperature controls. Hot water supply systems shall be equipped with automatic temperature controls capable of adjustments from the lowest to the highest acceptable temperature settings for the intended temperature operating range.

❖ The intent of this section is to prohibit manually controlled water-heating systems that might cause dangerous water temperatures to be generated. Automatic controls provide for reliable, consistent operation of the unit and most often have integrated safety devices to protect the equipment from unsafe operating conditions. Where systems consist of circulating-type heaters with separate hot water storage tanks, automatic controls connected to a remote temperature sensor in the storage tank(s) are required for control of the stored water temperature.

The user must be provided with a means for adjustment of the temperature of the hot water generated so that an appropriate water temperature for the intended purposes can be selected. Other than the requirement in Section 501.6 for tankless water heaters, the code does not specify a minimum or maximum temperature setting for a water heater temperature control because of the variety of applications requiring "hot" water. A few examples are: 1) 180°F (82°C) hot water might be necessary for the operation of a commercial dishwasher to provide for sterilization of dishware; 2) 120°F (49°C) hot water might be required for the potable water distribution system of a hotel not having a kitchen and 3) 140°F (60°C) hot water might be required to provide for space heating in addition to supplying water to the potable water distribution system. The user and installer of the water heating system are responsible for selecting the appropriate temperature setting for the water heater.

Although the need for adjustability of the hot water temperature setting of a water heater is understandable, this adjustability feature has drawbacks. Where a water heater serves fixtures in addition to the appliances or equipment that require a hot water temperature greater than 120°F (49°C), the potential for skin burns exists at those fixtures. Commentary Table 424.3 is a temperature burn chart illustrating the amount of exposure time at various hot water temperatures for third-degree burns of human skin. Because most water heater temperature controls can be easily adjusted by nearly anyone, the code requires certain fixtures to have temperature limiting devices to protect users from scalding hot water. These fixtures include showers, bathtubs, whirlpool tubs, bidets and faucets for public hand washing. The requirements for temperature limiting devices for these fixtures are located in Chapter 4. All other fixtures such as sinks, lavatories in sleeping and dwelling units, bar sinks and laundry tubs are not required to have temperature limiting devices and, as such, could be locations where scald injury could occur.

A water heater temperature setting that is too low could possibly cause health problems, specifically Legionnaires' disease. The disease was realized and named after an outbreak of pneumonia-like illness that was contracted by 221 persons attending an American Legion convention at a hotel in Philadelphia in July 1976. Two-thirds of the patients required hospitalization, of which 34 died. Although this type of bacteria had been in existence and causing disease for many years before that, this large-scale tragic event caused significant research to be performed to fully understand the origin and transmission of what is now called Legionella. Legionella are warm water organisms that are found primarily in surface waters of lakes, ponds, rivers and streams. Because Legionella are protected by being inside of host organisms, typical potable water supply disinfection methods do not eradicate the bacteria. Therefore, Legionella can be found in many potable water systems of buildings. The potential for Legionella to cause disease dramatically increases where conditions for growth (amplification) exist and actions/activities cause the bacteria to be inhaled (aspirated). Because the optimum temperature range for Legionella growth in potable water is 86°F to 104°F (30°C to 40°C), water heater temperature settings in this range should certainly be avoided. Temperatures of 122°F (50°C) for 2 hours or 140°F (60°C) for 2 minutes results in 90-percent destruction of Legionella bacteria. A temperature of approximately 158°F (70°C) results in instantaneous 100-percent destruction of the bacteria. Although not a requirement of the code, many hot water system designers require stored hot water to be at a temperature not less than 140°F (60°C) so as to reduce the potential for Legionella to exist and grow in hot water storage tanks. Regardless of the water heater temperature setting chosen, this code requires tempering valves to be installed at locations where the risk of scalding is especially high (see Chapter 4).

SECTION 502
INSTALLATION

502.1 General. Water heaters shall be installed in accordance with the manufacturer's instructions. Oil-fired water heaters shall conform to the requirements of this code and the *International Mechanical Code*. Electric water heaters shall conform to the requirements of this code and provisions of NFPA 70. Gas-fired water heaters shall conform to the requirements of the *International Fuel Gas Code*.

❖ The installation of a water heater relies heavily on the manufacturer's instructions, as the instructions are in integral part of the third-party certification for these appliances. However, in some cases, the referenced *International Codes* contain more restrictive requirements that must be followed as well. Fuel gas-fired water heater installation requirements are covered by the IFGC. The IMC covers installation requirements for oil-fired and electric water heaters.

Water heaters are appliances and, as such, the IMC or the IFGC contains additional requirements concerning them. For example, installation guidance for fuel pipe sizing and connections, appliance location and clearances, determination of the adequacy of combustion air, combustion air opening size and location and venting requirements are contained in the IFGC. NFPA 70 contains additional requirements for the installation of electric water heaters, such as the size of the electrical supply and disconnect provisions. In many situations, the installation requirements between the code and the manufacturer are identical. However, where there is a conflict between the code and the manufacturers' installation instructions, the requirements that provide the greatest level of safety apply.

Even if the water heating appliance appears to be in compliance with the manufacturer's instructions, the installation cannot be considered complete nor can it be approved until all associated components, connections and systems that serve the water heater are also in compliance with the applicable code provisions. For example, a fuel-fired water heater installation cannot be approved if the water heater is connected to a deteriorated, undersized or otherwise unsafe chimney or vent that does not comply with the IMC or the IFGC. Likewise, the same installation could not be approved if the existing fuel piping has insufficient capacity to supply the gas demand of the water heater, or the existing electrical supply conductors are inadequate or unsafe.

502.1.1 Elevation and protection. Elevation of water heater ignition sources and mechanical damage protection requirements for water heaters shall be in accordance with the *International Mechanical Code* and the *International Fuel Gas Code*.

❖ An ignition source could be an open flame such as a pilot light, an electrical switch, open resistance heating coils or an electrical igniter unit (such as hot surface or spark ignition types). Certain areas of a building are known to have a high potential for volatile liquids, such as gasoline and paint thinners, to be spilled or leak from their containers. Because the vapors of these liquids are heavier than air, they concentrate just above floor level, posing an explosion hazard. Testing has indicated that locating any possible ignition sources at some elevation above the floor in these areas provides a degree of safety due to the time delay for the vapors from a spilled combustible liquid to build up to ignitable concentrations at the elevation of the ignition source. However, given large enough spills and enough time, any ignition source in an enclosed area could cause an explosion.

Section 304.3 of the IMC requires that where equipment and appliances having an ignition source are located in hazardous locations, repair garages, private garages, motor fuel-dispensing facilities or parking garages, the ignition source must be located at least 18 inches (457 mm) above the floor of where the equipment rests. Equipment and appliances having an ignition source are prohibited in Group H occupancy or control areas where open use, handling or dispensing of combustible, flammable or explosive materials occurs. Note that rooms or spaces that are not part of the living space of a dwelling unit and that communicate directly with a private garage are considered to be part of the private garage. For example, a storage room on the same level of the garage having a door into the garage is in "direct communication" (in terms of the exchange of air) with the garage.

Most electric water heater thermostats have enclosed contacts but they are not sealed gas tight. Therefore, if an electric water heater having a thermostat located less than 18 inches (457 mm) from the bottom of the unit, the IMC requires that the unit be elevated so the ignition source (thermostat) is at least 18 inches (457 mm) above the garage floor. Electric water heaters having all switching controls located above 18 inches (457 mm) from the bottom of the water heater are not required to be elevated.

Sections 305.2 and 305.3 of the IFGC mirror the same ignition source requirements as the IMC except where the appliance is listed as flammable vapor ignition resistant (FVIR). Note that an FVIR-listed appliance is still prohibited from be located in a Group H area. Although FVIR technology has been proven to be reliable, some jurisdictions might still require elevation of FVIR fuel gas-fired water heaters. Therefore, always check with the authority having jurisdiction regarding elevation requirements for FVIR-listed appliances. (See the IFGC commentary for more information on FVIR appliances.)

A frequently asked question is how this section should be applied to a replacement water heater that originally was not required by the code at that time to be elevated or have protection against mechanical damage. The code makes no distinction between an installation in new construction and installation in existing construction. Where installers are confronted with obstacles that appear to prevent strict compliance with the code, they should consult with the code official before performing the replacement work.

502.2 Rooms used as a plenum. Water heaters using solid, liquid or gas fuel shall not be installed in a room containing air-handling machinery where such room is used as a plenum.

❖ Installation of fuel-fired water heaters is not permitted in rooms that are used as a plenum because of the potential for toxic products of combustion to be spread throughout the building in the event that the water heater or venting system malfunctions. Also, the negative and positive pressures developed within a plenum space can seriously affect the operation of a fuel-fired water heater located there.

For example, if a natural-draft fuel-fired water heater was installed in a room used as a return air plenum, the negative pressure of the plenum could overcome the draft of the water heater's venting system, and flue gases could be drawn into the plenum space. The improper operation of a water heater will greatly accelerate the deterioration of the water heater and its vent.

502.3 Water heaters installed in attics. Attics containing a water heater shall be provided with an opening and unobstructed passageway large enough to allow removal of the water heater. The passageway shall be not less than 30 inches (762 mm) in height and 22 inches (559 mm) in width and not more than 20 feet (6096 mm) in length when measured along the centerline of the passageway from the opening to the water heater. The passageway shall have continuous solid flooring not less than 24 inches (610 mm) in width. A level service space not less than 30 inches (762 mm) in length and 30 inches (762 mm) in width shall be present at the front or service side of the water heater. The clear access opening dimensions shall be not less than 20 inches by 30 inches (508 mm by 762 mm) where such dimensions are large enough to allow removal of the water heater.

❖ There is not always sufficient room for installations of water heaters in spaces such as basements, alcoves, utility rooms and furnace rooms. In an effort to save space or simplify an installation, water heaters are often installed in attics. Access to water heaters in attics is often difficult due to the lack of walking surfaces other than the edges of exposed ceiling joists. The intent of this section is to require a suitable access opening, passageway and work space that will provide for safe access by service personnel and consideration for water heater replacement without the need for removal of permanent portions of the building structure.

It is not uncommon for water heaters that are located in attics to be placed into position long before finish wall, ceiling and door/access opening trim materials are installed. A water heater of a size that "just fits" through the rough opening between ceiling joists or a water heater that was set prior to the gable ends of the attic space are completed might not be removable after the finish materials are in place. Coordination is required between the building designer and supplier/installer of the water heater to ensure that water heaters can be replaced without compromising the integrity of the building structure or requiring significant repair work to the building.

502.4 Seismic supports. Where earthquake loads are applicable in accordance with the *International Building Code*, water heater supports shall be designed and installed for the seismic forces in accordance with the *International Building Code*.

❖ Chapter 16 of the *International Building Code*® (IBC®) contains seismic maps that designate which areas of the country must design for earthquake loads. Section 308.2 of the code requires that piping supports be designed to resist seismic loads in designated zones. Similarly, this section requires that water heater supports be designed to resist the same seismic loads in accordance with the IBC. Failure of water heater supports or dislocation of a water heater has been shown to be a threat to health and safety of the building occupants. The requirements of this section are the minimum required criteria in view of life safety considerations.

502.5 Clearances for maintenance and replacement. Appliances shall be provided with *access* for inspection, service, repair and replacement without disabling the function of a fire-resistance-rated assembly or removing permanent construction, other appliances or any other piping or ducts not connected to the appliance being inspected, serviced, repaired or replaced. A level working space not less than 30 inches in length and 30 inches in width (762 mm by 762 mm) shall be provided in front of the control side to service an appliance.

❖ A water heater (an appliance) typically requires some inspection, service or repair over the life of the unit and eventually, the water heater requires replacement. Because previous editions of the code and many water heater manufacturers' installation instructions did not require a working space in front (control side) of the unit, some water heaters would be installed without consideration for providing ample access to the controls. With no prohibitions against installing piping, duct- work or other equipment not associated with the water heater such that access to the water heater was blocked, the code official had no code substantiation to require proper access to the water heater. Where the replacement of a "blocked" water heater was necessary, removal and replacement of those items as well as any permanent construction often resulted in improper reinstallation of those items because the persons performing water heater replacements are often unfamiliar with those other systems. Improper reinstallation (or worse, no reinstallation) of those removed items could lead to dangerous conditions.

The required size of the space in front (control side) of a water heater allows for ample kneeling area as well as provides a level area to set up a step ladder to access the top of tall water heaters, especially if the water heater is elevated above the floor. As this section prohibits removal of permanent construction even for replacement of the unit, building designers need to consider the size of the water heater with respect to its path of removal to the exterior of the building. During construction of a building, a water heater might be set before all of the walls, doorways and ceilings are completed, resulting in no possible way to move the old water heater out of building and move a new unit of similar size into position without removing permanent parts of the building. The removal and reinstallation of permanent portions of the building could cause problems with structural integrity, fire safety or operation of other building systems.

SECTION 503
CONNECTIONS

503.1 Cold water line valve. The cold water *branch* line from the main water supply line to each hot water storage tank or water heater shall be provided with a valve, located near the equipment and serving only the hot water storage tank or water heater. The valve shall not interfere or cause a disruption of the cold water supply to the remainder of the cold water system. The valve shall be provided with *access* on the same floor level as the water heater served.

❖ This section requires that a valve located near the water heater be installed in the cold water branch line from the main water supply. This section also provides requirements for separate water heater and storage tank hot water systems.

Valves are needed to isolate each hot water storage tank or water heater from the water distribution system to facilitate service, repair and replacement and to allow for emergency shut off in the event of a failure. The shutoff valve must be adjacent to the hot water storage tank or water heater, conspicuously located and within reach to permit it to be easily located and operated in the event of an emergency. The valve is to serve only the hot water storage tank or water heater, thereby not disrupting the flow of the cold water supply to other portions of the water distribution system when the hot water storage tank or water heater is taken out of service [see Commentary Figure 503.1(1)]. In accordance with Section 606.1, Item 8, the valve must be a full-open valve located on the water supply pipe to

Figure 503.1(1)
WATER HEATER VALVE

Figure 503.1(2)
PLASTIC PIPE TOO CLOSE TO DRAFT HOOD OF GAS-FIRED WATER HEATER

WATER HEATERS

every water heater (see commentary, Section 606.1).

Where plastic cold and hot water piping connections are made to a draft hood equipped gas- or oil-fired water heater, the Plastic Pipe and Fittings Association recommends maintaining at least a 6-inch (152 mm) clearance between the piping and the draft hood/vent pipe. Metallic pipe or metallic flexible connectors should be used in these locations [see Commentary Figure 503.1(2)].

This section does not prohibit the installation of a shutoff valve on the hot water line from the water heater. Some installations might have a water heater located below the system being served and placement of a valve in the hot water line enables heater servicing without a drain down of the hot water distribution system. The code is silent on what type of valve should be used if a valve is installed at this location.

Heat Traps

The *International Energy Conservation Code®* (IECC®) requires that for commercial building (as defined by the IECC) water heater applications, heat traps must be installed on storage-type water heaters and hot water storage tanks. The reader should refer to the IECC to determine when heat traps must be installed.

Heat traps reduce standby heat loss caused by convection carrying heated water up vertical piping from the water heater. The heat trap requirement only applies to storage tank water heaters and hot water storage tanks with vertical piping connecting to the top of the tank or to the side of tank. Heat traps can either

Figure 503.1(3)
SPECIAL HEAT TRAP PIPE NIPPLE

be of the "integral" type or be of an external piping configuration. An integral heat trap could be a specially designed pipe nipple having an internal elastomeric flapper to significantly block the natural flow of hot water caused by convection [see Commentary Figure 503.1(3)]. Other types of integral heat traps are: 1) a specially designed pipe nipple with an internal ball that either floats (for the cold water inlet) or sinks (for the hot water outlet) and 2) an elastomeric insert with multiple slits (creating flexible "blocking flaps") that are factory installed in the threaded opening of the water heater. Because all of these types of heat traps are installed in the inlet and outlet of the tank, they are considered to be "integral" to the tank.

In some cases, integral heat traps cannot be used because components are not available for the size of

Figure 503.1(4)
PIPE AND FITTING U-SHAPE HEAT TRAPS

Figure 503.1(5)
FLEXIBLE WATER CONNECTOR LOOP HEATER

the connection. In these cases, the piping to the cold water inlet and from the hot water outlet must be arranged to interrupt the convective action of the hot water. This interruption can be accomplished using an inverted U-shaped piping arrangement or a piping loop [see Commentary Figures 503.1(4) and 503.1(5)]. This code and the IECC are silent concerning the dimensions required to make external heat traps.

See the commentary to Section 607.5 concerning external insulation requirements for hot water piping.

503.2 Water circulation. The method of connecting a circulating water heater to the tank shall provide proper circulation of water through the water heater. The pipe or tubes required for the installation of appliances that will draw from the water heater or storage tank shall comply with the provisions of this code for material and installation.

❖ Occupancies such as restaurants, motels, hotels and laundries typically have high demands for hot water. In such structures, it is not uncommon to have a separate water heater and storage tank. When a separate water heater and storage tank are installed, they must be connected to provide proper circulation of water through the water heater (see Commentary Figure 503.2).

Gravity and forced circulation are two methods of providing circulation between the water heater and the storage tank. Gravity circulation is achieved by locating the water heating equipment below the storage tank. As water is heated, the hot water will rise out of the heat source, through the piping and into the storage tank. As the water in the storage tank cools, it will travel by gravity through the pipe back to the heat source. Forced circulation uses a pump or a circulator to send the water between the heat source and the storage tank. When there is high demand for hot water, forced circulation is generally used to meet the demand because it allows for a much faster recovery time (i.e., it can replace the hot water used faster than the slower gravity system).

The manufacturer's installation instructions must be followed for proper operation of the system. The manufacturer will specify the direction of flow, the pipe and pump size, necessary valves, safety flow switches and pressure relief valves necessary for the efficient operation of the system and to prevent damage to the water heater.

The materials used in connecting the water heater to the storage tank and for the hot water distribution system must meet the minimum quality standards prescribed in Section 605 (see commentary, Section 605).

SECTION 504
SAFETY DEVICES

504.1 Antisiphon devices. An *approved* means, such as a cold water "dip" tube with a hole at the top or a vacuum relief valve installed in the cold water supply line above the top of the heater or tank, shall be provided to prevent siphoning of any storage water heater or tank.

❖ Water heaters are designed to operate only when they are full of water. If some or all of the water is siphoned out of the tank during an interruption in the cold water supply, damage to the water heater from overheating could occur. Typically, water heater designs include a cold water "dip" tube, which directs the incoming cold water to the bottom of the tank. At the top of the tank, a hole is provided in the dip tube to prevent water from being siphoned from the tank through the tube. ANSI Z21.10.1 and UL 174 both require that the hole in the dip tube be located within 6 inches (152 mm) of the top of the tank.

If the water heater does not have an antisiphoning hole in the dip tube, a vacuum relief valve or other approved means must be installed to break any siphoning effect that might occur.

504.2 Vacuum relief valve. Bottom fed water heaters and bottom fed tanks connected to water heaters shall have a vacuum relief valve installed. The vacuum relief valve shall comply with ANSI Z21.22.

❖ A vacuum relief valve must be installed in bottom-fed water heaters and tanks connected to water heaters. The vacuum relief valve is intended to prevent a possible reduction within the tank to below atmospheric pressure (i.e., a partial vacuum). Many tanks are not

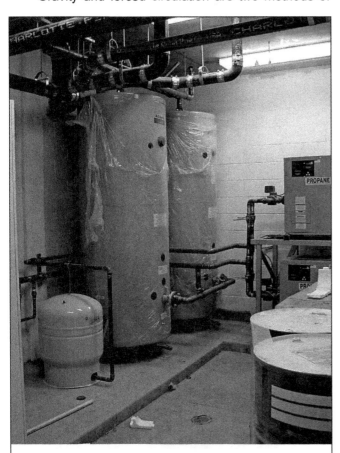

Figure 503.2
CIRCULATING-TYPE WATER HEATERS
CONNECTED TO STORAGE TANKS
(Photo courtesy of B&I Contractors, Inc.)

designed to resist external pressures exceeding internal pressures; therefore, a vacuum relief valve is necessary to prevent atmospheric pressure from possibly collapsing or otherwise damaging the tank or internal flueways.

The vacuum relief valve operates by automatically venting the system to the atmosphere when a partial vacuum is created, thereby permitting air to enter and maintaining the internal pressure (see Commentary Figure 504.2). Bottom-fed tanks would be susceptible to the creation of a vacuum by tank drainage resulting from negative supply pipe pressure or leakage.

Vacuum relief valves must be constructed and tested in accordance with ANSI Z21.22. This essentially requires the valve to be third-party certified as an indication to the installer, the inspector and the end user that it has been tested and evaluated by an approved agency and has been determined to perform in an acceptable manner. The presence of a certification mark is part of the information that the code official is to consider in the approval of the vacuum relief valve and the water heater.

504.3 Shutdown. A means for disconnecting an electric hot water supply system from its energy supply shall be provided in accordance with NFPA 70. A separate valve shall be provided to shut off the energy fuel supply to all other types of hot water supply systems.

❖ Just as the IMC and the IFGC require a shutoff valve (i.e., a disconnect means) for the energy supply, NFPA 70 requires electric water heaters have an electric disconnect switch. The shutoff valves for fuel-fired water heaters and disconnects for electric water heaters are necessary to allow for service, repairs and temporary and emergency shutdown (see Commentary Figure 504.3).

504.4 Relief valve. Storage water heaters operating above atmospheric pressure shall be provided with an *approved*, self-closing (levered) pressure relief valve and temperature relief valve or combination thereof. The relief valve shall conform to ANSI Z21.22. The relief valve shall not be used as a means of controlling thermal expansion.

❖ Every storage tank-type water heater must have both temperature and pressure relief protection to prevent tank failure. In many applications, the temperature and pressure relief valve safety functions are accomplished by a single combination-type pressure and temperature relief valve.

Where water is heated in a vessel open to atmospheric pressure (at sea level elevation), the water cannot be heated to a temperature any greater than 212°F (100°C) because water converts to steam at this temperature. However, if water is heated in a vessel that is under a pressure greater than atmospheric pressure, water temperatures greater than 212°F (100°C) are possible because the pressure in the vessel prevents the conversion of water into steam. Water at a temperature greater than its atmospheric pressure

Figure 504.2
WATER HEATER WITH BOTTOM FEED

Figure 504.3
ELECTRIC DISCONNECT ADJACENT TO WATER HEATER
(Photo courtesy of B&I Contractors, Inc.)

boiling point is said to be superheated. If the pressure in a vessel containing superheated water is rapidly lowered such as would occur when the vessel ruptures, the water will instantly and violently boil to create steam. The conversion of water to steam is an expansive process, that is, steam created from water occupies significantly more volume. Every gallon (3.8 L) of water converted to steam (at atmospheric pressure) will occupy approximately 224 cubic feet (6.3 m^3).

Similar conditions in a storage tank-type water heater under normal water distribution system pressures could occur if the heating control failed to cease heating the water at the time the thermostat setting was achieved and the temperature and pressure relief valve was inoperable. The sudden release of energy (the conversion of water into steam) due to tank rupture can have devastating results. Exploding water heaters have been known to penetrate through several stories of a building, knock down concrete walls and be propelled through roofs to land over 100 yards away. Human deaths and injuries have been attributed to water heater explosions.

A temperature relief valve prevents the water temperature in the tank from exceeding 210°F (99°C). If the water temperature should near 210°F (99°C), the valve opens to relieve any pressure buildup in the tank and to allow cooler water to enter the tank to lower the tank water temperature. With a properly installed and operating temperature relief valve, the water in the tank never reaches a temperature where steam could be generated under any pressure condition.

Many water heater tanks are designed for a working pressure of no greater than one half of the design pressure for the tank. A pressure relief valve is required to prevent the tank pressure from exceeding the working pressure of the tank. Although the tank is designed for a pressure twice that of the working pressure, other factors such as variations in manufacturing, in-service corrosion and weld joint fatigue due to pressure cycling necessitate limiting the operating pressure to no greater than the "working pressure" tank rating. Although pressure relief valves might be available with various pressure relief settings, at no time should a water heater tank be protected by a relief valve having a pressure relief setting greater than the working pressure rating of the tank. However, installing a pressure relief valve having a relief setting lower than the working pressure of the tank is not prohibited.

Obviously, the design, construction and quality of pressure and temperature relief valves must be controlled to ensure that these devices reliably perform the intended safety functions. The ANSI Z21.22 standard prescribes the necessary features, construction and testing of temperature and pressure relief valves.

As stated previously, when water is heated, the water will expand in volume. If the water being heated is in a closed vessel, the increase in volume will result in an increase in pressure. A storage tank-type water heater connected to a water distribution system could be a "closed vessel" such that when the water is heated, the system pressure will rise (see commentary, Section 607.3.2). Where thermal expansion-caused pressures must be controlled, the water heater pressure relief valve must not be used as the method to control the pressure increases. Relief valves are not designed to act as operating controls. A relief valve required to open frequently could become damaged or inoperative due to wear, corrosion or the valve might leak because of debris in the seat area. In any case, the intended safety function of the valve could become compromised.

Note that this section does not address the relief valve requirements for instantaneous and circulating water heaters. However, where these types of gas-fired water heaters are used, the applicable standards for these products require overtemperature protection as part of the burner control system and overpressure protection by installation of a pressure relief valve on one of the water connections to the unit. The UL 499 standard for electric instantaneous water heaters requires that overtemperature protection be built into the temperature control system, but does not require overpressure protection (see Commentary Figure 504.4).

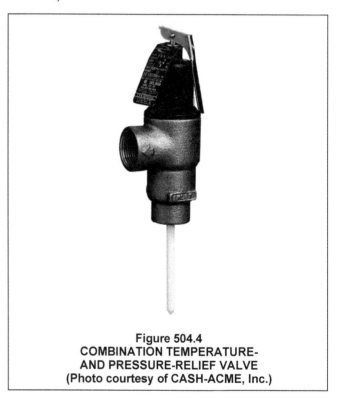

Figure 504.4
COMBINATION TEMPERATURE-
AND PRESSURE-RELIEF VALVE
(Photo courtesy of CASH-ACME, Inc.)

504.4.1 Installation. Such valves shall be installed in the shell of the water heater tank. Temperature relief valves shall be so located in the tank as to be actuated by the water in the top 6 inches (152 mm) of the tank served. For installations with separate storage tanks, the *approved*, self-closing (levered) pressure relief valve and temperature relief valve or combination thereof conforming to ANSI Z21.22 valves shall be installed on both the storage water heater and storage tank.

WATER HEATERS

There shall not be a check valve or shutoff valve between a relief valve and the heater or tank served.

❖ Storage tank water heaters commonly have one or more factory-installed threaded openings that are located to position the temperature-sensing element of a temperature relief valve within the top 6 inches (152 mm) of the tank. These openings can be on the side or the top of the tank and are typically marked "T&P Relief Valve" (see Commentary Figure 504.4.1). Because the highest water temperature in a storage tank water heater or hot water storage tank is at the top of the tank, the temperature sensing element of the relief valve must be positioned at that location to provide the earliest response to an over-temperature condition. As storage tank water heaters and hot water storage tanks are externally insulated, the temperature relief valve must be the proper length for the application. For example, it is not acceptable to install a "short-style" relief valve into a pipe coupling and then screw the coupling onto a pipe nipple that is screwed into the tank relief valve opening because the temperature-sensing element might not project sufficiently into the tank to properly sense the water temperature. Temperature relief valves are available with various body lengths to accommodate varying thicknesses of external tank insulation.

Where separate temperature and pressure relief valves are used for a storage tank water heater or a hot water storage tank, the pressure relief valve can be installed in any threaded opening in the tank. Or the pressure relief valve can be installed in the piping connected to the tank as long as there is not a shutoff or check valve installed between the pressure relief valve and the tank. For most applications, a combination temperature and pressure relief valve is used so that only one valve is needed.

For the 2012 code edition, this section was revised to clarify that each hot water storage tank and each storage tank water heater must have temperature and pressure relief valve protection. Even though a hot water system might have the storage water heater and a hot water storage tank connected by a pipe without any valves, both tanks require temperature and pressure relief valves (or a combination-type valve).

Shutoff or check valves must not be installed on either side (inlet or outlet) of a relief valve as this could render the relief valve inoperative, thereby creating a potentially dangerous condition.

504.5 Relief valve approval. Temperature and pressure relief valves, or combinations thereof, and energy cutoff devices shall bear the label of an *approved* agency and shall have a temperature setting of not more than 210°F (99°C) and a pressure setting not exceeding the tank or water heater manufacturer's rated working pressure or 150 psi (1035 kPa), whichever is less. The relieving capacity of each pressure relief valve and each temperature relief valve shall equal or exceed the heat input to the water heater or storage tank.

❖ Relief valves must be third-party certified by an approved agency as complying with ANSI Z21.22. The certification mark is the primary, if not the only, indication to the installer, the inspector and the end user that the relief valve has been tested and evaluated by an approved agency and has been determined to perform in an acceptable manner when installed and operated in accordance with its application, installation and operation instructions. The presence of a certification

FIGURE 504.5
COMBINATION TEMPERATURE-
AND PRESSURE-RELIEF VALVE LABEL
(Photo courtesy of CASH-ACME, Inc.)

For SI: 1 inch = 25.4 mm.

Figure 504.4.1
TEMPERATURE- AND PRESSURE-RELIEF VALVE INSTALLATION

mark is part of the information that the code official is to consider in the approval of the relief valve and the water heater.

Water heater temperature relief valves must be designed to open at a temperature not higher than 210°F (99°C). A relief valve device must be able to dissipate energy at a rate equal to or greater than the energy input rate of the water heater. Relief valve sizing information is provided by both water heater and relief device manufacturers (see Commentary Figure 504.5).

Water heaters must be marked with the maximum working pressure and the relief valve must be set to open at this pressure or 150 pounds per square inch (psi) (1035 kPa), whichever is less. Marking requirements for water heaters will also specify the minimum required relief-valve capacity. An undersized (insufficient capacity) relief valve would be unable to prevent the water heater pressure from exceeding the maximum capacity and a dangerous vessel failure could result. Relief-valve capacity is expressed in terms of British thermal units per hour (Btu/h).

A relief valve is designed to open in direct proportion to the water pressure and temperature acting on its closure disk. The higher the temperature or pressure, the greater the force and the more the valve opens.

504.6 Requirements for discharge piping. The discharge piping serving a pressure relief valve, temperature relief valve or combination thereof shall:

1. Not be directly connected to the drainage system.
2. Discharge through an *air gap* located in the same room as the water heater.
3. Not be smaller than the diameter of the outlet of the valve served and shall discharge full size to the *air gap*.
4. Serve a single relief device and shall not connect to piping serving any other relief device or equipment.
5. Discharge to the floor, to the pan serving the water heater or storage tank, to a waste receptor or to the outdoors.
6. Discharge in a manner that does not cause personal injury or structural damage.
7. Discharge to a termination point that is readily observable by the building occupants.
8. Not be trapped.
9. Be installed so as to flow by gravity.
10. Terminate not more than 6 inches (152 mm) above and not less than two times the discharge pipe diameter above the floor or *flood level rim* of the waste receptor.
11. Not have a threaded connection at the end of such piping.
12. Not have valves or tee fittings.
13. Be constructed of those materials listed in Section 605.4 or materials tested, rated and *approved* for such use in accordance with ASME A112.4.1.

❖ Because the discharge pipe from a water heater temperature and pressure relief valve is an extension of the potable water distribution system, the outlet of the discharge pipe must be treated as a potable water supply system outlet. Although Section 608.15.1 already requires protection of potable water supply outlets by requiring an air gap, installers of water heaters might not realize that the termination point of the relief valve discharge pipe must also provide for an air gap.

The requirement for an air gap at the end of the discharge pipe automatically provides for an indirect connection, should it be desired for the discharge to be captured by any type of drainage system. However, as water heater installers might not realize that the discharge pipe is not a waste pipe; the requirement for indirect connection is reiterated in this section.

A water heater temperature and pressure relief valve is an important safety device for the water heater as well as for the water distribution system that it is connected to (see commentary, Section 504.4). If the water heater is operating properly, the pressure in the water distribution system is no greater than is required by the code and the relief valve is in proper operating condition; there should not be any water discharge from the relief valve. Therefore, any water discharge from a relief valve is an indication of a water distribution system problem, a water heater problem or that the relief valve is malfunctioning. Because all of these problems are safety related and should be rectified as soon as possible, the discharge from the relief valve is an important "telltale" indicator to alert the occupants of the building to call for service. Therefore, it is important to terminate the relief valve discharge pipe within the same room as the water heater since most building occupants will usually associate evidence of leaking water at or near a water heater location as a potential water damage problem that deserves immediate attention [see Commentary Figure 504.6(1)].

An even more compelling reason for the relief valve discharge pipe to terminate at an air gap in the same room as the water heater is that a continuous routing of the discharge pipe from the relief valve through concealed spaces to a remote termination point can pose an extreme risk. Because the satisfactory condition of the discharge pipe is just as important as a properly operating relief valve, the condition of the discharge pipe must be observable by all who might need to perform service to the relief valve. Concealed discharge piping cannot be readily observed to verify that: 1) it has not been damaged (i.e., flattened, bent or cut); 2) the proper slope for gravity drainage still exists; 3) the pipe has not been tapped into, valved off or capped inadvertently and 4) the pipe length and number of elbows in the pipe are not in excess of the valve manufacturer's installation instructions. Since relief valve discharges are often just small amounts or continuous trickles of water, where freezing conditions might exist in spaces containing concealed piping or at a remote termination point, these small flows could easily freeze resulting in blockage of the pipe. The code requirement for the relief valve discharge pipe to terminate at an air gap in the same room as the water heater eliminates all uncertainties about the condition of the discharge pipe and its termination point.

Even though discharge of water from a relief valve should be a very rare occurrence, when discharge does occur, the flow of water must have a way to escape the area without causing damage to the building structure such as wood members, drywall and insulation. Note that the code is not concerned with damage to an occupant's property within the building. For example, if a water heater is located in a garage with a floor that slopes to the exterior through a garage door opening, and the perimeter walls of the garage at the floor level are of concrete block construction (without insulation or drywall applied), the water from the discharge pipe could flow to the outdoors without the need for a floor drain or waste receptor. The code is not concerned about damage to an occupant's property that might be stored on the floor of the garage. However, if the perimeter walls of the garage are of wood construction (with or without insulation or drywall) that rests directly upon the floor of the garage, the potential for damage to the building structure exists. In this situation, the discharge of water from the relief valve pipe must be captured and conveyed to a drain system or to the outdoors. Although it might be argued that the slope of the garage floor would convey the water out the garage door to the outdoors before contacting the wall, concrete floors are rarely perfectly sloped and an occupant's property within the space might cause diversion of the water towards the walls, thus causing damage to the structure.

The discharge pipe of a temperature and pressure relief valve must not be of a size that is smaller than the relief valve outlet size at any point from the outlet to the termination of the pipe. For example, a reduction in the size of the pipe such as from $^3/_4$-inch (19.1 mm) nominal to $^1/_2$-inch (12.7 mm) nominal size will impede the discharge flow, and reduce the discharge capacity of the relief valve. The discharge pipe must not have valves or tees installed. Obviously, a valve in the pipe that could be inadvertently closed would obstruct the discharge pipe and a tee fitting would allow connection of piping from other sources, both of which are dangerous conditions. The code is silent concerning the use of types of relief valve discharge piping that require the use of insert fittings. Common examples are PEX and PEX-AL-PEX tubing. Whether or not these insert fittings constitute enough of a restriction in flow to cause a reduction in maximum discharge capacity of the relief valve is unknown. The authority having jurisdiction must make the decision on whether the use of insert fittings in relief valve discharge piping violates the intent of "no reduction in size" of the discharge pipe.

The relief valve discharge pipe must terminate no more than 6 inches (152 mm) but not less than two times the pipe diameter above the floor surface or a waste receptor (which includes a floor drain) in order to prevent hot water discharge from being directed onto a building occupant [see Commentary Figure 504.6(1)]. The relief valve discharge pipe is also allowed to dis-

Figure 504.6(1)
RELIEF VALVE DISCHARGE

charge to the pan serving the water heater or hot water storage tank [see Commentary Figure 504.6(2)]. Obviously, for termination points to the floor, the floor must be a suitable location for water discharge as previously discussed. Otherwise, a floor drain or waste receptor must be provided to capture and direct discharges from the pipe. In multilevel buildings where water heaters are in the same location on each floor, it is common to provide an indirect piping system to capture any discharge and direct the flow to a single discharge location such as a floor drain on the first level [see Commentary Figure 504.6(3)] or to the outdoors. The code is silent as to the size/shape of waste receptors and the gravity drain line size for capturing and conveying the discharged water from relief valves.

Discharge of a relief valve to laundry trays/tubs, kitchen/utility sinks and shower floors are not suitable locations as this violates Item 6 of this code section. Persons using the fixture could be injured by hot water and steam that could come from the pipe.

The relief valve discharge pipe must drain by gravity and must not be trapped or have sags in the piping that would act as traps. Traps (or piping sags) would restrict the discharge flow and also cause water to be retained in the pipe. This retained water might cause a buildup of scale/corrosion in the pipe or, if the water heater is located in an area subject to freezing such as a garage or attic space, the water might freeze. Both situations have the potential to block the discharge flow from the relief valve, which would create a dangerous condition.

A threaded end on the termination of a relief valve discharge pipe is prohibited, as this would be an invitation for someone to screw on a cap or valve to stop water from dripping from the pipe. What might be perceived as a repair would, in reality, create a very dangerous condition [see Commentary Figure 504.6(4)].

Figure 504.6(2)
RELIEF VALVE INDIRECT WASTE

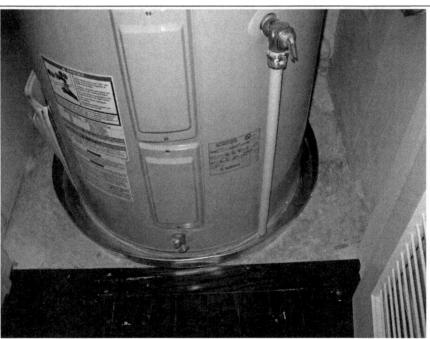

Figure 504.6(3)
T&P RELIEF VALVE PERMITTED TO DISCHARGE TO THE PAN

Figure 504.6(4)
CAPPED RELIEF VALVE TANK TAPPING AND PLUGGED RELIEF VALVE OPENING

Relief valve discharge piping must be one of the piping materials listed in Table 605.4. Although the pipe materials in this table must have a pressure rating of at least 100 psi (690 kPa) at 180°F (82°C) to serve as water distribution piping, all of these materials will most likely have the capability to survive limited exposure to higher temperatures at minimal (nearly atmospheric) pressure. Because a water heater relief valve is required to open at a water temperature of 210°F (99°C) and due to the short amount of time that this high temperature condition would usually exist (e.g., 15 minutes or less), any of the water distribution piping materials listed in Table 605.4 are deemed suitable for this application.

The code does not prohibit the use of the flexible types of water distribution piping listed in Table 605.4 for the relief valve discharge. However, Item 6 of this section would require that the end of the relief valve discharge pipe of flexible material be attached in some manner to direct the water flow to a "safe" location as well as prevent the end of the pipe from moving around so as to not expose a building occupant to a hot water discharge.

Relief valve discharge pipe assemblies manufactured in compliance with ASME A112.4.1 can also be used instead of one of the materials listed in Table 605.4. These assemblies are made of a material that can withstand 250°F (121°C) steam for 15 minutes without a change in the tube diameter and have a threaded fitting that allows hand tightening of the connection to the relief valve. These assemblies are not intended to have any additional fittings attached to or installed in the tubing or for the tubing to be lengthened with additional piping.

A frequently asked question is: How should this section (and even this chapter) be applied to the replacement of existing water heaters where the relief valve discharge pipe was originally allowed to terminate just above the floor, without a drain or waste receptor to capture the flow? The water heater might be in a basement where the building drain is above the elevation of the water heater temperature and pressure (T&P) valve or the water heater might be in the middle of a slab-on-grade building, without a nearby floor drain and not adjacent to an outside wall. Replacement appliances are subject to all of the requirements of the code in effect at the time of replacement. Where water heater replacement installers are confronted with obstacles that prevent strict compliance with this code section, they should consult with the code official before performing the replacement work.

504.7 Required pan. Where a storage tank-type water heater or a hot water storage tank is installed in a location where water leakage from the tank will cause damage, the tank shall be installed in a galvanized steel pan having a material thickness of not less than 0.0236 inch (0.6010 mm) (No. 24 gage), or other pans *approved* for such use.

❖ Where leaks from a storage tank-type water heater or hot water storage tank will cause damage to the building structure, the storage tank water heater or hot water storage tank must be installed in a pan. The pan must be made of galvanized steel of at least the thickness stated or be made of other materials as approved by the code official. Even though prefabricated aluminum and plastic pans are commonly available for many diameters of water heaters and are widely used by installers, these pans must be approved for use by the code official because they are not of galvanized steel. Note that the water heater manufacturer's instructions must also be reviewed for pan requirements because some manufacturers require the use of only a metal pan.

The required thickness of galvanized sheet metal coincides with the minimum thickness tolerance for 24-gage galvanized sheet metal so that a field measurement of thickness would allow for all thicknesses within the tolerance range of 24 gage to be acceptable.

504.7.1 Pan size and drain. The pan shall be not less than $1^1/_2$ inches (38 mm) in depth and shall be of sufficient size and shape to receive all dripping or condensate from the tank or water heater. The pan shall be drained by an indirect waste pipe having a diameter of not less than $^3/_4$ inch (19 mm). Piping for safety pan drains shall be of those materials listed in Table 605.4.

❖ The pan is required to capture water leakage from the water heater or hot water storage tank and convey the water to the pan drain. Therefore, the pan must be large enough to allow for the footprint of the tank as well as some drainage area around the tank so that the water can flow towards where the pan drain is connected to the pan. Generally, storage tank water heater manufacturers require that the pan minimum dimension be at least 2 inches (51 mm) greater than the diameter tank. The depth of the pan must be at least $1^1/_2$ (38 mm) inches to provide for the drain connection fitting, which is typically installed in the side wall of the pan. The pan drain must be at least a $^3/_4$-inch (19.1 mm) nominal pipe size and must be of one of the materials listed in Table 605.4. The pan drain pipe must not be directly connected to the sanitary drain system. Just as drain pipes downstream of the required air gaps at the relief valve discharge pipe termination can be combined together [see Commentary Figure 504.7(1)], pan drains can also be combined together in a similar manner to terminate at a single location at a floor drain, a waste receptor or to the outdoors.

Note that the pan drain pipe material must be of one of the materials listed for piping material suitable for water distribution system use. The concern is that a hot water storage tank or the water connections to the tank could leak hot water for extended periods of time until discovered. If the pan drain pipe material was not made of a material suitable for hot water, the pipe might sag between supports causing sluggish draining and clogging which could cause the drain pan to overflow.

A frequently asked question is: How should this section (and even this chapter) be applied to an existing water heater location that did not originally have a code requirement for installation of pan? The water heater

Figure 504.7(1)
WATER HEATER IN PAN

might be in a basement where the building drain is above the elevation of the bottom of the water heater or the water heater might be in the middle of a slab-on-grade building, without a nearby floor drain and not adjacent to an outside wall.

As stated in the commentary to Section 502.1.1, the code is silent on this topic, thus requiring the code official to determine to what degree the installation must comply with the code. Where water heater replacement installers are confronted with obstacles that prevent strict compliance with this code section, they should consult with the code official before performing the replacement work.

504.7.2 Pan drain termination. The pan drain shall extend full size and terminate over a suitably located indirect waste receptor or floor drain or extend to the exterior of the building and terminate not less than 6 inches (152 mm) and not more than 24 inches (610 mm) above the adjacent ground surface. Where a pan drain was not previously installed, a pan drain shall not be required for a replacement water heater installation.

❖ Although a pan drain is not required to follow the same rules as for gravity flow sanitary drainage, it should be obvious that any reductions in pan drain pipe size would restrict flow and be a potential clogging point, especially where pans have collected years of dust and possible rust flakes from corroding steel tanks. Since the drainage from the pan is basically clear water, the drain can terminate either outdoors or at a waste receptor or floor drain. Indirect connections at waste receptors or floor drains can be through either an air break or an air gap, since the only concern is to prevent a possible sewage backup into the pan. Where the pan drain terminates outdoors, a termination point of at least 6 inches (152 mm) above the ground protects the open end from being blocked by careless landscaping work and vegetation growth, which might also allow for insects to build nests in the opening. A maximum height of 24 inches (610 mm) provides an allowance for snow and ice build up on the ground but is low enough so as to not create a safety concern for passersby.

Replacement water heaters should be installed as if they are new water heaters. As such, the installation location might require a pan where one was not installed originally. Where the pan is required for the replacement installation, many times there is not a place for the pan drain to terminate. For example, the replacement of a water heater installed in a partially (or fully) finished basement would require a pan, but perhaps the building drain system is completely "overhead" or there is not a floor drain. Thus, there is not a drain for the pan drain to be routed to. In these situations, a pan drain is not required. Having a pan for the water heater will at least contain a small leak that might be noticed by the building occupants. Although the code does not require a water leak detection alarm, the building owner, occupants or even the water heater installer might be inclined to install a water leak detection alarm in the pan to provide a warning of the presence of water. Some water detection units have the capability to shut off the incoming water supply when a leak is detected.

SECTION 505
INSULATION

[E] 505.1 Unfired vessel insulation. Unfired hot water storage tanks shall be insulated to R-12.5 (h · ft^2 · °F)/Btu (R-2.2 m^2 · K/W).

❖ This section requires that unfired hot water storage tanks be insulated with not less than an *R*-value of 12.5 (h • ft^2 • °F)/Btu [R-2.2 m^2 • K/W]. External insulation is necessary for unfired hot water storage tanks to conserve energy. Such tanks are used in conjunction with circulating-type water heaters, hot water supply boilers and various heat exchangers. Unlike storage-type water heaters, unfired storage tanks can be purchased without an insulation jacket or covering.

Bibliography

The following resource materials are referenced in this chapter or are relevant to the subject matter addressed in this chapter.

Legionella 2003: Update and Statement by the Association of Water Technologies. Rockville, MD: Association of Water Technologies.

Chapter 6:
Water Supply and Distribution

General Comments

Potable water is viewed by the U.S. government as a precious commodity. The U.S. Environmental Protection Agency's (EPA) Office of Drinking Water specifies the quality requirements for water to be classified as potable. Water quality requirements regarding allowable levels of contaminants are becoming increasingly restrictive.

With this emphasis on water quality, plumbing equipment manufacturers have become increasingly concerned with the potential for introducing contaminants into the water through material selection. Faucets that supply drinking water are required to comply with National Sanitation Foundation (NSF) standards No. 61 and 372 to confirm that the product has a very low lead content. U.S. federal law requires drinking water faucets and components to contain not more than 0.25-percent lead content. As such, manufacturers of products that convey drinking water are paying greater attention to the selection of materials that are exposed to potable water. All materials used in the potable water supply system must be evaluated for their possible effect on the potable water with which they are in contact.

As the supplies of suitable raw water to make potable water become less available, some jurisdictions are seeing water utilities providing reclaimed water service for their customers. Reclaimed water is tertiary treated waste water that has been sufficiently disinfected to be safe for exposure to humans but not of drinking-water quality. In the majority of areas where reclaimed water is in use, local regulations only allow the water to be used for outside irrigation demands.

The shortage and rising cost of potable water have inspired designers, water utilities and water conservation activists to investigate other sources for water for use inside and outside of buildings. Gray water (that discharged from bathtubs, showers, handwash sinks and clothes washers) is, once properly treated, a viable water source for flushing water closets and urinals and for water supply to below-grade irrigation systems. Although this chapter was originally intended to only cover the potable water distribution system, the distribution of nonpotable water within buildings is also covered by some of the regulations within this chapter.

Purpose

Chapter 6 regulates the supply of potable water from both public and individual sources to every fixture and outlet so that it remains potable and uncontaminated through cross connection. Chapter 6 also regulates the design of the water distribution system, which will allow fixtures to function properly. The unique requirements of the water supply for health care facilities are addressed separately. It is critical that the potable water supply system remain free of actual or potential sanitary hazards by providing protection against backflow.

SECTION 601
GENERAL

601.1 Scope. This chapter shall govern the materials, design and installation of water supply systems, both hot and cold, for utilization in connection with human occupancy and habitation and shall govern the installation of individual water supply systems.

❖ Although this code section was written long ago when only potable water sources were used for the water supply for all fixtures, portions of this chapter are also applicable to the water distribution systems of nonpotable water supplies, such as treated gray water and rainwater. Where a supply of potable water is available from a public water utility or an individual source (e.g., a private well), all requirements in this chapter are applicable.

601.2 Solar energy utilization. Solar energy systems used for heating potable water or using an independent medium for heating potable water shall comply with the applicable requirements of this code. The use of solar energy shall not compromise the requirements for cross connection or protection of the potable water supply system required by this code.

❖ Solar heating systems consist of two basic types: direct connection and indirect connection. In a direct connection system, the heat transfer fluid is potable water. This type of system has potential safety problems that must be addressed in its design, inspection and maintenance. See the *International Mechanical Code*® (IMC®) for information on the design and installation of solar heating systems. The main concerns are contamination of the potable water source from piping and joint materials used in the system and from the inadvertent introduction of nonpotable or toxic transfer fluids at a future date. The use of direct connection systems is typically limited to solar water-heating systems where the potable water is heated directly by the solar collector and circulated through the system, and is suited for use only in areas where the water in the collectors is not subject to freezing.

In an indirect connection system, a freeze-protected heat transfer fluid is circulated through a closed loop to a heat exchanger. The heat is then transferred indirectly to the potable water. The fluids in these systems are not potable and are occasionally toxic. The type of heat exchanger used in these systems depends on the exact nature of the transfer fluid used. As required by Section 608.16.3, a single-wall heat exchanger may be used in systems where the fluid is nontoxic, and a double-wall heat exchanger must be used in systems using toxic fluids.

601.3 Existing piping used for grounding. Existing metallic water service piping used for electrical grounding shall not be replaced with nonmetallic pipe or tubing until other *approved* means of grounding is provided.

❖ In some older existing buildings, electrical equipment is grounded by connecting a grounding conductor to the interior metal water piping system. More commonly, a metal water pipe is used as a junction point for multiple grounding electrode conductors. The continuity of this electrical ground is essential to maintaining the safety of the electrical system. The replacement of a portion of that metal piping system with nonmetallic piping could interrupt the continuity of that electrical grounding system and create a potentially hazardous situation. When the replacement of any portion of an existing metal water piping system with nonmetallic piping is proposed, the method of maintaining electrical system continuity in the existing building must be approved by the code official.

Bonding of underground metallic water service piping is mandated in the *National Electric Code*® (NEC®) (NFPA 70) and the *International Residential Code*® (IRC®) as part of the grounding electrode system.

601.4 Tests. The potable water distribution system shall be tested in accordance with Section 312.5.

❖ Requirements for water supply system testing are described in Section 312.

601.5 Rehabilitation of piping systems. Where pressure piping systems are rehabilitated using an epoxy lining system, such lining system shall comply with ASTM F 2831.

❖ Metallic water-service piping and water distribution-piping can internally corrode, resulting in leaks. Replacing piping is often a very expensive option because of the process of getting to the piping, the extended labor time to replace the piping in confined areas, the process of repairing finished building materials that were removed and the time that the building or tenant space cannot be used during the replacement. Epoxy internal lining systems can be used to repair leaking piping without removal of the piping. Although some building materials might have to be removed for performing the lining work, the amount is often just a fraction of the materials removed to replace the piping. Piping is internally dried with hot air and grit-blasted to remove oxides and corrosion. The material and installation standard covers lining of piping $^1/_2$ inch (12.7 mm) to 36 inches (914 mm) in diameter by "blown through," spin cast or handsprayed methods. Curing of the coating is by either hot air or chemical reaction. For potable water service applications, the standard requires that the materials comply with NSF 14 and NSF 61. The standard requires pressure testing of the lined piping.

SECTION 602
WATER REQUIRED

602.1 General. Structures equipped with plumbing fixtures and utilized for human occupancy or habitation shall be provided with a potable supply of water in the amounts and at the pressures specified in this chapter.

❖ Table 403.1 specifies the minimum required number of fixtures for a building or tenant space. Even if there is only a single toilet room where the water closet uses gray water for flushing, the code requires a fixture for handwashing for which a potable water supply is required. If there is a plumbing fixture (requiring a supply of water) that is installed in a building that does not require any plumbing fixtures, a potable water supply is required for the building. The potable water supply must have the capacity to supply water at a pressure and flow rate that is required for the fixture to properly function.

602.2 Potable water required. Only potable water shall be supplied to plumbing fixtures that provide water for drinking, bathing or culinary purposes, or for the processing of food, medical or pharmaceutical products. Unless otherwise provided in this code, potable water shall be supplied to all plumbing fixtures.

❖ Potable water is required for all plumbing fixtures unless nonpotable water is used for purposes indicated in Chapter 13. Some common uses of nonpotable water are irrigation and the flushing of water closets and urinals.

The desire to use nonpotable water supplies is on the rise as a means of conserving water resources. Chapter 13 provides information on nonpotable water storage and handling.

602.3 Individual water supply. Where a potable public water supply is not available, individual sources of potable water supply shall be utilized.

❖ Where a public water system is unavailable at or near the location of the building, an individual water source must be provided. Note that an individual water supply must be determined to be potable before it can be used as a potable water source. The potability of water can only be determined by laboratory testing (see Section 602.3.3).

602.3.1 Sources. Dependent on geological and soil conditions and the amount of rainfall, individual water supplies are of the following types: drilled well, driven well, dug well, bored well, spring, stream or cistern. Surface bodies of water and land cisterns shall not be sources of individual water supply unless properly treated by *approved* means to prevent contamination.

❖ The sources for an individual water supply include wells, springs, cisterns, streams and surface water impoundments (man-made reservoirs). Wells are probably the most common source of individual water supply used today and consist of a vertical shaft that extends down into the earth to a water-bearing stratum. Wells may be bored, drilled, driven or dug. See

Section 202 for the definitions of each well type and the definition of "Cistern."

Water from surface water impoundments must be treated if used as a supply of potable water. This may include some form of chlorination or ozone treatment.

602.3.2 Minimum quantity. The combined capacity of the source and storage in an individual water supply system shall supply the fixtures with water at rates and pressures as required by this chapter.

❖ Some individual water supplies may not have the capacity to directly provide for the peak flow rate at the required pressure. A storage tank is often necessary to accumulate a volume of water to accommodate the demand on the building water distribution system. This section allows the combination of the source flow/pressure and a water storage tank to provide the required peak demand flow rate at the required pressure. For example, a well pump is typically used in combination with a water storage tank in order to provide for the peak water demand for a building.

602.3.3 Water quality. Water from an individual water supply shall be *approved* as potable by the authority having jurisdiction prior to connection to the plumbing system.

❖ An individual water supply source is always of unknown quality until the water from the source is tested under laboratory conditions. The only way the water supply can be approved is if a reputable test laboratory has determined that the water is potable (of drinking water quality).

602.3.4 Disinfection of system. After construction, the individual water supply system shall be purged of deleterious matter and disinfected in accordance with Section 610.

❖ A private water supply system must be disinfected after it has been constructed. Materials and equipment used for construction of wells could have been exposed to many kinds of pollutants and possibly, contaminants. Most well construction practices require "blowing" of the well (with compressed air) to determine the well's approximate water capacity. The test flushes water out of the well for a period of time, which helps to clear much of the debris from the well. However, as a precaution, the well must be disinfected before actual use of the water. The disinfection procedure must be as prescribed in Section 610 for all potable water systems (see commentary, Section 610). Typically, repairs do not introduce the higher potential for pollutants and contaminants. Additionally, when the well is in service, the need to get it back on-line does not typically allow time for the well to be disinfected.

602.3.5 Pumps. Pumps shall be rated for the transport of potable water. Pumps in an individual water supply system shall be constructed and installed so as to prevent contamination from entering a potable water supply through the pump units. Pumps shall be sealed to the well casing or covered with a water-tight seal. Pumps shall be designed to maintain a prime and installed such that ready *access* is provided to the pump parts of the entire assembly for repairs.

❖ Some materials used in the construction of pumps may not be suitable for contact with potable water. For example, a metal pump volute or impeller casting with a high-lead content would not be suitable for potable water service, as the lead could leach into the water system. Some pump designs could allow contamination of potable water by allowing undesirable materials to enter the flow stream. For example, shaft-bearing grease might be able to work its way into the water. Pump manufacturers offer specific pump models that are rated for pumping potable water.

Proper installation of a potable water pump is also very important towards keeping the condition of the water potable. Any water or contaminants outside of the pump, well casing or pump housing must be prevented from entering the pump or well. For example, potable water pumps must not be located in pits below grade, as ground water could flood the pit and possibly enter the pump. Pumps must also not require periodic priming (an initial filling with water) as part of their normal operation. As all pumps eventually require service or replacement, the installation must provide for ready access for performing those functions (see Chapter 2 definition of "Ready access").

602.3.5.1 Pump enclosure. The pump room or enclosure around a well pump shall be drained and protected from freezing by heating or other *approved* means. Where pumps are installed in basements, such pumps shall be mounted on a block or shelf not less than 18 inches (457 mm) above the basement floor. Well pits shall be prohibited.

❖ A well (pump) pit, being below grade, has the potential for collecting rainwater, surface water or ground water, which could enter the pump and contaminate the potable water. Where located in a basement, the potable water pump must be elevated in the event that the basement becomes partially flooded with ground water or a waste-water backup. Pumps located outside a building must be properly protected from freezing during the winter months. Insulated housings along with a heat source (a small electric heater or, sometimes, just an incandescent lightbulb) are typically used to protect the pump and piping from freezing. Insulation of the walls of a pump housing without the installation of a heat source is only suitable where climatic conditions are very mild (see commentary, Section 305.6).

SECTION 603
WATER SERVICE

603.1 Size of water service pipe. The water service pipe shall be sized to supply water to the structure in the quantities and at the pressures required in this code. The water service pipe shall be not less than $^3/_4$ inch (19.1 mm) in diameter.

❖ The water service system conveys water from a source of pressurized water supply to the beginning of the water distribution system in a building. The water service pipe is just one component of many in the water service system. The source of pressurized

WATER SUPPLY AND DISTRIBUTION

water supply could be a public water main, a private well pump or several other types of arrangements. Where a public water main is the source of supply, the water service system includes the tap or tee on the water main, the utility water service connector pipe, the water service pipe, pipe fittings, valves, a water meter and, more frequently than not, a backflow prevention valve [see Commentary Figure 603.1(1)]. Where a private well pump is the source of pressurized water supply, the water service system includes a check valve, the drop pipe down the well casing to the pump, the water service pipe, pipe fittings, valves and a pressure tank [see Commentary Figure 603.1(2)]. Note that the water service system ends inside the building at the required main water service valve (see Section 606.1, Item 2).

Although this code section is focused only on the sizing of the water service pipe, the sizes of the other components in the water service system can have an impact on determining the required size of the water service pipe. Thus, in order to properly size a water service pipe, the entire water service system must be considered. Because the design of a water service system requires a basic knowledge of pressure change caused by elevation difference, peak water demand, pipe friction loss, pipe velocity effects, pipe material selection and fitting/component pressure losses, the following is provided to briefly explain each of these subjects.

Pressure Change Caused by Elevation Difference

Consider a pipe between a public water main and a plumbing fixture in a building as shown in Commentary Figure 603.1(3). As the water must flow uphill to reach the fixture, energy is consumed, which will result in a loss of pressure. Conversely, as shown by Commentary Figure 603.1(4), as water flows downhill to reach the fixture, energy is gained, which will result in an increase in pressure. Thus, changes in elevation must be considered when determining the pressure at the discharge end of the water service system. Pressure gains are a bonus when designing a water service system. Pressure losses, however, are of great concern because a lack of adequate pressure for the water distribution system could cause operational

**Figure 603.1(1)
WATER SERVICE SYSTEM, PUBLIC WATER MAIN**

problems with plumbing fixtures. The following equation converts a change in elevation to pressure:

$$P = 0.433 \times h$$

where:

- P = Pressure created by a head of water (pound per square inch, abbreviated as "psi").
- h = Elevation difference (feet).

If the elevation of the outlet end of the water service system is higher than the elevation of the outlet of the pressurized water source, then pressure will be lost. If the elevation of the outlet end of the water service system is lower than the elevation of the outlet of the pressurized water source, then pressure will be gained.

Determining Peak Water Demand

Estimating the peak water demand (the maximum required water flow) for a building is the most difficult, and yet, the most important aspect of any water service system design problem. The term "peak water demand" is used to describe that flow, which is the maximum total water flow demand of the water distribution system at any given time. Peak water demand is not simply the addition of the maximum allowable flow rates of every fixture in the building because it is unrealistic, in most cases, to assume that all fixtures would be operating at the same time.

One commonly used and widely accepted method of estimating peak water demand is the "wsfu" method developed in 1923 by Dr. Roy B. Hunter. The "Hunter" method uses probability analysis to determine a frequency and duration of use for various types of fixtures. The method involves the use of unitless water demand factors called "water supply fixture units" (wsfu) for each common type of plumbing fixture. Actual studies of plumbing systems in federal buildings were performed to correlate the wsfu values to flow rates in gallons per minute (gpm). The results of these efforts were published in the National Bureau of Standards Report No. BMS 65, *Methods of Estimating Loads in Plumbing Systems*.

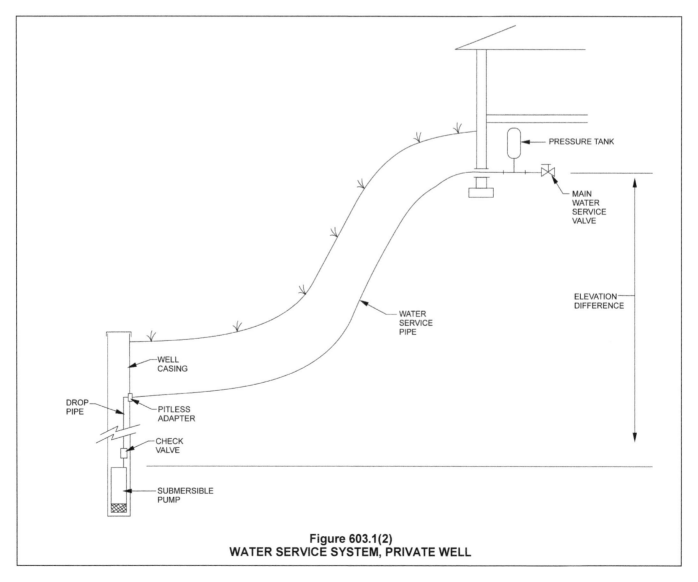

Figure 603.1(2)
WATER SERVICE SYSTEM, PRIVATE WELL

WATER SUPPLY AND DISTRIBUTION

Because water conversation measures have been mandated throughout the years since Dr. Hunter's original research, more recent studies by various parties have made adjustments to his original wsfu values for plumbing fixtures to account for the reduced flow rates of more modern fixtures. As these newer studies were performed independently from one another, plumbing system designers have available to them several different sets of plumbing fixture wsfu values and correlating flow rates. Caution must be exercised to make sure that wsfu values and correlating flow rates from the various studies are not intermixed as doing so might produce erroneous results.

A shortcoming of the various wsfu methods currently available is that they may not provide suitable results for projects that have extremely high water demands or very low water demands. For example, the peak water demand at halftime in a football stadium could approach a flow rate as if all fixtures were operating simultaneously. The use of most wsfu methods would probably result in undersized piping for that application. Conversely, the peak water demand in a very small building with only a few fixtures could be very small and, as such, most wsfu methods would probably oversize the piping for that application. While oversizing of piping will not affect the operation of the fixtures, larger than necessary piping is not economical. However, undersized piping can create operational problems that might be difficult or impossible to rectify after completion of a building. Ultimately, the system designer uses his or her experience to know when "adjustments" of calculated pipe sizes are needed for specific applications in order to reduce the potential liability of a system not working under peak water demand conditions.

Table E103.3(2) in Appendix E of the code contains one commonly used tabulation of wsfu values for various plumbing fixtures and groups of plumbing fixtures. Table E103.3(3) is the coordinating table for converting these wsfu values to gallons-per-minute flow. Note that Table E103.3(3) is actually two tables—one that covers systems having predominantly flush-tank-type water closets and the other that covers systems having predominantly flushometer-type water closets. Although Appendix E does not provide any guidelines that define what "predominantly" means, a conservative design practice might be to choose the flushometer column whenever the total number of flushometers is greater than 10 percent of the total number of fixtures in the building.

Figure 603.1(3)
DIFFERENCE IN ELEVATION, LOSS OF PRESSURE

Figure 603.1(4)
DIFFERENCE IN ELEVATION, GAIN OF PRESSURE

WATER SUPPLY AND DISTRIBUTION

Pipe Friction Loss

When water flows through a pipe, the friction of the water against the inside wall of the pipe resists the movement of the water. This resistance decreases the available flowing pressure at the discharge end of the pipe. As the water flows faster in the pipe, the resistance (and loss of pressure) becomes greater. This pressure loss is called "pressure drop caused by friction loss" or simply "friction loss."

One formula for determining the friction loss in a pipe is a variation of the Hazen-Williams equation:

$$f = 4.52 \times 100 \times (q^{1.85})/[(C^{1.85}) \times (d^{4.8655})]$$

where:

f = Friction loss in psi per 100 feet of pipe.

d = Inside diameter of pipe or tube, inches.

q = Fluid flow, gpm.

C = Surface roughness constant.

Typical values for C, which take into account some increase in roughness of the inside wall as the pipe ages, are:

- Asbestos-cement - 140
- Cast iron - 100
- Cement-mortar-lined ductile iron pipe - 140
- Concrete - 100
- Copper - 150
- Steel - 120
- Galvanized steel - 120
- Polyethylene - 150
- Polyvinyl chloride (PVC) - 150
- Fiber-reinforced plastic (FRP) - 150

Although this equation allows the designer to calculate the friction loss for any given flow, pipe material and diameter, the equation is difficult to solve without a scientific calculator or computer spreadsheet. As such, some designers use commonly available pipe friction loss graphs. Commentary Figure 603.1(5) illustrates a friction loss graph for PEX tubing having SDR 9 wall thickness. To use a friction loss graph, the graph is entered on the vertical axis and the flow rate that is required by the application is located. From that point, moving horizontally to the right, the intersection with one of the tube size lines is found. From that intersection, moving vertically down to the horizontal axis, the pressure drop (friction loss) per 100 feet (30 480 mm) of tube is read. For example, given a flow of 10 gpm and a 1-inch (25 mm) tube, the pressure drop in 100 feet (30 480 mm) of tube is approximately 5.5 psi. In other words, if the water source has a pressure of 50 psi and a flow of 10 gpm through a 100-foot-long (30 480 mm) section of tubing, the flowing pressure at the discharge end of the tube would be 50 - 5.5 = 44.5 psi.

Using the same friction loss graph and 10 gallons per minute, a $^3/_4$-inch (19.1 mm) nominal tube size could have been chosen, resulting in approximately 22 psi of pressure drop in 100 feet (30 480 mm) of tube. As pipe sizes become smaller, the same flow results in a higher velocity in the pipe, which causes significantly more pressure drop. Based upon a manufacturer's recommended maximum velocity for the pipe material, the size of the pipe is chosen to provide for a pressure loss that allows for an acceptable amount of flowing pressure at the discharge end of the pipe.

Flow Velocity Limitations

The velocity of water flowing in a pipe must be sufficient to keep suspended material in the water from building up. But more importantly, the velocity cannot be too great so as to avoid the destructive effects of erosion, cavitation and pressure surges. High velocities can also cause noise that might be a nuisance to building inhabitants. Much experience and testing involving the sizing of various water pipes have resulted in a "general engineering practice" of limiting the maximum velocity in water service pipes to not greater than 8 feet per second (2.4 m/s). However, manufacturers of specific types (materials) of pipe may allow or suggest higher velocities in order to permit smaller pipe sizes for a given flow. Additional discussion of recommended velocity limits is provided in the commentary to Section 604.

High velocities can result in erosion of the pipe wall caused by entrained air bubbles, sand particles and other suspended solid matter in the water. Another problem with excessive velocity in the pipe is cavitation. Where there is a change in direction of the flow, such as at an elbow, vapor bubbles can form (the water "boils") and then collapse (the vapor condenses). The formation and collapse of these bubbles can have a very destructive effect on the pipe wall material [see Commentary Figure 603.1(6)].

"Water hammer" is another problem that can occur because of high velocity in the pipe. The term "water hammer" describes the noise and vibrations produced by destructive forces known as hydraulic shock, which is the result of water flow at high velocity being abruptly stopped. See the commentary to Section 604.9 for further discussion of water hammer.

Pipe Material Selection

Knowledge of local water conditions is necessary to choose an appropriate pipe material for the water service system. The water conditions can also have an impact on the size of the pipes. Water containing dissolved mineral salts, such as sulfates, bicarbonates, chlorides of calcium, sodium and magnesium, can form a scale or buildup on the inner wall of the pipe. This phenomenon, known as "caking," can significantly increase the friction loss in a pipe. Corrosive conditions can also occur in metallic piping where the water has significant amounts of dissolved gases such as oxygen and carbon dioxide. A chemical analysis of the proposed water supply may be necessary to determine dissolved solid content, pH range, car-

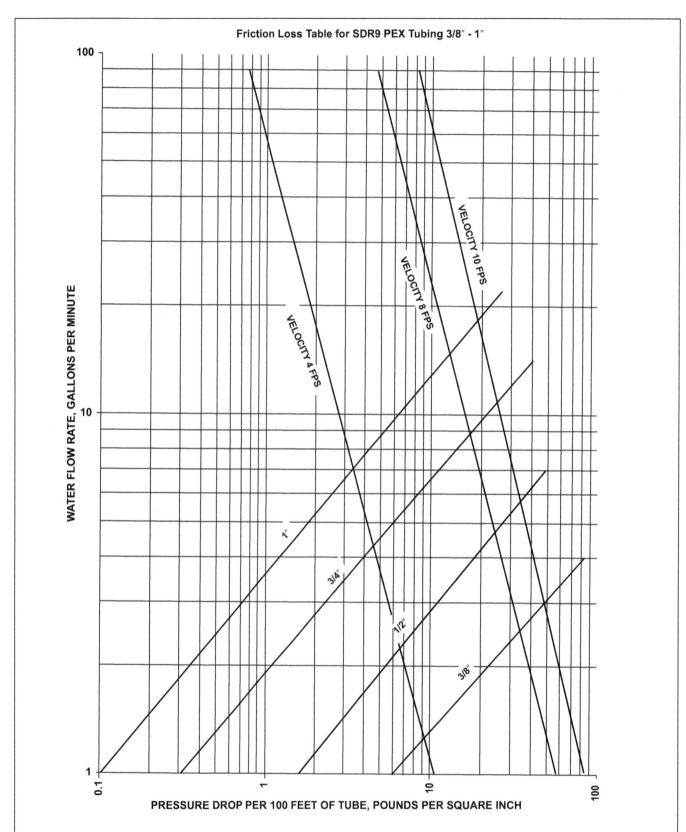

For SI: 1 inch = 25.4 mm, 1 foot = 304.8 mm, 1 gallon per minute = 3.785 L/m, 1 pound per square inch = 6.895 kPa, 1 foot per second = 0.3048 m/s.

Figure 603.1(5)
FRICTION LOSS IN SDR 9 PEX TUBING

**Figure 603.1(6)
CAVITATION**

bon dioxide content and the presence of dissolved oxygen. Local code officials, water utilities and piping manufacturers are often the best sources for information concerning the choice of pipe materials for the available potable water to be used.

As most water service system piping is buried, soil conditions could be a factor in choosing an appropriate pipe material. Soils that are especially aggressive to metallic piping, such as saltwater marshland and high sulphur mineral content soils, might require that the piping be made of corrosion-resistant materials such as plastic. Where soils have residual amounts of solvents, fuels or where insecticides or fertilizers are consistently used on the soils above, metallic pipe materials might be a more suitable pipe material choice. Local experienced soils engineers should be consulted wherever questionable soil conditions are present.

Fittings, Valves and Other Component Friction Losses

Pipe fittings in a piping system cause turbulence in the water flow, resulting in additional friction loss. One method to account for the friction losses caused by fittings is called the "equivalent length method." In other words, the friction caused by a fitting can be equated to the friction loss of a straight length of pipe. Tables E103.3(5) and E103.3(6) in Appendix E indicate the equivalent length of pipe and tube, respectively, for various fittings and valves.

A more commonly used method to estimate the friction loss caused by fittings and valves is to estimate an amount for the equivalent length for fittings and valves as a percentage of the actual developed length of straight piping. In general, 50 percent is frequently found to be a reasonably accurate estimate. For example, if the length of straight piping to be installed is 70 feet (21 336 mm), then the equivalent length of pipe used for the friction loss determination will be:

Equivalent length of pipe = actual length of pipe + [actual length of pipe × (0.50)]

Equivalent length of pipe = 70 + (70 × 0.50)

Equivalent length of pipe = 105 feet (32 004 mm)

Using this equivalent length estimating method requires careful judgment. For example, if the straight amount of piping to be installed is only 10 feet (3048 mm), but there are a number of fittings and valves within that length, then the use of 50 percent to determine the actual friction loss will probably underestimate the actual friction loss. Conversely, if a pipeline is 100 feet (30 480 mm) long with few fittings and full-open valves, then the use of 50 percent will probably result in overestimation of the actual friction loss. Designers should bear in mind that it is generally better to err on the side of computing slightly more friction loss in order to avoid underestimating the pressure loss for a water service system.

Other components such as taps (connection points to the water main), water meters, backflow prevention valves, strainers and filters cause restriction to the flow and, therefore, those pressure losses must be accounted for when calculating the flowing pressure available at the discharge end of the water service system. "Tap or tee loss" is the pressure loss caused by the entry of flow into the connection fitting on the public water main pipe. Table E103.3(4) in Appendix E indicates the tap or tee losses for various pipe sizes at various flow rates.

Water meters for metering a public water supply for water billing purposes are most often the responsibility of the public water utility. At the initial design stage of the water service supply system, the designer usually does not know the meter size, type or model in order to account for the water meter pressure loss. Initially, the designer can use the chart in Commentary Figure 603.1(7) to estimate the maximum allowable friction loss through various sizes of displacement-type (nutating disc) water meters. Displacement-type meters are the most common type of meter used for small-to-medium size residential and light commercial buildings. As the chart in Commentary Figure 603.1(7) is known to yield much higher than actual pressure losses, once the designer completes the preliminary design of the water service system, he or she can then consult with the public water utility to determine the exact brand/model of water meter to be used so that a more accurate pressure loss for the water meter can be entered into the calculations.

WATER SUPPLY AND DISTRIBUTION

Figure 603.1(7)
FRICTION LOSS FOR DISPLACEMENT-TYPE WATER METERS
(Source: Standard Plumbing Engineering Design, Nielsen, L. 1982)

Pressure losses for all other components in the water service system, such as backflow prevention valves, strainers and filters, must be obtained from the manufacturers of those components.

A Method for Sizing a Water Service Pipe

The code does not dictate a method for determining water service pipe size. The following sample problem illustrates one method based upon engineering principles:

Sample Problem

Consider the water service system arrangement illustrated in Commentary Figure 603.1(1). This public building will have eight flush tank water closets and six lavatories. The public water main pressure ranges from 50 to 55 psi. The utility water service connector is $1^1/_4$-inch (32 mm) Type L copper and is 10 feet (3048 mm) long. The length of the Type L copper water service pipe will be 160 feet. The water meter pressure loss will be 3 psi. The backflow prevention valve pressure loss will be 2 psi. The three valves in the water service system are full-open ball valves. There are not any angle-type pipe fittings (elbows) in the water service system. The elevation difference is 20 feet (uphill). The flowing pressure required at the main water service valve must be not less than 25 psi. Determine the smallest water service pipe that can be used and the resulting flowing pressure available at the main water service valve just downstream of the water meter.

Solution

1. Determine the peak water demand for the building by summing the wsfu values for all the fixtures. From Table E103.3(2), a public flush tank water closet is 5 wsfu and a public lavatory is 2.0 wsfu. Therefore, the total wsfu is (8 × 5) + (6 × 2) = 52 wsfu. From Table E103.3(3), under "Predominantly Flush Tanks," the interpolated flow rate for 52 wsfu is approximately 30 gpm.

2. Determine the pressure change caused by elevation difference. Because the water must flow uphill to the outlet end of the water service system, there will be a pressure loss. The amount of pressure loss is:

   $P = 0.433 \times h$

   $P = 0.433 \times 20$ feet

   $P = 8.66$ psi

   Because the water is flowing uphill, this pressure will be subtracted from the public water main pressure.

3. Determine the pressure loss caused by the tap on the water main. In Table E103.3(4), read down the left-hand column to find 30 gpm. Move across that row to the column for $1^1/_4$ inches (32 mm) and read a value of 0.69 psi as the pressure loss for the tap.

4. Determine the pressure loss through the utility water service connector. Using the friction loss

WATER SUPPLY AND DISTRIBUTION

chart in Commentary Figure 603.1(8), find the point for 30 gpm on the vertical scale on the left side of the chart. Move horizontally to the right until intersecting the "1^1/$_4$-inch tube" size line. Note that the intersecting point is midway between the 10-feet-per-second and 5-feet-per-second velocity lines, indicating the pipe velocity is slightly less than 8 feet per second (which is below the limit according to good engineering practice). Move vertically down from the intersecting point to the horizontal axis and read a pressure drop of 7.5 psi per 100 feet (30 480 mm). The pressure loss for the 10-foot (3048 mm) length of utility water service connector is 7.5 psi × (10 feet/100 feet) = 0.75 psi.

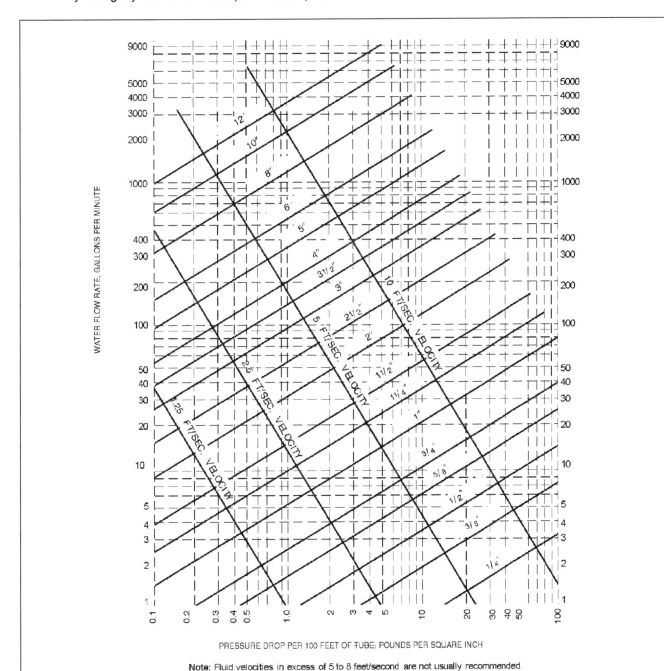

Figure 603.1(8)
FRICTION LOSS IN SMOOTH PIPE (Type L, ASTM B88 COPPER TUBING)

For SI: 1 inch = 25.4 mm, 1 foot = 304.8 mm, 1 gpm = 3.785 L/m, 1 psi = 6.895 kPa, 1 foot per second = 0.305 m/s.

a. This chart applies to smooth new copper tubing with recessed (streamline) soldered joints and to the actual sizes of types indicated on the diagram.

WATER SUPPLY AND DISTRIBUTION

5. Choose a water service pipe size and determine the pressure loss through that pipe size. Using the friction loss chart in Commentary Figure 603.1(8), find the point for 30 gpm on the vertical scale on the left side of the chart. Move horizontally to the right until intersecting the $1^1/_4$-inch (32 mm) tube size line. Note that the intersecting point is midway between the 10-feet-per-second and 5- feet-per-second velocity lines, indicating the pipe velocity is slightly less than 8 feet per second (again, within the velocity limit according to good engineering practice). Move vertically down from the intersecting point to the horizontal axis to read a pressure drop of 7.5 psi per 100 feet. The pressure loss for the 160 foot length of water service pipe is 7.5 psi × (160 feet/100 feet) = 12 psi.

6. Determine pressure loss for the valves. In Table E103.3(6), enter the left-hand column to find the $1^1/_4$-inch (32 mm) tube size. Move across that row to the "Ball Valve" column and read a value of 0.5 feet equivalent length of tube (per valve). The pressure loss for the three valves is 7.5 psi × [(3 × 0.5 feet)/100 feet)] = 0.11 psi.

7. Determine the flowing pressure at the end of the water service system. Using the minimum pressure of 50 psi as the public water main pressure, subtract the following pressures: 8.66 (elevation), 0.69 (tap), 0.75 (utility connector pipe), 12.0 (water service pipe), 0.11 (three valves), 3 (water meter) and 2 (backflow prevention valve): 50 - (8.66 + 0.69 + 0.75 + 12 + 0.11 + 3 + 2) = 22.79 psi.

    NOTE: The resulting flowing pressure is less than the 25 psi required; therefore, the pressure drop in the water service system must be reduced to yield a greater flowing pressure. Choosing the next larger size of water service pipe could be a solution. The following check is performed to verify the size is adequate.

8. Choose a $1^1/_2$-inch (38 mm) water service pipe size and determine the pressure loss through the water service pipe. Using the friction loss chart in Commentary Figure 603.1(8), find the point for 30 gpm on the vertical scale on the left side of the chart. Move horizontally to the right until intersecting the $1^1/_2$-inch (38 mm) tube size line. Move vertically down from the intersecting point to the horizontal axis and read a pressure drop of 3.5 psi per 100 feet (30 480 mm). The pressure loss for the 160-foot length of $1^1/_4$-inch (32 mm) water service pipe is 3.5 psi × (160 feet/ 100 feet) = 5.6 psi. The revised flowing pressure is 50 - (8.66 + 0.69 + 0.75 + 5.6 + 0.11 + 3 + 2) = 29.19 psi. The design provides an acceptable amount of flowing pressure.

    NOTE: It is not necessary to increase the size of the valves, the water meter or backflow prevention valve because the calculation already includes the pressure losses for the initially chosen sizes of those components. In actual practice, the ball valves would most likely be increased to a $1^1/_2$-inch (38 mm) size for "connection convenience" to the $1^1/_2$-inch (38 mm) water service pipe. However, because water meters and backflow prevention valves are substantially more costly in larger sizes, the size of those components can remain at the $1^1/_4$-inch (32 mm) size as the system design is acceptable with those sizes.

    For SI: 1 gallon per minute = 0.06309 L/s,
    1 pound per square inch = 6.895 kPa,
    1 foot per second = 0.3048 m/s,
    1 inch = 25.4 mm, 1 foot = 304.8 mm.

Fire Sprinkler Systems

Fire sprinkler systems can impact the design of the water service system if the sprinkler system and the water distribution system share a single water service system. In most cases, the sprinkler system water demand will require the components of the water service system to be much larger than what is required by just the demand of the water distribution system alone. It is generally assumed that the sprinkler system and the water distribution system will not be operating simultaneously.

Minimum Size of the Water Service Pipe

The water service pipe must be not smaller than $^3/_4$-inch (19.1 mm) nominal pipe or tube size. Even though the water demand of a very small building might only need the flow that could be supplied from a smaller water supply pipe, the $^3/_4$-inch (19.1 mm) minimum size provides for a level of durability for installation handling and shifting/settling ground conditions. Table 605.3 lists the pipe materials that can be used for water service pipe.

603.2 Separation of water service and building sewer. Where water service piping is located in the same trench with the *building sewer*, such *sewer* shall be constructed of materials listed in Table 702.2. Where the *building sewer* piping is not constructed of materials listed in Table 702.2, the water service pipe and the *building sewer* shall be horizontally separated by not less than 5 feet (1524 mm) of undisturbed or compacted earth. The required separation distance shall not apply where a water service pipe crosses a *sewer* pipe, provided the water service is sleeved to a point not less than 5 feet (1524 mm) horizontally from the *sewer* pipe centerline on both sides of such crossing. The sleeve shall be of pipe materials listed in Table 605.3, 702.2 or 702.3. The required separation distance shall not apply where the bottom of the water service pipe, located within 5 feet (1524 mm) of the *sewer*, is not less than 12 inches (305 mm) above the highest point of the top of the *building sewer*.

❖ This section was completely reworded for the 2015 edition; however, the requirements were not changed.

WATER SUPPLY AND DISTRIBUTION

The language in the previous editions was complex and seemed difficult to understand in an exception format.

The section begins with the most common method to install water service and sewer piping to a building: one trench is excavated and both water service and building sewer are installed in the same trench. In the majority of those situations, the building sewer is constructed of materials that are indicated in Table 702.2. Materials in that table are suitable for installation inside of a building (those materials also being indicated in Table 702.3 for building sewer piping). As Table 702.2 materials are more robust than the piping materials indicated only for building sewers, there is a low probability of sewage leaking from the building sewer.

The second sentence concerns building sewer piping that is of a material not indicated in Table 702.2. In other words, it is indicated only in Table 702.3. These piping materials are not as robust as those indicated in Table 702.2 and could leak sewage into the ground. Water service piping cannot be located in the same trench as these types of building sewer materials. Independent trenches must be excavated for the water service line and the building sewer so that not less than 5 feet of undisturbed earth are between the trenches. Compacted earth is considered to be the same as undisturbed earth.

The third sentence addresses the situation of a water service pipe crossing (either above or below) a building sewer that is of piping materials indicated only in Table 702.3. (Although the code doesn't directly state that the building sewer piping material referred to in this sentence is only one of those indicated in Table 702.3, it should be obvious because if the building sewer was of materials indicated in Table 702.2, the water service piping could be in the same trench or anywhere around a building sewer of a material indicated in Table 702.2.) The water service piping has to be sleeved for a length of not less than 5 feet on either side of the building sewer so that if the building sewer did leak sewage into the earth, the water service piping is not directly exposed. The sleeve material can be any of the materials indicated in Table 605.3, 702.2 or 702.3.

The last sentence concerns water service piping that is at an elevation that is not less than 12 inches above the highest part of the building sewer (where that sewer is of materials indicated only in Table 702.2). If sewage leaks from a building sewer, gravity will cause the sewage to seep downwards into the earth and not upwards. Therefore, if the water service piping is above the highest part of the building sewer, there is little probability of sewage contamination of the earth around the water service piping [see Commentary Figures 603.2(1) and 603.2(2)].

603.2.1 Water service near sources of pollution. Potable water service pipes shall not be located in, under or above cesspools, septic tanks, septic tank drainage fields or seepage pits (see Section 605.1 for soil and ground water conditions).

❖ Because private sewage disposal systems such as septic tanks, soil absorption systems (drainage fields) and cesspools rely on the infiltration of waste water into the soil as a method of treatment, locating a water service pipe in the area of these systems is prohibited. Commentary Figure 603.2.1 references the minimum horizontal separation distances between water source elements and components of soil absorption systems found in Sections 406 and 802 of the *International Private Sewage Disposal Code®* (IPSDC®).

For SI: 1 inch = 25.4 mm, 1 foot = 304.8 mm.

**Figure 603.2(1)
SEPARATION OF WATER SERVICE AND BUILDING SEWER**

WATER SUPPLY AND DISTRIBUTION

Figure 603.2(2)
WATER SERVICE CROSSING OVER BUILDING SEWER

ELEMENT	ABSORPTION FIELD	SEPTIC OR TREATMENT TANK
	DISTANCE (feet)	
Cistern	50	25
Lot line	5	2
Spring	100	50
Water main	50	—a
Water service	10	5
Water well	50	25

For SI: 1 foot = 304.8 mm.

a. No limitation is listed in Table 802.8 of the IPSDC.

Figure 603.2.1
MINIMUM HORIZONTAL SEPARATION DISTANCES
BETWEEN WATER SOURCE ELEMENTS AND SOIL
ABSORPTION SYSTEM COMPONENTS

SECTION 604
DESIGN OF BUILDING
WATER DISTRIBUTION SYSTEM

604.1 General. The design of the water distribution system shall conform to *accepted engineering practice*. Methods utilized to determine pipe sizes shall be *approved*.

❖ The water distribution system is the piping system within a building that conveys water from the main water service valve (at the termination of the water service system) to the connection points of plumbing fixtures and plumbing appliances. The sizes of the pipes in the water distribution system must be large enough to supply a sufficient volume of water at adequate pressure to allow the fixtures and equipment to properly function when operated during the period of peak water demand of the system.

There are many methods that can be used to determine pipe sizes for a water distribution system. The only code requirements concerning the method used are that: 1) the method is an accepted engineering practice and 2) the code official approves the method.

Sizing Individual Fixture Supply Pipes

Most water distribution pipe sizing methods are intended to be used for the sizing of only the branches, risers and mains of a system. The sizes of individual fixture supply pipes are determined from Table 604.5, unless the actual fixture requires larger pipe connections. For example, some designs of shower mixing valves have $^3/_4$-inch (19.1 mm) connections and, therefore, would require fixture supply pipes of $^3/_4$-inch (19.1 mm) and not the minimum size of $^1/_2$ inch (12.7 mm) indicated in the table. A basic assumption about the indicated pipe sizes in Table 604.5 is that the fixtures are within close proximity to the branches, risers and mains. For example, the length of fixture supply pipes should be limited to making the connection to a branch supply pipe that is either just below the floor or just above the ceiling for where the fixture is located.

Where a fixture is significantly remote from the branch supply pipe, the fixture supply pipe might have to be larger than the minimum size in order to provide adequate flow and pressure at the fixture. For example, a service sink could be located 150 feet (equivalent length) from a branch supply pipe. The minimum size fixture supply pipe, $^1/_2$ inch (12.7 mm), flowing 5 gpm, will have a pressure loss of 27 psi at the fixture end of the pipe. If the pressure in the branch, riser or main where the fixture pipe begins is low, then it is quite possible that a "no flow/low flow" condition could occur at the remote service sink. Pressure loss calculations should be performed to determine if a larger fixture supply pipe is necessary for remotely located fixtures.

Section C404.5.1 is newly added for the 2015 edition of the *International Energy Conservation Code*. The section requires that the size of hot (or tempered) water piping be limited, depending on the type of fixture being served, and the length of the piping be limited from the source of hot water. As the I-Codes are a family of integrated codes, these IECC requirements impact the design of the hot (or tempered) water distribution system. Section C404.5.1 was added to the IECC because of the (negative) energy implications of users not receiving hot (or tempered) water at fixtures in an amount of time that is considered reasonable. The simple explanation is the following. A user turns on a fixture in an attempt to receive hot (or tempered) water. The hot (or tempered) water starts moving from the hot (or tempered) water source towards the fixture. However, the user becomes impatient and doesn't wait long enough for the hot (or tempered) water to arrive and gives up. Some amount of time elapses and then the next user does the same thing. Between each event, when there is no flow, the water in the piping cools down. These events occur many, many times a day in many, many buildings. The result is that the energy to heat the water is "wasted" because it never arrives at the point of use. Moving the source of hot water closer to the fixtures is only a partial solution. IPC Section 607.2 requires that the source of water be not more than 50 feet (15 240 mm) of developed piping length away from the fixture. However, if the piping is large diameter and the flow from a single fixture is small, the result is the same: the hot (or tempered) water never reaches the fixture because the user gives up waiting. Certainly, there is a frustration factor for some users. For the users who are willing to wait or have no choice but to wait (as they might be attempting to shower), potable water is flowing down the drain. This is a related negative impact that is especially critical in regions where raw water supplies are dwindling. Also consider that if the pipe diameter is large and the user does wait for hot (or tempered) water to arrive, the entire length of large diameter piping has been filled with heated water, only to cool down until the next user arrives. In other words, large quantities of water are being heated, only to have that energy wasted by cooling when water is not being used.

Other than the requirement in Section 607.2, the code was silent concerning this very real energy problem. Therefore, IECC Section C404.5.1 provides the framework for designing hot (or tempered) water distribution systems that will be energy efficient and also save potable water.

IECC Section C404.5.1 classifies hot (or tempered) water use locations into two types: 1) public lavatories, and 2) all other fixtures and appliances. Public lavatories are in a separate category because these fixtures are required (see IPC Table 604.4) to have a very low flow rate (0.5 gpm). In IECC Table C404.5.1, for a public lavatory, the fixture supply pipe length from the source of tempered water to the lavatory is very short, for example, 2 feet (609 mm) of $^1/_2$-inch (12.7 mm) pipe. However, this table also introduces piping sizes that are smaller than $^1/_2$ inch (12.7 mm) in order to have more design flexibility for where the source of hot (or tempered) water can be located in relationship to the fixtures. The smaller pipe sizes might seem to be in conflict with the minimum fixture supply pipe sizes indicated in IPC Table 604.5. As the commentary for that table states, there are two purposes for the minimum pipe sizes: 1) to make the connections for shutoff valves (stops) of a consistent size and 2) to provide the designer with the latitude to not have to include the fixture supply piping in the pressure loss calculations for water distribution systems (provided that the fixtures are within close proximity to the branches, risers and mains). The designer's problem is that he/she doesn't really know how that last leg of the distribution system (the fixture supply pipe) will need to be routed. The sizes of the pipes in Table 604.5 are more than ample for any one fixture, providing that the piping lengths are relatively short. In other words, the designer need not be concerned about pressure losses in those short runs of piping.

However, the application of IECC Table C404.5.1 will make it necessary for the designer to scrutinize the pressure losses in hot (or tempered) water fixture supply pipes that are smaller than indicated in IPC Table 604.5. Although the designer may not have any more knowledge about the routing of the fixture supply pipe, decisions will need to be made about the where the source of hot water needs to located with respect to each fixture, the size of the fixture supply pipe (and the pressure loss in that pipe) and how that might impact the sizes of piping in the entire water distribution system. Does this mean that a smaller-than-normal required size of the hot water fixture supply pipe at its termination should violate the intent of IPC Table 604.5 to provide for consistency of size for connection (stop valve) purposes? Not currently, because there doesn't appear to be available hardware (stop valves) for piping (or tubing) for some of the smaller sizes, depending on the type of pipe or tubing material. Until the market responds with appropriate hardware for smaller pipe or tubing sizes, there will need to be a "conversion" or "adapter" fitting before the fixture supply pipe termination point so that currently available hardware can be utilized. And if appropriate hardware does become available, technically, the designer should submit this reduced-size fixture pipe termination hardware as an alternate method under Section 105.2 to receive approval from the code official.

The code is silent about the impact that smaller hot (or tempered) fixture supply pipes could have on the operation of the fixture that has "normal" (perhaps oversized) cold water fixture supply pipes. In other words, if the flowing pressures (of cold and hot water) at the fixture are significantly different, could there be

WATER SUPPLY AND DISTRIBUTION

operational problems with the fixture? The design professional will need to explore this topic and adjust the design of the water distribution system accordingly.

The following methods for sizing water distribution piping systems do not address the sizing of hot (or tempered) water piping in relationship to the requirements of IECC Section C404.5.1. These methods are only for understanding of general concepts and are not mandated or suggested by the code. The presentation of these methods is not an indication that a code official would or should approve these methods. The plumbing design industry uses many alternative methods. The design professional is responsible for using his or her expertise and experience to choose an appropriate method for a specific application, client and jurisdiction.

A Method for Pipe Sizing by Maximum Allowable Flow Velocity (Method No. 1)

This method for pipe sizing is generally adequate for limited area, "compact" buildings, three stories or fewer in height with the longest piping run not exceeding 150 feet equivalent length from the main water service valve where the residual flowing pressure (for overcoming pipe friction losses) is greater than 20 psi. The resulting short piping lengths in a "compact," "low-rise" building and the relatively high available flowing pressure allows the designer the freedom to ignore pipe friction losses, thus simplifying the calculations to obtain pipe sizes.

Residual flowing pressure is determined by subtracting elevation pressure loss and the largest minimum fixture flow pressure (from Table 604.3) from the amount of flowing pressure available at the main water service valve. For example, a flush tank water closet requires a flow pressure of 20 psi. If the highest fixture in the water distribution system is 25 feet above the main water service valve, then the elevation pressure loss will be 25 feet × 0.433 psi/foot = 10.8 psi. If the flowing pressure available at the main water service valve is 60 psi, then the residual flowing pressure available (to overcome pipe friction losses) will be 60 - 20 - 10.8 = 29.2 psi.

To begin the sizing process, a schematic of the water distribution system is drawn, showing each of the fixtures with the hot and cold wsfu values. Starting at the most remote fixtures and working back towards the main water service valve, sum the individual hot and cold wsfu values along the branches, risers and mains. Stop the summation of cold and hot water wsfus where the cold water supply for the water heater connects to the cold water distribution piping. Sum the combined wsfu values for all fixtures in the system and, starting at the main water service valve and working towards where the cold water supply for the water heater connects to the cold water distribution piping, subtract cold wsfu values until reaching the cold water pipe connection going to the water heater. Using Table E103.3(3), convert all branch, riser and main wsfu values to gpm values.

Choose the type of pipe to be used and determine the maximum allowable velocity for that pipe material from Commentary Figure 604.1(1) or other reliable sources, such as piping manufacturers. Using the inside diameters of the type of pipe selected, calculate the maximum allowable flow rates for the pipe sizes that are required for the project. The following equation (an adapted version of the basic flow formula, $Q = A \times V$) is used to calculate the maximum allowable flow rate in gpm for a given inside pipe diameter and maximum allowable flow velocity:

$$q = 2.448\, d^2\, v$$

where,

q = Flow, gpm.

d = Inside pipe diameter, inches.

v = Flow velocity, feet per second.

Using the "gpm flow rates" for the branches, risers and mains of the system, assign pipe sizes based upon the maximum allowable flow rates for each pipe size. Review the completed schematic and adjust pipe sizes based upon experience.

Sample Problem 1

Commentary Figure 604.1(2) shows a schematic for a public occupancy building of three floors having identical fixtures on each floor (letters have been added to the diagram for commentary discussion purposes). The water closets are flush-tank type. Hot and cold wsfu values for each fixture were determined from

MATERIAL	PIPE OR TUBE SIZE (inches)	MAXIMUM VELOCITY (feet per second)
All pipe and tube material	$^1/_2$ and smaller	5
Brass pipe	$^5/_8$ and larger	8
Copper or copper-alloy pipe and tubing (hot water systems)	$^5/_8$ and larger	5
Copper or copper-alloy pipe and tubing (cold water systems)	$^5/_8$ and larger	8
Chlorinated polyvinyl (CPVC) plastic pipe	$^5/_8$ - 1 $1^1/_4$ and larger	8 10
Cross-linked polyethylene (PEX) plastic tubing	$^5/_8$ and larger	8
Cross-linked polyethylene/ aluminum/cross-linked polyethylene (PEX-AL-PEX) pipe	$^5/_8$ and larger	8
Galvanized steel pipe	$^5/_8$ - 1 2 - 4 4 and larger	8 10 12
PE-AL-PE	$^5/_8$ and larger	8

For SI: 1 inch = 25.4 mm, 1 foot per second = 0.3048 m/s.

**Figure 604.1(1)
MAXIMUM FLOW VELOCITY FOR VARIOUS PIPE**

WATER SUPPLY AND DISTRIBUTION

Figure 604.1(2)
WATER SYSTEM SCHEMATIC, THREE-STORY COMPACT BUILDING

Table E103.3(2) and are noted on the diagram. The wsfu value for the hose bibb is assumed to be equivalent to cold wsfu value for a service sink. Type L copper piping will be used for the project. The water service system was designed to provide a flowing pressure of 60 psi at the main water service valve under conditions of peak water demand (89.25 wsfu, which is the sum of the combined wsfu values for all fixtures on the diagram).

Solution

Does the problem meet the criteria to use pipe sizing Method No. 1? Total equivalent length = (40 + 50) + (90 × 0.5) = 135 feet. Because this is less than 150 feet equivalent length, the first of the criteria for using Method No. 1 is satisfied. Is the residual flowing pressure available 20 psi or greater? The elevation pressure loss is (40 × 0.433) = 17.3 psi. The highest flowing pressure required by fixtures (from Table 604.3) is the water closet at 20 psi. Therefore, the residual flowing pressure available is 60 - 17.3 - 20 = 22.7 psi. Because this value exceeds 20 psi, the second criteria for using Method No. 1 is satisfied.

For Type L copper piping, the maximum allowable flow velocity [from Commentary Figure 604.1(1)] for cold water is 5 feet per second for $^1/_2$ inch (12.7 mm) and smaller and 8 feet per second for $^5/_8$ inch (15.9 mm) and larger. For hot water, the maximum allowable flow velocity is 5 feet per second. The maximum allowable flows for each pipe size are:

$^1/_2$ inch (ID = 0.545 inches) → $q = 2.448 \,(0.545^2)\,(5) = 3.6$ gpm cold and hot

$^3/_4$ inch (ID = 0.785 inches) → $q = 2.448 \,(0.785^2)\,(8) = 12$ gpm cold; 7.5 gpm hot

1 inch (ID = 1.025 inches) → $q = 2.448 \,(1.025^2)\,(8) = 20.6$ gpm cold; 12.9 gpm hot

$1^1/_4$ inch (ID = 1.265 inches) → $q = 2.448 \,(1.265^2)\,(8) = 31.3$ gpm cold; 19.6 gpm hot

$1^1/_2$ inch (ID = 1.505 inches) → $q = 2.448 \,(1.505^2)\,(8) = 44.4$ gpm cold; 27.8 gpm hot

2 inch (ID = 1.985 inches) → $q = 2.448 \,(1.985^2)\,(8) = 77.2$ gpm cold; 48.3 gpm hot

$2^1/_2$ inch (ID = 2.465 inches) → $2.448 \,(2.465^2)\,(8) = 118$ gpm cold; 73.8 gpm hot

Working back towards the water heater's cold water inlet connection to the cold water distribution system, sum the wsfu values for each branch, riser and main pipe section. Using Table E103.3(3), convert branch, riser and main pipe section wsfu values to gpm. Commentary Figure 604.1(3) shows the values determined and the pipe sizes assigned.

SECTION	WSFU	GPM	PIPE SIZE
A-B	10	14.6	1
B-C	15	17.5	1
C-D	20	19.6	1
D-E	21.5	20.2	1
E-F	23	20.7	1
F-G	24.5	21.3	1.25
G-H	26.75	22.1	1.25
H-I	52.5	29.9	1.25
I-J	79	37.7	1.5
M-N	3	6.5	.75
N-O	4.5	8.7	1
O-P	6.75	11.5	1
P-Q	13.5	16.8	1.25
Q-R	20.25	19.7	1.5
S-J	20.25	19.7	1
L-K	89.25	40.8	1.5
K-J	87	40.1	1.5

Figure 604.1(3)
SOLUTION FOR SAMPLE PROBLEM 1

Note that the sum of the wsfu values for I-J plus S-J (99.25) is greater than the wsfu value for Section J-K (87). This discrepancy will always exist because the sum of all of the individual hot and cold wsfu values will always be greater than the sum of the combined wsfu values for all the fixtures. However, once the wsfu values are converted to gpm values (43.3 and 40.1, respectively), the small difference in gpm will rarely require section I-J pipe size to be larger than Section J-K. In cases where there is a difference in pipe size, it should be intuitive that the pipe size for Section I-J does not need to be larger than Section J-K pipe size.

The size of individual fixture supply pipes can be as small as indicated in Table 604.5; however, common practice is to size most of the fixture supply pipes requiring fixture stop valves as $^1/_2$ inch (12.7 mm) in order to make it easier to install commonly available $^1/_2$-inch (12.7 mm) connection stop valves. Obviously, if the fixture supply pipe is required to be larger than $^1/_2$-inch (12.7 mm) pipe size (because of a large size fixture or faucet connection), fixture shutoff valves, where required, should be at least the same nominal size as the fixture supply pipe. For example, if a shower valve requires $^3/_4$-inch (19.1 mm) supply pipes, the fixture shutoff valves, where required, should be $^3/_4$-inch (19.1 mm) nominal size.

A Method for Pipe Sizing by Segment Friction Loss (Method No. 2)

This method for pipe sizing is generally adequate for buildings that are three stories or less in height, with piping runs of lengths over 150 feet or residual pressure less than 20 psi, or both. The reason why this method is limited to "low-rise" buildings is that as the water distribution system becomes taller than three stories, there can be problems with pressure availability at the highest fixtures during peak demand flow.

This method requires that the designer calculate a friction loss for each pipe segment of the water distribution system. The friction losses for each segment of pipe are summed and compared to the available flowing pressure for overcoming friction loss. If the sum of segment friction losses is larger than the available flowing pressure, the designer adjusts the pipe sizes to be larger so that the sum of the segment friction losses is less than the available flowing pressure. If the sum of the friction losses is significantly smaller than the available flowing pressure, the designer can decrease some pipe sizes to "use up" more flowing pressure.

Because of the number of calculations necessary for using this method, a computer spreadsheet is often the easiest way to perform the calculations and track the results. The design of smaller systems using this method can be accomplished by hand calculations and worksheet tabular forms; however, as trial and error calculation is always necessary and the potential for error is great, a spreadsheet is recommended.

Sample Problem 2

Consider the water distribution system for a 16-unit, single-story, extended-stay motel illustrated in Commentary Figure 604.1(4). The flowing pressure available at the main water service valve is 35 psi at the peak water demand flow of 91.2 wsfu. The highest plumbing fixture, a shower mixing valve, is 6 feet above the main water service valve elevation and requires a flow pressure of 20 psi. The pipe material is chosen to be Type L copper (flow velocity required to be 8 feet per second or less for cold and 5 feet per second or less for hot). What are the minimum pipe sizes for each pipe segment in the system?

Solution

The residual flowing pressure available to overcome pipe friction losses is 35 psi - (6 feet × 0.433 psi/foot) - 20 psi = 12.4 psi. A spreadsheet is developed for the cold side of the system [see Commentary Figure 604.1(5)].

Column B indicates the pipe section designation and Column C indicates the length of the pipe section in feet. Column D is the wsfu value for the segment of pipe. These values were obtained by starting at the most remote fixture connection points and working back towards Point Q, summing the wsfu demand of each apartment. The wsfu value for Section S-Q is the same as required at Point S.

Column E values will be obtained by using Table E103.3(3) to convert wsfu to gpm (interpolating as necessary). Column F will be the pipe size, in inches, that the designer chooses for the section. Column G will be the inside diameter, in inches, of the pipe chosen. Column H computes the velocity, V, in the pipe, in

Figure 604.1(4)
WATER SYSTEM SCHEMATIC, SINGLE-STORY, 16-UNIT MOTEL

For SI: 1 foot = 304.8 mm.

feet per second, using the equation V = Column E/ [2.448 (Column G)^2]. Column I is the equivalent length of fittings, in feet, that the designer inputs using Table E103.3(6) as a guide. Column J is the sum of the value in Column C and the value in Column J. Column K is the computation of loss, in psi, using a variation of the Hazen-Williams formula: Column K = Column J × 4.52 × (Column E)$^{1.85}$/[(150)$^{1.85}$ × (Column G)$^{4.8655}$]. Commentary Figure 604.1(6) shows the completed spreadsheet. The value at K21 is the sum of the pressure losses of all segments. This value must be less than the previously calculated residual flowing pressure of 12.4 psi.

Discussion

Initial selection of pipe sizes for each section begins with choosing pipe sizes that result in pipe velocities less than 8 fps. Fitting equivalent lengths (Column I) can be arbitrarily chosen as 1 foot (304.8 mm) in order to obtain a rough estimate of the actual pressure loss in the pipe section. The pipe sizes for each section are changed in a logical manner to result in a total pressure loss (the value at K21) that is less than the residual flowing pressure available. With the pipe sizes tentatively chosen, fitting equivalent lengths are adjusted based upon the tentative pipe sizes and the actual fittings used in each section. For example, Section S-Q flows into Section Q-P through a straight run of a tee fitting. From Table E103.3(6), for a 2-inch (51 mm) tube size, the equivalent length through a straight run of a tee is 0.5 feet. To be conservative, a value of 1 was used. In another example, Section B-A flows through a 90-degree elbow and from Table E103.3(6), for a 1^1/$_2$-inch (38 mm) tube size, the fitting equivalent length is 4 feet. If, after adjusting the fitting equivalent lengths, the total pressure loss exceeds the residual flowing pressure, one or more pipe sections will have to be increased in size to lower the total pressure loss. Fitting equivalent lengths are adjusted for the revised pipe sizes and the total pressure loss is compared to the residual flowing pressure loss. The process continues until the designer is satisfied with the design.

Could some pipe sections be smaller to "use up" the entire 12.4 psi of residual flowing pressure? Perhaps, but doing so would be based upon the designer's experience and his or her confidence in the project information at hand. While there is a tendency to focus on determining the smallest possible pipe size for each section, doing so might ignore the labor efficiencies of using fewer sizes of pipe and requiring fewer "set up" changes to work with smaller pipe. Again, designer experience is necessary to make educated decisions concerning pipe sizes versus installation costs.

WATER SUPPLY AND DISTRIBUTION

A	B	C	D	E	F	G	H	I	J	K
	COLD SECTION	LENGTH (feet)	WSFU	FLOW (gpm)	PIPE SIZE (inches)	PIPE INSIDE DIAMETER (inches)	VELOCITY (feet per second)	FTG LENGTH (feet)	TTL EQ LENGTH (feet)	LOSS (psi)
1	S-Q	20	91.2							
2	Q-P	20	75.2							
3	P-O	20	70.5							
4	O-N	20	65.8							
5	N-M	20	61.1							
6	M-L	20	56.4							
7	L-K	20	51.7							
8	K-J	20	47							
9	J-I	20	42.3							
10	I-H	20	37.6							
11	H-G	20	32.9							
12	G-F	20	28.2							
13	F-E	20	23.5							
14	E-D	20	18.8							
15	D-C	20	14.1							
16	C-B	20	9.4							
17	B-A	20	4.7							
18	A-A'	1	4.7							
19	A'-B'	2	3.2							
20	B'-C'	5	1							
21										

Figure 604.1(5)
SPREADSHEET SETUP FOR SAMPLE PROBLEM 2

To complete the design of the 16-unit, single-story, extended-stay motel, Commentary Figure 604.1(7) shows the completed spreadsheet for the hot side. Note that the flow velocity for all hot water pipes is kept below 5 fps in accordance with Commentary Figure 604.1(1).

Designing the hot side piping system follows the same process as was used for the cold side. Note the fitting lengths for Sections Q-R and T-U. Four 90-degree (1.57 rad) elbows have been anticipated (each 4 feet (1219 mm) in equivalent length), as heat trap piping arrangements are typically required for the inlet and outlet of large storage tank water heaters. Note the fitting length for Section R-T. As it is known that the water heater causes some pressure loss, the designer estimated this resistance to be equivalent to four 90-degree (1.57 rad) elbows and 5 feet (1524 mm) of (internal water heater dip tube) pipe.

A Method for Pipe Sizing by Uniform Friction Loss (Method No. 3)

This method can generally be used for any size building with any available supply pressure. The designer calculates friction losses [per 100 feet (30 480 mm)] for all pipe segments of a "basic design circuit" and compares those friction losses to a maximum allowable uniform friction loss for a basic design circuit. (The basic design circuit is the piping route from the main water service valve to the most remote fixture having the greatest demand at the highest elevation.) Pipe sizes are chosen based upon the maximum allowable velocity or maximum allowable friction loss [per 100 feet (30 480 mm)], whichever is smaller.

Sample Problem 3

Consider the cold water piping of a water distribution system for an apartment building of eight floors having 12 apartments on each floor. Each apartment has one bathroom group (flush tank water closet), a tankless water heater and a kitchen sink [see Commentary Figure 604.1(8)]. The wsfu values noted on the figure for each fixture and bathroom group are from Table E103.3(2). The piping will be Type L copper tube. The most remote fixture with the highest pressure requirement is the shower (20 psi minimum pressure required) on the eighth floor at the end of the branch run. The flowing pressure available at the main water

A	B	C	D	E	F	G	H	I	J	K
	COLD SECTION	LENGTH (inches)	WSFU	GPM	PIPE SIZE	PIPE ID	VELOCITY	FTG LGTH	TTL EQ LENGTH	LOSS
1	S-Q	20	91.2	64.7	2.00	1.985	6.7	1	21	0.7
2	Q-P	20	75.2	36.5	2.00	1.985	3.8	9	29	0.3
3	P-O	20	70.5	35	2.00	1.985	3.6	9	29	0.3
4	O-N	20	65.8	33.5	2.00	1.985	3.5	9	29	0.3
5	N-M	20	61.1	31.5	1.50	1.505	5.7	7	27	0.9
6	M-L	20	56.4	31	1.50	1.505	5.6	7	27	0.9
7	L-K	20	51.7	29.2	1.50	1.505	5.3	7	27	0.8
8	K-J	20	47	28.6	1.50	1.505	5.2	7	27	0.8
9	J-I	20	42.3	27	1.50	1.505	4.9	7	27	0.7
10	I-H	20	37.6	25	1.50	1.505	4.5	7	27	0.6
11	H-G	20	32.9	23.5	1.50	1.505	4.2	7	27	0.5
12	G-F	20	28.2	22.5	1.50	1.505	4.1	7	27	0.5
13	F-E	20	23.5	21	1.50	1.505	3.8	7	27	0.4
14	E-D	20	18.8	19	1.50	1.505	3.4	7	27	0.4
15	D-C	20	14.1	17	1.50	1.505	3.1	7	27	0.3
16	C-B	20	9.4	14	1.50	1.505	2.5	7	27	0.2
17	B-A	20	4.7	9	1.50	1.505	1.6	4	24	0.1
18	A-A'	1	4.7	9	0.75	0.785	6.0	3	4	0.3
19	A'-B'	2	3.2	6.8	0.75	0.785	4.5	1	3	0.1
20	B'-C'	5	1	3	0.50	0.545	4.1	2	7	0.4
21										9.7

Figure 604.1(6)
COMPLETED SPREADSHEET FOR SAMPLE PROBLEM 2—COLD SIDE

service valve is 75 psi. A normal amount of fittings will be used in the piping arrangements. Determine the minimum size of each of the piping segments using the uniform friction loss sizing method.

Solution

1. Calculate the length of piping in the basic design circuit: 5 feet + 5 feet + (6 × 20 feet) + (7 × 10 feet) + 5 feet + 5 feet. = 210 feet. Using an additional 50 percent of length to account for fittings: 210 + [210 × (50/100)] = 315 feet equivalent length.

2. Calculate the pressure available that can be used for friction loss in the basic design circuit: The pressure loss caused by elevation difference is (75 feet × 0.433) = 32.5 psi. The minimum pressure required at the shower fixture is 20 psi. Thus, the pressure remaining to overcome friction loss is (75 psi - 32.5 psi - 20 psi) = 22.5 psi.

3. Calculate the uniform friction loss per 100 feet for the basic design circuit: (22.5 psi/315 ft) × 100 = 7.1 psi per 100 feet.

4. Develop a table of pipe sizes and maximum allowable flow rates: Calculate the maximum flow at the 8 fps velocity limit from $q = 2.448\, d^2\, v$ [or read from the friction loss graph in Commentary Figure 603.1(8)] for each pipe size. NOTE: The flow rate associated with a friction loss of 7.1 psi per 100 feet (30 480 mm) for each size of pipe can be calculated from $q = 4.93\, p^{0.57} 1 d^{2.714}$ [or read from the friction loss chart in Commentary Figure 603.1(8)]. See Commentary Figure 604.1(9) for a summary of results. The smaller of the two gpm values for each pipe size is the maximum allowable flow rate for that pipe size.

5. Determine the wsfu for each pipe segment in the basic design circuit and list the values in a table. Using Table E103.3(3), convert wsfu values into flow rate of gpm for each section of pipe. Choose the pipe size for each pipe segment based upon the lesser of the flow rates (velocity or uniform friction loss) for each pipe size listed in Commentary Figure 604.1(9) [see Commentary Figure 604.1(10)].

WATER SUPPLY AND DISTRIBUTION

A	B	C	D	E	F	G	H	I	J	K
	HOT SECTION	LENGTH (inches)	WSFU	GPM	PIPE SIZE (inches)	PIPE ID (inches)	VELOCITY (fps)	FTG LENGTH (feet)	TTL EQ LENGTH (feet)	LOSS (psi)
1	S-Q	20	91.2	64.7	2.00	1.985	6.7	1	21.0	0.7
2	Q-R	10	40	26.3	1.50	1.505	4.7	16	26.0	0.6
3	R-T	5	40	26.3	1.50	1.505	4.7	16	21.0	0.5
4	T-U	10	40	26.3	1.50	1.505	4.7	16	26.0	0.6
5	U-V	20	40	26.3	1.50	1.505	4.7	1	21.0	0.5
6	V-W	20	37.5	25.6	1.50	1.505	4.6	1	21.0	0.5
7	W-X	20	35	24.9	1.50	1.505	4.5	1	21.0	0.5
8	X-Y	20	32.5	24.1	1.50	1.505	4.3	1	21.0	0.4
9	Y-Z	20	30	23.3	1.50	1.505	4.2	1	21.0	0.4
10	Z-AA	20	27.5	22.4	1.50	1.505	4.0	1	21.0	0.4
11	AA-BB	20	25	21.5	1.50	1.505	3.9	1	21.0	0.4
12	BB-CC	20	22.5	20.6	1.50	1.505	3.7	1	21.0	0.3
13	CC-DD	20	20	19.6	1.50	1.505	3.5	1	21.0	0.3
14	DD-EE	20	17.5	18.6	1.50	1.505	3.4	1	21.0	0.3
15	EE-FF	20	15	17.7	1.50	1.505	3.2	1	21.0	0.2
16	FF-GG	20	12.5	16.3	1.25	1.265	4.2	1	21.0	0.5
17	GG-HH	20	10	14.6	1.25	1.265	3.7	1	21.0	0.4
18	HH-II	20	7.5	12.3	1.25	1.265	3.1	1	21.0	0.3
19	II-JJ	1	5	9.4	1.00	1.025	3.7	5	6.0	0.1
20	JJ-KK	2	4.7	9	1.00	1.025	3.5	5	7.0	0.2
21	KK-LL	1	2.5	5.7	0.75	0.785	3.8	5	6.0	0.2
22	LL-MM	4	0.5	4	0.75	0.785	2.7	2	6.0	0.1
23	MM-NN	5	1	3	0.50	0.545	4.1	2	7.0	0.4
										8.3

Figure 604.1(7)
COMPLETED SPREADSHEET FOR SAMPLE PROBLEM 2—HOT SIDE

WATER SUPPLY AND DISTRIBUTION

Figure 604.1(8)
WATER SYSTEM SCHEMATIC, EIGHT-STORY, 96-UNIT APARTMENT BUILDING

For SI: 1 foot = 304.8 mm.

PIPE SIZE (inches)	MAXIMUM ALLOWABLE FLOW BASED ON 8 FPS VELOCITY LIMIT GPM	MAXIMUM ALLOWABLE FLOW BASED ON UNIFORM FRICTION LOSS LIMIT OF 7.1 PSI PER 100 FT GPM
1/2	5.8	2.9
3/4	12.1	7.8
1	20.6	16.1
1 1/4	31.3	28.6
1 1/2	44.4	45.8
2	77.2	97.1
2 1/2	118	174.7
3	169.8	283.1

Figure 604.1(9)
SAMPLE PROBLEM 3—PIPE SIZING TABLE

Water Distribution Pipe Sizing and Water Service Pipe Sizing Are Interrelated

Generally, a greater flowing pressure available at the termination point of the water service system (the main water service valve) will allow for smaller water distribution pipe sizes. Thus, the design of the water service system impacts the design of the water distribution system. If the pipe sizes of the water distribution system seem to be excessively large because of a lack of flowing pressure from the water service system, then the water service system might require design changes to make more flowing pressure available. Such changes might include choosing a larger

WATER SUPPLY AND DISTRIBUTION

size of water service pipe, increasing the size of water service system components or both. Economics play a significant role in designing both systems. For example, increasing the size of a pressure-reducing valve to gain more flowing pressure from a water service system might be more cost effective than increasing the sizes of numerous water distribution pipes.

Summary

The code official is responsible for determining that the pipe sizes for the water distribution system were chosen using an accepted engineering practice. In order for the code official to provide his or her approval, documentation in the form of hand calculations, computer program methodology, inputs and outputs and copies of any other relevant data might have to be supplied for review. The project drawings must reflect the pipe sizes determined by the approved sizing method so that the contractor installs the correct size of piping and the code official has a document to field verify that the design pipe sizes were installed.

604.2 System interconnection. At the points of interconnection between the hot and cold water supply piping systems and the individual fixtures, appliances or devices, provisions shall be made to prevent flow between such piping systems.

❖ Where a plumbing fixture, device or appliance is provided with both a hot and a cold water supply connection (other than terminal devices that blend hot and cold water, such as tempering valves, pressure-bal-

SECTION	WSFU	GPM	PIPE SIZE
SH-A	1.0	—	$1/2$ [a]
WC-A	2.2	—	$3/8$ [a]
LAV-B	0.5	—	$3/8$ [a]
KS-C	1.0	—	$1/2$ [a]
SH-A'	1.0	—	$1/2$ [a]
LAV-A'	0.5	—	$3/8$ [a]
KS-B'	1.0	—	$1/2$ [a]
A-B	3.2	6.8	$3/4$
A'-B'	1.5	4.0	$3/4$
WH-B'	2.5	5.8	$3/4$
WH-C	2.5	5.8	$3/4$
B-D	2.7	6.1	$3/4$
C-D	3.5	7.3	$3/4$
D-E	6.2	10.9	1
E-F	12.4	16.2	$1 1/4$
F-G	18.6	19.0	$1 1/4$
G-H	24.8	21.4	$1 1/4$
H-I	31	23.6	$1 1/4$
I-J	37.2	25.5	$1 1/4$
J-K	74.4	36.3	$1 1/2$
K-L	148.8	54.5	2
L-M	223.2	69.6	2
M-N	297.6	84.5	$2 1/2$
N-O	372	99.4	$2 1/2$
O-P	446.4	113.8	$2 1/2$
P-Q	520.8	127.8	3
Q-R	595.2	141.5	3

a. Minimum fixture supply pipe in accordance with Table 604.5.

Figure 604.1(10)
SAMPLE PROBLEM 3—SECTION PIPE SIZES

anced mixing valves, etc., to achieve a desired temperature), the water must be supplied in such a way that there will be no possibility for the flow of cold water to the hot water piping or vice versa. This is required so the hot water supply does not become cooled or tempered below the temperature required (see the definition of "Hot water" in Section 202) by the introduction of cold water, and that when a fixture is operated to provide cold water; hot water is not delivered. Another concern is a continuous flow of cold water into the hot water supply, could lead to increased operation of the water heater in an attempt to maintain the water temperature, resulting in an increased use of energy and a shortened heater life.

This section might also be applicable to "no return pipe" hot water recirculation system arrangements, which have become popular in recent years. Although originally intended as a retrofit method of providing for hot water recirculation where the installation of hot water return piping was impossible or too costly, these systems are also being installed in new construction for the labor and materials savings. Although system component designs vary between manufacturers, the system concept utilizes a pump in the hot water line to cause hot water flow through a special valve and into the cold water line near a fixture that is most remote from the water heater. When the valve senses that the temperature of the hot water flow becomes warm [approximately 90°F (32°C)], the valve automatically closes to stop flow so that the cold water line does not become filled with hot water. The valve also prevents flow of hot water to the cold supply line while cold water is flowing from the faucet. The minimal flow rate through the valve limits the amounts of warm water flowing into the cold water line such that when the user opens only the cold water control of the faucet, warm water flows from the faucet for only a few seconds before cooler flow begins. Typically, most users rarely sense the short period of warm water flow from the cold-only operation of the faucet. Even if they did have their hand under the faucet at the moment the cold water valve was opened, the flow is well within safe "tempered water" temperatures. Thus, the user is protected from the potential of being inadvertently scalded by hot water flow from cold water operation of the faucet. Because the recirculation flow is minimal, the operational cost of these systems is generally much less than a system utilizing a dedicated hot water return pipe. Thus, the energy "waste" for this type of hot water recirculation system is no greater than for a system using a dedicated hot water return pipe. Therefore, with respect to the apparent intents of this section for scald protection and preventing hot water storage depletion (and the resultant energy waste), these "no return pipe" systems do not appear to be a violation of this section.

However, if this section is read and interpreted in a literal manner, these "no return pipe" recirculation systems constitute an interconnection between hot and cold systems that does allow flow to occur between the systems and, therefore, these systems would violate this section. Ultimately, the code official must decide if these "no return pipe" systems comply with or violate this section based upon his or her interpretation of this section.

604.3 Water distribution system design criteria. The water distribution system shall be designed, and pipe sizes shall be selected such that under conditions of peak demand, the capacities at the fixture supply pipe outlets shall be not less than shown in Table 604.3. The minimum flow rate and flow pressure provided to fixtures and appliances not listed in Table 604.3 shall be in accordance with the manufacturer's installation instructions.

❖ Designers of water distribution systems must know the minimum flowing pressure that each fixture in the system must have so that the fixture will properly operate during periods of peak water demand. Table 604.3 provides the minimum flowing pressures for common fixtures and plumbing appliances. The designer must use this information for determining pipe sizes in the water distribution system.

Table 604.3 also provides a flow-rate demand for common fixtures and plumbing appliances. Knowledge of a fixture's flow rate is necessary to determine the amount of pressure loss through the piping in the water distribution system. Too small of pipe for a given flow rate can cause high flow velocities, which can result in excessive pressure loss at the discharge end of the pipe.

Where a fixture or appliance is not listed in Table 604.3, the designer must obtain the necessary flowing pressure and flow-rate information from the manufacturer of the fixture or appliance.

Does this section conflict with the commonly accepted Hunter method of using wsfu values to determine flow rates for fixtures? (See the commentary to Section 603.1 for details concerning the Hunter method for determining peak water demand for water distribution systems.) The answer is not straightforward because the flow rates of Table 604.3 are not coupled to a frequency of use such that a peak water demand on the water distribution system can be determined. As stated in the commentary to Section 603.1, in most cases, peak water demand is not simply the addition of the maximum allowable flow rates of every fixture in the building because it is usually unrealistic to assume that all fixtures would be operating at the same time. Therefore, the flow rates of Table 604.3 might be somewhat useful when evaluating water distribution systems for buildings such as stadiums that have a peak water demand during halftime that approaches the flow of all fixtures operating simultaneously. Otherwise, without application of frequency-of-use/duration-of-use information to the flow rates of Table 604.3, most pipe sizes determined for "normal use" building water distribution systems will be oversized.

WATER SUPPLY AND DISTRIBUTION

Note that some fixtures in Table 604.3 are also listed in Table 604.4 and that the actual maximum allowable flow rate for some fixtures is considerably less that the flow rate listed in Table 604.3 for the same fixture. This is another reason why determining pipe sizes using the flow rates in Table 604.3 could result in significant and unnecessary oversizing of piping.

Note a of Table 604.3 is a reminder to review Table 604.4 when reviewing the flow rate column.

Note b of Table 604.3 only requires that the capacity for piping serving a shower mixing valve be that of the lowest flow rating of the shower mixing valve. For many years since the showerhead flow rate was lowered to a maximum of 2.5 gallons per minute, product standards required that the mixing valve had to safely operate at a flow rate of 2.5 gallons per minute. In other words, at a flow rate of 2.5 gpm, the mixing valve must quickly self-adjust to prevent an outlet water temperature fluctuation exceeding plus or minus 3.6°F of the user set temperature. The required capacity (in Table 604.3) of the piping serving many fixtures has been reduced over several code editions so that this table now reflects a required piping flow rate of 2.5 gpm for shower mixing valves (matching the 2.5 gpm mixing valve scald-protection testing protocol). However, because of the strong emphasis on water use reduction, manufacturers now offer mixing valves that can safely operate at flows less than 2.5 gpm. Currently, most manufacturers are rating their pressure-balanced shower mixing valves at 1.5 gpm to comply with the scald-protection testing protocol of the standards. Note b allows the capacity of the piping serving those valves to be as low as the lowest rating of those mixing valves.

The table indicates a minimum required flow rate. Designers and installers could choose to install larger piping (as doing so would comply with the requirements of the table). However, larger piping than necessary is more costly to purchase, install and insulate (hot side only). Where the intent is to conserve water by choosing low flow-rated fixtures (and mixing valves), the installation of piping that is larger than necessary does not serve any purpose. The other added "benefit" of using the smallest size possible piping to serve a fixture is that the consumer (such as a condominium owner or an apartment dweller) cannot obtain a significantly higher flow rate by altering the fixture by changing the aerator or defeating the shower head flow restrictor. The smaller piping to the fixture might not be able to support the desired higher flow rate! See Section 607 for more information concerning smaller piping for the hot water supply to certain fixtures.

TABLE 604.3
WATER DISTRIBUTION SYSTEM DESIGN CRITERIA REQUIRED CAPACITY AT FIXTURE SUPPLY PIPE OUTLETS

FIXTURE SUPPLY OUTLET SERVING	FLOW RATE[a] (gpm)	FLOW PRESSURE (psi)
Bathtub, balanced-pressure, thermostatic or combination balanced-pressure/thermostatic mixing valve	4	20
Bidet, thermostatic mixing valve	2	20
Combination fixture	4	8
Dishwasher, residential	2.75	8
Drinking fountain	0.75	8
Laundry tray	4	8
Lavatory, private	0.8	8
Lavatory, private, mixing valve	0.8	8
Lavatory, public	0.4	8
Shower	2.5	8
Shower, balanced-pressure, thermostatic or combination balanced-pressure/thermostatic mixing valve	2.5[b]	20
Sillcock, hose bibb	5	8
Sink, residential	1.75	8
Sink, service	3	8
Urinal, valve	12	25
Water closet, blow out, flushometer valve	25	45
Water closet, flushometer tank	1.6	20
Water closet, siphonic, flushometer valve	25	35
Water closet, tank, close coupled	3	20
Water closet, tank, one piece	6	20

For SI: 1 pound per square inch = 6.895 kPa,
1 gallon per minute = 3.785 L/m.

a. For additional requirements for flow rates and quantities, see Section 604.4.
b. Where the shower mixing valve manufacturer indicates a lower flow rating for the mixing valve, the lower value shall be applied.

604.4 Maximum flow and water consumption. The maximum water consumption flow rates and quantities for all plumbing fixtures and fixture fittings shall be in accordance with Table 604.4.

Exceptions:

1. Blowout design water closets having a water consumption not greater than $3^1/_2$ gallons (13 L) per flushing cycle.

2. Vegetable sprays.

3. Clinical sinks having a water consumption not greater than $4^1/_2$ gallons (17 L) per flushing cycle.

4. Service sinks.

5. Emergency showers.

❖ Limiting water flow rates and volume per flush in certain fixtures is important to conserve not only our limited water supplies but also the energy required to pump, transport, treat, condition and heat water. Water conservation lessens the burden on water treatment facilities and private sewage disposal systems. The relationship between water and energy usage is evidenced by the fact that the federal government water conservation requirements are part of the U.S. Energy Policy Act (EPAct).

Exception 1 exempts blowout water closets from the requirements of Table 604.4 because such fixtures depend on high-volume and high-velocity water flow to evacuate the contents of the bowl. Blowout water closets may not be able to function at lower consumption rates. Blowout fixtures were frequently installed in assembly occupancies because they are considered to be less susceptible to the abuse and vandalism typically associated with public toilet facilities. However, the use of blowout fixtures is dwindling in favor of using water closets meeting the 1.6 gallons per flush [6.1 liters per flush (L/f)] limitation.

Exception 2 exempts vegetable sprays because they are not included in EPAct.

Exception 3 exempts clinical sinks because they are specialty fixtures used in institutional occupancies for the disposal of human waste and the cleaning of bedpans.

Exception 4 exempts service sinks because such fixtures are used for maintenance and janitorial operations. Because service sink faucets are usually used to fill mop buckets and similar containers or the sink itself, there is nothing to be gained by limiting the flow rate of such faucets. The same logic is extended to mop sinks, pots-and-pans sinks and bathtubs. Such fixtures are filled with water or are used to fill other containers, and limiting the flow rate of the faucet will not affect the amount of water needed by the user of the fixture, but will increase the filling time with no apparent benefit.

Exception 5 exempts emergency showers because they are used only in the event of an emergency. Emergency showers are provided for body and eye washing in occupancies where people are exposed to chemical hazards. When addressing life safety, water consumption is not a concern.

When 1.6-gpf (6.1 L/f) water closets were first introduced decades ago, their overall performance was lackluster at best. Over the years, designs for the 1.6 gpf (6.1 L/f) models have greatly improved and now high-performing designs are available that consume only 1.28 gpf (4.85 liters per flush). Although both designs might require a "double flush" from time to time, overall, these low consumption water closets have a substantial impact on the saving of water.

One concern involving low-water consumption water closets is the possible occurrence of pipe blockages caused by the lack of water to keep solids in suspension. This problem could become a greater issue where oversized drain piping has been installed and where gray-water-producing fixtures are not contributing waste water to the water closet drain piping. With respect to the 1.6 gpf (6.1 L/f) water closets, the "dry drain" problem has occurred only in isolated situations where other factors, such as fixture abuse and improperly sloped piping, have been involved. However, as the installation of 1.28 gpf (4.85 L/f) water closets becomes more prevalent, it is expected that pipe stoppages will occur with greater frequency even where drain piping has been carefully installed at not less than the minimum slope and has not been oversized. Because water conservation for plumbing fixtures is here to stay, it is foreseeable that drain and vent sizing methodology could change in response to the reduced fixture discharge rates.

The fixtures listed in the Table 604.4 are those targeted by the U.S. EPAct to be regulated so that water consumption can be controlled. Because the flow rate from a faucet or shower head is dependent upon pressure, pressures are indicated. However, the indicated pressures are not to be interpreted as maximum pressures for each fixture. The maximum pressure allowed in a water distribution system is 80 psi (552 kPa) (see Section 604.8).

TABLE 604.4
MAXIMUM FLOW RATES AND CONSUMPTION FOR PLUMBING FIXTURES AND FIXTURE FITTINGS

PLUMBING FIXTURE OR FIXTURE FITTING	MAXIMUM FLOW RATE OR QUANTITY[b]
Lavatory, private	2.2 gpm at 60 psi
Lavatory, public (metering)	0.25 gallon per metering cycle
Lavatory, public (other than metering)	0.5 gpm at 60 psi
Shower head[a]	2.5 gpm at 80 psi
Sink faucet	2.2 gpm at 60 psi
Urinal	1.0 gallon per flushing cycle
Water closet	1.6 gallons per flushing cycle

For SI: 1 gallon = 3.785 L, 1 gallon per minute = 3.785 L/m, 1 pound per square inch = 6.895 kPa.

a. A hand-held shower spray is a shower head.
b. Consumption tolerances shall be determined from referenced standards.

604.5 Size of fixture supply. The minimum size of a fixture supply pipe shall be as shown in Table 604.5. The fixture supply pipe shall terminate not more than 30 inches (762 mm) from the point of connection to the fixture. A reduced-size flexible water connector installed between the supply pipe and the fixture shall be of an *approved* type. The supply pipe shall extend to the floor or wall adjacent to the fixture. The minimum size of individual distribution lines utilized in gridded or parallel water distribution systems shall be as shown in Table 604.5.

❖ The fixture supply pipe, by definition, is the pipe from a fixture branch (which supplies water for two or more fixtures) to the individual fixture (see Commentary Figure 604.5). There are two reasons why the code specifies the minimum size of individual fixture supply pipes. First, water distribution system pipe sizing methods might indicate that a smaller pipe would provide the necessary flow and pressure for a fixture. This could result in several different sizes of small pipes being "stubbed out" of walls and floors, making it difficult to coordinate the connection of commonly available stop valves and other fixture fittings to those small pipes. Second, it is difficult to know at the design stage exactly how each fixture supply pipe will need to be routed so that an appropriate equivalent length of pipe can be determined for calculation of pressure loss in this final leg of the water distribution system. By specifying minimum pipe sizes, the code mostly eliminates the potential problem of having too small of pipe (resulting in too much pressure loss) for a fixture. Note the use of the phrase "mostly eliminates" in the previous sentence. The designer of the water distribution system should not assume that the minimize size for fixture supply pipes will be sufficient for all applications. For example, a shower mixing valve might have $^3/_4$-inch (19.1 mm) connection ports and, thus, the fixture supply pipe should be $^3/_4$ inch (19.1 mm) instead of the minimum size of $^1/_2$ inch (12.7 mm). Another example is where a fixture is located far away from the fixture branch pipe that supplies the fixture supply pipe. A long run of fixture supply pipe would result in significant pressure loss at the fixture end of the run. It may be necessary to increase the size of the fixture supply pipe for most of the length of the run in order to keep pressure loss to a minimum.

This section's requirement for locating the termination of the fixture supply pipe within 30 inches (762 mm) is for three reasons: 1) it limits the amount of friction loss caused by the small diameter connectors that are used; 2) it allows for a single length of commonly available connector lengths to make the connection (eliminating the need for couplings in the connector) and 3) it keeps the location of the fixture stop valves (where required) within close proximity of the fixture.

This section requires that the connector used between the termination of the fixture supply pipe and the fixture be approved (by the code official) so that "cobbled together" connection arrangements are not used. For example, it is unlikely that a code official would approve the use of rubber tubing forced onto barbed fittings as a connector in this application. Typically, connectors are soft (annealed) copper tubing (plain finish or chrome plated finish) with compression or flare end fittings, semirigid plastic tubing with compression or flare end fittings or flexible water connectors with braided reinforcement with flare or compression end fittings. Note that Section 605.6 requires that flexible water connectors conform to ASME A112.18.6.

The last sentence of Section 604.5 directs the code user to carefully look at Table 604.5 with respect to gridded and parallel water distribution systems. The intent is for the code user to notice Note a to realize that parallel water distribution systems (i.e., those having a central manifold) can have smaller water distribution (fixture supply) pipe sizes under certain conditions. (Note that Note a does not address gridded water distribution systems.) Because parallel water distribution systems generally have fewer fittings and the beginning of the fixture supply pipe starts at a central manifold that has significant flow and pressure availability, the fixture supply pipe can be smaller (by one pipe size) than what is necessary for a "mains, risers and branch"-type water distribution system design. The allowance for the one size smaller pipe is only applicable where the pressure at the meter location is not less than 35 psi (241 kPa) and the length of the distribution line (fixture supply pipe) is not greater than 50 feet (15 240 mm).

Although the code is silent about where the water meter location is in relationship to the central manifold for a parallel water distribution system, a reasonable assumption is that the two are most likely intended to be in very close proximity to one another. If this was not the case, and the meter was located at some distance away from the building (e.g., at the curb), there could be significant flowing pressure losses in the water service system to the building, resulting in less-than-required flowing pressure at the fixtures (see Table 604.3). However, even if the meter was located next to the central manifold and the available pressure was 35 psi, then a next size smaller tubing than what is required by Table 604.5 might not provide the required pressure at a remote fixture.

For SI: 1 inch = 25.4 mm.

Figure 604.5
WATER SUPPLY CONNECTION WITH REDUCED-SIZE FLEXIBLE SUPPLY

TABLE 604.3
MINIMUM SIZES OF FIXTURE WATER SUPPLY PIPES

FIXTURE	MINIMUM PIPE SIZE (inch)
Bathtubs[a] (60″ × 32″ and smaller)	1/2
Bathtubs[a] (larger than 60″ × 32″)	1/2
Bidet	3/8
Combination sink and tray	1/2
Dishwasher, domestic[a]	1/2
Drinking fountain	3/8
Hose bibbs	1/2
Kitchen sink[a]	1/2
Laundry, 1, 2 or 3 compartments[a]	1/2
Lavatory	3/8
Shower, single head[a]	1/2
Sinks, flushing rim	3/4
Sinks, service	1/2
Urinal, flush tank	1/2
Urinal, flushometer valve	3/4
Wall hydrant	1/2
Water closet, flush tank	3/8
Water closet, flushometer tank	3/8
Water closet, flushometer valve	1
Water closet, one piece[a]	1/2

For SI: 1 inch = 25.4 mm, 1 foot = 304.8 mm,
1 pound per square inch = 6.895 kPa.

a. Where the *developed length* of the distribution line is 50 feet or less, and the available pressure at the meter is 35 psi or greater, the minimum size of an individual distribution line supplied from a manifold and installed as part of a parallel water distribution system shall be one nominal tube size smaller than the sizes indicated.

604.6 Variable street pressures. Where street water main pressures fluctuate, the building water distribution system shall be designed for the minimum pressure available.

❖ In the design of a water distribution system, the lowest pressure in the public water main must be determined and used in the calculations to guarantee that adequate pressure water is available to the fixtures in the building during periods of peak water demand. In many areas of the country, the pressure in the public water main varies throughout the day and seasonally.

The lowest pressure in a public water main usually occurs in the summer because of the demands of lawn sprinkling and makeup water for air-conditioning cooling towers. The minimum pressure can usually be determined by a review of the records of pressures at different times of the day and year kept by the local water purveyor. The pressure can also be monitored in nearby buildings or ascertained from fire department hydrant inspections. However, future demands placed on a public water main as a result of community growth or expansion often result in a decrease of available pressure. While not required by the code, future public water main pressure losses caused by growth and expansion should be considered in determining the minimum available pressure for design purposes.

For individual water supply systems (e.g., wells), the lowest pressure for design pressure is the pump "cut in" pressure.

604.7 Inadequate water pressure. Wherever water pressure from the street main or other source of supply is insufficient to provide flow pressures at fixture outlets as required under Table 604.3, a water pressure booster system conforming to Section 606.5 shall be installed on the building water supply system.

❖ It might not be possible for a public water supply system to have the level of pressure required by some buildings. For example, a 500-foot-tall (152 500 mm) building with fixtures at the top floor would require a public water main pressure of at least 250 psi (1724 kPa) (accounting for pipe friction losses and elevation pressure loss). Also, some public water main systems are so antiquated that the public water main piping cannot withstand high pressures; therefore, water pressure boosting systems are necessary to raise the pressure of the incoming water supply so that adequate pressure is available at all fixtures in a building. Section 606.5 regulates the various methods for increasing the pressure of the water supply system (see commentary, Section 606.5).

604.8 Water pressure-reducing valve or regulator. Where water pressure within a building exceeds 80 psi (552 kPa) static, an *approved* water pressure-reducing valve conforming to ASSE 1003 or CSA B356 with strainer shall be installed to reduce the pressure in the building water distribution piping to not greater than 80 psi (552 kPa) static.

Exception: Service lines to sill cocks and outside hydrants, and main supply risers where pressure from the mains is reduced to 80 psi (552 kPa) or less at individual fixtures.

❖ The components and fixtures connected to a water distribution system are designed to properly operate and have reasonable life expectancy for pressures of not greater than 80 psi (552 kPa). Higher pressures can cause leaks from faucets, difficulty in operation of fixtures and reduced lifespan of other components such as water closet fill valves and electric solenoid water valves. Therefore, the code requires that in cases where the supply pressure is higher than 80 psig (552 kPa), a pressure-reducing valve be installed to reduce the pressure to 80 psi (552 kPa) or less.

Installation of a pressure-reducing valve requires that a strainer be installed along with the pressure-reducing valve to protect the valve's small passages and ports from dirt accumulation. A separate strainer is not always needed as some pressure-reducing valve designs have integral strainers. Strainers and pressure-reducing valves with integral strainers must be installed to enable access for removal of the strainer without removal of the strainer or valve (see Section 604.8.2).

The code provides an exception for allowing higher pressure for sillcocks and outside hydrants. It also allows higher pressure for water distribution system risers (in multistory buildings) where pressure-reducing valves are installed at points before branches serve fixtures.

There are two basic types of pressure-reducing valves: direct acting and pilot operated. Most applications use direct acting because of their ease of field adjustment and relatively low cost (as compared to pilot-operated valves). The direct-acting type uses spring force against a pressure diaphragm to accomplish pressure reduction. One shortcoming of direct-acting pressure-reducing valves is that of "fall-off" pressure. The fall-off pressure of a pressure-reducing valve varies by manufacturer, size and flow rate, but generally the fall-off pressure is in the 5- to 20-psi (34.5 to 137.8 kPa) range. For example, a direct-acting pressure-reducing valve set at 80 psi (552 kPa) static (no-flow condition) pressure and having a 20-psi (137.9 kPa) fall-off pressure at the design flow would only allow for 60-psi (414 kPa) pressure at design flow. Obviously, fall-off pressure has implications for the design of the water distribution system. Pilot-operated pressure-reducing valves do not have a fall-off pressure, but because of their relatively high cost, they are generally only used where precise control and absence of fall-off pressure are needed.

604.8.1 Valve design. The pressure-reducing valve shall be designed to remain open to permit uninterrupted water flow in case of valve failure.

❖ Water-pressure reducing valves must be designed so that if the pressure-reducing component of the valve fails to regulate the water pressure to the valve setting, the valve will remain in an open position to allow a continuous flow of water to the building. A valve without this fail-safe feature could close completely in event of a failure, thus depriving the building and its occupants of water service.

In the majority of cases where a pressure-reducing valve fails, the high side pressure is no longer reduced to acceptable downstream pressure levels. At the very least, this will only cause faucets to be more difficult to operate or drip from the spout. Pipes might thump or knock when a faucet is closed because of the higher flow rate at the elevated pressure condition. At the very worse, flexible fixture supply connectors and rubber washing machine hoses could blow off of their end connections and cause flooding. Where system pressures become greater than 150 psi (1024 kPa), water heater temperature and pressure relief valves begin leaking. Where the public water main pressure is 200 psi (1379 kPa) or greater, most manufacturers of pressure-reducing valves recommend the installation of two pressure-reducing valves, in series. While this arrangement provides for better pressure control and longer valve life, the effects of fall-off pressure during peak water demand could be significant.

Note that where public water main pressure exceeds 160 psi (1103 kPa), Section 605.3 requires that the water service piping pressure rating be at least that of the highest available pressure. Not all water service piping materials in Table 605.3 are suitable for pressures in excess of 160 psi (1103 kPa).

604.8.2 Repair and removal. Water pressure-reducing valves, regulators and strainers shall be so constructed and installed as to permit repair or removal of parts without breaking a pipeline or removing the valve and strainer from the pipeline.

❖ Water-pressure reducing valves, pressure regulators and strainers must be designed and installed to enable repairs (or cleaning, in the case of a strainer) without removal of the device from the pipe line. In the event of a valve malfunction, repairs can usually be made more quickly where removal of the device from the pipeline is not necessary. This limits the amount of time that a building water system is shut off, inconveniencing the building's occupants.

604.9 Water hammer. The flow velocity of the water distribution system shall be controlled to reduce the possibility of water hammer. A water-hammer arrestor shall be installed where *quick-closing valves* are utilized. Water-hammer arrestors shall be installed in accordance with the manufacturer's instructions. Water-hammer arrestors shall conform to ASSE 1010.

❖ Water hammer is a symptom of hydraulic shock occurring in a fluid-filled piping system. Hydraulic shock occurs when the velocity of fluid flowing through a pipe suddenly changes, such as when a valve closes. The change in velocity creates a pressure wave in the fluid upstream of the valve. The pressure wave travels at a velocity that is approximately equal to the speed of sound travel in the fluid. For water, that velocity (speed) is 4,660 feet per second (1421 meter per second), or in more familiar terms, over 3,000 miles per hour (4828 km/hr). The pressure wave travels upstream of the valve and bounces against turns (elbows and tees) in the piping. Some of the pressure wave continues upstream, bouncing off more piping turns while some of the wave is reflected back towards the closed valve. If the valve closes slowly, the reflected portions of the pressure wave dissipate through the still partially open valve. If the valve fully closes before the reflected portions of the pressure wave arrive at the valve, the wave will bounce against the closed valve and be re-reflected upstream. The pressure wave will continue bouncing back and forth until the friction forces in the fluid eventually cause the wave to dissipate. A strong pressure wave bouncing against a closed valve and turns (elbows and tees) in the piping can produce sounds similar to a hammer hitting a pipe, hence the term "water hammer." Note that hydraulic shock can be occurring and causing damage to the piping system even though the audible sounds (water hammer) of hydraulic shock are not present.

The speed of valve closure determines if hydraulic shock will occur in a pipe. The "critical" valve closure time is not a fixed value for all applications because it is dependent upon the length of piping leading up to the valve, the elasticity of the pipe material, the diameter of pipe and the wall thickness of the pipe. The general range for critical valve closure time, assuming a pipe length of 10 feet (3048 mm), is about 0.02 second for $^1/_2$-inch (12.7 mm) plastic tube (PEX SDR 9) to about 0.005 second for $^1/_2$-inch (12.7 mm) copper tube (Type L). Even if the pipe length was assumed to be 100 feet (30 480 mm), the critical valve closure times would be only 0.2 and 0.05 seconds, respectively. A few time reference points are: 1) the blink of a human eye: 0.3 to 0.4 second and 2) the closure time of a small electric solenoid water valve: 0.01 to 0.06 second.

This code section requires "water-hammer arrestors" to be used where quick-closing valves are utilized. From a theoretical point of view, with respect to hydraulic shock, a quick-closing valve would be a valve that closes in less than or equal to the critical time for a specific arrangement of piping. However, because critical valve closure time is difficult, perhaps impossible, to accurately determine in any application, the code cannot specify an exact time threshold for when a valve should be considered "quick closing." Therefore, even after much public hearing discussion on this subject, the code continues to rely upon the following simple definition for a quick-closing valve: "any valve that closes automatically when released manually or that is controlled by mechanical means for fast action closing" (see the definition of "Quick-closing valve" in Chapter 2).

This simple definition for a quick-closing valve can cause code enforcement questions, for there are a number of automatically closing valves (and faucets) that are commonly used in or connected to water distribution systems. Examples are flushometer valves, mechanical metering faucets, spring-return-to-close faucet controls, washing machine water solenoid valves, dishwasher water solenoid valves and ice-maker water solenoid valves. Which valves should be considered quick closing? Given that small size electric solenoid water valves have closure times less than the critical valve closure times estimated previously, it is a reasonable conclusion that piping connecting to such valves should have a water-hammer arrestor installed. Because automatically closing "mechanical-type" valves and faucets are unlikely to have as fast a closing time as that of small electric solenoid valves, it is doubtful that the mechanical-type valves would require water-hammer arrestors. Ultimately, the code official will have to make the final determination on whether or not other automatically closing valves and faucets require the installation of water-hammer arrestors on the piping to the valves.

Mechanical water-hammer arrestors (shock arrestors) help decrease hydraulic shock intensity by absorbing most of the energy of the pressure wave [see Commentary Figure 604.9(1)]. Mechanical arrestors should be located where they produce the maximum effect of controlling the shock intensity. The manufacturer typically provides guidelines for locating the arrestor. Because many water-hammer arrestors are now permanently charged and factory sealed, they require no maintenance and should last the life of the plumbing system. For this reason, the code official could consider allowing such devices to be installed in concealed locations that do not have access.

Note the distinction between water-hammer arrestors and simple air chambers. An air chamber, as shown in Commentary Figure 604.9(2) and typically constructed of a capped section of vertical pipe, is not a long-term/maintenance-free solution for controlling water hammer. An air chamber located on the fixture supply pipe to every fixture was assumed to help control water hammer; however, air chambers become waterlogged after a short period of time. This is because air chambers allow a direct interface between the trapped air and the water, which allows the air to be absorbed or displaced to other parts of the plumbing system.

Figure 604.9(1)
EXAMPLE OF MECHANICAL WATER-HAMMER ARRESTORS

WATER SUPPLY AND DISTRIBUTION

Figure 604.9(2)
AIR CHAMBER WATER-HAMMER ARRESTOR

Another way to reduce the intensity of hydraulic shock (and reduce the potential for water hammer) caused by quick-closing valves in the piping system is to choose larger pipe sizes, which will result in reduced flow velocities. The hydraulic shock pressure (in excess of the system pressure) is directly proportional to the flow velocity. While reducing the flow velocity does not eliminate the hydraulic shock, decreasing the normal flow velocity in a pipe by a factor of 2 will also decrease the hydraulic shock pressure by a factor of 2. Although reducing the flow velocity in a pipe connected to a quick-closing valve will go a long way towards reducing the potential for water hammer occurrence, this section still mandates the installation of a water-hammer arrestor where quick-closing valves are located.

604.10 Gridded and parallel water distribution system manifolds. Hot water and cold water manifolds installed with gridded or parallel connected individual distribution lines to each fixture or fixture fitting shall be designed in accordance with Sections 604.10.1 through 604.10.3.

❖ Traditionally, water distribution systems in buildings have been designed as "mains, risers and branch"-type systems (also known as "branch" systems). Gridded and parallel distribution systems differ from branch systems and use individual supply pipes that extend to each fixture or outlet from a central supply point. The central supply point is a multiple-outlet manifold to which the distribution lines connect. Manifold systems can be field-fabricated assemblies using conventional fittings or specialized fitting systems, or they could be constructed of factory-built units [see Commentary Figures 604.10(1) and 604.10(2)].

Parallel and gridded systems offer the advantages of simple design, small-size distribution lines and faster distribution of hot water. In branch distribution systems, the hot water contained in larger-diameter piping is often wasted by the user allowing the water to flow down the drain until it reaches the desired temperature. Where semirigid (bendable) tubing is used, gridded or parallel distribution systems can be installed with few, if any, fittings or connections in concealed spaces, thereby eliminating many potential leak points resulting in improved overall dependability of the system.

Table 604.10.1 dictates the minimum size of manifolds to be installed in parallel distribution systems. Table 604.3 dictates the minimum flow rates and flow pressures that must be provided at fixture outlets.

604.10.1 Manifold sizing. Hot water and cold water manifolds shall be sized in accordance with Table 604.10.1. The total gallons per minute is the demand of all outlets supplied.

❖ Manifolds serving gridded or parallel distribution systems must be sized in accordance with Table 604.10.1, based on both the total flow rate demand supplied by the manifold and the chosen design flow velocity of the system. Table 604.10.1 should be used by the designer to size manifold systems while considering the minimum and maximum flow rates required by Sections 604.3 and 604.4. The velocity rates given in Table 604.10.1 are not required minimum and maximum velocities for the design of the manifold system. As with all piping, the design flow velocity should not exceed that recommended by the pipe manufacturer. Generally, lower design flow velocities will prolong the life of the system and minimize noise [see Commentary Figure 604.1(1)].

The table provides sizing criteria for manifolds serving gridded or parallel distribution systems. Such systems are generally used in small buildings; therefore, the range of sizes is limited to those most likely to be encountered in such structures. The velocity columns represent the most common design velocities for the types of materials typically used in these systems. These do not represent required minimum and maximum velocities that must be used in the design of the manifold system.

TABLE 604.10.1
MANIFOLD SIZING

NOMINAL SIZE INTERNAL DIAMETER (inches)	MAXIMUM DEMAND (gpm)	
	Velocity at 4 feet per second	Velocity at 8 feet per second
$^1/_2$	2	5
$^3/_4$	6	11
1	10	20
$1^1/_4$	15	31
$1^1/_2$	22	44

For SI: 1 inch = 25.4 mm, 1 gallon per minute = 3.785 L/m,
1 foot per second = 0.305 m/s.

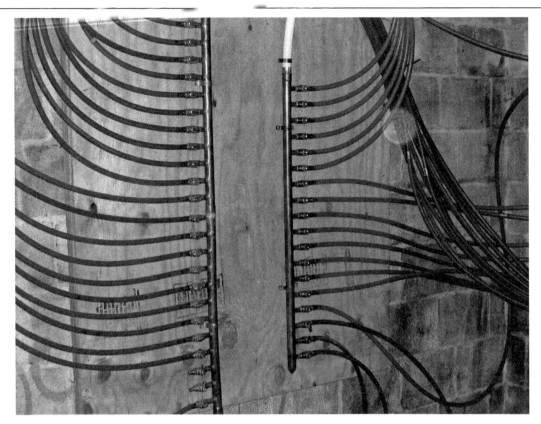

Figure 604.10(1)
FIELD-BUILT MANIFOLD FOR PARALLEL WATER DISTRIBUTION

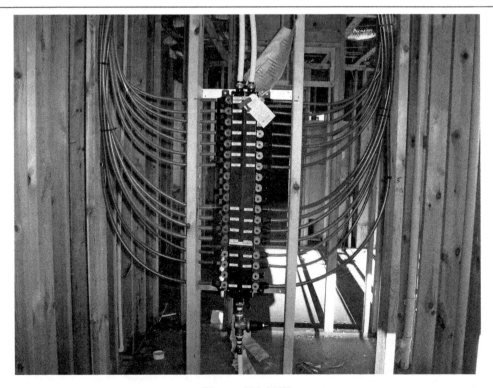

Figure 604.10(2)
FACTORY-BUILT MANIFOLD FOR PARALLEL WATER DISTRIBUTION

604.10.2 Valves. Individual fixture shutoff valves installed at the manifold shall be identified as to the fixture being supplied.

❖ Where parallel water distribution system manifolds are used, this section does not override the requirements of Item 1 in Section 606.2, which requires individual fixture supply shutoff valves. In most instances, the manifold valves are considered as branch control valves that could be used in isolating sections of the manifold system from the main water distribution supply. In the instance where a designer supplies one individual fixture directly from the manifold, such as a single-hose bibb opening, the manifold valve could be used to meet the requirements of Section 606.2 only if that valve is identified in accordance with Sections 604.10.2 and 606.4 and approved by the code official. The control function of valves located at the manifold would not be obvious; therefore, the valves must be labeled.

604.10.3 Access. *Access* shall be provided to manifolds with integral factory- or field-installed valves.

❖ Manifolds with valves, including those with field-installed valves, must be able to be accessed once the building is completed.

604.11 Individual pressure balancing in-line valves for individual fixture fittings. Where individual pressure balancing in-line valves for individual fixture fittings are installed, such valves shall comply with ASSE 1066. Such valves shall be installed in an accessible location and shall not be utilized alone as a substitute for the balanced pressure, thermostatic or combination shower valves required in Section 424.3.

❖ For faucets (fixture fittings) where the code does not require control of the temperature of the water discharge, there still may be a need (but not a code requirement) to keep the user-set, faucet-outlet temperature from fluctuating due to pressure changes in the water distribution system. For example, the water distribution system of a commercial kitchen having numerous pot and food preparation sinks could have significant pressure changes in both hot and cold piping systems caused by the simultaneous operation of two-handle, high-flow faucets typically used in these applications. The pressure changes in the systems could cause significant temperature changes at any one faucet, which would be annoying and perhaps dangerous. The ASSE 1066 pressure balancing valve has no adjustments and no upper temperature limit. The valve adjusts the flow rate through one side of the valve in order to compensate for a pressure differential between the sides (hot and cold) of up to 50 percent.

ASSE 1066 valves do not meet the requirements for balanced pressure, thermostatic or combination balanced-pressure/thermostatic mixing valves for showers (see Section 424.3). An ASSE 1066 valve is only intended for connection to a single-fixture fitting.

SECTION 605
MATERIALS, JOINTS AND CONNECTIONS

605.1 Soil and ground water. The installation of a water service or water distribution pipe shall be prohibited in soil and ground water contaminated with solvents, fuels, organic compounds or other detrimental materials causing permeation, corrosion, degradation or structural failure of the piping material. Where detrimental conditions are suspected, a chemical analysis of the soil and ground water conditions shall be required to ascertain the acceptability of the water service or water distribution piping material for the specific installation. Where detrimental conditions exist, *approved* alternative materials or routing shall be required.

❖ As most water service system piping is buried and some water distribution pipe is buried (e.g., slab-on-grade installations), soil conditions could be a factor in choosing an appropriate pipe material. Because soils are especially aggressive to metallic piping, such as saltwater marshland and high sulphur minerals, the piping might need to be made of nonmetallic materials. Or, if metallic piping must be used under these types of conditions, exterior protection such as plastic sleeving or specialized coatings might need to be used to ensure a reasonable life expectancy of the piping. Where soils have residual amounts of solvents, fuels or where insecticides or fertilizers are consistently used on the soils above, metallic pipe materials might be preferred over plastic piping. Local experienced soils engineers should be consulted wherever questionable soil conditions might be present.

605.2 Lead content of water supply pipe and fittings. Pipe and pipe fittings, including valves and faucets, utilized in the water supply system shall have not more than 8-percent lead content.

❖ A certain amount of lead in metallic materials makes the material softer and easier to machine (i.e., drill, turn, cut and mill); however, too high of a lead content in the internal "wetted" surfaces of the water-conveying products can allow significant leaching of lead into the potable water. Where the potable water is used for human consumption there is great concern about lead leaching because lead in drinking water was found to be a contributor of birth defects and childhood diseases. In June 1986, as authorized by the Safe Drinking Water Act (SWDA), the EPA required that the lead content in pipe, pipe fittings, faucets and valves conveying potable water be limited to not greater that 8 percent. (The same legislation also required that the lead content in solder and fluxes be reduced to not greater than 0.2 percent.) Further discussion regarding the SDWA can be found in the commentary to Section 424.1.

An amendment to the SDWA occurred in 2011 to further reduce the amount of lead in plumbing products. The amendment became law in January 2014. The amendment changes the meaning of "lead free" from 8 percent or less to 0.25 percent or less. However, the law also changes the determination of lead content to a "weighted average of wetted surfaces method." Therefore, the SDWA no longer requires laboratory testing of water exposed to products to determine the lead content. However, this section remains in the code.

605.2.1 Lead content of drinking water pipe and fittings. Pipe, pipe fittings, joints, valves, faucets and fixture fittings utilized to supply water for drinking or cooking purposes shall comply with NSF 372 and shall have a weighted average lead content of 0.25 percent or less.

❖ In recent years, additional health study data has been gathered to indicate that limiting the lead content to not greater than 8 percent might not be restrictive enough to avoid human health problems. Initially, two states (California and Vermont) enacted legislation to require the lead content of pipe, pipe fittings, plumbing fittings and fixture fittings (faucets) to be not greater than 0.25 percent (by wetted surface weighted average method). The U.S. EPA followed up with similar recommendation for all states and the "low lead" amendment to the SDWA became law in January 2014 (see commentary to Section 605.2).

Neither NSF 372 nor federal law requires low-lead compliant products be marked or identified in any particular manner. Identification markings of "low-lead" products are not standardized among manufacturers or third-party certification agencies. Some manufacturers might make both low-lead compliant products and products that are intended to be used in systems such as hydronic heating systems and nonpotable water systems that do not require low-lead compliant products. Contractors will need to be aware of the difference to avoid using the wrong product, such as a noncompliant valve, in an application where a low-lead compliant valve is required. Code officials will need to be knowledgeable about various product markings and vigilant for improper application errors when making inspections. Other manufacturers that make products such as kitchen faucets, lavatory faucets and drinking fountains that are intended for sale in areas covered by federal law will have only the low-lead compliant versions available for sale in those areas. However, with the increasing global access to plumbing products, contractors and code officials will need to be careful about the products that are being installed to ensure that federal law is not violated. This code's requirement for third-party certification to NSF 372 will make such verifications easier.

Evaluation of products for compliance with the low-lead federal law (and NSF 372) is by calculation method only.

605.3 Water service pipe. Water service pipe shall conform to NSF 61 and shall conform to one of the standards listed in Table 605.3. Water service pipe or tubing, installed underground and outside of the structure, shall have a working pressure rating of not less than 160 psi (1100 kPa) at 73.4°F (23°C). Where the water pressure exceeds 160 psi (1100 kPa), piping material shall have a working pressure rating not less than the highest available pressure. Water service piping materials not third-party certified for water distribution shall terminate at or before the full open valve located at the entrance to the structure. Ductile iron water service piping shall be cement mortar lined in accordance with AWWA C104.

❖ Piping material that comes in contact with potable water is required to conform to the requirements of NSF 61. The intent is to control the potential adverse health effects produced by indirect additives, products and materials that go into the making of the pipe so that those substances do not pollute or contaminate potable water. Plastic water pipe tested and labeled in accordance with NSF 14 as potable water pipe (see commentary, Section 303.3) also conforms to NSF 61 because NSF 14 makes reference to and requires compliance with NSF 61.

Not all water service piping materials are suitable for use as water distribution piping (in other words, are found in Table 605.4). For example, PVC pipe is not suitable for water distribution pipe because it cannot meet the pressure/temperature requirement for water distribution pipe (100 psi at 180°F) (689 kPa at 82.2°C). The concern is that, if PVC piping was allowed to be used inside of a structure for cold water piping, it might be inadvertently used for a hot water pipe. Therefore, water service piping that is not also suitable for use as water distribution piping must terminate at or before the main water shutoff valve (see Section 606.1, Item 2). Up to the 2003 code edition, this section specified an arbitrary 5 feet (1524 mm) from the point of entry as the point where the water service pipe had to end and the water distribution pipe had to begin. The code was revised in 2006 to allow more flexibility as to where the full-open main water valve can be located and, therefore, where the water service pipe ends and water distribution pipe begins [see Commentary Figures 605.3(1) and 605.3(2)].

The requirement for ductile iron piping to be internally cement coated according to AWWA C104 is to prevent the formation of rust on the inside of the pipe, which could flake off and enter the water distribution system.

TABLE 605.3. See page 6-37.

Acrylonitrile butadiene styrene (ABS) plastic pipe. ABS pipe is rarely manufactured for water service. ASTM D 1527 is the referenced standard for Schedule 40 and 80 ABS pipe. The schedule number indicates the wall thickness: the higher the schedule number, the thicker the pipe wall. ASTM D 2282 is the referenced standard for SDR-PR pipe. "SDR" is an abbreviation for standard pipe dimension ratio, which is the ratio of pipe diameter to wall thickness. The "PR" designation indicates that the pipe is pressure rated. The

WATER SUPPLY AND DISTRIBUTION

Figure 605.3(1)
PLASTIC WATER SERVICE EXTENDING INTO BUILDING

Figure 605.3(2)
PLASTIC WATER SERVICE EXTENDING UNDER SLAB AND THEN INTO BUILDING

pipe is black, must be rated for distribution of potable water and is marked at intervals of not more than 5 feet (1524 mm) with the following information:

- The manufacturer's name.
- The designation ASTM D 1527 or ASTM D 2282 SDR.
- Certification for conformance to NSF 61.
- Nominal pipe size.
- ABS 40 or 80 psi rating "ABS 1210."
- Coextruded ABS pipe "ABS Coex."

ASTM D 2468 is the standard for ABS plastic pipe fittings, Schedule 40. Fittings meeting the requirements of ASTM D 2468 are permitted with Schedule 40 pipe, and any fitting having a male thread must have an internal diameter not larger than Schedule 80 pipe of the same size.

Asbestos-cement pipe.

The 2015 IPC deleted asbestos cement pipe from Table 605.3 because of the concerns about working with this type of material (encapsulated asbestos) and because very little, if any, asbestos-cement pipe is

WATER SUPPLY AND DISTRIBUTION

TABLE 605.3
WATER SERVICE PIPE

MATERIAL	STANDARD
Acrylonitrile butadiene styrene (ABS) plastic pipe	ASTM D 1527; ASTM D 2282
Brass pipe	ASTM B 43
Chlorinated polyvinyl chloride (CPVC) plastic pipe	ASTM D 2846; ASTM F 441; ASTM F 442; CSA B137.6
Chlorinated polyvinyl chloride/aluminum/chlorinated polyvinyl chloride (CPVC/AL/CPVC)	ASTM F 2855
Copper or copper-alloy pipe	ASTM B 42; ASTM B 302
Copper or copper-alloy tubing (Type K, WK, L, WL, M or WM)	ASTM B 75; ASTM B 88; ASTM B 251; ASTM B 447
Cross-linked polyethylene (PEX) plastic pipe and tubing	ASTM F 876; ASTM F 877; AWWA C904; CSA B137.5
Cross-linked polyethylene/aluminum/cross-linked polyethylene (PEX-AL-PEX) pipe	ASTM F 1281; ASTM F 2262; CSA B137.10
Cross-linked polyethylene/aluminum/high-density polyethylene (PEX-AL-HDPE)	ASTM F 1986
Ductile iron water pipe	AWWA C151/A21.51; AWWA C115/A21.15
Galvanized steel pipe	ASTM A 53
Polyethylene (PE) plastic pipe	ASTM D 2239; ASTM D 3035; AWWA C901; CSA B137.11
Polyethylene (PE) plastic tubing	ASTM D 2737; AWWA C901; CSA B137.1
Polyethylene/aluminum/polethylene (PE-AL-PE) pipe	ASTM F 1282; CSA B137.9
Polyethylene of raised temperature (PE-RT) plastic tubing	ASTM F 2769
Polypropylene (PP) plastic pipe or tubing	ASTM F 2389; CSA B137.11
Polyvinyl chloride (PVC) plastic pipe	ASTM D 1785; ASTM D 2241; ASTM D 2672; CSA B137.3
Stainless steel pipe (Type 304/304L)	ASTM A 312; ASTM A 778
Stainless steel pipe (Type 316/316L)	ASTM A 312; ASTM A 778

being produced in the United States. This information is left in this commentary to assist persons who might come across this type of piping during remodeling and repairs.

In the United States, this material was predominantly used for public water mains and forced sewer mains that were installed many decades ago. Outside the United States, this type of pipe is still being used for such purposes. Asbestos-cement pipe is manufactured in 4-, 6-, 8-, 10-, 12-, 14-, 16-, 18-, 20-, 24-, 30- and 36-inch diameters. The pipe size indicates the nominal inside diameter of the pipe.

The pipe is made by forming a pipe section from a slurry of asbestos fibers and portland cement or portland slag cement. The product is autoclaved (heat and pressure) to substantially harden the mixture. The ends of the pipe section are accurately machined smooth so that asbestos cement couplings with internal rubber sealing rings can be used to join pipe sections at installation. Because the asbestos is encapsulated and bonded within the pipe, this type of pipe has not been considered a health hazard. However, very little, if any, asbestos-cement pipe is being produced in the United States because there is a hazard of breathing asbestos during production and cutting. Rubber rings to seal the joints must conform to ASTM D 1869. The product is marked with the following information:

- The manufacturer's name or trade name.
- Class (100, 150 or 200 psi).
- Hydrostaticproof pressure.
- Date of manufacture.
- The designation ASTM C 296.
- Nominal pipe size.

Brass pipe. This piping material is made from red brass that is composed of approximately 85-percent copper and 15-percent zinc. The pipe is available in both standard and extra-strong weight. Extra-strong weight has the same outside diameter as steel pipe, but with a thicker pipe wall.

Standard-weight pipe is often called by the expression "Schedule 40," and extra-strong pipe is named "Schedule 80." Brass pipe dimensions are very similar to steel pipe dimensions. ASTM B 43 is the referenced standard for brass pipe, available in diameters from $^1/_8$ inch (3.2 mm) to 12 inches (305 mm) and manufactured in 12-foot (3658 mm) lengths. The pipe must be rated for distribution of potable water and is marked with the following information:

- The manufacturer's name.
- The designation ASTM B 43.
- Certification for conformance to NSF 61.

- Nominal pipe size.
- Schedule 40 standard weight or Schedule 80 extra-strong weight.

Chlorinated polyvinyl chloride (CPVC) plastic pipe and tubing. The plastic material used to produce CPVC pipe is similar to polyvinyl chloride (PVC) material except that the molecular structure has additional chlorine elements to improve the material strength properties at elevated temperature. CPVC plastic pipe and tubing may be either a white, a milky white (cream), gray or black color or, for fire sprinkler service, an orange color. The raw material is designated CPVC 4120. The "41" designation of CPVC 4120 identifies the specific type and quality of the CPVC material. The "20" designation of CPVC 4120 indicates the hydrostatic design stress of the material in hundreds of psi. In other words, CPVC 4120 has a hydrostatic design stress of 2,000 psi (13 790 kPa). Note that this is a material design property and not a pressure rating of the pipe.

ASTM D 2846 is the referenced standard for SDR 11 CPVC pipe, tube and CPVC socket fittings. "SDR" means "standard dimension ratio," which is a ratio of the average outside diameter to the minimum wall thickness. Tube outside diameter (OD) dimensions for tube products included in this standard match copper water tube sizes $^1/_4$, $^3/_8$, $^1/_2$, $^3/_4$, 1, 1$^1/_4$, 1$^1/_2$, and 2 inch. The pipe OD dimensions for pipe products included in this standard match steel pipe sizes 1$^1/_2$ and 2 inch.

ASTM F 441 is the referenced standard for Schedule 40 and 80 plastic pipe, meaning that the OD dimensions of the pipe match those of steel pipe. Pipe sizes included in this standard are $^1/_4$, $^3/_8$, $^1/_2$, $^3/_4$, 1, 1$^1/_4$, 1$^1/_2$, 2, 2$^1/_2$, 3, 3$^1/_2$, 4, 5, 6, 8, 10, 12, 14, and 16 inch. The pipe wall thickness for Schedule 40 CPVC pipe is the same as for Schedule 40 steel pipe up to the 12-inch (305 mm) size. For the 14- and 16-inch sizes, the CPVC Schedule 40 pipe has thicker walls than steel pipe. The pipe wall thickness for Schedule 80 CPVC pipe is the same as for Schedule 80 steel pipe up to the 8-inch size. For sizes above 8-inch, CPVC Schedule 80 pipe has thicker walls than steel pipe. Pipe sizes included in this standard are $^1/_4$, $^3/_8$, $^1/_2$, $^3/_4$, 1, 1$^1/_4$, 1$^1/_2$, 2, 2$^1/_2$, 3, 3$^1/_2$, 4, 5, 6, 8, 10, 12, 14, and 16 inch.

ASTM F 442 is the referenced standard for SDR 13.5, 17, 21, 26 and 32.5 CPVC pipe. The pipe is also designated "PR," indicating that it is pressure rated for water. Pipe sizes included in this standard are $^1/_4$, $^3/_8$, $^1/_2$, $^3/_4$, 1, 1$^1/_4$, 1$^1/_2$, 2, 2$^1/_2$, 3, 3$^1/_2$, 4, 5, 6, 8, 10 and 12 inch. The OD dimensions of the pipes match those for steel pipe. The pipe wall thicknesses do not match steel pipe.

Section 303.3 requires identification of all plastic pipe and fittings with the mark of an approved agency establishing conformance to NSF 14 (see commentary, Section 303.4). Where used for drinking water, the product must be certified to NSF 61.

The product is marked at intervals of not more than 3 feet (914 mm) with the following information:

- The manufacturer's name.
- The standard designation (ASTM D 2846, ASTM D 441/442 or CSA B137.6).
- Certification for conformance to NSF 61.
- Nominal pipe size or nominal copper tube size.
- Type of plastic material (CPVC 4120).
- Schedule size and pressure rating (Schedule XX at 200 psi or SDR XX at 180 psi).

Copper or copper-alloy pipe. There are two types of copper pipe: threaded and threadless. Both pipes are seamless and have a chemical composition of 99.9-percent copper.

ASTM B 302 is the standard for threadless copper pipe for both regular and extra-strong weight, available in diameters from $^1/_4$ inch (6.4 mm) to 12 inches (305 mm) and manufactured in 20-foot (6096 mm) lengths. The pipe dimensions are similar to those of brass and steel pipe.

Threadless copper pipe has the same outside diameter as threaded copper pipe, steel pipe and brass pipe. The wall thickness of the pipe is less than that of threaded copper pipe; therefore, the inside diameter is greater. Threadless copper pipe is intended to be joined by brazing and is continuously marked in a gray color. The designation "TP" for threadless pipe is included in the marking. Section 303.4 also requires identification of copper pipe used for potable water with the mark of an approved third-party agency and notation that it conforms to the requirements of NSF 61 (see commentary, Section 303.4). The pipe must be rated for distribution of potable water and be marked with the following information:

- The manufacturer's name.
- The designation ASTM B 302 TP or ASTM B 42.
- Certification for conformance to NSF 61.
- Nominal pipe size.
- Schedule 40 standard weight or Schedule 80 extra-strong weight.

Copper or copper-alloy tubing (Type K, WK, L, WL, M or WM). The tubing is manufactured in two different tempers, which are identified as drawn (called "hard copper" with the designation "H") and annealed (called "soft copper" with the designation "O"). Hard (drawn) copper tubing comes in straight lengths, whereas soft (annealed) copper tubing comes in both straight lengths and coils.

Annealed copper and tempered drawn copper may be formed by bending. Bending tempered copper is identified on the pipe with the designation "BT." Only bending tempered copper tubing can be joined using flared connections.

Seamless copper water tubing is available in Type K, L or M. The tubing type indicates the wall thickness. Type K has the thickest wall, followed by Types L and M, respectively. ASTM B 75 is the referenced standard for annealed seamless copper tubing. ASTM B 88 and ASTM B 251 are the referenced standards for both drawn and annealed seamless copper water tubing. The outside diameter of the tubing is the same for all three types; only the inside diameter varies with wall thickness. Welded copper tubing has the designation "WK," "WL" or "WM." The "W" indicates that the tubing has a welded (longitudinal) seam. ASTM B 447 is the referenced standard for welded copper tubing. Welded tubing dimensions correspond to Type K, L or M seamless tubing.

The types of copper tubing are identified with a continuous colored marking for ease of identification. These colored markings are as follows: Type K (WK) is green, Type L (WL) is blue and Type M (WM) is red. All are available in diameters from $^1/_4$ inch (6.4 mm) to 12 inches (305 mm). Straight lengths are manufactured in 12-foot (3658 mm), 18-foot (5486 mm) and 20-foot (6096 mm) lengths and annealed copper coils are manufactured in 40-foot (12 192 mm), 60-foot (18 288 mm) and 100-foot (30 480 mm) lengths.

The tubing must be rated for distribution of potable water and is marked at intervals of not more than 3 feet (914 mm) with the following information:

- The manufacturer's name.
- The standard designation (ASTM Number).
- NSF 61 certification mark.
- Nominal size (inches, CTS).
- Color marking green, blue or red.

Cross-linked polyethylene (PEX) plastic pipe tubing. The material is available in nominal pipe sizes (NPS) of $^1/_4$-inch (6.4 mm) through 2 inches (51 mm) and a standard dimension ratio of SDR 9. Commentary Figure 605.3(3) shows a sample marking for PEX tubing.

PEX tubing can be produced by numerous different manufacturing processes, such as the Engel, Silane, Monosil, Point a'Mousson, Daoplas and electron beam methods. Although each manufacturing process yields tubing complying with the referenced standards, each produces a tube that possesses unique characteristics.

The cross-linked molecular structuring gives the pipe additional resistance to rupture over a wider range of temperatures and pressures than that of other polyolefin plastics [polyethylene (PE) and polypropylene (PP)]. Because of PEX pipe's unique molecular structure and resistance to heat, heat fusion is not possible as a joining method. Because PEX is a member of the polyolefin plastic family, it is resistant to solvents. Therefore, PEX pipe cannot be joined by solvent cementing.

PEX pipe is flexible, allowing it to be bent. The tubing can be bent using either hot or cold bending methods.

For hot bending, a hot-air gun with a diffuser nozzle is used. The pipe must not be exposed to an open flame. The hot air meeting the pipe surface must not exceed 338°F (170°C), and the heat-up time must not exceed 5 minutes. The pipe is bent using conventional methods and is allowed to cool to room temperature before removal from the bending tool. The minimum hot-bending radius is two and one-half times the outside diameter. PEX can be bent at room temperature (cold bending) to a minimum radius of six times the outside diameter.

Mechanical connectors and fittings for PEX pipe are proprietary and must be used only with the pipe for which they have been designed and tested. Fitting systems currently available include compression fittings with inserts and ferrules or O-rings, insert fittings using a metallic lock (crimp) ring and insert fittings with compression collars using a cold flaring tool. It is important to consult the manufacturer's installation instructions to identify the fittings specified for use with each brand of PEX piping (see commentary, Section 605.17.2).

Cross-linked polyethylene/aluminum/cross-linked polyethylene (PEX-AL-PEX) pipe. PEX-AL-PEX is a composite pipe made of an aluminum tube laminated to interior and exterior layers of cross-linked polyethylene. The layers are bonded together with an adhesive. The cross-linked molecular structuring gives the pipe additional resistance to rupture than that of polyethylene (see the commentary for cross-linked polyethylene tubing in this section). Therefore, the pipe is suitable for hot and cold water distribution and is pressure rated for 125 psi at 180°F (862 kPa at 82°C).

Although it is partially plastic, the PEX-AL-PEX pipe resembles metal tubing in that it can be bent by hand or with a suitable bending device while maintaining its shape without fittings or supports. The minimum bending radius specified by manufacturers of this type of pipe is five times the outside diameter.

Mechanical joints are the only method currently available to join PEX-AL-PEX pipe. A number of proprietary mechanical-compression-type connectors have been developed for use with the composite pipe to permit transition to other pipes and fittings. Such fittings must be installed in accordance with the manufacturer's instructions (see commentary, Section 605.20.3).

Cross-linked polyethylene/aluminum/high-density polyethylene (PE-AL-HDPE) plastic pipe. This pipe is similar to cross-linked polyethylene/aluminum/cross-linked polyethylene (PEX-AL-PEX) pipe except the outer layer is made of high-density polyethylene instead of PEX. The OD dimensions of this pipe are 0.630 (16 mm), 0.787 (20 mm), 1.024 (26 mm), 1.260 (32 mm), 1.575 (40 mm), 1.969 (50 mm) and 2.48 inches (63 mm). The respective inside diameter dimensions are 0.453 (11.5 mm), 0.591 (15 mm), 0.787 (20 mm), 1.024 (26 mm), 1.299 (33 mm), 1.654 (42 mm) and 2.126 inches (54 mm).

Ductile-iron water pipe. The piping material is identified as either cast iron or ductile iron. Both materials are within the classification of cast iron. Ductile iron is a cast iron that is higher in ductility, resulting in the name "ductile iron." It is not as brittle as cast-iron pipe. Section 605.3 requires that the inside surface of ductile-iron water pipe be cement-mortar lined in accordance with AWWA C104 [see Commentary Figure 605.3(1)]. This is necessary to provide a protective barrier between the potable water supply and the ductile-iron pipe to prevent impurities and contaminants from leaching into the water supply.

PIPE DIAMETER (inches)	MINIMUM LINING THICKNESS (inches)
3-12	$1/4$
14-24	$3/32$
30-54	$1/8$

For SI: 1 inch = 25.4 mm.

**Figure 605.3(1)
DUCTILE IRON PIPE LINING THICKNESS**

AWWA C151 and AWWA C115 are the referenced standards for ductile-iron water pipe. The pipe is manufactured with a bell hub end in lengths of 18 and 20 feet (5486 and 6096 mm). The pipe is available in diameters of 3 through 64 inches (76 through 1626 mm), and all required markings must be on or near the bell of the pipe with the following information:

- The manufacturer's name or mark.
- The country where cast.
- The weight, class or nominal thickness.
- The year in which the pipe is produced.
- The letters "DI" or the word "Ductile."

Galvanized steel pipe. A wide variety of pipe is classified as galvanized steel pipe. The grade of steel and the method of manufacturing can differ greatly.

The common pipe used in plumbing application is either continuous welded or electric-resistance welded (ERW). Continuous-welded pipe is heated and formed, whereas ERW pipe is cold rolled and welded.

Galvanized steel pipe is available in different wall thicknesses, the most common being standard weight, extra strong and double extra strong. These thicknesses are also referred to as Schedules 40, 80 and 160, respectively. As the schedule number increases, the wall thickness of the pipe increases. The outside diameter, however, remains the same for each pipe size. Steel pipe is hot dipped and zinc coated to produce a galvanized protective coating, both inside and outside. New galvanized pipe is a shiny silver, and, as the pipe ages, the coating oxidizes to protect the zinc. As a result, the color turns to a dull gray.

Galvanized steel pipe is sometimes called "galvanized pipe" or mistakenly called "wrought-iron pipe." Wrought-iron pipe has not been manufactured in the United States in more than 30 years. All pipe referred to as "wrought-iron pipe" is steel pipe.

The steel pipe industry claims to be responsible for nominal pipe dimensions, meaning the pipe is approximately the size by which it is identified. The dimensions date back to 1862 when Robert Briggs of Pascal Iron Works wrote the first pipe standards in the United States. It is believed that the dimensions were based on the pipe Briggs was manufacturing. That is why a $3/4$-inch (19.1 mm) galvanized steel pipe has an out-side diameter of 1.05 inches (27 mm) and an average inside diameter of 0.824 inch (21 mm) for standard-weight pipe. The pipe is available in diameters of $1/8$ inch through 26 inches (3.2 mm through 660 mm) and various lengths of 12 feet (3658 mm) to 22 feet (6706 mm). Galvanized steel pipe is marked with the following information:

- The manufacturer's name or mark.
- The type of pipe (i.e., A, B, XS, XXS).
- The specification number.
- The length of pipe.

See Commentary Figure 605.3(2) for dimensions of Schedule 40 steel pipe.

Polybutylene (PB) plastic pipe and tubing. Polybutylene pipe and tubing was the subject of much media attention decades ago. This was a result primarily of the settlement of a multimillion-dollar lawsuit and the announcement by the primary producer that it stopped manufacturing polybutylene resin for water supply piping. While some old stock of polybutylene may still be in warehouses, polybutylene pipe or tubing for potable water service or water distribution has not been produced in the United States for over 30 years. However, many homes built as late as the 1980s have had this product installed in water distribution systems. PB tubing for water service has been found installed as late as the 1990s as contractors and supply houses continued to deplete this obsolete stock. Because polybutylene pipe is no longer listed in the tables as an approved material for water service or water distribution piping, this material must not be installed in new construction, remodeling or repair work.

For reference and identification purposes only, PB pipe and tubing were marked "ASTM D 2662, D 2666, or D 3309." These materials were colored blue, black, or gray.

Polyethylene (PE) plastic pipe and tubing. Polyethylene is an inert polyolefin material. Because the material is highly resistant to chemical attack, solvent-cement joining of polyethylene pipe and tubing is not possible. Polyethylene pipe is blue or black when supplied for water service. The pipe is continuously marked with the manufacturer's name, the ASTM, AWWA or CSA standard and the grade of polyethylene.

Various grades of polyethylene resin can be used to produce pipe or tubing. In accordance with the ASTM standard, the material is identified by the term "PE," followed by four digits. The first digit indicates the type of material based on density. The second digit is the category identifying the extrusion flow rate. The last

WATER SUPPLY AND DISTRIBUTION

NOMINAL PIPE SIZE (inches)	OUTSIDE DIAMETER (inches)	AVERAGE WALL THICKNESS (inch)
$1^1/_4$	1.660	0.140
$1^1/_2$	1.9	0.145
2	2.375	0.154
$2^1/_2$	2.875	0.203
3	3.5	0.216
$3^1/_2$	4	0.226
4	4.5	0.226
5	5.563	0.258
6	6.625	0.280
8	8.625	0.322
10	10.75	0.365
12	12.75	0.375
14	14	0.437
16	16	0.5
18	18	0.562
20	20	.583
24	24	.687

For SI: 1 inch = 25.4 mm.

Figure 605.3(2)
STEEL PIPE DIMENSIONS
SCHEDULE 40

two digits show the hydrostatic design stress in hundreds of psi.

ASTM D 2239 is the referenced standard for SDIR-PR polyethylene pipe. The SDIRs of the pipe are 5.3, 7, 9, 11.5 and 15. The grades of polyethylene permitted include PE 1404, PE 2305, PE 2306, PE 3306, PE 3406 and PE 3408.

ASTM D 2737 is for polyethylene tubing with an SDR of 7.3, 9 and 11. The SDR for tubing is based on the average outside diameter.

Polyethylene/aluminum/polyethylene (PE-AL-PE) pipe. PE-AL-PE is identical to PEX-AL-PEX composite pipe except for the physical properties of the polyethylene. Polyethylene displays the same resistance to temperature and pressure as cross-linked polyethylene does; therefore, it is suitable for hot and cold water distribution and is pressure rated for 125 psi at 180°F (862 kPa at 82°C) (see the commentary to PEX-AL-PEX in this section).

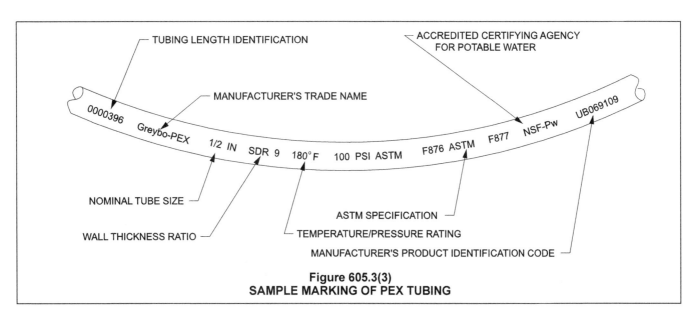

Figure 605.3(3)
SAMPLE MARKING OF PEX TUBING

Polyethylene of raised temperature (PE-RT) plastic tubing. Polyethylene of raised temperature is an inert polyolefin material. ASTM F 2769 covers tubing sizes $^1/_8$-, $^1/_4$-, $^5/_{16}$-, $^3/_8$-, $^1/_2$-, $^5/_8$-, $^3/_4$-, 1-, $1^1/_4$-, $1^1/_2$-, 2-, $2^1/_2$-, 3, $3^1/_2$-, 4-, $4^1/_2$-, 5- and 6-inch sizes. The wall thickness is based on an SDR of 9. Cold-bending radius must not be less than six times the outside diameter of the tubing. PE-RT must not be stored in locations or used in applications where there will be exposure to sunlight (UV light). PE-RT tubing is marked with the following information every 5 feet (1524 mm):

- The manufacturer's name or mark.
- The nominal tube size.
- Material specification in accordance with ASTM D 3350 (e.g., PE344373).
- Type of material (PE-RT).
- Standard Dimension Ratio SDR 9.
- Pressure rating (not less than 100 psi at 180°F and not less than 160 psi at 73°F).
- The specification number (ASTM F 2769).
- Chlorine-resistance classification code (e.g., CL1, CL3, or CL5).
- Standard designation for the fitting system(s) to be used with the tubing.
- Manufacturer's production code.
- Markings for NSF 14 and NSF 61.

Polypropylene (PP) pipe and tubing. Polypropylene is an inert polyolefin material. The pipe and tubing products are made of not more than 50 percent of other types of olefinic materials. ASTM F 2389 covers pipe sizes $^1/_2$, $^3/_4$, 1, $1^1/_2$, 2, 3, 4 and 6 inch. The pipe ODs match the dimensions of steel pipe. The pipe wall thicknesses match that of Schedule 80 steel pipe. The pipe is not to be externally threaded. ASTM F 2389 also covers metric size tubing in sizes 16, 20, 25, 32, 40, 50, 63, 75, 90, 110, 1125, 140 and 160 mm. The metric tubing wall thicknesses are according to the dimension ratio (DR) of the tubing. The available DRs are 11, 7.3, 6 and 5. The piping is joined using heat fusion methods. Polypropylene pipe is marked with the following information every 5 feet (1524 mm):

- The manufacturer's name or mark.
- The nominal pipe size.
- Pipe will be marked Schedule 80 or SCH 80.
- Type of material (PP-R).
- Pressure rating (100 psi at 180°F and 160 psi at 73°F).
- The specification number (ASTM F 2389 or CSA B137.11).
- Manufacturer's production code.
- Markings for NSF 14 and NSF 61.

Polyvinyl chloride (PVC) plastic pipe. PVC water service pipe is a different material than PVC drainage pipe, although both materials are white. The pipe must be continuously marked with the manufacturer's name, the ASTM or CSA standard and the grade of the PVC material.

A number of grades of PVC material are used to produce pipe. In accordance with the ASTM standards, the compounds are identified as PVC 12454-B, 12454-C and 14333-D. ASTM D 1785 is the referenced standard for Schedule 40, 80 and 160 PVC plastic pipe. These dimensions are the same as Schedule 40, 80 and 160 steel pipe.

ASTM D 2241 is the referenced standard for SDR-PR PVC plastic pipe. "SDR" means standard dimension ratio, which is the ratio between the average outside diameter of the pipe and the minimum wall thickness. "PR" indicates that the pipe is pressure rated. SDRs of the pipe are 13.5, 17, 21, 26, 32.5, 41 and 64.

ASTM D 2672 is the referenced standard for bell-end PVC plastic pipe conforming to either ASTM D 1785 or ASTM D 2241.

Stainless steel pipe (Types 304, 304L, 316 and 316L). Stainless steel is a family of iron-based alloys that must contain at least 10 percent or more chromium. The presence of chromium creates an invisible surface film that resists oxidation and makes the material corrosion resistant. Stainless steel is produced in an electric arc furnace using argon-oxygen degassing melting technology. This provides closely controlled chemistry in the final product. Most semifinished product receives a final annealing and pickling operation. The fact that stainless steel has a great resistance to corrosion means that using stainless will result in a very long pipe life compared to mild steel pipe. The benefits of stainless include corrosion resistance, fire and heat resistance, hygiene (easy cleaning ability), aesthetic appearance, strength-to-weight advantage, ease of fabrication, impact resistance and long-term value.

Wrought austenitic stainless steels unite a useful combination of outstanding corrosion and heat resistance with good mechanical properties over a wide temperature range. This type of pipe is typically used for chemicals, petroleum, pollution abatement, food processing, transportation equipment, water service and water distribution. There are several grades of stainless steel, such as ferritic, martensitic, austenitic, duplex and precipitation hardened. All grades other than the austenitic (300 series) stainless steels exhibit some degree of magnetism; the austenitic grades are nonmagnetic. Some austenitic welds may show some slight degree of magnetism because of the desired requirement to have a small amount of residual ferrite in the finished weld. Thus, one should not limit the selection of wrought austenitic stainless steels to those included herein, but rather should ask producers for the best grade—standard or specialty—for a particular use.

ASTM A 312 and ASTM A 778 are the referenced standards for water service and water distribution pipe. The material is suitable for both applications. Types 304/304L and 316/316L are referenced in those standards. Type 304/304L is the low-carbon grade, which provides the designer higher minimum design properties, while offering the benefits of the low-carbon grades of material suitable for welding. Type 316/316L stainless steels are characterized by addition of molybdenum and are more corrosion resistant than Type 304/304L.

Standard wrought austenitic stainless steels can be supplied in a variety of standard finishes, which include dull and smooth, pickled, ground finishes up to 500 grit, electropolished, rouge polished and tumbled.

Annealed tubing can be formed to minimum bend radii of two to six times OD. Tube ends can be flanged, flared or beaded. In addition to mechanical fastening, standard wrought austenitic stainless steels may be joined by brazing, soldering or welding.

Brazing can be readily accomplished, but should be well above the heat sensitizing range. Brazing without fluxes can be done in vacuum or in atmospheres of dry hydrogen, argon or helium.

Stainless steel can also be soldered using, for example, tin-lead filler metals. Fluxes are required to break down the protective oxide film and permit wetting. Chloride-free or phosphate fluxes should be used. Inert gas shielding is necessary to prevent porosity and oxidation of the weld metal.

605.3.1 Dual check-valve-type backflow preventer. Dual check-valve backflow preventers installed on the water supply system shall comply with ASSE 1024 or CSA B64.6.

❖ Operators of public water systems are required by the SDWA to protect the public water systems from possible sources of contamination. Every water service connection to a public water main is a possible entry point for contamination and pollutants to the public water supply system. For example, cleaning crews of an office building might use chemicals in such a manner that the chemicals could be introduced in the water distribution system of the building should a backflow event occur. The contamination could flow into the water service system and then into the public water system. In order to protect the public water supply, the water purveyor usually has legal right to protect the public system by installing a backflow preventer at the water service connection or request (demand) that the building owner install a backflow preventer in the water service system to the building. A dual check valve-type backflow preventer offers a minimum level of protection against backpressure and backsiphonage. In many cases, the water purveyor provides and installs a dual check valve backflow preventer as part of installing the water meter. Where dual check valve backflow preventers are installed, they must comply with ASSE 1024 or CSA B 64.6.

605.4 Water distribution pipe. Water distribution pipe shall conform to NSF 61 and shall conform to one of the standards listed in Table 605.4. Hot water distribution pipe and tubing shall have a pressure rating of not less than 100 psi (690 kPa) at 180°F (82°C).

❖ Piping material that comes in contact with potable water is required to conform to NSF 61 (see commentary, Section 605.3). Pipe installed in a water distribu-

**TABLE 605.4
WATER DISTRIBUTION PIPE**

MATERIAL	STANDARD
Brass pipe	ASTM B 43
Chlorinated polyvinyl chloride (CPVC) plastic pipe and tubing	ASTM D 2846; ASTM F 441; ASTM F 442; CSA B137.6
Chlorinated polyvinyl chloride/aluminum/chlorinated polyvinyl chloride (CPVC/AL/CPVC)	ASTM F 2855
Copper or copper-alloy pipe	ASTM B 42; ASTM B 302
Copper or copper-alloy tubing (Type K, WK, L, WL, M or WM)	ASTM B 75; ASTM B 88; ASTM B 251; ASTM B 447
Cross-linked polyethylene (PEX) plastic tubing	ASTM F 876; ASTM F 877; CSA B137.5
Cross-linked polyethylene/aluminum/cross-linked polyethylene (PEX-AL-PEX) pipe	ASTM F 1281; ASTM F 2262; CSA B137.10
Cross-linked polyethylene/aluminum/high-density polyethylene (PEX-AL-HDPE)	ASTM F 1986
Ductile iron pipe	AWWA C151/A21.51; AWWA C115/A21.15
Galvanized steel pipe	ASTM A 53
Polyethylene/aluminum/polyethylene (PE-AL-PE) composite pipe	ASTM F 1282
Polyethylene of raised temperature (PE-RT) plastic tubing	ASTM F 2769
Polypropylene (PP) plastic pipe or tubing	ASTM F 2389; CSA B137.11
Stainless steel pipe (Type 304/304L)	ASTM A 312; ASTM A 778
Stainless steel pipe (Type 316/316L)	ASTM A 312; ASTM A 778

WATER SUPPLY AND DISTRIBUTION

tion system must be capable of withstanding the working pressures of the system at the maximum hot water temperature that could be present in the system. The maximum allowable hot water temperature in a water distribution system is 180°F (82°C). The maximum allowable working pressure of a water distribution system is 80 psi (552 kPa) (see Section 604.8); however, as there can be pressure surges in a water distribution system, the maximum working pressure for the pipe is set at 100 psi (690 kPa).

605.5 Fittings. Pipe fittings shall be approved for installation with the piping material installed and shall comply with the applicable standards listed in Table 605.5. Pipe fittings utilized in water supply systems shall also comply with NSF 61. Ductile and gray iron pipe and pipe fittings utilized in water service piping systems shall be cement mortar lined in accordance with AWWA C104.

❖ Pipe (and tube) fittings must be appropriate for both the type of piping material and the service application. Even though a Schedule 40 PVC pipe fitting will fit Schedule 80 PVC pipe, the Schedule 40 fitting does not have the same pressure rating as the Schedule 80 pipe. Even though an insert fitting with crimp ring might appear to fit the PEX tubing of many manufacturers, only the fittings recommended by the manufacturer are to be used (see Section 605.17.2). Schedule 40 ABS pipe and fittings are dimensionally interchangeable with PVC Schedule 40 fittings; however, the code does not allow PVC and ABS to be solvent cemented together.

Compliance to NSF 61 for all water supply system components is paramount for protection of the public against contamination of the water supply (see commentary, Sections 605.2, 605.3 and 605.4).

Ductile and gray iron piping is frequently used for underground water service piping. Section 605.3 requires that such piping be cement mortar lined to protect the iron surface from rusting, giving the water a red tint. Fittings are no less subject to rusting so they too are required to be cement mortar lined. Note that the cement mortar lining requirement is only for water service piping applications. Although ductile iron and gray iron piping could be used as water distribution piping (inside of a building), the code does not require that the piping in those applications be cement mortar lined (see Section 605.4).

**TABLE 605.5
PIPE FITTINGS**

MATERIAL	STANDARD
Acrylonitrile butadiene styrene (ABS) plastic	ASTM D 2468
Cast-iron	ASME B16.4; ASME B16.12
Chlorinated polyvinyl chloride (CPVC) plastic	ASSE 1061; ASTM D 2846; ASTM F 437; ASTM F 438; ASTM F 439; CSA B137.6
Copper or copper alloy	ASSE 1061; ASME B16.15; ASME B16.18; ASME B16.22; ASME B16.23; ASME B16.26; ASME B16.29
Cross-linked polyethylene/aluminum/high-density polyethylene (PEX-AL-HDPE)	ASTM F 1986
Fittings for cross-linked polyethylene (PEX) plastic tubing	ASSE 1061, ASTM F 877; ASTM F 1807; ASTM F 1960; ASTM F 2080; ASTM F 2098, ASTM F 2159; ASTM F 2434; ASTM F 2735; CSA B137.5
Fittings for polyethylene of raised temperature (PE-RT) plastic tubing	ASTM F 1807; ASTM F 2098; ASTM F 2159; ASTM F 2735
Gray iron and ductile iron	AWWA C110/A21.10; AWWA C153/A21.53
Insert fittings for polyethylene/aluminum/polyethylene (PE-AL-PE) and cross-linked polyethylene/aluminum/cross-linked polyethylene (PEX-AL-PEX)	ASTM F 1974; ASTM F 1281; ASTM F 1282; CSA B137.9; CSA B137.10M
Malleable iron	ASME B16.3
Metal (brass) insert fittings for polyethylene/aluminum/polyethylene (PE-AL-PE) and cross-linked polyethylene/aluminum/cross-linked polyethylene (PEX-AL-PEX)	ASTM F 1974
Polyethylene (PE) plastic pipe	ASTM D 2609; ASTM D 2683; ASTM D 3261; ASTM F 1055; CSA B137.1
Polypropylene (PP) plastic pipe or tubing	ASTM F 2389; CSA B137.11
Polyvinyl chloride (PVC) plastic	ASTM D 2464; ASTM D 2466; ASTM D 2467; CSA B137.2; CSA B137.3
Stainless steel (Type 304/304L)	ASTM A 312; ASTM A 778
Stainless steel (Type 316/316L)	ASTM A 312; ASTM A 778
Steel	ASME B16.9; ASME B16.11; ASME B16.28

WATER SUPPLY AND DISTRIBUTION

605.5.1 Mechanically formed tee fittings. Mechanically extruded outlets shall have a height not less than three times the thickness of the branch tube wall.

❖ Constructing a manifold from copper tube and copper tee fittings is labor intensive and costly. Special equipment was developed to enable a skilled tradesperson to fabricate branch outlets in copper tube without the need for fittings. The branch outlets are formed in a continuous operation by drilling a pilot hole and drawing out the wall of the copper tube to form a branch outlet. To ensure that the branch outlet provides for an adequate connection for the branch tube, the height of the formed branch outlet tube must be not less than three times the thickness of the tube wall (see Commentary Figure 605.5.1).

Figure 605.5.1
DRILLING AND PULLING BRANCH OUTLETS
(Photo courtesy of T-Drill Industries, Inc.)

605.5.1.1 Full flow assurance. Branch tubes shall not restrict the flow in the run tube. A dimple serving as a depth stop shall be formed in the branch tube to ensure that penetration into the collar is of the correct depth. For inspection purposes, a second dimple shall be placed $^1/_4$ inch (6.4 mm) above the first dimple. Dimples shall be aligned with the tube run.

❖ The branch tube to be inserted into the formed branch outlet on the main run of tube must not interfere with the flow path in the main run of tube. To ensure that this interference does not occur, the end of the tube to be connected to the branch outlet must be notched to approximate the curvature of the main run of tube. In conjunction with the notching of the end of the branch tube, two dimples are pressed into the side of the tube to prevent the tube from being inserted too far into the main run of tube (see Commentary Figure 605.5.1.1). The upper dimple provides an indicator for inspection purposes as the lower dimple is often obscured by brazing filler material.

Figure 605.5.1.1
SPECIALLY PREPARED BRANCH PIPE END
(Photo courtesy of T-Drill Industries, Inc.)

605.5.1.2 Brazed joints. Mechanically formed tee fittings shall be brazed in accordance with Section 605.14.1.

❖ Because the height of the formed branch outlet on the main run of tube is very short and is not formed with the exacting tolerances of a factory-produced socket fitting, soft soldering of a branch tube in the formed branch outlet is inadequate to ensure the joint strength necessary for a dependable connection.

Brazing of the joint, in accordance with Section 605.14.1, is required (see Commentary Figure 605.5.1.2).

Figure 605.5.1.2
BRAZED CONNECTION
(Photo courtesy of T-Drill Industries, Inc.)

605.6 Flexible water connectors. Flexible water connectors exposed to continuous pressure shall conform to ASME A112.18.6/CSA B125.6. *Access* shall be provided to all flexible water connectors.

❖ ASTM A 112.18.6/CSA B125.6 is the referenced standard for flexible water connectors used in potable water systems. Compliance with the standard ensures that: 1) plastic materials coming into contact with potable water must comply with the applicable sections of NSF 14; 2) all other materials comply with the requirements of NSF 61; 3) solder and fluxes containing lead in excess of 0.2 percent are not used in the fabrication of the connector; 4) metal alloys exceeding 8-percent lead content are not used and 5) copper alloy components used contain not less than 58-percent copper (see commentary, Section 605.2). Flexible water connectors are designed to function at water pressure up to 125 psi (862 kPa) and supply temperature from 40°F to 180°F (4°C to 82°C).

605.7 Valves. Valves shall be compatible with the type of piping material installed in the system. Valves shall conform to one of the standards listed in Table 605.7 or shall be *approved*. Valves intended to supply drinking water shall meet the requirements of NSF 61.

❖ Valves must be of a material that is "compatible" with the pipe material so as to avoid valve and pipe failure caused by chemical or corrosive reactions. Table 605.7 contains a list of valve materials and the standards with which they must comply. For valve materials that are not indicated, such valves are required to be approved. Valves used for the supply of drinking water must comply with NSF 61; however, Sections 605.2 and 605.2.1 also apply in those situations.

Prior to the 2015 code, only ball valves, gate valves, butterfly valves, globe valves and plug valves that supplied drinking water were required to comply with NSF 61. Missing from this list were pressure-reducing valves, backflow-prevention valves, check valves and solenoid valves (which are typically a globe valve pattern). For the 2015 code, all types of valves that supply drinking water must comply with NSF 61. This increased coverage ensures that contaminants tested (especially lead) under NSF 61 are not being imparted to the drinking water supply for the building.

605.8 Manufactured pipe nipples. Manufactured pipe nipples shall conform to one of the standards listed in Table 605.8.

❖ Pipe nipples are short pieces [24 inches or less (610 mm)] of threaded pipe used in making up piping assemblies. The majority of brass, copper and galvanized steel pipe nipples are mass-produced in specific length increments in a factory as opposed to contractor- or supplier-made from pipe. Table 605.8 lists the referenced standards that apply to these manufactured products.

TABLE 605.8
MANUFACTURED PIPE NIPPLES

MATERIAL	STANDARD
Brass-, copper-, chromium-plated	ASTM B 687
Steel	ASTM A 733

❖ ASTM A 733 specifies material and dimensions for galvanized steel pipe nipples and stainless steel pipe nipples. Galvanized steel pipe nipples are made from ASTM A 53 or A 106 piping, which has been hot-dipped galvanized inside and out. Stainless steel nipples are made from ASTM A 312 piping. Steel pipe nipples are available in the range of nominal pipe sizes $^3/_8$ to 12 inches (3.2 mm to 305 mm) and in lengths up to 24 inches (610 mm).

ASTM B 687 requires brass- and chrome-plated nipples to be made from brass pipe conforming to ASTM B 43. Copper pipe nipples are made from copper pipe conforming to ASTM B 42. The nipples are available in pipe sizes $^1/_4$ inch to 8 inches (6.4 mm to 203 mm) and in lengths up to 12 inches (305 mm).

605.9 Prohibited joints and connections. The following types of joints and connections shall be prohibited:

1. Cement or concrete joints.
2. Joints made with fittings not *approved* for the specific installation.
3. Solvent-cement joints between different types of plastic pipe.
4. Saddle-type fittings.

❖ Pipe joints made with portland cement (such as brick mortar) or concrete are not effective in restraining and sealing a pipe joint because the materials are inflexi-

TABLE 605.7
VALVES

MATERIAL	STANDARD
Chlorinated polyvinyl chloride (CPVC) plastic	ASME A112.4.14; ASME A112.18.1/CSA B125.1; ASTM F 1970; CSA B125.3
Copper or copper alloy	ASME A112.4.14; ASME A112.18.1/CSA B125.1; ASME B16.34; CSA B125.3; MSS SP-67; MSS SP-80; MSS SP-110
Cross-linked polyethylene (PEX) plastic	ASME A112.4.14; ASME A112.18.1/CSA B125.1; CSA B125.3; NSF 359
Gray iron and ductile iron	AWWA C500; AWWA C504; AWWA C507; MSS SP-67; MSS SP-70; MSS SP-71; MSS SP-72; MSS SP-78
Polypropylene (PP) plastic	ASME A112.4.14; ASTM F 2389
Polyvinyl chloride (PVC) plastic	ASME A112.4.14; ASTM F 1970

ble and susceptible to cracking and displacement.

Fittings that are not approved for a specific installation can fail or be unacceptable because of a number of reasons, including: 1) incompatibility of materials; 2) not rated for the conditions of service or 3) contamination of the potable water supply.

A solvent-cement joint between a pipe and a fitting is a solvent "welding" process that melts the surfaces of the items to be joined. If the pipe and fitting material are not the same, it is questionable as to whether the fused zone of material in the joint has the same strength and durability as a joint made between identical materials. Note that Section 605.24.2 requires that joints between different grades of plastic pipe material must be made with an approved adapter fitting (see commentary, Section 605.24.2).

A saddle-type fitting is also known as a saddle tee or a saddle-tap tee. The installation of a saddle-type fitting essentially involves the drilling of a hole in the side of a pipe and installing a specially designed, rubber-gasketed, wide clamp around the pipe at the drilled hole location. The wide clamp has an internally threaded boss to enable connection of a threaded pipe or pipe fitting, forming a branch to the original pipe. Because of the gasket and the means of attachment of the clamp to the pipe, such saddle-type arrangements are not durable and, therefore, are undependable for long-term service.

605.10 ABS plastic. Joints between ABS plastic pipe and fittings shall comply with Sections 605.10.1 through 605.10.3.

❖ Acceptable joints for ABS plastic pipe are identified in the referenced sections.

605.10.1 Mechanical joints. Mechanical joints on water pipes shall be made with an elastomeric seal conforming to ASTM D 3139. Mechanical joints shall only be installed in underground systems, unless otherwise *approved*. Joints shall be installed only in accordance with the manufacturer's instructions.

❖ Mechanical joints are predominantly used in underground applications (e.g., bell or hub compression gasket joints) because earth burial provides axial restraint to keep the joint from separating. Because most mechanical joints are typically of the unrestrained type, above-ground mechanical joints are prohibited as most above-ground piping systems do not provide for the necessary axial restraint; however, there are situations where remodeling and system extension activities might make it impractical to install anything other than a mechanical joint. In these cases, the code official must evaluate the joint and determine if the arrangement provides for the required axial restraint to keep the joint intact during service. Note that the referenced standard regulates only the elastomeric material used to make the seal in a mechanical joint. The joint design and the installation details are the responsibility of the joint manufacturer.

605.10.2 Solvent cementing. Joint surfaces shall be clean and free from moisture. Solvent cement that conforms to ASTM D 2235 shall be applied to all joint surfaces. The joint shall be made while the cement is wet. Joints shall be made in accordance with ASTM D 2235. Solvent-cement joints shall be permitted above or below ground.

❖ Solvent welding ABS pipe and fittings together involves only the use of proper ABS solvent cement. No priming or solvent cleaning of the joint surfaces is required because ABS solvent cement is sufficiently aggressive enough to clean, prime and solvent weld all in one step. Solvent cement is not an adhesive or a filler as the term "cement" might imply. Solvent cement melts the surfaces to be joined together such that they merge together. As the solvent dissipates from the joint, the dissolved surfaces will harden as one and strengthen over time.

The keys to making consistent, high-quality solvent welded joints are:

- The pipe and fittings should be approximately the same temperature prior to solvent cementing.
- The pipe end and socket of the fitting must be clean, dry and free from grease, oil and other foreign materials.
- The solvent cement must be uniformly applied to the socket of the fitting and to the end of the pipe to the depth of the socket. Excessive solvent cement must be avoided to prevent a bead from forming on the inside of the pipe joint.
- The joint is made by twisting the pipe into the fitting socket immediately after applying the cement and while the cement is still wet. The pipe must be held together for a 5 to 10 seconds until the solvent cement begins to set as the pipe will tend to back out of the fitting.
- The small bead of solvent cement that forms between the pipe exterior and fitting should be removed with a clean cloth to avoid the possibility of weakening the pipe wall.
- There should be no rough handling of the joined pipe and fitting for at least 1 hour after the joint is assembled.
- The solvent welded joint should remain unpressurized for a period of 24 hours.

Because each type of solvent cement is specifically designed for a piping material, all-purpose solvent cement or universal solvent cement cannot be used to join ABS pipe or fittings unless the cement conforms to ASTM D 2235. Solvent cement must be identified on the can as complying with ASTM D 2235.

This section clarifies that solvent cementing is an acceptable ABS joining method for both above-ground and below-ground installations.

WATER SUPPLY AND DISTRIBUTION

605.10.3 Threaded joints. Threads shall conform to ASME B1.20.1. Schedule 80 or heavier pipe shall be permitted to be threaded with dies specifically designed for plastic pipe. *Approved* thread lubricant or tape shall be applied on the male threads only.

❖ Pipe thread dimensions can be historically traced to Robert Briggs, already mentioned in the commentary to Table 605.3 for galvanized steel pipe. The standard identifies pipe dimensions as "NPT." Often thought of as an abbreviation, some of the terms from which NPT has been shortened are: National Pipe Thread, Nominal Pipe Thread and National Pipe Tapered Thread.

NPT is a coded designation: N stands for USA Standard, P indicates pipe and T means that the threads are tapered.

Thread lubricant (pipe-joint compound) must be specifically designed for ABS plastic pipe. It is not the same lubricant that is commonly used for metallic pipe, such as steel. This section prohibits cutting threads into Schedule 40 ABS pipe. Threaded joints in Schedule 40 ABS pipe must be accomplished with factory-fabricated fittings made for that purpose. This section also recognizes that Schedule 80 pipe has sufficient pipe-wall thickness and strength to permit field threading.

Thread lubricant and tape are designed for two purposes: to lubricate the joint for proper thread mating and to fill in small imperfections on the surfaces of the threads.

605.11 Brass. Joints between brass pipe and fittings shall comply with Sections 605.11.1 through 605.11.4.

❖ Acceptable joints for brass pipe or fittings are identified in the referenced sections.

605.11.1 Brazed joints. All joint surfaces shall be cleaned. An *approved* flux shall be applied where required. The joint shall be brazed with a filler metal conforming to AWS A5.8.

❖ Although brazed joints are similar to soldered joints, brazed joints are joined at a higher temperature [in excess of 1,000°F (538°C)]. Because of a brazed joint's inherent resistance to elevated temperatures, it is much stronger than a soldered joint. Brazing is often referred to as "silver soldering." Silver soldering is more accurately described as silver brazing and uses high-silver-bearing alloys primarily composed of silver, copper and zinc. Silver soldering (brazing) typically requires temperatures in excess of 1,000°F (538°C), and such solders are classified as "hard" solders.

There has always been confusion over the distinction between "silver solder" and "silver-bearing solder." Silver solders are unique and can be further subdivided into soft and hard categories, which are determined by the percentages of silver and the other component elements of the particular alloy. The distinction must be made that silver-bearing solders [with a melting point below 600°F (315°C)] are used in soft-soldered joints, whereas silver solders [with a melting point above 1,000°F (538°C)] are used in silver-brazed joints.

The two common series of filler metals used for brazing are BCuP and BAg. BCuP contains phosphorus and copper alloys. BAg contains 30- to 60-percent silver with zinc and copper alloys. Some of the BAg filler metals also contain cadmium.

The joint surfaces to be brazed must be cleaned, typically by polishing the pipe with an emery cloth. The socket fitting is cleaned with a tube fitting brush.

Brazing flux is applied to the joint surfaces. Certain types of BCuP filler metals do not require flux for certain types of joints. The manufacturer's recommendations should be consulted to determine when flux is necessary.

The joints are typically heated with an oxyacetylene torch, though an air acetylene torch can be used on a smaller-diameter pipe. The pipe and fitting must be at approximately the same temperature when making the joint. The filler metal flows to the heat with the aid of the flux. The pipe should first be heated above the fitting. When the flux becomes clear, the heat should be directed to the base of the fitting socket. When the flux becomes clear and quiet, the filler metal is applied to the joint. The heat draws the filler metal into the socket (see Commentary Figure 605.11.1). Any remaining flux residue should be removed from the pipe and joint surfaces.

**Figure 605.11.1
HEAT APPLICATION OF BRAZED JOINT**

605.11.2 Mechanical joints. Mechanical joints shall be installed in accordance with the manufacturer's instructions.

❖ Mechanical joints are unique in that many acceptable types are available. Most mechanical joints use an elastomeric material to form a seal. Compression fittings that use a compression ring, which forms a metal-to-metal seal, are also considered as mechanical joints. In all cases, the joint must be carefully installed in accordance with the manufacturer's instructions.

605.11.3 Threaded joints. Threads shall conform to ASME B1.20.1. Pipe-joint compound or tape shall be applied on the male threads only.

❖ Pipe-joint compound is typically designed for metallic pipe. The pipe-joint compound used for brass pipe will also be applied for copper pipe and steel pipe (see commentary, Section 605.10.3).

605.11.4 Welded joints. All joint surfaces shall be cleaned. The joint shall be welded with an *approved* filler metal.

❖ A welded joint is similar to a brazed joint. The primary differences are the temperature at which the joint is made, the type of filler metals used and the fact that welding reaches the melting point of the base metal, whereas brazing temperatures are well below the melting point of the pipe (see commentary, Section 605.11.1).

605.12 Gray iron and ductile iron joints. Joints for gray and ductile iron pipe and fittings shall comply with AWWA C111/A21.11 and shall be installed in accordance with the manufacturer's instructions.

❖ This section requires rubber-gasket joints to connect ductile-iron water pipe. Although not specifically identified, the following are considered to be rubber-gasket joints:

Mechanical joints: a bottled joint of the stuffing type including a bell, an exterior flange with bolt openings, a socket, a sealing gasket, a follower gland with bolt holes and bolts.

Push-on joint: a single rubber-gasket joint consisting of a plain-end pipe forced into a socket and compressing the gasket around the outside of the pipe.

Flange joint: two flanged end pipes joined together by bolts that compress a ring gasket between the flange faces.

Modified push-on and mechanical joints: variations to the above-described joints that comply with the performance requirements of AWWA C111.

605.13 Copper pipe. Joints between copper or copper-alloy pipe and fittings shall comply with Sections 605.13.1 through 605.13.5.

❖ Acceptable joints for copper or copper-alloy pipe or fittings are identified in the referenced sections.

605.13.1 Brazed joints. All joint surfaces shall be cleaned. An *approved* flux shall be applied where required. The joint shall be brazed with a filler metal conforming to AWS A5.8.

❖ The installation technique and materials used to make a brazed joint for copper or copper-alloy pipe are identical to that of brass pipe and copper tubing (see commentary, Section 605.11.1).

605.13.2 Mechanical joints. Mechanical joints shall be installed in accordance with the manufacturer's instructions.

❖ Mechanical joints for copper or copper-alloy pipe are similar to those used to join brass pipe (see commentary, Section 605.11.2). Because the different mechanical joint couplings available are often unique to each manufacturer, the couplings must be installed in accordance with the manufacturer's instructions.

605.13.3 Solder joints. Solder joints shall be made in accordance with ASTM B 828. Cut tube ends shall be reamed to the full inside diameter of the tube end. Joint surfaces shall be cleaned. A flux conforming to ASTM B 813 shall be applied. The joint shall be soldered with a solder conforming to ASTM B 32. The joining of water supply piping shall be made with *lead-free solder and fluxes.* "Lead free" shall mean a chemical composition equal to or less than 0.2-percent lead.

❖ A soldered joint is the most common method of joining copper pipe and tubing.

Pipe must be cut square for proper alignment, adequate surface area for joining and an interior free from obstructions. When the pipe is cut, an edge (burr) is left protruding into the pipe. The pipe must be reamed to properly remove the burr. Chamfering is required to bevel the outer edge of a pipe cut end. Undercutting (excessive reaming) can reduce pipe wall thickness (see Commentary Figure 605.13.3).

ASTM B 32 covers many grades of solder, including tin-silver solders. Previously, the grades used in the plumbing industry were typically 40-60 (40-percent tin and 60-percent lead), 50-50 (50-percent tin and 50-percent lead) and 95-5 (95-percent tin and 5-percent antimony). Lead-based solders, however, are no longer permitted for joining copper tubing where used for the distribution of potable water.

Lead-based solders and fluxes have been implicated as a cause for an increased concentration of lead in drinking water. Potable water is exposed to the lead in the solder that commonly coats the inside surface of the pipe or fitting near the joint. The water can dissolve lead from the solder. The chemically combined or free lead in potable water can be ingested by drinking the water or by consuming food or beverages prepared with the water. In an effort to reduce the amount of lead in drinking water, lead-bearing solders and fluxes have been banned from use in joining water supply piping. After the code implemented requirements for lead-free solders and fluxes, the federal government followed suit. In June 1986, the EPA banned lead-bearing solders and fluxes throughout the country. Additionally, all solders and fluxes used on potable water lines must be tested and listed for compliance with NSF 61, which places further restrictions on the concentrations of lead permitted in these materials.

WATER SUPPLY AND DISTRIBUTION

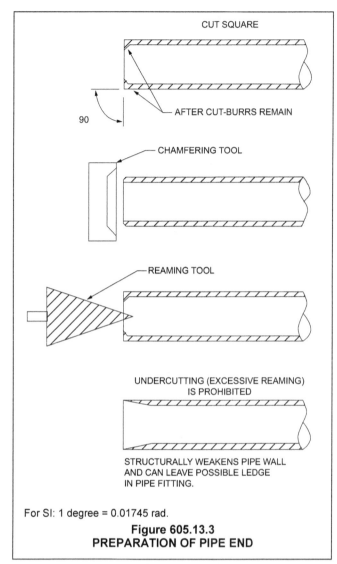

For SI: 1 degree = 0.01745 rad.

**Figure 605.13.3
PREPARATION OF PIPE END**

In addition to 95-5 (95-percent tin and 5-percent antimony), which is considered a viable alternative to lead solders in many applications, manufacturers have developed replacement solders for the plumbing industry.

Lead-bearing solder can sometimes be visually identified because the solder tends to darken with age. Replacement solders, tin-antimony solders and tin-silver solders tend to stay bright. Test kits are available that can be used to identify lead-bearing solders with absolute certainty.

ASTM B 828 governs the procedures for making capillary joints by soldering copper and copper alloy tubing and fittings. The joint surfaces for a soldered joint must be cleaned and shined bright, typically accomplished with emery cloth and specially designed fitting brushes. Once the copper is shined bright, the joint surfaces should not be touched by the human hand. The natural body oils on the hand will affect the making of the joint.

Flux must be applied to the joint surfaces. Flux is a chemically active material that removes and excludes oxides from the joint area during heating and allows the melted solder to spread out on the surfaces to be joined. The flux material must conform to ASTM B 813, which establishes minimum performance criteria, such as corrosivity and solubility, as well as a maximum lead content not to exceed 0.2 percent. The lead content is consistent with the 1986 amendments to the EPA Safe Drinking Water Act. The solder will flow by capillary action toward the heat. As such, the temperatures of the joint surfaces play a major role in making a properly soldered joint. The fitting and the pipe should be approximately the same temperature when the solder is applied. Heat causes the copper atoms to move further apart from each other and the solder atoms enter the spaces between the copper atoms, creating a strong bond when solidified.

If the temperatures are incorrect, either too high or uneven, the solder will run down the inside of the pipe and fitting or on the outside of the pipe. The result is a solder-lined pipe.

Care must also be exercised using wrought fittings because there is less of a ledge to prevent the flow of solder into the fitting socket. The same care must also be exercised when making a swage joint. A swage joint is where the end of one pipe is expanded to form a socket fitting. A swaging tool, similar to a hammer-type flaring tool, is used to expand the pipe. Because it is similar to a flared joint, only annealed and bending-tempered drawn copper can be swaged.

605.13.4 Threaded joints. Threads shall conform to ASME B1.20.1. Pipe-joint compound or tape shall be applied on the male threads only.

❖ Pipe-joint compound (also referred to as "pipe dope") is typically designed for metallic pipe. Pipe-joint compound used for copper pipe can also be used for brass pipe and steel pipe (see commentary, Section 605.10.3). Pipe-joint compound and tape must not be applied to female threads because this would allow the compound to enter and contaminate the piping system.

605.13.5 Welded joints. Joint surfaces shall be cleaned. The joint shall be welded with an *approved* filler metal.

❖ The installation technique and materials used to make a welded joint for copper or copper-alloy pipe are similar to those for brass pipe (see commentary, Section 605.11.4).

605.14 Copper tubing. Joints between copper or copper-alloy tubing and fittings shall comply with Sections 605.14.1 through 605.14.5.

❖ Acceptable joints for copper or copper-alloy tubing or fittings are identified in the referenced sections.

605.14.1 Brazed joints. Joint surfaces shall be cleaned. An *approved* flux shall be applied where required. The joint shall be brazed with a filler metal conforming to AWS A5.8.

❖ The brazing of a joint for copper or copper-alloy tubing is identical to that of copper, copper-alloy and brass pipe (see commentary, Section 605.13.1).

WATER SUPPLY AND DISTRIBUTION

605.14.2 Flared joints. Flared joints for water pipe shall be made by a tool designed for that operation.

❖ Because the pipe end is expanded in a flared joint, only annealed and bending-tempered drawn copper tubing may be flared. There are two types of flaring tools: hammer and screw yoke and block. The hammer-type flaring tool is inserted into the pipe and hammered to form a flare on the end of the pipe. With the screw yoke and block, the end of the pipe is firmly held by the block assembly. The screw yoke is attached to the block and a flared arbor is pressed into the pipe end, creating the flare.

The flared surface of the pipe is compressed against the mating surface of the fitting by the flare nut to form a water-tight seal. Commonly, pipe-joint compound is placed on flare-fitting threads to lubricate the assembly [see Commentary Figure 605.14.2(1)].

A flared joint can only be accomplished using soft (annealed) copper tubing because hard-drawn pipe is too easily split when flared. Because the joint can be readily dismantled and reassembled, it is in effect, a type of union connection [see Commentary Figure 605.14.2(2)].

Figure 605.14.2(1)
TYPICAL FLARED TUBE JOINT

Figure 605.14.2(2)
TYPICAL FLARED TUBE JOINT

605.14.3 Grooved and shouldered mechanical joints. Grooved and shouldered mechanical joints shall comply with ASTM F 1476, shall be made with an *approved* elastomeric seal and shall be installed in accordance with the manufacturer's instructions. Such joints shall be exposed or concealed.

❖ The grooved and shouldered joint has a one-piece elastomeric gasket that is "clamped" around the outside of a specially prepared pipe end or boss of a fitting. The pipe ends (or bosses of fittings) are prepared with a circumferential groove on the outside diameter. The groove is either cut or rolled depending on the type of material. The clamp (usually in two halves) is made with internal "shoulders," one for each side of the joint. When the clamp is assembled around the joint (with the gasket in place), the clamp presses on the outside of the gasket while at the same time, engaging the clamp's shoulders into the grooves on the pipe ends. As the clamp is tightened the gasket is firmly clamped onto the outside of the pipe ends.

This type of connection is most visible to the layperson where it is used in the construction of fire sprinkler system piping (usually visible in stairwells and parking garages). Because of the speed at which these joints can be assembled by semiskilled labor and the very low probability of joint leakage at test pressures sometimes exceeding 250 psi, this is the preferred type of joint that has been used for decades in sprinkler service. However, this type of joint has been used for more than 80 years in various industrial sectors such as the refinery, power, oil field and marine fields. Besides the advantages of fast assembly and low probability of leaks, grooved and shouldered joints have "flexibility" to accommodate some installation misalignments and absorb some piping movements caused by thermal changes, equipment shifting and earthquakes. In many industrial environments, required changes, upgrades and repairs are more easily facilitated by the ease of disassembly and reassembly using the same components except for a new gasket.

The clamps, fittings, and gaskets are available in a wide variety of materials depending on the application. These joints can be used in concealed locations and underground, provided that the manufacturer of the joint does not limit the use of the material chosen to above-ground service only.

605.14.4 Mechanical joints. Mechanical joints shall be installed in accordance with the manufacturer's instructions.

❖ Mechanical joints for copper tubing include a large array of fittings. The most common mechanical joint is a compression fitting. This type of fitting uses a compression ring (ferrule) to form the seal. The ring compresses around the pipe as the fitting is tightened. Commonly, pipe-joint compound is used on the fitting threads to lubricate the assembly.

605.14.5 Press-connect joints. Press-connect joints shall conform to one of the standards listed in Table 605.5, and shall be installed in accordance with the manufacturer's instructions. Cut tube ends shall be reamed to the full inside diameter of the tube end. Joint surfaces shall be cleaned. The tube shall be fully inserted into the press-connect fitting. Press-connect joints shall be pressed with a tool certified by the manufacturer.

❖ These fittings have an internal O-ring that provides the sealing. The metal fitting is crimped onto the pipe or tube with a special tool designed for that purpose. Typically, crimping tools are not interchangeable between different manufacturer's fittings. Once crimped in position, the fitting cannot be removed, and the pipe cannot be rotated in the fitting. Connecting copper tubing and pipe in this manner is useful, especially where water in the piping cannot be evacuated to allow soldering or brazing, or where open flames (torches) cannot be used.

605.14.6 Solder joints. Solder joints shall be made in accordance with the methods of ASTM B 828. All cut tube ends shall be reamed to the full inside diameter of the tube end. All joint surfaces shall be cleaned. A flux conforming to ASTM B 813 shall be applied. The joint shall be soldered with a solder conforming to ASTM B 32. The joining of water supply piping shall be made with *lead-free solders and fluxes*. "Lead free" shall mean a chemical composition equal to or less than 0.2-percent lead.

❖ The requirements and procedures for a soldered joint for copper and copper-alloy tubing are identical to those for copper and copper-alloy pipe (see commentary, Section 605.14.3).

605.15 CPVC plastic. Joints between CPVC plastic pipe and fittings shall comply with Sections 605.15.1 through 605.15.3.

❖ Acceptable joints for CPVC plastic pipe or fittings are identified in the referenced sections.

605.15.1 Mechanical joints. Mechanical joints shall be installed in accordance with the manufacturer's instructions.

❖ Many mechanical joints use an elastomeric material to form a seal. Because mechanical joints are typically unique to each particular manufacturer, they must be installed in accordance with the manufacturer's instructions (see commentary, Section 605.10.1).

605.15.2 Solvent cementing. Joint surfaces shall be clean and free from moisture. Joints shall be made in accordance with the pipe manufacturer's installation instructions. Where such instructions require that a primer be used, the primer shall be applied to the joint surfaces and a solvent cement orange in color and conforming to ASTM F 493 shall be applied to the joint surfaces. Where such instructions allow for a one-step solvent cement, yellow in color and conforming to ASTM F 493, to be used, the joint surfaces shall not require application of a primer before the solvent cement is applied. The joint shall be made while the cement is wet and in accordance with ASTM D 2846 or ASTM F 493. Solvent cemented joints shall be permitted above or below ground.

❖ Solvent cementing of CPVC piping must be performed in accordance with the pipe manufacturer's instructions using a solvent cement that complies with ASTM F 493. Some pipe manufacturers may require specific brands or types of solvent cement be used for certain types and sizes of piping. The pipe and fitting surfaces to be joined must be free of dirt, grease, oil and any other foreign substances. It is critical that joint surfaces are dry, as solvent cement quickly congeals in the presence of water.

Some pipe manufacturer's instructions may require that a primer liquid be used to prepare the joint surfaces. Primer liquids break down the glossy surface finish of the plastic, help remove any remaining foreign substances from the joint area and begin to soften the plastic surfaces. Because of its softening capability, primer liquids must not be allowed to puddle inside of fittings. Some pipe manufacturers may require specific brands or types of primer liquids be used for certain types and sizes of piping. This section does not require primer liquids to be tinted with a color (such as purple or blue); however, a colored primer could be useful for the installer for determining if all joint surfaces have been primed.

Where priming of joint surfaces is required by the pipe manufacturer, the color of the solvent cement to be used must be orange. In general, solvent cement is applied to all joint surfaces immediately after the joint surfaces have been primed. Immediately after application of the solvent cement, the joint is assembled. The assembled joint is held in the assembled position for a period of time as required by the pipe manufacturer. Different sizes of piping require very specific assembly procedures and holding times. The piping manufacturer's instructions must be followed in every detail to ensure a proper joint.

Where the pipe manufacturer allows for a one-step solvent cementing process (in other words, priming of the joint surfaces is not required), the color of the solvent cement to be used must be yellow. One-step solvent cementing can be in accordance with either ASTM D 2846 or ASTM F 493.

Assembled and cured joints can be located above or below ground.

605.15.3 Threaded joints. Threads shall conform to ASME B1.20.1. Schedule 80 or heavier pipe shall be permitted to be threaded with dies specifically designed for plastic pipe, but the pressure rating of the pipe shall be reduced by 50 percent. Thread by socket molded fittings shall be permitted. *Approved* thread lubricant or tape shall be applied on the male threads only.

❖ The requirements regarding CPVC pipe are similar to those of ABS plastic pipe (see Section 605.10.3). The key difference is that when CPVC plastic pipe is field threaded (Schedule 80 or heavier), the pressure rating of the pipe must be reduced by one-half because when the pipe is tested for a specific pressure, the sample of the pipe subjected to testing had a full wall thickness. Therefore, reducing pipe-wall thickness with the threading affects the pressure rating of the pipe.

The use of molded fittings for the joining of plastic pipe does not result in the 50-percent reduction in the pressure rating of the pipe because the molded fittings are fabricated so that the wall thickness of the material is maintained at the threads.

605.16 Chlorinated polyvinyl chloride/aluminum/chlorinated polyvinyl chloride (CPVC/AL/CPVC) pipe and tubing. Joints between CPVC/AL/CPVC plastic pipe or CPVC fittings shall comply with Sections 605.16.1 and 605.16.2.

❖ There are two methods for joining CPVC/AL/CPVC piping to fittings: mechanical and solvent cementing.

605.16.1 Mechanical joints. Mechanical joints shall be installed in accordance with the manufacturer's instructions.

❖ The manufacturer of the mechanical joint controls whether or not their mechanical joints can be used on CPVC/AL/CPVC pipe. Although many mechanical joints are suitable (as indicated by the manufacturer) for CPVC piping, CPVC/AL/CPVC piping is different in many respects. If a mechanical joint connection is used, the end of the CPVC/AL/CPVC piping must have the F2855 bushing solvent cemented into the end of the pipe (see commentary, Section 605.3 concerning CPVC/AL/CPVC piping). Although Table 605.5 includes the push-fit fitting standard ASSE 1061, the manufacturer of fittings complying with that standard must verify that their push-fit fitting is suitable for use on CPVC/AL/CPVC pipe.

605.16.2 Solvent cementing. Joint surfaces shall be clean and free from moisture, and an approved primer shall be applied. Solvent cement, orange in color and conforming to ASTM F 493, shall be applied to joint surfaces. The joint shall be made while the cement is wet, and in accordance with ASTM D 2846 or ASTM F 493. Solvent cement joints shall be permitted above or below ground.

Exception: A primer is not required where all of the following conditions apply:

1. The solvent cement used is third-party certified as conforming to ASTM F 493.

2. The solvent cement used is yellow in color.

3. The solvent cement is used only for joining $^1/_2$-inch (12.7 mm) through 2-inch-diameter (51 mm) CPVC/AL/CPVC pipe and CPVC fittings.

4. The CPVC fittings are manufactured in accordance with ASTM D 2846.

❖ Solvent cementing of CPVC/AL/CPVC piping must be performed in accordance with the pipe manufacturer's instructions using a solvent cement that complies with ASTM F 493. Some pipe manufacturers may require specific brands or types of solvent cement be used for certain types and sizes of piping. The pipe and fitting surfaces to be joined must be free of dirt, grease, oil and any other foreign substances. Especially critical is the need for the joint surfaces to be dry as solvent cement quickly congeals in the presence of water.

Priming of joint surfaces is required. Primer liquids break down the glossy surface finish of the plastic, help remove any remaining foreign substances from the joint area and begin to soften the plastic surfaces. Because of its softening capability, primer liquids must not be allowed to puddle inside of fittings. Some pipe manufacturers may require specific brands or types of primer liquids be used for certain types and sizes of piping. The color of the solvent cement to be used must be orange.

In general, solvent cement is applied to all joint surfaces immediately after they have been primed. Immediately after application of the solvent cement, the joint is assembled. The assembled joint is held in the assembled position for a period of time as required by the pipe manufacturer. Different sizes of piping require very specific assembly procedures and holding times. The piping manufacturer's instructions must be followed in every detail to ensure a proper joint.

A one-step solvent cementing process (in other words, priming of the joint surfaces is not required) can be used, provided that all of the conditions of the exceptions are met.

Assembled and cured joints can be located above or below ground.

605.17 PEX plastic. Joints between cross-linked polyethylene plastic tubing and fittings shall comply with Sections 605.17.1 and 605.17.2.

❖ Acceptable joints for cross-linked PEX plastic tubing or fittings are identified in the referenced sections.

605.17.1 Flared joints. Flared pipe ends shall be made by a tool designed for that operation.

❖ Flared joints must be made with tools specific to the design of the connector system. The assembly method must be in accordance with the manufacturer's instructions specific to the type of connectors being used. Flare configurations will vary depending on the tool employed. Using a cold flaring tool, which expands the tubing outside diameter mechanically, secures the flared end between the fitting. The flared surface serves as the sealing area between the tubing and fitting.

605.17.2 Mechanical joints. Mechanical joints shall be installed in accordance with the manufacturer's instructions. Fittings for cross-linked polyethylene (PEX) plastic tubing shall comply with the applicable standards listed in Table 605.5 and shall be installed in accordance with the manufacturer's instructions. PEX tubing shall be factory marked with the appropriate standards for the fittings that the PEX manufacturer specifies for use with the tubing.

❖ ASTM F 877, ASTM F 1807, ASTM F 1960, ASTM F 2098, ASTM F 2159, ASTM F 2434 and ASTM F 2080 are the standards for insert fittings, crimp rings and manifolds for cross-linked PEX plastic tubing systems. They are intended for use in 100 psi (690 kPa) cold and hot water distribution systems operating at temperatures up to and including 180°F (82°C). Insert fittings must be installed in accordance with the manufacturers' instructions. One type of system has insert fittings joined to PEX tubing by the compression of a copper crimp ring around the outer circumference of the tubing, forcing the tubing material into annular spaces formed by ribs on the fittings.

The crimping procedure involves sliding the crimp ring onto the tubing, then inserting the ribbed end of the fitting into the end of the tubing until the tubing contacts the shoulder of the fitting or tube stop. The crimp ring must then be positioned on the tubing so the edge of the crimp ring is $^1/_8$ to $^1/_4$ inch (3.2 to 6.4 mm) from the end of the tube. The jaws of the crimping tool must be centered over the crimp ring, and the tool must be held so that the crimping jaws are approximately perpendicular to the axis of the barb. The jaws of the crimping tool must be closed around the crimp ring, compressing the crimp ring onto the tubing. The crimp ring must not be crimped more than once.

Another type of system uses a special expander tool inserted into the end of the PEX tubing and the tube is expanded. The tool is removed and the insert fitting is quickly inserted before the tubing shrinks back to its original diameter. Depending on the proprietary design, a PEX or metal ring or collar is forced over the tube, locking it to the insert fitting (see Commentary Figure 605.17.2).

Insert fittings and crimp rings are not considered to be interchangeable between different manufacturers of PEX tubing. Therefore, PEX fittings specified by each manufacturer as indicated by the markings on the tube must be used.

Figure 605.17.2
PEX TUBING CONNECTION

605.18 Steel. Joints between galvanized steel pipe and fittings shall comply with Sections 605.18.1 through 605.18.3.

❖ Acceptable joints for galvanized steel pipe or fittings are identified in the referenced sections.

605.18.1 Threaded joints. Threads shall conform to ASME B1.20.1. Pipe-joint compound or tape shall be applied on the male threads only.

❖ Pipe-joint compound is typically designed for metallic pipe. The pipe-joint compound used for steel pipe will also be used for copper pipe and brass pipe (see commentary, Section 605.10.3). The application of pipe-joint compound to only male threads prevents the compound from entering the piping system.

605.18.2 Mechanical joints. Joints shall be made with an *approved* elastomeric seal. Mechanical joints shall be installed in accordance with the manufacturer's instructions.

❖ A wide variety of mechanical joints are available to join steel pipe, including rolled groove, cut groove and plain-end types.

A union, often mistakenly identified as a mechanical joint, qualifies as a pipe fitting in accordance with Section 605.5.

605.18.3 Grooved and shouldered mechanical joints. Grooved and shouldered mechanical joints shall comply with ASTM F 1476, shall be made with an *approved* elastomeric seal and shall be installed in accordance with the manufacturer's instructions. Such joints shall be exposed or concealed.

❖ See the commentary for Section 605.14.3.

605.19 PE plastic. Joints between polyethylene plastic pipe or tubing and fittings shall comply with Sections 605.19.1 through 605.19.4.

❖ Acceptable joints for PE plastic pipe and tubing or fittings are identified in the referenced sections. Sections 605.19.1 and 605.19.3 are also applicable to PEX, PEX-AL-PEX composite pipe, and PE-AL-PE composite pipe.

605.19.1 Flared joints. Flared joints shall be permitted where so indicated by the pipe manufacturer. Flared joints shall be made by a tool designed for that operation.

❖ Flared joints must be made with tools specific to the design of the connector system. The assembly method must be in accordance with the manufacturer's instructions specific to the type of connectors being used. A cold flaring tool is used to expand the end of the pipe into a cone shape. The cone-shaped end of the pipe is "clamped" between the cone-shaped end of a fitting and the internal cone shape of a nut with threads onto the fitting.

605.19.2 Heat-fusion joints. Joint surfaces shall be clean and free from moisture. All joint surfaces shall be heated to melt temperature and joined. The joint shall be undisturbed until cool. Joints shall be made in accordance with ASTM D 2657.

❖ The method of heat fusion of polyethylene pipe or tubing is similar to that of polypropylene (see commentary, Section 605.20.1).

605.19.3 Mechanical joints. Mechanical joints shall be installed in accordance with the manufacturer's instructions.

❖ This section addresses mechanical joining methods for polyethylene piping. This section also applies to the joining of PEX, PEX-AL-PEX and PE-AL-PE piping (see commentary, Table 605.4).

Mechanical joints for these materials include insert-type fittings, metallic lock-ring fittings, compression fittings and crimp-type fittings. Each fitting is intended for a specific pipe material and cannot be interchanged unless specified by the pipe manufacturer. In all cases, the fittings connecting any type of polyethylene pipe must be assembled and installed in accordance with the manufacturer's instructions.

605.19.4 Installation. Polyethylene pipe shall be cut square, with a cutter designed for plastic pipe. Except where joined by heat fusion, pipe ends shall be chamfered to remove sharp edges. Kinked pipe shall not be installed. The minimum pipe bending radius shall be not less than 30 pipe diameters, or the minimum coil radius, whichever is greater. Piping shall not be bent beyond straightening of the curvature of the coil. Bends shall be prohibited within 10 pipe diameters of any fitting or valve. Stiffener inserts installed with compression-type couplings and fittings shall not extend beyond the clamp or nut of the coupling or fitting.

❖ Any properly made pipe joint requires that the pipe ends be cut square for proper alignment, proper insertion depth and adequate surface area for joining (see Commentary Figure 605.13.3). Because polyethylene plastic pipe and tubing can be kinked and stress-weakened, the minimum pipe bending radius is specified to not exceed the structural capacity of the material. The minimum bending radius must be not less than the radius of the pipe coil as it came from the manufacturer, or the minimum radius must be not less than 30 pipe diameters, whichever of these two criteria is the largest (see Commentary Figure 605.19.4). The pipe can be straightened as it is taken from the coil, but it cannot be bent in a direction opposite of the bending direction of the coil (see Commentary Figure 605.19.4). To avoid stress on connected fittings or valves, the location of pipe bends is restricted to specific distances from such components.

605.20 Polypropylene (PP) plastic. Joints between PP plastic pipe and fittings shall comply with Section 605.20.1 or 605.20.2.

❖ Acceptable joints for PP plastic pipe and tubing or fittings are identified in the referenced sections.

605.20.1 Heat-fusion joints. Heat-fusion joints for polypropylene pipe and tubing joints shall be installed with socket-type heat-fused polypropylene fittings, butt-fusion polypropylene fittings or electrofusion polypropylene fittings. Joint surfaces shall be clean and free from moisture. The joint shall be undisturbed until cool. Joints shall be made in accordance with ASTM F 2389.

❖ The method of heat fusion of polypylene pipe or tubing is similar to that for polybutylene (see commentary, Section 605.19.2).

Figure 605.19.4
BENDING

605.20.2 Mechanical and compression sleeve joints. Mechanical and compression sleeve joints shall be installed in accordance with the manufacturer's instructions.

❖ This section addresses mechanical joining methods for polypropylene piping. Mechanical joints include insert-type fittings, metallic lock-ring fittings, compression fittings and crimp-type fittings.

Although an insert-type fitting reduces the inside diameter of the pipe, the reduction may be calculated as part of the design considerations of the water distribution system.

605.21 Polyethylene/aluminum/polyethylene (PE-AL-PE) and cross-linked polyethylene/aluminum/cross-linked polyethylene (PEX-AL-PEX). Joints between PE-AL-PE or PEX-AL-PEX pipe and fittings shall comply with Section 605.21.1.

❖ See the commentary to Section 605.21.1.

605.21.1 Mechanical joints. Mechanical joints shall be installed in accordance with the manufacturer's instructions. Fittings for PE-AL-PE and PEX-AL-PEX as described in ASTM F 1974, ASTM F 1281, ASTM F 1282, CSA B137.9 and CSA B137.10 shall be installed in accordance with the manufacturer's instructions.

❖ The ASTM standards listed are for insert fittings with either crimp rings or stainless steel clamps. Manufacturer's instructions must be followed for installation of the fittings to the pipe.

605.22 PVC plastic. Joints between PVC plastic pipe and fittings shall comply with Sections 605.22.1 through 605.22.3.

❖ Acceptable joints for PVC plastic pipe or fittings are identified in the referenced sections.

605.22.1 Mechanical joints. Mechanical joints on water pipe shall be made with an elastomeric seal conforming to ASTM D 3139. Mechanical joints shall not be installed in aboveground systems unless otherwise *approved*. Joints shall be installed in accordance with the manufacturer's instructions.

❖ PVC plastic hub pipe is connected using an elastomeric compression gasket (see commentary, Section 605.10.1).

605.22.2 Grooved and shouldered mechanical joints. Grooved and shouldered mechanical joints shall comply with ASTM F 1476, shall be made with an *approved* elastomeric seal and shall be installed in accordance with the manufacturer's instructions. Such joints shall be exposed or concealed.

❖ See the commentary for Section 605.14.3.

605.22.3 Solvent cementing. Joint surfaces shall be clean and free from moisture. A purple primer that conforms to ASTM F 656 shall be applied. Solvent cement not purple in color and conforming to ASTM D 2564 or CSA B137.3 shall be applied to all joint surfaces. The joint shall be made while the cement is wet and shall be in accordance with ASTM D 2855. Solvent-cement joints shall be permitted above or below ground.

❖ Like CPVC, PVC plastic pipe also requires a primer prior to solvent cementing. Although ASTM F 656 only recommends that PVC primer be purple, this is a code requirement. The purple coloring is intended to allow the installer and the inspector to verify that the required primer has been applied.

To distinguish the difference between primer and solvent cement, the solvent cement must be a color other than purple. Solvent cement for PVC is typically clear (see commentary, Sections 605.10.2 and 605.16.2). A PVC primer is first applied to both the pipe and the socket fitting, then PVC solvent cement is applied to both. The joint is made with a slight twisting until the pipe reaches the full depth of the socket fitting. It is then held in place until the solvent cement begins to set. The solvent cementing must be done in accordance with the referenced standard.

To minimize the health and fire hazards associated with solvent cements and primers, the manufacturers' instructions for the use and handling of these materials must be followed, in addition to the requirements in ASTM D 2855.

605.22.4 Threaded joints. Threads shall conform to ASME B1.20.1. Schedule 80 or heavier pipe shall be permitted to be threaded with dies specifically designed for plastic pipe, but the pressure rating of the pipe shall be reduced by 50 percent. Thread by socket molded fittings shall be permitted. *Approved* thread lubricant or tape shall be applied on the male threads only.

❖ The requirements regarding threaded joints for PVC plastic pipe are identical to those for CPVC plastic pipe (see commentary, Section 605.15.3).

605.23 Stainless steel. Joints between stainless steel pipe and fittings shall comply with Sections 605.23.1 and 605.23.3.

❖ Acceptable joints for stainless steel pipe or fittings are identified in the referenced sections.

605.23.1 Mechanical joints. Mechanical joints shall be installed in accordance with the manufacturer's instructions.

❖ Mechanical joints such as compression, grooved couplings, hydraulic pressed fittings or flange design can be used to join stainless steel pipe and fittings and are specified by the manufacturer. Such joints must be assembled and installed in compliance with the manufacturer's instructions.

605.23.2 Welded joints. All joint surfaces shall be cleaned. The joint shall be welded autogenously or with an *approved* filler metal as referenced in ASTM A 312.

❖ An important part of successful welding of the austenitic grades requires proper selection of alloy (for both the base metal and the filler rod) and corrects welding procedures. Two important objectives in making weld joints in austenitic stainless steel are: 1) preservation of corrosion resistance and 2) prevention of cracking. Welding naturally produces a temperature gradient in the metal being welded, ranging from the melting temperature of the fused weld metal to ambient temperature at some distance from the weld.

The two basic methods for welding stainless steels are fusion welding and resistance welding. In fusion welding, heat comes from an electric arc struck between an electrode and the metal to be welded. In resistance welding, bonding is the result of heat and pressure. Heat is produced by the resistance to the flow of electric current through the parts to be welded, and pressure is applied by the electrodes.

Welds and the surrounding area should be thoroughly cleaned to avoid impairment of corrosion resistance. Weld spatter, flux or scale may become focal points for corrosive attack if not properly removed, especially in aggressive environments.

605.23.3 Grooved and shouldered mechanical joints. Grooved and shouldered mechanical joints shall comply with ASTM F 1476, shall be made with an *approved* elastomeric seal and shall be installed in accordance with the manufacturer's instructions. Such joints shall be exposed or concealed.

❖ See the commentary for Section 605.14.3.

605.24 Joints between different materials. Joints between different piping materials shall be made with a mechanical joint of the compression or mechanical-sealing type, or as permitted in Sections 605.24.1, 605.24.2 and 605.24.3. Connectors or adapters shall have an elastomeric seal conforming to ASTM F 477. Joints shall be installed in accordance with the manufacturer's instructions.

❖ Mechanical joints can be used to join dissimilar piping materials. Because many piping materials have the same outside diameters, many fittings are designed by manufacturers for use with a number of different piping materials [see Commentary Figures 605.24.1(1) and (2)].

605.24.1 Copper or copper-alloy tubing to galvanized steel pipe. Joints between copper or copper-alloy tubing and galvanized steel pipe shall be made with a brass fitting or dielectric fitting or a dielectric union conforming to ASSE 1079. The copper tubing shall be soldered to the fitting in an *approved* manner, and the fitting shall be screwed to the threaded pipe.

❖ When copper or copper-alloy tubing is joined with galvanized steel pipe, protecting against galvanic corrosion is required. Galvanic corrosion occurs when two different metals come in contact in the presence of an electrolyte, such as water. Galvanic corrosion accelerates the natural corrosion process that occurs in all metals, and, because certain metals corrode faster than others, they have been placed in a hierarchy in order of rate of corrosion [see Commentary Figure 605.24.1(1)]. The more reactive metal at a connection is called the "anode" (galvanized steel pipe), and the less reactive metal is called the "cathode" (copper tubing). When the copper tubing and galvanized steel pipe are coupled, the galvanized steel pipe will tend to dissolve in the electrolyte, thereby generating electric current flow between the metals.

This section prescribes three methods of protection against galvanic corrosion: 1) a brass fitting such as a short section of brass pipe or a brass valve body; 2) a dielectric fitting such as a plastic-lined short pipe section or 3) a dielectric union either flange type or insulated-nut type. While the code text seems to imply that all of these types must conform to ASSE 1079, only the union type is covered by the referenced standard.

A variety of dielectric fitting designs is available. Commentary Figures 605.24.1(2) and 605.24.1(3) illustrate two of the most commonly used types. Other dielectric fittings of varying designs are available. Dielectric fittings provide an electrical barrier to prevent the passage of electric current either between dissimilar metals or between the piping and the electrolyte. Because the code does not define the term "dielectric fitting" or prescribe a standard for gauging the design or performance of a dielectric fitting, the code official must evaluate and approve the type of dielectric fitting and its application.

The joint between the copper tube and the threaded galvanized steel pipe can also be made with a cast-brass fitting. A wrought-copper fitting is not permitted for this transition. A cast-brass fitting is easily recognized by its yellowish color. The brass fitting serves as a buffer unlike the dielectric fitting, which provides an electrically insulating barrier. The brass fitting greatly retards the rate of corrosion. Because brass, copper and galvanized steel are dissimilar metals, there will be a minimal amount of corrosion among such materials.

Anode end (least noble, the reactive or corroding end)
   Magnesium
   Magnesium alloys
   Zinc
   Galvanized steel
   Aluminum
   Aluminum alloys
   Carbon steel
   Cast iron
   Stainless steel (active)
   Tin
   Lead
   Nickel (active)
   Brass
   Copper
   Bronze
   Nickel-copper alloys
   Nickel (passive)
   Titanium
   Silver
   Graphite
   Gold
   Platinum
Cathode end (most noble, the protected or less-reactive end)

**Figure 605.24.1(1)
CORROSIVE HIERARCHY OF METALS**

WATER SUPPLY AND DISTRIBUTION

Because plastic is usually the material that provides the electrical barrier in a dielectric fitting, special care is necessary when applying heat to the joint to avoid damaging the plastic sleeve. This is usually accomplished by sliding the plastic sleeve and steel nut away from the area to be heated. The soldered joint must conform to the requirements of Section 605.13.3. Commentary Figure 605.24.1(4) illustrates the result of connecting copper piping directly to a steel water heater tank. Welds on the top of this water heater tank began to leak in less than six years of service.

Figure 605.24.1(4)
CORROSION CAUSED BY ABSENCE OF DIELECTRIC FITTINGS

Figure 605.24.1(2)
DIELECTRIC UNION

605.24.2 Plastic pipe or tubing to other piping material. Joints between different types of plastic pipe or between plastic pipe and other piping material shall be made with *approved* adapters or transition fittings.

❖ The joining of differing types of plastic pipe (e.g., CPVC to PVC, ABS to PVC) and plastic pipe to other piping materials such as copper, brass and galvanized steel is allowed only through the use of adapter fittings. The adapter fittings must be evaluated individually and approved by the code official. Some key items to review when evaluating adapter fittings are joint tightness and compatibility of the adapter to the dissimilar materials being joined.

605.24.3 Stainless steel. Joints between stainless steel and different piping materials shall be made with a mechanical joint of the compression or mechanical sealing type or a dielectric fitting or a dielectric union conforming to ASSE 1079.

❖ Where two dissimilar metals are in contact in the presence of an electrolyte a battery effect is created, current flows and one of the metals corrodes. Any metal in this sense will tend to have corrosion accelerated when it is coupled in the presence of an electrolyte. A

Figure 605.24.1(3)
DIELECTRIC FLANGE FITTING

very important factor to consider in evaluating the potential for galvanic corrosion is the relative surface area of the two different metals in contact.

In theory all metals immersed in a electrolyte, such as water, have a voltage potential. The relative activity of the metal is determined by voltage potential, with those metals with negative voltages being most active and more likely to corrode. For example, magnesium and iron have a negative voltage and corrode very rapidly when exposed to a moist atmosphere. Copper and stainless steel have a positive voltage and are less likely to corrode.

Dielectric sleeves may be used with flange bolting or as gaskets for connections between different types of materials to prevent such corrosion. The plastic or rubber gasket engages the thread of the fitting attached, thereby insulating the metal from the water. This prevents corrosion on either part.

While the code text seems to imply that all of these types must conform to ASSE 1079, only the union type is covered by the referenced standard [see Commentary Figures 605.24.1(1) and 605.24.1(2)].

605.25 PE-RT plastic. Joints between polyethylene of raised temperature plastic tubing and fittings shall be in accordance with Section 605.25.1.

❖ Polyethylene of raised temperature tubing can only be connected by mechanical joining methods in accordance with Section 605.25.1.

605.25.1 Mechanical joints. Mechanical joints shall be installed in accordance with the manufacturer's instructions. Fittings for polyethylene of raised temperature plastic tubing shall comply with the applicable standards listed in Table 605.5 and shall be installed in accordance with the manufacturer's instructions. Polyethylene of raised temperature plastic tubing shall be factory marked with the applicable standards for the fittings that the manufacturer of the tubing specifies for use with the tubing.

❖ Mechanical joints for these materials include insert-type fittings, metallic lock-ring fittings, compression fittings and crimp-type fittings. Each fitting is intended for a specific pipe material and cannot be interchanged unless specified by the pipe manufacturer. In all cases, the fittings connecting any type of polyethylene pipe must be assembled and installed in accordance with the manufacturer's instructions.

SECTION 606
INSTALLATION OF THE BUILDING WATER DISTRIBUTION SYSTEM

606.1 Location of full-open valves. Full-open valves shall be installed in the following locations:

1. On the building water service pipe from the public water supply near the curb.

2. On the water distribution supply pipe at the entrance into the structure.

3. On the discharge side of every water meter.

4. On the base of every water riser pipe in occupancies other than multiple-family residential *occupancies* that are two stories or less in height and in one- and two-family residential *occupancies*.

5. On the top of every water down-feed pipe in *occupancies* other than one- and two-family residential *occupancies*.

6. On the entrance to every water supply pipe to a dwelling unit, except where supplying a single fixture equipped with individual stops.

7. On the water supply pipe to a gravity or pressurized water tank.

8. On the water supply pipe to every water heater.

❖ Valves are needed to isolate components and portions of a water distribution system to facilitate service, repair and replacement and to allow emergency shutoff in the event of catastrophic leak. A full-open valve, as defined in the legacy codes, is a shutoff valve that, in the full-open position, has a straight-through flow passageway with a diameter of not less than one nominal pipe size smaller than the nominal pipe size of the connecting pipe, or with an area not less than 85 percent of the cross-sectional area of the connecting pipe. Unlike shutoff valves and stops, full-open valves cause very little resistance to flow and, therefore, cause minimal pressure loss downstream of the valve. Full-open valves are also referred to as "water service valves" and include gate, ball and butterfly valves.

The following discusses each one of this section's list items:

1. A curb valve (also known as a "curb stop") is typically installed by the water utility company at the end of the water utility service connector pipe [see Commentary Figure 606.1(1)]. Decades ago, a curb valve was typically a lubricated rotor (plug)-type valve. However, newer installations often use ball valve-type curb valves. Where the water meter is located near the curb, a valve in the water meter vault or crock could serve the same purpose as the curb valve.

2. Where the water meter is located inside the building, the building entrance valve is located before the water meter. Where the water meter is located outside of the building (such as in a vault or crock), the building entrance valve is the valve between the end of the water service system and the beginning of the water distribution system.

3. The location of a valve before and after the water meter allows meter replacement without having to drain down the building water distribution system or empty the water service system [see Commentary Figure 606.1(2)]. Where the water meter is in a vault or crock outside the building, a valve immediately downstream of

the water meter is not required since the building entrance valve can serve as the valve after the water meter.

4. A water pipe extending vertically through one or more stories and conveying water vertically upward is considered a riser. Valves at the base of risers are very helpful for servicing a multistory building so that the entire water distribution system does not require draining in order to facilitate repairs in one area [see Commentary Figure 606.1(3)]. One- and two-family residential buildings are exempt from the water riser valve requirement as are multiple-family dwellings that are not over two stories in height.

5. A water pipe extending vertically through one or more stories and conveying water vertically downward is considered a downfeed. Valves at the top of downfeeds are very helpful for servicing a multistory building so that only small areas of the building will be without water in order to facilitate repairs in one area [see Commentary Figure 606.1(3)]. One- and two-family residential buildings are exempt from the water downfeed valve requirement. Note, however, that multiple-family dwellings that are not over two stories in height are not exempt from the requirement for valves on downfeed pipes.

6. Where a building contains more than one dwelling unit, each dwelling unit must have a valve located at the entrance of every water distribution pipe to the dwelling unit. For example, consider a multiunit apartment building where each apartment is provided with a hot and a cold water supply pipe for delivery of water to the fixtures in the apartment. Both hot and cold water

Figure 606.1(1)
CURB VALVE ON WATER SUPPLY

Figure 606.1(2)
VALVE AT ENTRANCE TO THE BUILDING AND ON THE DISCHARGE SIDE OF METER

supply pipes must have a full-open valve so that the water can be shut off to just that apartment's water distribution system. This allows for extended maintenance or remodeling activities without interruption of the water supplies for adjacent apartments [see Commentary Figure 606.1(4)].

The exception portion of this list item, "except where supplying a single fixture equipped with individual stops," might have several meanings. One possible interpretation is that where a single fixture in a dwelling unit is supplied from piping that is not shut off by the dwelling unit's required full-open valve(s) previously mentioned, that fixture's supply pipes are not required to have full-open valves if the fixture has individual stops (valves). For example, consider risers or downfeeds that supply only the kitchen sinks in a "stack" of dwelling units in an apartment building. The individual stop valves for the kitchen sink can serve as the water shut-off valves for the fixture supply pipes. Another possible interpretation might be that if every fixture in a dwelling unit has individual stops (valves), then full-open valves are not required on the water supply pipes to the dwelling unit. The code official must determine the final meaning of this item's exception statement.

7. Gravity water tanks and pressurized water tanks require service and replacement. The provision of a full-open valve on the water supply to such tanks allows for those activities without the need to shut off pumping systems and water supply systems [see Commentary Figure 606.1(5)].

8. Water heaters require periodic servicing such as temperature and pressure relief valve replacement, sediment removal and element/thermostat replacement. Eventual replacement of the water heater is often necessary. The provision of a full-open valve on the water supply to a water heater allows for those activities without interrupting the supply of cold water to the building

Figure 606.1(3)
VALVES FOR EACH RISER AND DOWNFEED

Figure 606.1(4)
DWELLING UNIT VALVE

WATER SUPPLY AND DISTRIBUTION

[see Commentary Figure 606.1(6)]. Note that the code does not prohibit the installation of a valve on the hot water pipe from the water heater. Installation of such valves is sometimes prudent where portions of the hot water distribution system are located above the elevation of the water heater so that extensive draining of the hot water distribution system is not required for water heater service. Although the code is silent about the type of valve that should be used for this application, a full-open valve would be the best choice to avoid significant pressure losses in the hot water distribution system.

606.2 Location of shutoff valves. Shutoff valves shall be installed in the following locations:

1. On the fixture supply to each plumbing fixture other than bathtubs and showers in one- and two-family residential *occupancies*, and other than in individual sleeping units that are provided with unit shutoff valves in hotels, motels, boarding houses and similar *occupancies*.

2. On the water supply pipe to each sillcock.

3. On the water supply pipe to each appliance or mechanical equipment.

❖ Shutoff valves can be any type of valve, such as a globe valve, a ball valve (full open or reduced port), a gate valve or angle valve. A shutoff valve has no minimum requirements for the cross-sectional area of flow. The following discusses each of this section's list items:

1. Fixtures require a shutoff valve on each of the fixture supply pipes supplying water to the fixture so that if the fixture requires repair or replacement, the building's water distribution system does not have to be shut down and

**Figure 606.1(5)
VALVES ON SUPPLY TO EACH TANK**

**Figure 606.1(6)
WATER HEATER VALVE**

drained. There are two exceptions within this list item. The first exception is that bathtub and shower faucets/mixing valves in one- and two-family residences are not required to have shutoff valves. The reason for this is that the fixture supply pipes to these faucets/mixing valves are concealed, making it difficult, if not impossible, to install shutoff valves with access. The second exception is where individual sleeping units in hotels, motels, boarding houses and similar occupancies have a unit shutoff valve (i.e., a valve that shuts off the water supply to only that unit), fixture shutoff valves are not required. The reason for this exception is to reduce the number of shutoff valves required in a building of that type since fixture shutoff valves can be a source of leaks.

The code does not directly specify the location for fixture shutoff valves. Where fixture supply pipes originate from a "mains, risers and branch"-type water distribution system, fixture shutoff valves are usually located adjacent to the fixture and, more commonly, attached to the end of the fixture supply pipes [see Commentary Figure 606.2(1)]. Note that Section 604.5 requires that fixture supply pipes must terminate not more than 30 inches (762 mm) from the point of connection to the fixture. Thus, fixture shutoff valves attached to the end of the fixture supply pipes will be located within 30 inches (762 mm) from the point of connection to the fixture. The shutoff valves that are typically installed at the end of fixture supply pipes are specialized fixture fittings that are called "stops" or "stop valves." Stop valves are available in straight pattern and angle pattern, are available with a variety of inlet and outlet connection sizes/configurations and, because, in some cases the fittings are exposed, are available in a variety of finishes such as chrome, brushed nickel and polished brass.

Where fixture supply pipes originate from a parallel water distribution central manifold with valves, the fixture shutoff valves could be the valves on the manifold. As long as the manifold shutoff valves can shut off the water supply to only an individual fixture, the manifold valves could serve as the fixture shutoff valves and stop valves would not be required adjacent to the fixture. Note Section 606.4 requires marking of valves that are not adjacent to the fixture.

2. A sillcock is a hose connection faucet located on the outside of a building. Sillcocks include hose bibbs and freezeproof hose faucets. Such faucets are often vandalized, abused (left running) and freeze damaged such that the water supply to the faucet must be shut off for extended periods of time. The shutoff valve is necessary to isolate the faucet from the remainder of the water distribution system [see Commentary Figure 606.2(2)]. The code is silent with respect to the location of the shutoff valve but if it is remote from the hose faucet, Section 606.4 requires labeling of the shutoff valve. Even if the shutoff valve is not remote from the hose faucet, Section 606.4 requires marking of the valve.

3. Appliances and equipment that require a water supply must have shutoff valves on the water supply pipes so that repairs and replacement can be performed without affecting the remainder of the water distribution system. The code is silent with respect to the location of the shutoff valve but if it is remote from the appliance or equipment, Section 606.4 requires labeling of the shutoff valve(s).

For SI: 1 inch = 25.4 mm.

Figure 606.2(1)
SHUTOFF VALVE FOR EACH FIXTURE

**Figure 606.2(2)
SILL FAUCET VALVE**

606.3 Access to valves. *Access* shall be provided to all full-open valves and shutoff valves.

❖ To serve their intended purpose, full-open and shutoff valves must be able to be accessed for operation (see the definition of "Access" in Chapter 2).

606.4 Valve identification. Service and hose bibb valves shall be identified. All other valves installed in locations that are not adjacent to the fixture or appliance shall be identified, indicating the fixture or appliance served.

❖ The term "service valve" in this section is intended to mean the main water service valve for the building. This valve will be the full-open valve between the end of the water service supply system and the beginning of the water distribution system. The code requirement for marking this valve should be obvious as locating this valve during an emergency would be critical.

The term "hose bibb valves" in this section refers to sillcocks, freezeproof outdoor faucets, yard hydrants and wall hydrants. Section 606.2 requires shutoff valves for these types of valves. The required shutoff valve for such hose connection faucets must have a tag or similar method of identification showing the valve's purpose. All other valves must be labeled if the valve's purpose is not obvious because of the proximity of the fixture or appliance served by the valve (see commentary, Section 604.10.2). Identification of these "remote" valves is necessary since service personnel can only see the valve; they will not have any clue as to what the valve is connected to because they are unable to trace concealed piping. Obviously, locating specific valves in an emergency is also important. Where a parallel water distribution system with a central manifold is installed, the valves on the manifold must be marked for the individual fixtures that the valves serve.

606.5 Water pressure booster systems. Water pressure booster systems shall be provided as required by Sections 606.5.1 through 606.5.10.

❖ Requirements for water pressure booster systems are described in Sections 606.5.1 through 606.5.10.

606.5.1 Water pressure booster systems required. Where the water pressure in the public water main or individual water supply system is insufficient to supply the minimum pressures and quantities specified in this code, the supply shall be supplemented by an elevated water tank, a hydro-pneumatic pressure booster system or a water pressure booster pump installed in accordance with Section 606.5.5.

❖ There is no guarantee that a public water main will have the required pressure for the water distribution system of a proposed new building. Some public water supply systems are so antiquated that the public water main piping cannot withstand pressures much over 30 psi (207 kPa). In other situations, a public water supply system might be installed in hilly terrain where the pressure will be low at the highest point in the system because pressures might be too great at the lowest point in the system. In an individual water supply system (well system), the depth of the well might limit the amount of pressure that the well pump can provide at grade level. In these types of situations, a method of boosting the existing water pressure must be provided so that a building's water distribution system has the necessary pressure to overcome piping friction losses and elevation pressure loss.

Generally, there are three types of water pressure boosting methods. The first and perhaps the oldest is that of a water tank (or tanks) placed on the roof of a building. Pumps in the basement of the building pump water from the source (the public main or well system) up to the tank to keep it filled. The tank gravity feeds the water distribution system of the building. The top of the tank is vented to the outdoors. A second method for boosting pressure is the use of a closed tank so that when water is pumped into the tank, the displaced air becomes pressurized. The air pressure in the tank provides the pressure onto the surface of the water. The third method is the use of variable speed pumps to push water into the water distribution system. These pumps automatically speed up and slow down to maintain a constant pressure based on the variations in demand (flow).

Any pump that draws directly from the public water

main or a well pump discharge must be protected against cavitation damage. Therefore, this section requires that the pump installation be in accordance with Section 606.5.5.

606.5.2 Support. All water supply tanks shall be supported in accordance with the *International Building Code*.

❖ The weight of a full water tank must be considered when calculating the dead loads in the building's structural design. Water weighs 62.4 pounds per cubic foot (1000 kg/m^3) or 8.34 pounds per gallon (1 kg/L).

606.5.3 Covers. All water supply tanks shall be covered to keep out unauthorized persons, dirt and vermin. The covers of gravity tanks shall be vented with a return bend vent pipe with an area not less than the area of the down-feed riser pipe, and the vent shall be screened with a corrosion-resistant screen of not less than 16 by 20 mesh per inch (630 by 787 mesh per m).

❖ This section refers to gravity water tanks that must be open to the outdoors. Because the water tank contains potable water, a cover provides protection against contamination and unauthorized entry. The vent is necessary to prevent pressures from developing in the tank resulting from water being withdrawn or pumped into the tank and to allow for thermal expansion.

606.5.4 Overflows for water supply tanks. A gravity or suction water supply tank shall be provided with an overflow with a diameter not less than that shown in Table 606.5.4. The overflow outlet shall discharge at a point not less than 6 inches (152 mm) above the roof or roof drain; floor or floor drain; or over an open water-supplied fixture. The overflow outlet shall be covered with a corrosion-resistant screen of not less than 16 by 20 mesh per inch (630 by 787 mesh per m) and by $^1/_4$-inch (6.4 mm) hardware cloth or shall terminate in a horizontal angle seat check valve. Drainage from overflow pipes shall be directed so as not to freeze on roof walks.

❖ Gravity or suction water tanks are not designed to be pressurized. They are designed to withstand the pressure exerted by the water being contained in the tank. As such, these tanks must have an overflow pipe installed on the top cover of the tank or on the side of the tank near the top. The pipe is routed to a roof drain, floor drain or, in some cases, to a water-supplied fixture having a drain. The end of the pipe must terminate above the drain or flood level rim of the fixture by at least 6 inches (152 mm) to prevent the possibility of the backsphonage of waste water into the overflow pipe. The pipe must be protected against entry by insects or small animals with a screen on the end of the open pipe or by a swing check valve installed in a horizontal pipe section. The size of the overflow pipe is based on its capacity to gravity discharge water at a rate that is commensurate with the pressurized filling rate.

TABLE 606.5.4
SIZES FOR OVERFLOW PIPES FOR WATER SUPPLY TANKS

MAXIMUM CAPACITY OF WATER SUPPLY LINE TO TANK (gpm)	DIAMETER OF OVERFLOW PIPE (inches)
0 – 50	2
50 – 150	2$^1/_2$
150 – 200	3
200 – 400	4
400 – 700	5
700 – 1,000	6
Over 1,000	8

For SI: 1 inch = 25.4 mm, 1 gallon per minute = 3.785 L/m.

606.5.5 Low-pressure cutoff required on booster pumps. A low-pressure cutoff shall be installed on all booster pumps in a water pressure booster system to prevent creation of a vacuum or negative pressure on the suction side of the pump when a positive pressure of 10 psi (68.94 kPa) or less occurs on the suction side of the pump.

❖ The creation of low pressures on the upstream (suction) side of pumps increases the possibility of backflow in the portion of the supply system that precedes the pumping equipment. Additionally, where there is inadequate pressure and volume of water on the suction side of the pump, cavitation can occur in the pump housing. Cavitation, which is the formation of partial vacuums within a flowing liquid (e.g., air bubbles), can rapidly deteriorate a pump impeller. Net positive suction head (NPSH) requirements vary with impeller design and must be checked for each pump. A low-pressure cutoff prevents the pump from operating when the supply pressure at the pump inlet is 10 psi (69 kPa) or less.

606.5.6 Potable water inlet control and location. Potable water inlets to gravity tanks shall be controlled by a fill valve or other automatic supply valve installed so as to prevent the tank from overflowing. The inlet shall be terminated so as to provide an *air gap* not less than 4 inches (102 mm) above the overflow.

❖ Float switches, float valves, electrode controllers and electronic level devices are used for controlling the flow of water into a tank. The water level in gravity tanks must be maintained by automatic means so that the supply of potable water is maintained. An air gap must be installed between the tank filling pipe termination and highest water level possible in the tank. The highest water level possible in the tank is the elevation of the overflow pipe connection to the tank.

606.5.7 Tank drain pipes. A valved pipe shall be provided at the lowest point of each tank to permit emptying of the tank. The tank drain pipe shall discharge as required for overflow pipes and shall not be smaller in size than specified in Table 606.5.7.

❖ A valve is required on every tank so that it can be emptied for service, repair or replacement of the tank.

TABLE 606.5.7
SIZE OF DRAIN PIPES FOR WATER TANKS

TANK CAPACITY (gallons)	DRAIN PIPE (inches)
Up to 750	1
751 to 1,500	$1^1/_2$
1,501 to 3,000	2
3,001 to 5,000	$2^1/_2$
5,000 to 7,500	3
Over 7,500	4

For SI: 1 inch = 25.4 mm, 1 gallon = 3.785 L.

❖ The valve at the base of the water tank is sized to empty the tank quickly.

606.5.8 Prohibited location of potable supply tanks. Potable water gravity tanks or manholes of potable water pressure tanks shall not be located directly under any soil or waste piping or any source of contamination.

❖ Any opening on a water tank must be protected against contamination. If drainage piping is located above a tank, there is a possibility of a leak from the piping or from a fixture served by such piping dripping onto the tank. Similar occurrences in the past have led to major outbreaks of disease.

606.5.9 Pressure tanks, vacuum relief. All water pressure tanks shall be provided with a vacuum relief valve at the top of the tank that will operate up to a maximum water pressure of 200 psi (1380 kPa) and up to a maximum temperature of 200°F (93°C). The size of such vacuum relief valve shall be not less than $^1/_2$ inch (12.7 mm).

Exception: This section shall not apply to pressurized captive air diaphragm/bladder tanks.

❖ This section addresses pressure tanks that are part of a water pressure booster system. There is always the possibility of a pressure tank being subjected to a vacuum, especially elevated tanks supplying downfeed piping. Many tanks are not designed for a vacuum condition; therefore, a vacuum relief valve is necessary to prevent atmospheric pressure from collapsing or otherwise damaging the tank. A vacuum relief valve is a type of check valve having a shielded inlet.

606.5.10 Pressure relief for tanks. Every pressure tank in a hydropneumatic pressure booster system shall be protected with a pressure relief valve. The pressure relief valve shall be set at a maximum pressure equal to the rating of the tank. The relief valve shall be installed on the supply pipe to the tank or on the tank. The relief valve shall discharge by gravity to a safe place of disposal.

❖ Hydropneumatic tanks are rated for a given pressure. When connected in line with pumping equipment, there is a possibility of exceeding the pressure rating of the tank. The pressure relief valve installed on the tank prevents the buildup of excessive pressure, thereby preventing the possibility of tank rupture or failure.

606.6 Water supply system test. Upon completion of a section of or the entire water supply system, the system, or portion completed, shall be tested in accordance with Section 312.

❖ It is advantageous to conduct a water test before the system is concealed. Any leaks are easily repaired at this time without damaging or replacing existing materials. The test pressure required by Section 312.5 using water must be equal to or greater than the working pressure of the water supply system that will serve the building water service and water distribution system. When using air, the test pressure must not be less than 50 psi (344 kPa) (see Section 312.5).

606.7 Labeling of water distribution pipes in bundles. Where water distribution piping is bundled at installation, each pipe in the bundle shall be indentified using stenciling or commercially available pipe labels. The identification shall indicate the pipe contents and the direction of flow in the pipe. The interval of the identification markings on the pipe shall not exceed 25 feet (7620 mm). There shall be not less than one identification label on each pipe in each room, space or story.

❖ Unlabeled bundled water piping presents major problems during renovations and repairs. With gray-water systems now being introduced or retrofitted into buildings, pipe identification becomes more critical since not all water piping will be potable water piping.

SECTION 607
HOT WATER SUPPLY SYSTEM

607.1 Where required. In residential *occupancies*, *hot water* shall be supplied to plumbing fixtures and equipment utilized for bathing, washing, culinary purposes, cleansing, laundry or building maintenance. In nonresidential *occupancies*, *hot water* shall be supplied for culinary purposes, cleansing, laundry or building maintenance purposes. In nonresidential *occupancies*, *hot water* or *tempered water* shall be supplied for bathing and washing purposes.

❖ Hot water is almost always necessary to achieve thorough cleaning of just about anything within a building. This section requires hot water to be supplied to plumbing fixtures and equipment that are used for culinary purposes (cleaning of pots, pans, dishes, flatware, cups and glasses), general cleansing (cleaning of fixtures, furniture, walls and floors), laundry (washing of clothing in machines and laundry trays) and building maintenance (cleaning of tools, paint brushes and dirty equipment).

Where bathing or washing (of the human body) is concerned, this section requires that plumbing fixtures used for those purposes in residential occupancies be supplied with hot water. In nonresidential occupancies, those same fixtures must be provided with either hot or tempered water. Chapter 2 defines "Hot water" as water having a temperature of 110°F (43°C) or greater. "Tempered water" is defined in Chapter 2 as water having a temperature between (but not including) 85°F and 110°F (29°C and 43°C).

The decision to supply hot or tempered water to fixtures used for bathing or washing in buildings of nonresidential occupancy is the responsibility of the building designer with the exception of public hand-washing facilities. Section 416.5 requires the designer to provide only tempered water to be delivered from public hand-washing facilities. Providing tempered water for these applications offers greater energy savings as well as protects users from water temperatures that could be dangerous because of the potential for scald injury.

607.1.1 Temperature limiting means. A thermostat control for a water heater shall not serve as the temperature limiting means for the purposes of complying with the requirements of this code for maximum allowable *hot* or *tempered water* delivery temperature at fixtures.

❖ Because it is relatively easy to adjust a thermostat on a water heater and the fact that water heater thermostats do not hold water temperatures within a repeatable narrow range, water heater thermostats must not be used to control temperatures where maximum water temperatures must not be exceeded. Thermostatic water temperature control valves offer precise temperature control and repeatability. They are also a type of device that is unfamiliar to most laypersons such that it is unlikely that the device will be tampered with.

607.1.2 Tempered water temperature control. *Tempered water* shall be supplied through a water temperature limiting device that conforms to ASSE 1070 and shall limit the *tempered water* to a maximum of 110°F (43°C). This provision shall not supersede the requirement for protective shower valves in accordance with Section 424.3.

❖ Where tempered water is to be supplied to a tempered water system or to one or more fixtures, the temperature of the water must be controlled by a device that complies with ASSE 1070. These devices are thermostatic temperature control valves that precisely control the temperature of the water discharged from the valve so that the temperature is relatively constant. Where an ASSE 1070 device is used to supply water to one or more individual shower valves, Section 424.3 requiring the shower valves to comply with ASSE 1016 or ASME A112.18.1/CSA B125.1 still applies. This is because the ASSE 1070 device does not have the fast-response time that is required for individual shower valves.

607.2 Hot or tempered water supply to fixtures. The *developed length* of *hot* or *tempered water* piping, from the source of hot water to the fixtures that require *hot* or *tempered water*, shall not exceed 50 feet (15 240 mm). Recirculating system piping and heat-traced piping shall be considered to be sources of *hot* or *tempered water*.

❖ A significant amount of water can be wasted by a fixture user while waiting for hot water to arrive at the fixture. In addition to being an inconvenience, the costs for the water, for the energy to heat the water and for the disposal of the water are literally going down the drain. There will be less waste and inconvenience at the fixture where the source of hot water is located near to the fixture.

Hot water must be available within 50 feet (15 240 mm) of developed pipe length of the fixture requiring hot water. Where the developed length of piping is greater than 50 feet (15 240 mm), the distance must be reduced to be not greater than 50 feet (15 240 mm). The following example illustrates the impact of this 50-foot (30 480 mm) maximum length. A 50-foot (15 240 mm) length of $^1/_2$-inch (12.7 mm) copper tube contains approximately 0.6 gallon (2.3 liters) of water. A public lavatory is limited to a flow rate of 0.5 gpm (1.9 L/m). Therefore, if the entire 50-foot (15 240 mm) length of tube is at ambient temperature, the fixture user would have to wait approximately 1.2 minutes for hot water to arrive at the fixture. Under these conditions, most public lavatory users would have given up long before hot water arrived. In reality, where fixtures are being used frequently and the hot water piping system is well insulated, the wait time (and water waste) is substantially reduced.

The code is silent on the methods that can be used to ensure that a source of hot water is within the required distance limit. For example, if a lavatory is located 63 feet (19 202 mm) of developed piping length from a water heater, several methods could be used to reduce the distance to no greater than 50 feet (15 240 mm). The most common methods are: 1) locate the water heater closer to the fixture: 2) install a hot water recirculation system such that the piping of that system provides a connection point for hot water that is closer to the fixture or 3) install a pipe heating system on the piping beginning at the source of the hot water to a point that is closer to the fixture.

In small buildings where all the fixtures are clustered together in one location, it is often possible to locate a central water heater within the specified distance. Where a building is large and the fixtures requiring hot water are "spread out," multiple water heating units can be strategically located to service groups of fixtures. Single "point of use" water heaters are frequently installed to provide hot water to just one fixture. In larger buildings where all of the water heating appliances are located in a mechanical room, it is often impossible to locate the mechanical room within 50 feet (15 240 mm) of developed piping length to all fixtures requiring hot water. Where water heating units cannot be located near to the fixtures, pumped hot water recirculation piping systems are typically installed to convey the hot water to a point near to all fixtures.

A hot water recirculating system circulates hot water from the water heater or hot water storage tank through a loop of piping that returns the water to the water heater or hot water storage tank. A continuous flow of hot water in the piping loop maintains a desired water

WATER SUPPLY AND DISTRIBUTION

temperature at the point of connection for the piping to each fixture. Ideally, the location of the recirculation piping would pass in very close proximity to all fixtures requiring hot water. This code section only requires that the recirculation piping be within 50 feet (15 240 mm) of developed piping length to each fixture. There are also "no-return pipe" hot water recirculation systems available (see Commentary Figure 607.2).

Electric cable or steam tracing lines can also be used to maintain water temperatures in hot water piping.

Operating costs for the methods previously discussed are a concern for building designers, especially if they are designing for maximum energy efficiency. Recirculating pumps, heating cables and steam tracing all require energy. There is also the concern about heat loss from the piping which requires additional energy for maintaining the temperature of the hot water in the system.

607.2.1 Circulation systems and heat trace systems for maintaining heated water temperature in distribution systems. For Group R2, R3 and R4 occupancies that are three stories or less in height above grade plane, the installation of heated water circulation and temperature maintenance systems shall be in accordance with Section R403.5.1 of the *International Energy Conservation Code*. For other than Group R2, R3 and R4 occupancies that are three stories or less in height above grade plane, the installation of heated water circulation and heat trace systems shall be in accordance with Section C404.6 of the *International Energy Conservation Code*.

❖ The IECC has requirements for heated water circulation systems and heated water temperature maintenance systems (heat trace) because these systems can waste a significant amount of energy if not designed and installed to minimize energy consumption. This code section is somewhat complicated by the fact that the IECC is split into two parts: commercial buildings (IECC-CE) and residential buildings (IECC-RE). This code covers residential buildings that are classified by the IBC as Group R2, R3 or R4. However, where those buildings are "low rise" (i.e., three stories or less in height above grade plane), the IECC-RE requirements apply. For all other buildings covered by the IPC, the IECC-CE requirements apply.

This section is a pointer to alert the user of the IPC that requirements affecting heated water circulation systems and heated water temperature maintenance systems (heat trace) for water distribution systems exist in the IECC.

IECC C404.6 (and its subsections)

IECC C404.6 (and its subsections) prohibit gravity and thermo-syphon circulation systems. Automatic controls, temperature sensors and pumps must have access. Manual controls must have ready access. The controls for circulating hot water system pumps must start the pump when there is a demand for hot water within the occupancy. The controls must automatically turn off the pumps when the water in the circulation loop is either at the desired temperature or when there is not demand for hot water.

Electric heat trace systems must comply with IEEE 515.1. The controls these systems must be able to automatically adjust the energy input to the heat tracing to maintain the desired water temperature in the piping in accordance with the times when heated water is used in the occupancy. The heat trace system must be arranged to be turned off automatically when there is not a demand for hot water.

For SI: 1 foot = 304.8 mm.

**Figure 607.2
AUTOMATIC TEMPERATURE MAINTENANCE SYSTEM BY HOT WATER (PUMPED) CIRCULATION LOOP**

WATER SUPPLY AND DISTRIBUTION

IECC R403.5.1 (and its subsections)

These sections have the same requirements as IECC C404.6 with one difference. The heat trace systems can comply with either IEEE 515.1 or UL 151.

607.2.1.1 Pump controls for hot water storage systems. The controls on pumps that circulate water between a water heater and a storage tank for heated water shall limit operation of the pump from heating cycle startup to not greater than 5 minutes after the end of the cycle.

❖ This section addresses water-heating systems that utilize a separate water-heating unit to generate hot water for storage in tanks. These systems have a circulating pump to move hot water into the storage tank (and cooler water back to the heating unit). This section requires that the circulation pump not continue to operate after the hot water storage tank is at the desired temperature. The run time of up to 5 minutes after the end of the heating cycle is to allow for the heating unit's heat exchanger to "cool down" and equalize temperature with the stored hot water. This section is duplicated in IECC-CE, but not in IECC-RE.

607.2.1.2 Demand recirculation controls for distribution systems. A water distribution system having one or more recirculation pumps that pump water from a heated water supply pipe back to the heated water source through a cold water supply pipe shall be a demand recirculation water system. Pumps shall have controls that comply with both of the following:

1. The control shall start the pump upon receiving a signal from the action of a user of a fixture or appliance, sensing the presence of a user of a fixture, or sensing the flow of hot or tempered water to a fixture fitting or appliance.

2. The control shall limit the temperature of the water entering the cold water piping to 104°F (40°C).

❖ This section applies only to heated water (hot or tempered water) recirculation systems that pump water back to the heated water source through a cold water supply pipe. Specialized equipment for recirculating hot water to remote fixture systems and back to the heated water source through a cold water supply pipe has been used for several decades. They were most often used as a retrofit for buildings where routing of dedicated return piping was too costly or impossible. See the commentary for Section 604.2 for additional understanding of "no dedicated return line" recirculation systems.

Extensive research by many organizations, including the Oak Ridge National Laboratory and the California Energy Commission, has determined that having hot water "ready and waiting" at fixtures is a significant energy waste. On the other hand, where a hot (or tempered) water circulation system is not provided, there is a significant waste of energy of heating water that never arrives at the intended point of use except where a large volume of cooler water is flushed down the drain in order to have the heated water arrive at the point of use. Certainly, such water waste is of concern (because of rising water and sewer charges) but more so, the waste of "embedded energy" (utility treatment and pumping energy) in that water is of equal or greater concern. The focus of the research was to determine how to best install and operate hot (or tempered) water recirculation systems for the lowest possible operating cost.

Section 607.2 has been in the code for many editions, primarily because of the inconvenience that users experience when trying to obtain hot (or tempered) water from fixtures. This requires that many building arrangements have a hot (or tempered) water recirculation system for fixtures. The code doesn't specify which type of recirculation system to use (dedicated return pipe or "cold water supply" return pipe). If a "cold water supply" return pipe system is chosen, then Section 607.2.1.2 applies.

This section introduces a new defined term of "Demand recirculation water system." See Chapter 2 for the definition and commentary to understand how this system works. Section 607.2.1.2 requires that if the recirculation system is of the "cold water supply" return pipe type of system, then the recirculation system is required to be a "demand recirculation water system." Items 1 and 2 of Section 607.2.1.2 specify how the pump controls must operate.

This code section is duplicated in both the IECC-CE and IECC-RE.

607.2.2 Piping for recirculation systems having master thermostatic valves. Where a thermostatic mixing valve is used in a system with a hot water recirculating pump, the *hot water* or *tempered water* return line shall be routed to the cold water inlet pipe of the water heater and the cold water inlet pipe or the hot water return connection of the thermostatic mixing valve.

❖ Where hot water systems have a master thermostatic mixing valve and a circulating pump downstream of the mixing valve to maintain temperatures at remote areas of the system, the hot water or tempered water return line from the mixing valve must be split and routed to the cold water inlet pipe of the water heater and the cold water inlet of the mixing valve or hot water return connection of the mixing valve [see Commentary Figure 607.2.3(1)]. If the hot water return pipe is routed only to the water heater [see Commentary Figure 607.2.3(2)], the mixed water system temperature will rise when there is no fixture use in the system. Where the hot water return pipe is routed only to the mixing valve [see Commentary Figure 607.2.3(3)], the mixed water system temperature will drop to ambient temperature when there is no fixture use in the system.

607.3 Thermal expansion control. Where a storage water heater is supplied with cold water that passes through a check valve, pressure reducing valve or backflow preventer, a thermal expansion tank shall be connected to the water heater

WATER SUPPLY AND DISTRIBUTION

For SI: °C = [32 - (°F)]/1.8.

**Figure 607.2.3(1)
PROPER RECIRCULATION RETURN**

**Figure 607.2.3(2)
PROHIBITED RECIRCULATION RETURN**

WATER SUPPLY AND DISTRIBUTION

Figure 607.2.3(3)
PROHIBITED RECIRCULATION RETURN

cold water supply pipe at a point that is downstream of all check valves, pressure reducing valves and backflow preventers. Thermal expansion tanks shall be sized in accordance with the tank manufacturer's instructions and shall be sized such that the pressure in the water distribution system shall not exceed that required by Section 604.8.

❖ Liquids expand in volume when heated. If the liquid being heated completely fills a vessel of fixed volume, the pressure in the vessel will increase in proportion to the rise in temperature of the liquid. As long as sufficient heat is applied to cause a rise in liquid temperature, the pressure in the vessel will continue to rise until liquid is allowed to escape from the vessel (relieving the pressure) or the vessel ruptures due to overpressurization. Because a hot water distribution system is necessarily connected to a water heating device, there is the potential for thermal expansion to cause a rise in system pressure. The intent of this section is to require the control of any pressure increase caused by thermal expansion of water being heated.

Uncontrolled pressure increases caused by thermal expansion can have a detrimental effect on all components in a water distribution system. Faucets, typically designed for long-life operation at pressures not greater than 80 psig (552 kPa), will wear out rapidly and leak at higher pressures. Other system components, such as supply shutoff valves and flexible water supply connectors, may leak or fail at pressures exceeding 125 psig (862 kPa). Leaks might develop in water distribution piping that was tested to only the "working pressure" of the system. Water heater tank welds may fatigue and fail due to frequent pressure fluctuations caused by uncontrolled thermal expansion.

This begs the question: Is there any water distribution system where control of pressure due to thermal expansion is not required? The following examples illustrate the few cases where pressure control is not required or necessary:

Case A: Consider a building where the main water supply line pressure is always below 80 psig (therefore, no pressure-reducing valve is installed) and there are not any backflow preventers, check valves or other one-way flow devices in the system or in the connections to the main water supply line (the public water main). When water in the water distribution system is heated, the increased volume is easily absorbed by the massive volume of the water main and, thus, no pressure increase in the water distribution system can be detected. Since there is no pressure increase, no pressure control is required.

Case B: Consider a building where a private water source (e.g., a well system with an accumulator tank) supplies the water distribution system and no one-way flow devices (check valves) are between the accumulator tank and the water distribution system. The cut-off pressure setting of the pump is 60 psi. When water in the water distribution system is heated, the increased volume is easily absorbed by the relatively large volume of the accumulator tank (also known as a pressure tank) and, thus, only a small pressure increase in the water distribution system can be detected. Therefore, no thermal expansion pressure control is required.

With respect to Case A, low-pressure [80 psig (552 kPa) or less] public water mains are becoming a rarity as fire service demands, new building construction demands, and better water main pipe materials all contribute towards new and replacement public water mains being designed for pressures much higher than 80 psig. Therefore, most buildings will require a pressure-reducing valve and, thus, a device to control pressures due to thermal expansion of water being heated is required.

Regardless of the public water main pressure (and the need for a pressure-reducing valve), it is becoming commonplace for water purveyors to install a backflow prevention device at the water meter or to require the building owner to provide a backflow prevention device in the water service piping to the building. The water purveyor requires these devices in order to comply with federal regulations and maintain certification as a safe drinking water provider. As such, most water providers have necessarily established backflow prevention programs in order to ensure that the public water system is protected from possible backflow contamination from buildings connected to the water system. At a minimum, most water purveyors install a dual check valve (in accordance with ASSE 1024 or CSA B64.6) on the upstream side of the water meter for a building. However, certain building occupancies, such as restaurants and factories, are required by the water purveyor to install backflow devices with a higher degree of protection. The installation of any backflow devices in the water service line will require the control of water distribution system pressures caused by thermal expansion of water being heated.

Note that existing buildings are typically not exempt from having to comply with a water purveyor's demands to install backflow protection. As water purveyors ramp up their backflow prevention programs, existing buildings (having no or limited backflow protection) are identified by their backflow hazard potential and a schedule for compliance is established. For example, the first year of a program might target all existing high-hazard factory and industrial occupancies. The next year might target all restaurants and smaller "medium-hazard" commercial establishments. The following year might be all residential irrigation systems. In some cases, periodic replacement of worn out, inaccurate outdated water meters often triggers the automatic installation of a backflow prevention device (dual check valve) with the new water meter. While most water providers will inform the building owner, by letter, that a backflow device was installed with the new water meter, these letters are often tossed aside without due respect for the need to install thermal expansion pressure control devices.

Beyond the requirements for backflow prevention devices, the installation of check valves (a one way flow device) is often necessary for proper operation of a plumbing system. A check valve is sometimes installed in the cold supply pipe to a water heater to prevent hot water migration into the cold water distribution piping. Some hot water recirculation systems require the installation of check valves that could result in a closed system. Water pressure booster pumps might also have internal check valves that could create a closed system.

Note that this section identifies only "storage water heating equipment" as the generator of hot water. While this appears to imply that thermal expansion-induced pressure increases are only created by storage tank water heaters, this is not the case. A question that is frequently asked is, "Does a water distribution system with a tankless water heater require control of pressure increases caused by thermal expansion?" The answer is provided in the following two system examples:

System A: Water distribution system with a tankless, on-demand water heater (electric or gas) without a storage tank. In this system, water is heated only when a hot water outlet is open. Because the water distribution system is now an "open" system (i.e., a hot water outlet is open), any thermal expansion is relieved through the open hot water outlet and no pressure increase can occur. When the outlet is closed, the system becomes a closed system but, since the flow has ceased, the tankless water heater stops heating the water. Therefore, methods for controlling pressure increases due to thermal expansion are not required.

System B: Water distribution system with tankless, on demand water heater (electric or gas) with a storage tank. The water service line has a backflow preventer (such as a dual check valve at the water meter). In this system, an unfired hot water storage tank may be required to provide for hot water recirculation to lessen the "wait time" for hot water to arrive at any fixture. Or the storage tank might be required to provide for large simultaneous hot water demands (that the tankless water heater cannot produce without the storage). For either reason, as the hot water storage tank cools off, a circulation pump is switched on to cause flow out of the storage tank, through the tankless water heater and back into the storage tank until the water in the storage tank is at an acceptable temperature level. Because there will be times where no hot water outlet is opened during the time that the tankless water heater is heating water, the water distribution system is a closed system. Therefore, a means for controlling pressure increases caused by the thermal expansion must be provided.

The method for controlling thermal expansion is the installation of a thermal expansion tank (see Commentary Figure 607.3). Prior to the 2015 edition, the code was open to any method that could control an

**Figure 607.3
THERMAL EXPANSION CONTROL**

increase in pressure caused by thermal expansion. One method was a pressure relief valve in the cold water line to the water heater that wasted a small amount of water to a drain or to the outdoors when the system pressure exceeded a set value such as 75, 80 or 125 psi. Another method was a specially designed water closet (tank type) fill valve with a pressure relief valve that would waste water to the flush tank when the system pressure exceeded a set value such as 80 psi. There are several reasons for this change:

1. Any device that "wastes" potable water is not in the spirit of water conservation that the code and the plumbing industry are moving towards. Although the amount/frequency of the water waste may be insignificant when the relief devices are new, constant operation can wear out the device such that it leaks constantly. This might occur for years before the building owner, occupant or astute service provider notices the problem.

2. The relief devices are not always where one would expect them. The replacement of a water closet (having the special fill valve) would remove the relief device without anyone discovering the impact to the system. Or a relief valve located far away from the water heater could be plugged to stop "nuisance" leaking.

3. The code did not indicate a relief pressure setting for these relief devices. Because there is a pressure relief valve on the storage water heater (and hot water storage), some have (incorrectly) considered this as the thermal expansion relief device, not realizing that there is a prohibition in Section 504.4 for the water heater pressure relief valve to serve as a thermal expansion device.

Section 604.8 requires that the pressure in the water distribution system not exceed 80 psi (552 kPa). Where a thermal expansion tank is required to control thermal expansion-caused pressure increases, the pressure in the water distribution system must be set at less than 80 psi. Although a thermal expansion tank does "control" the increase in pressure caused by thermal expansion, the size of the thermal expansion tank determines how much the water distribution system pressure will increase after the thermal expansion event. This is because a thermal expansion tank has an air-pressurized bladder in the tank. As the water in the system expands, the water moves into the thermal expansion tank to compress the air bladder. The pressure in the air bladder increases and the system (water distribution system and the air bladder) will be at a new pressure greater than before the thermal expansion event. If the initial air bladder volume is very small, the water volume expanding into the thermal expansion tank will cause a significant increase in air bladder pressure and thus, the final system pressure will be significantly greater than it was before the thermal expansion event. If the initial air bladder volume is very large, that same water volume expanding into the thermal expansion tank will cause very little increase in air bladder pressure and thus, the final system pressure will increase very little. The question is not whether the pressure will increase (because it always will), but how much will it increase based on the size of the thermal expansion tank chosen. The goal is to not allow the water distribution system pressure to exceed 80 psi. Therefore, the initial system pressure needs to be less than 80 psi before the thermal expansion event occurs. Sizing of the thermal expansion tank must be in accordance with the manufacturer's instructions.

Is a thermal expansion tank required to be at the water heater? The thermal expansion tank is only required to be downstream of any check valves, pressure-reducing valves or backflow preventers before the cold water supply reaches the water heater. Section 606.1 requires shutoff valves before the inlet to a storage water heater. However, the thermal expansion tank is not required to be located after these shutoff valves. Although having the thermal expansion tank at the water heater is often convenient for servicing, it is assumed that if the shutoff valve is closed, it is to remove the water heater from service and the energy source to the heater is turned off. Some reasons for the thermal expansion tank not to be located at the water heater are: 1) no available space and 2) one thermal expansion tank serves a multiheater system.

The code section does not prohibit the use of methods other than a thermal expansion tank (such as pressure relief valves) to control thermal expansion. Perhaps a pressure relief valve of robust design (for frequent operation) is located in a mechanical room where its purpose is well known and any discharges are routed to a water recycling system. However, those other methods would require code official approval under Section 105.2.

There are situations where installing a thermal expansion tank for every storage water heater might not make sense. For example, consider a water distribution system that has an initial pressure of 40 psi. This is one way to conserve water in a building, provided that the water distribution piping system is designed such that at this pressure and at periods of peak demand, the minimum pressures for proper operation of the fixtures are provided. When a thermal expansion event occurs, the volume of the entire system (both cold and hot) might be large enough such that the thermal expansion results in a final system pressure of 80 psi or less. This intent of the code section is to prevent the water distribution system pressure from exceeding 80 psi. The intent is not to prevent pressure fluctuations as those will always occur during the normal operation of the system. Another example is a multiple heater system. This could be a number of water heaters in a mechanical room serving a building or a water heater in every tenant space (or apartment) of a building with many spaces (or apartments). If there is not a check valve, backflow preventer or pressure-reducing valve between the heaters and one main thermal expansion tank, and the tank is sized appropriately, then one tank can work just as well as many smaller tanks for each water heater. However, the arrangements in these two examples would require code official approval under Section 105.2.

One topic that was discussed during the code change hearings for the proposal that made the changes to this section was whether a thermal expansion-caused pressure increase would ever occur in a large building such as an apartment building or a multistory condominium building where each unit had a water heater (provided that there were not any check valves, backflow preventers or pressure-reducing valves between the building's water service entrance and all of the water heaters). Even though the total volume of water that could be heated is significant, not all of those heaters will be heating at the same time. There only needs to be a few leaky water outlets in the entire water distribution system to relieve the thermal expansion-caused pressure increase from several water heaters heating simultaneously. Perhaps the only time that all water heaters in the building could be heating at the same time would be after utility services (gas or electric) were turned back on after being cut off for days. Even if that rare event happened, chances are good that many water outlets in the building would be frequently opened during the heating of water (i.e., the system would not be closed). This is a topic that the design professional and the code official will need to work out under Section 105.2.

607.4 Flow of hot water to fixtures. Fixture fittings, faucets and diverters shall be installed and adjusted so that the flow of hot water from the fittings corresponds to the left-hand side of the fixture fitting.

Exception: Shower and tub/shower mixing valves conforming to ASSE 1016/ASME A112.1016/CSA B125.16 or ASME A112.18.1/CSA B125.1, where the flow of hot water corresponds to the markings on the device.

❖ One of the oldest expressions in plumbing is: "Hot on the left, cold on the right." The code mandates that "hot" correspond to the left side of the fixture fitting for safety reasons. It has become an accepted practice to equate the left side of the faucet with hot water. The intent is to protect an individual from potential scalding when turning on what is believed to be cold water and get scalded by hot water.

Some manufacturers design a fixture fitting that can be reversed inside the faucet if a plumber inadvertently (or intentionally in the case of back-to-back faucets) pipes the water supplies in reverse. This type of installation would be acceptable. The exception provides for single-handle shower valves that comply with ASSE 1016/ASME A112.18.16/CSA B125.16 and that have identifiable hot designations on the control valve. These devices operate by rotating a single handle or knob away from the "off" position to initiate water flow. The handle is moved farther in the same direction to increase the temperature of flowing water. Because the rotating operation is limited to one direction, the user cannot mistakenly activate water flow at the maximum hot-water setting without first rotating the handle or knob through the cold-water setting. Also, ASSE 1016/ASME A112.18.16/CSA B125.16 requires control valves to have settings that indicate clearly the direction or means of adjustment to change temperature.

[E] 607.5 Insulation of piping. For other than Group R2, R3 and R4 occupancies that are three stories or less in height above grade plane, piping to the inlet of a water heater and piping conveying water heated by a water heater shall be insulated in accordance with Section C404.4 of the *International Energy Conservation Code*. For Group R2, R3 and R4 occupancies that are three stories or less in height above grade plane, piping to the inlet of a water heater and piping conveying water heated by a water heater shall be insulated in accordance with Section R403.5.3 of the *International Energy Conservation Code*.

❖ The *International Energy Conservation Code* (IECC) has requirements for insulation of heated water piping because without insulation, these systems can waste a significant amount of energy and add to the cooling loads of buildings. This code section is somewhat complicated by the fact that the IECC is split into a commercial buildings part (IECC-CE) and a residential buildings part (IECC-RE). This code covers residential buildings that are classified by the IBC as Group R2, R3 or R4, however, where those buildings are "low-rise" buildings (i.e., three stories or less in height above grade plane), the IECC-RE requirements apply. For all other buildings covered by the IPC, the IECC-CE requirements apply.

This section is controlled by the IECC as indicated by the [E] at the beginning of the section number.

Why is a section on insulating heated water piping in the IPC? On most large projects, insulation installation is performed by an insulation contractor and those large projects often have insulation details on the drawings so that the plumbing designers/installers understand what allowances to make around the piping to accommodate the insulation thicknesses. But not all projects are large nor have such insulation details on the drawings. The plumbing designer and installer need to understand what the IECC's minimum insulation requirements are so allowances can be made for the insulation in accordance with the code (the IECC).

IECC Section C404.4

IECC Section C404.4 requires that hot water piping be insulated in accordance with Table C403.2.10. This table indicates the required insulation wall thicknesses and thermal conductivity rating, k, of the insulating material based upon the fluid temperature in the piping.

IECC Section C404.4 also requires that the inlet piping to a storage water heater or a hot water storage tank be insulated for not less than 8 feet of developed piping length or to the heat trap, whichever is less. See the commentary on heat traps in Section 503.1 for an understanding of heat traps for water heaters. The purpose of insulating the inlet piping is to reduce the "standby heat losses" because the heat travels a distance up the inlet pipe. If the inlet pipe was not insulated (because it is thought of as a cold water inlet pipe), during periods where hot water is not be used, the heat from the storage tank would radiate out from the inlet piping and be lost. The installation of a heat trap significantly reduces this heat loss. Note that if the heat trap is not integral to the water heater, then it is external to the tank and the piping to the heat trap must be insulated. Heat traps on water heaters are only required where the water heater is not part of a hot (or tempered) water recirculation system (See IECC Section 404.3).

One method for providing heated water to locations that are remote from the heated water source is to heat trace the hot water piping. Obviously, this piping needs to also be insulated. IECC Section C404.4 requires that the piping be insulated in accordance with Table C403.2.10 or the heat trace manufacturer's instructions.

For tubular-type insulation, insulation manufacturers' instructions require that all seams and joints in the tube be sealed. The insulation must be continuous around any corners (elbows and tees). Gaps and openings in the piping insulation system result in poor insulation performance.

Pipe insulation must be continuous except where the piping passes through a framing member. The hot water piping must be completely covered by the required insulation. The tubular pipe insulation manufacturer's instructions imply this but for inspection purposes this obvious requirement is stated. However, note the phrase "except where the piping passes through a framing member." Although the code requires continuous piping insulation, there could be significant structural ramifications if this allowance was not made. Consider a $3/4$-inch pipe ($7/8$-inch actual outside diameter) with $1/2$-inch wall tubular insulation. The outside diameter of the assembly is about 2 inches. Requiring 2-inch-diameter holes in structural members creates a lot of extra work and might not be structurally possible without making the structural member deeper or thicker. The code doesn't intend for insulation thickness to be driving the design of structural members. However, if an installer wants to install a hole large enough for the insulated pipe and the hole doesn't violate any structural member cutting, notching or boring limitation, that is not a violation of this code section. Although the larger hole might be more effort, routing a continuous-insulated piping assembly though the member might be easier in the long run (noting that many factory-insulated piping products are available).

The piping insulation required by the section is not associated with insulation requirements that might be needed for the protection of the piping from freezing temperatures (see Section 305.4). This section is only concerned with the loss of heat from the piping to "within the building thermal envelope." Although routing of hot water piping in the exterior walls of a building is not commonly done in large buildings but on smaller buildings it is sometimes installed in this manner. Thus, the heated water piping is not completely "within the building thermal envelope" and is not totally "protected" from freezing.

External surface temperatures of the insulation of hot water piping could be an issue in some locations where personnel could touch the pipe insulation. The insulation thicknesses in Table C403.2.10 do increase for higher fluid temperatures in the piping, however, it is unknown if those values ensure an insulation surface temperature that is below what is needed for personnel protection.

IECC Section C404.4 has a list of exceptions for where piping insulation is **not** required to be installed. Many of these are common sense.

1. Insulation is not required from the connection at the end of the fixture supply pipe to the fixture of the plumbing appliance. The end of the fixture supply pipe is where the shutoff valve for the fixture or appliance is attached. It is too cumbersome to accomplish this and would most likely be discarded the first time a repair on the fixture was needed. Also, some fixtures have exposed and visible supply piping (e.g., pedestal lavatories) and certainly the occupant would not appreciate what might be the best attempt at insulating these connections. Note that this allowance for not installing insulation does not have anything to do with the requirements for supply piping "covers" where the fixture is intended for use by persons with disabilities (see Section 404).

2. Insulation is not required to be installed on valves, pumps, strainers and threaded unions in heated water piping that is 1 inch and smaller. These small components are difficult to insulate. There would be minimal energy benefit gained.

3. Insulation is not required to be installed on the piping from shower and bath "user controlled" mixing valves to the water outlets. This piping is of short length and the water is warm for only a short period of time. There would be minimal energy benefit gained.

4. Insulation is not required to be installed on the cold water piping that receives heated water as part of a water recirculation system that does not have a dedicated return pipe to the water heater. See Section 607.2.1.2 for an understanding of "no dedicated return pipe" recirculation systems. Circulating hot water back to the hot water source through a cold water supply pipe could cause someone to consider the cold water supply pipe as conveying heated water. That is not the intent of this section. The water in the cold water supply pipe will only be warm for a short distance away from the fixture. There is no energy savings by insulating the pipe, as the water in that pipe is intended to be used as cold water.

5. Insulation is not required to be installed on tubing from hot drinking-water heating units to the water outlet. This piping is of short length and the water in the tubing is hot for a short period of time. There would be minimal energy benefit gained.

6. Insulation is not required to be installed on piping at locations where a vertical support of the piping is installed. Vertical support of water piping requires that the support grip the pipe to be able to carry the weight of the pipe. Insulating the pipe through this type of support is impossible.

   For horizontal piping, the "hangers," "clamps," "hooks" or "straps" need to be large enough to accommodate the outside diameter of the insulation. This brings up the topic of saddles at horizontal support points. The insulation manufacturer must be consulted to determine if saddles are needed and how to accomplish support with the saddle.

7. Insulation is not required for heated water piping that is surrounded by insulation having a thermal resistance (R-value) of not less than R-3. There could be piping arrangements in buildings where the heated water piping is routed through an area that will have "building thermal envelope"-type insulation installed. In other words, insulation material that has published R-values and that is intended to be used in "flat" assemblies such as walls, floors and ceilings. Some common insulation materials are mineral wool, fiberglass and spray foams where an installed thickness provides for a specific insulation R-value. The exception does not necessarily mean that the insulation location is what actually defines the building thermal envelope (although it could be in that location). R-value-rated insulation can be used in other building locations that are within the space that the building thermal envelope surrounds. For example, insulation installed in walls for the reduction of noise or for the separation of conditioned spaces from semiconditioned spaces. This exception allows for R-value-rated insulation to do "double duty" by serving as the insulation for hot water piping and the insulation for the flat area.

   Does this exception mean that R-value insulation can be used to "wrap a pipe" and be the only function that the insulation serves? As stated previously, R-value insulation is intended to be used in flat assemblies such as walls, floors and ceilings. Installing such insulation around a pipe at the minimum thickness that provides R-3 will not provide for the insulating effect that Table C403.2.10 intends. A circular profile is different than a flat surface. Could "overinsulating" with R-value insulation (where "wrapping a pipe" is performed) provide for the intended insulation effect? That might be possible; however, it would need to be approved by the code official under the alternate methods and materials section (Section 105.2) of the code. One example of this might be the use of spray foam to insulate hot water piping in a trench. Although extra insulation coverage (thickness) for an R-value greater than R-3 is planned for around the pipe, does that meet or exceed the insulating effect that Table C403.2.10 intends? Only a design professional could determine this.

For piping conveying fluids in the temperature range of 104 °F to 140 °F, Table C403.2.10 indicates a tubular insulation wall thickness of 1 inch (25 mm). Larger pipe sizes require greater wall thickness [unless Note "a" is used to reduce wall thickness back to 1 inch (25 mm)]. How can "burying" a hot water pipe in R-value rated wall (ceiling or floor) insulation "equate" to the required 1-inch (25 mm) thickness of tubular insulation on the pipe? One possible explanation is that in most cases, there is significantly more (thickness) of R-value insulation to the sides and "back" of the pipe as compared to the "front" side of the pipe where the minimum thickness might be in order to "just attain" an R-3 coverage. The average of the heat losses around the pipe is thought to be somewhat "equivalent" in the temperature range under consideration.

The code does not provide any details for the use of R-rated materials in order for this exception to be used. However, the discussion in the previous two paragraphs does provide a few general concepts that might be used to develop criteria where the exception can be used. For example, is it acceptable to strap a $1^1/_2$-inch diameter (38 mm) hot water pipe directly to a $3^1/_2$-inch-deep (89 mm) stud and place 1 inch (25 mm) (R-3) of fiberglass batt insulation behind and in front of the pipe? Or would it be better to route the pipe in the middle of the stud space so that there are many more inches of insulation on either side of the pipe? The designer, installer and code official will have to work out the answers to such questions.

IECC Section R403.5.3

IECC Section R403.5.3 recognizes two ways to insulate heated water piping: 1) put the piping in insulation having an R-value rating or 2) apply tubular pipe insulation having a thermal conductivity rating, k. Or in more simple terms, bury the pipe in "building thermal envelope" insulation (mineral wool, fiberglass batt, spray foam, etc.) or put insulation tubing (rubber, polyethylene, mineral wool, fiberglass, etc.) on the piping. Where R-value insulation is used, the installed R-value completely around the pipe cannot be less than R-3. Where tubular insulation is used, the wall thickness of tube cannot be less than $^1/_2$ inch (12.7 mm). The thermal conductivity rating, k, of the tube material cannot be greater than 0.31 Btu per inch/h $\times$ ft^2 $\times$ °F [0.45 W/(m $\times$ K)] at 200 °F (93.3 °C). Thermal conductivity, k, of a material is the ability of the material to transfer thermal energy. The lower the k value, the better the material is for insulating (keeping the heat in the pipe). The k $\leq$ 0.31 covers most of the commonly available materials.

Why is not tubular insulation "R-value rated"? Because the R-value is different for different pipe sizes and insulation wall thicknesses. There is a mathematical conversion that can approximate the R-value for a particular size of tubular insulation; however, it is only an estimate that is not verified by testing in accordance with a standard for such products. Tubular insulation is specified by wall thickness and k value.

The following example will illustrate the intent of IECC Section R403.5.3. A hot water pipe is routed from the water heater in the middle of an unheated basement, up through an exterior wall to a tub/shower fixture located on the second floor. The mixing valve control for the tub/shower is not located on the exterior wall but is located on an interior wall between the bathroom and a bedroom. Starting at the hot water outlet of the water heater, the piping runs straight up into the building thermal envelope insulation on the underside of the first floor. The piping between the water heater and underside of the floor above requires tubular insulation. Once the piping is in the insulation on the underside of the floor, the tubular insulation is stopped and the piping is "nested" into the fiberglass batt insulation so that the piping has not less than R-3 around the piping. Generally, standard fiberglass batt insulation has an R-value of about R-3 for 1-inch (25 mm) thickness. (High-performance fiberglass batt insulation has a slightly higher R-value per inch.) Proper nesting of the piping in the insulation will require that the piping installer coordinate with the insulation installer. After the piping reaches the exterior wall of the building and turns up through the sill plate, the exterior wall will be insulated with spray foam (closed cell) having an R-value of about R-6 per inch of thickness. As only R-3 is required for the piping insulation, the piping installer needs to locate the piping so that not less than $^1/_2$ inch (12.7 m) of space is available all around the pipe as the pipe is routed up through the exterior wall. Again, the piping installer will need to coordinate with the insulation installer to understand how to locate and keep the piping in place during the spraying process so that the required thickness of insulation is maintained around the piping. Once the piping reaches the joist or truss space above the first floor, the piping exits the exterior wall and runs through the joist or truss area, over to and up into the interior wall to the mixing valve. Because those areas do not have building thermal insulation installed, the piping is insulated with tubular insulation.

IECC Section R403.5.3 has a few "rules and allowances" for the installation of insulation. The overall requirement is that the insulation be installed in accordance with the manufacturer's instructions.

For batt-type insulation, such instructions require that piping be "nested" into the insulation. All too commonly, batt insulation is installed with the entire batt thickness "behind" piping, cables and conduits in walls and ceilings. Compressing the insulation thickness to less than the intended installed thickness reduces the insulating value of the insulation product. And certainly, if the hot water piping is to take advantage of the insulation qualities of the building thermal envelope insulation, the piping needs to be carefully nested into the insulation and not have batt material packed in tightly around the piping. Compressing batt insulation reduces its insulating value. Lack of coordination between the insulation installer and the piping installer is the primary cause of poor piping insulation practices.

For tubular-type insulation, insulation manufacturers' instructions require that all seams and joints in the tube be sealed, and that the insulation be continuous around any corners (elbows and tees). All too commonly, length-way slit tubular insulation is installed on piping without sealing the slit or end to end joints. Insulation of elbows and tees is often poorly accomplished or not done at all. Gaps and openings in the piping insulation system result in poor insulation performance.

This leads into the next requirement of IECC Section R403.5.3, which is that pipe insulation must be continuous except where the piping passes through a framing member. The hot water piping must be completely covered by the required insulation. Even though the tubular pipe insulation manufacturer's instructions imply this, the "building thermal envelope" manufacturer's instructions do not directly address their product being used as piping insulation. Therefore, this is why this obvious requirement is stated. However, note the phrase "except where the piping passes through a framing member." Although the code requires continuous piping insulation, there could be significant structural ramifications if this allowance was not made. Consider a $^3/_4$-inch (19.1 m) pipe $^7/_8$-inch actual outside diameter) with $^1/_2$-inch (12.7 m) wall tubular insulation. The outside diameter of the assembly is about 2 inches (51 mm). Requiring 2-inch-diameter (51 mm) holes in structural members creates a lot of extra work and might not be structurally possible without making the structural member deeper or thicker. The code doesn't intend for insulation thickness to be driving the design of structural members. However, if an installer wants to install a hole large enough for the insulated pipe and the hole doesn't violate any structural member cutting, notching or boring limitation, that is not a violation of this code section. Although the larger hole might be more effort, routing a continuous-insulated piping assembly through the member might be easier in the long run (noting that many factory-insulated piping products are available).

IECC Section R403.5.3 also warns the reader that the piping insulation required by the section is not, in any way, associated with insulation requirements that might be needed for the protection of the piping from freezing temperatures (see Section 305.4). This section is only concerned with the loss of heat from the piping to "within the building thermal envelope." A routing of hot water piping in exterior walls is not completely "within the building thermal envelope." However, it is an accommodation because the residential building industry has been doing this "successfully" for decades. Note that this requirement for hot water piping insulation could work *against* the efforts for keeping piping routed in exterior walls from freezing. The more insulation between the "heat" of the interior of the building and the cold on the exterior of the building will make the piping colder (see commentary, Section 305.4).

IECC Section R403.5.3 also warns the reader about external surface temperatures of the insulation of hot water piping. This probably will not be an issue in a residential environment because hot water piping temperatures would rarely be great enough such that the temperature on the exterior of the insulation would be so extreme that someone could be burned by touching it. However, with the resurgence of solar thermal heating systems and the potentially high temperatures that those systems can generate, the insulation requirements of this section are intended to keep the temperature on the outside of the insulation at a level where people could not be burned.

IECC Section R403.5.3 has a list of exceptions for where piping insulation is **not** required to be installed. Many of these are common sense.

1. Insulation is not required on flexible connectors or reduced-size fixture supply tubing from the connection at the end of the fixture supply piping to a fixture fitting. It is too cumbersome to accomplish this and would most likely be discarded the first time a repair on the fixture was needed. Also, some fixtures have exposed and visible supply piping (e.g., pedestal lavatories) and certainly the occupant would not appreciate such attempts for insulating. Note that this allowance for not installing insulation does not have anything to do with the requirements for supply piping "covers" where the fixture is intended to be a fixture for use by persons with disabilities (see Section 404).

2. Insulation is not required to be installed on valves, pumps and threaded unions in heated water piping. In a residential environment,

these components are relatively small, making them difficult to insulate. There would be minimal energy benefit gained.

3. Insulation is not required to be installed on the piping from shower and bath mixing valves to the water outlets. This piping is of short length, and the water is warm for only a short period of time. There would be minimal energy benefit gained.

4. Insulation is not required to be installed on the cold water piping that receives heated water as part of a water recirculation system that does not have a dedicated return pipe to the water heater. See Section 607.2.1.2 for an understanding of "no dedicated return pipe" recirculation systems. Circulating hot water back to the hot water source through a cold water supply pipe could cause someone to consider the cold water supply pipe as conveying heated water. That is not the intent of this section. The water in the cold water supply pipe will only be warm for a short distance away from the fixture. There is no energy savings by insulating the pipe, as the water in that pipe is intended to be used as cold water.

5. Insulation is not required to be installed on tubing from hot drinking-water heating units to the water outlet. This piping is of short length and the water in the tubing is hot for a short period of time. There would be minimal energy benefit gained.

6. Insulation is not required to be installed on piping at locations where a vertical support of the piping is installed. Vertical support of water piping requires that the support grip the pipe to be able support the weight of the pipe. Insulating the pipe through this type of support is impossible. However, recognize that this exception applies only to vertical support of piping. It is intended that piping insulation be continuous "through" horizontal piping supports. This point deserves further explanation.

Where batt insulation is used to insulate horizontal piping, insulating the back side of the support (and over the pipe) with the required thickness of insulation meets the requirement of the insulation being continuous through the horizontal piping support. For example, consider hot water piping routed in a ceiling joist space (parallel to the joists) in an attic. A 2 x 4 block is positioned across the joist space at a point about midway of the depth of the joists. The piping is laid on top of the 2 x 4 blocking. The insulator cuts open the insulation batt, places a portion of the batt thickness under the 2 x 4 and places the remainder of the batt thickness on top of the piping. As long as the required insulation thickness is below the blocking and on top of the piping, the insulation requirement is met.

The same concept applies to blown-in or sprayed-in insulation. Again, coordination between the insulator and the piping installer is paramount to making this happen with minimal difficulty.

Where tubular insulation is used for horizontal piping, the "hangers" clamps," "hooks" or "straps" need to be large enough to accommodate the outside diameter of the insulation. This brings up the topic of saddles at horizontal support points. The insulation manufacturer must be consulted to determine if saddles are needed and how to accomplish support with the saddle.

7. Insulation is not required to be installed on piping or tubing from a tankless water heater serving only one fixture. The intent of this exception is to cover the small tankless water heater that is surface mounted just below a lavatory. Similar to item 1, this application does not require insulation on the hot water tubing.

SECTION 608
PROTECTION OF POTABLE WATER SUPPLY

608.1 General. A potable water supply system shall be designed, installed and maintained in such a manner so as to prevent contamination from nonpotable liquids, solids or gases being introduced into the potable water supply through cross connections or any other piping connections to the system. Backflow preventer applications shall conform to Table 608.1, except as specifically stated in Sections 608.2 through 608.16.10.

❖ This section contains the requirements for protecting the potable water supply from contamination. The most important aspect of a plumbing code is the protection of the potable water system. History is filled with local and widespread occurrences of sickness and disease caused by not safeguarding the water supply. It is imperative that the potable water supply be maintained in a safe-for-drinking condition at all times and at all actual or potential connections and outlets.

To understand the importance of protecting the potable water supply, identifying potential threats and determining the appropriate backflow prevention method, an understanding of how backflow prevention can occur is necessary.

Backflow depends on atmospheric pressure, water pressures and variations in the water supply at any moment. The following is a discussion on pressure and backflow.

Pressure

Atmospheric pressure (psi): This is pressure exerted by the weight of the atmosphere above the earth. Atmospheric pressure at sea level is 14.7 psi (101 kPa), however, it varies with altitude [see Commentary Figure 608.1(1)].

WATER SUPPLY AND DISTRIBUTION

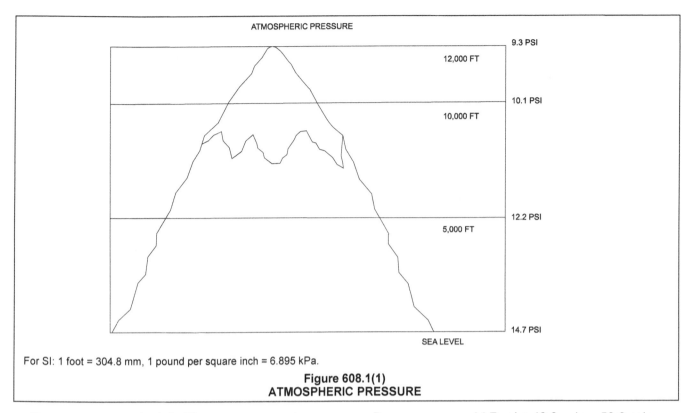

For SI: 1 foot = 304.8 mm, 1 pound per square inch = 6.895 kPa.

**Figure 608.1(1)
ATMOSPHERIC PRESSURE**

Gauge pressure (psig): The pressure read on a gauge. The reference point for the gauge is atmospheric pressure equal to zero. In other words, gauge pressure is the amount of pressure above or below atmospheric pressure.

Absolute pressure (psia): The sum of atmospheric pressure and gauge pressure. For example, if a gauge reads 20 psig (138 kPa), the absolute pressure would be determined as follows:

$P_{ABSOLUTE}$ = 14.7 psi (atmospheric) + 20 psi (gauge reading)

$P_{ABSOLUTE}$ = 34.7 psia

For SI: 1 pound per square inch = 6.895 kPa.

Water pressure: This is a function of the weight and height of a column of water. A cubic foot of water [see Commentary Figure 608.1(2)] weighs 62.4 pounds (28.3 kg); the pressure exerted at the base of the cube is 62.4 pounds per square foot (psf) (0.433 lb/in.^{2}). Therefore, the height of a column of water determines the pressure that is exerted at its base [see Commentary Figure 608.1(3)].

The diameter of the pipe or vessel has no bearing on this pressure. For example, a column of water 100 feet (30 480 mm) high and 20 feet (6096 mm) in diameter has the same pressure at its base as a 100-foot-high (30 480 mm) column of water, 1 inch (25 mm) in diameter [see Commentary Figure 608.1(4)]. Columns of water with a free surface (exposed to atmosphere) will have atmospheric pressure in addition to pressure caused by the weight of water exerted at its base [see Commentary Figure 608.1(5)].

Base pressure = 14.7 psi + 43.3 psig = 58.0 psia

Water pressure is sometimes referred to as "head pressure," "head" or "feet of head."

1 foot of head = 0.433 psig
100 feet of head = 43.3 psig

For SI: 1 pound per square inch = 6.895 kPa.

Backflow

Backflow is the flow of liquids in potable water distribution piping in reverse of their intended path. There are two types of pressure conditions that cause backflow: backsiphonage and backpressure.

Backsiphonage: The backflow of water caused by system pressure falling below atmospheric pressure. Atmosphere supplies the force that reverses flow. Any gauge reading of pressure below atmospheric pressure will be negative. Commentary Figures 608.1(6) and 608.1(7) illustrate the effect atmospheric pressure has on water. An open tube is inserted into a container of water in Commentary Figure 608.1(6). Atmospheric pressure acts equally on the surface of the water within the tube and on the outside of the tube.

If, as shown in Commentary Figure 608.1(7), the tube is sealed and a vacuum pump is used to evacuate all the air from the tube, a vacuum with a pressure of 0 psia (0 kPa) is created within the tube.

Because liquids will flow toward the point of lowest pressure, water will rise to 33.9 feet (10 333 mm). In other words, the weight of atmosphere at sea level balances the weight of a column of water 33.9 feet (10 333 mm) in height.

Figure 608.1(2)
PRESSURE EXERTED BY 1 CUBIC FOOT OF WATER

Backsiphonage occurs in the water distribution system when supply pressure falls below atmospheric pressure. The following are practical examples of conditions that can cause backsiphonage.

A simple siphon occurs when pressure is reduced by a difference in water levels at two separate points within a continuous fluid system [see Commentary Figure 608.1(8)].

Aspiration or venturi principle is achieved when reduced pressure is created within a fluid system as a result of fluid motion. Intentional or unintentional constriction of flow through an opening or pipe increases velocity and decreases pressure. If this point of reduction is linked to a source of pollution, backsiphonage of the pollutant can occur [see Commentary Figure 608.1(9)].

Supply pressure can be reduced at the pump intake or suction side. Insufficient water pressure or undersized piping supplying the pump can cause a negative pressure in the system, allowing contaminated water to be drawn into the system [see Commentary Figure 608.1(10)].

Backpressure: Pressure created in a nonpotable system in excess of the water supply mains causing backflow. Backpressure can be created by mechanical means, such as a pump; by static head pressure including an elevated tank or by thermal expansion from a heat source, such as a water heater.

TABLE 608.1. See page 6-83.

❖ Where the code does not specifically address backflow prevention methods or devices for a specific application in Sections 608.2 through 608.16.10, this table must be used to select a suitable method or device for the application. To use the table, first identify whether the potential hazard substance that might enter the potable water system is a low or high hazard to the potable water supply. Note a directs the reader to the Chapter 2 definitions of "Pollution" and "Contamination" for determining whether the hazard is low or high, respectively. Once the hazard is selected, this should eliminate some entries in the table as being viable for the application. Next, determine if the application will be subject to backpressure conditions. If so, this will eliminate additional table entries as being viable. (If not, then no table entries will be eliminated because all methods and devices are suitable for backsiphonage conditions.) Of the remaining entries, the code user must then select a method or device

WATER SUPPLY AND DISTRIBUTION

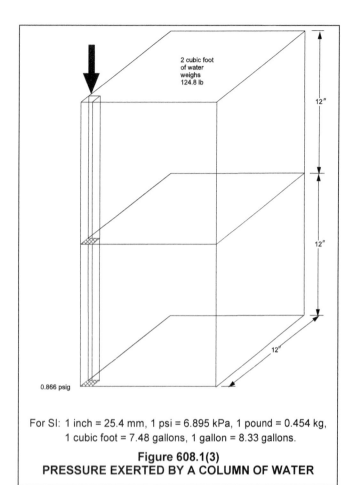

For SI: 1 inch = 25.4 mm, 1 psi = 6.895 kPa, 1 pound = 0.454 kg, 1 cubic foot = 7.48 gallons, 1 gallon = 8.33 gallons.

**Figure 608.1(3)
PRESSURE EXERTED BY A COLUMN OF WATER**

BASE PRESSURE = 14.7 psi + 43.3 psig
= 58.0 psia

WATER PRESSURE IS SOMETIMES REFERRED TO AS HEAD PRESSURE, HEAD, OR FEET OF HEAD.

1 FOOT OF HEAD = 0.433 psig
100 FEET OF HEAD = 43.3 psig

For SI: 1 inch = 25.4 mm, 1 pound per square inch = 6.896 kPa, 1 pound = 0.454 kg.

**Figure 608.1(5)
COLUMN OF WATER EXPOSED TO ATMOSPHERE**

BASE PRESSURE = 43.3 psi   BASE PRESSURE = 43.3 psi

PIPE DIAMETER DOES NOT INFLUENCE THE PRESSURE AT THE BASE OF A COLUMN

For SI: 1 inch = 25.4 mm, 1 pound per square inch = 6.896 kPa, 1 pound = 0.454 kg.

**Figure 608.1(4)
BASE PRESSURE**

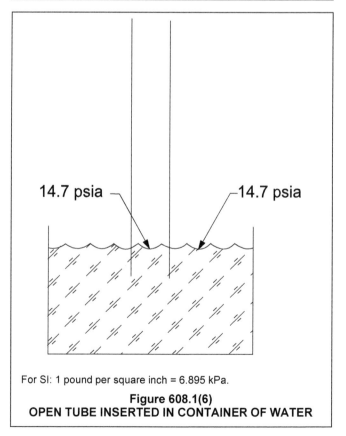

For SI: 1 pound per square inch = 6.895 kPa.

**Figure 608.1(6)
OPEN TUBE INSERTED IN CONTAINER OF WATER**

WATER SUPPLY AND DISTRIBUTION

TABLE 608.1
APPLICATION OF BACKFLOW PREVENTERS

DEVICE	DEGREE OF HAZARD[a]	APPLICATION[b]	APPLICABLE STANDARDS
Backflow prevention assemblies:			
Double check backflow prevention assembly and double check fire protection backflow prevention assembly	Low hazard	Backpressure or backsiphonage Sizes $3/8''$–$16''$	ASSE 1015, AWWA C510, CSA B64.5, CSA B64.5.1
Double check detector fire protection backflow prevention assemblies	Low hazard	Backpressure or backsiphonage Sizes $2''$–$16''$	ASSE 1048
Pressure vacuum breaker assembly	High or low hazard	Backsiphonage only Sizes $1/2''$–$2''$	ASSE 1020, CSA B64.1.2
Reduced pressure principle backflow prevention assembly and reduced pressure principle fire protection backflow assembly	High or low hazard	Backpressure or backsiphonage Sizes $3/8''$–$16''$	ASSE 1013, AWWA C511, CSA B64.4, CSA B64.4.1
Reduced pressure detector fire protection backflow prevention assemblies	High or low hazard	Backsiphonage or backpressure (Fire sprinkler systems)	ASSE 1047
Spill-resistant vacuum breaker assembly	High or low hazard	Backsiphonage only Sizes $1/4''$–$2''$	ASSE 1056
Backflow preventer plumbing devices:			
Antisiphon-type fill valves for gravity water closet flush tanks	High hazard	Backsiphonage only	ASSE 1002, CSA B125.3
Backflow preventer for carbonated beverage machines	Low hazard	Backpressure or backsiphonage Sizes $1/4''$–$3/8''$	ASSE 1022
Backflow preventer with intermediate atmospheric vents	Low hazard	Backpressure or backsiphonage Sizes $1/4''$–$3/4''$	ASSE 1012, CSA B64.3
Dual-check-valve-type backflow preventer	Low hazard	Backpressure or backsiphonage Sizes $1/4''$–$1''$	ASSE 1024, CSA B64.6
Hose connection backflow preventer	High or low hazard	Low head backpressure, rated working pressure, backpressure or backsiphonage Sizes $1/2''$–$1''$	ASME A112.21.3, ASSE 1052, CSA B64.2.1.1
Hose connection vacuum breaker	High or low hazard	Low head backpressure or backsiphonage Sizes $1/2''$, $3/4''$, $1''$	ASME A112.21.3, ASSE 1011, CSA B64.2, CSA B64.2.1
Laboratory faucet backflow preventer	High or low hazard	Low head backpressure and backsiphonage	ASSE 1035, CSA B64.7
Pipe-applied atmospheric-type vacuum breaker	High or low hazard	Backsiphonage ony Sizes $1/4''$–$4''$	ASSE 1001, CSA B64.1.1
Vacuum breaker wall hydrants, frost-resistant, automatic-draining-type	High or low hazard	Low head backpressure or backsiphonage Sizes $3/4''$, $1''$	ASME A112.21.3, ASSE 1019, CSA B64.2.2
Other means or methods:			
Air gap	High or low hazard	Backsiphonage or backpressure	ASME A112.1.2
Air gap fittings for use with plumbing fixtures, appliances and appurtenances	High or low hazard	Backsiphonage or backpressure	ASME A112.1.3
Barometric loop	High or low hazard	Backsiphonage only	(See Section 608.13.4)

For SI: 1 inch = 25.4 mm.

a. Low hazard—See Pollution (Section 202).
   High hazard—See Contamination (Section 202).
b. See Backpressure, low head (Section 202).
   See Backsiphonage (Section 202).

based upon the specific application for which the description fits. For example, a backflow preventer for a carbonated beverage machine would be an inappropriate selection where a 4-inch (102 mm) backflow prevention device is needed for the application. Of the final list of viable entries, the best selection is typically the one that is most economical to purchase and install.

For SI: 1 foot = 304.8 mm, 1 pound per square inch = 6.895 kPa, 1 pound = 0.454 kg.

Figure 608.1(7)
VACUUM APPLIED TO A SEALED TUBE INSERTED IN A CONTAINER OF WATER

Figure 608.1(8)
EXAMPLE OF SIMPLE SIPHON

For SI: 1 pound per square inch = 6.895 kPa.

Figure 608.1(9)
EXAMPLE OF ASPIRATION OR VENTURI PRINCIPLE

Figure 608.1(10)
EXAMPLE OF PUMP CAUSING NEGATIVE PRESSURE

608.2 Plumbing fixtures. The supply lines and fittings for plumbing fixtures shall be installed so as to prevent backflow. Plumbing fixture fittings shall provide backflow protection in accordance with ASME A112.18.1/CSA B125.1.

❖ To prevent the potable water supply system from being contaminated, plumbing fixtures must be installed in a manner that will prevent backflow. Backflow devices must be in accordance with ASME A112.18.1/CSA B125.1.

608.3 Devices, appurtenances, appliances and apparatus. Devices, appurtenances, appliances and apparatus intended to serve some special function, such as sterilization, distillation, processing, cooling, or storage of ice or foods, and that connect to the water supply system, shall be provided with protection against backflow and contamination of the water supply system. Water pumps, filters, softeners, tanks and other appliances and devices that handle or treat potable water shall be protected against contamination.

❖ This section requires backflow protection for fixtures, appliances and equipment that serve some special function and are connected to the potable water supply system. The specialty equipment being regulated is typically associated with sanitation, food service and health care facilities. This section combines the intents of Sections 608, 609 and 802.

Generally, one backflow prevention device is required for each fixture, appliance and appurtenance. The water downstream of a backflow prevention device is considered to be potable water until a backflow event occurs to cause the water to be nonpotable. Where a single fixture, piece of equipment or individual appliance is connected downstream of the backflow preventer, a malfunction of that appurtenance that causes the water to become nonpotable only affects the appurtenance causing the contamination. If multiple fixtures, appliances or appliances are supplied with water from a single backflow prevention device, the water contamination caused by one fixture could cause the other fixture to be supplied with nonpotable water. Thus, a backflow protection device is required for each item to prevent "cross contamination" between items.

Only where all of the connected fixtures do not require potable water could one backflow preventer serve all of those items. One example is a warehouse having multiple standard frostproof yard hydrants to supply water for floor and lot cleaning. Because a standard frostproof yard hydrant has a stop-and-waste valve below grade, the potable water supply to the hydrant must be protected from possible contamination from any one of those yard hydrants. One backflow preventer could serve this multiple hydrant system as long as the piping identification requirements of Section 608.8 are met (noting that the riser of the yard hydrant is part of the piping). Another example might be a series of machines requiring water (but not necessarily potable water) for cooling purposes. One backflow preventer could supply water for this equipment line. Again, note the piping identification requirements of Section 608.8.

608.3.1 Special equipment, water supply protection. The water supply for hospital fixtures shall be protected against backflow with a reduced pressure principle backflow prevention assembly, an atmospheric or spill-resistant vacuum breaker assembly, or an *air gap*. Vacuum breakers for bedpan washer hoses shall not be located less than 5 feet (1524 mm) above the floor. Vacuum breakers for hose connections in health care or laboratory areas shall not be less than 6 feet (1829 mm) above the floor.

❖ This section establishes the minimum level of protection required for the water supply to various health care fixtures. In health care facilities, the risk of contamination to the potable water supply is greater because of the biological hazards present in many of the procedures conducted. Therefore, to reduce the potential for contamination, procedures or processes that use a hose must have their vacuum breakers

WATER SUPPLY AND DISTRIBUTION

installed at an elevation higher than that where the procedure is taking place.

608.4 Water service piping. Water service piping shall be protected in accordance with Sections 603.2 and 603.2.1.

❖ To reduce the risk of contaminating the potable water supply, water service piping must be installed in accordance with Sections 603.2 and 603.2.1.

608.5 Chemicals and other substances. Chemicals and other substances that produce either toxic conditions, taste, odor or discoloration in a potable water system shall not be introduced into, or utilized in, such systems.

❖ The prohibition of chemicals, regardless of hazard classification, includes the use of substances that clean the inside of the piping system or that de-lime various components, all of which could affect the potability of the water and, therefore, must never be added to the system. The generalization can be made that nothing should be introduced into a potable water supply after it leaves the water treatment facility.

608.6 Cross connection control. Cross connections shall be prohibited, except where *approved* backflow prevention assemblies, backflow prevention devices or other means or methods are installed to protect the potable water supply.

❖ A cross connection is where a system, other than a potable water system, is connected to or has the potential of being connected to a potable water system. If the cross connection is established and certain pressure conditions occur, pollutants or contaminants will enter the potable water supply. Systems other than potable water systems include nonpotable water, industrial fluid, gas or other substance-conveying systems. A cross connection could be established as the result of any of the following:

- Bypass arrangements;
- Jumper connections;
- Removable sections;
- Swivel or changeover devices;
- Direct connections and
- Submerged inlets.

Where it is necessary to establish a cross connection, the potable water supply must be protected by an approved means of backflow prevention assemblies, backflow prevention devices or other means and methods (see commentary, Section 608.1).

608.6.1 Private water supplies. Cross connections between a private water supply and a potable public supply shall be prohibited.

❖ A private water supply (auxiliary water supply) is a supply or source not under control or management of a public water purveyor. These sources typically include wells, springs and surface supply sources, such as lakes and reservoirs. These sources are subject to contamination by industrial or animal wastes, or other natural or man-made materials (see Commentary Figure 608.6.1).

608.7 Valves and outlets prohibited below grade. Potable water outlets and combination stop-and-waste valves shall not be installed underground or below grade. Freezeproof yard hydrants that drain the riser into the ground are considered to be stop-and-waste valves.

Exception: Freezeproof yard hydrants that drain the riser into the ground shall be permitted to be installed, provided that the potable water supply to such hydrants is protected upstream of the hydrants in accordance with Section 608 and the hydrants are permanently identified as nonpotable outlets by *approved* signage that reads as follows: "Caution, Nonpotable Water. Do Not Drink."

❖ Any fixture or device that incorporates a stop-and-waste feature is prohibited if the waste opening is underground or in any location where waste water or water-borne contaminants could enter the device or water supply from the ground or other source by reversal of flow.

Many water utilities routinely use stop-and-waste valves on curb stops and fire hydrants. Another common location for stop-and-waste valves is in yard hydrants equipped with a below-grade self-draining weep hole [see Commentary Figures 608.7(1) and 608.7(2)].

Sanitary yard hydrants are available that protect against ground-water contamination by eliminating

Figure 608.6.1
CONNECTION BETWEEN TWO SYSTEMS

WATER SUPPLY AND DISTRIBUTION

Figure 608.7(1)
YARD HYDRANT WITH STOP AND WASTE VALVE

the waste opening below grade. These devices use a sealed reservoir to collect water draining from the hydrant head and riser and empty the reservoir upon opening of the valve the next time the hydrant is in use.

The exception allows for installations of frostproof yard hydrants (having a combination stop-and-waste valve below grade) for applications where nonpotable water is acceptable for the use intended. Examples of nonpotable water use applications are hydrants used for farm area wash down, public park area cleaning and garbage area cleanup. Yard hydrants having stop and waste valves below grade must have backflow protection for the potable water supply to the hydrant, as specified in Section 608. In addition, the hydrants must be marked permanently with a sign having the code-specified words. The sign must meet the approval of the code official.

608.8 Identification of nonpotable water systems. Where nonpotable water systems are installed, the piping conveying the nonpotable water shall be identified either by color marking, metal tags or tape in accordance with Sections 608.8.1 through 608.8.2.3.

❖ Where nonpotable water supply systems are installed in a building, the piping of the nonpotable system must be identified so as to prevent advertent cross connection between different systems, especially the potable water system, during repairs or renovations. For example, a building could be provided with a recycled gray-water supply system that supplies nonpotable water for flushing urinals and water closets. Without identification of the recycled gray-water piping, the piping system might be assumed to convey potable

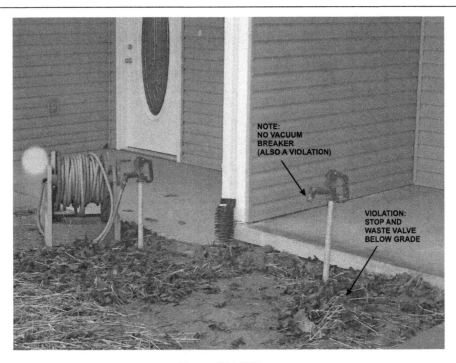

Figure 608.7(2)
PROHIBITED INSTALLATION OF YARD HYDRANTS CONNECTED TO WATER SERVICE LINES

water since it is commonplace for urinals and water closets to be supplied with potable water. A connection to an unmarked nonpotable piping system for the purposes of obtaining potable water could lead to significant health risks for the building's occupants.

608.8.1 Signage required. Nonpotable water outlets, such as hose connections, open ended pipes and faucets, shall be identified with signage that reads as follows: "Nonpotable water is utilized for [application name]. CAUTION: NONPOTABLE WATER – DO NOT DRINK." The words shall be legibly and indelibly printed on a tag or sign constructed of corrosion-resistant waterproof material or shall be indelibly printed on the fixture. The letters of the words shall be not less than 0.5 inch (12.7 mm) in height and in colors in contrast to the background on which they are applied. In addition to the required wordage, the pictograph shown in Figure 608.8.1 shall appear on the required signage.

❖ Some nonpotable water systems have outlets such as hose connections, open-ended pipes and faucets that could be mistaken for potable water outlets. Therefore, at these locations, the code requires signage to warn the building's occupants that water obtained from these outlets is not safe for drinking. The pictograph in Figure 608.8.1 conveys the intent in a visual manner.

**Figure 608.8.1
"PICTOGRAPH—DO NOT DRINK"**

608.8.2 Distribution pipe labeling and marking. Nonpotable distribution piping shall be purple in color and shall be embossed, or integrally stamped or marked, with the words: "CAUTION: NONPOTABLE WATER – DO NOT DRINK" or the piping shall be installed with a purple identification tape or wrap. Pipe identification shall include the contents of the piping system and an arrow indicating the direction of flow. Hazardous piping systems shall also contain information addressing the nature of the hazard. Pipe identification shall be repeated at intervals not exceeding 25 feet (7620 mm) and at each point where the piping passes through a wall, floor or roof. Lettering shall be readily observable within the room or space where the piping is located.

❖ Distribution piping conveying nonpotable water must be of purple color and have the caution lettering as indicated. Many types of water distribution pipe materials are offered with the external purple color and caution lettering. For those materials that are not offered with purple color or lettering, or in cases where the installer wishes to use standard colored pipe, the code allows for pipe identification using tapes or wraps. The tapes and wraps would need to be of purple color (primarily) and have the cautionary lettering. The intent of this section is not to require that elbows, tees and valves in the piping be purple colored or have the cautionary lettering. Identification of the pipe is all that is required, as the pipe is located where someone would most likely attempt to "tap into."

Piping identification must indicate the contents and the direction of flow, as well as any hazardous nature of the flow such as temperature, pressure and flammability and toxicity. The repetition of the identification locations is necessary so that those performing maintenance and modifications can easily find the pipe identification to determine the contents of the piping. Obviously, the pipe identification must be in a position such that a person within the room or space can easily see the identification markings.

608.8.2.1 Color. The color of the pipe identification shall be discernable and consistent throughout the building. The color purple shall be used to identify reclaimed, rain and gray water distribution systems.

❖ Where color is used for pipe identification, it must be clearly recognizable from any other pipe identification colors and be the same color for the same type of piping in the entire building. Where identifying reclaimed, rain and gray water distribution systems with color, the color purple must be used.

608.8.2.2 Lettering size. The size of the background color field and lettering shall comply with Table 608.8.2.2.

❖ See Table 608.8.2.2 for the required size of pipe identification.

**TABLE 608.8.2.2
SIZE OF PIPE IDENTIFICATION**

PIPE DIAMETER (inches)	LENGTH BACKGROUND COLOR FIELD (inches)	SIZE OF LETTERS (inches)
3/4 to 1 1/4	8	0.5
1 1/2 to 2	8	0.75
2 1/2 to 6	12	1.25
8 to 10	24	2.5
over 10	32	3.5

For SI 1 inch = 25.4 mm.

608.8.2.3 Identification tape. Where used, identification tape shall be at least 3 inches (76 mm) wide and have white or black lettering on a purple field stating "CAUTION: NONPOTABLE WATER – DO NOT DRINK." Identification tape shall be installed on top of nonpotable rainwater distribution pipes, fastened at least every 10 feet (3010 mm) to each pipe length and run continuously the entire length of the pipe.

❖ This section provides requirements for identification tapes for nonpotable water distribution piping. Also, this section requires that identification tape used for rainwater distribution piping be located on "top" of the piping and fastened at 10-foot (3048 mm) intervals (and that the tape be continuous throughout the entire length of the pipe).

It is believed that this text comes from one of the many rainwater re-use "best practices" manuals that have been created over the last decade and that the text was selected primarily for the requirements for the identification tape. The part about use of the tape for identifying rainwater distribution piping, while important for buried rainwater distribution piping, doesn't seem to be logical to apply to rainwater piping inside of buildings (unless that piping is buried under the building).

The code doesn't specify how to use/apply identification tapes, other than the buried application for rainwater piping. Where tape is used for identifying buried piping for any nonpotable water distribution piping, the requirements for buried rainwater distribution piping appear to be reasonable, although this is not a code requirement. Where nonpotable water distribution piping is in a building (and not buried below the building), nonself-adhering tapes could be (spiral) wrapped around nonpotable water distribution piping, as simply fastening at 10-foot (3048 mm) intervals might not result in a "permanent" marking. For example, if the tape were fastened at 10-foot (3048 mm) intervals on a horizontal pipe, the tape could sag and eventually be torn off during other work. Spiral wrapping of tape is not a code requirement but seems to be a logical method.

Identification tapes could also be self-adhering. However, application of such tapes should take into consideration whether the adhesive will have a negative impact on the pipe material, if the adhesive will hold up in the expected environmental conditions (heat, condensation, etc.) and if the adhesive can be removed without damage to the piping, should a repair or alteration be necessary.

608.9 Reutilization prohibited. Water utilized for the cooling of equipment or other processes shall not be returned to the potable water system. Such water shall be discharged into a drainage system through an *air gap* or shall be utilized for nonpotable purposes.

❖ Water used for cooling or processing could become contaminated and, therefore, must not be returned to the potable water system. Water usage that has no effect on the potability of the water could be acceptable in accordance with Section 501.2 or under the alternative approval provisions of Section 105.2. Systems affecting the potability of the water are prohibited.

It is not the intent of this section to prohibit potable water heaters from supplying hot water to heat exchangers that are constructed of materials approved for potable water distribution piping. Such systems are used to provide space heating and are designed to have no effect on the potability of the water (see Section 501.2).

608.10 Reuse of piping. Piping that has been utilized for any purpose other than conveying potable water shall not be utilized for conveying potable water.

❖ When a fluid or gas other than potable water is conveyed through piping, it can react with the interior pipe wall. The interior of the pipe could develop deposits or a coating that could contaminate the potable water if that pipe is used in a potable water supply system. Therefore, piping used to convey potable water must either be new (unused) or have been previously used only to convey potable water.

608.11 Painting of water tanks. The interior surface of a potable water tank shall not be lined, painted or repaired with any material that changes the taste, odor, color or potability of the water supply when the tank is placed in, or returned to, service.

❖ Only coatings that do not affect the potability of the water can be applied to the inside of a potable water tank.

608.12 Pumps and other appliances. Water pumps, filters, softeners, tanks and other devices that handle or treat potable water shall be protected against contamination.

❖ Pumps, filters, tanks and apparatus for use with potable water in a water supply system must be designed and approved for this specific use. Such devices and equipment must be installed to prevent contamination. For example, a water softener regeneration cycle drain must be protected against backflow.

608.13 Backflow protection. Means of protection against backflow shall be provided in accordance with Sections 608.13.1 through 608.13.10.

❖ Sections 608.13.1 through 608.13.10 list the nine methods of protection against backflow that are recognized by the code. Any other method would require compliance with Table 608.1 or an alternative approval in accordance with Section 105.2.

608.13.1 Air gap. The minimum required *air gap* shall be measured vertically from the lowest end of a potable water outlet to the *flood level rim* of the fixture or receptacle into which such potable water outlet discharges. *Air gaps* shall comply with ASME A112.1.2 and *air gap* fittings shall comply with ASME A112.1.3.

❖ ASME A112.1.2 is the standard for air gaps, which require specific dimensions as illustrated in Table 608.15.1. An air gap is not a device and has no moving parts. It is, therefore, the most effective and dependable method of preventing backflow and should be used where feasible. However, when this method of separation is used, there must be an assurance or at least a level of awareness by the user that the air gap cannot be bypassed or defeated by adding a hose or extension from the end of the potable water outlet to the source of contamination. The potable water opening or outlet must terminate at an elevation above the level of the source of contamination. The potable water could be contaminated only if the entire area or room becomes flooded to a depth that would submerge the potable water opening or outlet. An air gap installation must always be constructed so that the air gap remains permanently fixed. Air gaps can be fabricated from commercially available plumbing components or obtained as separate units and integrated into plumbing and piping systems.

This section requires air gaps to comply with Table 608.15.1 based on two variables:

- The size of the effective opening of the outlet or supply and
- The distance between the outlet and a wall or similar obstruction (see Table 608.15.1).

ASME A112.1.3 is the standard for air gap fittings for use with plumbing fixtures, appliances and appurtenances. This standard was developed for applications where an air gap fitting is manufactured for appliance applications.

608.13.2 Reduced pressure principle backflow prevention assemblies. Reduced pressure principle backflow prevention assemblies shall conform to ASSE 1013, AWWA C511, CSA B64.4 or CSA B64.4.1. Reduced pressure detector assembly backflow preventers shall conform to ASSE 1047. These devices shall be permitted to be installed where subject to continuous pressure conditions. The relief opening shall discharge by *air gap* and shall be prevented from being submerged.

❖ A reduced pressure principle backflow preventer is a device considered to be the most reliable mechanical method to prevent backflow. These devices consist of dual independently acting, spring-loaded check valves that are separated by a chamber or "zone" equipped with a relief valve. The pressure downstream of the device and in the central chamber between the check valves is maintained at a minimum of 2 psi (13.8 kPa) less than the potable water supply pressure at the device inlet, hence the name "reduced pressure principle." The relief valve located in the central chamber is held closed by the pressure differential between the inlet supply pressure and the central chamber pressure. If the inlet supply pressure decreases, the device will discharge water from the central chamber, thus maintaining a lower pressure in the chamber. In the event of backpressure or negative supply pressure, the relief valve will open to the atmosphere and drain any backflow that has leaked through the check valve. The relief vent will also allow air to enter to prevent any siphonage.

During a static or no-flow condition, both check valves and the relief vent remain closed. When there is normal flow, the two check valves open and the relief vent port remains closed. The supply pressure drops across the valve because of the force required to open the spring-loaded check valves. If backsiphonage or backpressure occurs, both check valves close, and the relief vent opens to the atmosphere, discharging the water in the intermediate zone. If both check valves are fouled and held in the open position, and there is either a backsiphonage or backpressure condition or both, the relief vent port opens, discharging the possibly contaminated water to the atmosphere and allowing air to enter to break the siphon. This gives a visual indication of a malfunction of one or both of the check valves or relief valve.

ASSE 1013 is the standard for two types of backflow prevention assemblies, identified as "reduced pressure principle backflow preventers" (RP) and "reduced pressure principle fire protection backflow preventers" (RPF). An RP device is designed for a working pressure of 150 psi (1024 kPa), and an RPF device is designed for a working pressure of 175 psi (1207 kPa). The RP and RPF are identical in their backflow protection. The RPF has specific performance requirements relating to its use on fire protection systems.

ASSE 1047 is the standard for devices known as "reduced pressure detector fire protection backflow prevention assemblies" (RPDF). This device is primarily used to protect the public water supply from contaminants found in fire sprinkler systems and to detect low rates of flow up to 0.126 L/s (2 gpm) within the sprinkler system caused by leakage or unauthorized use. This standard also allows provisions for alarm signaling devices to be included in the assembly. The application and installation of a reduced pressure principle backflow preventer is as follows:

- Provides protection against low- or high-hazard contamination.
- Provides protection against backsiphonage and backpressure backflow.
- The device can be used where it is under continuous pressure from the water supply.
- The relief vent opening must discharge through an air gap.
- The device must be accessible for testing and inspection.
- The device cannot be installed below grade where it may be subject to submersion.

Provisions must be made at or near the location of the installation to prevent drainage from the relief vent opening causing damage to the structure when the device is installed in a building or structure.

608.13.3 Backflow preventer with intermediate atmospheric vent. Backflow preventers with intermediate atmospheric vents shall conform to ASSE 1012 or CSA B64.3. These devices shall be permitted to be installed where subject to continuous pressure conditions. The relief opening shall discharge by *air gap* and shall be prevented from being submerged.

❖ A backflow preventer with intermediate atmospheric vent is a nontestable device consisting of two spring-loaded check valves in the closed position and a vent to atmosphere between the two check valves. The design and principle of operation of these backflow preventers are similar to those of a reduced pressure principle backflow preventer. Supply pressure keeps the atmospheric vent closed, but zero supply pressure or backsiphonage will open the inner chamber to the atmosphere.

The application and installation of a backflow preventer with intermediate atmospheric vent is as follows:

- Provides protection against low-hazard contamination.
- Provides protection against backsiphonage and backpressure backflow.
- The relief vent opening must discharge through an air gap.
- The device can be used where it is under continuous pressure from the water supply.
- The device cannot be installed below grade where it may be subject to submersion.
- The device must be accessible for inspection and replacement.

Provisions must be made at or near the location of the installation to prevent drainage from the relief vent opening causing damage to the structure when the device is installed in a building or structure.

608.13.4 Barometric loop. Barometric loops shall precede the point of connection and shall extend vertically to a height of 35 feet (10 668 mm). A barometric loop shall only be utilized as an atmospheric-type or pressure-type vacuum breaker.

❖ The code considers a barometric loop to be equivalent to a vacuum breaker. A barometric loop may be installed in any location where the code would otherwise require a vacuum breaker.

The principle of a barometric loop is that siphon action is limited to the height that a column of water can be drawn. The 35-foot (10 668 mm) dimension is greater than the theoretical head of a pump suction. A perfect vacuum is equivalent to 0 psi (0 kPa) absolute (gauge pressure minus atmospheric pressure). Atmospheric pressure cannot support a water column in excess of 34 feet (10 363 mm) in height. Because siphonage is dependent on atmospheric pressure, siphonage cannot occur if the vertical distance exceeds 34 feet (10 363 mm).

A barometric loop would have control valves located in the vertical pipe or downstream of the connection (see Commentary Figure 608.13.4). A barometric loop cannot prevent backflow caused by backpressure.

608.13.5 Pressure vacuum breaker assemblies. Pressure vacuum breaker assemblies shall conform to ASSE 1020 or CSA B64.1.2. Spill-resistant vacuum breaker assemblies shall comply with ASSE 1056. These assemblies are designed for installation under continuous pressure conditions where the critical level is installed at the required height. Pressure vacuum breaker assemblies shall not be installed in locations where spillage could cause damage to the structure.

❖ A pressure vacuum breaker assembly contains one or two independently operating spring-loaded check valves and an independently operating spring-loaded air inlet valve located on the discharge side of the

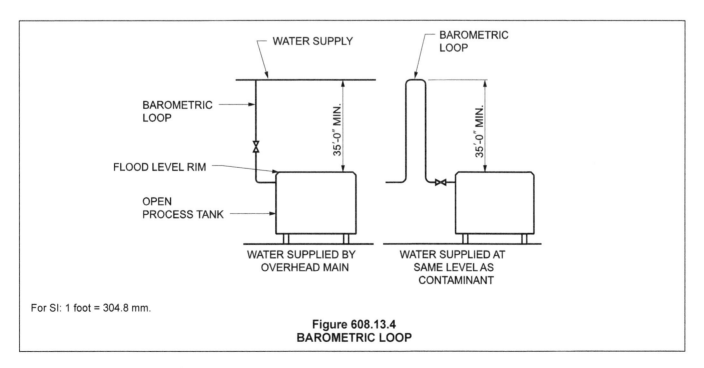

For SI: 1 foot = 304.8 mm.

**Figure 608.13.4
BAROMETRIC LOOP**

check valve. This device is testable because it has a shutoff valve and test cock at each end (see Commentary Figure 608.13.5). While a pressure vacuum breaker assembly operates by the same principle as an atmospheric vacuum breaker, a pressure vacuum breaker assembly is spring activated. The water pressure acting against the spring closes off the air opening. Because of the spring activation, a valve can be located downstream of a pressure vacuum breaker assembly. A spill-resistant vacuum breaker assembly is considered a pressure vacuum breaker assembly (see Section 608.13.8).

The application and installation of a pressure vacuum breaker assembly is as follows:

- Provides protection against low- and high-hazard contamination.
- Provides protection against backsiphonage only.
- Can be used where it is under continuous pressure from the water supply (valving permitted downstream).
- The device has a critical level of installation of 6 inches (152 mm) above flood level rim.
- The device must be accessible for testing and inspection.

When the device is installed in a building or structure, provisions must be made at or near the location of the installation to prevent damage caused by possible leakage from the air vent opening.

Figure 608.13.5
PRESSURE VACUUM BREAKER

608.13.6 Atmospheric-type vacuum breakers. Pipe-applied atmospheric-type vacuum breakers shall conform to ASSE 1001 or CSA B64.1.1. Hose-connection vacuum breakers shall conform to ASME A112.21.3, ASSE 1011, ASSE 1019, ASSE 1035, ASSE 1052, CSA B64.2, CSA B64.2.1, CSA B64.2.1.1, CSA B64.2.2 or CSA B64.7. These devices shall operate under normal atmospheric pressure when the critical level is installed at the required height.

❖ An atmospheric vacuum breaker is designed to prevent siphonic action from occurring downstream of the device. When siphon action or a vacuum is applied to the water supply, the vacuum breaker opens to the atmosphere, allowing air to enter the piping system, thus breaking the vacuum. This prevents contamination of the potable water system by stopping the backsiphonage backflow. When a siphon occurs on the supply inlet of the vacuum breaker, the disc float drops over the opening. Air enters the atmospheric port opening, allowing the remaining water in the piping downstream from the vacuum breaker to drain.

These devices are not effective in preventing backflow resulting from backpressure. The outlet of atmospheric vacuum breakers must remain open to the atmosphere by terminating with a pipe, spout or similar unobstructed opening. Valves must not be installed downstream of this device because this would subject the device to supply pressure, thereby rendering it inoperative.

Application and installation of an atmospheric-type vacuum breaker is as follows:

- Provides protection against low- or high-hazard contamination.
- Provides protection against backsiphonage only.
- The device cannot be installed where it is under continuous pressure from the water supply (12-hour or less intervals).
- Critical level installed 6 inches (152 mm) above highest point of use downstream.
- The device must be accessible for inspection and replacement.

Hose connection backflow preventers are designed to be installed on the discharge side of a hose-threaded outlet on a potable water system. This two-check device protects against backflow resulting from backsiphonage and low-head backpressure, under the high-hazard conditions present at a hose-threaded outlet. This device must be used only on systems where sources of backpressure are not introduced and where low-head backpressure does not exceed that generated by an elevated hose equal to or less than 10 feet (3048 mm) in height.

608.13.7 Double check backflow prevention assemblies. Double check backflow prevention assemblies shall conform to ASSE 1015, CSA B64.5, CSA B64.5.1 or AWWA C510. Double check detector fire protection backflow prevention assemblies shall conform to ASSE 1048. These assemblies shall be capable of operating under continuous pressure conditions.

❖ These assemblies are designed for low-hazard applications subject to backpressure and backsiphonage.

The devices consist of two independent spring-loaded check valves in series. Test cocks are provided to allow testing of the devices (see Commentary Figure 608.13.7). Note that these devices must not be confused with dual check-valve devices or two single check valves placed in series.

Double-detector fire protection check-valve assemblies are always allowed as an option where double check-valve backflow prevention assemblies are required because they offer identical protection. Double-detector assemblies provide the ability to monitor water flow through the device to determine whether there is a system leak or water is being intentionally withdrawn from the system downstream of the assembly.

608.13.8 Spill-resistant pressure vacuum breaker assemblies. Spill-resistant pressure vacuum breaker assemblies shall conform to ASSE 1056 or CSA B64.1.3. These assemblies are designed for installation under continuous-pressure conditions where the critical level is installed at the required height.

❖ A spill-resistant vacuum breaker assembly is sometimes referred to as a backsiphonage backflow vacuum breaker assembly. A spill-resistant vacuum breaker assembly can be installed in any location where the code requires a pressure vacuum breaker assembly. A spill-resistant vacuum breaker assembly is identical to a pressure vacuum breaker assembly except for an additional feature that significantly reduces the potential for water discharging through the air inlet vent each time the device is pressurized. As the name implies, spill resistant is not spillproof; therefore, careful consideration of the location of these devices is necessary to avoid the potential for water damage caused by leakage from the air inlet vent.

A spill-resistant vacuum breaker assembly is a testable device consisting of one check valve, spring loaded to the closed position, and an air inlet vent valve, spring loaded to the open position. The air inlet valve is located downstream from the check valve. The check valve and air inlet vent are located between two shutoff valves and two test cocks.

The application and installation of a spill-resistant vacuum breaker are as follows:

- Protects against low- and high-hazard contamination.
- Protects against backsiphonage backflow only.
- Can be subjected to continuous pressure (closed valves downstream is allowed).
- The critical level must be installed 6 inches (152 mm) above the highest outlet served.

The device must be located to allow for testing and inspection.

608.13.9 Chemical dispenser backflow devices. Back-flow devices for chemical dispensers shall comply with ASSE 1055 or shall be equipped with an *air gap* fitting.

❖ Chemical dispensing systems mix potable water with chemicals to provide the user with a ready-to-use chemical solution. These systems must be equipped with a backflow prevention device complying with ASSE 1055 or an air gap fitting, which is a device with an integral air gap. ASSE 1055 applies to devices classified as chemical dispensing systems having self-contained means of backflow protection, as follows:

Type A: These devices have the chemical(s) pressurized above atmospheric pressure.

Type B: These devices do not pressurize the chemical(s) above atmospheric pressure, with the only source of backpressure coming from an elevated hose.

Figure 608.13.7
DOUBLE CHECK-VALVE ASSEMBLY

608.13.10 Dual check backflow preventer. Dual check backflow preventers shall conform to ASSE 1024 or CSA B64.6.

❖ This type of backflow preventer is not field testable. This device is distinctly different than a double check backflow prevention assembly, which is field testable. A dual check device is sometimes installed by the water utility in conjunction with the installation of the water meter for small buildings and residential homes where the water service is 1 inch (25 mm) and smaller.

608.14 Location of backflow preventers. Access shall be provided to backflow preventers as specified by the manufacturer's instructions.

❖ Every backflow preventer must be located so that it can be reached for inspection, testing, repair and replacement without removing a wall, floor, ceiling or other permanent portion of the building structure. It can be located behind access panels or in utility-type spaces. Backflow preventer devices, other than double-check-valve assemblies, have an atmospheric opening, and the location must take this into consideration. The devices must be located not only in ventilated spaces to allow air to enter them, but also where water discharge from them will not cause damage or create a hazard.

Backflow preventers must not be installed in locations subject to flooding, such as underground vaults. If the atmospheric opening on a backflow preventer is submerged, the protection provided by the device is lost.

608.14.1 Outdoor enclosures for backflow prevention devices. Outdoor enclosures for backflow prevention devices shall comply with ASSE 1060.

❖ ASSE 1060 is the standard that details the requirements of an outside enclosure for various types of backflow prevention assemblies. It includes enclosure types for freezing and nonfreezing locations. The enclosures incorporate features to provide for freeze protection, positive drainage to prevent submergence of the assembly, security and accessibility for testing and repair.

Freeze protection enclosures are divided into the following classes:

1. Class I enclosures are designed for all devices other than pressure and atmospheric vacuum breakers.
2. Class I-V enclosures are designed for pressure and atmospheric vacuum breakers.

Freeze-retardant enclosures are divided into the following classes:

1. Class II enclosures are designed for all devices other than pressure and atmospheric vacuum breakers.
2. Class II-V enclosures are designed for pressure and atmospheric vacuum breakers.

Enclosures without freeze protection are divided into the following classes:

1. Class III enclosures are designed for all devices other than pressure and atmospheric vacuum breakers.
2. Class III-V enclosures are designed for pressure and atmospheric vacuum breakers.

Freeze-protection enclosures are designed and constructed to maintain a minimum internal temperature of 40°F (4°C) when the external temperature is as low as -30°F (-34°C).

608.14.2 Protection of backflow preventers. Backflow preventers shall not be located in areas subject to freezing except where they can be removed by means of unions or are protected from freezing by heat, insulation or both.

❖ Frequently, backflow prevention valves are required to be located outside of buildings for a variety of reasons. If a valve will be exposed to freezing temperatures, the valve must be either protected during the freezing condition or removed during freezing weather. Where valves are necessary to function during freezing weather, removal is not possible, thus requiring that insulation or heat or both be used to keep the valve above freezing temperatures.

There is usually no question as to the need for an insulated enclosure and heat source for backflow preventers exposed to severe freezing weather. Valves exposed to milder winter conditions might only need insulating or where freezing conditions rarely exist, no protection may be an appropriate solution. The code is silent on the level of protection needed for these milder climates. Enclosure vendors and code officials could be sources for information to decide the degree of freeze protection required.

For those backflow prevention valve applications where use of the valve is seasonal such as for lawn sprinklers, or swimming pool water service, removal is usually the easiest solution. If removal is intended, unions must be provided to enable the valve to be easily removed without disturbing permanent piping. Even though the valve is removed, removal of the water in the remaining permanent piping should not be overlooked to prevent freeze damage to the piping.

608.14.2.1 Relief port piping. The termination of the piping from the relief port or *air gap* fitting of a backflow preventer shall discharge to an *approved* indirect waste receptor or to the outdoors where it will not cause damage or create a nuisance.

❖ In particular, reduced pressure zone (RPZ) backflow prevention valves will discharge water during a backflow event, a drop in supply pressure or where an obstruction (dirt) fouls the check valves. If the valve is located outdoors, an air gap fitting is usually not needed as the discharge dumps to the ground. If the valve is indoors or the discharge in an outdoor location would cause damage or a nuisance, the valve must be located over an appropriate waste receptor (floor drain, sump with drain, etc.) to capture the flow from the relief port. If the relief port discharge must be piped to a suitable discharge point (either outdoors or to a nearby waste receptor), the relief port must be fitted with an air gap fitting (as supplied by the valve manufacturer) for connection of the relief discharge piping. Where the piping discharges to a waste receptor, it must connect indirectly.

In most situations, relief port discharges are a trickle of water that is easily accommodated by the pipe size dictated by the air gap connection. However, large obstructions (dirt, rocks, etc.) or sudden large drops in supply pressure may cause an extreme and prolonged discharge event such that the air gap fitting and relief pipe size will be ineffective in capturing the discharge. Designers should consider this type of event and provide for a means to capture and direct the flow from an extreme event to a safe location in order to avoid nearby equipment and property damage.

608.15 Protection of potable water outlets. All potable water openings and outlets shall be protected against backflow in accordance with Section 608.15.1, 608.15.2, 608.15.3, 608.15.4, 608.15.4.1 or 608.15.4.2.

❖ An opening or outlet is where the potable water system comes in contact with the atmosphere. This includes, but is not limited to, faucets, hose bibbs, plumbing fixtures, plumbing appliances and other points where water is withdrawn from the potable water supply system. The openings should be protected by one of four methods: air gap, backflow preventer with intermediate atmospheric vent, vacuum breaker or reduced pressure principle backflow preventer.

There are two types of cross-connection control philosophies: containment and isolation.

Containment: The practice of confining potential contamination within the facility where it arises by installing a backflow preventer at the meter or curb stop.

Isolation: The practice of confining potential contamination by installing a backflow preventer at each outlet or connection to a fixture or nonpotable system.

❖ The code requires isolation only for cross connections. The practice of containment is optional and widely used in waste-water treatment plants and marinas where potable water is supplied to boats and buildings with numerous nonpotable piping systems.

Many water supplies to commercial or industrial facilities have containment protection due in part to the complexity or diversity of use of the potable water supply within the facility. Although containment alone will protect the public water system, it does not protect the potable water within the facility.

608.15.1 Protection by air gap. Openings and outlets shall be protected by an *air gap* between the opening and the fixture *flood level rim* as specified in Table 608.15.1. Openings and outlets equipped for hose connection shall be protected by means other than an *air gap*.

❖ An air gap is the simplest, most common and most reliable form of backflow protection for openings and outlets. Most sink and lavatory faucets are designed with a built-in air gap. A hose connection outlet cannot rely on an air gap to provide backflow protection. Once a hose is added to the outlet, the air gap no longer exists. Therefore, an alternative method of backflow prevention is required for any hose outlet.

Although an air gap is an extremely effective backflow preventer, it does interrupt the piping flow with corresponding loss of pressure for subsequent use.

Consequently, air gaps are used primarily at end-of-the-line service where reservoirs or storage tanks are desired.

Additionally, it exposes the potable water supply to the surrounding air with its inherent bacteria, dust particles and other air-borne pollutants or contaminants. This presents a backflow potential and must be considered when selecting appropriate backflow methods.

An air gap must be inspected as frequently as mechanical backflow preventers in accordance with Section 312.9 [see Commentary Figures 608.15.1(1) and 608.15.1(2)].

For SI: 1 inch = 25.4 mm.

**Figure 608.15.1(1)
AIR GAP FOR A LAVATORY FAUCET**

WATER SUPPLY AND DISTRIBUTION

**Figure 608.15.1(2)
AIR GAP PIPING SYSTEM**

608.15.2 Protection by reduced pressure principle backflow prevention assembly. Openings and outlets shall be protected by a reduced pressure principle backflow prevention assembly or a reduced pressure principle fire protection backflow prevention assembly on potable water supplies.

❖ A reduced pressure principle backflow prevention assembly or a reduced pressure principle fire-protection backflow prevention assembly can also be used to protect outlets and openings, although it is rarely used for this purpose (see commentary, Section 608.13.2).

TABLE 608.15.1. See below.

This table has been a mainstay in the plumbing industry since 1940. The values specified have not changed in more than 50 years.

The middle column of the table shows minimum air gap dimensions for outlets located away from walls or obstructions. The third column contains more stringent air gap dimensions when the outlet is located near a wall or obstruction [see Commentary Figures 608.15.1(3) and 608.15.1(4)].

608.15.3 Protection by a backflow preventer with intermediate atmospheric vent. Openings and outlets shall be protected by a backflow preventer with an intermediate atmospheric vent.

❖ The devices addressed in Section 608.13.3 are acceptable means of backflow prevention for openings and outlets.

608.15.4 Protection by a vacuum breaker. Openings and outlets shall be protected by atmospheric-type or pressure-type vacuum breakers. The critical level of the vacuum breaker shall be set not less than 6 inches (152 mm) above the *flood level rim* of the fixture or device. Fill valves shall be set in accordance with Section 425.3.1. Vacuum breakers shall not be installed under exhaust hoods or similar locations that will contain toxic fumes or vapors. Pipe-applied vacuum breakers shall be installed not less than 6 inches (152 mm) above the *flood level rim* of the fixture, receptor or device served.

❖ For a vacuum breaker to operate properly, it must be located above the flood level rim of the fixture or device that it is protecting.

**TABLE 608.15.1
MINIMUM REQUIRED AIR GAPS**

FIXTURE	MINIMUM AIR GAP	
	Away from a wall[a] (inches)	Close to a wall (inches)
Lavatories and other fixtures with effective openings not greater than $^1/_2$ inch in diameter	1	$1^1/_2$
Sinks, laundry trays, gooseneck back faucets and other fixtures with effective openings not greater than $^3/_4$ inch in diameter	$1^1/_2$	$2^1/_2$
Over-rim bath fillers and other fixtures with effective openings not greater than 1 inch in diameter	2	3
Drinking water fountains, single orifice not greater than $^7/_{16}$ inch in diameter or multiple orifices with a total area of 0.150 square inch (area of circle $^7/_{16}$ inch in diameter)	1	$1^1/_2$
Effective openings greater than 1 inch	Two times the diameter of the effective opening	Three times the diameter of the effective opening

For SI: 1 inch = 25.4 mm, 1 square inch = 645 mm^2.

a. Applicable where walls or obstructions are spaced from the nearest inside-edge of the spout opening a distance greater than three times the diameter of the *effective opening* for a single wall, or a distance greater than four times the diameter of the *effective opening* for two intersecting walls.

Figure 608.15.1(3)
AIR GAPS—LOCATED NEAR SINGLE WALL

Figure 608.15.1(4)
AIR GAPS—LOCATED NEAR INTERSECTING WALLS

If the vacuum breaker is below the flood level rim, it is subject to pressure from the fixture or outlet piping because of the static head (see commentary, Section 608.13.6). Valves cannot be located downstream from an atmospheric vacuum breaker. A downstream valve will subject the device to backpressure, thereby causing damage or impaired operation. All valves or controlling devices must be located upstream of the device. Valves may be downstream only where a pressure vacuum breaker is installed (see Commentary Figure 608.15.4).

608.15.4.1 Deck-mounted and integral vacuum breakers. *Approved* deck-mounted or equipment-mounted vacuum breakers and faucets with integral atmospheric vacuum breakers or spill-resistant vacuum breaker assemblies shall be installed in accordance with the manufacturer's instructions and the requirements for labeling with the critical level not less than 1 inch (25 mm) above the *flood level rim*.

❖ This section recognizes a type of vacuum breaker that is installed on the deck (horizontal surface at rim) of plumbing fixtures such as bathtubs, whirlpools, bidets and sinks. Such devices must be tested for installation with lesser distances between the critical level and the fixture flood level rim. Such devices are intended to be more aesthetically pleasing and would be built into fixture trim hardware, such as fill spouts or diverter controls.

WATER SUPPLY AND DISTRIBUTION

For SI: 1 inch = 25.4 mm.

**Figure 608.15.4
SUBMERGED INLET PROTECTED BY VACUUM BREAKER**

608.15.4.2 Hose connections. Sillcocks, hose bibbs, wall hydrants and other openings with a hose connection shall be protected by an atmospheric-type or pressure-type vacuum breaker or a permanently attached hose connection vacuum breaker.

Exceptions:

1. This section shall not apply to water heater and boiler drain valves that are provided with hose connection threads and that are intended only for tank or vessel draining.

2. This section shall not apply to water supply valves intended for connection of clothes washing machines where backflow prevention is otherwise provided or is integral with the machine.

❖ In accordance with this section, outlets and openings to which a hose can be connected must be protected by some type of vacuum breaker device. Commentary Figure 608.15.4.2(1) shows a hose connection in violation of this requirement. Commentary Figure 608.15.4.2(2) shows one type of hose faucet that has one type of integral vacuum breaker. Atmospheric vacuum breakers screwed onto the threads of the hose faucet may also be acceptable as long as the connection is permanent (locking type) or the threads between the hose faucet and the vacuum breaker are nonstandard (i.e., neither pipe threads nor garden hose threads). The devices addressed in Sections 608.13.2 and 608.13.3 would also be permitted because they would afford a greater level of protection. Where a hose can be elevated above the critical level of a vacuum breaker, the vacuum breaker must be designed to function under such conditions. For example, hose connection vacuum breakers incorporate a check valve in conjunction with an atmospheric port. Such devices can prevent siphonage and can also resist backflow caused by the head pressure developed in an elevated hose. In accordance with Table 608.1 and the definition for "Backflow, low head backpressure" in Section 202, this backpressure is limited to 10 feet (3048 mm) of water column or 4.33 psi (29.9 kPa). If the possibility exists for backpressure to exceed 10 feet (3048 mm) of water column [the equivalent of raising a hose filled with water 10 feet (3048 mm) above the vacuum breaker], another backflow prevention device, designed to handle greater backpressures, must be installed. If a hose is equipped with a control valve, an atmospheric vacuum breaker cannot be used to protect that outlet [see commentary, Section 608.15.4 and Commentary Figure 608.15.4.2(3)].

**Figure 608.15.4.2(1)
HOSE (SILL) FAUCET WITHOUT VACUUM BREAKER**

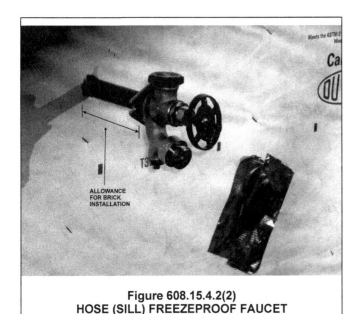

**Figure 608.15.4.2(2)
HOSE (SILL) FREEZEPROOF FAUCET
WITH INTEGRAL VACUUM BREAKER**

**Figure 608.15.4.2(3)
HOSE CONNECTION VACUUM BREAKER**

In Commentary Figure 608.15.4.2(3), the supply valve shutoff disc (1) seals against the diaphragm (2) when there is no flow in the pipe. The atmospheric port (3) is open. As flow occurs in the pipe, the atmospheric ports are closed before the supply valve disc opens. Note that this type of vacuum breaker can be acceptable only if the installation is made permanent (e.g., a locking break-off set screw).

The exceptions are intended to address situations where hose threads are commonly provided but do not require the installation of a vacuum breaker. The two exceptions are not exempting these outlets from any form of backflow prevention. In Exception 1, water heater and boiler drain valves (with or without hose threads) must be provided with an air gap, and in Exception 2, the clothes washing machine must be provided with an integral form of backflow prevention (typically an air gap between the fill pipe and the overflow rim of the washing machine tub).

Table 608.1 contains a reference to a relatively new device called a "hose connection backflow preventer." The device is testable and consists of two check valves spring loaded in the closed position with an atmospheric vent between them. The backflow preventer can be either an add-on device to or integral with a wall hydrant or hose bibb. Because the add-on device is designed to attach to hose threads, it would be appropriate for these devices to serve as the required method of backflow protection for hose connections.

608.16 Connections to the potable water system. Connections to the potable water system shall conform to Sections 608.16.1 through 608.16.10.

❖ Sections 608.16.1 through 608.16.10 regulate backflow prevention for direct connections between a potable and nonpotable system. A cross connection exists where there is no atmospheric opening (air gap) between the potable water supply and the source of contamination.

This section addresses cross connections that result from the direct connection of the potable water supply to a closed system such as boilers, vessels, tanks, heat exchangers and nonpotable piping systems.

608.16.1 Beverage dispensers. The water supply connection to beverage dispensers shall be protected against backflow by a backflow preventer conforming to ASSE 1022 or by an *air gap*. The portion of the backflow preventer device downstream from the second check valve and the piping downstream therefrom shall not be affected by carbon dioxide gas.

❖ The potable water supply to a carbonated beverage dispenser must be protected by a backflow preventer with an intermediate atmospheric vent conforming to ASSE 1022 or by an air gap. Concerns have been raised that a dual check valve alone will not show any visible indication of failure. The backflow preventer for beverage dispensing equipment includes two check valves and an atmospheric vent. If there is failure of the downstream check and the backpressure exceeds the supply pressure, the vent will discharge, giving a visual indication of the check valve's failure. Piping downstream of the backflow preventer with intermediate atmospheric vent must not be copper. These requirements are intended to prevent CO_2 backflow into copper water supply lines, which causes leaching of copper into the potable water supply (see Commentary Figure 608.16.1).

Prior to the 2009 edition of the code, a carbonated beverage dispenser backflow preventer that met the CSA B64.3.1 standard was acceptable. An in depth analysis comparing the test protocols between the ASSE and CSA standards suggested that the CSA standard was not equal to the requirements of ASSE. Because the use of ASSE 1022 devices has a long history of providing safe conditions, and the majority of beverage dispenser manufacturers only specify the ASSE 1022 device for use with their equipment, the CSA standard was removed from this section.

608.16.2 Connections to boilers. The potable supply to the boiler shall be equipped with a backflow preventer with an intermediate atmospheric vent complying with ASSE 1012 or CSA B64.3. Where conditioning chemicals are introduced into the system, the potable water connection shall be protected by an *air gap* or a reduced pressure principle backflow preventer, complying with ASSE 1013, CSA B64.4 or AWWA C511.

❖ Because boilers are pressurized vessels, the potential for backflow caused by backpressure is quite high. Any time the potable water supply pressure drops below boiler pressure, backflow can occur.

If a boiler system contains only untreated, unconditioned water supplied from a potable source, a backflow preventer with an intermediate atmospheric vent can be installed (see Commentary Figure 608.16.2).

If any chemicals are added to the boiler system, the potable water supply connection must be protected with a reduced pressure principle backflow preventer.

608.16.3 Heat exchangers. Heat exchangers utilizing an essentially toxic transfer fluid shall be separated from the potable water by double-wall construction. An *air gap* open

Figure 608.16.1
BEVERAGE DISPENSER

Figure 608.16.2
FOR BOILER WITHOUT CONDITIONING CHEMICALS

to the atmosphere shall be provided between the two walls. Heat exchangers utilizing an essentially nontoxic transfer fluid shall be permitted to be of single-wall construction.

❖ The extent of isolation required for a heat exchanger depends on the type of fluid used on the exchanger's nonpotable side. From the definition of "Essentially nontoxic transfer fluid," the Gosselin rating of nonpotable fluid must be evaluated. If the fluid has a Gosselin rating of 1, a single-wall heat exchanger is permitted.

The Gosselin rating is a measure of the toxicity of a substance. The name originates from one of the prime developers of the rating system, Dr. Robert E. Gosselin, professor of Pharmacology at Dartmouth Medical School in New Hampshire.

Gosselin toxicity ratings are the values used by medical personnel to analyze poison victims. The ratings are based on the probable lethal dose for a human. The six levels of Gosselin ratings are outlined in Commentary Figure 608.16.3(1).

Some of the commercially available transfer fluids with a Gosselin rating of 1 are identified in Commentary Figure 608.16.3(2).

If the heat transfer fluid has a Gosselin rating of 2 or more, a double-wall heat exchanger is required. The double-wall heat exchanger must have an intermediate space between the walls that is open to the atmosphere. This type of construction would allow any leakage of fluid through the walls of the heat exchanger to discharge externally to the heat exchanger where it would be observable [see Commentary Figure 608.16.3(3)].

TRADE NAME	MANUFACTURER
Caloria Ht-43	Exxon Co.
DowFrost	Dow
Drewsol	Drew Chemical
Freeze Proof	Commonwealth
Mobiltherm 603	Mobile Oil Co.
Mobiltherm Light	Mobile Oil Co.
Nutek 835	Nuclear Technology
Nutek 876	Nuclear Technology
Process Oil 3029	Exxon Co.
Propylene Glycol U.S.P.	Union Carbide
Solargard G.	Daystar Corp.
Solar Winter Ban	Solar Alternative Inc.
Sunsol 60	Sunworks
Suntemp	Resource Technology Corp.
Syltherm 444	
Therminal 66	Dow Corning Corp.
UCAR Food Freeze 35	Monsanto Co.
FDA Approved Boiler Additives	Union Carbide
Freon 12	
Freon 22	
Freon 112	
Freon 114	

Figure 608.16.3(2)
HEAT TRANSFER FLUIDS (ESSENTIALLY NONTOXIC) GOSSELIN RATING OF 1

TOXICITY RATING OR CLASS	PROBABLE ORAL LETHAL DOSE (HUMAN)	
	Dose	For 7 kg person (150 pounds)
6 Super toxic	Less than 5 mg/kg	A taste (less than 7 drops)
5 Extremely toxic	5-50 mg/kg	Between 7 drops and 1 teaspoon
4 Very toxic	50-500 mg/kg	Between 1 teaspoon and 1 ounce
3 Moderately toxic	0.5-5 gm/kg	Between 1 ounce and 1 pint (pound)
2 Slightly toxic	5-15 gm/kg	Between 1 pint and 1 quart
1 Practically nontixic	Above 16 gm/kg	More than 1 quart (2.2 pounds)

Figure 608.16.3(1)
GOSSELIN RATINGS

Figure 608.16.3(3)
DOUBLE WALL HEAT EXCHANGER

608.16.4 Connections to automatic fire sprinkler systems and standpipe systems. The potable water supply to automatic fire sprinkler and standpipe systems shall be protected against backflow by a double check backflow prevention assembly, a double check fire protection backflow prevention assembly or a reduced pressure principle fire protection backflow prevention assembly.

Exceptions:

1. Where systems are installed as a portion of the water distribution system in accordance with the requirements of this code and are not provided with a fire department connection, isolation of the water supply system shall not be required.

2. Isolation of the water distribution system is not required for deluge, preaction or dry pipe systems.

❖ One of the most common arrangements for supplying water to an automatic fire sprinkler system or a fire service standpipe system is through a common water service system that also provides potable water to a building's water distribution system [see Commentary Figure 608.16.4(2)]. Because the water that initially fills the piping of the automatic fire sprinkler system is potable water, the degree of hazard to the potable water supply should a backflow event occur is low. In other words, the potable water supply might become polluted, but not contaminated (see Chapter 2 definitions of "Pollution" and "Contamination"). Therefore, a double check valve backflow protection assembly (or double check fire-protection backflow prevention assembly) is the minimum acceptable form of backflow prevention required between the potable water supply and the automatic fire sprinkler system.

However, if antifreeze or other chemicals are added to a sprinkler or standpipe system or if a nonpotable secondary water supply is or could be connected to the sprinkler or standpipe system, the potable water supply must be protected with a reduced pressure principle fire-protection backflow prevention assembly to protect the potable water supply (see commentary, Section 609.16.4.1). An example of a nonpotable secondary water supply would be the connection of a fire truck pumper unit through a fire department connection on the building where the pump is conveying nonpotable water such as from a pond or lake.

Exception 1 addresses the installation of dual-purpose (combination) water distribution and fire sprinkler piping systems. Where only spot protection is provided in limited-area sprinkler systems and in residential applications, it is common for fire sprinklers to be supplied directly from the building's potable water distribution system. In such systems, there is not a fire department connection through which water can be pumped into the system, and the entire piping system, including the sprinkler piping, is constructed of materials that are approved for potable water distribution. Because there is not a potential source of pollution or contamination in such systems and it would be nearly impossible to install, maintain and test a backflow preventer at each take-off point for a sprinkler, a backflow preventer is not required [see Commentary Figure 608.16.4(1)].

Note that Exception 1 is not intended to apply to separate (independent) water distribution and fire sprinkler systems that share only a common water service system [see Commentary Figure 608.16.4(2)]. Where the fire sprinkler and water distribution system do not share piping other than the water service system and even if the sprinkler system was constructed of materials approved for water distribution, isolation of the fire sprinkler system from the potable water supply system is still required because it keeps the sprinkler system or standpipe filled with water should a public water main lose pressure during a water main break or there is a heavy fire demand that might occur elsewhere.

Discussion of Exception 2 requires a brief explanation of wet-pipe, deluge, preaction and dry-pipe sprinkler systems. A wet-pipe sprinkler system is the most common type of system in most buildings. A network of pipes terminate at a number of closed sprinklers. Each sprinkler has a fuse-like device, that when heated above a certain temperature, melts or breaks to allow flow through the sprinkler. The entire sprinkler network is filled with water and kept under pressure. The only sprinklers that discharge water are the ones where the fusible device melts or breaks.

A deluge sprinkler system is comprised of a network of pipes terminating in a number of open "high-flow"-type sprinklers. There is no water in the system until a main valve is triggered to open by thermal sensors in the protected area. Once the valve opens, water flows from all sprinklers to deluge the entire protected area with water, even though only a portion of the area might be experiencing a fire condition.

A dry-pipe sprinkler system consists of a network of pipes terminating in a number of closed sprinklers. Just as in a wet-pipe system, each sprinkler has a fuse-like device, that when heated above a certain temperature, melts or breaks to allow flow through the sprinkler. However, instead of water filling the system, compressed air or compressed nitrogen is used to fill the system and is maintained by a small air compressor (or high-pressure tank) in order to keep a special check valve closed. The special check valve is connected to the potable water supply. Once a sprinkler activates (i.e., the fusible device melts or breaks), the compressed air (or gas) is released from the system, allowing the special check valve to open and water to flow into the network of piping and to the sprinklers. However, only the sprinklers that are activated will be discharging water into the area. Dry pipe systems are often used where areas to be sprinkler protected are subject to freezing temperatures.

A preaction sprinkler system is a cross between a deluge system and a dry-pipe system. The valve holding back water from entering the sprinkler system is electrically controlled by thermal sensors in the protected area. The sprinklers are also of the closed-with-fusible-device type. This type of system is typically used where it is desirable to not have water flow on just the independent action of a valve opening or a sprinkler opening. Both events must occur before water will flow from a sprinkler.

WATER SUPPLY AND DISTRIBUTION

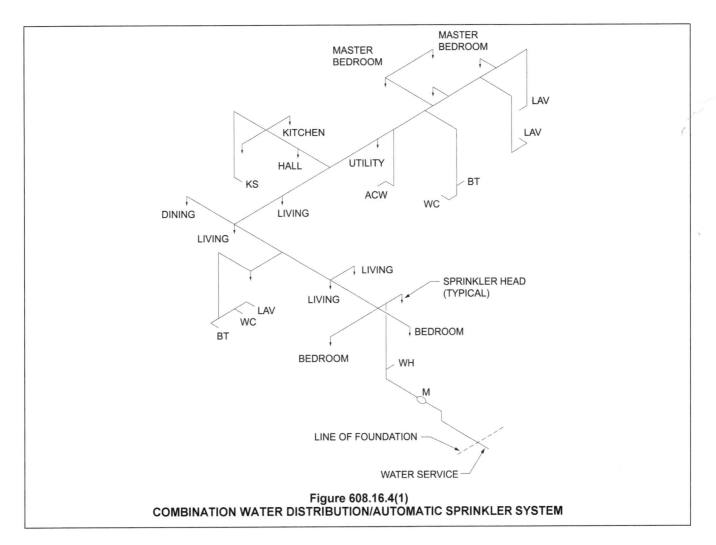

Figure 608.16.4(1)
COMBINATION WATER DISTRIBUTION/AUTOMATIC SPRINKLER SYSTEM

Figure 608.16.4(2)
COMMON FIRE SPRINKLER AND DOMESTIC WATER SUPPLY

Because deluge, dry-pipe and preaction sprinkler systems are not filled with any liquid that could pollute or contaminate the potable water supply, Exception 2 allows these systems to be installed without a backflow preventer. However, where a fire department connection or auxiliary source of water supply could introduce nonpotable water into a sprinkler system, a backflow preventer is required.

The following seven cases illustrate how to determine when a backflow preventer must be installed and which type (either double check or reduced pressure principle) is required to be installed:

Case 1: The suction piping system for pumping from a secondary source of potable water has underground piping that is in close proximity to a sanitary sewer, septic tank, seepage pit, privy, cesspool or a piping system carrying high-hazard materials. Table 608.17.1 and Section 703.1 can be used to establish acceptable vertical and horizontal separations from sources of contamination. If the underground piping is within any of the required separation distances, a reduced pressure principle fire-protection backflow prevention assembly is required at the connection of the sprinkler system to the potable water supply.

Case 2: The potability of the water entering through the fire department connection to the sprinkler system cannot be guaranteed. Tanks of fire trucks often contain corrosion inhibitors, antifreeze, extinguishing agents or other chemicals. Pumper trucks sometimes draw from lakes, ponds, rivers, reservoirs, nonpotable water wells, mill water systems or other industrial water sources. [AWWA Manual 14 indicates 1,700 feet (518 m) as the maximum distance for an auxiliary water source to be considered a potential suction supply.] If any of these conditions are anticipated, a reduced pressure principle fire-protection backflow prevention assembly is required at the connection of the sprinkler system to the potable water supply.

Case 3: The sprinkler system is directly connected to a nonpotable secondary water supply. A reduced pressure principle fire-protection backflow prevention assembly is required at the connection of the sprinkler system to the potable water supply.

Case 4: The sprinkler system contains chemical additives, antifreeze or liquid foam fire-extinguishing concentrates. A reduced pressure principle fire-protection backflow prevention assembly is required at the connection of the sprinkler system to the potable water supply.

Case 5: A wet sprinkler system where none of the conditions in Case 1 or 2 exist. A double check valve assembly is the minimum form of backflow protection at the connection of the sprinkler system to the potable water supply.

Case 6: A dry, preaction or deluge sprinkler system of any type of pipe material, where none of the conditions of Case 1 or 2 exist. No backflow preventer is required at the connection of the sprinkler system to the potable water supply.

Case 7: The sprinkler system is a multipurpose type system where all the piping in the system is suitable for use in water distribution systems. A backflow preventer is not required at the take-off points in the water distribution system for the sprinkler piping.

608.16.4.1 Additives or nonpotable source. Where systems under continuous pressure contain chemical additives or antifreeze, or where systems are connected to a nonpotable secondary water supply, the potable water supply shall be protected against backflow by a reduced pressure principle backflow prevention assembly or a reduced pressure principle fire protection backflow prevention assembly. Where chemical additives or antifreeze are added to only a portion of an automatic fire sprinkler or standpipe system, the reduced pressure principle backflow prevention assembly or the reduced pressure principle fire protection backflow prevention assembly shall be permitted to be located so as to isolate that portion of the system. Where systems are not under continuous pressure, the potable water supply shall be protected against backflow by an air gap or an atmospheric vacuum breaker conforming to ASSE 1001 or CSA B64.1.1.

❖ A nonpotable secondary water supply could be water from above- or below-ground tanks, private wells, ponds, reservoirs and lakes. In the event of loss or inadequacy of the potable water supply, the secondary nonpotable water source is pumped into the system, thereby contaminating the potable supply in the event backflow occurs. In some localities, fire departments will use mobile equipment to pump water from nonpotable sources through the fire department connection into the sprinkler system. Where chemicals, additives, antifreeze or nonpotable water are introduced into a sprinkler system, a reduced pressure principle backflow prevention assembly is required. It is not uncommon for a reduced pressure principle backflow prevention assembly to be installed on a branch of the sprinkler system to isolate only that portion of the system containing antifreeze.

Fire sprinkler systems that are not under continuous pressure are classified as indirect cross connections. When subject to only backsiphonage, these systems are to be protected by an atmospheric vacuum breaker conforming to ASSE 1001 or CSA B64.1.1. These devices protect the potable water supply against pollutants or contaminants that enter the system as a result of backsiphonage through the outlet. Under backsiphonage conditions, a small amount of water may exit through the air ports. These devices must be installed downstream of the last control valve and not be subjected to backpressure for more than 12 hours of continuous water pressure.

608.16.5 Connections to lawn irrigation systems. The potable water supply to lawn irrigation systems shall be protected against backflow by an atmospheric vacuum breaker, a pressure vacuum breaker assembly or a reduced pressure principle backflow prevention assembly. Valves shall not be installed downstream from an atmospheric vacuum breaker. Where chemicals are introduced into the system, the potable water supply shall be protected against backflow by a reduced pressure principle backflow prevention assembly.

❖ Lawn irrigation systems are not considered plumbing systems and begin at the point of connection to the potable water distribution (the point of connection being the backflow preventer interface). Irrigation system outlets can be below grade level, exposing them to contamination. The piping in irrigation systems is not regulated by the code. For these reasons, the minimum form of backflow prevention is an atmospheric vacuum breaker.

However, many irrigation systems use zone valve controls that could render atmospheric vacuum breakers inoperative; therefore, in many cases, atmospheric vacuum breakers are not permitted. Pressure vacuum breaker assemblies are designed to operate under pressure at the device outlet. These devices are permitted for all irrigation systems except those having chemical injection. Where fertilizers, herbicides, pesticides or any other chemicals can be introduced into an irrigation system, the potable water supply must be protected by a reduced pressure principle backflow prevention assembly.

It is not uncommon for lawn maintenance personnel to pump chemicals directly into an irrigation system, thereby creating a hazardous cross connection subject to backpressure. Installation of pumps, connections for pumping equipment and auxiliary air tanks on irrigation systems are also common to improve pressure. These types of installations produce backpressure within the system. Under these circumstances, the only acceptable means of backflow prevention is a reduced pressure principle backflow prevention assembly.

Because of the varying nature and unpredictable usage of irrigation systems, most installations are equipped with pressure vacuum breaker assemblies or reduced pressure principle backflow prevention assemblies.

608.16.6 Connections subject to backpressure. Where a potable water connection is made to a nonpotable line, fixture, tank, vat, pump or other equipment subject to high-hazard backpressure, the potable water connection shall be protected by a reduced pressure principle backflow prevention assembly.

❖ A reduced pressure principle backflow prevention assembly is the only acceptable means of protecting a direct connection between the potable water supply and a source of high-hazard contamination that can create a backpressure on the potable water supply. The backpressure can be created by mechanical means, such as a pump, or by static pressure head, including an elevated tank or pipe arrangement.

A direct connection to a pressurized source of contamination is the most dangerous form of cross connection; therefore, the most stringent means of backflow prevention is required, which is a reduced pressure principle backflow prevention assembly (see Commentary Figure 608.16.6).

This section is not intended to supersede the requirements of Section 608.16.4, which permits the use of a double check valve assembly for automatic fire sprinkler systems. Although a connection by the fire department during a fire will likely elevate the automatic sprinkler pressure beyond the domestic water supply pressure (resulting in backpressure), the double check valve backflow prevention assembly is tested and rated for backpressure applications typically associated with automatic sprinkler systems. However, a double check valve assembly is not rated for high-hazard applications; therefore, the potential sources of contamination must be analyzed before a backflow prevention device is selected (see commentary, Section 608.16.4).

**Figure 608.16.6
REDUCED PRESSURE PRINCIPLE
BACKFLOW PREVENTER
(Photo courtesy of Watts Regulator Company)**

608.16.7 Chemical dispensers. Where chemical dispensers connect to the potable water distribution system, the water supply system shall be protected against backflow in accordance with Section 608.13.1, 608.13.2, 608.13.5, 608.13.6, 608.13.8 or 608.13.9.

❖ Chemical dispensers are typically cleaning-related devices having multiple chambers that hold sanitizing solvents. They are stationary self-contained units that, when connected directly to the potable water supply, allow the user to mix the chemicals it contains with water, producing the desired type and strength of cleaning agent. Chemical dispensers are commonly found in a janitor closet or cleaning room in large commercial operations that require a higher-than-average level of sanitation, such as a school or hospital. A connection between the potable water supply and a chemical dispenser creates a situation that is susceptible to backsiphonage with the potential for backpressure. Because of these sources for contamination, chemical dispensers must be protected using an air gap, a reduced principle backflow preventer, a backflow preventer with intermediate vents, a pressure-type vacuum breaker or an atmospheric-type vacuum breaker. After identifying the degree of hazard of the potential contaminants and establishing whether backsiphonage or backpressure can occur, the appropriate device can be selected using Table 608.1.

608.16.8 Portable cleaning equipment. Where the portable cleaning equipment connects to the water distribution system, the water supply system shall be protected against backflow in accordance with Section 608.13.1, 608.13.2, 608.13.3, 608.13.7 or 608.13.8.

❖ Portable cleaning equipment, similar to the chemical dispensers described in Section 608.16.7, are typically cleaning-related devices having multiple chambers that hold sanitizing solvents. These are self-contained mobile units that transport the chemicals with the unit. They can be connected directly to the potable water supply with a hose, which allows the user to mix the chemicals with water, producing the desired type and strength of cleaning agent. Because the location (elevation) of the hose is not restricted, any connection between the potable water supply and portable cleaning equipment is susceptible to backsiphonage with the potential for backpressure. Because of the sources for contamination, portable cleaning equipment must be protected using an air gap, a reduced pressure principle backflow preventer, a backflow preventer with intermediate vents or a double check valve assembly.

Because this cleaning equipment is portable, it is imperative that potential points of connection to the potable water supply be properly protected. Some manufacturers of portable cleaning equipment have tried to address this issue by installing a backflow preventer integral with the equipment. When evaluating this equipment, it is important to verify that the integral devices are rated for the type of hazard and installed in a fixed position with the critical level above the equipment outlet. See Section 608.13 for additional discussion on the proper selection of backflow prevention methods.

608.16.9 Dental pump equipment. Where dental pumping equipment connects to the water distribution system, the water supply system shall be protected against backflow in accordance with Section 608.13.1, 608.13.2, 608.13.5, 608.13.6 or 608.13.8.

❖ Dental equipment that is connected to the potable water supply and operates through the use of a pump is susceptible to backsiphonage with the potential for backpressure and therefore must be protected using an air gap, a pressure-type vacuum breaker or an atmospheric-type vacuum breaker.

This equipment could introduce pathogens or become reservoirs of water-borne disease because of contact with the microorganisms in a patient's mouth. These devices include water-flushed cuspidors, water/air syringes, water-powered aspirators/suction systems, water-flushed ultrasonic cleaning apparatus and water-cooled handpieces.

Some of this equipment has built-in air gaps and does not require external protection or is not provided with water power. However, water pumped into water/air syringes and drill handpieces connected to waterlines are still a significant backflow concern. Given the high-hazard nature of dental unit water lines, a backflow preventer must be installed to isolate them from the rest of the plumbing system.

The code requirements of Section 608.16.9 can be found to conflict with dental industry literature; therefore, the code official must consider the intent of these requirements. Biohazard contamination of the potable water supply has some of the most severe implications, which justifies thorough consideration and the necessity for protection against actual or potential sources of contamination. Therefore, each application of water supply use in conjunction with dental equipment must be scrutinized carefully, considering the design of the equipment and how the water supply integrates into the function of the equipment. The slightest potential for cross contamination must be addressed where potable water supply use is intergrated into the function of any medical equipment.

608.16.10 Coffee machines and noncarbonated beverage dispensers. The water supply connection to coffee machines and noncarbonated beverage dispensers shall be protected against backflow by a backflow preventer conforming to ASSE 1022 or by an *air gap*.

❖ The same protection must be provided for these types of devices as one must provide for carbonated drink dispensers.

608.17 Protection of individual water supplies. An individual water supply shall be located and constructed so as to be safeguarded against contamination in accordance with Sections 608.17.1 through 608.17.8.

❖ The important concern with an individual water supply is that the water remains potable and free from any contamination.

608.17.1 Well locations. A potable ground water source or pump suction line shall not be located closer to potential sources of contamination than the distances shown in Table 608.17.1. In the event the underlying rock structure is limestone or fragmented shale, the local or state health department shall be consulted on well site location. The distances in Table 608.17.1 constitute minimum separation and shall be increased in areas of creviced rock or limestone, or where the direction of movement of the ground water is from sources of contamination toward the well.

❖ A well must be separated from sources of contamination to reduce the possibility of affecting the potability of the water. An analysis of the well location must also include an evaluation of soil and drainage conditions of the area.

WATER SUPPLY AND DISTRIBUTION

TABLE 608.17.1
DISTANCE FROM CONTAMINATION TO PRIVATE WATER SUPPLIES AND PUMP SUCTION LINES

SOURCE OF CONTAMINATION	DISTANCE (feet)
Barnyard	100
Farm silo	25
Pasture	100
Pumphouse floor drain of cast iron draining to ground surface	2
Seepage pits	50
Septic tank	25
Sewer	10
Subsurface disposal fields	50
Subsurface pits	50

For SI: 1 foot = 304.8 mm.

❖ The distances specified in the table have been established by soil scientists and engineers. They take into consideration the variety of soil conditions that may exist and have a built-in factor of safety to drastically reduce the possibility of a well becoming contaminated.

608.17.2 Elevation. Well sites shall be positively drained and shall be at higher elevations than potential sources of contamination.

❖ It is always preferable to have a well located at a higher elevation than any potential source of contamination. This is not always possible, however, especially when the source of contamination is located on an adjacent lot.

608.17.3 Depth. Private potable well supplies shall not be developed from a water table less than 10 feet (3048 mm) below the ground surface.

❖ So the water will remain potable, the well must be at least 10 feet (3048 mm) below grade. It is preferable, however, to have a well a minimum of 20-feet (6096 mm) deep. High water tables and shallow wells are more susceptible to contamination because of their close proximity to the ground surface (see Commentary Figure 608.17.3).

608.17.4 Water-tight casings. Each well shall be provided with a water-tight casing extending to not less than 10 feet (3048 mm) below the ground surface. Casings shall extend not less than 6 inches (152 mm) above the well platform. Casings shall be large enough to permit installation of a separate drop pipe. Casings shall be sealed at the bottom in an impermeable stratum or extend several feet into the water-bearing stratum.

❖ Any contaminant could enter the well with relative ease if the well does not have a proper casing. High water tables are more susceptible to contamination because of their close proximity to the ground surface (see Commentary Figure 608.17.4).

608.17.5 Drilled or driven well casings. Drilled or driven well casings shall be of steel or other *approved* material. Where drilled wells extend into a rock formation, the well casing shall extend to and set firmly in the formation. The annular space between the earth and the outside of the casing shall be filled with cement grout to a depth of not less than 10 feet (3048 mm) below the ground surface. In an instance of casing to rock installation, the grout shall extend to the rock surface.

❖ The most common type of well, a drilled well, can be drilled through rock. This is considered the best all-around method for constructing a well. A driven well is most practical when the water table is relatively shallow [not deeper than 50 feet (15 240 mm)].

The well casing must be installed to prohibit the entrance of contaminants. The grout between the earth and the well casing provides the well with an additional barrier of protection. Use of a drive shoe is recommended to prevent casing distortion (see Commentary Figure 608.17.5).

For SI: 1 inch = 25.4 mm, 1 foot = 304.8 mm.

Figure 608.17.3
DEPTH OF WELL

WATER SUPPLY AND DISTRIBUTION

**Figure 608.17.4
WELL CASING**

**Figure 608.17.5
CASING FOR DRILLED OR DRIVEN WELL**

608.17.6 Dug or bored well casings. Dug or bored well casings shall be of water-tight concrete, tile or galvanized or corrugated metal pipe extending to not less than 10 feet (3048 mm) below the ground surface. Where the water table is more than 10 feet (3048 mm) below the ground surface, the water-tight casing shall extend below the table surface. Well casings for dug wells or bored wells constructed with sections of concrete, tile or galvanized or corrugated metal pipe shall be surrounded by 6 inches (152 mm) of grout poured into the hole between the outside of the casing and the ground and extending not less than 10 feet (3048 mm) below the ground surface.

❖ The opening is much larger with a dug or bored well than with a drilled well. A water-tight casing designed to resist cave-in is required to protect such a well from the entrance of a contaminant.

608.17.7 Cover. Potable water wells shall be equipped with an overlapping water-tight cover at the top of the well casing or pipe sleeve such that contaminated water or other substances are prevented from entering the well through the annular opening at the top of the well casing, wall or pipe sleeve. Covers shall extend downward not less than 2 inches (51 mm) over the outside of the well casing or wall. A dug well cover shall be provided with a pipe sleeve permitting the withdrawal of the pump suction pipe, cylinder or jet body without disturbing the cover. Where pump sections or discharge pipes enter or leave a well through the side of the casing, the circle of contact shall be water tight.

❖ A cover is important to protect the well water from contamination from an outside source. It must fit firmly to the casing with cracks or openings filled tightly to create a water-tight seal (see Commentary Figure 608.17.7).

**Figure 608.17.7
DUG WELL COVER**

608.17.8 Drainage. Potable water wells and springs shall be constructed such that surface drainage will be diverted away from the well or spring.

❖ One of the greatest threats to a well is the risk of contamination from surface drainage. To protect against this, surface drainage must be prevented from draining to the well area and must be diverted away from the casing. In flood-hazard and high-hazard zones, well casings must extend at least 1 foot (305 mm) above the base flood elevation, protecting the well from being contaminated by floodwaters.

SECTION 609
HEALTH CARE PLUMBING

609.1 Scope. This section shall govern those aspects of health care plumbing systems that differ from plumbing systems in other structures. Health care plumbing systems shall conform to the requirements of this section in addition to the other requirements of this code. The provisions of this section shall apply to the special devices and equipment installed and maintained in the following occupancies: nursing homes, homes for the aged, orphanages, infirmaries, first aid stations, psychiatric facilities, clinics, professional offices of dentists and doctors, mortuaries, educational facilities, surgery, dentistry, research and testing laboratories, establishments manufacturing pharmaceutical drugs and medicines and other structures with similar apparatus and equipment classified as plumbing.

❖ This section governs the water supply to health care facilities and fixtures. Hospitals are required to have dual water services so that in the event of a water main or water service pipe break, the hospital will not experience a complete loss of water supply. In health care facilities there are unique plumbing fixtures with special requirements. Although typically found in hospitals, these fixtures could be located in other health care facilities. Note that the list of occupancies is not all-inclusive; it is intended only to be representative of the occupancies in which special plumbing fixtures are regulated by this chapter.

609.2 Water service. Hospitals shall have two water service pipes installed in such a manner so as to minimize the potential for an interruption of the supply of water in the event of a water main or water service pipe failure.

❖ This section applies only to hospitals. It is not applicable to other health care facilities, such as urgent care centers or doctors' offices. The intent of the section is to require two remote water services; however, they can connect to the same water main. Where a single water main is provided, it is desirable to connect the water services on opposite sides of the water main control valve. In the event that the water main control valve is closed, one of the two water service taps will remain in service (see Commentary Figure 609.2).

609.3 Hot water. *Hot water* shall be provided to supply all of the hospital fixture, kitchen and laundry requirements. Special fixtures and equipment shall have hot water supplied at a temperature specified by the manufacturer. The hot water system shall be installed in accordance with Section 607.

❖ The hot water demands of a hospital vary greatly depending on its type of equipment and the services provided. When a laundry is installed, the newer systems have a separate hot water supply system independent of the hospital system. The systems are designed for uninterrupted laundry service.

Kitchen equipment demands also vary depending on the type of kitchen and the accompanying equipment. The design professional must determine the needs of hospital systems when designing the hot water supply system.

609.4 Vacuum breaker installation. Vacuum breakers shall be installed not less than 6 inches (152 mm) above the *flood level rim* of the fixture or device in accordance with Section 608. The *flood level rim* of hose connections shall be the maximum height at which any hose is utilized.

❖ Because a vacuum breaker has a critical level and must not be subjected to any backpressure, vacuum breakers with hose connections must be elevated above the highest level that the hose can be raised (see commentary, Section 608.13). Vacuum breakers serving hose connections are typically located 7 feet (2134 mm) or more above the floor.

609.5 Prohibited water closet and clinical sink supply. Jet- or water-supplied orifices, except those supplied by the flush

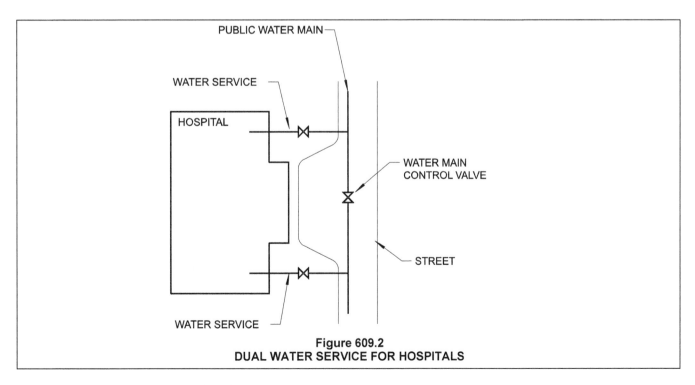

**Figure 609.2
DUAL WATER SERVICE FOR HOSPITALS**

connections, shall not be located in or connected with a water closet bowl or clinical sink. This section shall not prohibit an *approved* bidet installation.

❖ The concern with water closets and clinical sinks is that any submerged water supply outlet must be protected against backflow. If the water supply is through the flush connection, it is protected by the vacuum breaker of the flushing device.

609.6 Clinical, hydrotherapeutic and radiological equipment. Clinical, hydrotherapeutic, radiological or any equipment that is supplied with water or that discharges to the waste system shall conform to the requirements of this section and Section 608.

❖ This section parallels Section 423 in that it addresses any and all specialty-type fixtures, equipment and appliances that would not generally be classified as plumbing fixtures. The intent is to protect the potable water supply from backflow contamination that could be caused by the variety of special health care-related equipment supplied by the potable water distribution system. All such fixtures and equipment must comply with the provisions in Section 609 and with the water supply protection requirements of Section 608.

Because such fixtures and equipment are not normally viewed as plumbing fixtures, they are commonly overlooked during the design or plumbing inspection stages. Specialty fixtures and equipment could pose a backflow hazard equal to or greater than any typical plumbing fixture (see commentary, Section 608)

609.7 Condensate drain trap seal. A water supply shall be provided for cleaning, flushing and resealing the condensate trap, and the trap shall discharge through an *air gap* in accordance with Section 608.

❖ An indirect waste receptor or floor drain receiving the discharge of condensate in a health care facility must have a water supply for flushing the receptor and resealing the trap should the sealing water evaporate. There must be an air gap between the water supply opening and the waste receptor to protect against backflow.

609.8 Valve leakage diverter. Each water sterilizer filled with water through directly connected piping shall be equipped with an *approved* leakage diverter or bleed line on the water supply control valve to indicate and conduct any leakage of unsterile water away from the sterile zone.

❖ Sterilized water must be protected against leakage from the unsterile potable water supply.

SECTION 610
DISINFECTION OF POTABLE WATER SYSTEM

610.1 General. New potable water systems shall be purged of deleterious matter and disinfected prior to utilization. The method to be followed shall be that prescribed by the health authority or water purveyor having jurisdiction or, in the absence of a prescribed method, the procedure described in either AWWA C651 or AWWA C652, or as described in this section. This requirement shall apply to "on-site" or "in-plant" fabrication of a system or to a modular portion of a system.

1. The pipe system shall be flushed with clean, potable water until dirty water does not appear at the points of outlet.

2. The system or part thereof shall be filled with a water/chlorine solution containing not less than 50 parts per million (50 mg/L) of chlorine, and the system or part thereof shall be valved off and allowed to stand for 24 hours; or the system or part thereof shall be filled with a water/chlorine solution containing not less than 200 parts per million (200 mg/L) of chlorine and allowed to stand for 3 hours.

3. Following the required standing time, the system shall be flushed with clean potable water until the chlorine is purged from the system.

4. The procedure shall be repeated where shown by a bacteriological examination that contamination remains present in the system.

❖ Pipe and pipe fittings are potentially exposed to a variety of pollutants and contaminants from manufacturing, handling and installation. Extensive flushing of a completed potable water system with potable water is usually not adequate to ensure that bacteria or microorganisms that might be present will be flushed away. Disinfection is the only way to kill bacteria and microorganisms living within potable water systems.

The most common disinfection method is the exposure of the internal surfaces of the piping system to a chlorine solution. Chlorine breaks down the cell walls of bacteria and microorganisms and destroys the enzymes and structures inside the cell. This process is called "oxidation" and renders the remainder of the cells harmless.

Chlorine is available in water-based solutions having various concentrations of "free chlorine." Free chlorine is the portion of the solution that is available for disinfecting purposes. Free chlorine is typically indicated by the percentage of sodium hypochlorite in the solution. For example, the majority of common household bleaches have 5.25-percent NaOCl (sodium hypochlorite). Commonly available liquid pool chlorine has approximately 12-percent NaOCl. Dry chlorine tablets are also available that will, when dissolved in a specific volume of water, make a water-based solution of a specific concentration of free chlorine. Although pure chlorine is available in gaseous form, it is very dangerous if not handled with extreme care. Chlorine gas is colorless, spreads quickly and can result in death if inhaled only momentarily.

The true test of whether the piping system has been properly disinfected is for a bacteriological analysis to be performed. As the sampling procedures and testing protocols are critical, specialized testing agencies are used to determine if any bacteria or microorganisms are still present in the potable water system.

For the 2015 code, the requirement for disinfection

of a repaired water supply system was removed because 1) repairs are often minor as compared to the size of the entire system 2) there is better control of new/repair materials to avoid contaminants and 3) the required system downtime for disinfection cannot be tolerated.

SECTION 611
DRINKING WATER TREATMENT UNITS

611.1 Design. Drinking water treatment units shall meet the requirements of NSF 42, NSF 44, NSF 53, NSF 62 or CSA B483.1.

❖ This section contains the requirements for point of use and point of entry drinking water treatment units. These units are intended to augment the treatment of the drinking water by the municipality and to provide user treatment for private wells. Water delivered to a building is expected to be within established guidelines for health, odor and chemical content. These guidelines are established by the U.S. Department of Health and are applicable to potable water sources, whether they are public municipal water treatment facilities or individual private wells. Even though the guidelines make certain that the water is safe to consume, there are times when the water supply has an objectionable odor or dissolved materials within it are noticeable to the occupants of the building, particularly in drinking water.

NSF 42 contains requirements for units that are designed to improve the perceived quality of drinking water supplied to the occupants of the building. These units are typically installed either at the final distribution point of the drinking water, referred to as "point of use" units, or at a central location where several fixtures can be served from the same treatment unit, referred to as "point of entry" units.

Regardless of the location of the treatment units, their function is the same. These units are designed to reduce the amount of specific contaminants present in the water supply. The contaminants affect the aesthetic quality of the water, causing occupants to smell an offensive odor or to notice some chemical or particulate in the water. Most water treatment units are designed to remove certain odors or dissolved materials in the drinking water supply.

The user or installer of the treatment unit is responsible for the selection of the proper unit based on the offensive characteristics of the water and the unit's reported performance capabilities. Treatment units require periodic maintenance so they will continue to function as designed. Because of this requirement, the treatment unit must be installed in a location where it is accessible for maintenance.

NSF 53, similar to the NSF 42 standard discussed above, contains the standards for units designed to improve the quality of drinking water by removing dissolved materials or substances that might prove to have a negative effect on the health of the building occupants. These materials or substances could be microbiological, chemical or particulate and are often considered either an established or potential health hazard. Although the exact source of the potentially harmful material or substances is not known, the intent of these treatment units is to remove these materials or substances from the water just prior to its consumption by the occupants.

611.2 Reverse osmosis systems. The discharge from a reverse osmosis drinking water treatment unit shall enter the drainage system through an *air gap* or an *air gap* device that meets the requirements of NSF 58 or CSA B483.1.

❖ Similar to the drinking water treatment units discussed in Section 611.1, the reverse osmosis systems manufactured in accordance with NSF 58 are intended to remove specific contaminants from public or private drinking water supplies considered to be microbiologically safe. The treatment units designed in conformance with NSF 58 are intended to reduce the total dissolved solids (TDS) and other contaminants present in the drinking water, many of which are considered to be either established or potential health hazards.

A reverse osmosis system reduces the objectionable dissolved materials by the use of pressure to force the water through a filter membrane or medium in such a manner that the concentration level of the discharged water is less than the concentration of the incoming water. Reverse osmosis treatment systems are intended for installation as a point of use system. This means that the treatment unit serves the drinking water fixture located immediately adjacent to the treatment system.

Treatment units manufactured in accordance with this standard are to be clearly identified as conforming to this standard, including the specific performance claims of the unit as verified and substantiated by test data. The waste discharge from a reverse osmosis drinking water treatment unit must be designed, constructed and located so that the discharge orifice is directed downward and the lower edge of the outlet must be at an elevation not less than 2 inches (51 mm) above the flood rim of the waste receptacle.

611.3 Connection tubing. The tubing to and from drinking water treatment units shall be of a size and material as recommended by the manufacturer. The tubing shall comply with NSF 14, NSF 42, NSF 44, NSF 53, NSF 58 or NSF 61.

❖ The tubing used for drinking water treatment units must be of a size and material recommended by the manufacturer. The tubing materials are certified during the evaluation of the equipment for compliance with materials or products that come into contact with drinking water.

SECTION 612
SOLAR SYSTEMS

612.1 Solar systems. The construction, installation, alterations and repair of systems, equipment and appliances intended to utilize solar energy for space heating or cooling, domestic hot water heating, swimming pool heating or process heating shall be in accordance with the *International Mechanical Code*.

❖ Solar energy can be used for a variety of purposes including space heating, space cooling, domestic water heating, swimming pool heating and process heating. Therefore, the scope encompasses all of these potential solar energy applications and includes the construction, installation, alteration and repair of all systems and equipment.

SECTION 613
TEMPERATURE CONTROL DEVICES AND VALVES

613.1 Temperature-actuated mixing valves. Temperature-actuated mixing valves, which are installed to reduce water temperatures to defined limits, shall comply with ASSE 1017. Such valves shall be installed at the hot water source.

❖ ASSE 1017 is the standard for temperature-actuated mixing valves for hot water distribution systems. Water mixing (also defined as "tempering" or "blending") valves are used extensively in applications for domestic service to mix hot and cold water to reduce high service water temperature to the building distribution piping system.

These devices are not intended for final temperature control at the fixtures and appliances (see ASSE 1016). Final control must be provided by suitable individual mixing devices, which should be supplemented by appropriate high-limit alarms or temperature-limiting devices at the point of use. These valves are designed for automatic control of hot water temperature and to maintain this temperature within a reasonable degree of uniformity.

Bibliography

The following resource materials are referenced in this chapter or are relevant to the subject matter addressed in this chapter.

Hunter, Roy B. *Water-Distribution Systems for Buildings*. Washington, DC: National Bureau of Standards, NBS BMS, 1941.

Chapter 7:
Sanitary Drainage

General Comments

The conventional method of sizing a sanitary drainage system is by drainage fixture unit (dfu) load values. The drainage fixture unit approach takes into consideration the probability of a load on a drainage system.

The probability method of sizing drainage systems was developed largely by Dr. Roy Hunter. Through his research, Dr. Hunter attempted to standardize and simplify design principles, while reducing the cost of plumbing systems.

Fixture unit values were determined based on the average rate of discharge by a fixture, the time of a single operation and the frequency of use or interval between operations. The theoretical approach considers a large group of fixtures being connected to the plumbing system with only a small fraction of the total number of fixtures in use simultaneously. The probability method has also been effective in the design of smaller plumbing systems because of the excessive design factors added by Dr. Hunter.

Dr. Hunter sought an adequate design methodology to provide satisfactory service without interruption or inconvenience to the user.

Because the fixture unit values have a built-in probability factor, they cannot be directly translated into flow or discharge rates. For example, one dfu is equivalent to a discharge rate of 1 cubic foot per minute (0.4719 L/s) [approximately 7.5 gallons per minute (gpm) (28 L/min)]. Two independent fixtures, however, each having a value of one dfu, cannot be considered as having a combined discharge rate of 15 gpm (57 L/min) because the dfu value incorporates the element of probability.

Purpose

The purpose of Chapter 7 is to regulate the materials, design and installation of sanitary drainage piping systems as well as the connections made to the system. The intent is to design and install sanitary drainage systems that will function reliably, are neither undersized nor oversized and are constructed from materials, fittings and connections whose quality is regulated by this section.

In the 19th century, typhoid fever, cholera and dysentery were a threat to survival. The modern plumbing system, with proper drainage piping, has been one of the main reasons for the elimination of these diseases.

Medical professionals give much of the credit to the plumbing profession for improvements in health and longevity. Medicine alone would have only a marginal effect without improved sanitation practices.

SECTION 701
GENERAL

701.1 Scope. The provisions of this chapter shall govern the materials, design, construction and installation of sanitary drainage systems.

❖ This section establishes that design and installation provisions for sanitary drainage systems are located in Chapters 3, 7 and 10.

Chapter 7 regulates the design, sizing, components and installation of a sanitary drainage system. Many other chapters relate to Chapter 7 because they regulate components of sanitary drainage systems. For example, Chapters 3, 9 and 10 regulate pipe support, venting and traps, respectively, all of which are combined to create a sanitary drainage system.

701.2 Sewer required. Buildings in which plumbing fixtures are installed and premises having drainage piping shall be connected to a public *sewer*, where available, or an *approved* private sewage disposal system in accordance with the *International Private Sewage Disposal Code*.

❖ Plumbing fixtures installed in a building or structure must be directly connected to a drainage system. The sanitary drainage systems must connect to the public systems provided by the local jurisdiction or public utility if they are available. If a public system is not available, the drainage system must connect to an approved private sewage disposal system in accordance with the *International Private Sewage Disposal Code*® (IPSDC®), which contains requirements for the design and installation of private sewage disposal systems. This section must not be construed to prevent the indirect waste systems required by Chapter 8.

701.3 Separate sewer connection. A building having plumbing fixtures installed and intended for human habitation, occupancy or use on premises abutting on a street, alley or easement in which there is a public *sewer* shall have a separate connection with the *sewer*. Where located on the same lot, multiple buildings shall not be prohibited from connecting to a common *building sewer* that connects to the public *sewer*.

❖ This requirement intends to prohibit the combining of sewers serving different buildings prior to connection to the public sewer. The only exception is where the sewers to be combined are serving buildings on the same lot or parcel of land. This section does not prohibit the use of adjoining properties that have been included in a dedicated easement approved by the administrative authority. Commentary Figure 701.3(1) illustrates the use of a dedicated public sewer easement to accommodate Buildings A and B. The common building sewer is actually an extension of the public sewer and under control of the public authority.

When the sanitary discharge from more than one building is connected to the public sewer by means of a shared or common sewer line, the responsibility for the maintenance of the common sewer line can become a problem. Should a stoppage develop in the common portion of the drain, all parties connected to the drain could disclaim responsibility for having the stoppage cleared. This presents a problem because there is no one party to hold responsible for the maintenance of the drain. Also, replacement or repairs of a sewer are the responsibility of all parties sharing a common drain connection, and the process of determining the point of origin of the wastes into the common drain is more difficult. The first sentence of this section is intended to apply to the connection of more than one building where each building is under separate ownership, and the second sentence applies in instances where connection of more than one structure is proposed where all the buildings are under the same ownership. The commercial property owner, such as a shopping complex owner, may want to build another building on his or her property, and connecting the new building sewer to existing sewers on the same property would be convenient and less expensive than making a new connection to the public sewer [see Commentary Figure 701.3(2)].

Figure 701.3(1)
COMMON BUILDING SEWER LOCATED WITHIN AN APPROVED PUBLIC SEWER EASEMENT

Figure 701.3(2)
BUILDINGS ON SAME LOT

701.4 Sewage treatment. Sewage or other waste from a plumbing system that is deleterious to surface or subsurface waters shall not be discharged into the ground or into any waterway unless it has first been rendered innocuous through subjection to an *approved* form of treatment.

❖ The discharge of untreated waste and sewage from a sanitary drainage system creates an environmental and human health hazard. This section requires treatment of sewage that poses a hazard before discharging to ponds, rivers, streams, lakes, aquifers or similar bodies of water. The form of treatment for the waste must be approved by the code official and is often governed by the local health department or state or U.S. Environmental Protection Agency (EPA) having jurisdiction. These requirements are also applicable to gray-water recycling systems, as stated in Appendix C of the code.

701.5 Damage to drainage system or public sewer. Waste detrimental to the public *sewer* system or to the functioning of the sewage-treatment plant shall be treated and disposed of in accordance with Section 1003 as directed by the code official.

❖ Materials that are detrimental to the drainage system must not discharge into such system. Any substances that will clog, produce explosive mixtures, degrade the pipes or their joints or any wastes that interfere with the sewage disposal or treatment process must not be allowed to enter the building drainage system. Sections 803.2 and 1003.1 contain requirements for collecting harmful materials and the neutralization of corrosive wastes (see commentary, Sections 803.2 and 1003.1).

701.6 Tests. The sanitary drainage system shall be tested in accordance with Section 312.

❖ Requirements for sanitary drainage and vent system testing are described in Section 312.

701.7 Engineered systems. Engineered sanitary drainage systems shall conform to the provisions of Sections 316 and 714.

❖ Engineered sanitary drainage systems are designed using approved computer program methods and may not comply with all the provisions prescribed in this chapter. The system design must be approved by the code official in accordance with Section 316 and must conform to the provisions of Section 714.

701.8 Drainage piping in food service areas. Exposed soil or waste piping shall not be installed above any working, storage or eating surfaces in food service establishments.

❖ A potential for contamination exists in areas where piping is exposed. This protection is necessary for health, safety and sanitation of food service areas. Many design alternatives are available, such as secondary containment, ceilings and perimeter installations that avoid these areas.

SECTION 702
MATERIALS

702.1 Above-ground sanitary drainage and vent pipe. Above-ground soil, waste and vent pipe shall conform to one of the standards listed in Table 702.1.

❖ This section contains material requirements for all sanitary drainage, waste and vent (DWV) systems. The referenced table specifies allowable piping materials for above-ground DWV systems. In selecting pipe and fittings, it is important to consider their reliability for installation in a drainage system. The corrosion resistance of the pipe and fittings is a measure of the ability of the materials used to resist both the internal corrosive effects of the effluent likely to flow through them and the effects of soils on their exterior. Corrosion can be reduced or eliminated by the application of coating, lining and cathodic protection. Physical strength of the pipe and fittings is another consideration because of the physical damage that may occur either during installation or after being placed in service. Fire-resistance rating of the pipe and fittings is a measure of their ability to remain intact and not fail, or to carry water, during a fire. Additionally, pipe, fittings, joints, hangers and supports should be selected taking this into consideration.

SANITARY DRAINAGE

TABLE 702.1
ABOVE-GROUND DRAINAGE AND VENT PIPE

MATERIAL	STANDARD
Acrylonitrile butadiene styrene (ABS) plastic pipe in IPS diameters, including Schedule 40, DR 22 (PS 200) and DR 24 (PS 140); with a solid, cellular core or composite wall	ASTM D 2661; ASTM F 628; ASTM F 1488; CSA B181.1
Brass pipe	ASTM B 43
Cast-iron pipe	ASTM A 74; ASTM A 888; CISPI 301
Copper or copper-alloy pipe	ASTM B 42; ASTM B 302
Copper or copper-alloy tubing (Type K, L, M or DWV)	ASTM B 75; ASTM B 88; ASTM B 251; ASTM B 306
Galvanized steel pipe	ASTM A 53
Glass pipe	ASTM C 1053
Polyolefin pipe	ASTM F 1412; CSA B181.3
Polyvinyl chloride (PVC) plastic pipe in IPS diameters, including schedule 40, DR 22 (PS 200), and DR 24 (PS 140); with a solid, cellular core or composite wall	ASTM D 2665; ASTM F 891; ASTM F 1488; CSA B181.2
Polyvinyl chloride (PVC) plastic pipe with a 3.25-inch O.D. and a solid, cellular core or composite wall	ASTM D 2949, ASTM F 1488
Polyvinylidene fluoride (PVDF) plastic pipe	ASTM F 1673; CSA B181.3
Stainless steel drainage systems, Types 304 and 316L	ASME A112.3.1

❖ Table 702.1 identifies the materials that are approved and the standards with which these materials must comply.

Acrylonitrile butadiene styrene (ABS) plastic pipe.

The exterior of ABS plastic pipe is black in color. Depending on which standard the pipe is made to, the pipe can be solid wall or cellular core (frequently called "foam core"). The solid-wall pipe is produced by either a single extrusion or a simultaneous multiple coextrusion. The cellular core pipe is produced by a coextrusion process with concentric inner and outer solid ABS layers and the core consisting of closed-cell foamed plastic.

The pipe is marked with not less than the following information:

- The manufacturer's name.
- One or more standard designations: ASTM D 2661, ASTM F 628, ASTM F 1488, CSA B181.1.
- Nominal pipe size.
- "ABS DWV" for single extrusion pipe.
- "COEX ABS DWV" for coextruded pipe.
- "COEX ABS CELLULAR CORE DWV" for coextruded pipe with cellular core.

CSA B181.1: This standard covers ABS pipe and fittings for drain waste and vent applications. The available pipe sizes are $1^1/_4$ to 6 inches (32 to 152 mm) with the outside diameters and wall thicknesses matching Schedule 40 steel pipe.

ASTM D 2661: This standard covers ABS pipe and fittings for drain waste and vent applications. The available pipe sizes are $1^1/_4$ to 6 inches (32 to 152 mm) with the outside diameters and wall thicknesses matching Schedule 40 steel pipe. The pipe is made from virgin material with no reprocessed or recycled plastic. The standard allows single extruded and coextruded forms. Piping and fittings covered by this standard do not have a pressure rating.

ASTM F 628: This standard covers ABS pipe and fittings for drain waste and vent applications. The available pipe sizes are $1^1/_4$ to 6 inches (32 to 152 mm) with the outside diameters and wall thicknesses matching Schedule 40 steel pipe. The pipe is made from virgin material with no reprocessed or recycled plastic. The standard covers the cellular core form of pipe. Piping and fittings covered by this standard do not have a pressure rating.

ASTM F 1488: This standard covers composite pipe, which means that layers of the pipe could be of different materials such as PVC or CPVC. The pipe external layer is made from virgin ABS material with no reprocessed or recycled plastic. The standard covers coextruded forms, including the cellular core form. Piping covered by this standard does not have a pressure rating.

The available pipe sizes for the IPS-SCH 40 series are $1^1/_4$ to 12 inches (32 to 305 mm) including $2^1/_2$-, $3^1/_2$- and 5-inch (64, 89 and 127 mm) sizes with the outside diameters and wall thicknesses matching Schedule 40 steel pipe [except for the 12-inch (305 mm) size, which has an 0.460-inch (12 mm) wall thickness].

The available pipe sizes for the IPS-DR-PS series are $1^1/_4$ to 12 inches (32 to 305 mm) including $2^1/_2$-, $3^1/_2$-, and 5-inch (64, 89 and 127 mm) sizes with the outside diameters matching Schedule 40 steel pipe. The wall thickness is based upon a dimension ratio (DR) of 24 or 22 [corresponding to a pipe stiffness (PS) of PS140 and PS200, respectively].

Brass pipe

This piping material is made from red brass, which is composed of approximately 85-percent copper and 15-percent zinc. The pipe is available in both standard and extra-strong weight. Extra-strong weight has the same outside diameter as steel pipe, but with a thicker pipe wall.

Standard-weight pipe is named "Schedule 40," and extra-strong pipe is named "Schedule 80." Brass pipe dimensions are very similar to steel pipe dimensions. ASTM B 43 is the referenced standard for brass pipe, available in diameters from $1/_8$ inch (3.2 mm) to 12 inches (305 mm) and manufactured in 12-foot (3658 mm) lengths. The pipe is marked with the following information:

- The manufacturer's name.
- The designation of ASTM B 43.
- Nominal pipe size.
- Schedule 40 or Schedule 80.

Cast-iron pipe

Cast-iron is manufactured in three classifications: service (standard) weight, extra-heavy weight and hubless. The pipe is usually lined internally with cement or coal-tar enamel and is coated externally with a variety of materials to reduce corrosion by soils. Two types of pipe ends are manufactured: hub-and-spigot and hubless. ASTM A 74 is the referenced standard for hub-and-spigot and is available in diameters from 3 to 15 inches (76 to 381 mm) and manufactured in 5- and 10-foot (1524 mm and 3048 mm) lengths. CISPI 301 and ASTM A 888 are the referenced standards for hubless pipe, which is available in diameters from $1^1/_2$ to 15 inches (38 to 381 mm) and manufactured in 5-, 10- and 18-foot (1524, 3048 and 5486 mm) lengths. The hub and spigot ends may be joined by either caulking or an elastomeric compression gasket. Hubless ends can be joined only by external compression couplings.

Cast-iron pipe is marked continuously on the barrel with the following information:

- The manufacturer's name or registered trademark.
- The country of origin.
- Letters to indicate the proper classification, extra heavy (XH) or service weight (SV).

Copper or copper-alloy pipe

See the commentary to Table 605.3, Copper or copper-alloy pipe.

Copper or copper-alloy tubing (Type K, L, M or DWV)

ASTM B 306 is the referenced standard for Type DWV copper tubing, which has a thinner pipe wall than Type M copper. The tubing is produced only in a drawn (hard) temper (see commentary, Table 605.3, Copper or copper-alloy tubing, Type K, L, M or DWV). Type DWV copper tubing is identified with a continuous yellow marking with the following information:

- The manufacturer's name or registered trademark and "DWV."
- The country of origin.

Galvanized steel pipe

The ASTM A 53 standard covers continuous welded and electric-resistance welded (ERW) pipe. Continuous-welded pipe is made from a flat strip of steel that is heated, formed into a cylinder and forged welded along the longitudinal seam. ERW pipe is made from a flat strip of steel that is cold formed into a cylinder and electric-resistance welded along the longitudinal seam.

Galvanized steel pipe is available in different wall thicknesses, the most common being standard, extra-strong and double-extra-strong weight. These thicknesses are also referred to as Schedules 40, 80 and 160, respectively. As the schedule number increases, the wall thickness of the pipe increases. The outside diameter, however, remains the same for each pipe size. Steel pipe is hot dipped and zinc coated to produce a galvanized protective coating, both inside and outside. New galvanized pipe is a shiny silver, and, as the pipe ages, the coating oxidizes to protect the zinc. As a result, the color turns to a dull gray.

Galvanized steel pipe is sometimes called "galvanized pipe" or mistakenly called "wrought-iron pipe." Wrought-iron pipe has not been manufactured in the United States in more than 30 years. All pipe referred to as "wrought-iron pipe" is steel pipe.

The steel pipe industry claims to be responsible for nominal pipe dimensions, meaning the pipe is approximately the size by which it is identified. The dimensions date back to 1862 when Robert Briggs of Pascal Iron Works wrote the first pipe standards in the United States. It is believed that the dimensions were based on the pipe Briggs was manufacturing. That is why a $^3/_4$-inch (19.1 mm) galvanized steel pipe has an outside diameter of 1.05 inches (27 mm) and an average inside diameter of 0.824 inch (21 mm) for standard-weight pipe. The pipe is available in diameters of $^1/_8$ inch through 26 inches (3.2 mm through 660 mm) and various lengths of 12 feet (3658 mm) to 22 feet (6706 mm).

Galvanized steel pipe is marked with the following information:

- The manufacturer's name or mark.
- The type of pipe (i.e., A, B, XS, XXS).
- The specification number.
- The length of pipe.

See Commentary Figure 702.1 for dimensions of Schedule 40 steel pipe.

Glass pipe

Borosilicate glass pipe is low-expansion, low-alkali-content glass pipe that is extremely chemical resistant. Glass pipe is used primarily for above-ground chemical waste drainage systems. ASTM C 1053 is the referenced standard for borosilicate glass pipe and fittings used for drainage, waste and vent systems and governs the chemical requirements, operating temperatures, pressure rating and dimensions for glass pipe and fittings. Glass piping has a minimum rated operating temperature of -40°F (-40°C) and a maximum operating temperature of 212°F (100°C). The maximum rated working pressure of glass DWV piping systems is the pressure caused by gravity flow. Glass pipe is manufactured in $1^1/_2$-, 2-, 3-, 4- and 6-inch (38, 51, 76, 102 and 152 mm) diameters and standard lengths of 5 and 10 feet (1524 and 3048 mm).

SANITARY DRAINAGE

NOMINAL PIPE SIZE (inches)	OUTSIDE DIAMETER (inches)	AVERAGE WALL THICKNESS (inch)
1 1/4	1.660	0.140
1 1/2	1.9	0.145
2	2.375	0.154
2 1/2	2.875	0.203
3	3.5	0.216
3 1/2	4	0.226
4	4.5	0.226
5	5.563	0.258
6	6.625	0.280
8	8.625	0.322
10	10.75	0.365
12	12.75	0.375
14	14	0.437
16	16	0.5
18	18	0.562
20	20	.583
24	24	.687

For SI: 1 inch = 25.4 mm.

**Figure 702.1
STEEL PIPE DIMENSIONS**

Polyolefin pipe

Polyolefin is a term used to describe a family of plastics that includes polyethylene (PE) and polypropylene (PP). ASTM F 1412 and CSA B181.3 are the referenced standards for polyolefin pipe and fittings used in laboratory drainage systems for corrosive applications. This material may not be compatible with all chemicals, making consultation with the manufacturer necessary to determine the suitability of the pipe for a particular application (see commentary, Section 702.5). Polyolefin is identified with the following information:

- The manufacturer's name or trademark.
- The designation CSA B181.3 or the wording "Lab Drainage."
- Nominal pipe size.
- Raw material designation.

Polyvinyl chloride (PVC) plastic pipe

PVC plastic pipe for drainage and vent systems is available in solid wall, coextruded with either a cellular core or composite pipe. The pipe is either white or a milky white (cream) and is continuously marked with the manufacturer's name, the ASTM International (ASTM) or Canadian Standards Association (CSA) standard and Type DWV. Cellular core pipe is also marked "COEX CELLULAR CORE PVC PIPE," and composite pipe is marked with the applicable DR and PS, for example, DR35/PS50 or sewer and drain series DR35/PS50. If the pipe also conforms to the requirements for water service pipe, it will be double marked. ASTM D 2665 and CSA B 181.2 are the referenced standards for PVC plastic pipe and fittings. The pipe has the same dimensions as Schedule 40 steel pipe. For this reason, it is often called "Schedule 40 PVC drain pipe." ASTM F 891 is the referenced standard for coextruded cellular core PVC pipe. The standard specifies various grades of pipe. Only Schedule 40 pipe, also designated as Type DWV, is approved for above-ground use. ASTM F 1488 is the referenced standard for coextruded composite PVC pipe. The standard specifies various grades of pipe based on different DR and PS applications. IPS Schedule 40 and IPS DR-PS Series are used for above- or below-ground installation for drain, waste and vent pipe. Sewer and Drain DR-PS Series is used for gravity flow sewer and drain pipe. ASTM D 2949 is the referenced standard for 3-inch (76 mm) PVC plastic pipe and fittings only. The pipe has an outside diameter of $3^1/_4$ inches (83 mm), as compared to $3^1/_2$ inches (89 mm) for other 3-inch (76 mm) PVC plastic pipe. The inside diameter is the same as a 3-inch (76 mm) pipe conforming to ASTM D 2665. The resulting pipe wall is $^1/_8$ inch (3.2 mm) less in thickness.

The $3^1/_4$-inch (83 mm) outside diameter pipe was developed to permit installation of PVC plastic pipe in a nominal 2-inch by 4-inch (51 mm by 102 mm) stud wall cavity. The pipe is often incorrectly identified as Schedule 30 PVC plastic pipe or "thin" wall pipe.

CSA 181.2: This standard covers PVC pipe and fittings for drain waste and vent applications. OD and wall match Schedule 40 Steel up to 12 inches (305 mm). Over 12 inches (305 mm), walls are thicker than steel pipe sizes. Up to 24-inch (610 mm) size.

ASTM D 2665: This standard covers PVC pipe and fittings. The pipe is made in a single extrusion with the OD and wall matching Schedule 40 steel pipe up to 12 inches (305 mm). Over 12 inches (305 mm) and up to 24-inch (610 mm) size, the walls of the pipe are thicker than steel pipe sizes. Although this pipe is pressure tested, it is only indicated for DWV service.

ASTM D 2949: This standard covers PVC pipe and fittings for one size of pipe: $3^1/_4$ inches (83 mm) outside diameter. The pipe has a wall thickness of $^1/_8$ inch (3.2 mm).

ASTM F 891: This standard covers PVC coextruded cellular core pipe. It is suitable for nonpressure use. The OD and wall matches IPS Schedule 40 steel pipe except that the 12-inch (305 mm) size has a wall thickness of 0.406 inch (10.3 mm).

ASTM F 1488: This standard covers composite pipe, which means that layers of the pipe could be of different materials such as ABS or CPVC. The pipe external layer is made from virgin PVC material with no reprocessed or recycled plastic. The standard covers coextruded forms, including the cellular core form. Piping covered by this standard does not have a pressure rating.

The available pipe sizes for the IPS-SCH 40 series are 1^1/$_4$ to 12 inches (32 to 305 mm) including 2^1/$_2$-, 3^1/$_2$- and 5-inch (66, 89 and 127 mm) sizes with the outside diameters and wall thicknesses matching Schedule 40 steel pipe [except for the 12-inch (305 mm) size, which has an 0.460-inch (11.7 mm) wall thickness].

The available pipe sizes for the IPS-DR-PS series are 1^1/$_4$ to 12 inches (32 to 305 mm), including 2^1/$_2$-, 3^1/$_2$-, 5-inch (66, 89 and 127 mm) sizes with the outside diameters matching steel pipe. The wall thickness is based upon a DR of 24 or 22 (corresponding to PS 140 and PS 200, respectively).

Only the 3^1/$_4$-inch (83 mm) OD size of the sewer and drain-DR-PS series is listed for above-ground drainage and vent applications pipe.

Polyvinylidene fluoride (PVDF) pipe

The ASTM F 1673 standard covers pipe and fittings for corrosive applications. The available pipe sizes are 1^1/$_4$ to 12 inches (32 to 305 mm) with outside diameters matching steel pipe. The pipe is available in wall thicknesses matching Schedule 40 and Schedule 80 steel pipe dimensions. The pipe is also available in a wall thickness corresponding to a standard dimension ratio (SDR) of 21.

Stainless steel drainage systems

Stainless steel drainage and vent systems are made of stainless steel alloy of Type 304 or Type 316L. The thickness of the pipe wall is 0.040 inch (1 mm) except for 6-inch (152 mm) size, which has a pipe wall thickness of 0.060 inch (1.5 mm).

Stainless steel systems are used for both sanitary waste and chemical waste systems. ASME A112.3.1 is the referenced standard for stainless steel drainage system pipe and fittings. For building drainage and vent piping, Type 304 is approved for use only in above-ground applications and Type 316L is approved for use only in below-ground applications. Either type may be used for building sewers. Stainless steel drainage system pipe is manufactured in 2-, 3-, 4- and 6-inch (51, 76, 102 and 152 mm) diameters and in lengths of 3.3, 4.9, 6.6, 9.8 and 16.4 feet (1006, 1494, 2012, 2987 and 4999 mm). Stainless steel pipe is identified with the following information:

- Manufacturer's name or trademark.
- Designation ASTM A 112.3.1.
- Nominal pipe size.
- Date of manufacture.

702.2 Underground building sanitary drainage and vent pipe. Underground building sanitary drainage and vent pipe shall conform to one of the standards listed in Table 702.2.

❖ The referenced table specifies the acceptable piping materials for underground drainage and vent systems.

TABLE 702.2. See next column.

The materials listed in this table are for installation underground within a building.

Acrylonitrile butadiene styrene (ABS) plastic pipe

See the commentary to Table 702.1, ABS pipe.

Cast-iron pipe

See the commentary to Table 702.1, Cast-iron pipe.

Copper or copper-alloy tubing (Type K, L, M or DWV)

See the commentary to Table 702.1, Copper or copper-alloy tubing (Type K, L, M or DWV).

Polyolefin pipe

See the commentary to Table 702.1, Polyolefin pipe.

Polyvinyl chloride (PVC) plastic pipe

See the commentary to Table 702.1, PVC pipe.

Polyvinylidene fluoride (PVDF) pipe

See commentary for Table 702.1, PVDF pipe.

Stainless steel drainage systems

See the commentary to Table 702.1, Stainless steel drainage systems. Note that only Type 316L must be used underground within a building.

TABLE 702.2
UNDERGROUND BUILDING DRAINAGE AND VENT PIPE

MATERIAL	STANDARD
Acrylonitrile butadiene styrene (ABS) plastic pipe in IPS diameters, including Schedule 40, DR 22 (PS 200) and DR 24 (PS 140); with a solid, cellular core or composite wall	ASTM D 2661; ASTM F 628; ASTM F 1488; CSA B181.1
Cast-iron pipe	ASTM A 74; ASTM A 888; CISPI 301
Copper or copper-alloy tubing (Type K, L, M or DWV)	ASTM B 75; ASTM B 88; ASTM B 251; ASTM B 306
Polyolefin pipe	ASTM F 1412; CSA B181.3
Polyvinyl chloride (PVC) plastic pipe in IPS diameters, including Schedule 40, DR 22 (PS 200) and DR 24 (PS 140); with a solid, cellular core or composite wall	ASTM D 2665; ASTM F 891; ASTM F 1488; CSA B181.2
Polyvinyl chloride (PVC) plastic pipe with a 3.25-inch O.D. and a solid, cellular core or composite wall	ASTM D 2949, ASTM F 1488
Polyvinylidene fluoride (PVDF) plastic pipe	ASTM F 1673; CSA B181.3
Stainless steel drainage systems, Type 316L	ASME A 112.3.1

For SI: 1 inch = 25.4 mm.

SANITARY DRAINAGE

702.3 Building sewer pipe. *Building sewer* pipe shall conform to one of the standards listed in Table 702.3.

❖ The referenced table specifies the acceptable piping materials for sanitary sewer systems.

TABLE 702.3
BUILDING SEWER PIPE

MATERIAL	STANDARD
Acrylonitrile butadiene styrene (ABS) plastic pipe in IPS diameters, including Schedule 40, DR 22 (PS 200) and DR 24 (PS 140); with a solid, cellular core or composite wall	ASTM D 2661; ASTM F 628; ASTM F 1488; CSA B181.1
Acrylonitrile butadiene styrene (ABS) plastic pipe in sewer and drain diameters, including SDR 42 (PS 20), PS 35, SDR 35 (PS 45), PS 50, PS 100, PS 140, SDR 23.5 (PS 150) and PS 200; with a solid, cellular core or composite wall	ASTM F 1488; ASTM D 2751
Cast-iron pipe	ASTM A 74; ASTM A 888; CISPI 301
Concrete pipe	ASTM C 14; ASTM C 76; CSA A257.1M; CSA A257.2M
Copper or copper-alloy tubing (Type K or L)	ASTM B 75; ASTM B 88; ASTM B 251
Polyethylene (PE) plastic pipe (SDR-PR)	ASTM F 714
Polyvinyl chloride (PVC) plastic pipe in IPS diameters, including Schedule 40, DR 22 (PS 200) and DR 24 (PS 140); with a solid, cellular core or composite wall	ASTM D 2665; ASTM F 891; ASTM F 1488
Polyvinyl chloride (PVC) plastic pipe in sewer and drain diameters, including PS 25, SDR 41 (PS 28), PS 35, SDR 35 (PS 46), PS 50, PS 100, SDR 26 (PS 115), PS 140 and PS 200; with a solid, cellular core or composite wall	ASTM F 891; ASTM F 1488; ASTM D 3034; CSA B182.2; CSA B182.4
Polyvinyl chloride (PVC) plastic pipe with a 3.25-inch O.D. and a solid, cellular core or composite wall	ASTM D 2949, ASTM F 1488
Polyvinylidene fluoride (PVDF) plastic pipe	ASTM F 1673; CSA B181.3
Stainless steel drainage systems, Types 304 and 316L	ASME A112.3.1
Vitrified clay pipe	ASTM C 4; ASTM C 700

For SI: 1 inch = 25.4 mm.

❖ Table 702.3 identifies the materials that are permitted and the standards with which these materials must comply.

Acrylonitrile butadiene styrene (ABS) plastic pipe

See the commentary to Table 702.1, ABS pipe. In addition to the pipe sizes described in the commentary to Table 702.1, the following piping is allowed to be used:

ASTM F 1488: The available pipe sizes for ABS Sewer and Drain-DR-PS series are 2, 3, 4, 6, 8, 9, 10 and 12 inches (51, 76, 102, 152, 203, 229, 250 and 305 mm) with the outside diameters of 2.25, 3.25, 4.215, 6.275, 8.4, 9.44, 10.5 and 12.5 inches (57, 83, 107, 159, 213, 240, 267 and 318 mm), respectively. The wall thicknesses are based upon DR of 38, 34, 27, 24 or 22 (corresponding to a PS of 35, 50, 100, 140 and 200, respectively).

ASTM D 2751: The available pipe sizes for ABS sewer pipe are 3, 4, 6, 8, 10 and 12 inches (76, 102, 152, 203, 250 and 305 mm) with the outside diameters of 3.25, 4.215, 6.275, 8.4, 10.5 and 12.5 inches (83, 107, 159, 213, 267 and 318 mm), respectively. The wall thicknesses are based upon SDR of 23.5, 35 and 42 (corresponding to a PS of 35, 50, 100, 140 and 200, respectively).

Cast-iron pipe

See the commentary to Table 702.1, Cast-iron pipe.

Concrete pipe

The pipe is made from portland cement, blast-furnace slag cement or portland-pozzolan cement-and-aggregate mixture. The pipe is produced both reinforced and nonreinforced. ASTM C 14 and CSA A 257.1 are the referenced standards for nonreinforced concrete pipe. ASTM C 14 concrete pipe is available in three classes: 1, 2 or 3. The pipe classification relates to the concrete strength based on diameter: Class 1 has the lowest strength requirements and Class 3 has the highest. The pipe is manufactured in diameters of 4 through 36 inches (102 through 914 mm). ASTM C 76 and CSA A 257.2 are the referenced standards for reinforced concrete pipe, which is strengthened with steel wire, hot rod or welded wire fabric. ASTM C 76 pipe is produced in five classes: 1, 2, 3, 4 or 5. The class determines the strength of the concrete, pipe and reinforcement geometry: Class 1 has the lowest strength requirements and Class 5 has the highest. The pipe is manufactured in sizes 12 inches (305 mm) and larger.

Copper or copper-alloy tubing (Type K, L, M or DWV)

See the commentary to Table 702.1, Copper or copper-alloy tubing (Type K, L, M or DWV)

Polyethylene (PE) plastic pipe

ASTM F 714 is the referenced standard for polyethylene plastic pipe based on outside diameters of $3^1/_2$ inches (90 mm) and larger. This piping material is intended for new construction and insertion renewal of old piping systems used for the transport of municipal sewage, domestic sewage, industrial process liquids and effluents in both pressure and nonpressure systems. All pipes produced under ASTM F 714 are pressure rated. Polyethylene plastics are defined as plastics or resins prepared by the polymerization of no less than 85-percent ethylene and no less than 95 percent of total olefins with additional compounding ingredients.

Polyolefin pipe

See the commentary to Table 702.1, Polyolefin pipe.

Polyvinyl chloride (PVC) plastic pipe

See the commentary to Table 702.1, PVC pipe.

ASTM F 891 specifies sewer-grade pipe by a PS designation, in addition to Schedule 40 (Type DWV). PS indicates pipe stiffness. The acceptable grades for sewers are PS50 and PS100. The numerical designation indicates pipe stiffness in pounds per square inch (psi). The higher value indicates the thicker pipe wall. ASTM D 2665 and D 3034 cover sewer-grade PVC pipe and fittings. The materials may be used interchangeably. Both pipes are produced in an SDR of 35 and 41, with a hub end. The PVC material used to produce the pipe is classified as 12454-B, 12454-C or 13364-B. The pipe is produced in diameters of 4 through 15 inches (102 through 381 mm). CSA B182.2 is the referenced standard for SDR 28, 35 and 41 PVC sewer pipe and fittings, and CSA B 182.4 is the referenced standard for PVC ribbed sewer pipe and fittings (see commentary, Table 702.1).

Polyvinylidene fluoride (PVDF) pipe

See the commentary to Table 702.1, PVDF pipe.

Stainless steel drainage systems

See the commentary to Table 702.1, Stainless steel drainage systems.

Vitrified clay pipe

The pipe material is surface clay, fire clay, shale or a combination of these materials. The pipe is classified as standard strength, extra strength, extra quality and heavy duty. The classification designates the strength of the pipe. Vitrified clay pipe is produced in diameters of 3 inches (76 mm) and larger. Each length of pipe is marked with the manufacturer's name and the classification of the pipe.

702.4 Fittings. Pipe fittings shall be *approved* for installation with the piping material installed and shall comply with the applicable standards listed in Table 702.4.

❖ Fittings are designed to be installed in a particular system with a given material or combination of materials. Drainage pattern fittings must be installed in drainage systems. Vent fittings are limited to the venting system. Many fittings are intended for use only for water distribution systems. There are also fittings available that may be used in any type of plumbing system. Fittings must be of the same material as, or compatible with, the pipe. This prevents any chemical or corrosive action between dissimilar materials. Many pipe standards also include fittings. There are a number of other standards that strictly regulate pipe fittings. For example, ASTM D 3311 addresses fitting patterns for ABS and PVC complying with ASTM D 2661, D 2665, F 628 and F 891. It does not address material or performance requirements, only fitting geometric and laying lengths. So in some instances, several referenced standards may regulate a single fitting.

**TABLE 702.4
PIPE FITTINGS**

MATERIAL	STANDARD
Acrylonitrile butadiene styrene (ABS) plastic pipe in IPS diameters	ASTM D 2661; ASTM F 628; CSA B181.1
Acrylonotrile butadiene styrene (ABS) plastic pipe in sewer and drain diameters	ASTM D 2751
Cast iron	ASME B 16.4; ASME B 16.12; ASTM A 74; ASTM A 888; CISPI 301
Copper or copper alloy	ASME B 16.15; ASME B 16.18; ASME B 16.22; ASME B 16.23; ASME B 16.26; ASME B 16.29
Glass	ASTM C 1053
Gray iron and ductile iron	AWWA C 110/A21.10
Malleable iron	ASME B 16.3
Polyolefin	ASTM F 1412; CSA B181.3
Polyvinyl chloride (PVC) plastic in IPS diameters	ASTM D 2665; ASTM F 1866
Polyvinyl chloride (PVC) plastic pipe in sewer and drain diameters	ASTM D 3034
Polyvinyl chloride (PVC) plastic pipe with a 3.25-inch O.D.	ASTM D 2949
Polyvinylidene fluoride (PVDF) plastic pipe	ASTM F 1673; CSA B181.3
Stainless steel drainage systems, Types 304 and 316L	ASME A 112.3.1
Steel	ASME B 16.9; ASME B 16.11; ASME B 16.28
Vitrified clay	ASTM C 700

For SI: 1 inch = 25.4 mm.

❖ The table specifies acceptable materials for pipe fittings, along with the appropriate standards with which the various fittings must comply.

Acrylonitrile butadiene styrene (ABS) plastic pipe fittings

See the commentary to Table 702.1, ABS pipe. In addition to that information, the following is provided:

The ASTM D 2751 standard covers solvent cement-type drainage fittings for ABS sewer pipe.

Cast-iron pipe fittings

See the commentary to Table 702.1, Cast iron pipe. In addition to that information, the following is provided:

The ASME B 16.4 standard covers threaded fittings made of gray iron. These are not drainage pattern fittings. These fittings are not suitable for use with cast-iron pipe used for DWV systems. The standard covers Class 125 and 250 fittings.

The ASME B 16.12 standard covers threaded fittings made of cast iron. These are special drainage pattern fittings for use with galvanized steel pipe. These fittings are not suitable for use with cast-iron pipe used for DWV systems.

Copper or copper-alloy tubing fittings

The standard ASME B 16.23 covers cast copper-alloy drainage fittings. The standard ASME B 16.29 covers wrought copper and wrought-copper drainage fittings. The standard ASME B 16.15 covers cast bronze-threaded fittings. The standard ASME B 16.22 covers wrought-copper and copper-alloy pressure fittings. The standard ASME B 16.18 covers cast copper-alloy pressure fittings. The standard ASME B 16.26 covers cast copper-alloy fittings for flared copper tubes.

Glass pipe fittings

See the commentary to Table 702.1, Glass pipe.

Gray Iron and ductile iron fittings

The standard AWWA C 110 covers gray iron and ductile iron fittings suitable for use with ductile iron pipe. However, the code does not recognize ductile iron pipe as an acceptable material for conveying waste water.

Malleable iron fittings

The ASME B 16.3 standard covers threaded fittings made of malleable iron. These are not drainage pattern fittings so these fittings cannot be used in drainage applications. Also, because these fittings are not recessed thread fittings, they cannot be used for vent piping.

Polyolefin pipe fittings

See the commentary to Table 702.1, Polyolefin pipe.

Polyvinyl chloride (PVC) plastic pipe fittings

See the commentary to Table 702.3, PVC pipe. In addition to that information, the following information is provided:

The standard ASTM F 1866 covers DWV fittings for PVC plastic piping that has an outside diameter matching Schedule 40 steel pipe. The standard also covers fabricated DWV fittings (as opposed to injection-molded, one-piece fittings).

Polyvinylidene fluoride pipe fittings

See the commentary to Table 702.1, PVDF pipe.

Stainless steel drainage system fittings

See the commentary to Table 702.1, Stainless steel drainage systems.

Steel pipe fittings

The standard ASME B 16.9 is for factory-made butt welded fittings ASME B16.11 is for forged fittings, socket and threaded types. ASME B 16.28 is for wrought steel butt-welding short radius elbows and returns.

Vitrified clay pipe

See the commentary to Table 702.3, Vitrified clay pipe.

702.5 Temperature rating. Where the waste water temperature will be greater than 140°F (60°C), the sanitary drainage piping material shall be rated for the highest temperature of the waste water.

❖ The pipe materials indicated in Tables 702.1 through 702.3 are suitable for waste water temperatures up to 140°F (60°C). Where waste water temperatures are expected to be greater than 140°F, the design professional needs to make sure that the selected pipe material can handle the higher temperature.

702.6 Chemical waste system. A chemical waste system shall be completely separated from the sanitary drainage system. The chemical waste shall be treated in accordance with Section 803.2 before discharging to the sanitary drainage system. Separate drainage systems for chemical wastes and vent pipes shall be of an *approved* material that is resistant to corrosion and degradation for the concentrations of chemicals involved.

❖ A measure of the degree of acidity or alkalinity of a liquid is its pH value, which defines the concentration of free hydrogen ions on a scale of 0 to 14. A pH value of 7 is considered neutral. As the value decreases, the liquid becomes more acidic; as the value increases, it becomes more alkaline. Mildly alkaline liquids are not harmful to a piping system. When the pH of a liquid is less than 6.8, it is considered to be corrosive. A drainage and vent system that handles waste with a pH lower than 5.5 is called an "acid waste system." This type of system is usually found in hospitals, research laboratories, food processing plants, photo labs, battery rooms and similar facilities.

There is no universal material available to resist all chemical action. Each pipe has qualities that make it resistant to certain chemical waste systems. An individual analysis of compatibility is required for the installation of a chemical waste and vent system. Piping material such as borosilicate glass and polyolefin plastic is often used for special waste applications and is listed in Table 702.1; however, not all materials commonly used for chemical waste and vent systems are listed in the code. Materials such as high-silicon iron pipe, lead pipe and ceramic glazed or unglazed vitrified clay are commonly used. Some of those materials are proprietary products and cannot be listed in the code.

Acid wastes in the vapor state are more corrosive than in the liquid state; therefore, it is important to provide material that resists chemical action for the vent piping as well as for the drainage piping. The code official must evaluate the material or rely on an evaluation by authoritative sources, and approve the product as an alternative material. Section 803 governs the requirements for discharge of special wastes into the sanitary drainage system.

702.7 Lead bends and traps. The wall thickness of lead bends and traps shall be not less than $^1/_8$ inch (3.2 mm).

❖ The walls of lead bends and traps must be at least $^1/_8$ inch (3.2 mm). They are typically installed in chemical waste systems.

SECTION 703
BUILDING SEWER

703.1 Building sewer pipe near the water service. The proximity of a *sewer* to a water service shall comply with Section 603.2.

❖ Where the building sewer is installed in close proximity to the water service, there is a concern that the water service could be subjected to contamination from a sewer leak or failure. To minimize the possibility of contamination, only materials with high reliability are allowed to be installed in a common trench. See Section 603.2 for additional requirements for water service and building sewer separation.

703.2 Drainage pipe in filled ground. Where a *building sewer* or *building drain* is installed on filled or unstable ground, the drainage pipe shall conform to one of the standards for ABS plastic pipe, cast-iron pipe, copper or copper-alloy tubing, or PVC plastic pipe listed in Table 702.3.

❖ Filled or unstable ground is considered to be a marshy or landfill area. Because of the increased possibility of movement, only the strongest materials with the greatest ability to maintain pipe alignment are permitted to be installed for the building sewer.

703.3 Sanitary and storm sewers. Where separate systems of sanitary drainage and storm drainage are installed in the same property, the sanitary and storm building sewers or drains shall be permitted to be laid side by side in one trench.

❖ The concern for contamination of the storm sewer from a sewer leak or failure is not as critical as that of contamination of the water service; therefore, it is permissible to place the storm and sanitary sewers side by side in the same trench.

703.4 Existing building sewers and drains. Existing *building sewers* and *drains* shall connect with new *building sewer* and drainage systems only where found by examination and test to conform to the new system in quality of material. The code official shall notify the owner to make the changes necessary to conform to this code.

❖ Use of an existing drainage system may continue to be used if the piping material is found to be in satisfactory condition. Some existing materials are no longer manufactured; however, the piping may be more than adequate. Materials of this type that may be encountered include wrought-iron pipe, cast-iron medium-weight pipe and cast-iron "victory" pipe.

703.5 Cleanouts on building sewers. Cleanouts on *building sewers* shall be located as set forth in Section 708.

❖ Improperly located cleanouts may create difficulties in the cleaning of stoppages in the drainage system. Cleanout locations and spacing requirements are stated in Section 708.

703.6 Combined sanitary and storm public sewer. Where the public sewer is a combined system for both sanitary and storm water, the sanitary sewer shall be connected independently to the public sewer.

❖ There are numerous major (older) cities where the public sewer system conveys both sanitary drainage and storm water drainage. Because storm water flows often create overflows of the public sewer system (and the state environmental authorities fine the cities for these overflows), cities are working to separate the public systems by installing a storm (only) sewer system. The sewers for new buildings must be separate to accommodate the future installation of public storm (only) sewers.

SECTION 704
DRAINAGE PIPING INSTALLATION

704.1 Slope of horizontal drainage piping. Horizontal drainage piping shall be installed in uniform alignment at uniform slopes. The slope of a horizontal drainage pipe shall be not less than that indicated in Table 704.1.

❖ This section contains installation requirements for sanitary drainage systems. The minimum desired velocity in a horizontal drain pipe is approximately 2 feet per second (0.61 m/s). This velocity is often referred to as the "scouring velocity," even though it may not always clean the walls of the drainage pipe. The scouring velocity is intended to keep solids in suspension. For example, if the velocity is too low where a drain pipe is excessively oversized, the solids tend to drop out of suspension, settling to the bottom of the pipe. This may eventually result in a drain stoppage. A greater velocity is required to move solids at rest than is required to keep moving solids in suspension. A true scouring velocity would then be higher than the velocity initially used to transport the solid.

TABLE 704.1
SLOPE OF HORIZONTAL DRAINAGE PIPE

SIZE (inches)	MINIMUM SLOPE (inch per foot)
$2^1/_2$ or less	$^1/_4$
3 to 6	$^1/_8$
8 or larger	$^1/_{16}$

For SI: 1 inch = 25.4 mm, 1 inch per foot = 83.33 mm/m.

❖ A simplified method of determining the minimum slope of a drainage pipe is provided. The slope listed for each given pipe size is intended to produce the minimum required drainage flow velocity. A drainage pipe can be installed with greater slopes because a maximum slope is not prescribed. The slope is commonly expressed in percentages as total fall per length of pipe. For example, a 1-percent slope is approximately equivalent to $^1/_8$ inch per foot (10 mm per m) and a 2-percent slope is approximately equal to $^1/_4$ inch per foot (20 mm per m).

704.2 Change in size. The size of the drainage piping shall not be reduced in size in the direction of the flow. A 4-inch by 3-inch (102 mm by 76 mm) water closet connection shall not be considered as a reduction in size.

❖ One of the fundamental requirements of a drainage system is that the size of piping cannot be reduced in the direction of drainage flow. A size reduction would create an obstruction to flow, possibly resulting in a backup of flow, an interruption of service in the drainage system or stoppage in the pipe. A 4-inch by 3-inch (102 mm by 76 mm) water closet connection is not considered to be a reduction in pipe size.

704.3 Connections to offsets and bases of stacks. Horizontal *branches* shall connect to the bases of stacks at a point located not less than 10 times the diameter of the drainage *stack* downstream from the *stack*. Horizontal *branches* shall connect to horizontal *stack* offsets at a point located not less than 10 times the diameter of the drainage *stack* downstream from the upper *stack*.

❖ At the base of every stack, a phenomenon of flow may occur that is commonly called "hydraulic jump." Hydraulic jump is the rising in the depth of flow above a half full pipe. A horizontal drain designed in accordance with the code is expected to have a normal flow depth of not greater than half full. The rise in flow may be great enough to close off the opening of the pipe, thus creating pressure fluctuations in the system that may affect trap seals. Hydraulic jump occurs within a distance of 10 times the diameter of the drainage stack downstream of the stack connection. To avoid flow interference, backup of flow and extreme pressure fluctuations, horizontal branch connections are prohibited in this area. Hydraulic jump also results from the high-velocity vertical flow in the stack making a transition to a slower horizontal flow in the building drain. The initial velocity remains high in the horizontal building drain until flow levels off, achieving a uniform flow condition (see Commentary Figure 704.3). A horizontal stack offset creates flow conditions similar to that created at the base of any other stack. The stack above the offset connects to horizontal piping, thereby causing hydraulic jump.

704.4 Future fixtures. Drainage piping for future fixtures shall terminate with an *approved* cap or plug.

❖ If piping is installed for future fixtures, it must be closed with caps or plugs to prevent the escape of sewer gas and waste and the entrance of foreign substances. Section 905.6 also requires a vent to be roughed in for drainage pipe that will serve fixtures to be installed in the future.

SECTION 705
JOINTS

705.1 General. This section contains provisions applicable to joints specific to sanitary drainage piping.

❖ This section contains requirements that apply to materials used for the joining of pipes and the installation of those materials. Acceptable joints and connections for the sanitary drainage piping materials addressed in Section 702 are listed in alphabetical order in this chapter. The main section title identifies the piping material to which the section applies. Subsequent sections list acceptable joints or connections for each particular material. Joints and connections between dissimilar materials are regulated in Section 705.19.

705.2 ABS plastic. Joints between ABS plastic pipe or fittings shall comply with Sections 705.2.1 through 705.2.3.

❖ Acceptable joints for ABS plastic pipe are identified in the referenced sections.

705.2.1 Mechanical joints. Mechanical joints on drainage pipes shall be made with an elastomeric seal conforming to ASTM C 1173, ASTM D 3212 or CSA B602. Mechanical joints shall be installed only in underground systems unless otherwise *approved*. Joints shall be installed in accordance with the manufacturer's instructions.

❖ Mechanical joints are predominantly used in underground applications (e.g., bell or hub compression gasket joints and elastomeric transition couplings). Mechanical joints are typically of the unrestrained type. Above-ground mechanical joints are limited primarily to rehabilitation installations and connections to existing systems when it is impractical to make another type of joint. Mechanical joints are sometimes used where other joining means cannot be used. For example, underground PVC pipe may be joined with an elastomeric coupling where, because of water in the trench, solvent cementing is not possible. The referenced standards are addressed only with respect to

For SI: 1 inch = 25.4 mm.

**Figure 704.3
HYDRAULIC JUMP**

the elastomeric material used to make the seal in a mechanical joint, not the entire coupling or fitting assembly.

705.2.2 Solvent cementing. Joint surfaces shall be clean and free from moisture. Solvent cement that conforms to ASTM D 2235 or CSA B181.1 shall be applied to all joint surfaces. The joint shall be made while the cement is wet. Joints shall be made in accordance with ASTM D 2235, ASTM D 2661, ASTM F 628 or CSA B181.1. Solvent-cement joints shall be permitted above or below ground.

❖ Solvent cementing is the most common method of joining ABS plastic pipe and fittings. This section clarifies that solvent cementing is an acceptable joining method for both above-ground and below-ground installations. Although it is the most common method, solvent cementing is one of the most misunderstood methods of joining pipe.

The ASTM and CSA referenced standards indicate how to handle solvent cement safely and recommend a practice for making joints. ABS solvent cement is a flammable liquid and a source of flammable vapors. It must be kept a safe distance away from any source of ignition. Solvent cement should also be used in a well-ventilated area. Prolonged breathing of solvent cement vapors can have adverse health effects. Solvent cement is also an eye and skin irritant. If contact with the eyes is possible, protective goggles should be worn. When frequent contact with the skin is likely, impervious gloves should be worn.

The pipe and fittings should be approximately the same temperature prior to solvent cementing. The pipe should be cut square and chamfered. The pipe end and socket fitting must be clean, dry and free from grease, oil and other foreign materials. The solvent cement must be uniformly applied to the socket of the fitting and to the end of the pipe to the depth of the socket. Excessive solvent cement should be avoided to prevent a puddle of cement from forming on the inside of the pipe joint, which may deform or structurally weaken the fitting or pipe. The joint is made by twisting the pipe into the fitting socket immediately after applying the cement. The pipe will tend to back out unless it is restrained; therefore, it must be held in place for a few seconds until the solvent cement begins to set.

A small bead of solvent cement forms between the pipe exterior and fitting. The bead must be removed with a clean cloth to avoid the possibility of weakening the pipe wall. There should be no rough handling of the pipe for at least 1 hour after the joint is made. A solvent-cement joint should remain for a period of 24 hours before being exposed to working pressure. Because each type of solvent cement is specifically designed for a given piping material, all-purpose solvent cement or universal solvent cement cannot be used to join ABS pipe or fittings unless it conforms to ASTM D 2235 or CSA B181.1.

Solvent cement for each plastic pipe material requires a unique mixture or combination of solvents and dissolved plastic resins to conform to the applicable standard. Although all-purpose cement may dissolve the surfaces of ABS, PVC and CPVC, it may be too aggressive or not aggressive enough to make a joint strong enough to meet the minimum strength requirements of the referenced standards. Solvent cement must be identified on its container as complying with the applicable ASTM or CSA standard.

705.2.3 Threaded joints. Threads shall conform to ASME B1.20.1. Schedule 80 or heavier pipe shall be permitted to be threaded with dies specifically designed for plastic pipe. *Approved* thread lubricant or tape shall be applied on the male threads only.

❖ Pipe thread dimensions can be historically traced to Robert Briggs, mentioned in the commentary to Table 605.3 for galvanized steel pipe. ASME B 1.20.1 identifies pipe dimensions as "NPT," which is often thought of as an abbreviation. Some of the terms from which NPT has been shortened are: National Pipe Thread, Nominal Pipe Thread and National Pipe Tapered Thread. NPT is a coded designation: "N" stands for USA Standard, "P" indicates pipe and "T" means that the threads are tapered.

Thread lubricant (pipe-joint compound) must be specifically designed for ABS plastic pipe. It is not the same lubricant that is commonly used for metallic pipe, such as steel. Thread lubricant and tape are designed for two purposes: to lubricate the joint for proper thread mating and to fill in small imperfections on the surfaces of the threads. Field threading of ABS is limited to solid-core Schedule 80 pipe because the wall thickness is sufficient to accommodate threads without affecting pipe strength.

705.3 Brass. Joints between brass pipe or fittings shall comply with Sections 705.3.1 through 705.3.4.

❖ Acceptable joints for brass pipe or fittings are identified in the referenced sections.

705.3.1 Brazed joints. All joint surfaces shall be cleaned. An *approved* flux shall be applied where required. The joint shall be brazed with a filler metal conforming to AWS A5.8.

❖ Although brazed joints are similar to soldered joints, brazed joints are joined at a higher temperature [in excess of 1,000°F (538°C)]. Because of the brazed joint's inherent resistance to elevated temperatures, it is much stronger than a soldered joint. Brazing is often referred to as "silver soldering." Silver soldering is more accurately described as silver brazing and uses high silver-bearing alloys primarily composed of silver, copper and zinc. Silver soldering (brazing) typically requires temperatures in excess of 1,000°F (538°C), and such solders are classified as "hard" solders.

There has always been confusion over the distinction between "silver solder" and "silver-bearing solder." Silver solders are unique and may be further subdivided into soft and hard categories, which are determined by the percentages of silver and the other component elements of the particular alloy. The distinction must be made that silver-bearing solders [with

SANITARY DRAINAGE

a melting point less than 600°F (316°C)] are used in soft-soldered joints, whereas silver solders [with a melting point above 1,000°F (538°C)] are used in silver-brazed joints. The two common series of filler metals used for brazing are BCuP and BAg. BCuP contains phosphorus and copper alloys. BAg contains 30- to 60-percent silver with zinc and copper alloys. Some of the BAg filler metals also contain cadmium.

During brazing, such filler metal produces toxic fumes. Brazing should always be done in well-ventilated areas, and protective eye gear must also be worn. The joint surfaces to be brazed must be cleaned, which is typically accomplished by polishing the pipe with an emery cloth. The socket fitting is cleaned with a specially designed brush. Brazing flux is applied to the joint surfaces. Certain types of BCuP filler metals do not require flux for certain types of joints. The manufacturer's recommendations should be consulted to determine when flux is necessary. The joints are typically heated with an oxyacetylene torch, though an air acetylene torch may be used on a smaller-diameter pipe. The pipe and fitting must be at approximately the same temperature when making the joint. The filler metal flows to the heat with the aid of the flux. The pipe must first be heated above the fitting. When the flux becomes clear, the heat must be directed to the base of the fitting socket. When the flux becomes clear and quiet, the filler metal is applied to the joint. The heat draws the filler metal into the socket. Any remaining flux residue must be removed from the pipe and joint surfaces (see Commentary Figure 705.3.1).

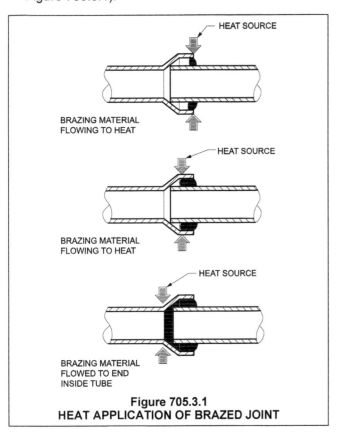

Figure 705.3.1
HEAT APPLICATION OF BRAZED JOINT

705.3.2 Mechanical joints. Mechanical joints shall be installed in accordance with the manufacturer's instructions.

❖ Mechanical joints are unique in that there are many acceptable types available. Most mechanical joints use an elastomeric material to form a seal. Compression fittings that use a compression ring (ferrule) which forms a metal-to-metal seal are also considered mechanical joints. In all cases, the joint must be carefully installed in accordance with the manufacturer's instructions.

705.3.3 Threaded joints. Threads shall conform to ASME B1.20.1. Pipe-joint compound or tape shall be applied on the male threads only.

❖ The pipe-joint compound is typically designed for all metallic pipe. Generally, the pipe-joint compound used for brass pipe may also be applied to copper and steel pipe (see commentary, Section 705.2.3).

705.3.4 Welded joints. All joint surfaces shall be cleaned. The joint shall be welded with an *approved* filler metal.

❖ A welded joint is similar to a brazed joint. The primary differences are the temperature at which the joint is made, the type of filler metal used and the fact that welding reaches the melting point of the base metal, whereas brazing temperatures are well below the melting point of the pipe (see commentary, Section 705.4.1).

705.4 Cast iron. Joints between cast-iron pipe or fittings shall comply with Sections 705.4.1 through 705.4.3.

❖ Acceptable joints for cast-iron pipe or fittings are identified in the referenced sections.

705.4.1 Caulked joints. Joints for hub and spigot pipe shall be firmly packed with oakum or hemp. Molten lead shall be poured in one operation to a depth of not less than 1 inch (25 mm). The lead shall not recede more than $^{1}/_{8}$ inch (3.2 mm) below the rim of the hub and shall be caulked tight. Paint, varnish or other coatings shall not be permitted on the jointing material until after the joint has been tested and *approved*. Lead shall be run in one pouring and shall be caulked tight. Acid-resistant rope and acidproof cement shall be permitted.

❖ A caulked joint is the oldest form of joining cast-iron pipe (see Commentary Figure 705.4.1). The joint is sealed with tightly packed oakum or hemp. The oakum or hemp expands the first time it becomes wet, thus sealing the joint. Lead is poured in place behind the oakum to prevent the oakum from expanding out of the joint. Quite often, a caulked joint will leak when it is first tested. This is a result of the oakum not yet expanding in place. Once the oakum has expanded, the joint will not leak.

The joint is made by packing the oakum or hemp into the hub fitting around the pipe to be joined. The surfaces must be dry to prevent the oakum from expanding prematurely. Once in place, the oakum is packed tight with a packing iron and hammer. Molten lead is poured in one continuous operation, securing the oakum. If the joint surfaces or oakum is wet when pouring the lead, rapidly expanding steam causes the lead to pop or explode. After the lead solidifies, it must be packed tight. A properly made caulked joint will bear marks on the lead caused by a finishing iron used to pack the lead.

Because caulked joints are labor intensive and involve the risk of worker injury, and because lead is a health hazard, such joints are becoming increasingly rare. The code does not recognize a substitute for lead in caulked joints. In the case of remodeling and repair work where one or more caulked joints must be made to allow transitions to different types of piping, code officials had on occasion permitted the use of approved lead substitutes, such as cement-based compounds. Elastomeric seals are also available for making transition joints with hubbed cast-iron pipe (see commentary, Section 705.4.2).

705.4.2 Compression gasket joints. Compression gaskets for hub and spigot pipe and fittings shall conform to ASTM C 564 and shall be tested to ASTM C 1563. Gaskets shall be compressed when the pipe is fully inserted.

❖ A mechanical compression joint is used with hubbed cast-iron drainage pipe. The elastomeric gasket is often called a "rubber gasket" or "rubber joint." The gasket is inserted into the hub of the pipe. A lubricant is used to coat the end of the joining pipe and the gasket. The pipe is inserted into the hub using a special tool that pulls the pipe and hub together, compressing the gasket in the annular space between the pipe and the hub. The pipes must be completely pulled together to make the joint properly. A typical compression joint has two main compression points in the seal (see Commentary Figure 705.4.2).

705.4.3 Mechanical joint coupling. Mechanical joint couplings for hubless pipe and fittings shall consist of an elastomeric sealing sleeve and a metallic shield that comply with CISPI 310, ASTM C 1277 or ASTM C 1540. The elastomeric sealing sleeve shall conform to ASTM C 564 or CSA B602 and shall be provided with a center stop. Mechanical joint couplings shall be installed in accordance with the manufacturer's instructions.

❖ Mechanical joint couplings are devices used with hubless cast-iron drainage, waste and vent pipe and fittings. Mechanical joint couplings are manufactured in a variety of types, all of which include a metallic shield or sleeve elastomeric seal and a clamp assembly that compresses the seal around the outside circumfer-

Figure 705.4.1
CAULKED JOINT FOR HUB AND SPIGOT (CAST-IRON)

Figure 705.4.2
CAST-IRON COMPRESSION GASKET

ence of the pipe or fitting. Mechanical joint couplings are typically designed to resist the shear forces that can cause misalignment of pipe and fittings. ASTM C 564 and CSA CAN/CSA-B602 apply only to the elastomeric seal component of mechanical joint couplings. CISPI 310 and ASTM C 1277 address the entire coupling assembly. The center stop prevents the pipe or fitting from penetrating too far into the coupling. The center stop is designed to ensure proper penetration depth for both of the pipes or fittings being joined.

The code regulates the design and quality of the elastomeric seals and the overall design of mechanical joint coupling assemblies used to join cast-iron DWV pipe and fittings. CISPI 310 and ASTM C 1277 address couplings that use a metal shield. The shield protects the elastomeric seal and contributes to the shear force resistance and rigidity to the assembled joint. The use of coupling designs that do not use a metal shield may have to be approved in accordance with Section 105.2. Mechanical joints and mechanical joint couplings must be installed in accordance with the manufacturer's instructions (see Commentary Figure 705.4.3).

705.5 Concrete joints. Joints between concrete pipe and fittings shall be made with an elastomeric seal conforming to ASTM C 443, ASTM C 1173, CSA A257.3M or CSA B602.

❖ The only acceptable method of joining concrete pipe is an elastomeric gasket that fits in the hub of the pipe. The gasket is compressed between the hub and the pipe to form a seal. The ASTM and CSA standards regulate the design, material and performance of the elastomeric gasket (see Commentary Figure 705.5).

705.6 Copper pipe. Joints between copper or copper-alloy pipe or fittings shall comply with Sections 705.6.1 through 705.6.5.

❖ Acceptable joints for copper or copper-alloy pipe or fittings are identified in the referenced sections.

705.6.1 Brazed joints. All joint surfaces shall be cleaned. An *approved* flux shall be applied where required. The joint shall be brazed with a filler metal conforming to AWS A5.8.

❖ The installation technique and materials used to make a brazed joint for copper or copper-alloy pipe are identical to that of brass pipe and copper tubing (see commentary, Section 705.4.1).

705.6.2 Mechanical joints. Mechanical joints shall be installed in accordance with the manufacturer's instructions.

❖ Mechanical joints for copper or copper alloy pipe are similar to those used to join brass pipe (see commentary, Section 705.4.2). Because many different types of mechanical joints are available and most couplings are unique to one manufacturer, the joining method must be in accordance with the manufacturer's installation instructions. Such joints are highly resistant to being pulled apart.

705.6.3 Solder joints. Solder joints shall be made in accordance with the methods of ASTM B 828. Cut tube ends shall be reamed to the full inside diameter of the tube end. All joint

Figure 705.4.3
HUBLESS CAST-IRON DRAINAGE PIPE JOINT

Figure 705.5
CONCRETE MECHANICAL JOINT

surfaces shall be cleaned. A flux conforming to ASTM B 813 shall be applied. The joint shall be soldered with a solder conforming to ASTM B 32.

❖ Soldering is the most common method of joining copper pipe and tubing. ASTM B 828 governs the procedures for making capillary joints by soldering copper and copper alloy tube and fittings. To consistently make satisfactory joints, the following sequence of joint preparation and operations must be followed: measuring and cutting, reaming, cleaning, fluxing, assembly and support, heating, applying the solder and cooling and cleaning. The joint surfaces must be accurately measured because if the tube is too short, it will not reach all the way into the fitting and a proper joint cannot be made. If the tube is too long, the system strain may affect service life.

Tubing may be cut in a number of different ways to produce a square end. The tubing may be cut with a tubing cutter, a hacksaw, an abrasive wheel or a band saw. Regardless of the method, the cut must be square with the run of the tube so it will seat properly in the fitting. The tubing ends, outside of the tube ends to the inside diameter of the tube, must be reamed to remove the burrs created by the cutting operation (see Commentary Figure 705.6.3).

The joint surfaces for a soldered joint must be cleaned to remove all oxides and surface soil from the tube ends and fittings. Cleaning may be accomplished using a sand cloth, an abrasive pad and a fitting brush. The space between the tube and the fitting is where the solder metal fills this gap by capillary action. The spacing is critical for the solder metal to flow into the gap and form a strong joint. Uniformity of capillary space will ensure metal capillary flow.

Once the joint surfaces are cleaned, do not touch the cleaned area with bare hands or oily gloves. Skin oils, lubricating oils or grease could interfere with the process. Apply a thin, even coating of flux to both the tubing and fitting that will dissolve and remove traces of oxide from the cleaned surfaces. Care must be taken in applying flux because excessive amounts will cause corrosion that may perforate the wall of the tubing, the fitting or both. The flux material must conform to ASTM B 813, which establishes minimum performance criteria. This also limits the maximum lead content to 0.2 percent.

Begin heating with the flame alternating from the fitting cup back onto the tube. Touch the solder to the joint and if the solder does not melt, remove it and continue the heating process. Once the melting temperatures are reached, apply heat to the base of the fitting to aid capillary action in drawing the molten solder into the cup toward the heat source. Molten solder metal is drawn into the joint by capillary action regardless of the direction that the metal is being fed (horizontal or vertical). If excessive heat is used, the flux will burn out and char the pipe, and where the temperature is too uneven, the solder will run down either the inside or the outside of the pipe. Either condition will prevent capillary action.

Allow the completed joint to cool naturally; cooling it with water will cause unnecessary shock and stress on the joint. When the joint is cool, clean off any remaining flux.

For SI: 1 degree = 0.0175 rad.

**Figure 705.6.3
PREPARATION OF PIPE END**

705.6.4 Threaded joints. Threads shall conform to ASME B1.20.1. Pipe-joint compound or tape shall be applied on the male threads only.

❖ Before assembly, the threads must be cleaned with a wire brush to remove burrs or chips resulting from the cutting process. Pipe-joint compound or Teflon tape is applied on the male thread to ensure a tight seal. If pipe-joint compound is applied to the female thread, it will enter the system. The male thread is placed inside the female thread and the joint is tightened using a pipe wrench. Threaded joints are generally limited to smaller pipe diameters because of the amount of effort required to turn a pipe of larger size in making the joint.

705.6.5 Welded joints. All joint surfaces shall be cleaned. The joint shall be welded with an *approved* filler metal.

❖ The installation technique and materials used to make a welded joint for copper or copper-alloy tubing are similar to that of brass pipe (see commentary, Section 705.3.4).

705.7 Copper tubing. Joints between copper or copper-alloy tubing or fittings shall comply with Sections 705.7.1 through 705.7.3.

❖ Acceptable joints for copper or copper-alloy tubing or fittings are identified in the referenced sections.

705.7.1 Brazed joints. All joint surfaces shall be cleaned. An *approved* flux shall be applied where required. The joint shall be brazed with a filler metal conforming to AWS A5.8.

❖ The brazing of a joint for copper or copper-alloy tubing is identical to that of copper, copper-alloy and brass pipe (see commentary, Section 705.4.1).

705.7.2 Mechanical joints. Mechanical joints shall be installed in accordance with the manufacturer's instructions.

❖ There is a variety of joining methods for mechanical joints for copper or copper-alloy tubing. The most common is the compression-type joint. This type of joint is made by placing a compression nut over the outside of the pipe followed by a compression ring. The pipe is inserted into the fitting and tightened. As the compression nut is tightened, the pipe and fitting are brought together, thereby forming a leakproof joint. Mechanical joints have some flexibility, allowing the pipes to be installed with an angle of deflection. Such joints are highly resistant to being pulled apart.

705.7.3 Solder joints. Solder joints shall be made in accordance with the methods of ASTM B 828. Cut tube ends shall be reamed to the full inside diameter of the tube end. All joint surfaces shall be cleaned. A flux conforming to ASTM B 813 shall be applied. The joint shall be soldered with a solder conforming to ASTM B 32.

❖ The requirements and procedures for a soldered joint for copper and copper-alloy tubing are identical to those for copper and copper-alloy pipe (see commentary, Section 705.6.3 and Commentary Figure 705.6.3).

705.8 Borosilicate glass joints. Glass-to-glass connections shall be made with a bolted compression-type, 300 series stainless steel coupling with contoured acid-resistant elastomeric compression ring and a fluorocarbon polymer inner seal ring; or with caulked joints in accordance with Section 705.8.1.

❖ Two methods are recognized for joining glass pipe: one is a mechanical compression joint that is used to join bead-end or plain-end glass pipe; the other is a caulked joint that is used to join bead-end glass pipe to hubbed pipe. The compression-type mechanical joint is available in two standard types: one type is used to join beaded-end glass pipe to beaded-end glass pipe, and the other one is used to join beaded-end glass pipe to plain-end glass pipe.

Both couplings consist of an elastomeric compression ring (typically rubber) with a chemically inert tetrafluoroethylene (TFE) inner seal ring and a stainless steel outer band with bolt. The bead-end to bead-end pipe connection is made by slipping the compression liner and inner seal ring over one section of pipe, aligning the pipes to be joined and sliding the compression liner over both bead ends. The stainless steel outer band is then slid over the compression liner and the joint is secured by tightening the bolt on the outer band. The bead-end to plain-end joint is made in a similar manner. The methods for joining glass pipe to other pipe materials are listed in Section 705.16.6.

705.8.1 Caulked joints. Lead-caulked joints for hub and spigot soil pipe shall be firmly packed with oakum or hemp and filled with molten lead not less than 1 inch (25 mm) in depth and not to recede more than $^1/_8$ inch (3.2 mm) below the rim of the hub. Paint, varnish or other coatings shall not be permitted on the jointing material until after the joint has been tested and *approved*. Lead shall be run in one pouring and shall be caulked tight. Acid-resistant rope and acidproof cement shall be permitted.

❖ The method recognized for joining glass pipe to hubbed pipe is a poured-lead caulk joint. This joint method is used to join beaded-end glass pipe to glass or metal hubbed pipe. The bead end of the glass pipe is inserted into the bell of the other pipe and oakum, hemp or acid-resistant rope is then packed around the glass pipe to prevent direct contact between the two pipes. The installation of the joint is identical to that of cast-iron pipe (see commentary, Section 705.4.1).

705.9 Steel. Joints between galvanized steel pipe or fittings shall comply with Sections 705.9.1 and 705.9.2.

❖ Acceptable joints for galvanized steel pipe or fittings are identified in the referenced sections.

705.9.1 Threaded joints. Threads shall conform to ASME B1.20.1. Pipe-joint compound or tape shall be applied on the male threads only.

❖ The pipe-joint compound used for steel pipe will also be used for both copper and brass pipe (see commentary, Section 705.2.3). The application of pipe-joint compound or Teflon tape to only male threads prevents the compound from entering the piping system.

705.9.2 Mechanical joints. Joints shall be made with an *approved* elastomeric seal. Mechanical joints shall be installed in accordance with the manufacturer's instructions.

❖ A wide variety of mechanical joints are available to join steel pipe, including plain or beveled ends, cut or rolled grooves. Each requires different pipe-end preparation in making a joint. Rolled grooves (instead of cut grooves) are used where the wall of the pipe is not thick enough to have a groove formed around the pipe near the ends. The coupling assembly for both types of grooving is the same. The joint consists of an inner elastomeric gasket and an outer two-piece (split) collar with integral bolts used for tightening. The assembly is placed over both ends and set into the grooves in the pipe. The bolts are tightened to the torque requirements listed in the manufacturer's installation instructions.

705.10 Lead. Joints between lead pipe or fittings shall comply with Sections 705.10.1 and 705.10.2.

❖ Acceptable joints for lead pipe or fittings are identified in the referenced sections.

705.10.1 Burned. Burned joints shall be uniformly fused together into one continuous piece. The thickness of the joint shall be at least as thick as the lead being joined. The filler metal shall be of the same material as the pipe.

❖ A burned joint is formed by fitting the end of one lead pipe into the flared end of another lead pipe. Heat is then applied evenly around the joint, melting the overlapping edges and fusing them together. Lead burning is a method used for joining sheet lead or making lead pans.

705.10.2 Wiped. Joints shall be fully wiped, with an exposed surface on each side of the joint not less than $^3/_4$ inch (19.1 mm). The joint shall be not less than $^3/_8$ inch (9.5 mm) thick at the thickest point.

❖ A wiped joint is formed by fitting the end of one pipe into the flared end of the other. Molten lead is poured onto the joint and wiped with a hand-held pad, providing a minimum dimension of $^3/_4$ inch (19.1 mm) and a maximum thickness of $^3/_8$ inch (9.5 mm) (see Commentary Figure 705.10.2). With the aid of modern technology and introduction of new materials, lead pipe is rarely used in drainage systems.

705.11 PVC plastic. Joints between PVC plastic pipe or fittings shall comply with Sections 705.11.1 through 705.11.3.

❖ Acceptable joints for PVC plastic pipe or fittings are identified in the referenced sections.

705.11.1 Mechanical joints. Mechanical joints on drainage pipe shall be made with an elastomeric seal conforming to ASTM C 1173, ASTM D 3212 or CSA B602. Mechanical joints shall not be installed in above-ground systems, unless otherwise *approved*. Joints shall be installed in accordance with the manufacturer's instructions.

❖ A typical underground sewer line is plastic hub pipe with an elastomeric compression gasket inside the hub. The pipe must be inserted to the full depth of the hub to make a proper joint (see commentary, Section 705.2.1). The typical mechanical joint must be restrained to prevent separation; therefore, such joints are limited to underground installations where restraint is accomplished by burial.

705.11.2 Solvent cementing. Joint surfaces shall be clean and free from moisture. A purple primer that conforms to ASTM F 656 shall be applied. Solvent cement not purple in color and conforming to ASTM D 2564, CSA B137.3, CSA B181.2 or CSA B182.1 shall be applied to all joint surfaces. The joint shall be made while the cement is wet and shall be in accordance with ASTM D 2855. Solvent-cement joints shall be permitted above or below ground.

Exception: A primer is not required where both of the following conditions apply:

1. The solvent cement used is third-party certified as conforming to ASTM D 2564.

2. The solvent cement is used only for joining PVC drain, waste and vent pipe and fittings in nonpressure applications in sizes up to and including 4 inches (102 mm) in diameter.

❖ PVC plastic pipe requires a primer prior to solvent cementing. Although ASTM F 656 only recommends that PVC primer be purple, this is a code requirement. The purple coloring is intended to allow the installer and the inspector to verify that the required primer has been applied. To distinguish the difference between primer and solvent cement, the solvent cement must be a color other than purple. Solvent cement for PVC is typically clear (see commentary, Section 705.2.2). A PVC primer is first applied to both the pipe and the socket fitting. The PVC solvent cement is then applied to the pipe and socket fitting. The joint is made with a slight twisting until the pipe reaches the full depth of the socket fitting. It is then held in place until the solvent cement begins to set. It is wise to wear impervious gloves to prevent solvent cement from coming into contact with the skin. The solvent-cementing procedure must be done in accordance with the referenced standard.

To minimize the health and fire hazards associated with solvent cements and primers, the manufacturers' instructions for the use and handling of these materials must be followed, in addition to the requirements found in ASTM D 2855.

For SI: 1 inch = 25.4 mm.

**Figure 705.10.2
LEAD-WIPED JOINT**

The exception allows for primer to not be used for joints of piping that is 4 inches (102 mm) and smaller. Testing by NSF revealed that for nonpressure applications, a solvent cement-only joint provides adequate strength.

705.11.3 Threaded joints. Threads shall conform to ASME B1.20.1. Schedule 80 or heavier pipe shall be permitted to be threaded with dies specifically designed for plastic pipe. *Approved* thread lubricant or tape shall be applied on the male threads only.

❖ The requirements for threaded joints in PVC plastic pipe are identical to those of ABS and CPVC plastic pipe (see commentary, Section 705.2.3).

705.12 Vitrified clay. Joints between vitrified clay pipe or fittings shall be made with an elastomeric seal conforming to ASTM C 425, ASTM C 1173 or CSA B602.

❖ An acceptable joint for vitrified clay pipe is a mechanical joint sealed with an elastomeric material. This is a type of compression joint similar to that used with other hubbed pipe and fittings.

705.13 Polyethylene plastic pipe. Joints between polyethylene plastic pipe and fittings shall be underground and shall comply with Section 705.13.1 or 705.13.2.

❖ Acceptable joints for polyethylene plastic pipe or fittings are identified in the referenced sections. Note that joints can be located only underground.

705.13.1 Heat-fusion joints. Joint surfaces shall be clean and free from moisture. All joint surfaces shall be cut, heated to melting temperature and joined using tools specifically designed for the operation. Joints shall be undisturbed until cool. Joints shall be made in accordance with ASTM D 2657 and the manufacturer's instructions.

❖ Three types of processes are used: 1) hot plate welding for butt welding of pipes; 2) heat fusion for joining pipe to fittings and 3) electrofusion welding for joining pipe to fittings. Special clamping fixtures are used to hold the pipe ends and fittings in alignment. Some fixtures for small pipe diameters are small enough to hand carry; larger pipes require fixtures to be on a wheeled cart or even a motorized/tracked machine.

In the hot plate welding process, the ends of pipe are faced and squared with the centerline of the pipe using a rotating-blade planer block that accurately rotates around the pipe end while planing.

After cleaning debris from the pipes, the two pipe ends are clamped in precise alignment in the fixture so that no perceptible gap will exist between the pipe ends once they are pushed together. A plate with internal heating elements and a nonstick surface is inserted between and in contact with the two pipe ends. After sufficient heating to soften the ends of the pipe, the heating plate is removed and the ends of the pipe are mated with a specified force, "fusing" the softened pipe ends together. After an appropriate cooling time, the pipe is removed from the fixture and is ready for installation. A completed quality butt weld has: 1) complete and uniform beads; 2) bead rollback onto pipe and 3) proper alignment between the two pipes with centerlines parallel and, more importantly, outside wall surfaces in line (no offset).

In the heat fusion and electrofusion processes, a pipe end is joined to a fitting such as coupling, tee or valve having a recessed socket.

In the heat fusion method, the exterior of the pipe and interior of the fitting socket are electrically heated independently with specialized electric heating devices. Once the materials are sufficiently softened, the heating devices are removed and the pipe is precisely inserted in the socket fitting and allowed to cool.

In the electrofusion method, the socket fitting has a heating element "implant" that is connected to an appropriate power source. The pipe is inserted into the fitting and the heating element implant is energized for a specific period of time, causing the pipe and fitting interior to be softened and fused together. The joint is allowed to cool for an appropriate amount of time before the joint is disturbed.

All three processes require trained and qualified personnel as well as strict adherence to procedures to make fusion joints with integrity. Cleanliness and protection against weather elements during the process are essential to produce high-quality joints.

705.13.2 Mechanical joints. Mechanical joints in drainage piping shall be made with an elastomeric seal conforming to ASTM C 1173, ASTM D 3212 or CSA B602. Mechanical joints shall be installed in accordance with the manufacturer's instructions.

❖ Three types of mechanical joints can be used depending on the size of the pipe: 1) compression nut fittings [small diameters up to 2 inches (51 mm)]; 2) stab-type fittings [small diameters up to 2 inches (51 mm)], bolt-type compression fittings [from $1^1/_4$ inches (32 mm) and up] and 3) a variety of flanged ends. Each requires following the manufacturer's instructions for proper application and installation. Note that only joints using an elastomeric seal conforming to ASTM C 1173, D 3212 or CSA B602 comply with the code.

705.14 Polyolefin plastic. Joints between polyolefin plastic pipe and fittings shall comply with Sections 705.14.1 and 705.14.2.

❖ Acceptable joints for polyethylene plastic pipe or fittings are identified in the referenced sections.

705.14.1 Heat-fusion joints. Heat-fusion joints for polyolefin pipe and tubing joints shall be installed with socket-type heat-fused polyolefin fittings or electrofusion polyolefin fittings. Joint surfaces shall be clean and free from moisture. The joint shall be undisturbed until cool. Joints shall be made in accordance with ASTM F 1412 or CSA B181.3.

❖ Only the heat fusion and electrofusion process using socket-type fitting is allowed (see commentary, Section 705.16.1).

705.14.2 Mechanical and compression sleeve joints. Mechanical and compression sleeve joints shall be installed in accordance with the manufacturer's instructions.

❖ Three types of mechanical joints can be used depending upon the size of the pipe: 1) compression nut fittings [small diameters up to 2 inches (51 mm)]; 2) stab-type fittings [small diameters up to 2 inches (51 mm)] and 3) bolt-type compression fittings [from 1 1/4 inches (32 mm) and up], and a variety of flanged ends. Each requires following the manufacturer's instructions for proper application and installation.

705.15 Polyvinylidene fluoride plastic. Joints between polyvinylidene plastic pipe and fittings shall comply with Sections 705.15.1 and 705.15.2.

❖ This pipe material (pronounced poly-vy-NIL-ih-DEAN floor-IDE) is a highly nonreactive and pure thermoplastic fluoropolymer. Several different companies manufacture this material under trademarked trade names but the material is the same. Because of the multiple strong carbon-fluorine molecular bonds, the material has high purity, high strength, and an exceptional resistance to solvents, acids and bases, making it an excellent material for conveying corrosive wastes. It has a melting point of about 351°F (177°C) and is a low generator of heat and smoke in a fire event.

705.15.1 Heat-fusion joints. Heat-fusion joints for polyvinylidene fluoride pipe and tubing joints shall be installed with socket-type heat-fused polyvinylidene fluoride fittings or electrofusion polyvinylidene fittings and couplings. Joint surfaces shall be clean and free from moisture. The joint shall be undisturbed until cool. Joints shall be made in accordance with ASTM F 1673.

❖ Polyvinylidene fluoride pipe and tubing joints can be heat fused or electrofused in a manner similar to that used for polyolefin pipe.

705.15.2 Mechanical and compression sleeve joints. Mechanical and compression sleeve joints shall be installed in accordance with the manufacturer's instructions.

❖ Polyvinylidene fluoride pipe and tubing can be joined with compression fittings or mechanical joints that are specifically intended for use with the piping material. The manufacturer's instructions for installation of the fittings must be followed.

705.16 Joints between different materials. Joints between different piping materials shall be made with a mechanical joint of the compression or mechanical-sealing type conforming to ASTM C 1173, ASTM C 1460 or ASTM C 1461. Connectors and adapters shall be *approved* for the application and such joints shall have an elastomeric seal conforming to ASTM C 425, ASTM C 443, ASTM C 564, ASTM C 1440, ASTM F 477, CSA A257.3M or CSA B602, or as required in Sections 705.16.1 through 705.16.7. Joints between glass pipe and other types of materials shall be made with adapters having a TFE seal. Joints shall be installed in accordance with the manufacturer's instructions.

❖ Mechanical joints are typically used to join dissimilar piping materials. Because many piping materials have the same outside diameters, many fittings are designed by manufacturers for use with a number of different piping materials. Some mechanical joints are manufactured specifically for joining pipes having different outside diameters. Transition couplings use bushings (inserts) to match the outside diameters of the pipes or are simply designed to fit over different outside diameter pipe ends. ASTM C 1173, ASTM C 1460 or ASTM C 1461 are the standards for the complete coupling assembly, which includes an elastomeric gasket, a clamp assembly or both.

705.16.1 Copper or copper-alloy tubing to cast-iron hub pipe. Joints between copper or copper-alloy tubing and cast-iron hub pipe shall be made with a brass ferrule or compression joint. The copper or copper-alloy tubing shall be soldered to the ferrule in an *approved* manner, and the ferrule shall be joined to the cast-iron hub by a caulked joint or a mechanical compression joint.

❖ An adapter must be used when joining copper tubing to a cast-iron hub. The adapter is called by a wide variety of names, the most common being "soil adapter" or "caulk adapter." The adapter creates a spigot end similar to that of cast-iron pipe and fittings.

705.16.2 Copper or copper-alloy tubing to galvanized steel pipe. Joints between copper or copper-alloy tubing and galvanized steel pipe shall be made with a brass converter fitting or dielectric fitting. The copper tubing shall be soldered to the fitting in an *approved* manner, and the fitting shall be screwed to the threaded pipe.

❖ When joining copper or copper-alloy tubing with galvanized steel pipe, a method of protecting against galvanic corrosion is required. Galvanic corrosion occurs when two different metals come in contact in the presence of an electrolyte, such as water. Galvanic corrosion accelerates the natural corrosion process that occurs in metals, and, because certain metals corrode faster than others, they have been placed in a hierarchy in order of rate of corrosion. The more reactive metal at a connection is called the "anode" (galvanized steel pipe), and the less reactive metal is called the "cathode" (copper tubing). When the copper tubing and galvanized steel pipe are coupled, the galvanized steel pipe will tend to dissolve in the electrolyte, thereby generating electric current flow between the metals. This section prescribes two methods of protection against galvanic corrosion: dielectric fittings and brass fittings. The fittings are designed to provide a barrier or buffer between the two materials.

705.16.3 Cast-iron pipe to galvanized steel or brass pipe. Joints between cast-iron and galvanized steel or brass pipe

SANITARY DRAINAGE

shall be made by either caulked or threaded joints or with an *approved* adapter fitting.

❖ Galvanized steel pipe and brass pipe may be caulked directly into the hub of cast-iron pipe without an adapter fitting. Certain cast-iron fittings have a female tapped inlet to permit a direct threaded connection (see Commentary Figure 705.16.3).

Figure 705.16.3
CAST-IRON ADAPTER FITTING

705.16.4 Plastic pipe or tubing to other piping material. Joints between different types of plastic pipe or between plastic pipe and other piping material shall be made with an *approved* adapter fitting. Joints between plastic pipe and cast-iron hub pipe shall be made by a caulked joint or a mechanical compression joint.

❖ ABS and PVC plastic pipe may connect directly into the hub of cast-iron pipe with a caulked joint or by using a compression gasket without the use of an adapter fitting. Special caulk-adapter plastic fittings are available that provide a spigot end for use with hubbed pipe and fittings. Where different grades of plastic pipe are joined or where plastic pipe and cast-iron spigot end pipe, Schedule 40 steel pipe or copper pipe are joined, an adapter fitting must be used.

705.16.5 Lead pipe to other piping material. Joints between lead pipe and other piping material shall be made by a wiped joint to a caulking ferrule, soldering nipple or bushing or shall be made with an *approved* adapter fitting.

❖ The common method of adapting to lead is by a wiped joint to a ferrule, nipple or bushing; however, a mechanical adapter fitting using an elastomeric material may also be used.

705.16.6 Borosilicate glass to other materials. Joints between glass pipe and other types of materials shall be made with adapters having a TFE seal and shall be installed in accordance with the manufacturer's instructions.

❖ Glass pipe may be joined to plain-end or threaded-end metal or plastic pipe with adapter couplings that incorporate a TFE (tetrafluoroethylene) seal. This seal isolates the glass pipe from contact with the pipe it is being joined to and helps provide a positive seal around the perimeter of the glass pipe. Typically, a manufacturer of glass pipe also makes the couplings that are designed for use with the pipe; therefore, the coupling manufacturer's installation instructions must be followed. The TFE seal is resistant to chemical attack, as is glass pipe. TFE is commonly known by the registered trademark name "Teflon."

705.16.7 Stainless steel drainage systems to other materials. Joints between stainless steel drainage systems and other piping materials shall be made with *approved* mechanical couplings.

❖ Joints between stainless steel drainage systems must be made with a mechanical coupling and have an elastomeric seal conforming to ASTM C 425, ASTM C 443, ASTM C 564, ASTM C 1173, ASTM D 1869, ASTM F 477, CAN/CSA A257.3M or CAN/CSA B602. Because many different types of mechanical joints are available and most couplings are unique to each manufacturer, the joining method must be in accordance with the manufacturer's installation instructions.

705.17 Drainage slip joints. Slip joints shall comply with Section 405.8.

❖ An elastomeric gasket must be used to make the seal for drainage slip joints [see Commentary Figures 705.17(1) and 705.17(2) and the definition of "Slip joint" in Chapter 2]. Slip joints cannot be concealed and are limited to those connections that occur between a fixture waste outlet and its connection to the fixture drain, including the trap adapter on the trap outlet arm. In locations where a slip joint connection is concealed, an access panel must be provided so that the connection can be serviced. This would include tub waste and overflow connections (see commentary, Section 405.8).

705.18 Caulking ferrules. Ferrules shall be of red brass and shall be in accordance with Table 705.18.

❖ Caulking ferrules are used to adapt cast-iron hub pipe to either copper tubing or threaded pipe. The ferrule is installed in the hub of the cast-iron pipe by either a caulked joint or an elastomeric gasket compression joint. The copper tubing is soldered to, or the threaded pipe is screwed into, the ferrule.

TABLE 705.18
CAULKING FERRULE SPECIFICATIONS

PIPE SIZES (inches)	INSIDE DIAMETER (inches)	LENGTH (inches)	MINIMUM WEIGHT EACH
2	$2^{1}/_{4}$	$4^{1}/_{2}$	1 pound
3	$3^{1}/_{4}$	$4^{1}/_{2}$	1 pound 12 ounces
4	$4^{1}/_{4}$	$4^{1}/_{2}$	2 pounds 8 ounces

For SI: 1 inch = 25.4 mm, 1 ounce = 28.35 g, 1 pound = 0.454 kg.

❖ The table specifies the acceptable dimensions and weights for caulking ferrules.

SANITARY DRAINAGE

**Figure 705.17(1)
SLIP JOINT**

**Figure 705.17(2)
SLIP JOINTS**

705.19 Soldering bushings. Soldering bushings shall be of red brass and shall be in accordance with Table 705.19.

❖ A soldering bushing is used to adapt lead pipe to other piping materials.

**TABLE 705.19
SOLDERING BUSHING SPECIFICATIONS**

PIPE SIZES (inches)	MINIMUM WEIGHT EACH
$1^1/_4$	6 ounces
$1^1/_2$	8 ounces
2	14 ounces
$2^1/_2$	1 pound 6 ounces
3	2 pounds
4	3 pounds 8 ounces

For SI: 1 inch = 25.4 mm, 1 ounce = 28.35 g, 1 pound = 0.454 kg.

❖ The table specifies the acceptable weights for soldering bushing sizes.

705.20 Stainless steel drainage systems. O-ring joints for stainless steel drainage systems shall be made with an *approved* elastomeric seal.

❖ O-ring joints for stainless steel drainage systems are identical to those between stainless steel drainage systems and other piping materials (see commentary, Section 705.16.7).

SECTION 706
CONNECTIONS BETWEEN DRAINAGE PIPING AND FITTINGS

706.1 Connections and changes in direction. All connections and changes in direction of the sanitary drainage system shall be made with *approved* drainage fittings. Connections between drainage piping and fixtures shall conform to Section 405.

❖ This section contains the requirements for the proper use of fittings and connections used to accomplish changes in direction. The design of a drainage system is based on uniform flow in the piping. Special drainage fittings have been developed with design patterns providing the least resistance to flow. All types of connections and drainage fittings are subject to the approval of the code official. Section 405 regulates the installation of fixtures connected to the sanitary drainage system.

706.2 Obstructions. The fittings shall not have ledges, shoulders or reductions capable of retarding or obstructing flow in the piping. Threaded drainage pipe fittings shall be of the recessed drainage type. This section shall not be applicable to tubular waste fittings used to convey vertical flow upstream of the trap seal liquid level of a fixture trap.

❖ Because the flow of waste within the drainage system is slow and induced by gravity, any shoulder edge or obstruction on the inside surface of the pipe may inhibit or retard flow and likely cause a stoppage. Therefore, joints, connections and fittings must be of the drainage type, free from flow-obstructing elements. Recessed drainage-type threaded fittings are designed so that the inside surface of the pipe will be flush with the inside surface of the fitting. An offset closet flange is not considered an obstruction. Tubular waste fittings upstream of the trap seal sometimes have internal baffles that might appear to block flow; however, this section does not pertain to those fittings.

706.3 Installation of fittings. Fittings shall be installed to guide sewage and waste in the direction of flow. Change in direction shall be made by fittings installed in accordance with Table 706.3. Change in direction by combination fittings, side inlets or increasers shall be installed in accordance with Table 706.3 based on the pattern of flow created by the fitting. Double sanitary tee patterns shall not receive the discharge of back-to-back water closets and fixtures or appliances with pumping action discharge.

> **Exception:** Back-to-back water closet connections to double sanitary tees shall be permitted where the horizontal *developed length* between the outlet of the water closet and the connection to the double sanitary tee pattern is 18 inches (457 mm) or greater.

❖ Drainage fittings and connections must provide a smooth transition of flow without creating obstructions or causing interference. The use of proper fittings helps to maintain the required flow velocities and reduces the possibility of stoppage in the drainage system. Combination fittings are commonly used in drainage systems and must be evaluated for their pattern of flow. Combination fittings include two or more fittings, such as combination wye and eighth bends or tee-wyes. A double sanitary tee (cross) cannot be used for connections to fixtures and appliances with pumping action because it has a short pattern for change of direction. In addition to blowout-type water closets, 1.6-gallon-per-flush (6.1 L/flush) flushometer tank (pressure assisted) and gravity tank-type fixtures may also produce crossover flow into the opposite fixture connection. The high-velocity flow may discharge through the opposing openings of a double sanitary tee, causing an interruption or blowback at the other fixture. Back-to-back water closets are fixtures that are installed directly opposite of each other and discharge through a short distance into a double pattern fitting. If back-to-back fixtures discharge through several feet of pipe or multiple changes of direction, the flow velocity in the fixture drains would decrease to the point where cross flow would not occur. The exception provides for back-to-back water closet connections to double sanitary tee patterns where the horizontal developed length between the outlet of the water closet and connection to the double sanitary tee pattern is 18 inches (457 mm) or more. The critical distance at which there is potential for crossover is a horizontal developed length of less than 10 times the diameter of the horizontal pipe, for example, 30 inches (762 mm) for a 3-inch-diameter (76 mm) pipe. Once the developed length is over 30 inches (762 mm), the surging flow conditions will no longer exist, and the frictional resistance of the pipe retards the velocity to that of uniform flow conditions.

It is mistakenly thought that the use of a double combination fitting as shown in Commentary Figure 706.3(5) is not permitted for the water closet arrangement because it will not permit the vent to connect above the trap weir. However, this arrangement is permitted because the vent connection limitation is designed to prevent self-siphonage, and water closets must self-siphon to operate properly. Use of fittings such as those having side inlets, heel inlets and specialty opening configurations is permitted only in applications that comply with the intent of Table 706.3 [see Commentary Figures 706.3(1) through (5)].

TABLE 706.3
FITTINGS FOR CHANGE IN DIRECTION

TYPE OF FITTING PATTERN	CHANGE IN DIRECTION		
	Horizontal to vertical	Vertical to horizontal	Horizontal to horizontal
Sixteenth bend	X	X	X
Eighth bend	X	X	X
Sixth bend	X	X	X
Quarter bend	X	X[a]	X[a]
Short sweep	X	X[a,b]	X[a]
Long sweep	X	X	X
Sanitary tee	X[c]	—	—
Wye	X	X	X
Combination wye and eighth bend	X	X	X

For SI: 1 inch = 25.4 mm.
a. The fittings shall only be permitted for a 2-inch or smaller *fixture drain*.
b. Three inches or larger.
c. For a limitation on double sanitary tees, see Section 706.3.

❖ When the flow is in the horizontal plane only, long-pattern fittings are necessary to prevent excessive reduction in flow velocity. When flow is from the horizontal to the vertical or from the vertical to the horizontal, a shorter pattern fitting may be used because the acceleration of flow caused by gravity will maintain higher velocities.

An "X" in a column indicates that a particular fitting is acceptable for the described change in direction. For example, the only "X" shown for a sanitary tee is in the column for horizontal to vertical change in direction. The table has been developed to identify where use of a particular drainage pattern fitting is allowed.

The terminology used to identify drainage pattern fittings was originally developed by the cast-iron soil pipe industry. The pattern of the fittings is regulated by the fitting standards referenced in Table 702.4. A fitting of one material will not necessarily look the same as a fitting of another material with the same name. Only cast-iron pipe fittings have a separate quarter-bend fitting and short-sweep fitting. For other drainage piping materials, a quarter bend and a short sweep are the same fitting.

Copper tubing fittings are unique by having only one sweep-type fitting identified as a DWV 90-degree (1.6 rad) elbow. This is considered a long-sweep drainage fitting. Some manufacturers make a copper fitting identified as a DWV 90-degree (1.6 rad) long radius elbow. This is also not included in the fitting standards, but could be used as a long sweep.

Note b permits the use of short-sweep fittings of 3 inches (76 mm) in diameter or larger where the change in direction is from the vertical to the horizontal.

Note a permits the use of 2-inch (51 mm) or smaller short-sweep and quarter-bend fittings for drains serving a single fixture where the change in direction is from the vertical to the horizontal or from the horizontal to the horizontal.

The intent of Note c is to allow short-pattern fittings for fixture drains, which are generally difficult to install using long-pattern fittings in frame construction. For example, a long-sweep PVC 90-degree (1.6 rad) elbow would not fit within a nominal 2-inch by 4-inch (51 mm by 102 mm) stud wall cavity where used to make a horizontal-to-horizontal change in direction. A quarter bend with a very short radius, less than that of a short sweep, is identified as a vent elbow or vent quarter bend. These particular fittings are designed for vent system applications only.

Double sanitary tees must not be installed where water closets or appliances with pump action are back to back. Appliances with pump action are typically washing machines or dishwashers (see commentary, Section 706.3).

Side inlet fittings are acceptable for drainage connections. The typical application for side inlet fittings is horizontal to vertical changes in direction involving side inlet sanitary tees, side inlet elbows and side inlet wyes. When used to connect a vent, however, the requirements for venting as specified in Section 905 must be met [see Commentary Figures 706.3(1) through 706.3(8)].

Low- and high-heel inlet elbows are acceptable depending on the application. A low-heel branch inlet must not be used for a drainage connection where solid waste enters the run of the elbow from the vertical. In such a case, solid waste could enter the branch inlet opening, causing a blockage. Because there is no waste flow in vent piping, and the airflow in vent piping may be in either direction, the types of fittings used in vent piping are not regulated by this section. For example, a sanitary tee placed on its back with the branch inlet connected to a vertical dry vent riser is a common installation practice and is not prohibited. In this application, the only waste flow is through the straight run of the fitting, and the branch inlet conducts only air into or out of the horizontal drain pipe. Note that the tee must be oriented to allow compliance with Section 905.3. The restrictions placed on the use of a sanitary tee are based on its inability to guide waste flow entering through the branch inlet. Such restrictions do not apply where the branch inlet is used for a dry vent connection in compliance with Section 905. In all cases, it is the intent of the code to require that fittings be installed in a manner that will encourage high-velocity drainage flow and minimize the chances of blockage.

The details of the patterns for drainage, waste and vent fittings made of PVC and ABS are covered by ASTM D 3311. This standard requires 90-degree (1.6 rad) fittings (except vent patterns) to have sockets aligned in the fitting to cause horizontal drainage pipes to have a slope of $^1/_4$ inch per foot (21 mm per m) [see Commentary Figure 706.3(9)]. In some applications, the use of these $^1/_4$-bend fittings, having an angle of 91.2 degrees, results in an improper alignment of the connecting pipes resulting in a poor joint. Commentary Figure 706.3(10) shows the correct orientation for the sanitary tee to prevent this piping alignment problem. Many plumber apprentices learned to install sanitary tees in vent piping in this manner after being told that the orientation promoted better airflow or allowed the condensate to flow better back to the drain; however, there is no technical substantiation for these explanations.

Sanitary tees may be used for horizontal drainage flow through the run of the tee where the branch of the tee is for venting only and is oriented within 45 degrees (0.79 rad) of the vertical. A sanitary tee can be installed on its back for venting applications in sanitary drainage.

Other bend patterns for DWV fittings of PVC and ABS do not have this special socket alignment; thus, an $^1/_8$ bend fitting is 45 degrees (0.79 rad), a $^1/_6$-bend fitting is 60 degrees (1 rad) and so forth. In some applications, use of these fittings might a create pipe slope problem [see Commentary Figure 706.3(10)]. Although the code is silent on the use of a close combination of fittings to serve the purpose of one fitting, full and square engagement of the pipes into a close combination "made up" 90-degree (1.57 rad) bend might produce undesirable results.

706.4 Heel- or side-inlet quarter bends. Heel-inlet quarter bends shall be an acceptable means of connection, except where the quarter bend serves a water closet. A low-heel inlet shall not be used as a wet-vented connection. Side-inlet quarter bends shall be an acceptable means of connection for drainage, wet venting and *stack* venting arrangements.

❖ Because of the possibility of waste from a water closet clogging a heel-inlet (either high or low heel) quarter bend, this fitting must not be used in a closet bend application. Similarly, low-heel-inlet quarter bends must not receive flow from drainage pipes that serve as wet vents for a fixture. Only side-inlet quarter bends can be used for these applications [see Commentary Figures 706.4(1), (2) and 706.3(6)]. Note the obvious cut floor joist violation in Commentary Figure 706.4(2).

SANITARY DRAINAGE

For SI: 1 degree = 0.0175 rad, 1 inch = 25.4 mm.

Figure 706.3(1)
COMPARISON OF DIFFERENT 1½-INCH-LONG SWEEPS FOR VARIOUS PIPING MATERIALS

For SI: 1 degree = 0.0175 rad, 1 inch = 25.4 mm.

Figure 706.3(2)
COMPARISON OF 3-INCH COMBINATION WYE AND EIGHTH BEND FOR VARIOUS PIPING MATERIALS

For SI: 1 degree = 0.0175 rad.

Figure 706.3(3)
DRAINAGE PATTERN FITTINGS
(continued)

7-26   2015 INTERNATIONAL PLUMBING CODE® COMMENTARY

SANITARY DRAINAGE

Figure 706.3(3)—continued
DRAINAGE PATTERN FITTINGS

For SI: 1 degree = 0.0175 rad.

Figure 706.3(4)
DIFFERENCE BETWEEN PVC PLASTIC $^1/_4$-BEND SHORT SWEEP AND $^1/_4$-BEND VENT

SANITARY DRAINAGE

For SI: 1 inch = 25.4 mm.

Figure 706.3(5)
BACK-TO-BACK FIXTURES

Figure 706.3(6)
HEEL INLET FITTINGS

Figure 706.3(7)
COMBINATION TEES

SANITARY DRAINAGE

For SI: 1 inch = 25.4 mm, 1 degree = 0.0175 rad.

**Figure 706.3(8)
FITTING USES**

**Figure 706.3(9)
IMPROPER USE OF DWV-PVC/ABS 90-DEGREE FITTINGS**

SANITARY DRAINAGE

Figure 706.3(10)
PROPER USE OF DWV-PVC/ABS 90-DEGREE FITTINGS

Figure 706.4(1)
HEEL- AND SIDE-INLET QUARTER BEND APPLICATION RESTRICTIONS

Figure 706.4(2)
SIDE-INLET QUARTER BEND FOR WATER CLOSET

SECTION 707
PROHIBITED JOINTS AND CONNECTIONS

707.1 Prohibited joints. The following types of joints and connections shall be prohibited:

1. Cement or concrete joints.
2. Mastic or hot-pour bituminous joints.
3. Joints made with fittings not *approved* for the specific installation.
4. Joints between different diameter pipes made with elastomeric rolling O-rings.
5. Solvent-cement joints between different types of plastic pipe.
6. Saddle-type fittings.

❖ This section contains requirements for joints between pipe and fittings of different materials.

Cement and concrete are not effective in sealing a pipe joint because such materials are inflexible and susceptible to cracking and displacement. A rolling O-ring has no resistance to being pushed or rolled out of a joint when exposed to pressure or the movement of pipe caused by expansion and contraction. A solvent-cement joint is a homogeneous chemical bond made between a pipe and a fitting. The bond is accomplished because the chemical composition of the joint surfaces is the same. If the materials are different, the joint may not form a proper chemical bond, and the strength and integrity of the joint will be adversely affected. For example, solvent cementing is prohibited between ABS and PVC, PVC and CPVC or ABS and CPVC (see commentary, Section 705.2.2). Saddle-type fittings that are moved out of alignment can weaken the pipe because of the pipe wall penetration and typically do not form a drainage pattern connection. "Saddle-type" fittings are commonly used for connecting a building sewer to an existing public sewer.

SECTION 708
CLEANOUTS

708.1 Cleanouts required. Cleanouts shall be provided for drainage piping in accordance with Sections 708.1.1 through 708.1.11.

❖ There must be points of access to drainage systems for clearing blockages. These access points are called cleanouts (see the definition for "Cleanout" in Chapter 2). Blockages in vertical piping rarely, if ever, occur because gravity will cause the blockage to move downward. Therefore, only horizontal piping is required to have cleanouts. In locating cleanouts, the designer or installer should consider the type of equipment that might be necessary for clearing blockages. Typical drain clearing/cleaning equipment includes hand-held cable augers, flat tapes, power-driven cables from $^1/_4$ diameter to $1^1/_8$-inch diameter (or larger), and water jetting hoses from $^1/_4$-inch (6.4 mm) diameter to 1 inch (25 mm) diameter (or larger). This equipment, especially power driven units, can be sizable and heavy. A large machine could be difficult to maneuver into smaller areas, let alone be safe to operate in a confined area. This section does provide a few limitations for where cleanouts can be located. Otherwise, the experience of the designer or installer is leaned upon heavily to determine the best, most logical locations for cleanouts.

A frequently asked question is whether mechanical joints such as elastomeric connections (hubless joint connectors) can be considered as cleanouts. A significant amount of drain cleaning is performed by personnel that, although trained on drain cleaning methods and equipment, may not have the necessary training on the proper removal and reinstallation of piping. Hangers could be left out, couplings reused or couplings overtightened such that the plumbing system might not work as intended or become a hazard (either environmentally or structurally). In the author's opinion, appropriate cleanout access such as removal of a cleanout plug is a far superior and a safer approach to provide access to the drainage system. However, some code officials might believe that removal of piping connectors would provide the intended access. The code is silent on this topic.

708.1.1 Horizontal drains and building drains. Horizontal drainage pipes in buildings shall have cleanouts located at intervals of not more than 100 feet (30 480 mm). *Building drains* shall have cleanouts located at intervals of not more than 100 feet (30 480 mm) except where manholes are used instead of cleanouts, the manholes shall be located at intervals of not more than 400 feet (122 m). The interval length shall be measured from the cleanout or manhole opening, along the *developed length* of the piping to the next drainage fitting providing access for cleaning, the end of the horizontal drain or the end of the *building drain*.

Exception: Horizontal *fixture drain* piping serving a non-removable trap shall not be required to have a cleanout for the section of piping between the trap and the vent connection for such trap.

❖ Modern drain-cleaning equipment is available with cable (or jetter hose) length to clear blockages in horizontal drainage piping that is 100 feet (30 480 mm) (or less) in developed piping length from the cleanout opening. Manholes are sometimes used for cleanouts for larger diameter piping. The larger piping diameter allows for larger diameter cables (or larger diameter/higher pressure jetter hoses) to be used that are capable of clearing blockages in horizontal drainage piping that is 400 feet (122 m) in length (or less). A key point to consider in locating cleanouts is that the lengths indicated in this section are measured from the cleanout opening to the farthest point where the next cleanout allows access to the horizontal drain piping. For example, if the cleanout is on a stack and the cleanout opening is 10 feet (30 480 mm) above the horizontal drain pipe (that the stack connects to), then the length of horizontal drainage piping that can be cleared is only 90 feet. The lengths in this section are for drainage piping that does not have changes in direction any

SANITARY DRAINAGE

greater than 45 degrees (0.79 rad). See Section 708.1.4 for horizontal changes in direction that are greater than 45 degrees (0.79 rad).

The exception covers the special circumstance for horizontal drain piping that begins with a nonremovable trap such as for a floor drain or a shower drain. It is understood that a cable or jetter hose can be negotiated through a trap. However, the convoluted path of the trap creates additional drag on a cable such that running the cable for too far of a distance from the trap can cause high stresses in the cable resulting in a greater potential for cable breakage. Therefore, the cleanout access for these horizontal drains can be located downstream of these traps for a distance of up to the maximum allowable pipe length between the trap and the vent connection for the trap (see Table 909.1 for trap-to-vent distances).

708.1.2 Building sewers. *Building sewers* smaller than 8 inches (203 mm) shall have cleanouts located at intervals of not more than 100 feet (30 480 mm). *Building sewers* 8 inches (203 mm) and larger shall have a manhole located not more than 200 feet (60 960 mm) from the junction of the *building drain* and *building sewer* and at intervals of not more than 400 feet (122 m). The interval length shall be measured from the cleanout or manhole opening, along the *developed length* of the piping to the next drainage fitting providing access for cleaning, a manhole or the end of the *building sewer*.

❖ The commentary for Section 708.1.1 explains most of the information in this section. Sewers that are 8 inches (203 mm) or larger are required to have manholes for cleanout access because the large cables (or jetter hoses) needed for these pipelines cannot be easily negotiated through the standard pattern wye and $^1/_8$-bend fitting that would be needed to bring the cleanout access opening to grade level. Building sewers could be broken at wye and $^1/_8$-bend entry into the horizontal drainage piping resulting in earth or stone backfill later migrating into the sewer. Manholes for these larger cables and hoses provide the needed space for access to the piping. Note that the first manhole for an 8-inch (203 mm) or larger diameter sewer is not required to be at the building. A manhole located up to 200 feet (60 960 mm) away from the building can be used to clear blockages in the sewer line back towards the building (see Commentary Figure 708.1.2).

708.1.3 Building drain and building sewer junction. The junction of the *building drain* and the *building sewer* shall be served by a cleanout that is located at the junction or within 10 feet (3048 mm) of the *developed length* of piping upstream of the junction. For the requirements of this section, the removal of the water closet shall not be required to provide cleanout access.

❖ The building sewer begins at the termination of the building drain that ends, by definition, at a point that is 30 inches (762 mm) from the outside wall of the building. Access to the building sewer is very important because blockages can frequently occur as the result of low-flow velocities being unable to transport solids and tree roots that can penetrate the sewer. Blockages might also be the result of other underground utilities that were installed by underground horizontal directional boring methods or "slit" trenching, both of which can sever/penetrate a building sewer without the knowledge of the equipment operator.

The allowance to have the cleanout located upstream of the building drain/building sewer juncture but up to 10 feet (3048 mm) provides for building sewer cleanouts to be in basements, crawl spaces, interior rooms or corridors. The removal of a water closet to provide cleanout access has a twofold purpose: 1) water closets are either in a water closet compartment or a small single-user toilet room. Although removal of the water closet could be a cleanout (see "Cleanout" definition in Chapter 2), such spaces are usually not large enough to facilitate the safe operation of cable machines that are powerful enough to have root cutting capability and 2) the location of a wall (stack) or floor cleanout that requires removal of the water closet to gain safe access to the cleanout defeats the purpose of having installed the cleanout— to facilitate drain clearing. For example, a cleanout might be located directly behind a water closet, on the (close) wall side of a water closet or on the floor next to the water closet such that the water closet would need to be removed to enable positioning of a large drain-cleaning machine near the cleanout opening. A heavy machine near or on the closet flange could cause damage to the flange that might not be properly repaired. The location of a building sewer cleanout on the interior of a building needs to be well thought out (see Commentary Figure 708.1.3).

It is important to maintain safety and situational awareness when attempting to clear building sewer blockages. As stated previously, underground hori-

Figure 708.1.2
MANHOLES INSIDE THE BUILDING

zontal directional boring or slit trenching are frequently being used to install many types of underground utilities. Because building sewers are usually not marked on drawings, are frequently made of nonmetallic materials and are rarely installed with detectable tapes or wires such that detection equipment can locate them, a building sewer could have been inadvertently penetrated by a utility service. And that may be the reason for the sewer blockage! Thus, drain cleaning personnel need to be aware of any evidence (including information from the building occupant/owner) of recent construction activities in the area that could have impacted the sewer line. Especially if those activities involve natural gas lines or electric power cabling. There have been injuries and deaths associated with sewer cleaning work where gas lines or electric cables had penetrated a sewer and were cut by the sewer cleaning operations.

708.1.4 Changes of direction. Where a horizontal drainage pipe, a *building drain* or a *building sewer* has a change of horizontal direction greater than 45 degrees (0.79 rad), a cleanout shall be installed at the change of direction. Where more than one change of horizontal direction greater than 45 degrees (0.79 rad) occurs within 40 feet (12 192 mm) of *developed length* of piping, the cleanout installed for the first change of direction shall serve as the cleanout for all changes in direction within that 40 feet (12 192 mm) of *developed length* of piping.

❖ Table 706.3 covers what types of fittings can be used for changes in direction of drainage piping. However, where there are many changes in direction greater than 45 degrees (0.79 rad), it becomes more difficult for the drain-cleaning equipment (cables or jetter hoses) to negotiate the turns. The length of 40 feet (12 192 mm) is based on the ability of the drain-cleaning equipment to negotiate all of the turns without significant problems.

Where the length of drainage piping to be cleared is longer than 40 feet (12 192), then accommodations are necessary to ease the way for the cable or jetter hose for these longer pipelines. The accommodation is to not use single fitting patterns for changes in direction greater than 45 degrees (0.79 rad).

A single-pattern fitting or a fitting made up from "combining" fittings to make a pattern that is virtually identical to a single fitting must be avoided. Examples are: 1) a $^1/_4$-bend long sweep fitting can be duplicated by assembling a $^1/_8$-bend spigot end elbow into the hub end of a $^1/_8$-bend elbow and 2) a combination wye and $^1/_8$-bend fitting can be duplicated by assembling a $^1/_8$-bend spigot end elbow into the branch of a wye fitting. A frequent question is whether two $^1/_8$-bend pattern fittings can be connected together by a short length of pipe and not be considered duplicating a single pattern fitting (and not requiring a cleanout for that change in direction). For example, making a 90-degree (1.57 rad) change in direction using two $^1/_8$-bend fittings (hubs on both ends) connected by a "make-up" length of pipe (such that the hub end of each fitting touches the other). The code does allow 45-degree (0.79 rad) changes in direction and does not state how far apart those changes are required to be. The logical conclusion is that the length of pipe between the two fittings could be minimal and the made-up fitting still be in compliance such that a cleanout would not be required. This might not necessarily be the same conclusion arrived at by the local code official. Additionally, the type of pipe material (and how the fittings are made for that type of pipe such as hubless cast-iron fittings versus plastic fittings) might alter the code official's viewpoint in the matter. In the author's opinion, if a length of pipe needs to be used to make up a fitting to accomplish a change of direc-

For SI: 1 inch = 25.4 mm, 1 foot = 304 mm.

**Figure 708.1.3
CLEANOUTS AT SEWER JUNCTION**

SANITARY DRAINAGE

tion greater than 45 degrees (0.79 rad), then it would be better to plan for and use a length that clearly separates the two fittings into two separate changes in direction rather than be frowned upon for causing the question to be raised in the first place (see Commentary Figure 708.1.4).

708.1.5 Cleanout size. Cleanouts shall be the same size as the piping served by the cleanout, except that cleanouts for piping larger than 4 inches (102 mm) need not be larger than 4 inches (102 mm).

Exceptions:

1. A removable P-trap with slip or ground joint connections can serve as a cleanout for drain piping that is one size larger than the P-trap size.
2. Cleanouts located on *stacks* can be one size smaller than the *stack* size.
3. The size of cleanouts for cast-iron piping can be in accordance with the referenced standards for cast-iron fittings as indicated in Table 702.4.

❖ Cleanout openings for 1¹/₄-, 1¹/₂-, 2-, 2¹/₂-, 3- and 4-inch pipes must not be smaller than the pipe served, thus allowing use of full-bore cleaning equipment. For larger piping, a 4-inch (102 mm) opening is considered adequate for today's cleaning equipment, and full-bore cleaning equipment is typically not used for larger piping [see Commentary Figure 708.1.5(1)].

Manufactured cleanout fittings, like pipe, are sized by nominal dimensions. An industry rule of thumb for pipes larger than 4 inches (102 mm) is that the cleanout is sized in accordance with Commentary Figure 708.1.5(2). The requirements contained in this figure, although not required by the code, are intended to facilitate the use of drain-cleaning attachments (heads) and equipment of the proper size to accomplish full-bore or near full-bore pipe cleaning.

Commentary Figure 708.1.5(2) contains minimum cleanout sizes for larger-diameter drains and sewers. The cleanout sizing is based solely on the need to provide a large enough access opening into the drain pipe for adequate service with larger drain-cleaning equipment. The figure provides sizing for drainage pipe that is 10 inches (254 mm) or larger in diameter. The 8-inch (203 mm) minimum cleanout size for a 10-inch (254 mm) or larger drain, however, applies only to cleanouts serving drains other than building sewers because Section 708.1.4 requires a manhole for all building sewers 8 inches (203 mm) or larger.

Exception 1 allows a smaller nominal size fitting to serve as a cleanout for a larger nominal pipe size. This exception is very specific and limited to removable "P" traps. Exception 1 is intended to permit the removal of fixture traps to provide access to the connecting drains. Exception 2 accommodates some stack fitting designs that have always had the "one pipe size smaller" cleanout opening.

Exception 3 references Table 702.4 where the cast-iron fitting standards ASTM A 74, ASTM A 888 or CISPI 301 are indicated. Cleanout sizes for those fittings vary depending on the fitting chosen.

For SI: 1 degree = 0.0175 rad, 1 foot = 304.8 mm.

**Figure 708.1.4
CLEANOUTS**

SANITARY DRAINAGE

**Figure 708.1.5(1)
ACCEPTABLE CLEANOUTS**

NOMINAL PIPE SIZE (inches)	NOMINAL SIZE CLEANOUT (inches)
5	4
6	4
8	6
10 and above	8

**Figure 708.1.5(2)
CLEANOUT SIZES FOR CAST-IRON PIPING**

708.1.6 Cleanout plugs. Cleanout plugs shall be of brass, plastic or other *approved* materials. Cleanout plugs for borosilicate glass piping systems shall be of borosilicate glass. Brass cleanout plugs shall conform to ASTM A 74 and shall be limited for use only on metallic piping systems. Plastic cleanout plugs shall conform to the referenced standards for plastic pipe fittings, as indicated in Table 702.4. Cleanout plugs shall have a raised square head, a countersunk square head or a countersunk slot head. Where a cleanout plug will have a trim cover screw installed into the plug, the plug shall be manufactured with a blind end threaded hole for such purpose.

❖ Metallic cleanout plugs must be brass to provide for easy removal. If the cleanout plug is corroded in place, the brass plug is a soft enough material to be chiseled out. Brass cleanout plugs are limited to metallic fittings because the metal plug threads may damage the softer plastic threads of a fitting. Plastic plugs are intended for use with plastic fittings; however, this section does not prohibit the use of plastic plugs with metallic fittings. Like brass plugs, plastic plugs are less likely to seize in metallic fittings. The cleanout plug must have a square turning surface to allow for ease of removal while minimizing the possibility of stripping the surface during removal.

708.1.7 Manholes. Manholes and manhole covers shall be of an *approved* type. Manholes located inside of a building shall have gas-tight covers that require tools for removal.

❖ There many designs for manholes and manhole covers. Frequently, jurisdictions have specifications on manhole design and require specially embossed covers to be used. The code official must approve the design.

Where manholes are installed inside of buildings, the manhole covers must be of gas-tight design to prevent sewer gases from entering the building. Covers need to be bolted down or otherwise require tools to remove so access is limited.

The designer specifying the installation of manholes in areas subject to flooding must consider the protection of manholes from the effects of floodwaters. Floodwaters may damage or completely displace manholes and covers if they are not properly sealed and secured. Additionally, submerged manhole openings (covers) may allow floodwaters to enter the drainage system. The manhole cover may be located above the base flood level elevation or designed to resist hydrostatic forces when submerged in water and dynamic forces resulting from wave action or high-velocity water.

SANITARY DRAINAGE

708.1.8 Installation arrangement. The installation arrangement of a cleanout shall enable cleaning of drainage piping only in the direction of drainage flow.

Exceptions:

1. Test tees serving as cleanouts.

2. A two-way cleanout installation that is *approved* for meeting the requirements of Section 708.1.3.

❖ Many fixtures and buildings have been damaged by drain cleaning work that allowed the cable or jetter hose to be routed in an upstream direction. Drainage piping frequently smaller in diameter at upstream locations and augers can break or permanently jam in the piping. Almost every plumbing contractor has heard of stories or has personally witnessed the aftermath of a cleaning cable that found its way out through a water closet, breaking the water closet and causing significant wall, floor and ceiling damage in a toilet room. Cleanouts must be well thought out so the preferred cleanout arrangement is for cleaning in a downstream direction.

Exception 1 recognizes that a test tee might be required to be installed for a cleanout. The intent is not for enabling drain cleaning in the upstream direction. The drainage piping upstream of the test tee (serving as a cleanout) must have the required cleanout. Though not required by the code, some permanent indication of flow should be indicated on test tees serving as cleanouts, except those where the direction of flow is obvious, such as on a vertical pipe.

Exception 2 recognizes a specialized two-way cleanout fitting that is constructed with a branch throat that curves toward both the upstream and downstream directions of the run of the fitting. This fitting is only allowed to be used for the building drain/building sewer cleanout requirement of Section 708.1.3 (see Commentary Figure 708.1.8). This fitting is sometimes used as a retrofit for existing buildings where the building drain was not provided with an upstream cleanout. A question that is often asked is how far below grade can a two-way cleanout fitting be located? The concern is about being able to reliably guide the cleaning cable (or jetter hose) in the intended direction (and not in an unintended direction). Or it should be obvious that the cleanout is a specialized two-way cleanout fitting so that a drain cleaner doesn't just assume that the fitting will always direct the cable downstream. The code is silent on this issue.

**Figure 708.1.8
TWO-WAY CLEANOUT FITTING**

708.1.9 Required clearance. Cleanouts for 6-inch (153 mm) and smaller piping shall be provided with a clearance of not less than 18 inches (457 mm) from, and perpendicular to, the face of the opening to any obstruction. Cleanouts for 8-inch (203 mm) and larger piping shall be provided with a clearance of not less than 36 inches (914 mm) from, and perpendicular to, the face of the opening to any obstruction.

❖ The usability of cleanouts is not to be compromised by their placement in the building. The required clearances in this section are the *minimum* conditions that will allow the cleanout to be used for the intended purpose. The designer is encouraged to provide additional clearance whenever possible and practical. The required clearances are to be measured perpendicular to the face of the cleanout opening to any obstruction. Obstructions could mean fixed equipment, plumbing fixtures and equipment, appliances that cannot be moved by hand only (such as ice machines), or permanent building construction such as water closet compartment panels that are fixed to the building (see Commentary Figure 708.1.9).

MINIMUM CLEARANCE:
X = MINIMUM 18″ FOR 6″ AND SMALLER PIPES
X = MINIMUM 36″ FOR 8″ AND LARGER PIPES

For SI: 1 inch = 25.4 mm.

**Figure 708.1.9
CLEARANCE FOR CLEANOUTS**

708.1.10 Cleanout access. Required cleanouts shall not be installed in concealed locations. For the purposes of this section, concealed locations include, but are not limited to, the inside of plenums, within walls, within floor/ceiling assemblies, below grade and in crawl spaces where the height from the crawl space floor to the nearest obstruction along the path from the crawl space opening to the clean-out location is less than 24 inches (610 mm). Cleanouts with openings at a finished wall shall have the face of the opening located within $1^1/_2$ inches (38 mm) of the finished wall surface. Cleanouts located below grade shall be extended to grade level so that the top of the cleanout plug is at or above grade. A cleanout installed in a floor or walkway that will not have a trim cover installed shall have a countersunk plug installed so the top surface of the plug is flush with the finished surface of the floor or walkway.

❖ A cleanout cannot be located in permanently concealed areas, be concealed by permanent finish materials or be located in an area that cleaning equipment could not be obviously moved to perform the work. A cleanout that cannot be found is a useless cleanout. A cleanout that is poorly located (such as in floor/ceiling area) such that it presents difficulties resulting in building damage is not an intelligent choice (see Commentary Figure 708.1.10).

Figure 708.1.10
CLEANOUT INSTALLATIONS

708.1.10.1 Cleanout plug trim covers. Trim covers and access doors for cleanout plugs shall be designed for such purposes and shall be *approved*. Trim cover fasteners that thread into cleanout plugs shall be corrosion resistant. Cleanout plugs shall not be covered with mortar, plaster or any other permanent material.

❖ Trim covers and access panels are required to be designed for such purpose so that they are recognizable as a potential location for a cleanout. Such covers and panels must be approved by the code official so that obscure and inappropriate methods for covering up cleanouts are not chosen (see Commentary Figure 708.1.10.1).

708.1.10.2 Floor cleanout assemblies. Where it is necessary to protect a cleanout plug from the loads of vehicular traffic, cleanout assemblies in accordance with ASME A112.36.2M shall be installed.

❖ Load-rated cleanout assemblies are required for vehicular traffic areas.

708.1.11 Prohibited use. The use of a threaded cleanout opening to add a fixture or to extend piping shall be prohibited except where another cleanout of equal size is installed with the required access and clearance.

❖ If a cleanout opening is eliminated by a permanent piping connection, a new approved cleanout of equal access and capacity must be installed. In existing structures, cleanouts provide a convenient opening for the connection of new piping in remodeling, addition and alteration work. The cleanout fitting is commonly removed to allow a new connection; however, a substitute cleanout must be provided to serve in the same capacity as the cleanout that was eliminated. Many threaded cleanout openings have only a few threads and, therefore, are not intended to receive threaded pipe or male adapters.

Figure 708.1.10.1
CLEANOUTS ON CONCEALED PIPING

SECTION 709
FIXTURE UNITS

709.1 Values for fixtures. *Drainage fixture unit* values as given in Table 709.1 designate the relative load weight of different kinds of fixtures that shall be employed in estimating the total load carried by a soil or waste pipe, and shall be used in connection with Tables 710.1(1) and 710.1(2) of sizes for soil, waste and vent pipes for which the permissible load is given in terms of fixture units.

❖ This section contains the requirements to establish fixture unit amounts for plumbing fixtures for use in determining drainage system piping size. For a discussion on the determination of fixture units, see the General Comments at the beginning of this chapter. The term "fixture unit" refers to "drainage fixture units." Refer to the commentary on the definitions of "Fixture units" and "Drainage fixture units" in Chapter 2.

The method of sizing a sanitary drainage system is based on the dfu values listed in Table 709.1.

TABLE 709.1. See page 7-39.

❖ This table identifies common plumbing fixtures with their corresponding dfu values. The dfu values are used to determine the load on a drain, which, in turn, is used to determine the minimum drainage pipe size requirement for the load served. In determining the fixture unit values, each fixture was evaluated for its impact on the sanitary drainage system. The original development of fixture unit values was based on an arbitrary scale established by Dr. Hunter. Dr. Hunter conceived the idea of assigning a fixture unit value to represent the degree to which a fixture loads a system when used at the maximum assumed frequency. The scale was based on two fixtures: the lavatory and water closet (public). Dr. Hunter determined that a lavatory load on a drainage system was one-sixth of that for a water closet. Hence, the lavatory was assigned the value of 1 while the water closet was 6.

The sole purpose of the fixture unit concept is to make it possible to calculate the design load on the system directly when the system is composed of different kinds of fixtures, each having a different loading characteristic. The last column of the table lists the minimum trap size required for the corresponding fixture. Minimum trap size is also the minimum drainage pipe size because the size of a drainage pipe cannot be reduced in the direction of flow (see commentary, Section 1002.5). Although the code does not regulate the maximum size of a fixture trap, traps should not be oversized because the reduced flow volume and velocity will impair the trap's self-scouring ability.

Note a references Table 709.2 for determining dfu values for commercial washing machines with traps larger than 3 inches (76 mm) in diameter.

Note b allows for shower heads and whirlpool attachments to a bathtub fixture without increasing the fixture unit value.

Note c references Sections 709.2 through 709.4 to determine fixture unit values of plumbing fixtures not found in the table and fixtures with intermittent flows.

Note d requires consistency between the fixture outlet size and trap size of urinals and water closets. Both the trap size and the design of a water closet are regulated by ANSI Z124.4 and ASME A112.19.2.

Note e permits assigning a water closet or urinal a lower dfu value than that listed in the table when appropriate testing justifies the reduction. This can be significant when installing 1.6-gallon-per-flush (6.1 L per flush) water closets because some manufacturers have conducted tests and determined that such fixtures have a dfu value of 2. An emergency floor drain is installed in areas such as toilet rooms, boiler rooms and laundry rooms and is used only in cases of fixture overflow, water pipe failure and vandalism (see commentary, Section 412.4). Such floor drains do not add to the normal load of the drainage system.

Note f permits adding fixtures to a dwelling unit bathroom group to the dfu value of the bathroom group fixture count.

Note g references Section 406.3 for sizing requirements for the trap and fixture drain for an automatic clothes washer and the branch drain or drainage stack.

Note h addresses waste receptors and the appropriate determination of dfu loads based upon what is being discharged to the waste receptor. Section 709.4 requires a minimum dfu load to be assumed for a trap size if the sum of indirectly connected fixtures to the waste receptor is less than the minimum dfu load for the trap size. Section 709.4.1 allows for the special application of condensate drains for refrigerated cases and a reduction of dfu load (see commentary, Section 704.4.1)

709.2 Fixtures not listed in Table 709.1. Fixtures not listed in Table 709.1 shall have a *drainage fixture unit* load based on the outlet size of the fixture in accordance with Table 709.2. The minimum trap size for unlisted fixtures shall be the size of the drainage outlet but not less than 1^1/$_4$ inches (32 mm).

❖ When a specific plumbing fixture is not listed in Table 709.1, the dfu value is based on the outlet size of the fixture. For example, floor sinks (receptors) are not listed in Table 709.1; therefore, Table 709.2 would be used. A 1^1/$_4$-inch (32 mm) drain is the minimum acceptable size to permit proper open channel flow in the pipe for sanitary drainage. The minimum trap size is based on this pipe size (see commentary, Section 709.4).

SANITARY DRAINAGE

**TABLE 709.1
DRAINAGE FIXTURE UNITS FOR FIXTURES AND GROUPS**

FIXTURE TYPE	DRAINAGE FIXTURE UNIT VALUE AS LOAD FACTORS	MINIMUM SIZE OF TRAP (inches)
Automatic clothes washers, commercial[a,g]	3	2
Automatic clothes washers, residential[g]	2	2
Bathroom group as defined in Section 202 (1.6 gpf water closet)[f]	5	—
Bathroom group as defined in Section 202 (water closet flushing greater than 1.6 gpf)[f]	6	—
Bathtub[b] (with or without overhead shower or whirlpool attachments)	2	$1^1/_2$
Bidet	1	$1^1/_4$
Combination sink and tray	2	$1^1/_2$
Dental lavatory	1	$1^1/_4$
Dental unit or cuspidor	1	$1^1/_4$
Dishwashing machine[c], domestic	2	$1^1/_2$
Drinking fountain	$^1/_2$	$1^1/_4$
Emergency floor drain	0	2
Floor drains[h]	2[h]	2
Floor sinks	Note h	2
Kitchen sink, domestic	2	$1^1/_2$
Kitchen sink, domestic with food waste disposer and/or dishwasher	2	$1^1/_2$
Laundry tray (1 or 2 compartments)	2	$1^1/_2$
Lavatory	1	$1^1/_4$
Shower (based on the total flow rate through showerheads and body sprays) Flow rate:		
5.7 gpm or less	2	$1^1/_2$
Greater than 5.7 gpm to 12.3 gpm	3	2
Greater than 12.3 gpm to 25.8 gpm	5	3
Greater than 25.8 gpm to 55.6 gpm	6	4
Service sink	2	$1^1/_2$
Sink	2	$1^1/_2$
Urinal	4	Note d
Urinal, 1 gallon per flush or less	2[e]	Note d
Urinal, nonwater supplied	$^1/_2$	Note d
Wash sink (circular or multiple) each set of faucets	2	$1^1/_2$
Water closet, flushometer tank, public or private	4[e]	Note d
Water closet, private (1.6 gpf)	3[e]	Note d
Water closet, private (flushing greater than 1.6 gpf)	4[e]	Note d
Water closet, public (1.6 gpf)	4[e]	Note d
Water closet, public (flushing greater than 1.6 gpf)	6[e]	Note d

For SI: 1 inch = 25.4 mm, 1 gallon = 3.785 L, gpf = gallon per flushing cycle, gpm = gallon per minute.
a. For traps larger than 3 inches, use Table 709.2.
b. A showerhead over a bathtub or whirlpool bathtub attachment does not increase the *drainage fixture unit* value.
c. See Sections 709.2 through 709.4.1 for methods of computing unit value of fixtures not listed in this table or for rating of devices with intermittent flows.
d. Trap size shall be consistent with the fixture outlet size.
e. For the purpose of computing loads on *building drains* and *sewers*, water closets and urinals shall not be rated at a lower *drainage fixture unit* unless the lower values are confirmed by testing.
f. For fixtures added to a *bathroom group*, add the *dfu* value of those additional fixtures to the *bathroom group* fixture count.
g. See Section 406.3 for sizing requirements for *fixture drain*, branch drain and drainage *stack* for an automatic clothes washer standpipe.
h. See Sections 709.4 and 709.4.1.

SANITARY DRAINAGE

TABLE 709.2
DRAINAGE FIXTURE UNITS FOR FIXTURE DRAINS OR TRAPS

FIXTURE DRAIN OR TRAP SIZE (inches)	DRAINAGE FIXTURE UNIT VALUE
$1^1/_4$	1
$1^1/_2$	2
2	3
$2^1/_2$	4
3	5
4	6

For SI: 1 inch = 25.4 mm.

❖ If a fixture is not listed in the table, the dfu value is based on the outlet or trap size of the fixture. This method of assigning fixture unit values is consistent with the historical development of sizing methodology because the outlet size restricts the discharge flow rate of the fixture. Special consideration should be given to the sizing of receptors, such as floor sinks that receive the discharge of multiple indirectly connected fixtures. For example, the table may be inappropriate for determining the fixture unit value for an indirect waste receptor with a 4-inch (102 mm) trap that serves five or six $1^1/_2$-inch-diameter (38 mm) culinary fixture drains (see commentary, Section 709.4).

709.3 Values for continuous and semicontinuous flow. *Drainage fixture unit* values for continuous and semicontinuous flow into a drainage system shall be computed on the basis that 1 gpm (0.06 L/s) of flow is equivalent to two fixture units.

❖ Equipment that discharges either continuously or semicontinuously, such as pumps and ejectors, is assigned two fixture units for each gallon per minute (gpm) of discharge rate. For example, a continuously operating pump with a discharge rate of 20 gpm (78 L/min) would have a dfu value of 40. This conversion applies only to the type of continuous flow addressed in this section. Because a dfu is based on probability, only whole numbers are used to express fixture unit values. A fractional fixture unit value is always rounded up to the next whole number. This calculation does not work in reverse. The relationship of dfu to gpm is not a constant ratio that allows direct conversion of units. Thus, it cannot be determined that a dfu value of 40 would yield a continuous flow of 20 gpm (78 L/min).

709.4 Values for indirect waste receptor. The *drainage fixture unit* load of an indirect waste receptor receiving the discharge of indirectly connected fixtures shall be the sum of the *drainage fixture unit* values of the fixtures that discharge to the receptor, but not less than the *drainage fixture unit* value given for the indirect waste receptor in Table 709.1 or 709.2.

❖ Waste receptors such as hub drains and floor sinks are typically used to receive waste from multiple indirectly connected fixtures (see Section 802). The simultaneous discharge of indirectly connected fixtures may result in a greater load on the drainage system than accounted for by Table 709.2. For example, a 3-inch (76 mm) hub drain could be serving four $1^1/_2$-inch (38 mm) indirect wastes, which may represent a total load of 8 dfu (4 by 2 dfu); however, Table 709.2 would assign a total load of only 5 dfu. If the total dfu load determined by this section is less than that given by Table 709.1 or 709.2, as applicable, the table values must prevail.

709.4.1 Clear-water waste receptors. Where waste receptors such as floor drains, floor sinks and hub drains receive only clear-water waste from display cases, refrigerated display cases, ice bins, coolers and freezers, such receptors shall have a *drainage fixture unit* value of one-half.

❖ In large food stores and warehouses having many refrigerated display cases and freezer units, many traps are installed with either hub drains or waste receptors to collect the condensate from the units. Even though the condensate flow from each unit is only a trickle, the trap sizes dictated the size of the drain system connecting all the traps. For example, a 3-inch (76 mm) trap with hub drain has a drainage fixture unit value (derived Table 709.2.) of 5 dfu. In situations with several hundred traps, the total dfu load results in unnecessarily large pipe sizes compared to the actual flow in the pipe. This section allows the designer to use a smaller more realistic dfu loading for these applications.

SECTION 710
DRAINAGE SYSTEM SIZING

710.1 Maximum fixture unit load. The maximum number of *drainage fixture units* connected to a given size of *building sewer*, *building drain* or horizontal *branch* of the *building drain* shall be determined using Table 710.1(1). The maximum number of drainage fixture units connected to a given size of horizontal *branch* or vertical soil or waste *stack* shall be determined using Table 710.1(2).

❖ This section contains the requirements for the sizing of gravity drainage piping.
  The maximum number of dfu that can be connected to a drainage pipe is given in Tables 710.1(1) and 710.1(2) and is dependent on whether the pipe is a building drain, sewer, stack or horizontal branch.

710.1.1 Horizontal stack offsets. Horizontal *stack* offsets shall be sized as required for building drains in accordance with Table 710.1(1), except as required by Section 711.3.

❖ Horizontal offsets (see the definition of "Horizontal pipe" in Chapter 2) cause a disruption of flow in the stack, and horizontal piping has less flow capacity than vertical piping. The vertical flow must convert to a horizontal flow in horizontal offsets; therefore, the increased turbulence and decreased flow capacity of the horizontal offset must be considered.
  In many cases, Table 710.1(1) will require that the diameter of the offset be larger than the stack piping above it. The reference to Section 711.3 in the last

phrase of this section is of no consequence because Section 711.3 would require the same sizing as this section.

TABLE 710.1(1)
BUILDING DRAINS AND SEWERS

DIAMETER OF PIPE (inches)	MAXIMUM NUMBER OF DRAINAGE FIXTURE UNITS CONNECTED TO ANY PORTION OF THE BUILDING DRAIN OR THE BUILDING SEWER, INCLUDING BRANCHES OF THE BUILDING DRAIN[a]			
	Slope per foot			
	$1/16$ inch	$1/8$ inch	$1/4$ inch	$1/2$ inch
$1 1/4$	—	—	1	1
$1 1/2$	—	—	3	3
2	—	—	21	26
$2 1/2$	—	—	24	31
3	—	36	42	50
4	—	180	216	250
5	—	390	480	575
6	—	700	840	1,000
8	1,400	1,600	1,920	2,300
10	2,500	2,900	3,500	4,200
12	3,900	4,600	5,600	6,700
15	7,000	8,300	10,000	12,000

For SI: 1 inch = 25.4 mm, 1 inch per foot = 83.3 mm/m.

a. The minimum size of any *building drain* serving a water closet shall be 3 inches.

❖ This table gives the maximum fixture units that can be connected to building drains, sewers and every horizontal branch connecting to a building drain. Such drains are the lowest portions of the system that drain horizontally by gravity (see the definition of "Building drain" in Chapter 2). The pipe diameters indicated are the nominal dimension designations for piping material. These dimensions are regulated by the piping material standards referenced in this chapter.

The minimum size of each pipe section is based on the slope of the pipe (pitch or grade) and the number of dfu that connect to that section. Drainage fixture units are additive in the direction of flow. The dfu values listed in the table are the maximum value for the given pipe size and slope.

The principle of a building drain flowing approximately half full at a maximum capacity design load was used by Dr. Hunter to develop a sizing method. The sizing is based further on a steady-state flow condition in the pipe (level flow).

The building drain actually has a combination of a steady-state and surging (highly fluctuating) flow. Surging flow results from the high entrance velocity from the discharge of a fixture, horizontal branch or stack.

The dfu load values in this table are conservative, and half-full flow will rarely, if ever, occur [see Commentary Figure 710.1(1)].

For SI: 1 inch = 25.4 mm, 1 inch per foot = 83.3 mm/m.

**Figure 710.1(1)
BUILDING DRAIN SIZING**

SANITARY DRAINAGE

Figure 710.1(2)
HORIZONTAL BRANCH
(Complete venting not shown)

For SI: 1 inch = 25.4 mm.

710.1.2 Vertical stack offsets. Vertical *stack* offsets shall be sized as required for straight stacks in accordance with Table 710.1(2), except where required to be sized as a *building drain* in accordance with Section 711.1.1.

❖ Because vertical offsets (see the definition of "Vertical pipe" in Chapter 2) do not have as much impact on flow in the stack, such offsets are sized no differently than the stack itself (see commentary, Section 711.1.1).

TABLE 710.1(2)
HORIZONTAL FIXTURE BRANCHES AND STACKS[a]

DIAMETER OF PIPE (inches)	MAXIMUM NUMBER OF DRAINAGE FIXTURE UNITS (dfu)			
	Total for horizontal branch	Stacks[b]		
		Total discharge into one branch interval	Total for stack of three branch Intervals or less	Total for stack greater than three branch intervals
1¹⁄₂	3	2	4	8
2	6	6	10	24
2¹⁄₂	12	9	20	42
3	20	20	48	72
4	160	90	240	500
5	360	200	540	1,100
6	620	350	960	1,900
8	1,400	600	2,200	3,600
10	2,500	1,000	3,800	5,600
12	3,900	1,500	6,000	8,400
15	7,000	Note c	Note c	Note c

For SI: 1 inch = 25.4 mm.
a. Does not include branches of the building drain. Refer to Table 710.1(1).
b. Stacks shall be sized based on the total accumulated connected load at each story or branch interval. As the total accumulated connected load decreases, stacks are permitted to be reduced in size. Stack diameters shall not be reduced to less than one-half of the diameter of the largest stack size required.
c. Sizing load based on design criteria.

❖ The table may be considered as four separate tables: one for sizing horizontal branches connecting to a stack, one for limiting the discharge per branch interval for any height stack, one for sizing stacks three branch intervals or less in height and one for sizing stacks more than three branch intervals in height. The pipe dimensions listed in the first column are again the nominal dimension designations identified in the pipe standards.

To size a horizontal branch, only the first two columns of the table are required. The second column relates only to horizontal branches; it does not affect the sizing of stacks. A horizontal branch, however, may have a vertical section of pipe.

The table lists the maximum number of fixture units that can be discharged into a given size pipe. As the number of dfu increases in the direction of flow, the pipe may also have to be increased in size.

The dfu load value for a horizontal branch is typically lower than the permitted load for a building drain, a branch of a building drain or a sewer the same size as listed in Table 710.1(1). When sizing a horizontal branch, consideration must be given to the flow entering a stack. If there is simultaneous flow in the stack and horizontal branch, the flow in the horizontal branch may adversely affect the flow in the stack, and the flow in the stack may impede the flow in the horizontal branch. A significant flow in the stack may cause a backup of flow in the horizontal branch connecting to the stack. The table accounts for the effect on the stack by limiting the load capacity of the horizontal branch [see Commentary Figure 710.1(3)].

The table does not have an entry for a 1¼-inch-diameter (32 mm) pipe, and the most upstream lavatory in Commentary Figure 710.1(2) is served by a 1¼-inch (32 mm) horizontal fixture branch. This is intended to illustrate that the omission of a 1¼-inch (32 mm) pipe from the table does not prohibit its use as a horizontal drain. A 1¼-inch (32 mm) pipe is limited to 1 dfu.

When sizing a stack, there are three main criteria that must be considered: the discharge into a branch interval of a stack, the total discharge into a stack and the total number of branch intervals on the stack.

For sizing a stack of three branch intervals or less, the third and fourth columns of the table are used [see Commentary Figure 710.1(3)]. Within any given branch interval, the maximum amount of discharge a stack may receive is limited. If the maximum discharge per branch interval is exceeded, the stack would have to be increased in size. The discharge may be received from more than one horizontal branch.

Commentary Figure 710.1(4), Example 1, depicts two 3-inch (76 mm) horizontal branches connecting within one branch interval. The total fixture unit load is 30, meaning that the stack diameter must be a minimum of 4 inches (102 mm). The example demonstrates a situation where the branches must be only 3 inches (76 mm) and, because of the location of their connection to the stack, a larger-size stack is required. In this example, the total permitted capacity of a 3-inch (76 mm) stack was not exceeded; only the limitation for a given branch interval was exceeded.

The flow, discharging within a branch interval, is limited to avoid interference with the flow in the stack. The collision of horizontal flow in a branch with the vertical flow in a stack may disrupt the annular flow pattern in the stack. This disruption may cause the cross-sectional area of the stack to be partially or completely filled with waste, which in turn causes the formation of slugs or pistons of waste in the stack. As stated previously, the flow from the horizontal branches may greatly affect the pressure differentials in the drainage system.

Another example of the limitations for a given branch interval, as shown in Commentary Figure 710.1(4), Example 2, is a single 4-inch (102 mm) pipe connecting to a stack.

If a total of 110 dfu discharged into a horizontal branch, the pipe size is a minimum of 4 inches (102 mm). The stack receiving the discharge of the branch must be a minimum of 5 inches (127 mm). The horizontal branch exceeds the 90 dfu limitation for a single branch interval of a 4-inch (102 mm) stack.

The stack is also sized based on the total dfu discharge to the stack. The discharge of each horizontal branch is added to compute the stack's total dfu load.

For sizing stacks over three branch intervals, the third and fifth columns are used [see Commentary Figure 710.1(3)]. The difference in sizing the taller stacks is the total dfu load permitted in the stack. Taller stacks are allowed a much greater capacity than stacks three branch intervals or less [see Commentary Figures 710.1(5) and 710.1(6)]. The flow in the stack, based on the table, will occupy approximately one-quarter of the cross-sectional area of the pipe under the maximum design condition. Dr. Hunter discovered in his early research that this capacity of flow results in the drainage spiraling down the pipe wall with a center core of air. Only small volumes of waste tend to fall down the inner core of the stack. As the volume increases, any slug of water is forced by air frictional factors against the pipe wall [see Commentary Figure 710.1(7)].

After a short distance of fall [approximately 15 feet (4572 mm)], terminal velocity is reached by the drainage flow in the stack. Terminal velocity is the maximum speed of drainage flow in the stack, resulting from a balancing of air and pipe wall friction with gravitational force.

Although terminal velocity always occurs in a tall stack, it rarely occurs in shorter stacks of three branch intervals or less. This phenomenon is reflected in the table by permitting a greater capacity of flow in stacks with more than three branch intervals.

Note a of the table is a reminder that Table 710.1(1) is to be used for sizing horizontal branches of the building drain.

Note b permits reducing the size of the stack in the direction opposite of flow as the dfu load decreases. This is commonly referred to as "telescoping the stack." A stack cannot be reduced to less than half the diameter of its base where it terminates as a stack vent [see Commentary Figure 710.1(8)].

DIAMETER OF PIPE (inches)	TOTAL DISCHARGE INTO ONE BRANCH INTERVAL	TOTAL FOR STACK GREATER THAN THREE BRANCH INTERVALS
1½	2	8
2	6	24
2½	9	42
3	20	72
4	90	500
5	2000	1,100
6	350	1,900
8	600	3,600
10	1,000	5,600
12	1,500	8,400
15	—	—

For SI: 1 inch = 25.4 mm.

Figure 710.1(3)
ALLOWABLE STACK DRAINAGE FIXTURE UNITS

SANITARY DRAINAGE

For SI: 1 inch = 25.4 mm.

**Figure 710.1(4)
STACK SIZING**

For SI: 1 inch = 25.4 mm.

**Figure 710.1(5)
EXAMPLE OF SIZING STACK GREATER THAN THREE BRANCH INTERVALS**

For SI: 1 inch = 25.4 mm.

**Figure 710.1(6)
SIZING OF STACK THREE BRANCH INTERVALS OR LESS**

**Figure 710.1(7)
FLOW IN A STACK**

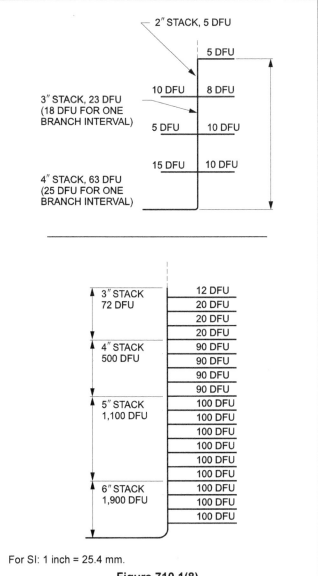

For SI: 1 inch = 25.4 mm.

**Figure 710.1(8)
REDUCING SIZE OF DRAINAGE STACK**

710.2 Future fixtures. Where provision is made for the future installation of fixtures, those provided for shall be considered in determining the required sizes of drain pipes.

❖ Anticipated dfu loads for future plumbing fixtures must be considered in the design of the spacing. Piping installed for future plumbing fixtures must be closed to prevent the escape of sewer gas and waste and to prevent the entrance of foreign substances. Section 905.6 also requires that a vent be roughed in for drainage pipe that will serve fixtures to be installed in the future.

SECTION 711
OFFSETS IN DRAINAGE PIPING IN BUILDINGS OF FIVE STORIES OR MORE

711.1 Horizontal branch connections above or below vertical stack offsets. If a horizontal *branch* connects to the *stack* within 2 feet (610 mm) above or below a vertical *stack* offset, and the offset is located more than four *branch intervals* below the top of the *stack*, the offset shall be vented in accordance with Section 906.

❖ Offsets in stacks produce turbulent flow that can create pressure fluctuations in the stack. Stacks having five or more branch intervals above a vertical offset have been found to have significant pressure fluctuations in the stack at locations within 2 feet (610 mm) above and below the vertical offset. These pressure fluctuations can adversely affect the liquid seal in traps connected to horizontal branches that connect to the stack within 2 feet (610 mm) above or below the offset. If the horizontal branch connections to the stack are more than 2 feet (610 mm) above and below the vertical offset, the pressure fluctuations near the offset will not disrupt the liquid seals of the traps connected to the horizontal branches. Therefore, venting of the offset is not required [see Commentary Figure 711.1(1)]. However, if a horizontal branch does connect within 2 feet (610 mm) of a vertical offset, either above or below, the traps connected to that branch will experience significant pressure fluctuations that could cause the liquid seal in the traps to be siphoned out or blown out. Therefore, venting of the vertical offset in accordance with Section 907 is required to neutralize the pressure fluctuations [see Commentary Figure 711.1(2)].

711.1.1 Omission of vents for vertical stack offsets. Vents for vertical offsets required by Section 711.1 shall not be required where the *stack* and its offset are sized as a *building drain* [see Table 710.1(1)].

❖ This section presents an alternative to the venting requirements of Section 711.1 for stacks having vertical offsets and a horizontal branch connecting within 2 feet (610 mm) above or below the offset. If the stack and the offset are sized as if they were building drains, the additional airspace created by this pipe oversizing provides the necessary venting. Note that this does not necessarily mean that the stack and offset size will always need to be increased in size. For example,

SANITARY DRAINAGE

from the fifth column of Table 710.1(2), a load of 50 dfu requires a stack and offset size of 3 inches (76 mm). If this size is checked against the fifth column of Table 710.1(1), a load of 50 dfu on a building drain requires a pipe size of 3 inches (76 mm). Thus, no increase in the size of the stack or the offset is necessary because the stack and offset are already sufficiently underloaded (or oversized) to provide the necessary airspace for venting without installing the relief venting required by Section 711.1. In other words, oversizing (or underutilization) of the stack and the offset sufficiently controls the pressure fluctuations in the stack without the need for additional venting. However, in another example, if the upper stack load was 400 dfu, the required stack size in accordance with the fifth column of Table 710.1(2) is 4 inches (102 mm). Compare this to the fifth column of Table 710.1(1) where a building drain size of 5 inches (127 mm) is required to carry the 400 dfu. Thus, for a modest size increase of the drainage stack and the offset, the offset vent piping system can be eliminated [see Commentary Figure 711.1(1)].

711.2 Horizontal stack offsets. A *stack* with a horizontal offset located more than four *branch intervals* below the top of the *stack* shall be vented in accordance with Section 907 and sized as follows:

1. The portion of the *stack* above the offset shall be sized as for a vertical *stack* based on the total number of *drainage fixture units* above the offset.

2. The offset shall be sized in accordance with Section 710.1.1.

3. The portion of the *stack* below the offset shall be sized as for the offset or based on the total number of *drainage fixture units* on the entire *stack*, whichever is larger [see Table 710.1(2), Column 5].

❖ Offsets in stacks produce turbulent flow that can create pressure fluctuations in the stack. Stacks having five or more branch intervals above a horizontal offset have significant pressure fluctuations. These pressure fluctuations can adversely affect the liquid seal in traps connected to branches that connect to the stack nearest to the offset. Therefore, in order to control the pressure fluctuations, the horizontal offset must be

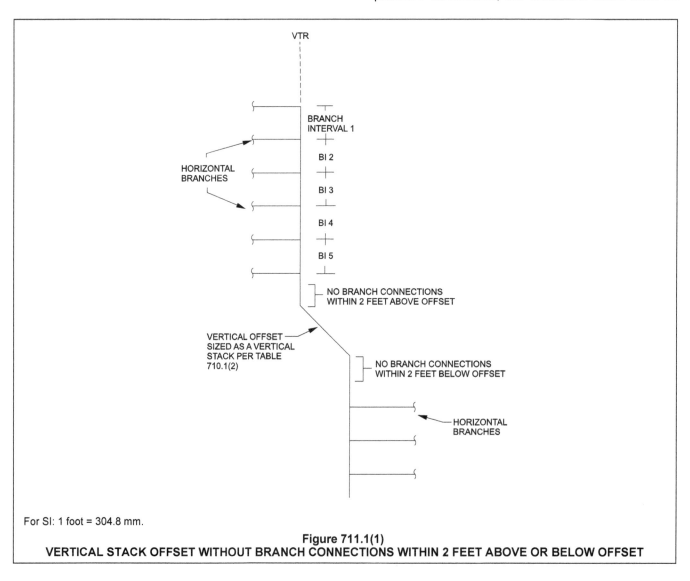

For SI: 1 foot = 304.8 mm.

Figure 711.1(1)
VERTICAL STACK OFFSET WITHOUT BRANCH CONNECTIONS WITHIN 2 FEET ABOVE OR BELOW OFFSET

vented in accordance with Section 907, the offset piping must be sized as if it were a building drain and the lower stack must be at least as big as the horizontal offset piping or larger if the total (upper plus lower) stack load requires a larger size.

The upper section of the stack (i.e., above the offset) is sized as if it were a separate stack by using Table 710.1(2). The horizontal offset is sized as if it was a building drain by using Table 710.1(1). Note that the correct slope column of Table 710.1(2) must be used. [Where slope of the offset is greater than $^1/_2$ inch per foot (12.7 mm), the $^1/_2$-inch-per-foot column is used for sizing]. The stack below the offset is sized by using the fifth column of Table 710.1(2) and the number of dfu for the entire stack. At a minimum, the lower stack must be not less than the offset piping [see Commentary Figure 711.2(1)].

Example

Consider the stack with a horizontal offset as shown in Commentary Figure 711.2(2). The upper stack has a load of 360 dfu. The fifth column of Table 710.1(2) requires the stack to be 4 inches (102 mm). The stack cannot be reduced in diameter (telescoped) since the cumulative load on the stack decreases (going up the stack) because a 3-inch (76 mm) stack cannot accept a single branch discharge over 20 dfu. The load on the offset is 360 dfu and, when sizing this offset as if it were a building drain [Table 710.1(1)], the offset piping must be not less than 5 inches (127 mm). Because the offset piping is required to be 5 inches (127 mm), the lower stack must be at least 5 inches (127 mm). However, for the entire stack, the load is 1,160 dfu. The lower stack section is sized by the fifth column of Table 710.1(2), which requires a stack size

For SI: 1 foot = 304.8 mm.

Figure 711.1(2)
VERTICAL STACK OFFSET WITH BRANCH CONNECTIONS WITHIN 2 FEET ABOVE OR BELOW OFFSET

SANITARY DRAINAGE

of 6 inches (152 mm) to carry 1,160 dfu. Thus, the upper stack would be 4 inches (102 mm), the offset would be 5 inches (127 mm) and the lower stack would be 6 inches (152 mm). The vent piping would be installed in accordance with Section 907 (sizing in accordance with Section 906.1).

711.2.1 Omission of vents for horizontal stack offsets. Vents for horizontal stack offsets required by Section 711.2 shall not be required where the stack and its offset are one pipe size larger than required for a building drain [see Table 710.1(1)] and the entire stack and offset are not less in cross-sectional area than that required for a straight stack plus the area of an offset vent as provided for in Section 907.

❖ This section provides an alternative to the venting requirements of Section 711.2 for stacks having horizontal offsets. If the stack above and below the offset and the horizontal offset are sized one pipe size larger than the pipe size required for a building drain and the resulting stack and offset piping size provides the same or larger cross-sectional area than would be required for a straight stack plus the area of the required offset vent piping as required by Section 907, then venting of the offset in accordance is not required (see Commentary Figure 711.2.1). This generous stack and offset oversizing requirement provides for plenty of airspace to neutralize pressure fluctuation caused by the horizontal offset.

711.3 Offsets below lowest branch. Where a vertical offset occurs in a soil or waste *stack* below the lowest horizontal *branch*, a change in diameter of the *stack* because of the offset shall not be required. If a horizontal offset occurs in a soil or waste *stack* below the lowest horizontal *branch*, the required diameter of the offset and the *stack* below it shall be determined as for a *building drain* in accordance with Table 710.1(1).

❖ No size increase is required in vertical offsets that are below the lowest horizontal branch connection to the stack [see Commentary Figure 711.3(1)]. The pressure differences created by such offsets mainly affect the behavior of horizontal branches downstream of the offset, and if there are no horizontal branches downstream, there is nothing to protect by sizing increases. Horizontal offsets that are below the lowest horizontal branch connection must be sized as a building drain in accordance with Table 710.1(1) [see Commentary Figure 711.3(2)]. Because of the greater offset angle, oversizing is required to minimize pressure fluctuations; however, offset venting is not required.

Figure 711.2(1)
HORIZONTAL STACK OFFSET WITH VENTING

SANITARY DRAINAGE

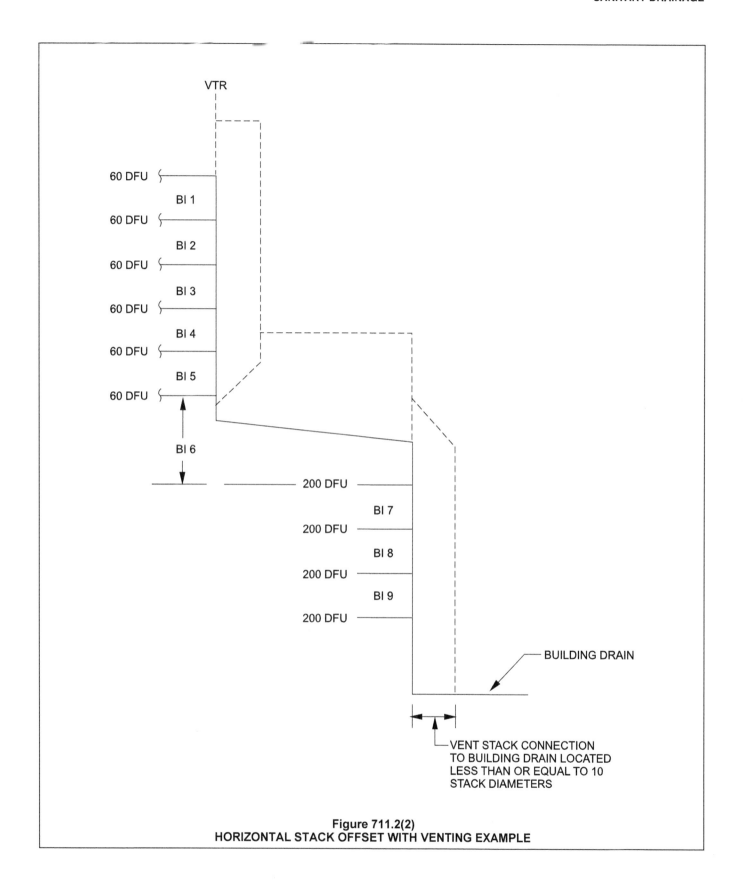

Figure 711.2(2)
HORIZONTAL STACK OFFSET WITH VENTING EXAMPLE

SANITARY DRAINAGE

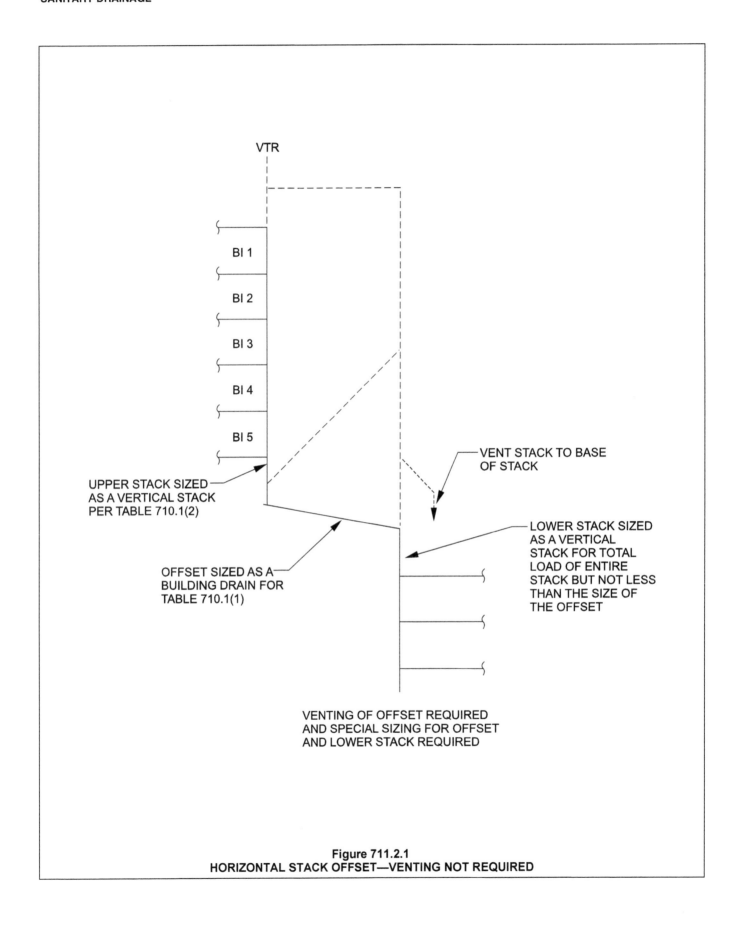

Figure 711.2.1
HORIZONTAL STACK OFFSET—VENTING NOT REQUIRED

SANITARY DRAINAGE

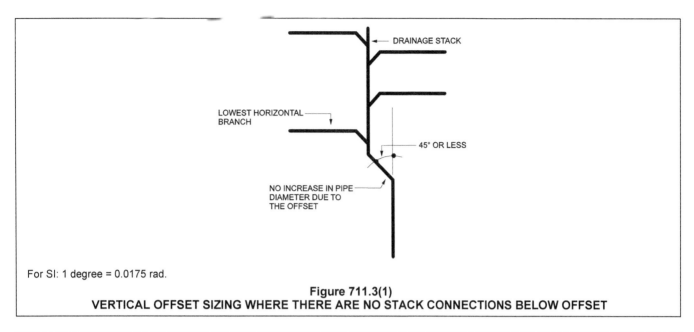

For SI: 1 degree = 0.0175 rad.

Figure 711.3(1)
VERTICAL OFFSET SIZING WHERE THERE ARE NO STACK CONNECTIONS BELOW OFFSET

For SI: 1 degree = 0.0175 rad.

Figure 711.3(2)
HORIZONTAL OFFSET SIZING WHERE THERE ARE NO STACK CONNECTIONS BELOW OFFSET

SECTION 712
SUMPS AND EJECTORS

712.1 Building subdrains. *Building subdrains* that cannot be discharged to the *sewer* by gravity flow shall be discharged into a tightly covered and vented sump from which the liquid shall be lifted and discharged into the building gravity drainage system by automatic pumping equipment or other *approved* method. In other than existing structures, the sump shall not receive drainage from any piping within the building capable of being discharged by gravity to the *building sewer*.

❖ Where the drainage system or portions of it cannot discharge by gravity to the sewer, the drainage is collected in a tightly sealed and vented sump and pumped to a gravity sewer or drain. The sump must be sized to provide adequate holding capacity and to limit the retention period of the waste. Though not specifically required by the code, the intent of this section is that the capacity of the sump must not exceed one-half of a day's (12 hours) discharge load from the piping system connected to the sump under normal use. The intent is to keep the waste retention period short to prevent the sump from acting as a waste decomposition (septic) tank.

The minimum capacity of the sump must be such that the pumping equipment operates for at least 15 seconds per pumping cycle to prevent short cycling, thereby extending the life of the equipment.

The cover for the sump must be gas tight to prevent the escape of sewer gas into the building (see commentary, Section 916.5). Sumps, other than pneumatic ejectors, must be vented in accordance with Section 916.5.1.

SANITARY DRAINAGE

Sumps receiving water closet and urinal discharge use a sewage ejector to pump the soil and waste to a gravity drain. Quite often, the pumping equipment is referred to as a sewage pump by technical standards. A pneumatic ejector is a special type of sewage "pump" that operates by air pressure, which forces the sewage from the pressurized receiver instead of mechanically pumping it. Pneumatic ejectors require special relief vents to instantaneously relieve the pressure in the receiver [see commentary, Section 916.5.2 as well as Commentary Figures 712.1(1) and 712.1(2) for two types of pumps].

712.2 Valves required. A check valve and a full open valve located on the discharge side of the check valve shall be installed in the pump or ejector discharge piping between the pump or ejector and the gravity drainage system. *Access* shall be provided to such valves. Such valves shall be located above the sump cover required by Section 712.1 or, where the discharge pipe from the ejector is below grade, the valves shall be accessibly located outside the sump below grade in an access pit with a removable *access* cover.

❖ The discharge piping must have a check valve to prevent the previously pumped waste from returning to

Figure 712.1(1)
PNEUMATIC SEWAGE EJECTOR

Figure 712.1(2)
VERTICAL SUSPENDED WET PIT SEWAGE PUMP

the sump pit when the pump or ejector shuts off. This prolongs the life of the pumping equipment and conserves energy (see Commentary Figure 712.2). The check valve will also prevent the gravity drainage system from backing up into the sump or receiver.

The requirement for installation of a full-open valve on the discharge side of the check valve indicates a similar concern. There will be some point over the service life of the sump or check valve when replacement or repairs are necessary. This full-open valve is there to prevent waste from running back into the discharge piping when maintenance is performed on the check valve or pump/ejector. It also allows isolation of the drainage system to prevent backups and sewer gas leakage during pump replacement and maintenance. Both valves must be accessible for maintenance on the pump and be installed on the discharge pipe between the pump or ejector and the gravity drainage system.

712.3 Sump design. The sump pump, pit and discharge piping shall conform to the requirements of Sections 712.3.1 through 712.3.5.

❖ This is the main section, under which Sections 712.3.1 to 712.3.5 will give the specific requirements for the construction of the sump pump, the pit and the discharge piping related to sumps and ejectors. Section 712.3.1 relates to sump pump construction. Section 712.3.2 relates to pit construction. Section 712.3.3 relates to discharge piping. Section 712.3.4 contains the amount of space to be maintained between the inlet piping into the sump pit and the top of the effluent level. Section 712.3.5 contains requirements for the connection of the discharge piping to the drainage system.

712.3.1 Sump pump. The sump pump capacity and head shall be appropriate to anticipated use requirements.

❖ The sewage pump or pneumatic ejector must be sized properly to accommodate the peak flow into the receiver or pit, to provide the head pressure required to lift or eject waste, to prevent excessive cycling of the pumping equipment and to handle the type of waste that discharges to the sump.

Sewage pumps and ejectors must be able to handle, without creating blockages, the types of solid waste discharged into the sump pit or receiver. Pumps and ejectors serving water closets must be able to handle the solids associated with soil drainage.

Grinder pumps and grinder ejectors that serve water closets are exempt from the 2-inch (51 mm) spherical solids-handling requirements of this section. These pumps and ejectors reduce solids to a near-liquid state and pump the slurry to the drainage system. The liquidized solids pass readily through the pump and into the drainage system, thus reducing the possibility of creating a stoppage in either the pump or the receiving drainage line.

In certain instances, the design professional may choose to install duplex pumping equipment. Duplex pumping installations consist of two pumps with special controls installed to alternate pump duty. If the inflow into the sump is at a high rate or one pump fails to operate, the second pump provides assistance.

For SI: 1 foot = 304.8 mm.

Figure 712.2
SEWAGE EJECTOR INSTALLATION

The intent behind duplex pumping equipment is to reduce the possibility of disruption of service of the plumbing system. If a pump fails, it may be repaired while the drainage system functions normally with the remaining pump. Duplex pumping equipment is not required by the code.

The sewage pump or ejector must have a minimum capacity that is based on the size of the discharge pipe. The required capacity provides a full-flow velocity of at least 2 feet per second (0.61 m/s) in the discharge pipe (see Table 712.4.2).

712.3.2 Sump pit. The sump pit shall be not less than 18 inches (457 mm) in diameter and not less than 24 inches (610 mm) in depth, unless otherwise *approved*. The pit shall be accessible and located such that all drainage flows into the pit by gravity. The sump pit shall be constructed of tile, concrete, steel, plastic or other *approved* materials. The pit bottom shall be solid and provide permanent support for the pump. The sump pit shall be fitted with a gas-tight removable cover that is installed flush with grade or floor level, or above grade or floor level. The cover shall be adequate to support anticipated loads in the area of use. The sump pit shall be vented in accordance with Chapter 9.

❖ As discussed in the commentary to Section 712.1, the sump pit must be properly sized to receive waste and sewage from the plumbing fixtures that it will serve, as well as proper operation of the pump.

In this section, the minimum dimensions of 18 inches (457 mm) in diameter and 24 inches (610 mm) in depth are given, unless the designer or manufacturer presents calculations and other supporting data that will allow the code official to determine that other dimensions work for the situation being addressed. This approval is strictly on a case-by-case basis by the code official. This section states that the sump pit is to be accessible for maintenance and requires that all drainage flow into the pit must be by the force of gravity.

The sump pit must be constructed of a durable material, such as tile, concrete, steel, plastic or other approved materials. The bottom of the sump pit must be solid and structurally capable of supporting the sump pump. The sump pit must have a gas-tight removable cover to prevent the escape of sewer gas. This cover is to be structurally capable of supporting the weight of loads it will receive based on the location of the sump pit. The sump pit must be vented in accordance with Chapter 9 (see commentary, Section 916.5). The sump pit cover must be readily removable; therefore, the method used to connect the vent to the sump pit must be designed for disassembly.

712.3.3 Discharge pipe and fittings. Discharge pipe and fittings serving sump pumps and ejectors shall be constructed of materials in accordance with Sections 712.3.3.1 and 712.3.3.2 and shall be *approved*.

❖ The discharge pipe and fitting materials must be approved by the code official (see commentary, Section 712.2).

712.3.3.1 Materials. Pipe and fitting materials shall be constructed of brass, copper, CPVC, ductile iron, PE, or PVC.

❖ Previous editions of the code were silent on the type of pipe materials used for pump discharge. The materials listed are determined to be suitable for handling clear water or sewage.

712.3.3.2 Ratings. Pipe and fittings shall be rated for the maximum system operating pressure and temperature. Pipe fitting materials shall be compatible with the pipe material. Where pipe and fittings are buried in the earth, they shall be suitable for burial.

❖ The pump discharge piping and fittings must be suitable for the internal pressure and fluid temperatures to be encountered. If buried in the earth, the materials must be able to withstand earth loads from above.

Refer to Commentary Figures 712.1(1) and 712.1(2) for illustrations of sumps and related components.

712.3.4 Maximum effluent level. The effluent level control shall be adjusted and maintained to at all times prevent the effluent in the sump from rising to within 2 inches (51 mm) of the invert of the gravity drain inlet into the sump.

❖ This section states that the control mechanism that starts and stops the pump must be maintained and adjusted to limit the level of effluent in the sump to not higher than 2 inches (51 mm) below the invert level of the gravity drainage pipe(s) entering the sump. This requirement is to reduce the chance that the gravity drainage piping will become flooded or clogged because of standing effluent in the sump that may back up into the drain piping.

712.3.5. Pump connection to the drainage system. Pumps connected to the drainage system shall connect to a *building sewer*, *building drain*, soil *stack*, waste *stack* or *horizontal branch drain*. Where the discharge line connects into horizontal drainage piping, the connection shall be made through a wye fitting into the top of the drainage piping and such wye fitting shall be located not less than 10 pipe diameters from the base of any soil *stack*, waste *stack* or *fixture drain*.

❖ This is part of the main section, titled "Sumps and ejectors." As such, it may be interpreted to be applicable to both sewage pumps (including the "grinder type") and pneumatic sewage ejectors. The terms "sewage pump" and "ejector" are sometimes used interchangeably to describe the same device. This section imposes certain restrictions on how and where sewage pumps and pneumatic sewage ejectors connect to the gravity drainage system.

In previous editions of the code, the connection of the ejector or pump discharge piping was required to be only to the building sewer or to the building drain. This severe limitation for the connection point was found to be unnecessary since the pump or ejector discharge flow can be accommodated by most any drain pipe (horizontal or vertical) that is sized large enough to accept both the gravity flow in the drain plus the pumped flow. The 10-foot (3048 mm) distance between the connection point and any upstream vertical stack connection was also found to be severely limiting. The connection of the discharge to a horizontal drain needs to avoid sections of the horizontal drain where gravity flow in the pipe might be turbulent so pressure surges or backups do not occur. Thus, a pump or ejector discharge pipe connection to a horizontal drain must be located at least 10 (stack or fixture drain) pipe diameters away from a soil stack, waste stack or fixture drain connection to the horizontal drain. Note that the connection arrangement to horizontal piping using a wye fitting in the top half of the pipe is still required to allow the pumped waste flow to blend into the gravity flow in the pipe with the least amount of disruption.

The code is silent on the connection arrangement of pump or ejector discharge piping to stacks. Although use of a wye fitting might create less disruption of the flow in the stack, a sanitary tee connection might not be any more disruptive to stack flow than the discharge of a fully loaded horizontal branch drain would cause. Even though this code section does not prohibit connecting the pumped discharge to one side of a double sanitary tee in a stack, a gravity flow pipe connection should probably not be located directly across from a pumped discharge connection unless the stack size is significantly larger than the pumped discharge connection.

A question that is often asked about pumps or ejectors discharging to gravity drain piping is how to size a gravity drain pipe for a "gallon-per-minute" flow. While Section 709.3 states that the dfu load for semicontinuous flows is 2 dfu for 1 gpm, it is unclear whether this should apply to the intermittent discharge of a pump or ejector. As an example, consider a 2-inch (51 mm) discharge pipe having a 21-gpm (79 L/m) flow (refer to Table 712.4.2). Using the 2-dfu-to-1-gpm (3.8 L/m) equivalency yields a 42-dfu load on the gravity flow pipe that the discharge pipe connects to. A review of Table 710.1(2) indicates that at least a 4-inch (102 mm) horizontal drain or at least a 3-inch (76 mm) stack would be necessary to accommodate this load. While these required sizes of drain or stack in this example might seem reasonable, the problem is that, typically, the smallest 2-inch (51 mm) discharge sewage pump is capable of at least 50 gpm (1894 m) at low head [e.g., 10 foot (3053 mm)] conditions. Using the 2-dfu-to-1-gpm (3.8 L/m) conversion now yields a 100 dfu load! That dfu load combined with the gravity flow dfu design load for the pipe could require significantly larger drain or stack sizes. The designer of the system has the responsibility to decide how to size the gravity flow pipe to accommodate the pumped discharge flow.

712.4 Sewage pumps and sewage ejectors. A sewage pump or sewage ejector shall automatically discharge the contents of the sump to the building drainage system.

❖ This section is redundant with Section 712.1 and requires such equipment to operate automatically.

712.4.1 Macerating toilet systems. Macerating toilet systems shall comply with ASME A112.3.4/CSA B45.9 and shall be installed in accordance with the manufacturer's instructions.

❖ Macerating toilet systems are unique in that there are many acceptable types available. CSA B45.9 and ASME A112.3.4 are the standards for macerating toilet systems. Such systems must be installed in accordance with the manufacturer's instructions for their proper use and function. A macerating/pumping system encompasses three components: the container that houses the operating mechanisms, the pressure chamber that automatically activates and deactivates the macerating pump and an induction motor that drives the shredder blades and pump assembly. Such systems collect waste from a single water closet, lavatories, bathtub or any combination thereof located in the same room, and grind and pump these macerated wastes to some point in the sanitary drainage system [see Commentary Figure 712.4.1(1)]. This system is designed for above-the-floor installation and incorporates either a directly affixed macerating tank and integral grinder/pump mechanism or an integral system consisting of a single water closet with a built-in grinder/pump mechanism [see Commentary Figure 712.4.1(2)].

712.4.2 Capacity. A sewage pump or sewage ejector shall have the capacity and head for the application requirements. Pumps or ejectors that receive the discharge of water closets shall be capable of handling spherical solids with a diameter of up to and including 2 inches (51 mm). Other pumps or ejectors shall be capable of handling spherical solids with a diameter of up to and including 1 inch (25 mm). The capacity of a pump or ejector based on the diameter of the discharge pipe shall be not less than that indicated in Table 712.4.2.

Exceptions:

1. Grinder pumps or grinder ejectors that receive the discharge of water closets shall have a discharge opening of not less than $1^1/_4$ inches (32 mm).

2. Macerating toilet assemblies that serve single water closets shall have a discharge opening of not less than $^3/_4$ inch (19.1 mm).

❖ The sewage pump or ejector must be sized properly to accommodate the peak flow into the receiver, to provide the head pressure required for the elevation of lift, to prevent excessive cycling of the pumping equipment and to handle the type of waste that discharges to the sump. Sewage pumps and ejectors must be able to handle, without creating blockages, the types of solid waste discharged into the sump or receiver. Pumps and ejectors serving water closets must be able to handle the solids associated with soil drainage. Grinder pumps/ejectors and macerating toilet assemblies that serve water closets are exempt from the 2-inch (51 mm) spherical solids-handling requirements of this section. These pumps and ejectors pulverize solids to a near liquid state and pump the slurry

SANITARY DRAINAGE

**Figure 712.4.1(1)
MACERATING TOILET SYSTEMS
(Photo courtesy of SFA Saniflo)**

**Figure 712.4.1(2)
MACERATING TOILET SYSTEMS**

to the drainage system. The 1$^1/_4$-inch (32 mm) minimum-size discharge for grinder pumps and ejectors and $^3/_4$-inch (19.1 mm) minimum-size discharge for macerating grinder/pump systems will permit solids that are completely reduced to pass without creating a stoppage.

**TABLE 712.4.2
MINIMUM CAPACITY OF SEWAGE PUMP OR SEWAGE EJECTOR**

DIAMETER OF THE DISCHARGE PIPE (inches)	CAPACITY OF PUMP OR EJECTOR (gpm)
2	21
2$^1/_2$	30
3	46

For SI: 1 inch = 25.4 mm, 1 gallon per minute = 3.785 L/m.

❖ Table 712.4.2 specifies sewage pump capacity in gallons per minute with respect to the pump's discharge opening size. This relationship between pipe size and pump capacity will maintain flow velocity in the piping to help prevent blockage and restriction.

SECTION 713
HEALTH CARE PLUMBING

713.1 Scope. This section shall govern those aspects of health care plumbing systems that differ from plumbing systems in other structures. Health care plumbing systems shall conform to this section in addition to the other requirements of this code. The provisions of this section shall apply to the special devices and equipment installed and maintained in the following occupancies: nursing homes; homes for the aged; orphanages; infirmaries; first aid stations; psychiatric facilities; clinics; professional offices of dentists and doctors; mortuaries; educational facilities; surgery, dentistry, research and testing laboratories; establishments manufacturing pharmaceutical drugs and medicines; and other structures with similar apparatus and equipment classified as plumbing.

❖ Section 713 regulates the installation of special fixtures and the related waste and vent connections found in health care facilities. The special plumbing fixtures addressed in this section include clinical sinks, bedpan washers, bedpan steamers, vacuum fluid suction systems, sterilizers and hydrotherapeutic baths.

Section 713.1 defines the scope of this chapter and describes the occupancies involved. Section 202 contains the definitions of key terms and phrases used throughout this section.

Section 422 regulates the use of special fixtures in the health care industry.

Section 609 governs the water supply to health care facilities and fixtures. Hospitals are required to have dual water services so that in the event of a water main or water service pipe break, the hospital will not experience a complete loss of water supply.

The unique plumbing fixtures in health care facilities are regulated by the special requirements of this section as well as those previously mentioned such as Sections 202, 422 and 609. Health care fixtures must be installed to maintain the same level of sanitation and protection against backflow as other plumbing fixtures.

Some of the requirements for health care fixtures in this section may appear archaic. As technology improves, the fixtures change. Requirements addressing older fixtures will no longer be applicable to certain newer fixtures.

This is most evident with bedpan washers and sterilizers. Water closets are often equipped with an add-on flushing mechanism that may function as a bedpan washer. Installation of a separate fixture may not be needed.

Sterilizers may have a local or vapor vent connection. The new units are designed to condense the vapor rather than discharge it through a vent.

Although the code lists many applications, some are bound to be missing. The requirements for plumbing installations in health care facilities apply to facilities whether or not they are specifically listed. These requirements are to be considered the minimum special attention given to health care plumbing. The fixture installer must carefully follow any other requirements contained in the manufacturer's instructions.

This section regulates the waste and vent connections for health care plumbing fixtures, appliances and equipment. The requirements of this section are in addition to other applicable code requirements (see commentary, Sections 202, 422 and 609).

713.2 Bedpan washers and clinical sinks. Bedpan washers and clinical sinks shall connect to the drainage and vent system in accordance with the requirements for a water closet. Bedpan washers shall also connect to a local vent.

❖ These fixtures are to be vented and connected to the drainage system like a water closet because they receive similar human waste and function in a similar manner. A local vent is required because a bedpan washer uses high-temperature hot water to clean the bedpan. A clinical sink is not equipped with a local vent (see commentary, Section 713.3).

713.3 Indirect waste. Sterilizers, steamers and condensers shall discharge to the drainage through an indirect waste pipe by means of an *air gap*. Where a battery of not more than three sterilizers discharges to an individual receptor, the distance between the receptor and a sterilizer shall not exceed 8 feet (2438 mm). The indirect waste pipe on a bedpan steamer shall be trapped.

❖ An indirect waste pipe protects sterilizers, steamers and condensers from contamination from the drainage system through the use of the required air gap. An indirect waste pipe is typically piped individually to the indirect waste receptor. This section permits a common indirect waste pipe for two or three sterilizers installed together (see Commentary Figure 713.3). Additionally, the indirect waste pipe on the bedpan steamer must be trapped to prevent the escape of steam vapors and odor.

SANITARY DRAINAGE

Figure 713.3
COMMON INDIRECT WASTE PIPE FOR STERILIZERS

713.4 Vacuum system station. Ready *access* shall be provided to vacuum system station receptacles. Such receptacles shall be built into cabinets or recesses and shall be visible.

❖ Vacuum system stations are typically installed in recesses in locations where medical professionals have ready access to them. The person locating the station must take into consideration the movements of personnel during an operation or emergency procedure.

These stations are to be visible to allow for ready supervision of the equipment to make sure it is functioning as required.

713.5 Bottle system. Vacuum (fluid suction) systems intended for collecting, removing and disposing of blood, pus or other fluids by the bottle system shall be provided with receptacles equipped with an overflow prevention device at each vacuum outlet station.

❖ A vacuum or fluid suction system is a central system of piping in a health care facility, with terminals located throughout the building. The terminals are typically found in operating rooms, emergency rooms, delivery rooms, recovery rooms, intensive care rooms and similar locations.

Typically, a bottle collector connects to the terminal when the suction system is used. Fluids collected must be discharged to the bottle while the vacuum system remains free of fluids; therefore, a means of preventing the overflow of the bottle is required. This is commonly done by having a separate vacuum trap with a shutoff installed between the suction collection bottle and the vacuum system terminal inlet.

713.6 Central disposal system equipment. Central vacuum (fluid suction) systems shall provide continuous service. Systems equipped with collecting or control tanks shall provide for draining and cleaning of the tanks while the system is in operation. In hospitals, the system shall be connected to the emergency power system. The exhausts from a vacuum pump serving a vacuum (fluid suction) system shall discharge separately to open air above the roof.

❖ The vacuum system is essential for a health care facility and must operate under all conditions. To provide continuous service, at least two vacuum pumps are installed. These systems must also connect to the building's emergency power system (see Commentary Figure 713.6).

Many systems operate with a receiver tank to prevent the constant cycling of the vacuum pumps. The tanks must be equipped with a drain and must be arranged for the cleaning of the tank while the system is still in service. This is accomplished with a bypass piping arrangement located around the receiver tank. This section also requires the exhaust from the vacuum pump to discharge to the outdoors at an elevation that is above the roof of the building in which the equipment is located. The exhaust may be considered a biohazard and must be discharged in a manner similar to plumbing vent terminals.

713.7 Central vacuum or disposal systems. Where the waste from a central vacuum (fluid suction) system of the barometric-lag, collection-tank or bottle-disposal type is connected to the drainage system, the waste shall be directly connected to the sanitary drainage system through a trapped waste.

❖ Certain vacuum systems are designed to connect directly to the drainage system. These types of installations cannot be connected by an indirect waste pipe. Rather, they must be directly connected. This is intended to minimize possible contamination by preventing the waste from being exposed to the atmosphere. The connection must be trapped to protect the system from sewer gases.

713.7.1 Piping. The piping of a central vacuum (fluid suction) system shall be of corrosion-resistant material with a smooth interior surface. A *branch* shall be not less than $^{1}/_{2}$-inch (12.7 mm) nominal pipe size for one outlet and shall be sized in accordance with the number of vacuum outlets. A main shall be not less than 1-inch (25 mm) nominal pipe size. The pipe sizing shall be increased in accordance with the manufacturer's instructions as stations are increased.

❖ The common piping materials used in vacuum systems are copper tubing, galvanized steel pipe and stainless steel pipe. These materials are corrosion resistant and have smooth interior surfaces. The sizing, cleanout and access requirements are considered the very minimum acceptable.

**Figure 713.6
CENTRAL VACUUM SYSTEM
(Shown for illustrative purposes only)**

713.7.2 Velocity. The velocity of airflow in a central vacuum (fluid suction) system shall be less than 5,000 feet per minute (25 m/s).

❖ The air velocity of a vacuum system is limited to 5,000 feet per minute (25 m/s) to reduce the noise factor and erosion in the piping.

713.8 Vent connections prohibited. Connections between local vents serving bedpan washers or sterilizer vents serving sterilizing apparatus and normal sanitary plumbing systems are prohibited. Only one type of apparatus shall be served by a local vent.

❖ Both a local vent and a sterilizer vent are used for the removal of condensate and odor. Both vents function differently than a vent system designed to protect the trap seals in a drainage system. Local vents must be completely independent of the sanitary drain and waste venting system. An interconnection between local vents and sanitary drain and waste vents would allow sewer gases to enter the local vents, thereby resulting in equipment contamination and the escape of sewer gases into the building interior.

Each local or sterilizer vent must be designed to serve only one type of equipment to prevent possible cross contamination. A local vent connecting to equipment used to clean or sterilize bedpans must not connect to any other sterilizer vents. Although the equipment might be sterilizing the bedpan, there could be bacterial growth in the local vent originating from human waste. The installer must carefully read and follow the fixture manufacturer's instructions.

713.9 Local vents and stacks for bedpan washers. Bedpan washers shall be vented to open air above the roof by means of one or more local vents. The local vent for a bedpan washer shall be not less than a 2-inch-diameter (51 mm) pipe. A local vent serving a single bedpan washer is permitted to drain to the fixture served.

❖ A local vent is occasionally referred to as a "vapor vent." It connects to the fixture or equipment and is designed to vent steam or hot water vapor and foul odors. The local vent prevents water vapor from escaping into the room or space. Besides increasing the humidity or forming condensation, escaping water vapor may also contain biological contaminants. Local vents are disappearing with the use of modern equipment designed to contain and condense the water vapor in the equipment. This type of equipment may not provide a connection for, nor does it require, a local vent. This requirement applies only to bedpan washers having a connection for a local vent.

713.9.1 Multiple installations. Where bedpan washers are located above each other on more than one floor, a local vent *stack* is permitted to be installed to receive the local vent on the various floors. Not more than three bedpan washers shall be connected to a 2-inch (51 mm) local vent *stack*, not more than six to a 3-inch (76 mm) local vent *stack* and not more than 12 to a 4-inch (102 mm) local vent *stack*. In multiple installations, the connections between a bedpan washer local vent and a local vent *stack* shall be made with tee or tee-wye sanitary pattern drainage fittings installed in an upright position.

❖ The requirements for this section may be simplified into a table form [see Commentary Figure 713.9.1(2)]. The stack may be increased in size as additional bedpan washers connect to the local vent stack [see Commentary Figure 713.9.1(1)].

SANITARY DRAINAGE

For SI: 1 inch = 25.4 mm.

**Figure 713.9.1(1)
LOCAL VENT STACK**

LOCAL VENT STACK SIZE (inches)	MAXIMUM NUMBER OF BEDPAN WASHERS
2	3
3	6
4	12

For SI: 1 inch = 25.4 mm.

**Figure 713.9.1(2)
MULTIPLE CONNECTION FOR BEDPAN
WASHERS TO LOCAL VENT STACK**

713.9.2 Trap required. The bottom of the local vent *stack*, except where serving only one bedpan washer, shall be drained by means of a trapped and vented waste connection to the sanitary drainage system. The trap and waste shall be the same size as the local vent *stack*.

❖ Because a local vent is designed to remove water vapor, there will be a certain amount of water condensing inside the pipe. The condensate must be drained to the sanitary drainage system to prevent the local vent stack from accumulating water. The connection to the drainage system must be by means of a trapped and vented waste connection (see Commentary Figure 713.9.2).

713.9.3 Trap seal maintenance. A water supply pipe not less than $\frac{1}{4}$ inch (6.4 mm) in diameter shall be taken from the flush supply of each bedpan washer on the discharge or fixture side of the vacuum breaker, shall be trapped to form not less than a 3-inch (76 mm) water seal and shall be connected to the local vent *stack* on each floor. The water supply shall be installed so as to provide a supply of water to the local vent *stack* for cleansing and drain trap seal maintenance each time a bedpan washer is flushed.

❖ A $\frac{1}{4}$-inch (6.4 mm) tube is looped to form a trap and is connected to the local vent. With each use of the bedpan washer, water is discharged to maintain the trap seal of the drainage connection. The looped tube must be connected downstream of the fixture's vacuum breaker to prevent a cross connection.

713.10 Sterilizer vents and stacks. Multiple installations of pressure and nonpressure sterilizers shall have the vent connections to the sterilizer vent *stack* made by means of inverted wye fittings. *Access* shall be provided to vent connections for the purpose of inspection and maintenance.

❖ A sterilizer vent is similar to a local vent. The vent is designed to remove the steam vapor developed in sterilizer equipment (see Commentary Figure 713.10).

Like local vents, sterilizer vents are not used because current equipment condenses the steam vapor in the equipment.

713.10.1 Drainage. The connection between sterilizer vent or exhaust openings and the sterilizer vent *stack* shall be designed and installed to drain to the funnel or basket-type waste fitting. In multiple installations, the sterilizer vent *stack* shall be drained separately to the lowest sterilizer funnel or basket-type waste fitting or receptor.

❖ The sterilizer vent exhausts the steam created in the sterilizer. As a result, condensation forms inside the piping. To maintain proper operation of the system, the sterilizer vent must be designed to drain the condensate to an indirect waste receptor at the base of the vent.

**Figure 713.9.2
BOTTOM CONNECTION TO LOCAL VENT STACK**

SANITARY DRAINAGE

Figure 713.10
STERILIZER PIPING

713.11 Sterilizer vent stack sizes. Sterilizer vent *stack* sizes shall comply with Sections 713.11.1 through 713.11.4.

❖ The requirements for sterilizer vent stack sizes are described in Sections 713.11.1 through 713.11.4.

713.11.1 Bedpan steamers. The minimum size of a sterilizer vent serving a bedpan steamer shall be 1¹/₂ inches (38 mm) in diameter. Multiple installations shall be sized in accordance with Table 713.11.1.

❖ The minimum size sterilizer vent is 1¹/₂ inches (38 mm). When multiple sterilizer vents are interconnected, the piping must be sized in accordance with Table 713.11.1.

TABLE 713.11.1
STACK SIZES FOR BEDPAN STEAMERS AND BOILING-TYPE STERILIZERS (Number of Connections of Various Sizes Permitted to Various-sized Sterilizer Vent Stacks)

STACK SIZE (inches)	CONNECTION SIZE	
	1¹/₂"	2"
1¹/₂ª	1 or	0
2ª	2 or	1
2ᵇ	1 and	1
3ª	4 or	2
3ᵇ	2 and	2
4ª	8 or	4
4ᵇ	4 and	4

For SI: 1 inch = 25.4 mm.
a. Total of each size.
b. Combination of sizes.

❖ This table often appears confusing because of the use of the terms "or" and "and." When the term "or" is used under the permitted connection size, it means that either one or the other listed connection is permitted. When the term "and" is used, it means that both of the listed connections are permitted together.

713.11.2 Boiling-type sterilizers. The size of a sterilizer vent *stack* shall be not less than 2 inches (51 mm) in diameter where serving a utensil sterilizer and not less than 1¹/₂ inches (38 mm) in diameter where serving an instrument sterilizer. Combinations of boiling-type sterilizer vent connections shall be sized in accordance with Table 713.11.1.

❖ Table 713.11.1 is also used for sizing boiling-type sterilizer vent stacks. Refer to Section 713.11.1 for commentary on the use of the table.

713.11.3 Pressure sterilizers. Pressure sterilizer vent *stacks* shall be 2¹/₂ inches (64 mm) minimum. Those serving combinations of pressure sterilizer exhaust connections shall be sized in accordance with Table 713.11.3.

❖ Pressure sterilizer vent stacks must not be less than 2¹/₂ inches (64 mm) in diameter; however, where such stacks serve multiple pressure sterilizers, the vent stacks must be sized in accordance with Table 713.11.3.

TABLE 713.11.3
STACK SIZES FOR PRESSURE STERILIZERS (Number of Connections of Various Sizes Permitted To Various-sized Vent Stacks)

STACK SIZE (inches)	CONNECTION SIZE			
	³/₄"	1"	1¹/₄"	1¹/₂"
1¹/₂ª	3 or	2 or	1	—
1¹/₂ᵇ	2 and	1	—	—
2ª	6 or	3 or	2 or	1
2ᵇ	3 and	2	—	—
2ᵇ	2 and	1 and	1	—
2ᵇ	1 and	1 and	—	1
3ª	15 or	7 or	5 or	3
3ᵇ	1 and	1 and 5 and	2 and	2 1

For SI: 1 inch = 25.4 mm.
a. Total of each size.
b. Combination of sizes.

❖ The table dictates the allowable number of individual sterilizer vent connections to a vent stack. Connections are listed as numbers and combinations of various sizes (see commentary, Table 713.11.1). The rows designated by Note a allow one or more connections of a single sterilizer exhaust connection size. Note a rows do not permit commingling of different sterilizer exhaust connection sizes. The rows designated by Note b allow multiple connections of sterilizer exhausts that are combinations of sizes.

713.11.4 Pressure instrument washer sterilizer sizes. The diameter of a sterilizer vent *stack* serving an instrument washer sterilizer shall be not less than 2 inches (51 mm). Not more than two sterilizers shall be installed on a 2-inch (51 mm) *stack*, and not more than four sterilizers shall be installed on a 3-inch (76 mm) *stack*.

❖ The requirements of this section are illustrated in Commentary Figure 713.11.4.

STACK SIZE (inches)	NUMBER OF STERILIZERS CONNECTED
2	2
3	4

For SI: 1 inch = 25.4 mm.

**Figure 713.11.4
PRESSURE INSTRUMENT WASHER STERILIZER SIZES**

SECTION 714
COMPUTERIZED DRAINAGE DESIGN

714.1 Design of drainage system. The sizing, design and layout of the drainage system shall be permitted to be designed by *approved* computer design methods.

❖ This section establishes a parameter for sizing a drainage system using a computerized method. Various computer programs have been developed using the same technical guidelines used to develop the sizing tables in Sections 703, 706, 710 and 711. One of the major differences is that the computerized methods are more exact, knowing the exact discharge considerations or load on the drainage system fixture profiles, material specifications and slope of drainage piping.

714.2 Load on drainage system. The load shall be computed from the simultaneous or sequential discharge conditions from fixtures, appurtenances and appliances or the peak usage design condition.

❖ When sizing to exact parameters, the discharge conditions for the particular system must be known. This is similar to determining the peak demand of the water distribution system, as indicated in the commentary to Section 604.3.

714.2.1 Fixture discharge profiles. The discharge profiles for flow rates versus time from fixtures and appliances shall be in accordance with the manufacturer's specifications.

❖ Each fixture has a specific discharge profile, plotting flow rates versus time. When designing by a computerized method, the actual discharge profile of the individual fixtures being installed may be used. The table in Section 604.3 applies a conservative, general discharge profile for each category of fixtures.

714.3 Selections of drainage pipe sizes. Pipe shall be sized to prevent full-bore flow.

❖ The drainage system must be designed to maintain open channel flow (gravity). If full-bore flow is achieved, the piping may have pressure flow occurring. This may be detrimental to the operation of the system because the venting design provisions are based on the hydraulics of a gravity system.

714.3.1 Selecting pipe wall roughness. Pipe size calculations shall be conducted with the pipe wall roughness factor (ks), in accordance with the manufacturer's specifications and as modified for aging roughness factors with deposits and corrosion.

❖ Exact sizing uses the parameters of the given material to be installed. The roughness coefficient of the interior pipe wall is used in sizing. Pipe wall roughness is a factor used to determine the flow losses attributable to friction. These factors must contain modifying factors that reflect that changes occur as the pipe ages and develops deposits and corrosion.

714.3.2 Slope of horizontal drainage piping. Horizontal drainage piping shall be designed and installed at slopes in accordance with Table 704.1.

❖ Piping sized by a computerized method is subject to the same slope requirements as conventionally sized piping. The minimum slope must be as specified in Table 704.1. Refer to that section and its commentary for related issues, requirements and concerns.

SECTION 715
BACKWATER VALVES

715.1 Sewage backflow. Where plumbing fixtures are installed on a floor with a finished floor elevation below the elevation of the manhole cover of the next upstream manhole in the public *sewer*, such fixtures shall be protected by a backwater valve installed in the *building drain*, or horizontal *branch* serving such fixtures. Plumbing fixtures installed on a floor with a finished floor elevation above the elevation of the manhole cover of the next upstream manhole in the public *sewer* shall not discharge through a backwater valve.

Exception: In existing buildings, fixtures above the elevation of the manhole cover of the next upstream manhole in the public *sewer* shall not be prohibited from discharging through a backwater valve.

❖ A backwater valve is required in areas where the public sewer might back up into the building through the sanitary drainage system [see Commentary Figure 715.1(1)]. Where plumbing fixtures are located on a floor that is above the next upstream manhole cover from the building sewer connection to the public sewer, the public sewer will back up through the street manhole before entering the building.

Public sewers might become blocked or overloaded, which will result in sewage backing up into the manholes and any laterals (taps) connected to the sewer system. The point of overflow for the public sewer will be the top of the manholes in the backed-up portion of the system.

Fixtures or drains located on a floor that is at an elevation below that of the tops of the manholes for the relative portion of the sewer system are subject to backflow and must be protected by backwater valves [see Commentary Figure 715.1(2)]. Plumbing fixtures that are not subject to backflow are not permitted to discharge through a backwater valve. In theory, limiting the fixtures that discharge through a backwater valve will prevent waste from upstream fixtures from backing up through downstream fixtures because it cannot pass through the backwater valve when the public sewer is blocked or overloaded. Additionally, the valve will be protected from excess wear and potential failure resulting from debris and accumulations.

715.2 Material. Bearing parts of backwater valves shall be of corrosion-resistant material. Backwater valves shall comply with ASME A112.14.1, CSA B181.1 or CSA B181.2.

❖ This section requires that bearing parts of the backwater valve be made of corrosion-resistant materials so that the valve remains operable for its expected life, thus protecting the building it serves. This section also gives the industry standards to which the backwater valve must be manufactured. As required in Section 107, these backwater valves are to be inspected and verified by an independent third-party quality-assurance facility.

715.3 Seal. Backwater valves shall be so constructed as to provide a mechanical seal against backflow.

❖ A backwater valve is designed to prevent the backflow of drainage in a piping system. The valve incorporates a swing check on the inlet side of the body. This section requires that the swing check have a positive seal against the backflow of drainage. Some backwater valves are also equipped with a gate valve that is manually operated to close the outlet side of the device. Backwater valves equipped with gate valves are typically used in areas subject to severe flooding conditions [see Commentary Figure 715.1(2)].

715.4 Diameter. Backwater valves, when fully opened, shall have a capacity not less than that of the pipes in which they are installed.

❖ This section simply requires that the capacity of the pipe in which the backwater valve is located not be reduced by the valve. This would include components of the backwater valve, including any optional manual gate valves. Backwater valves are designed so as to not create a restriction in the piping.

715.5 Location. Backwater valves shall be installed so that *access* is provided to the working parts for service and repair.

❖ Because a backwater valve has movable parts, there is a possibility of stoppage or malfunction. This section requires that backwater valves be located to be accessible to permit the necessary maintenance or repairs.

SECTION 716
VACUUM DRAINAGE SYSTEMS

716.1 Scope. Vacuum drainage systems shall be in accordance with Sections 716.2 through 716.4.

❖ This section regulates the design and installation methods for vacuum drainage systems. Plumbing systems are designed to drain by gravity whenever possible; however, there are many instances where the discharge from plumbing fixtures must be lifted by a vacuum to an approved point of disposal. This type of system is commonly specified in railcars, aircraft, buses, marine vessels, recreational vehicles and land-based projects.

716.2 System design. Vacuum drainage systems shall be designed in accordance with the vacuum drainage system manufacturer's instructions. The system layout, including piping layout, tank assemblies, vacuum pump assembly and other components necessary for proper function of the system shall be in accordance with the manufacturer's instructions. Plans, specifications and other data for such systems shall be submitted to the code official for review and approval prior to installation.

❖ The vacuum system connections are made on the inlet rather than on the outlet as is the case with sewage ejectors or pumps. The wastes in these types of systems are being pulled, not pushed, through the piping system. Typically, the waste is introduced into the system from plumbing fixtures, conveyed to a collection tank and then to an approved point of disposal. A grinding pump that macerates waste is often installed prior to the collection tank, thereby providing additional holding capacity. The negative pressure is maintained in the system by a vacuum pump. A con-

Figure 715.1(1)
BACKWATER VALVE

SANITARY DRAINAGE

Figure 715.1(2)
BACKWATER VALVES

trol panel links the system and typically includes a timing valve, discharge valve and water valve. For example, when the flush mechanism is activated for a water closet, the discharge valve opens, allowing atmospheric pressure to force waste through the piping to the vacuum tank. At the same time, the water valve injects water to rinse the bowl. A lavatory or sink waste flows by gravity to an interface valve, which then opens and allows the waste to enter the vacuum piping. The discharge from showers or floor drains is typically collected in a holding tank, and when the tank is filled, a pressure sensor triggers an actuator to open a valve, allowing the waste to move into the vacuum piping. A control system automatically regulates the operation of the vacuum and sewage ejector pumps. The vacuum pump creates a vacuum in the system by pumping air from the vacuum tank and piping. Once a predetermined level of sewage is reached in the vacuum tank, the sewage is automatically pumped out by sewage ejector pumps to a treatment plant or sewer main. Manufacturers must provide detailed installation and servicing instructions with each type of system.

716.2.1 Fixtures. Gravity-type fixtures installed in vacuum drainage systems shall comply with Chapter 4.

❖ Fixtures utilized in a vacuum system must adhere to the same standards for plumbing fixtures in Chapter 4 of the code. Plumbing fixtures regulated in this type of system must be free from deficiencies that affect the fixture's intended purpose. The referenced standards specify quality, dimensions, construction requirements and the testing of such fixtures.

716.2.2 Drainage fixture units. *Drainage fixture units* for gravity drainage systems that discharge into, or receive discharge from, vacuum drainage systems shall be based on the values in this chapter.

❖ Table 709.1 of the code provides the drainage fixture unit (dfu) value for various type of fixtures. A water closet installed in a vacuum system typically discharges 1.1 liters or 1 quart of water per flush. Note e in this table indicates that water closets and urinals may have a lower dfu value confirmed by test results. Many vacuum-type water closets have a dfu value lower than the same fixture type indicated in Table 709.1.

716.2.3 Water supply fixture units. Water supply fixture units shall be based on the values in Chapter 6 of this code, except that the water supply fixture unit for a vacuum-type water closet shall be 1.

❖ The demand on the water supply system for vacuum-drained fixtures will be the same as if the fixtures were gravity drained. Methods for determining water supply pipe sizes usually involve some type of dimensionless value assigned to a fixture so that flow rates can be determined (and pipe sizes determined), based on the total number of fixtures (of all types) in a building. The dimensionless values take into account flow rates and the *probability of use* of fixtures. One common name for this dimensionless value is the "water supply fixture unit." Although Chapter 6 does not directly address water supply fixture units, the typical approved methods that conform to accepted engineering practice commonly utilize water supply fixture units for determining water supply piping systems.

The code requirement for the water supply fixture unit for a water closet on a vacuum drainage system to be 1 is an indication that the water closet will require minimal water requirements for flushing. The vacuum system evacuates the bowl. Water is only necessary for bowl refill so that there is a water surface.

716.2.4 Traps and cleanouts. Gravity drainage fixtures shall be provided with traps and cleanouts in accordance with this chapter and Chapter 10.

❖ Drainage piping in a vacuum system must have cleanouts installed in accordance with Section 708. Fixture traps must be provided in accordance with Section 1002.1. Many vacuum systems can eliminate fixture venting because of the positive seal between discharge waste water and the fixture; hence, the necessity for traps. The code official must evaluate each system individually based on the manufacturer's recommended installation instructions and the requirements of the code.

716.2.5 Materials. Vacuum drainage pipe, fitting and valve materials shall be in accordance with the vacuum drainage system manufacturer's instructions and the requirements of this chapter.

❖ Before installation of drainage pipes and fittings, the manufacturer's recommendations must be consulted in order to evaluate the type of material to be used. Table 702.1 provides the acceptable material to be used in vacuum systems.

716.3 Testing and demonstrations. After completion of the entire system installation, the system shall be subjected to a vacuum test of 19 inches (483 mm) of mercury and shall be operated to function as required by the code official and the manufacturer of the vacuum drainage system. Recorded proof of all tests shall be submitted to the code official.

❖ Vacuum systems must be subjected to testing to determine potential defects and leaks. Vacuum pumps maintain an operating range of 16 to 20 inches (406 to 508 mm) of mercury vacuum. Differential air pressure propels the sewage at velocities of 15 to 18 feet per second (4.5 to 5.5 m/s). Such systems are subjected to a vacuum test of 19 inches (483 mm) of mercury, which equals 9.3 pounds per square inch (64 kPa) of air pressure.

716.4 Written instructions. Written instructions for the operation, maintenance, safety and emergency procedures shall be provided to the building owner. The code official shall verify that the building owner is in receipt of such instructions.

❖ Manufacturer's installation instructions must be provided because they are an enforceable extension of the code for determining whether a system has been properly installed. A detailed description of operating maintenance, safety and emergency procedures is required to reflect the scope of information needed to determine compliance.

SECTION 717
REPLACEMENT OF UNDERGROUND SEWERS BY PIPE-BURSTING METHODS

717.1 General. This section shall govern the replacement of existing *building sewer* piping by pipe-bursting methods.

❖ This is a method for replacing existing building piping that does not require complete excavation of the existing piping. This methods uses an expanding mandrel that is attached to the end of the replacement pipe. The mandrel (with the replacement pipe in tow) is pulled by a cable through the existing sewer line. The mandrel has a hydraulically activated expanding shell that breaks (bursts) the existing pipe to make room for the new pipe as it is pulled into position.

The method requires that launching and receiving pits be excavated at the beginning and ending points of the pipe to be replaced. If there are any lateral connections that need to be made between the beginning and the end, excavated pits are required at those locations as well.

717.2 Applicability. The replacement of *building sewer* piping by pipe-bursting methods shall be limited to gravity drainage piping of sizes 6 inches (152 mm) and smaller. The replacement piping shall be of the same nominal size as the existing piping.

❖ Although the pipe-bursting method of underground pipe replacement is used for large pipe sizes, the code limits its use to the replacement of pipes that are 6 inches (152 mm) and smaller. The code also limits this method to use on gravity sewers and a replacement using the same size as existing piping.

717.3 Pre-installation inspection. The existing piping sections to be replaced shall be inspected internally by a recorded video camera survey. The survey shall include notations of the position of cleanouts and the depth of connections to the existing piping.

❖ An internal video camera survey of the existing installation is required to identify problem areas and location of connections and cleanouts.

717.4 Pipe. The replacement piping shall be manufactured with a standard dimension ratio (SDR) of 17 and in compliance with ASTM F 714.

❖ The only material that the code allows for sewer replacement is polyethylene plastic pipe (ASTM F714) having an SDR 17 wall thickness. This pipe is an outer-diameter- and wall-thickness-controlled product. SDR 17 provides for an inside diameter that is very close to the inside diameter of PVC and cast-iron sewer piping of the same nominal pipe size

717.5 Pipe fittings. Pipe fittings to be connected to the replacement piping shall be of extra-high molecular weight PE3408 material and shall be manufactured with an SDR of 17 and in compliance with ASTM D 2683.

❖ Fittings for this material must be compatible with the pipe. The requirement for compliance to ASTM D2683 avoids the use of butt welded fittings. A butt-welded fitting could have internal edges that could catch and restrict flow. Use of a socket fitting allows for pipe ends to be deburred/chamfered before assembly.

717.6 Cleanouts. Where the existing *building sewer* did not have cleanouts meeting the requirements of this code, cleanout fittings shall be installed as required by this code.

❖ Cleanouts are required in accordance with Section 708, whether the existing sewer had them or not.

717.7 Post-installation inspection. The completed replacement piping section shall be inspected internally by a recorded video camera survey. The video survey shall be reviewed and *approved* by the code official prior to pressure testing of the replacement piping system.

❖ An internal video camera survey of the completed installation is required to be reviewed by the code official prior to pressure testing of replacement pipe, including all connections to the replacement pipe.

717.8 Pressure testing. The replacement piping system as well as the connections to the replacement piping shall be tested in accordance with Section 312.

❖ As with any installation of a new sewer, pressure testing is required.

Chapter 8:
Indirect/special Waste

General Comments

An indirect waste pipe is a method of discharging waste from a fixture or appliance into the sanitary drainage system to prevent backflow and possible contamination. The waste pipe discharges through an air gap or air break into an open receptor or standpipe. Section 801 outlines the scope of the chapter, which is to indicate how to protect fixtures from possible contamination and to indicate how to dispose of special wastes properly. Section 802 requires most plumbing fixtures and appliances in food-handling establishments to discharge through an indirect waste. This eliminates a possible cause of food contamination from the sanitary drainage system. Section 802 also regulates the location of floor drains in food storage areas and waste water from swimming pools, pool filters and deck drains. Section 803 regulates special wastes that require treatment before being discharged to the sanitary drainage system, such as waste from a chemical laboratory.

Commonly referred to as "open site" or "safe waste," an indirect waste pipe is a means of protecting a fixture or appliance in the event of a backup of the sanitary drainage system. Sewage cannot back up into a fixture or appliance if a stoppage occurs downstream of the waste receptor receiving the discharge of an indirect waste pipe. Instead, sewage will spill onto the surrounding floor area. This protection is necessary when the contamination presents a health hazard or creates a potential cross connection with the potable water system. Note that an indirect waste pipe conveying waste from a fixture such as a sink, floor drain or swimming pool is part of the drainage system and, therefore, must conform to other applicable provisions of the code, including those found in Chapters 3 and 7. Condensate disposal piping must comply with Section 314 of the code, which is identical to and controlled by Section 307 of the *International Mechanical Code*® (IMC®). This section requires condensate pipe to be cast iron, galvanized steel, copper, cross-linked polyethylene (PEX), polyethylene (PE), polybutylene (PB), acrylonitrile butadiene styrene (ABS), chlorinated polyvinyl (CPVC) or polyvinyl chloride (PVC). Condensate piping must be not less than $^3/_4$-inch (19.1 mm) internal diameter and sized in accordance with an approved method. Where condensate is generated by multiple refrigeration units, the minimum size of the piping carrying the combined flows must be in accordance with Table 314.2.2 of this code. A condensate drain is required to have a trap installed if required by the equipment or appliance manufacturer. Condensate drain traps are not required to be vented as is required for sanitary drainage traps.

Purpose

Chapter 8 regulates drainage installations requiring an indirect connection to the sanitary drainage system. Fixtures and plumbing appliances, such as those associated with food preparation or handling, health care facilities and potable liquids, must be protected from contamination that can result from connection to the drainage system. An indirect connection prevents sewage from backing up into a fixture or appliance, thus providing protection against potential health hazards. The chapter also regulates special wastes containing hazardous chemicals. The waste must be treated to prevent any damage to the piping or sewage treatment process.

SECTION 801
GENERAL

801.1 Scope. This chapter shall govern matters concerning indirect waste piping and special wastes. This chapter shall further control matters concerning food-handling establishments, sterilizers, clear-water waste, swimming pools, methods of providing *air breaks* or *air gaps*, and neutralizing devices for corrosive wastes.

❖ This section outlines the scope of the chapter, which is to protect fixtures and appliances from possible contamination and to dispose of special wastes properly.

Indirect waste connections are necessary where a higher degree of health protection from sewage backup into the fixture or appliance is required because of the nature of its use. The chapter also has performance requirements for the neutralizing of special waste.

801.2 Protection. Devices, appurtenances, appliances and apparatus intended to serve some special function, such as sterilization, distillation, processing, cooling, or storage of ice or foods, and that discharge to the drainage system, shall be provided with protection against backflow, flooding, fouling, contamination and stoppage of the drain.

❖ This section requires protection of appliances and specialized equipment against contamination resulting from backflow. This section combines the intent of Sections 608, 609 and 802.

SECTION 802
INDIRECT WASTES

802.1 Where required. Food-handling equipment, in other than dwelling units, clear-water waste, dishwashing machines and utensils, pots, pans and dishwashing sinks shall discharge through an indirect waste pipe as specified in Sections

INDIRECT/SPECIAL WASTE

802.1.1 through 802.1.8. Health-care related fixtures, devices and equipment shall discharge to the drainage system through an indirect waste pipe by means of an *air gap* in accordance with this chapter and Section 713.3. Fixtures not required by this section to be indirectly connected shall be directly connected to the plumbing system in accordance with Chapter 7.

❖ This section establishes where an indirect waste is required, as well as appropriate indirect waste methods and installation procedures. The requirements for indirect wastes are described in Sections 802.1.1 through 802.1.8. Health care plumbing fixtures, appliances and equipment that must discharge through an indirect waste are included in Section 713.3. Fixtures that need not be indirectly connected by Chapter 8 must be directly connected to the plumbing system in accordance with Section 301.3.

802.1.1 Food handling. Equipment and fixtures utilized for the storage, preparation and handling of food shall discharge through an indirect waste pipe by means of an air gap. Each well of a multiple-compartment sink shall discharge independently to a waste receptor.

❖ All food must be protected from possible contamination caused by the sanitary drainage system. The requirement for indirect waste connections extends to all storage, cooking and preparation equipment, including vegetable sinks; food-washing sinks; refrigerated cases and cabinets; ice boxes; ice-making machines; steam kettles; steam tables; potato peelers; egg boilers; coffee urns and brewers; drink dispensers and similar types of equipment and fixtures [see Commentary Figures 802.1.1(1), 802.1.1(2) and 802.1.1(3)]. Food stored in walk-in coolers and freezers is specifically addressed in Section 802.1.2. Note that sinks used for washing, rinsing and sanitizing utensils, dishes, pots, pans or service ware are also considered as if they were food preparation fixtures because of the connection requirements of Section 802.1.8. Domestic and commercial dishwashing machines must discharge indirectly through an air gap or air break into a standpipe or waste receptor or as allowed by Section 802.1.6 (see Commentary Figure 802.1.6 or 802.1.7).

Occasionally, commercial kitchen designers will desire to mount a food waste grinder to a food preparation sink for ease of disposal of food prep wastes such as leaves, skins, spoiled produce, bones and meat trimmings. Unfortunately, this creates several code conflicts. A food waste grinder is not a plumbing "fixture" that is required to be indirect connected; therefore, it must be direct connected to the drain system. However, because the discharge of the food waste grinder is also the outlet for the food-prep sink, this section requires the drain to be indirectly connected. The concern of the code for indirectly connecting drains of food prep sinks is to avoid

Figure 802.1.1(1)
INDIRECTLY CONNECTED THREE-COMPARTMENT FOOD PREPARATION SINK

**Figure 802.1.1(2)
REFRIGERATED FOOD CASE DISCHARGING TO HUB DRAIN**

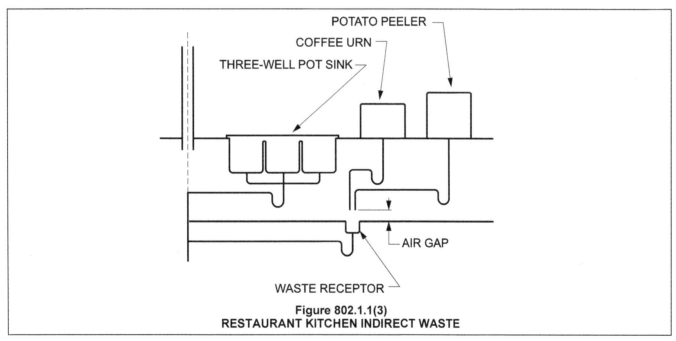

**Figure 802.1.1(3)
RESTAURANT KITCHEN INDIRECT WASTE**

contamination of food from a backup of the drainage system. Therefore, the code's intent is to protect food from bacterial contamination. Without question, the grinding chamber of a food waste grinder will always contain significant bacteria regardless of any cleaning or sanitation practices. Although, not directly stated in the code, the requirement for food waste grinders to be directly connected and the code's intent to protect food from contamination suggests that food waste grinders must not be installed on food-prep sinks.

802.1.2 Floor drains in food storage areas. Floor drains located within walk-in refrigerators or freezers in food service and food establishments shall be indirectly connected to the sanitary drainage system by means of an *air gap*. Where a floor drain is located within an area subject to freezing, the waste line serving the floor drain shall not be trapped and shall indirectly discharge into a waste receptor located outside of the area subject to freezing.

Exception: Where protected against backflow by a backwater valve, such floor drains shall be indirectly connected

INDIRECT/SPECIAL WASTE

to the sanitary drainage system by means of an *air break* or an *air gap*.

❖ This section is consistent with the intent of Section 802.1.1, which is to protect food stored in walk-in refrigerators and freezers from possible contamination caused by the sanitary drainage system. Floor drains located in these areas must be indirectly connected to the sanitary drainage system through an air gap. The exception allows for floor drains in refrigerators and freezers to connect by means of an air break. This maintains backflow protection while providing an indirect waste connection option for large food storage areas where it is impractical to elevate the cooler/freezer floor to a level necessary to achieve an air gap (see Commentary Figure 802.1.2). Section 305.6 prohibits locating traps in areas subject to freezing, and Section 802.2 requires indirect waste piping to be trapped where the length limit is exceeded.

802.1.3 Potable clear-water waste. Where devices and equipment, such as sterilizers and relief valves, discharge potable water to the building drainage system, the discharge shall be through an indirect waste pipe by means of an *air gap*.

❖ Clear-water waste includes all of the various condensate drains plus the other devices and systems listed. Clear-water wastes are not required to discharge to the sanitary drainage system. However, where they

For SI: 1 foot = 304.8 mm.

Figure 802.1.2
FLOOR DRAIN IN FOOD STORAGE FREEZER

Figure 802.1.3
INDIRECT WASTE

do, an indirect waste connection by means of an air gap is required. An air gap prevents possible cross connection between potable water source and the drain system (see Commentary Figure 802.1.3). Indirect waste pipes from potable water sources should not be connected together (manifolded) as the discharge from one source might possibly contaminate the other source of potable water. For example, the connection of two separate ice storage bin indirect waste pipes would be prohibited. Each must discharge independently to the waste receptor.

802.1.4 Swimming pools. Where waste water from swimming pools, backwash from filters and water from pool deck drains discharge to the building drainage system, the discharge shall be through an indirect waste pipe by means of an *air gap*.

❖ This section addresses the method of discharge where a swimming pool drain connects to the drain-

Figure 802.1.5
INDIRECT WASTE—CLEAR-WATER WASTE

INDIRECT/SPECIAL WASTE

age system. An indirect waste connection by means of an air gap is required.

802.1.5 Nonpotable clear-water waste. Where devices and equipment such as process tanks, filters, drips and boilers discharge nonpotable water to the building drainage system, the discharge shall be through an indirect waste pipe by means of an *air break* or an *air gap*.

❖ Where there is no possibility of contaminating the potable water (nonpotable discharge), the indirect waste may be by means of an air gap or air break. An air break is often preferred to reduce any splashing that may occur. This section does not require the devices and equipment listed to discharge to the drainage system; it only indicates the method of discharge if they do connect (see Commentary Figure 802.1.5).

802.1.6 Domestic dishwashing machines. Domestic dishwashing machines shall discharge indirectly through an *air gap* or *air break* into a waste receptor in accordance with Section 802.2, or discharge into a wye branch fitting on the tailpiece of the kitchen sink or the dishwasher connection of a food waste disposer. The waste line of a domestic dishwashing machine discharging into a kitchen sink tailpiece or food waste disposer shall connect to a deck-mounted air gap or the waste line shall rise and be securely fastened to the underside of the sink rim or counter.

❖ Dishwashing machines must discharge indirectly to the drainage system. An indirect connection by air gap

Figure 802.1.6
RESIDENTIAL DISHWASHER WASTE CONNECTIONS

Figure 802.1.7
COMMERCIAL DISHWASHER WASTE CONNECTION

or air break is required. The waste line must be looped as high as possible and securely fastened to the underside of the sink rim or countertop to minimize the potential for waste backflow into the dishwasher cabinet (see Commentary Figure 802.1.6).

802.1.7 Commercial dishwashing machines. The discharge from a commercial dishwashing machine shall be through an *air gap* or *air break* into a waste receptor in accordance with Section 802.2.

❖ The waste from a commercial dishwashing machine must discharge by means of an air gap or air break to prevent sewage from backing up into the dishwashing compartment in the event of a stoppage in the drainage system (see Commentary Figure 802.1.7). In accordance with ASSE 1004, the minimum drain connection is 1^1/$_2$ inches (38 mm) nominal size.

802.1.8 Food utensils, dishes, pots and pans sinks. Sinks, in other than dwelling units, used for the washing, rinsing or sanitizing of utensils, dishes, pots, pans or service ware used in the preparation, serving or eating of food shall discharge indirectly through an *air gap* or an *air break* to the drainage system.

❖ Because commercial kitchen pot-and-pan sinks are typically required by the local health department regulations to indirectly connect to the sanitary drainage system, the code requires these sinks to be indirectly connected. The logic for this is that, even though such sinks are said to be used only for washing dishes, pots and pans, the truth is that, many times, the sink is often used as a food-preparation sink.

802.2 Installation. Indirect waste piping shall discharge through an *air gap* or *air break* into a waste receptor. Waste receptors shall be trapped and vented and shall connect to the building drainage system. Indirect waste piping that exceeds 30 inches (762 mm) in *developed length* measured horizontally, or 54 inches (1372 mm) in total *developed length*, shall be trapped.

Exception: Where a waste receptor receives only clear-water waste and does not directly connect to a sanitary drainage system, the receptor shall not require a trap.

❖ To minimize bacterial growth and odor in indirect waste piping, a trap is required when the piping exceeds 30 inches (762 mm) in developed length measured horizontally or 54 inches (1372 mm) in total developed length. The trap requirement is intended to apply only to fixtures or devices that are open to the room atmosphere. It does not apply to an indirect

For SI: 1 inch = 25.4 mm, 1 foot = 304.8 mm.

**Figure 802.2
TRAPS FOR INDIRECT WASTE PIPES**

INDIRECT/SPECIAL WASTE

waste pipe making a direct connection to a device, such as a water heater relief valve. A trapped relief valve discharge pipe is prohibited because it creates a restriction, causes backpressure to develop and could possibly freeze, creating a blockage. The trap prescribed by this section does not require a vent because such trap does not serve the same purpose as other traps in the drainage system (see Commentary Figure 802.2).

802.2.1 Air gap. The *air gap* between the indirect waste pipe and the *flood level rim* of the waste receptor shall be not less than twice the effective opening of the indirect waste pipe.

❖ An air gap is a drainage pipe that terminates above the flood level rim of the waste receptor. Any stoppage downstream of the waste receptor will result in waste overflowing the receptor without submerging the indirect waste pipe outlet. This type of indirect waste connection offers the highest level of protection by not allowing any possibility for backsiphoning of waste or sewage (see Commentary Figure 802.2.1). An air gap is always acceptable where an air break is specified.

802.2.2 Air break. An *air break* shall be provided between the indirect waste pipe and the trap seal of the waste receptor.

❖ The indirect waste pipe is allowed to extend below the flood level rim, but it must terminate above the seal of the trap serving the waste receptor [see Commentary Figures 802.2.2(1) and 802.2.2(2)]. An air gap is always acceptable where an air break is specified.

Figure 802.2.1
AIR GAP

Figure 802.2.2(1)
AIR BREAK

INDIRECT/SPECIAL WASTE

802.3 Waste receptors. For other than hub drains that receive only clear-water waste and standpipes, a removable strainer or basket shall cover the outlet of waste receptors. Waste receptors shall not be installed in concealed spaces. Waste receptors shall not be installed in plenums, crawl spaces, attics, interstitial spaces above ceilings and below floors. Ready *access* shall be provided to waste receptors.

❖ There are no specific standards regulating waste receptors. They may be identified by various names, including open hub drain, floor sink and waste sink. A floor drain can be considered to be a waste receptor where an indirect waste pipe is directed into it or where it is fitted with a means (such as a funnel attachment) to receive an indirect waste pipe. Otherwise, a floor drain is a plumbing fixture intended to drain a floor surface and is not subject to the code requirements for waste receptors.

Floor sinks are specialized sinks designed to be installed in the floor. Manufacturers produce a special line of fixtures designed to be used as waste receptors, some of which look like kitchen sinks. A laundry sink is commonly used as a waste receptor to receive the discharge of an automatic clothes washer [see Commentary Figures 802.3(1) and 802.3(2)]. Waste receptors must be located to allow adequate ventilation and air circulation, minimize misuse, prevent a tripping or walking hazard and allow maintenance. [see Commentary Figure 802.3(3)].

**Figure 802.2.2(2)
AIR BREAK**

**Figure 802.3(1)
WASTE RECEPTORS**

INDIRECT/SPECIAL WASTE

Figure 802.3(2)
WASTE RECEPTORS

Figure 802.3(3)
PROHIBITED LOCATIONS FOR WASTE RECEPTORS

802.3.1 Size of receptors. A waste receptor shall be sized for the maximum discharge of all indirect waste pipes served by the receptor. Receptors shall be installed to prevent splashing or flooding.

❖ Sizing a waste receptor is as difficult as sizing a drainage system. There must be a determination of the probability of simultaneous discharge of the various fixtures connecting to the same waste receptor. The receptor must be capable of accepting all of the discharge without overflowing. The code does not specify a method for sizing receptors; however, it would be logical to size the drain for the receptor based on the total drainage fixture units (dfus) that discharge to the receptor. While accepting the discharge of an indirect waste pipe, the receptor is simultaneously discharging to the drainage system at a rate based on the size of its drain pipe.

Two major problems in the design and sizing of a waste receptor are splashing and flooding. Both must be minimized to avoid creating an insanitary condition. Both the shape of the waste receptor and the discharge rate of the indirect waste pipe affect the amount of splashing and flooding involved. A shallow, flat bottom receptor receiving a high rate of discharge will splash to the exterior of the receptor.

The depth and bowl shape may have to be selected to reduce splashing. Another means of reducing splashing into a waste receptor is by terminating the indirect waste pipe with a diagonally cut end. The lower section of the diagonal cut is faced to the closest exterior wall of the waste receptor. If the waste receptor receives only the discharge of emergency drains, such as a temperature and pressure relief valve, splashing is of little concern. If the receptor is used only during an emergency, not on an everyday basis, such limited use will not create any major health hazard [see Commentary Figures 802.3.1(1) and 802.3.1(2)].

A high rate of discharge also increases a waste receptor's susceptibility to flooding. The designer should pay particular attention to receptor depth, trap sizing and the free area of the drain's protective grating. As is often the case, the waste receptor trap size restricts the discharge flow rate of the fixture. A close coordination of the sizing of the trap, rate of discharge, receptor depth and drain selection will circumvent an insanitary flooding condition.

802.3.2 Hub drains. A hub drain shall be in the form of a hub or a pipe extending not less than 1 inch (25 mm) above a water-impervious floor.

❖ An open hub drain is a common method of providing a waste receptor in water-impervious floors. It eliminates the need to provide a separate fixture with a strainer. A hubbed drain pipe or plain-end section of pipe serves as the receptor for receiving indirect waste. An open hub drain is restricted to clear-water waste without solids, unless the hub drain has a strainer. The pipe or hub is required to be a minimum of 1 inch (25 mm) above the floor to prevent the drain from serving as a floor drain (see Commentary Figure 802.3.2).

802.3.3 Standpipes. Standpipes shall be individually trapped. Standpipes shall extend not less than 18 inches (457

Figure 802.3.1(1)
SHALLOW, FLAT BOTTOM WASTE RECEPTOR

Figure 802.3.1(2)
SIZE AND SHAPE OF WASTE RECEPTOR

For SI: 1 inch = 25.4 mm.

Figure 802.3.2
CLEAR-WATER WASTE RECEPTOR

**Figure 802.3.3
STANDPIPES**

mm) but not greater than 42 inches (1066 mm) above the trap weir. *Access* shall be provided to standpipes and drains for rodding.

❖ A standpipe can serve as a waste receptor. The minimum dimension for the height of the receptor is intended to provide minimal retention capacity and head pressure to prevent an overflow. This is especially necessary when the standpipe receives a high rate of discharge, such as the pumped discharge of a clothes washer (see Commentary Figure 802.3.3). The maximum height requirement is intended to control the waste flow velocity at the trap inlet. Excessive inlet velocity can promote trap siphonage.

SECTION 803
SPECIAL WASTES

803.1 Neutralizing device required for corrosive wastes. Corrosive liquids, spent acids or other harmful chemicals that destroy or injure a drain, *sewer*, soil or waste pipe, or create noxious or toxic fumes or interfere with sewage treatment processes shall not be discharged into the plumbing system without being thoroughly diluted, neutralized or treated by passing through an *approved* dilution or neutralizing device. Such devices shall be automatically provided with a sufficient supply of diluting water or neutralizing medium so as to make the contents noninjurious before discharge into the drainage system. The nature of the corrosive or harmful waste and the method of its treatment or dilution shall be *approved* prior to installation.

❖ When deleterious waste is suspected, it must be completely analyzed for any detrimental effects on the sanitary drainage system, public sewer or sewage treatment facility. A special waste system must treat any deleterious waste and reduce the threat to an acceptable level before it is discharged to the sanitary drainage system. The code does not specify what an "acceptable level" is because it varies by pipe material and type of waste. Special waste systems are associated with atypical plumbing systems, including laboratories, chemical plants, processing plants and hospitals. They may be necessary, however, when emergency shower systems are installed for use by persons exposed to harmful chemicals or nuclear material.

803.2 System design. A chemical drainage and vent system shall be designed and installed in accordance with this code. Chemical drainage and vent systems shall be completely separated from the sanitary systems. Chemical waste shall not discharge to a sanitary drainage system until such waste has been treated in accordance with Section 803.1.

❖ When a chemical drainage system is installed, the drainage and vent piping must still conform to all code requirements. The pipe sizing and installation would have to conform to the requirements of Chapters 7 and 9 and other chapters as applicable. Both the drainage and vent piping must be completely separate systems. Although it is understood that the drainage piping must be separate until the waste is treated or neutralized, it is always questionable whether a separate system is necessary for a vent. The reason for a completely separate vent system is the uncertainty of the nature of fumes and vapors in a chemical waste system. The gases in the chemical drain, waste and vent (DWV) system could be incompatible with the piping materials and gases in the normal DWV system. Because the effect of combining the systems is unknown, such venting systems are kept separate.

SECTION 804
MATERIALS, JOINTS AND CONNECTIONS

804.1 General. The materials and methods utilized for the construction and installation of indirect waste pipes and systems shall comply with the applicable provisions of Chapter 7.

❖ This section references all of the acceptable materials and methods described in Chapter 7 that are used specifically for the installation and joining of indirect waste pipes and systems.

Chapter 9: Vents

General Comments

In the most common method of venting, each trap or trapped fixture is provided with a separate or individual vent, which then connects to the building's main venting system. The minimum size of any vent is one-half of the required size of the drain pipe, but not less than $1^1/_4$ inches (32 mm) (see commentary, Section 906.2).

Other methods of venting include common venting (Section 911), wet venting (Section 912), waste stack venting (Section 913), circuit venting (Section 914), combination waste and vent (Section 915), island fixture venting (Section 916) and single stack venting (Section 917).

A common vent is a single vent pipe that vents more than one fixture. It functions as an individual vent for each fixture. Wet venting is a method of venting single or double bathroom groups or combinations thereof. One vent pipe may function as a vent for all of the fixtures connected to the wet vent. Circuit venting is the venting of a battery of up to eight fixtures with a single vent pipe. A combination waste and vent system is restricted to floor drains, sinks, drinking fountains and lavatories. The system relies on the oversizing of the drain pipe. Island fixture venting is a method of individual or common venting in which the vent is installed below the flood level rim of the fixture before rising to connect to another vent or vent terminal. A waste stack vent is a method of venting individual fixtures through a drainage stack. The oversized stack functions as a vent.

Other areas addressed in Chapter 9 include outdoor vent extension requirements (Section 904), vent terminals and frost closure (Section 903), vent grade (Section 905), trap to vent distance limitations (Section 909) and relief vents (Section 908).

Some have compared a vent system to the open/closed straw phenomenon. For example, imagine that a soda straw is inserted into a glass of water. A thumb is placed over the open end of the straw, and the straw is raised above the glass. Water remains in the straw until the thumb is removed. The claim is made that without a vent, the drainage system would not properly operate—just as the straw did not empty with the thumb over the top. (This analogy is not completely accurate because a drainage system is not flowing completely full as is the soda straw; therefore, a column of water could not be suspended in this manner.) Drainage systems are designed to flow partially filled.

The vent system reduces the pressure differentials in the drainage system, thus protecting the trap seals from positive pressures, self-siphonage and induced siphonage that may result in complete or partial loss of the seals.

To better understand venting methods and terminology, all venting methods may be conceptually categorized as either dry vents or wet vents. Vent stacks, stack vents, branch vents, relief vents, island vents and individual vents are dry vents. Wet vents, circuit vents, combination drain and vents, waste stack vents and different level common vents are versions of "wet venting," in which the vent is wetted by drainage flow.

Purpose

The purpose of venting is to protect the trap seal of each trap. The vents are designed to maintain maximum differential pressures at each trap of not more than 1 inch of water column (249 Pa). Pressure fluctuations are created by waste flow in the drainage system. Venting is not intended to provide for the circulation of air within the drainage system. If there were no traps in a drainage system, venting would not be required. The system would function adequately because it would be open to the atmosphere at the fixture connections, thereby allowing airflow.

Chapter 9 covers the requirements for vents and venting. Knowing why venting is required makes it easier to understand the intent of this chapter. Venting protects every trap against the loss of its seal. Provisions set forth in this chapter are geared toward limiting the air pressure differentials in the drainage system to a maximum of 1 inch of water column (249 Pa) above or below atmospheric pressure (i.e., positive or negative pressures)

SECTION 901
GENERAL

901.1 Scope. The provisions of this chapter shall govern the materials, design, construction and installation of vent systems.

❖ Chapter 9 describes a variety of methods to vent plumbing fixtures and traps. Some methods have been laboratory tested to determine sizing and installation requirements that provide proper venting to a drainage system. Other venting methods have been field tested, establishing a long history of satisfactory service.

901.2 Trap seal protection. The plumbing system shall be provided with a system of vent piping that will permit the admission or emission of air so that the seal of any fixture trap shall not be subjected to a pressure differential of more than 1 inch of water column (249 Pa).

❖ Protection of the trap seal means that during the normal operation of the plumbing system, the water seal remains in the trap. Each trap has a minimum of a 2-inch (51 mm) trap seal depth translating to a hydrostatic pressure equal to a 2-inch water column (498 Pa). If exposed to a 1-inch water column (249 Pa) pressure differential, a 1-inch (25 mm) water seal remains in the trap. The vent methods identified in this chapter are intended to limit the air pressure differential at trap seals to 1 inch of water column (249 Pa) or less (see commentary, Sections 1002 through 1002.5).

Air pressure fluctuations downstream of the trap caused by certain flow conditions can occur. High positive pressures could force the trap seal back into the fixture, possibly resulting in water spilling onto the floor from fixtures such as floor drains. This will also result in sewer gases being pushed through the trap seal. For example, bubbles rising through the trap seal of a water closet are an indication of excessive positive pressure on the drainage side of the trap.

When a trap seal is exposed to a lower pressure on the drainage side of the trap, the water seal will rise and flow over the trap weir and into the drainage system. This will result in a complete or partial loss of the trap water seal. The lower pressure on the drainage side of the trap (also called "negative pressure" or "partial vacuum") causes loss of trap seal that is proportional to the amount of negative pressure. Such a pressure differential must never exceed 1 inch of water column (249 Pa) [see Commentary Figures 901.2(1) and 901.2(2)].

Section 1002.4 contains additional provisions to protect a trap seal that is subject to loss by evaporation, which might occur where trap seals do not regularly receive waste.

Figure 901.2(2)
MAXIMUM TRAP SEAL RESISTANCE

For SI: 1 inch = 25.4 mm.

Figure 901.2(1)
LOSS OF TRAP SEAL

901.2.1 Venting required. Traps and trapped fixtures shall be vented in accordance with one of the venting methods specified in this chapter.

❖ This section establishes that traps and trapped fixtures must be vented. This section also indicates that the method of venting can be any one of the applicable methods described in Chapter 9. Proper application of these venting options can substantially reduce the amount of pipe and fittings while still providing proper venting. Inspectors and plumbers often overlook the opportunities afforded by the different venting methods. For example, a particular installation might not be in compliance with the code for a certain type of vent; however, that same installation might comply if it were considered as a different type of vent system or slightly modified to comply with the rules of a different type of venting system.

901.3 Chemical waste vent systems. The vent system for a chemical waste system shall be independent of the sanitary vent system and shall terminate separately through the roof to the outdoors or to an air admittance valve that complies with ASSE 1049. Air admittance valves for chemical waste systems shall be constructed of materials *approved* in accordance with Section 702.5 and shall be tested for chemical resistance in accordance with ASTM F 1412.

❖ Where a chemical waste system is installed, the drainage system must be separate from the sanitary system (see commentary, Section 803.2). The vent system is also required to be separate. The fumes and vapors in a chemical vent system may adversely affect a sanitary vent system. In accordance with Section 702.5, vent materials must be resistant to corrosion and degradation from the chemical waste and vapors involved.

For the 2012 code edition, air-admittance valves are allowed to be used on chemical waste vent systems. The air-admittance valves must comply with ASSE 1049, which requires that testing of the valve materials comply with ASTM F 1412.

901.4 Use limitations. The plumbing vent system shall not be utilized for purposes other than the venting of the plumbing system.

❖ The venting system is designed for a specific function and cannot be used for any other purpose because that might interfere with the proper operation of the plumbing system or create an extreme hazard. For example, vent piping must not serve as a drain for condensate from heating, ventilating and air-conditioning (HVAC) equipment. Such use introduces waste into piping that is not intended to convey waste and might allow sewer gases to enter the buildings via the condensate drain connection.

901.5 Tests. The vent system shall be tested in accordance with Section 312.

❖ This section requires that vent systems be tested in accordance with Section 312, which describes the tests and inspections necessary to verify a vent system performs as required by the code. Although visual inspections play a vital part in the integrity of the system, it is impossible for the code official to identify defects without testing.

901.6 Engineered systems. Engineered venting systems shall conform to the provisions of Section 919.

❖ This section requires that the design of alternative vent systems be in accordance with Section 919. Section 919 describes an alternative sizing method for individual fixture vents, which may not comply with the requirements of this chapter. Specific requirements are applicable, however, with respect to design criteria, submittals, technical data, construction documents, design approval and inspection and tests.

SECTION 902
MATERIALS

902.1 Vents. The materials and methods utilized for the construction and installation of venting systems shall comply with the applicable provisions of Section 702.

❖ This section requires the construction and installation of venting systems, which are a necessary part of the complete sanitary drainage system, to comply with Section 702, which includes requirements for materials used in the sanitary drainage system (see commentary, Section 702).

902.2 Sheet copper. Sheet copper for vent pipe flashings shall conform to ASTM B 152 and shall weigh not less than 8 ounces per square foot (2.5 kg/m^2).

❖ Sheet copper was a predominant material used as a lining in plumbing fixtures. Today, it is rarely used because of the high cost of material and labor. Often in today's roofing systems, copper is replaced with rubber or asphaltic materials because they are less costly and labor intensive. The ASTM International (ASTM) standard regulates cold-rolled tempered sheet copper, hot-rolled tempered sheet copper and annealed sheet copper. When sheet copper is used for flashing around vents, the sheeting must not weigh less than 8 ounces per square foot (2.5 kg/m^2). This provides better durability and protection for the area being penetrated by the vent. Sheet copper is still used for flashing on long-life roof systems, such as slate or tile. Copper can cause corrosion to occur when used with or contacting other metals; therefore, the designer who specifies copper flashing must use only copper or carefully isolate it from other metals.

902.3 Sheet lead. Sheet lead for vent pipe flashings shall weigh not less than 3 pounds per square foot (15 kg/m^2) for field-constructed flashings and not less than $2^1/_2$ pounds per square foot (12 kg/m^2) for prefabricated flashings.

❖ Where sheet lead is used in flashing installations around vent pipe, it must weigh at least 3 pounds per square foot (15 kg/m^2) where constructed at the job site and $2^1/_2$ pounds per square foot (12 kg/m^2) for prefabricated units.

Lead flashing is usually wrapped over the exposed

rim of the vent terminal. If the flashing is not properly formed around the internal surface of the vent pipe, it will reduce the effective cross-sectional area of the vent terminal and restrict the flow of air into and out of the drainage system. This may compromise the effectiveness of the vent system to protect trap seals and may promote frost closure at the vent terminal.

SECTION 903
VENT TERMINALS

903.1 Roof extension. Open vent pipes that extend through a roof shall be terminated not less than [**NUMBER**] inches (mm) above the roof. Where a roof is to be used for assembly or as a promenade, observation deck, sunbathing deck or similar purposes, open vent pipes shall terminate not less than 7 feet (2134 mm) above the roof.

❖ This section contains requirements for the location, installation and protection of vent terminals.

Where the roof is occupied for any purpose, such as a sunning or promenade deck, plumbing vents must extend at least 7 feet (2134 mm) above the roof to prevent harmful sewer gases from polluting the area. The sewer gases will tend to disperse into the air, rather than accumulate near the roof surface. The minimum termination height must be determined by the local jurisdiction based on its snowfall rates. The termination height should be no less than 6 inches (152 mm) above the roof to protect the terminal from being blocked by snow and to allow sufficient length of pipe for proper roof flashing.

The vent terminal through the roof must be designed to tolerate movement resulting from expansion and contraction of the piping material. This is extremely important where the vent is a piping material with a high rate of expansion and contraction, such as acrylonitrile butadiene styrene (ABS) and polyvinyl chloride (PVC) plastic pipe. Vent terminals extending higher than the distances specified in Table 308.5 for vertical pipe hanger spacing must have intermediate supports (see Commentary Figure 903.1).

903.2 Frost closure. Where the 97.5-percent value for outside design temperature is 0°F (-18°C) or less, vent extensions through a roof or wall shall be not less than 3 inches (76 mm) in diameter. Any increase in the size of the vent shall be made not less than 1 foot (305 mm) inside the thermal envelope of the building.

❖ The possibility of frost closure occurs only in areas of the country having cold climates with winter outside $97^1/_2$-percent design temperatures of 0°F (-18°C) or less. Temperatures for many cities in the United States are shown in Appendix D. For example, Pueblo, Colorado, has a winter $97^1/_2$-percent temperature of 0°F (-18°C) and would require a minimum of 3 inches (76 mm) in diameter vent.

A vent 3 inches (76 mm) or larger in size is not likely to completely close up from frost. However, vents smaller than 3 inches (76 mm) in diameter must be increased in size where frost closure is possible. Any increase must occur within the structure to protect the smaller-diameter pipe from direct exposure to the outside atmosphere. The vent pipes must increase diameter starting at not less than 1 foot (305 mm) inside the building's thermal envelope. In other words, the vent enlargement occurs inside the heated zone of the building.

See the "Degree Day and Design Temperature" table (located in Appendix D) and the ASHRAE *Fundamentals Handbook* for temperature information.

903.3 Flashings. The juncture of each vent pipe with the roof line shall be made water tight by an *approved* flashing.

For SI: 1 foot = 304.8 mm.

**Figure 903.1
VENT TERMINAL**

❖ Water that is allowed to enter between the intersections of roofing and vent piping will cause decay of the wood and degradation of other building materials; therefore, flashings, including those of materials described in Sections 902.2 and 902.3, are required at these intersections.

903.4 Prohibited use. A vent terminal shall not be used for any purpose other than a vent terminal.

❖ The use of the vent terminal to support imposed loads is prohibited.

903.5 Location of vent terminal. An open vent terminal from a drainage system shall not be located directly beneath any door, openable window, or other air intake opening of the building or of an adjacent building, and any such vent terminal shall not be within 10 feet (3048 mm) horizontally of such an opening unless it is 3 feet (914 mm) or more above the top of such opening.

❖ The vent terminal must be located away from any building air intake opening (gravity or mechanical) to reduce the possibility of sewer gases entering a building (see Commentary Figure 903.5).

903.6 Extension through the wall. Vent terminals extending through the wall shall terminate at a point not less than 10 feet (3048 mm) from a lot line and not less than 10 feet (3048 mm) above average ground level. Vent terminals shall not terminate under the overhang of a structure with soffit vents. Sidewall vent terminals shall be protected to prevent birds or rodents from entering or blocking the vent opening.

❖ Sidewall vent terminations are an alternative to roof penetrations that may result in significant cost savings and a more aesthetically pleasing installation. For example, a sidewall vent may be preferred to penetrating membrane built-up, slate or tile roofs. Such roof penetrations are difficult to make leakproof, may be expensive and are often considered unsightly.

In multistory buildings where remodeling work is being done, a sidewall vent termination can serve as a welcome alternative to running vent piping through finished or occupied stories above.

Where a sidewall vent is installed, the vent opening must be protected with a screen or louver to prevent birds from building a nest in the pipe. This also prevents rodents from entering the vent system (see Commentary Figure 903.6). Such vents must not terminate where the emissions from the vent can either cause structural damage or enter the building envelope through soffit vents. Such vents must also comply with the location requirements of Section 903.5.

903.7 Extension outside a structure. In climates where the 97.5-percent value for outside design temperature is less than 0°F (-18°C), vent pipes installed on the exterior of the structure shall be protected against freezing by insulation, heat or both.

❖ A vent pipe located outside a building increases the likelihood of frost closure in colder climates. To prevent a buildup of frost on the inside of the vent pipe, it must be protected. If provided with adequate insula-

For SI: 1 foot = 304.8 mm.

**Figure 903.5
SEPARATION OF VENT TERMINAL**

VENTS

For SI: 1 foot = 304.8 mm.

Figure 903.6
SIDEWALL VENT

tion, the heat of the vent gases and vapors may be relied upon to prevent frost from developing in the vent pipe.

SECTION 904
OUTDOOR VENT EXTENSIONS

904.1 Required vent extension. The vent system serving each *building drain* shall have not less than one vent pipe that extends to the outdoors.

❖ The intent of this section of the code is to make sure that the building drain is vented to the outdoors. The required vent connecting to the outdoors can be satisfied in a number of ways including using a: 1) soil stack vent; 2) waste stack vent; 3) individual fixture vent; 4) common vent; 5) vent stack or 6) relief vent.

The code does not specify where the required vent must be located or the required routing except that it must eventually connect to the building drain either directly or through an extension of a drain. Therefore, any vent connecting to a soil or waste stack, any vent stack, any relief vent or any fixture vent may be considered as the required vent extension to outdoors [see Section 904.1.1 and Commentary Figures 904.1(1) through (3)]. However, in no case can the required vent extension to outdoors be less than one-half of the required size of the building drain (see Sections 904.1.2 and 906.1).

The following examples illustrate the required vent extension for building drains with a soil stack:

Example A:

A 3-inch (76 mm) soil stack with a stack vent serving as the required vent extension to the outdoors, connecting to a 3-inch (76 mm) building drain, must have at least a 1^1/$_2$-inch (38 mm) stack vent [maximum of 102 fixture units served and a maximum 25-foot (7620 mm) developed length] in accordance with Section 906.1 and Table 906.1. This 1^1/$_2$-inch (38 mm) stack vent is the minimum size required to comply with this section because it is at least one-half the size of the building drain.

Example B:

A 3-inch (76 mm) soil stack with a stack vent serving as the required vent extension to the outdoors, connecting to a 4-inch (102 mm) building drain, must have at least a 2-inch (51 mm) stack vent [maximum of 102 fixture units served and a maximum 86-foot (26 213 mm) developed length] in accordance with Section 906.1 and Table 906.1. The 2-inch (51 mm) stack vent is the minimum size required for the purpose of complying with this section because the 4-inch (102 mm) building drain requires a vent extension to outdoors of not less than one-half the building drain size.

For SI: 1 inch = 25.4 mm.

Figure 904.1(1)
REDUCTION OF STACK VENT CONNECTION

VENTS

Example C:

A 1¹/₂-inch (38 mm) waste stack with a 1¹/₂-inch (38 mm) stack vent serving as the required vent extension to the outdoors, connecting to a 4-inch (102 mm) building drain, must have at least a 2-inch (51 mm) stack vent in accordance with Section 906.1. Therefore, the waste stack must also be at least 2 inches (51 mm) from the dry vent connection down to the building drain. If the waste stack with stack vent was not serving as the required vent extension to the outdoors, the stack could remain at 1¹/₂ inches (38 mm) and the stack vent 1¹/₂ inches (38 mm) [maximum of two fixture units served in accordance with Table 913.4 and a maximum 150-foot (45 720 mm) developed length in accordance with Table 906.1].

Example D:

A 3-inch (76 mm) building drain with a 1¹/₂-inch (38 mm) vented individual lavatory drain serving as the required vent extension to the outdoors must have at least a 1¹/₂-inch (38 mm) vent [up to 40-feet (12 192 mm) developed length] in accordance with Section 906.2 because the vent must be at least one-half the size of the drain that it serves. If this fixture drain with vent was not serving as the required vent extension to the outdoors, the vent size minimum would be 1¹/₄ inches (32 mm) [up to 40-feet (12 192 mm) developed length].

Example E:

A 4-inch (102 mm) building drain with a 1¹/₂-inch (38 mm) individual vented lavatory drain serving as the required vent extension to the outdoors must have at least a 2-inch (51 mm) vent [up to 40-feet (12 192 mm) developed length] in accordance with Section 906.2 because the vent must be at least one-half the size of the drain it serves. Therefore, any portion of the fixture drain downstream of the vent connection must also be at least 2 inches (51 mm). If this fixture drain with vent was not serving as the required vent extension to the outdoors, the vent size minimum would be 1¹/₄ inches (32 mm) [up to 40-feet (12 192 mm) developed length to outdoors].

904.1.1 Installation. The required vent shall be a dry vent that connects to the *building drain* or an extension of a drain that connects to the *building drain*. Such vent shall not be an island fixture vent as allowed by Section 916.

❖ A vent leading to the outdoors that connects directly to the building drain or to a drain connecting to the building drain can be used for the required vent extension. Island fixture vents cannot be used for this function because of the high potential for clogging of the vent piping that has horizontal components below the flood rim level of the island fixture (see commentary, Section 904.1).

904.1.2 Size. The required vent shall be sized in accordance with Section 906.2 based on the required size of the *building drain*.

❖ The minimum size is based on Section 906.2 unless required to have a 3-inch (76 mm) minimum diameter vent terminal as required by Section 903.2 (see commentary, Section 903.2).

904.2 Vent stack required. A vent *stack* shall be required for every drainage *stack* that has five *branch intervals* or more.

Exception: Drainage stacks installed in accordance with Section 913.

❖ If a drainage stack is five or more branch intervals in height, a vent stack is required. If the drainage stack is less than five branch intervals in height, a vent stack is not required because the pressures in the drainage stack are not likely to create an air pressure differential at the trap seals in excess of 1 inch of water column (249 Pa). Vent stacks are dry and required to connect at or near the base of the stack served to act

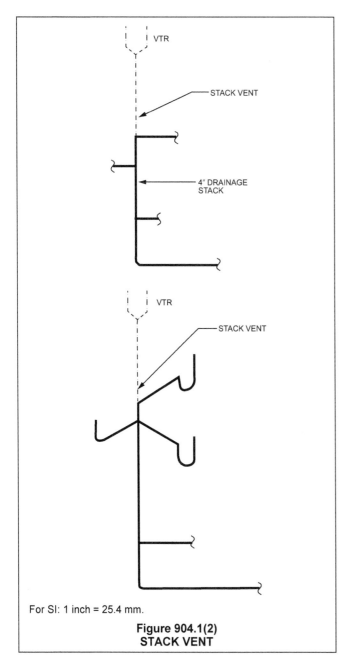

For SI: 1 inch = 25.4 mm.

Figure 904.1(2)
STACK VENT

VENTS

**Figure 904.1(3)
REQUIRED VENT TO OUTDOORS**

as a relief vent for the pressures that develop in the lowest portion of stacks (see commentary, Section 904.4). The code does not require stack vents for drainage stacks except as required by Section 913.3 because the minimum required venting of each fixture has been accomplished when the system complies with the venting methods outlined in this chapter. A stack vent is typically used as a collection point for vent pipes so that a single roof penetration can be made. Keep in mind that "vent stacks" and "stack vents" are distinct (see Section 202).

The exception allows waste stack vented systems to not to have a vent stack when the stack has five or more branch intervals connected. Waste stack vented systems (see Section 913) are already oversized to provide for adequate venting without the need for additional venting.

904.3 Vent termination. Vent *stacks* or *stack vents* shall terminate outdoors to the open air or to a stack-type air admittance valve in accordance with Section 918.

❖ No vent is allowed to terminate as an open-ended pipe in an attic, cupola, open parking structure, screened-in porch or any other space within a building envelope, regardless of how well such space is ventilated. The vent must extend to the outdoors if it is the required vent extension for the building drain. If the vent is not the required vent extension for the building drain, it may terminate outdoors or terminate with an air admittance valve in accordance with Section 918.

904.4 Vent connection at base. Vent *stacks* shall connect to the base of the drainage *stack*. The vent *stack* shall connect at or below the lowest horizontal *branch*. Where the vent *stack* connects to the *building drain*, the connection shall be located downstream of the drainage *stack* and within a distance of 10 times the diameter of the drainage *stack*.

❖ A drainage stack of five or more branch intervals must be vented at or below the lowest branch connection to relieve the positive pressure developed in the stack. The connection can be to the stack or to the building drain downstream of the stack and within a distance of 10 times the drainage stack diameter.

The size of the vent connection to the drainage system must be not less than the size required for the vent stack. Vent stacks are sized in accordance with Section 906.1, based on the size of the drainage stack at its base. Where the vent stack is connected to the building drain at the base of the drainage stack, the point of connection must be within the section of horizontal drain where the greatest air pressure fluctuations and turbulence occur (see commentary, Section 704.3 and Commentary Figure 904.4).

904.5 Vent headers. *Stack vents* and vent stacks connected into a common vent header at the top of the *stacks* and extending to the open air at one point shall be sized in accordance with the requirements of Section 906.1. The number of fixture units shall be the sum of all fixture units on all *stacks* connected thereto, and the *developed length* shall be the longest vent length from the intersection at the base of the most distant *stack* to the vent terminal in the open air, as a direct extension of one *stack*.

❖ It is quite often economically or aesthetically desirable to connect a number of vent stacks or stack vents to a common header to minimize the number of penetrations of the roof. The header must be adequately sized to both supply and relieve air in every vent stack simultaneously.

Section 906.1 requires that the header be sized to carry the total fixture unit load served by the vent stacks connected to the header. In determining the developed length, the longest vent lengths connected thereto are added to the overall length of the header.

Example:

Commentary Figure 904.5 illustrates how to determine the size of a vent header connecting four vent stacks serving soil stacks more than three branch intervals in height.

The fixture unit served by each vent stack is added as it connects to the header. Each section of the header is sized as a vent stack serving the accumulated load being vented by the individual vent stacks.

The section of the vent header between vent stack A and vent stack B is a horizontal offset of vent stack A.

For SI: 1 inch = 25.4 mm.

Figure 904.4
VENT CONNECTION AT BASE OF STACK

For SI: 1 inch = 25.4 mm, 1 foot = 304.8 mm.

Figure 904.5
VENT HEADERS

Therefore, it is the same size as vent stack A. The developed length for vent stack A is 240 feet (220 feet + 20 feet) (73 152 mm). The number of drainage fixture units (dfu) served by vent stack A is 200.

The total fixture unit load served by the vent header between stacks B and C is 350 drainage fixture unit (dfu) (200 dfu + 150 dfu). The section of the header between vent stacks B and C is sized based on the accumulated load of drainage stacks A and B and the longest developed length of either stack A or B. The developed length is the longest vent length from the base of the most distant stack (vent stack A) to the vent terminal in the open air. In this example, the developed length is 220 feet + 20 feet + 20 feet + 20 feet + 20 feet = 300 feet (91 440 mm).

From Table 906.1, the stack and the vent must be 4 inches (102 mm), which can serve up to 540 fixture units at a maximum developed length of 580 feet (176.8 m); therefore, the size required for the vent stack header is 4 inches (102 mm).

The total fixture unit load served by the vent header between vent stacks C and D is 600 dfu, which exceeds 540 dfu listed as the maximum total for a 4-inch (102 mm) drainage stack; therefore, a 5-inch (127 mm) drainage stack and vent would be required, which may serve up to 1,400 dfu at a maximum developed length of 590 feet (179.8 m). Thus, the required size of the vent stack header is 5 inches (127 mm).

The total fixture unit load between vent stack D and the termination is 730 dfu. This value is still within the capacity of a 5-inch (127 mm) drainage stack and vent stack; therefore, the size of the vent stack header between stack D and the termination is still 5 inches (127 mm).

SECTION 905
VENT CONNECTIONS AND GRADES

905.1 Connection. Individual, *branch* and circuit vents shall connect to a vent *stack*, *stack vent*, air admittance valve or extend to the open air.

❖ This section further emphasizes the requirement that vents are to extend to the outdoors except where allowed to terminate with an air admittance valve. This might be accomplished by connecting to a vent stack or a stack vent that extends outdoors and terminates to the open air (see commentary, Section 904.3) or by extending these vents independently to the outdoors. This prevents sewer gases from escaping into the building. The use of air admittance valves must comply with the provisions of Section 918.

905.2 Grade. Vent and *branch* vent pipes shall be so graded and connected as to drain back to the drainage pipe by gravity.

❖ Vents convey water vapor from the drainage system that might condense in the vent piping. Also, rainwater can enter a vent system at the vent termination; therefore, the vent must be graded (sloped) to the drainage system, thus preventing any accumulation of condensate or rainwater. The slope of a vent pipe does not affect the movement of air inside the pipe. It makes no

Figure 905.2
GRADE OF VENTS

difference whether the vent piping is sloped to drain back to the drain piping served by the vent or the vent piping slopes to drain to a vent stack or stack vent, provided that water will not stand in any portion of the venting system (see Commentary Figure 905.2).

Vent piping must be supported according to Table 308.5 to maintain slope and to prevent sags where condensate could collect and restrict airflow.

905.3 Vent connection to drainage system. Every dry vent connecting to a horizontal drain shall connect above the centerline of the horizontal drain pipe.

❖ When drainage enters a pipe, it is naturally assumed that the liquid proceeds down the pipe in the direction of flow. Being a liquid, however, the drainage seeks its own level and may move against the direction of flow before reversing and draining down the pipe. When this occurs, the solids may drop out of suspension to the bottom of the pipe. In a drain, the solids will again move down the pipe with the next discharge of liquid into the drain.

If the horizontal pipe serves only as a dry vent, the solids have no possibility of being brought back into suspension and, over a period of time, the horizontal vent pipe may be completely obstructed by waste. To avoid these blockages, the vent must connect to horizontal drains above the centerline (see Commentary Figure 905.3). The intent is to prevent waste from entering a dry vent by connecting the vent above the flow line of the drain. Such connections will result in vent piping that forms an angle of 45-degrees (0.79 rad) or more with the horizontal.

905.4 Vertical rise of vent. Every dry vent shall rise vertically to a point not less than 6 inches (152 mm) above the *flood level rim* of the highest trap or trapped fixture being vented.

Exception: Vents for interceptors located outdoors.

❖ This code provision is very important for the proper design, longevity and maintenance-free operation of a plumbing system. Surprisingly, this is also one of the most frequently violated code provisions.

The intent of this section is to prohibit horizontal vent piping from occurring where it will be subjected to waste flow. A vertical rise of the vent pipe substantially reduces the possibility of it having a blockage or stoppage. Most pipe stoppages occur in horizontal pipes. The vertical rise of the vent to a point above the flood level rim is necessary to protect the vent from a backup of waste in the event of blockage in the drainage system. A blockage causes the drainage to rise in the vent pipe and in the plumbing fixture. Any drainage system backup will flow over the fixture's flood level rim, thereby preventing it from entering the horizontal vent pipe.

A stoppage in the drain may be easily cleared by rodding through a cleanout; however, this is not the case with a vent blockage, which is seldom, if ever, accessible (see Commentary Figure 905.4).

A vent blockage will rarely produce any noticeable or recognizable symptoms of a problem; therefore, extraordinary precautions are taken to protect vents. Note that a drain obstruction will always be apparent, but a vent obstruction will almost never be apparent to the building occupants. It is often argued that horizontal vent runs below the fixture rim are necessary in order to place the vent connection within the distance limits of Section 909.1. This argument may be valid if the code specified individual fixture vents as the only venting option; however, this is not the case. The code allows many venting options; thus, there is always an option to avoid the installation of horizontal vents below the fixture flood rim.

The exception provides a measure of design flexibility and acknowledges where a gravity grease interceptor is located in an open parking area, it is impossible to install a vent that rises 6 inches (152 mm) above the top of the interceptor.

905.5 Height above fixtures. A connection between a vent pipe and a vent *stack* or *stack vent* shall be made at not less than 6 inches (152 mm) above the *flood level rim* of the highest fixture served by the vent. Horizontal vent pipes forming *branch* vents, relief vents or loop vents shall be located not less than 6 inches (152 mm) above the *flood level rim* of the highest fixture served.

❖ This section expresses the same concern as the previous section. In the event of drain stoppage, waste

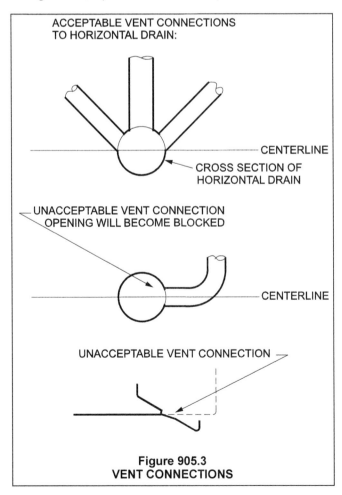

Figure 905.3
VENT CONNECTIONS

could rise into the vent piping. Over a period of time, the vent piping may become blocked with solids that have settled out of the waste liquids. Additionally, individual, common or branch vents that improperly connect to vent stacks or stack vents may become drains if the fixture drain or drains served are blocked or restricted. It is entirely possible for a vent to serve unintentionally as a drain, and this condition may go unnoticed for a long period of time [see Commentary Figures 905.5(1) and 905.5(2)].

Figure 905.4
VERTICAL RISE OF VENT

VENTS

Figure 905.5(1)
IMPROPERLY CONNECTED VENT SERVING A DRAIN

For SI: 1 inch = 25.4 mm.

Figure 905.5(2)
PROPERLY CONNECTED VENT SERVING A DRAIN

For SI: 1 inch = 25.4 mm.

Figure 905.5(3)
HORIZONTAL BRANCH VENT ELEVATION

A properly installed horizontal dry vent pipe will never contain waste liquids because the fixture or fixtures served will overflow before the waste can enter the vent piping [see Commentary Figure 905.5(3) and commentary, Section 905.4].

905.6 Vent for future fixtures. Where the drainage piping has been roughed-in for future fixtures, a rough-in connection for a vent shall be installed. The vent size shall be not less than one-half the diameter of the rough-in drain to be served.

The vent rough-in shall connect to the vent system, or shall be vented by other means as provided for in this chapter. The connection shall be identified to indicate that it is a vent.

❖ Section 710.2 addresses the installation of drainage pipe for future plumbing fixtures. Where future fixture rough-in piping occurs, vent piping must also be installed. This section does not necessarily require complete connection of the rough-in vent to the future fixture drainage pipe. The vent pipe is simply roughed-in in a manner to remain accessible for future connection when plumbing fixtures are installed. The vent rough-in must be tied in to the vent system, as required by Section 905.1, or must extend to a vent terminal in the outdoors.

The vent rough-in requirement of this section is intended to make sure that a venting means will be available in advance of the future fixture installation. It also eliminates the cost and hardship associated with the installation of vent piping at the time fixtures are installed. It is common to find that fixtures have been installed without vents because a vent was not provided for future use.

Note that a future fixture rough-in may be served by a wet vent, common vent, waste stack vent, circuit vent, combination drain and vent or an air admittance valve; therefore, a dry vent rough-in may not be required. Venting configurations must be sized and installed to accommodate future fixture installations.

VENTS

Where the installation of a future fixture will require making a connection to the vent rough-in, such connection must be identified as a vent so that the purpose of the original installation is evident when the future fixture is installed.

SECTION 906
VENT PIPE SIZING

906.1 Size of stack vents and vent stacks. The minimum required diameter of *stack vents* and vent *stacks* shall be determined from the *developed length* and the total of *drainage fixture units* connected thereto in accordance with Table 906.1, but in no case shall the diameter be less than one-half the diameter of the drain served or less than $1^1/_4$ inches (32 mm).

❖ The sizing criteria for stack vents and vent stacks are based on three variables: the developed length of the vent, the size of the stack served by the vent and the total dfu connected to the stack.

The vent is sized in accordance with Table 906.1, but in no case is the vent pipe diameter to be less than one-half the required diameter of the drain served or less than $1^1/_4$ inches (32 mm).

The developed length of the vent pipe must also be considered when sizing the vent system. The flow of air that a given pipe size can convey with a limited friction loss depends on the pipe length. The longer the pipe, the greater the effect friction will have on the flow of air. If the resistance to airflow is too high, the pressure drop will be excessive, and the vent will be unable to prevent air pressure differentials from exceeding 1 inch of water column (249 Pa). For example, if the pressure loss in a vent caused by pipe friction exceeds 2 inches of water column (498 Pa), pressures in the drainage system would exceed 2 inches of water column (498 Pa) before any air would move through the vent to neutralize pressure fluctuations. The vent pipe is often increased in size to compensate for this effect. Refer to the description of "Developed length" in Section 906.3. Commentary Figures 906.1(1), 906.1(2) and 906.1(3) contain examples for proper sizing of vents.

TABLE 906.1. See page 9-16.

This table contains the requirements for sizing the piping serving as a stack vent or a vent stack. Included in this table are the size of the soil or waste stack to be vented, the number of (drainage) fixture units connected to the drain line being vented and the developed length of vent pipe. Since the sizing limitations are in one table, the user can easily determine that the requirements for stack vents and vent stacks have been satisfied.

906.2 Vents other than stack vents or vent stacks. The diameter of individual vents, *branch* vents, circuit vents and relief vents shall be not less than one-half the required diameter of the drain served. The required size of the drain shall be determined in accordance with Table 710.1(2). Vent pipes shall not be less than $1^1/_4$ inches (32 mm) in diameter. Vents exceeding 40 feet (12 192 mm) in *developed length* shall be increased by one nominal pipe size for the entire *developed length* of the vent pipe. Relief vents for soil and waste *stacks* in buildings having more than 10 *branch intervals* shall be sized in accordance with Section 908.2.

❖ This section contains the sizing criteria for dry portions of a conventional venting system, including individual vents, branch vents, circuit vents, relief vents, common vents and dry vent connections to wet vents. The size of the vent is based on two variables: the required size of the drain served by the vent and the developed length of the vent.

The sizing of vent pipes other than vent stacks and stack vents is very similar to the method used to size the vent stacks and the stack vents discussed in Section 906.1. The drainage piping is sized in accordance with Table 710.1(2) based on the drainage fixture load determined by Tables 709.1 and 709.2 and Sections 709.3 and 709.4. The vent size is then determined to be a minimum of one-half the required diameter of the drain served, but not less than $1^1/_4$ inches (32 mm).

Note that this section ties vent sizing to the drain sizing criteria of Table 710.1(2) for horizontal fixture branches, which results in more conservative vent sizing. For example, a circuit-vented branch serving 30 dfu may consist of a 3-inch (76 mm) drain and a $1^1/_2$-inch (38 mm) circuit vent in accordance with Table 710.1(1), but would have to consist of a 4-inch (102 mm) drain and a 2-inch (51 mm) circuit vent in accordance with Table 710.1(2). The developed length of the vent pipe must also be considered when sizing the vent system. The amount of air that a given pipe size can convey depends on its length. The effect of friction

EXAMPLE 1: TO DETERMINE THE MINIMUM VENT STACK SIZE FROM TABLE 906.1, USE THE ROW FOR A 4-INCH SOIL OR WASTE STACK AND A FIXTURE UNIT LOAD OF 540:

DRAINAGE STACK = 4″
FIXTURE UNIT LOAD/BRANCH = 35 DFU
FIXTURE UNIT LOAD = 455 DFU
DEVELOPED LENGTH = 140′
MINIMUM VENT DIAMETER = 3″

For SI: 1 inch = 25.4 mm, 1 foot = 304.8 mm.

Figure 906.1(1)
VENT STACK SIZING

increases as pipe length increases and has a greater effect on the flow of air. If the friction increases to such a level that the vent is not able to prevent pressure differentials from exceeding 1 inch of water column (249 Pa), the vent pipe size must be increased to compensate for this effect. Refer to the definition of "Developed length" in Section 906.3. Commentary Figures 906.2(1) through 906.2(4) contain several examples for proper sizing of vents. This section makes the assumption that a developed length exceeding 40 feet (12 192 mm) will result in excessive friction losses; therefore, such vents must be increased by one nominal pipe size. A vent increased in size in accordance with this section would be allowed to have an unlimited developed length; however, there will always be some length at which the friction loss would again be excessive. It is assumed that the natural constraints of construction practice will prevent vent lengths that would exceed the friction losses of a vent increased by one nominal pipe size. Note that the increase in pipe size caused by the 40-foot (12 192 mm) length limitation will increase the size of the entire developed length of the vent, not only the portion that extends beyond 40 feet (12 192 mm).

906.3 Developed length. The *developed length* of individual, *branch*, circuit and relief vents shall be measured from the farthest point of vent connection to the drainage system to the point of connection to the vent *stack, stack vent* or termination outside of the building.

❖ Section 906.2 requires that a vent pipe be increased one nominal pipe size when its developed length exceeds 40 feet (12 192 mm). This section establishes how to determine the developed length for individual, branch, circuit and relief vents [see Commentary Figures 906.3(1) and 906.3(2)].

For SI: 1 inch = 25.4 mm, 1 foot = 304.8 mm.

**Figure 906.1(2)
VENT STACK SIZING**

For SI: 1 inch = 25.4 mm, 1 foot = 304.8 mm.

**Figure 906.1(3)
STACK VENT SIZING**

VENTS

TABLE 906.1
SIZE AND DEVELOPED LENGTH OF STACK VENTS AND VENT STACKS

DIAMETER OF SOIL OR WASTE STACK (inches)	TOTAL FIXTURE UNITS BEING VENTED (dfu)	MAXIMUM DEVELOPED LENGTH OF VENT (feet)[a] DIAMETER OF VENT (inches)									
		1¼	1½	2	2½	3	4	5	6	8	10	12
1¼	2	30	—	—	—	—	—	—	—	—	—	—
1½	8	50	150	—	—	—	—	—	—	—	—	—
1½	10	30	100	—	—	—	—	—	—	—	—	—
2	12	—	75	200	—	—	—	—	—	—	—	—
2	20	30	50	150	—	—	—	—	—	—	—	—
2½	42	26	30	100	300	—	—	—	—	—	—	—
3	10	—	42	150	360	1,040	—	—	—	—	—	—
3	21	—	32	110	270	810	—	—	—	—	—	—
3	53	—	27	94	230	680	—	—	—	—	—	—
3	102	—	25	86	210	620	—	—	—	—	—	—
4	43	—	—	35	85	250	980	—	—	—	—	—
4	140	—	—	27	65	200	750	—	—	—	—	—
4	320	—	—	23	55	170	640	—	—	—	—	—
4	540	—	—	21	50	150	580	—	—	—	—	—
5	190	—	—	—	28	82	320	990	—	—	—	—
5	490	—	—	—	21	63	250	760	—	—	—	—
5	940	—	—	—	18	53	210	670	—	—	—	—
5	1,400	—	—	—	16	49	190	590	—	—	—	—
6	500	—	—	—	—	33	130	400	1,000	—	—	—
6	1,100	—	—	—	—	26	100	310	780	—	—	—
6	2,000	—	—	—	—	22	84	260	660	—	—	—
6	2,900	—	—	—	—	20	77	240	600	—	—	—
8	1,800	—	—	—	—	—	31	95	240	940	—	—
8	3,400	—	—	—	—	—	24	73	190	729	—	—
8	5,600	—	—	—	—	—	20	62	160	610	—	—
8	7,600	—	—	—	—	—	18	56	140	560	—	—
10	4,000	—	—	—	—	—	—	31	78	310	960	—
10	7,200	—	—	—	—	—	—	24	60	240	740	—
10	11,000	—	—	—	—	—	—	20	51	200	630	—
10	15,000	—	—	—	—	—	—	18	46	180	571	—
12	7,300	—	—	—	—	—	—	—	31	120	380	940
12	13,000	—	—	—	—	—	—	—	24	94	300	720
12	20,000	—	—	—	—	—	—	—	20	79	250	610
12	26,000	—	—	—	—	—	—	—	18	72	230	500
15	15,000	—	—	—	—	—	—	—	—	40	130	310
15	25,000	—	—	—	—	—	—	—	—	31	96	240
15	38,000	—	—	—	—	—	—	—	—	26	81	200
15	50,000	—	—	—	—	—	—	—	—	24	74	180

For SI: 1 inch = 25.4 mm, 1 foot = 304.8 mm.

a. The *developed length* shall be measured from the vent connection to the open air.

The developed length of a vent is measured from the farthest point of connection to the drainage system to a vent stack, stack vent or termination outdoors. Where connecting to a vent stack or stack vent, either one must be sized to supply adequate air to each vent connection. For example, the circumstance may arise where the size of a stack vent or vent stack would have to be increased to serve a connected, individual, branch, circuit or relief vent. Consideration must be given to the fact that the developed length of a stack vent will also contribute to the overall friction loss of any vent connected to the stack vent. For example, a 2-inch (51 mm) branch vent connecting to a 2-inch (51 mm) stack vent will have an overall friction loss that is based on the developed length of the branch vent plus the developed length of the stack vent.

If a vent does not connect to a vent stack or stack vent, the developed length is measured to the point of termination in the outdoor air.

VENTS

For SI: 1 inch = 25.4 mm, 1 foot = 304.8 mm.

**Figure 906.2(1)
VENT SIZING**

NOTE: ALL VENTS SHOWN LESS THAN 40'-0" IN DEVELOPED LENGTH

SECTION	LAV SOURCE OF WET VENT	SHOWER SOURCE OF WET VENT
A-B	1¹/₂"	1¹/₄"
B-D	2"	2"
C-F	1¹/₂"	1¹/₂"
B-F	1¹/₂"	2"
A-E	1¹/₂"	1¹/₄"
E-F	1¹/₄"	1¹/₂"

For SI: 1 inch = 25.4 mm, 1 foot = 304.8 mm.

**Figure 906.2(2)
VENT SIZING**

906.4 Multiple branch vents. Where multiple *branch* vents are connected to a common *branch* vent, the common *branch* vent shall be sized in accordance with this section based on the size of the common horizontal drainage *branch* that is or would be required to serve the total *drainage fixture unit* load being vented.

❖ A common branch vent with multiple branch vent connections must be sized as though it were serving a branch drain sized to carry the total fixture load of all drains served by the branch vents.

Example:

Determine the proper sizes for the vent system shown in Commentary Figure 906.4. Note that vent developed lengths are assumed to be less than 40 feet (12 192 mm) up to point Z.

The total dfu load of the horizontal drain branches served by branch vents A, B and C at point W is 16 dfu. In accordance with Table 710.1(2), the minimum required size of a horizontal drain branch with that load is 3 inches (76 mm); therefore, the size of the common branch vent between points W and X is 1¹/₂ inches (38 mm).

The horizontal drain branch served by the common branch vent at point X is 19 dfu. Table 710.1(2) requires the horizontal branch drain to be 3 inches (76 mm). The common branch vent between points X and Y is then required to be 1¹/₂ inches (38 mm). Because the horizontal branch drain size is 3 inches (76 mm), the common branch vent between points Y and Z must also be 1¹/₂ inches (38 mm) in diameter. Because the dfu load of the horizontal drain branch served by the common branch vent at point Z is 22 dfu, the vertical vent at point Z must be at least 2 inches (51 mm) in diameter. If the developed length of the vertical vent from point Z to the vent terminal plus the developed length of any branch vent exceeds 40 feet (12 192 mm), the vertical vent and such branch vent must be increased in diameter by one pipe size.

VENTS

For SI: 1 inch = 25.4 mm, 1 foot = 304.8 mm.

**Figure 906.2(3)
VENT SIZING**

For SI: 1 inch = 25.4, 1 foot = 304.8 mm.

**Figure 906.2(4)
VENT SIZING**

VENTS

**Figure 906.3(1)
DEVELOPED LENGTH OF VENTS**

906.5 Sump vents. Sump vent sizes shall be determined in accordance with Sections 906.5.1 and 906.5.2.

❖ The following sections specify the method of sizing sump vents and sewage ejector relief pipes.

906.5.1 Sewage pumps and sewage ejectors other than pneumatic. Drainage piping below *sewer* level shall be vented in the same manner as that of a gravity system. Building sump vent sizes for sumps with sewage pumps or sewage ejectors, other than pneumatic, shall be determined in accordance with Table 906.5.1.

❖ Sumps receiving sanitary drainage must be vented. Where a sewage pump is used, the vent is allowed to connect to the building venting system. Table 906.5.1 specifies the minimum vent pipe size based on the discharge capacity of the pump and the developed length of vent pipe. The sump vent must be a minimum of $1^1/_4$ inches (32 mm) (see Commentary Figure 906.5.1).

Building subdrain systems, including their venting systems, are designed no differently than gravity drainage systems, except that a sewage pump or ejector is required. The vents for both gravity and subdrain systems may be combined if pneumatic ejector vents are independently terminated to the outdoors (see Section 906.5.2).

906.5.2 Pneumatic sewage ejectors. The air pressure relief pipe from a pneumatic sewage ejector shall be connected to an independent vent *stack* terminating as required for vent extensions through the roof. The relief pipe shall be sized to relieve air pressure inside the ejector to atmospheric pressure, but shall be not less than $1^1/_4$ inches (32 mm) in size.

❖ Pneumatic sewage ejectors operate at high air pressures, forcing the sewage from the ejector sump to the gravity drainage system. An independent vent is necessary to relieve the pressure in the sump at the com-

**Figure 906.3(2)
DEVELOPED LENGTH OF VENTS**

VENTS

**Figure 906.4
COMMON BRANCH VENT**

pletion of the ejector cycle. If such a vent is connected to the building venting system, tremendous pressure differentials would be created, disrupting the normal operation of the drainage system.

**Figure 906.5.1
SUMP VENT**

SECTION 907
VENTS FOR STACK OFFSETS

907.1 Vent for horizontal offset of drainage stack. Horizontal offsets of drainage *stacks* shall be vented where five or more *branch intervals* are located above the offset. The offset shall be vented by venting the upper section of the drainage *stack* and the lower section of the drainage *stack*.

❖ This section requires venting offsets to reduce air pressure differentials. Offsets involve changes in the direction of flow from true vertical to greater than 45 degrees (0.79 rad) from vertical. Such changes upset the nonturbulent annular flow in the stack, thus creating turbulent flow with the possibility for full bore flow in sections of the offset. Significant pressure fluctuations may result, which are stabilized by the offset vents. Refer to Section 704.3 for information on hydraulic jump.

An offset is vented to reduce the air pressure differentials that occur in the drainage stack and offset (see the commentary to Section 704.3 for flow in an offset). Venting is necessary when the drainage stack above the offset is five or more branch intervals in height.

An offset in a drainage stack is vented as two separate stacks with a vent connection for both the upper stack and the lower stack (see commentary, Sections 907.2 and 907.3). The offset vents must be no less than the size of the vent stack, which is determined in accordance with Section 906.

TABLE 906.5.1. See page 9-21.

The vent for a sump provides a supply of air into the sump while the pumping equipment is operating. The supply of air prevents a pressure differential from occurring on the drainage supply pipe.

The sump vent also allows air to be displaced when drainage is filling the sump and air to be admitted when the sump is emptied. Because the air requirements are minimal, the resulting pipe size is small.

VENTS

TABLE 906.5.1
SIZE AND LENGTH OF SUMP VENTS

DISCHARGE CAPACITY OF PUMP (gpm)	MAXIMUM DEVELOPED LENGTH OF VENT (feet)[a]					
	Diameter of vent (inches)					
	1 1/4	1 1/2	2	2 1/2	3	4
10	No limit[b]	No limit	No limit	No limit	No limit	No limit
20	270	No limit	No limit	No limit	No limit	No limit
40	72	160	No limit	No limit	No limit	No limit
60	31	75	270	No limit	No limit	No limit
80	16	41	150	380	No limit	No limit
100	10[c]	25	97	250	No limit	No limit
150	Not permitted	10[c]	44	110	370	No limit
200	Not permitted	Not permitted	20	60	210	No limit
250	Not permitted	Not permitted	10	36	132	No limit
300	Not permitted	Not permitted	10[c]	22	88	380
400	Not permitted	Not permitted	Not permitted	10[c]	44	210
500	Not permitted	Not permitted	Not permitted	Not permitted	24	130

For SI: 1 inch = 25.4 mm, 1 foot = 304.8 mm, 1 gallon per minute = 3.785 L/m.

a. *Developed length* plus an appropriate allowance for entrance losses and friction due to fittings, changes in direction and diameter. Suggested allowances shall be obtained from NBS Monograph 31 or other *approved* sources. An allowance of 50 percent of the *developed length* shall be assumed if a more precise value is not available.
b. Actual values greater than 500 feet.
c. Less than 10 feet.

The developed length of a sump vent is measured from the sump connection to a vent stack or stack vent connection. If the sump vent does not connect to a vent stack, the developed length is measured to the point of termination outside. In accordance with Note a, the overall (equivalent) length of a vent is the actual developed length plus an allowance for the equivalent length of fittings. For example, using the rule-of-thumb 50-percent allowance, a sump vent with an actual developed length of 90 feet (27 432 mm) would have an overall developed length of 135 feet (41 148 mm) because of the addition of the equivalent lengths of fittings.

907.2 Upper section. The upper section of the drainage *stack* shall be vented as a separate *stack* with a vent *stack* connection installed in accordance with Section 904.4. The offset shall be considered the base of the *stack*.

❖ The vent for the upper section of the stack must connect to the stack in accordance with Section 904.4. The venting would be equivalent to venting the upper section as if it were a separate stack. If the vent stack for the upper section does not connect to the vent stack for the lower section (see commentary, Section 907.3), the vent need be sized only for the upper section (see Commentary Figure 907.2).

Figure 907.2
VENT CONNECTIONS FOR HORIZONTAL OFFSET

907.3 Lower section. The lower section of the drainage *stack* shall be vented by a yoke vent connecting between the offset and the next lower horizontal *branch*. The yoke vent connection shall be permitted to be a vertical extension of the drainage *stack*. The size of the yoke vent and connection shall be a minimum of the size required for the vent *stack* of the drainage *stack*.

❖ The vent for the lower section of the stack may connect as an extension of the drainage stack (similar to a stack vent) or as a yoke vent. A yoke vent is an upright (vertical) wye connection to the vertical stack. The vent system for the lower section must be sized for the total load on the drainage stack. The size of the vent connection must not be less than the size of the vent stack that serves the drainage stack (see Commentary Figure 907.2).

SECTION 908
RELIEF VENTS—STACKS OF
MORE THAN 10 BRANCH INTERVALS

908.1 Where required. Soil and waste *stacks* in buildings having more than 10 *branch intervals* shall be provided with a relief vent at each tenth interval installed, beginning with the top floor.

❖ This section requires a method of relieving pressure conditions in drainage stacks more than 10 branch intervals in height.

The flow in a drainage stack creates both negative and positive pressures. The vent system is designed to equalize the air pressures at the trap seal. A relief vent is required to assist the fixture venting and the venting at the base of the stack by providing midpoint connections to the stack. The relief vent prevents excessive pressure from being created by emitting or admitting air at specified points within the stack. A relief vent must be located every 10 branch intervals, measured from the highest horizontal drainage branch, which is typically the top floor. The interval breaks are then calculated downward to the base of the stack (see Commentary Figure 908.1).

908.2 Size and connection. The size of the relief vent shall be equal to the size of the vent *stack* to which it connects. The lower end of each relief vent shall connect to the soil or waste *stack* through a wye below the horizontal *branch* serving the floor, and the upper end shall connect to the vent *stack* through a wye not less than 3 feet (914 mm) above the floor.

❖ This section requires the size of the relief vent to be at least equal to the size of the vent stack to which it connects. This is considered the minimum necessary to provide sufficient venting capacity in both the relief vent and the stack vent to which it is connected.

The lower end of the relief vent is connected to the soil or waste stack below the level of the horizontal branch that serves the floor level within the branch interval required to have the relief vent. The location of this connection is intended to allow waste that might get into the relief vent, including condensation, to reach a waste line. This connection is made using a wye fitting installed as a drainage fitting in order to not impair the flow. The upper connection of the relief vent is made to the vent stack and is to be located a minimum of 3 feet (914 mm) above the floor level of the same horizontal branch. This connection is made using a wye fitting installed in an inverted position. The 3-foot (914 mm) minimum height required is a common theme in this chapter and is intended to prevent waste flow from entering the vent stack.

**Figure 908.1
RELIEF VENT CONNECTION**

VENTS

SECTION 909
FIXTURE VENTS

909.1 Distance of trap from vent. Each fixture trap shall have a protecting vent located so that the slope and the *developed length* in the *fixture drain* from the trap weir to the vent fitting are within the requirements set forth in Table 909.1.

Exception: The *developed length* of the *fixture drain* from the trap weir to the vent fitting for self-siphoning fixtures, such as water closets, shall not be limited.

❖ This section contains the location and installation requirements for vent pipes serving as fixture vents. The distance from every trap to its vent is limited to reduce the possibility of the trap self-siphoning. Self-siphoning of a trap is the siphoning caused by the discharge from the fixture the trap serves. The intent is to locate the vent within the hydraulic gradient of the fixture drain so as to prevent the opening to the vent from being occluded by waste flow [see commentary, Section 909.2 and Commentary Figure 909.1(1)].

For SI: 1 inch = 25.4 mm.

Figure 909.1(1)
DISTANCE FROM TRAP TO VENT

Self-siphoning fixtures, such as water closets and siphon action urinals, can have an unlimited distance from the trap weir to the vent. The maximum distance between the trap weir and the vent is to protect the trap against self-siphoning. Fixtures that depend on siphon action for proper operation will operate properly no matter how far the vent is located from the fixture.

Studies have been done addressing the trap-to-vent distance for a water closet. Because the water closet relies on self-siphonage to operate properly, and the trap is resealed after each use, there is no need for limiting the distance from the water closet to a vent. The water closet fixture drain must be vented in all cases, but limiting the distance between the vent connection and the water closet appears to serve no purpose. A water closet or other self-siphoning fixture drain can have a total fall greater than the diameter of the fixture drain. This is reinforced by Section 909.2.

Commentary Figure 909.1(2) illustrates the slope required to achieve the maximum length of a 1$^1/_2$-inch (38 mm) fixture drain between the trap and the vent serving the trap. For shorter distances between the trap weir, the fixture drain could be sloped greater that $^1/_4$ inch per foot (6 mm per 300 mm) as long as the elevation of the trap weir is not higher than the elevation of the top of the vent opening. Because some drainage fittings might not accommodate the connection of a fixture drain having a significant slope, it would be prudent to set all fixture drain pipes from the trap weir to the vent at a slope of $^1/_4$ inch per foot (6 mm per 300 mm). Note that the illustration shows the measurement of the pipe length to be from the bottom of each fitting's hub and not necessarily from the trap weir to the top of the opening of the fitting in the stack. There are two reasons for this: 1) the depth of flow slightly decreases as the liquid flows towards the vent opening because the flow velocity increases slightly due to the slope of the fixture drain and 2) the indicated points are easily identified in the field whereas the exact points of the trap weir and the vent opening locations are difficult to determine.

TABLE 909.1
MAXIMUM DISTANCE OF FIXTURE TRAP FROM VENT

SIZE OF TRAP (inches)	SLOPE (inch per foot)	DISTANCE FROM TRAP (feet)
1$^1/_4$	$^1/_4$	5
1$^1/_2$	$^1/_4$	6
2	$^1/_4$	8
3	$^1/_8$	12
4	$^1/_8$	16

For SI: 1 inch = 25.4 mm, 1 foot = 304.8 mm, 1 inch per foot = 83.3 mm/m.

❖ The distances listed in Table 909.1 are intended to prevent the weir of the trap from being located above the highest inlet to the vent. As the total change in elevation of a fixture drain (resulting from the fixture drain's slope) approaches the inside diameter of the drain, the weir of the trap approaches the highest level of the vent connection and self-siphoning is more likely.

When the trap weir rises above the highest inlet of the vent opening, it is commonly referred to as either a full or half S trap. This slang expression is derived from the concept that such a piping arrangement creates a self-siphoning potential similar to that of an S trap [see Commentary Figures 909.1(1) and 909.1(2)]. The intent of Section 909.2 is to prevent creating an S-trap effect, which increases the possibility of self-siphonage.

For SI: 1 inch = 25.4 mm.

**Figure 909.1(2)
MAXIMUM FALL OF FIXTURE DRAIN BETWEEN TRAP AND VENT**

**Figure 909.1(3)
TRAP TO VENT DISTANCE FOR WATER CLOSETS**

909.2 Venting of fixture drains. The total fall in a *fixture drain* due to pipe slope shall not exceed the diameter of the *fixture drain*, nor shall the vent connection to a *fixture drain*, except for water closets, be below the weir of the trap.

❖ This section reinforces the requirement in Section 909.1. The trap weir must stay below the highest inlet to the vent except for fixtures with integral traps (see commentary, Section 909.1). The fixture drain cannot offset vertically or drop vertically, other than the required slope, between the trap and its vent connection.

The type of fitting used to connect a fixture drain to the fixture branch is also limited. Where the fixture drain pipe runs horizontally and then turns vertically at the point of connection to the vent, a short-pattern fitting is necessary to keep the trap weir below the highest inlet to the vent. The type of transition fitting will necessarily be a sanitary tee except where the vent connects to the horizontal portion of the fixture drain. Although a long-pattern fitting (including a tee-wye or a combination wye-eighth bend) is recommended for creating the proper flow into the vertical drain, such a fitting greatly increases the possibility of self-siphonage of the trap seal and is, therefore, prohibited by this section [see Commentary Figures 909.2(1) and (2)].

Water closets are not regulated by this section. This section specifies a requirement to reduce the possibility of self-siphoning of the trap seal for a given fixture. Water closets (and siphon-action urinals), however, are designed for siphon action of the fixture. The fixture traps must self-siphon during each flushing cycle to accomplish total evacuation of the fixture contents. The flushing device is further designed to refill the trap seal at the completion of the flushing cycle. This operation may be observed during each flushing cycle of a water closet when the water in the bowl is almost completely removed. This self-siphoning action is what the requirements of Sections 909.1 and 909.2 attempt to prevent in other fixtures [see Commentary Figure 909.1(3)].

909.3 Crown vent. A vent shall not be installed within two pipe diameters of the trap weir.

❖ Crown venting is any arrangement in which a vent connects at the top of the weir (crown) of a trap. Crown vented "P" and "S" traps used to be manufactured with vent connections as an integral part of the trap.

The problem with this type of connection is that the vent opening becomes blocked, thus closing the vent. The blockage is a result of the action of the drainage flowing through the trap. The flow direction and velocity will force waste up into the vent connection, eventually clogging it with debris. The vent connection must therefore be a minimum of two pipe diameters downstream from the trap weir to prevent the vent from becoming blocked [see Commentary Figures 909.3(1) and 909.3(2)].

For SI: 1 inch = 25.4 mm, 1 inch per foot = 83.33 mm/m.

Figure 909.2(1)
VENT CONNECTION

VENTS

Figure 909.2(2)
SELF-SIPHONING PHENOMENON

For SI: 1 inch = 25.4 mm.

Figure 909.3(1)
CROWN VENTING

SECTION 910
INDIVIDUAL VENT

910.1 Individual vent permitted. Each trap and trapped fixture is permitted to be provided with an individual vent. The

Figure 909.3(2)
MINIMUM TRAP TO VENT DISTANCE

individual vent shall connect to the *fixture drain* of the trap or trapped fixture being vented.

❖ The simplest form of venting any trap or trapped fixture is an individual vent for each trap. A single vent pipe is connected between the trap or fixture and the branch connection to the drainage system. The protecting vent must be between the fixture trap and the source of pressure fluctuations. If drainage flows from another fixture between the vent connection and the fixture trap, the vent is not installed as an individual vent. The only drainage flowing past a properly installed individual vent is from the fixture being vented.

An individual vent was previously called a "revent." This expression dates back to the original concepts in venting when it was believed that the only vent necessary was an extension of the drainage stack through the roof. It was later determined that a stack vent was inadequate for protecting the trap seals. Individual vents were added to assist the stack vent in protecting the trap seals. The individual vent was called a "revent" because it was considered to be additional venting to an already-vented system. The definition of the prefix "re-" is "repetition of a previous action." The name was changed to "individual vent" when it was determined that a stack vent was unnecessary if each trap was protected by a vent [see Commentary Figures 910.1(1) and 910.1(2)].

SECTION 911
COMMON VENT

911.1 Individual vent as common vent. An individual vent is permitted to vent two traps or trapped fixtures as a common vent. The traps or trapped fixtures being common vented shall be located on the same floor level.

❖ For common venting, the vent is classified and sized as an individual vent. The fixture drains being common vented are allowed to connect to either a vertical or horizontal drainage pipe. Any two fixtures can be common vented. Where pump-action discharge fixtures or water closets are common vented, Section 706.3 requires directional drainage fittings to prevent the discharge action of one fixture from interfering with the other fixture or causing a blowback into another fixture.

If two water closets are installed back to back at the same level and are common vented, the connection to the vertical drainage branch must be through a double combination wye and eighth bend. An exception for the installation of back-to-back water closet connections allows the use of a double sanitary tee where the distance between the outlet of the water closet and the

Figure 910.1(2)
IMPROPERLY INSTALLED INDIVIDUAL VENTS

Figure 910.1(1)
INDIVIDUAL VENTS

connection to the double sanitary tee pattern is 18 inches (457 mm) or more. The fixtures being common vented must be located on the same floor level so that the flow velocities are kept low by the limiting distance of the vertical drop of waste [see Commentary Figures 911.1 and 706.3(5)].

911.2 Connection at the same level. Where the fixture drains being common vented connect at the same level, the vent connection shall be at the interconnection of the fixture drains or downstream of the interconnection.

❖ The most popular form of common venting is connecting two fixtures at the same level. A typical installation is two horizontal drains connecting to a vertical drain through a double-pattern fitting. The extension of the vertical pipe serves as the vent; however, two drains connecting horizontally to a horizontal drain through a double-pattern fitting is also a form of common venting. The vent is allowed to connect at the interconnection of the fixture drains or downstream along the horizontal drain.

The requirements of Section 905 regarding vent connection would apply, as well as those of Section 909. The trap-to-vent distances for each of the fixtures must comply with Table 909.1 (see commentary, Section 911.1 and Commentary Figure 911.2).

911.3 Connection at different levels. Where the fixture drains connect at different levels, the vent shall connect as a vertical extension of the vertical drain. The vertical drain pipe connecting the two *fixture drains* shall be considered the vent for the lower *fixture drain*, and shall be sized in accordance with Table 911.3. The upper fixture shall not be a water closet.

❖ This section applies only to vertical drain applications. Two fixtures within a single story are allowed to connect at different levels to a vertical drain and still be common vented. The vent remains an individual vent serving as a common vent. The drain pipe between the upper fixture and the lower fixture, however, must meet the oversizing requirement specified in Table 911.3. Although drainage flows in the pipe between the upper and lower fixture connections, the section of pipe is not classified as a wet vent despite the fact that it functions in the same manner as a wet vent. The

**Figure 911.1
COMMON VENTS**

For SI: 1 inch = 25.4 mm.

**Figure 911.2
COMMON VENTS WITH FIXTURE DRAINS CONNECTING AT SAME LEVEL**

pipe is wet and does serve as a vent; however, it is classified as a common vent for purposes of the code (see Commentary Figure 911.3).

The fixture at the higher level is not allowed to be a water closet because the pressures created by the high-volume rapid discharge of a water closet could adversely affect the venting of lower fixtures.

TABLE 911.3
COMMON VENT SIZES

PIPE SIZE (inches)	MAXIMUM DISCHARGE FROM UPPER FIXTURE DRAIN (dfu)
$1^1/_2$	1
2	4
$2^1/_2$ to 3	6

For SI: 1 inch = 25.4 mm.

❖ The sizing of the common vent serving as a drain for the upper fixture and vent for the lower fixture is the same as the sizing for a wet vent. The same concepts of venting apply, hence the same sizing. The dry vent extension above the upper fixture connection is sized as an individual vent and is based on the largest fixture drain served. For example, the size of a dry vent serving a lavatory and water closet connecting at different levels is $1^1/_2$ inches (38 mm) because the water closet fixture drain cannot be less than 3 inches (76 mm).

For SI: 1 inch = 25.4 mm.

Figure 911.3
COMMON VENTS WITH FIXTURE DRAINS CONNECTING AT DIFFERENT LEVELS

SECTION 912
WET VENTING

912.1 Horizontal wet vent permitted. Any combination of fixtures within two *bathroom groups* located on the same floor level is permitted to be vented by a horizontal wet vent. The wet vent shall be considered the vent for the fixtures and shall extend from the connection of the dry vent along the direction of the flow in the drain pipe to the most downstream *fixture drain* connection to the *horizontal branch drain*. Each wet-vented *fixture drain* shall connect independently to the horizontal wet vent. Only the fixtures within the *bathroom groups* shall connect to the wet-vented *horizontal branch drain*. Any additional fixtures shall discharge downstream of the horizontal wet vent.

❖ A horizontal wet vent is a horizontal branch drain pipe that is sized larger than is required by Section 710. The oversizing of the pipe allows for a large airspace above the maximum probable waste flow level in the pipe so as to provide for adequate venting air in the same pipe as the waste flow. In other words, the horizontal wet vent is a special type of vent pipe that is "wet" because it carries waste flow. Commentary Figure 912.1(1) shows a commonly installed horizontal wet vent for a single bathroom group. A horizontal wet vent can be used only for a limited number and type of fixtures contained in a maximum of two bathroom groups (see the definition of "Bathroom group" in Section 202).

The acceptance of horizontal wet venting for bathroom groups is based on the low probability of simultaneous discharge from the types of fixtures within the group, significant oversizing of the drain pipe and the limited number of drainage fixtures units allowed. Years of satisfactory performance of horizontal wet venting of bathroom groups in single-family residential dwellings [as covered by the *International Residential Code®* (IRC®)] has proven this method of venting. This method is similar to the long-proven methods of circuit venting and combination waste and vent arrangements that also use airspace above the waste flow level for venting fixtures connected to a horizontal drain pipe.

The maximum number and type of fixtures that can be accommodated by a horizontal wet vent are: two water closets, two lavatories, two bidets, two emergency floor drains (see definition of "Emergency floor drain" in Section 202) and two bathing fixtures (either bathtubs or showers). Note that a urinal is not on the list of bathroom group fixtures that can be served by a horizontal wet vent. A horizontal wet vent could have as few as two fixtures or as many as 10 fixtures, but no more than two fixtures of any type can be connected to the system. For example, three lavatories or three water closets cannot be connected to a horizontal wet vent. No other types of fixtures are allowed to discharge into the horizontal wet vent system. All fixtures connecting to a horizontal wet vent must be within the same floor level to avoid high flow velocities, which could cause excessive turbulence that might block the airspace (venting area) above the waste flow level.

As with other horizontal wet venting methods (e.g., combination waste and vent and circuit vent), the main section of the horizontal wet vent pipe is intended to be installed horizontally without vertical offsets with each fixture drain connected "independently" to the horizontal drain in order to keep flow turbulence at a minimum. Note that fixtures connecting though a double pattern fitting are considered to be independently connected to the horizontal wet vent. The individual fixture drains vented by horizontal wet vent must enter the horizontal wet vent in the horizontal plane so that the connection is not below the trap weir of the connected fixture. Water closets, having integral traps that depend on self-siphoning, are exempt from this requirement. Other fixtures that are not vented by the horizontal wet vent (i.e., vented by a dry vent) can enter the horizontal wet vent horizontally, vertically or at any angle in between. The minimum slopes of the piping and the required fittings must be in accordance with Chapter 7. While there is no specified maximum slope for the horizontal sections of horizontal wet vent piping, the intent is to avoid high flow velocities, which might disrupt the airspace for venting. Thus, the horizontal wet vent pipe is intended to be installed at or near the minimum slope requirements for horizontal branch drains.

Section 912.2.1 regulates the dry vent connection to the horizontal wet vent, but is discussed in this commentary section for continuity. The dry vent for a horizontal wet vent must connect to a vent stack, stack vent, air admittance valve or terminate outdoors.

The required vent could be a dry vent directly connected to the main section of the horizontal wet vent pipe downstream of the most upstream fixture [see Commentary Figure 912.1(2)]. In this arrangement, the dry vent is classified as an individual vent for the most upstream fixture. The required vent could be a dry vent for common vented fixtures [see Commentary Figure 912.1(3)].

The last sentence of Section 912.2.1 attempts to reinforce the definition of a horizontal wet vent so that improper fixture connections are avoided. A horizontal wet vent begins at the dry vent connection and extends downstream to the last fixture being wet vented. The horizontal wet vent does not extend upstream of the

Figure 912.1(1)
HORIZONTAL WET VENTING OF BATHROOM FIXTURES

dry vented fixture connection. Thus, multiple interconnected fixture drains that connect upstream of the dry vented fixture connection or the dry vent connection are not vented by the horizontal wet vent. Commentary Figures 912.1(4) and (5) are examples of prohibited arrangements.

Not all fixtures connected to a horizontal wet vent are required to be vented by the horizontal wet vent [see Commentary Figure 912.1(6)]. In this arrangement the shower trap was required to be located at a distance that exceeded the maximum trap-to-vent distance requirements of Table 909.1, thus requiring the trap to be independently vented. Any number of individually vented fixtures of the types allowed for a bathroom group can connect to the horizontal wet vent as long as the total number of fixtures allowed for the bathroom groups is not exceeded [see Commentary Figure 912.1(7)].

The sizing for both the horizontal and vertical "wet" sections of a horizontal wet vent is regulated by Section 912.3 and Table 912.3. A unique feature of a horizontal wet vent that is different from other horizontal "wet-type" venting methods described in this chapter is that the pipe does not have to be of uniform size throughout. The horizontal wet vent pipe size need only increase so as to accommodate the connected load and provide adequate airspace above the flow line for venting. Note that the last fixture downstream, vented by the horizontal wet vent, does not contribute to the dfu loading and thus does not contribute to the sizing of the wet vent. Where a water closet is connected to a horizontal wet vent, the piping must be at least 3 inches (76 mm) in diameter from the fixture, downstream to the end of the horizontal wet vent. Pipe sizes larger than the minimum required size might be necessary in order to satisfy clean-out access provisions as required by Chapter 7. The code does not specify a maximum

Figure 912.1(2)
HORIZONTAL WET VENT WITH INDIVIDUAL VENT

Figure 912.1(3)
HORIZONTAL WET VENT
WITH COMMON VENTED FIXTURES UPSTREAM

Figure 912.1(4)
PROHIBITED CONNECTION OF TWO FIXTURES
UPSTREAM OF DRY-VENTED FIXTURE CONNECTION

Figure 912.1(5)
PROHIBITED CONNECTION OF TWO FIXTURES
UPSTREAM OF DRY-VENTED FIXTURE CONNECTION

length for a horizontal wet vent. However, since this method was originally developed for applications where two bathroom groups were adjacent or back to back, the intent for the use of this method is that the fixtures to be horizontally wet vented are to be in the same vicinity and not in remote groups.

Various acceptable horizontal wet vent arrangements, complete with the required pipe sizes, are shown in Commentary Figures 912.1(8) through (12). Horizontal offsets are shown, as well as pipe size increases as required by the increase in dfu loading in accordance with Table 912.3. Using pipe sizes greater than those required by Table 912.3 is acceptable. However, as with any horizontal drain installed at near minimum slope, excessive oversizing could result in sluggish flow and clogging problems.

Commentary Figure 912.1(11) is a good example to illustrate the dry vent sizing requirements for horizontal wet vents. The size of the required dry vent for a horizontal wet vent is regulated by Section 912.3. For dry vents that are not the "required" dry vent for the horizontal wet vent, sizing is to be in accordance with the appropriate sections covering those venting methods. For example, if the dry vent of an individually vented lavatory fixture is not serving as the required vent for the horizontal wet vent, then the minimum size of the dry vent is $1^1/_4$ inches (32 mm) in accordance with Section 906.2. However, if this dry vent is the "required" dry vent for a horizontal wet vent that serves a 3-inch (76 mm) drain, then the required dry vent must be at least $1^1/_2$ inches (38 mm) since the required vent must be at least one-half the size of the "largest required diameter of pipe within the wet-vented system." This is the reason why the dry vent shown in Commentary Figure 912.1(11) is $1^1/_2$ inches (38 mm) and not $1^1/_4$ inches (32 mm).

Prohibited horizontal wet vented arrangements are illustrated in Commentary Figures 912.1(13) through (18). Notes in each figure indicate the reason for noncompliance.

Commentary Figures 912.1(19) through (24) show examples of proper horizontal wet venting.

Figure 912.1(6)
HORIZONTAL WET VENT WITH COMMON VENT WITH AN ADDITIONAL INDIVIDUALLY DRY-VENTED FIXTURE

Figure 912.1(7)
HORIZONTAL WET VENT WITH THREE ADDITIONAL INDEPENDENTLY VENTED FIXTURES

For SI: 1 inch = 25.4 mm.

Figure 912.1(8)
HORIZONTAL WET VENT SIZING (PRIVATE BATH GROUP)

For SI: 1 inch = 25.4 mm.

Figure 912.1(9)
HORIZONTAL WET VENT (PUBLIC BATH GROUP)

For SI: 1 inch = 25.4 mm.

Figure 912.1(10)
HORIZONTAL WET VENT

VENTS

For SI: 1 inch = 25.4 mm.

**Figure 912.1(11)
HORIZONTAL WET VENT**

For SI: 1 inch = 25.4 mm.

**Figure 912.1(12)
HORIZONTAL WET VENT**

**Figure 912.1(13)
NONCOMPLIANT HORIZONTAL WET VENT**

VENTS

Figure 912.1(14)
NONCOMPLIANT HORIZONTAL WET VENT

Figure 912.1(15)
NONCOMPLIANT HORIZONTAL WET VENT

VENTS

Figure 912.1(16)
NONCOMPLIANT HORIZONTAL WET VENT

Figure 912.1(17)
NONCOMPLIANT HORIZONTAL WET VENT

Figure 912.1(18)
NONCOMPLIANT HORIZONTAL WET VENT

VENTS

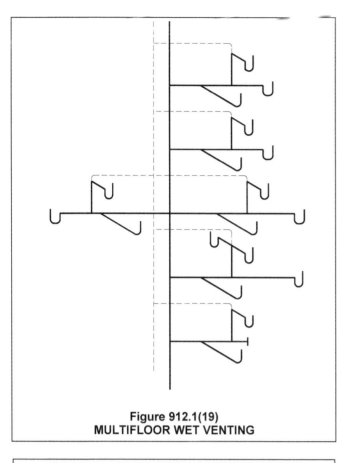

Figure 912.1(19)
MULTIFLOOR WET VENTING

THE DISTANCE FROM EACH TUB TO POINT C MUST BE WITHIN THE DISTANCE SPECIFIED IN TABLE 906.1

DRAIN COMPONENT	MIN. SIZE	DFU (TABLE 709.1)
LAV DRAIN	1 1/4"	1
BT DRAIN	1 1/2"	2
WC DRAIN	3"	3
A-B[a]	2"[a]	2
B-C[a]	2"[a]	2
C-D[a]	2 1/2"[a]	6
D-E[a]	3"[a]	9
A-X	1 1/2"	NA

a. FROM TABLE 912.3
NA. NOT APPLICABLE

For SI: 1 inch = 25.4 mm.

Figure 912.1(21)
DOUBLE BATHROOM GROUP (PRIVATE) WET VENT

DRAIN COMPONENT	MIN. SIZE	DFU (TABLE 709.1)
LAV FIXTURE DRAIN	1 1/4"	1
SH FIXTURE DRAIN	1 1/2"	2
WC FIXTURE DRAIN	3"	3
A-B[a]	1 1/2"[a]	1
B-C[a]	2"[a]	3
A-X	1 1/2"	NA

a. FROM TABLE 912.3
NA. NOT APPLICABLE

THE WET VENT EXTENDS FROM C TO A

SIZING OF THE SYSTEM IS ACCORDING TO TABLE

For SI: 1 inch = 25.4 mm.

Figure 912.1(20)
SINGLE BATHROOM GROUP (PRIVATE) WET VENT

SECTION	SIZE WHEN WET VENT STARTS AT A	SIZE WHEN WET VENT STARTS AT C
A-B	1 1/2"	1 1/4"
B-C	1 1/2"	2"
B-D	2"	2"

WHEN THE BATHTUB FIXTURE DRAIN EXCEEDS THE DISTANCE FROM TRAP TO VENT REQUIREMENTS, AN ADDITIONAL INDIVIDUAL VENT IS ADDED FOR THE FIXTURE. THE WET VENT MAY BE CONSIDERED AS EITHER SECTION A-D BY USE OF THE LAV CONNECTION, OR SECTION C-D BY USE OF THE BATHTUB CONNECTION.

For SI: 1 inch = 25.4 mm.

Figure 912.1(22)
WET VENT VARIATIONS FOR A SINGLE BATHROOM GROUP

VENTS

For SI: 1 inch = 25.4 mm.

**Figure 912.1(23)
HORIZONTAL WET VENTING**

**Figure 912.1(24)
MULTIPLE INDIVIDUAL VENTS**

912.1.1 Vertical wet vent permitted. Any combination of fixtures within two bathroom groups located on the same floor level is permitted to be vented by a vertical wet vent. The vertical wet vent shall be considered the vent for the fixtures and shall extend from the connection of the dry vent down to the lowest *fixture drain* connection. Each wet-vented fixture shall connect independently to the vertical wet vent. Water closet drains shall connect at the same elevation. Other *fixture drains* shall connect above or at the same elevation as the water closet fixture drains. The dry-vent connection to the vertical wet vent shall be an individual or common vent serving one or two fixtures.

❖ Vertical wet vents used to be referred to as "stack venting." The vertical wet vent extends from the connection to the dry vent down to the lowest fixture drain connection. A common example of vertical wet venting of a single bathroom group is shown in Commentary Figure 912.1.1(1). The two most important requirements of vertical wet venting are that all fixtures must connect above or at the water closet connection elevation and all fixtures must connect independently to the vertical wet vent. Note that fixtures connected through a double pattern fitting are considered to be independently connected. If more than one water closet is to be connected, both must connect at the same elevation [see Commentary Figure 912.1.1(2)]. Note that in this figure, a double sanitary tee is only acceptable where the water closet drain connection to the sanitary tee is 18 inches (457 mm) or more in horizontal developed length distance from the outlet of the water closet (see Section 706.3 Exception). Otherwise, the water closets must connect through double wye and eighth-bend fitting (also known as a "double T-wye") [see Commentary Figure 912.1.1(3)].

Any combination of fixtures within two bathroom groups located on the same floor level can be installed without individual fixture vents (see the definition of "Bathroom group" in Section 202). However, each fixture drain must connect to the wet vent within the maximum length for trap-to-vent distance as specified in Table 909.1. Note that a bathroom group does not include a urinal fixture. Without being specifically stated, the vertical wet vent must be vertical without

For SI: 1 inch = 25.4 mm.

Figure 912.1.1(2)
VERTICAL WET VENT OF TWO BATHROOM GROUPS

For SI: 1 inch = 25.4 mm.

Figure 912.1.1(3)
VERTICAL WET VENT OF TWO BATHROOM GROUPS

For SI: 1 inch = 25.4 mm.

Figure 912.1.1(1)
VERTICAL WET VENT OF SINGLE BATHROOM GROUP

vertical or horizontal offsets. Commentary Figure 912.1.1(4) illustrates various vertical wet vent arrangements.

Section 912.2.2 regulates how the dry vent connection is made to the vertical wet vent and Section 912.3 regulates the size of the dry vent. Both are discussed in this commentary section for continuity. The dry vent is an extension of the vertical wet vent and is connected to a vent stack, stack vent, air admittance valve or point of termination outside of the building. The dry vent size must be at least one-half of the largest required pipe size served by the wet vent. For example, in Commentary Figure 912.1.1(2), the largest pipe served by the vertical wet vent is 3 inches (76 mm), requiring the dry vent to be a minimum of $1^1/_2$ inches (38 mm). If the run of the dry vent from the start of the dry vent fixture connection to the termination of the vent (to a vent stack, stack vent, air admittance valve or termination outdoors) exceeds 40-feet (12 192 mm) developed length, Section 906.2 requires the dry vent size to be increased by one nominal pipe size for its entire length.

912.2 Dry vent connection. The required dry-vent connection for wet-vented systems shall comply with Sections 912.2.1 and 912.2.2.

❖ See the commentary to Sections 912.2.1 and 912.2.2.

912.2.1 Horizontal wet vent. The dry-vent connection for a horizontal wet-vent system shall be an individual vent or a common vent for any *bathroom group* fixture, except an *emergency floor drain*. Where the dry-vent connects to a water closet *fixture drain*, the drain shall connect horizontally to the horizontal wet-vent system. Not more than one wet-vented *fixture drain* shall discharge upstream of the dry-vented *fixture drain* connection.

❖ See the commentary to Section 912.1. Note that previous code editions did not allow a water closet to be the upstream dry-vented fixture. Because a water closet can be dry vented in a nonwet-vented system, there is not any technical reason why this venting arrangement will not work in a wet-vented system, Thus, this edition of the code allows a dry-vented water closet to be the dry-vented fixture that vents the horizontal wet vent.

Another change for this edition is that two (or more) fixture drains cannot connect upstream of the dry-vented fixture drain connection for a horizontal wet-vented system. Commentary Figure 912.1(5) illustrates one common violation. As the figure shows, the wet vent does not continue horizontally upstream towards the bathtub. The section of pipe between the first bathtub upstream and the dry-vented fixture connection is simply an individual fixture drain that con-

For SI: 1 inch = 25.4 mm.

**Figure 912.1.1(4)
VERTICAL WET VENTING**

nects to the wet vent. Because the code requires that each fixture trap be provided with its own vent (see Section 909.1) (i.e., each fixture does not connect individually to the horizontal wet vent), the most upstream bathtub fixture drain in Commentary Figure 912.1(5) does not have a vent. The violation illustrated in Commentary Figure 912.1(4) follows the same reasoning.

912.2.2 Vertical wet vent. The dry-vent connection for a vertical wet-vent system shall be an individual vent or common vent for the most upstream *fixture drain*.

❖ See the commentary to Section 912.1.1.

912.3 Size. The dry vent serving the wet vent shall be sized based on the largest required diameter of pipe within the wet-vent system served by the dry vent. The wet vent shall be of a size not less than that specified in Table 912.3, based on the fixture unit discharge to the wet vent.

❖ See the commentary to Sections 912.1 and 912.1.1, which specifically reference this section for horizontal wet vents and vertical wet vents, respectively.

Each section of the wet-vented system is individually sized, depending on the dfu load carried by that section. In other words, wet vent piping size increases as the load increases. As opposed to circuit venting, a wet-vented horizontal branch drain need not be uniformly sized. The fixture unit loads shown in Table 912.3 range from 1 dfu (single lavatory or bidet) to 12 dfu (two bathroom groups) (see the definition of "Bathroom group" in Section 202).

TABLE 912.3
WET VENT SIZE

WET VENT PIPE SIZE (inches)	DRAINAGE FIXTURE UNIT LOAD (dfu)
$1^1/_2$	1
2	4
$2^1/_2$	6
3	12

For SI: 1 inch = 25.4 mm.

❖ The table states the drain pipe oversizing that is fundamental to wet venting. The sizing of wet vent piping is based on the fixture unit load discharging through such piping. When comparing Table 710.1(2) to Table 912.3, the degree of wet-vent oversizing is evident. The individual fixture drains are sized in accordance with Section 710.1.

The developed length of a fixture drain is measured from the fixture to the point of connection to the wet vent. The length of the fixture drain is limited to the distance specified in Table 906.1 (see commentary, Sections 909.1 and 909.1.1).

SECTION 913
WASTE STACK VENT

913.1 Waste stack vent permitted. A waste *stack* shall be considered a vent for all of the fixtures discharging to the *stack* where installed in accordance with the requirements of this section.

❖ A waste stack vent uses the waste stack as the vent for fixtures other than urinals and water closets. The principles of use are based on some of the original research that was done in development of the first plumbing codes. The system has been identified by a variety of names, including vertical wet vent, Philadelphia single-stack and multifloor stack venting. Note that due to the oversized design of a waste stack vent system, no relief vents are needed for stacks greater than 10 branch intervals.

913.2 Stack installation. The waste *stack* shall be vertical, and both horizontal and vertical offsets shall be prohibited between the lowest *fixture drain* connection and the highest *fixture drain* connection. *Fixture drains* shall connect separately to the waste *stack*. The *stack* shall not receive the discharge of water closets or urinals.

❖ Because the drainage stack serves as the vent, there are certain limitations on the design to prevent pressures in the system from exceeding plus or minus 1 inch of water column (249 Pa). The system is identified as a "waste" stack vent because it prohibits the connection of water closets and urinals. Only "waste" may discharge to the stack.

Water closets and some urinals have a surging discharge that results in too great a pressure fluctuation in the system. The system can serve sinks, lavatories, bathtubs, bidets, showers, floor drains, drinking fountains and standpipes.

To preserve the desirable annular flow in the stack, offsets (of any degree) are prohibited and the stack must be vertical [90 degrees (1.57 rad) from horizontal] from the lowest fixture connection to the highest fixture connection. Offsets are permitted in the drainage stack below the lowest branch connection to the waste stack vent and within the dry stack vent portion (see commentary, Section 913.3).

Each fixture must connect independently through a single or double sanitary tee to the drainage stack. Independent connections will prevent large discharges into the stack at any one point and will direct the flow into the stack with minimum interference with any vertical flow already occurring, thus minimizing pressure fluctuations [see Commentary Figures 913.2(1) and 913.2(2)].

913.3 Stack vent. A *stack vent* shall be provided for the waste *stack*. The size of the *stack vent* shall be not less than the size of the waste *stack*. Offsets shall be permitted in the *stack vent*, shall be located not less than 6 inches (152 mm)

VENTS

For SI: 1 inch = 25.4 mm.

**Figure 913.2(1)
WASTE STACK VENT**

above the flood level of the highest fixture and shall be in accordance with Section 905.2. The *stack vent* shall be permitted to connect with other *stack vents* and vent *stacks* in accordance with Section 904.5.

❖ A full-size stack vent provides the vent opening to the outdoors, allowing the stack to remain at neutral pressures. This is the only type of stack addressed in the code that is required to have a stack vent. Offsets are allowed in the stack vent because such offsets are dry and have no effect on the annular flow in the waste stack. As always, dry vents must not run horizontally below the flood level rim of fixtures because this may allow waste and solids to enter the vent and impair its function. The stack vent can be terminated at a stack type air admittance valve, extended to the outdoors or tied into another vent stack, stack vent or vent header (see Commentary Figure 913.3).

913.4 Waste stack size. The waste *stack* shall be sized based on the total discharge to the *stack* and the discharge within a *branch* interval in accordance with Table 913.4. The waste *stack* shall be the same size throughout its length.

❖ Two factors govern sizing of the waste stack vent: the total discharge into the stack and the total discharge within any one branch interval. The size of the stack must be uniform from the base of the stack to the stack vent terminal to allow unimpeded air movement to control pressure fluctuations.

For SI: 1 inch = 25.4 mm.

**Figure 913.2(2)
WASTE STACK VENT**

Figure 913.3
WASTE STACK VENT OFFSETS AND STACK VENT CONNECTIONS

TABLE 913.4
WASTE STACK VENT SIZE

STACK SIZE (inches)	MAXIMUM NUMBER OF DRAINAGE FIXTURE UNITS (dfu)	
	Total discharge into one branch interval	Total discharge for stack
1 1/2	1	2
2	2	4
2 1/2	No limit	8
3	No limit	24
4	No limit	50
5	No limit	75
6	No limit	100

For SI: 1 inch = 25.4 mm.

❖ To reduce the possibility of exceeding plus or minus 1 inch of water column (249 Pa) pressure, the stack is greatly oversized. Although Chapter 7 permits a drainage stack to be fully loaded, a waste stack vent may be loaded only to a fraction of its total capacity. Full loading of a stack will result in pressure fluctuations that will necessitate additional venting for fixtures connecting to the stack.

SECTION 914
CIRCUIT VENTING

914.1 Circuit vent permitted. A maximum of eight fixtures connected to a horizontal *branch* drain shall be permitted to be circuit vented. Each *fixture drain* shall connect horizontally to the horizontal *branch* being circuit vented. The *horizontal branch drain* shall be classified as a vent from the most downstream *fixture drain* connection to the most upstream *fixture drain* connection to the horizontal *branch*.

❖ Circuit venting involves up to eight fixtures and a single vent. A dry vent connects between the two most upstream fixtures on the horizontal branch. The horizontal branch serves as a "wet vent" for the battery of fixtures. The piping arrangement has the appearance of multiple fixtures connecting to a horizontal branch, with only the most upstream fixture being individually vented.

The principle of circuit venting is that the flow of drainage never exceeds a half-full flow condition. The air for venting the fixtures circulates in the top half of the horizontal branch drain pipe. The flow velocity in the horizontal branch is slow and nonturbulent, thereby preventing pressure differentials from affecting the connecting fixtures [see Commentary Figure 914.1(1)]. The circuit-vented fixtures must connect to the circuit-vented branch in the horizontal plane to limit the amount of turbulence created by fixture discharge [see Commentary Figure 914.1(2)].

914.1.1 Multiple circuit-vented branches. Circuit-vented horizontal *branch* drains are permitted to be connected together. Each group of a maximum of eight fixtures shall be considered a separate circuit vent and shall conform to the requirements of this section.

❖ Circuit-vented branches may connect either in series or in parallel. Each group of up to eight fixtures must be considered as a separate circuit-vented branch and be installed accordingly.

The concept of multiple circuit-vented branches is that horizontal flow in a drain pipe does not require a large quantity of air to balance the pressure. The slower flow velocity in a horizontal drain also prevents pressure fluctuations that may affect the traps of the connecting fixtures (see Commentary Figure 914.1.1).

914.2 Vent connection. The circuit vent connection shall be located between the two most upstream fixture drains. The vent shall connect to the horizontal *branch* and shall be installed in accordance with Section 905. The circuit vent pipe shall not receive the discharge of any soil or waste.

❖ The circuit vent must connect between the two most upstream fixture drains to allow proper circulation of air. The vent is located so that the most upstream fixture discharges past the vent connection, thereby washing that section of pipe and preventing the buildup of solids. For this reason, the most upstream fixture should not be a fixture that is seldom used (such as a floor drain). When back-to-back fixtures are installed, the vent is connected between the most upstream pairs (see commentary, Section 914.1 and Commentary Figure 914.2).

Section 905 requires that the vent connect vertically to the drain and rise vertically to 6 inches (152 mm) above the flood level rim before offsetting horizontally.

914.3 Slope and size of horizontal branch. The slope of the vent section of the *horizontal branch drain* shall be not greater than one unit vertical in 12 units horizontal (8.3-percent slope). The entire length of the vent section of the *horizontal branch drain* shall be sized for the total drainage discharge to the *branch*.

❖ The principle of a circuit vent is that the drainage branch is entirely horizontal. A maximum pitch of 1 inch per foot (83 mm/m) ensures that the branch remains horizontal without any vertical offsets.

The horizontal drainage branch must also be uniformly sized to establish consistent flow characteristics and the free movement of air. Because it is uniformly sized for its full length, the horizontal branch will be oversized for all but the most downstream portion. Table 710.1(2) is used to size the horizontal branch (see Commentary Figure 914.3).

914.3.1 Size of multiple circuit vent. Each separate circuit-vented horizontal *branch* that is interconnected shall be sized independently in accordance with Section 914.3. The downstream circuit-vented horizontal *branch* shall be sized for the total discharge into the *branch*, including the upstream *branches* and the fixtures within the *branch*.

❖ Each circuit-vented branch must be uniformly sized based on the total drainage load on that branch, including any drainage load from upstream branches. If an increase in pipe size is required, the location must be at the interconnection of the circuit-vented branches (see Commentary Figure 914.3.1).

914.4 Relief vent. A relief vent shall be provided for circuit-vented horizontal *branches* receiving the discharge of four or more water closets and connecting to a drainage *stack* that receives the discharge of soil or waste from upper horizontal *branches*.

❖ If a circuit-vented horizontal branch connects to a drainage stack that is receiving drainage discharge from floors above, a relief vent is required if more than three water closets connect to the circuit-vented horizontal branch. The relief vent will prevent any pressure differential in the drainage stack from affecting the horizontal branch by relieving the pressures.

For SI: 1 inch = 25.4 mm.

**Figure 914.1(1)
CIRCUIT VENTING**

VENTS

Figure 914.1(2)
ROUGH-IN FOR EIGHT CIRCUIT-VENTED FIXTURES
(Photo courtesy of B&I Contractors)

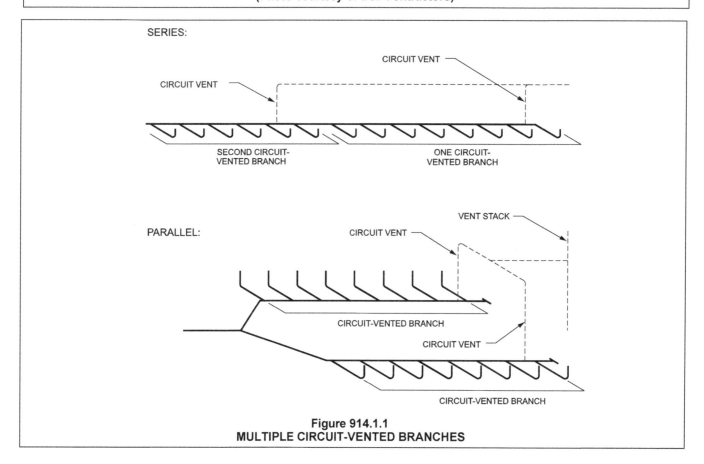

Figure 914.1.1
MULTIPLE CIRCUIT-VENTED BRANCHES

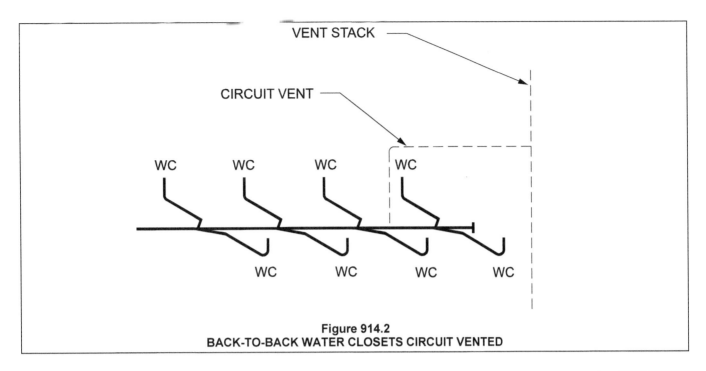

Figure 914.2
BACK-TO-BACK WATER CLOSETS CIRCUIT VENTED

For SI: 1 inch = 25.4 mm.

Figure 914.3
SIZING A CIRCUIT VENT SYSTEM

VENTS

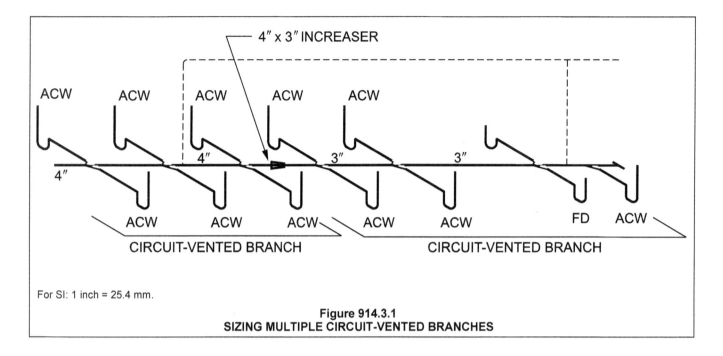

Figure 914.3.1
SIZING MULTIPLE CIRCUIT-VENTED BRANCHES

914.4.1 Connection and installation. The relief vent shall connect to the horizontal *branch* drain between the *stack* and the most downstream *fixture drain* of the circuit vent. The relief vent shall be installed in accordance with Section 905.

❖ The relief vent must connect between the stack and the most downstream fixture on the horizontal branch. The relief vent is sized in accordance with Section 906.2 (see Commentary Figure 914.4.1).

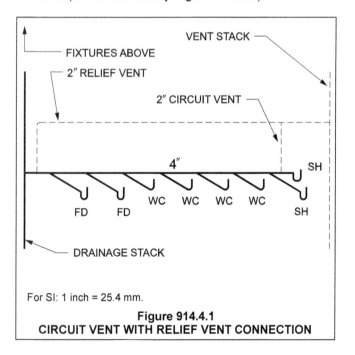

Figure 914.4.1
CIRCUIT VENT WITH RELIEF VENT CONNECTION

914.4.2 Fixture drain or branch. The relief vent is permitted to be a *fixture drain* or fixture *branch* for fixtures located within the same *branch interval* as the circuit-vented horizontal *branch*. The maximum discharge to a relief vent shall be four fixture units.

❖ Unlike the circuit vent, which must be dry, the relief vent is permitted to serve as a drain for fixtures installed within the same circuit-vented branch. When a relief vent serves as a fixture branch, that vent is subject to the same sizing requirements as a common vent or wet vent. A $1^1/_2$-inch (38 mm) relief vent may receive 1 dfu and a 2-inch (51 mm) or larger relief vent is limited to 4 dfu.

914.5 Additional fixtures. Fixtures, other than the circuit-vented fixtures, are permitted to discharge to the *horizontal branch drain*. Such fixtures shall be located on the same floor as the circuit-vented fixtures and shall be either individually or common vented.

❖ Where additional fixtures (other than the circuit-vented fixtures) are located on the same floor as a circuit-vented branch, such fixtures are allowed to connect to the circuit-vented branch. The additional fixtures must be separately vented, and the additional drainage load of those fixtures must be considered in sizing the circuit-vented branch [see Commentary Figures 914.5(1) and 914.5(2)]. This provision is intended to allow the connection of lavatories, urinals and similar fixtures that are located in the same area as the circuit-vented fixtures. This is common practice for toilet room installations that use circuit vents.

VENTS

For SI: 1 inch = 25.4 mm.

Figure 914.5(1)
CIRCUIT-VENTED BRANCH WITH ADDITIONAL FIXTURE CONNECTIONS

For SI: 1 inch = 25.4 mm.

Figure 914.5(2)
CIRCUIT-VENTED BRANCH WITH ADDITIONAL FIXTURE CONNECTIONS

VENTS

SECTION 915
COMBINATION WASTE AND VENT SYSTEM

915.1 Type of fixtures. A *combination waste and vent system* shall not serve fixtures other than floor drains, sinks, lavatories and drinking fountains. *Combination waste and vent systems* shall not receive the discharge from a food waste disposer or clinical sink.

❖ This section contains requirements for a venting method serving floor drains, sinks, lavatories and drinking fountains. Like wet vents and circuit vents, a combination waste and vent system is another variation of "wet" venting. Although the specific requirements for each are unique, all of the "wet" venting means (circuit, wet and combination) utilize the same basic principle of oversized drains with ample space for airflow above the waste flow.

The number of fixtures connecting to a combination waste and vent system is unlimited, provided that the fixtures are floor drains, sinks, lavatories or drinking fountains. The system is commonly used where floor drains are installed in large open areas that cannot accommodate vertical vent risers from the floor drains. A combination waste and vent is also practical to use where island sinks are installed in dwelling units, laboratories and classrooms.

To prevent blockages, such systems are intended to receive only clear and gray-water wastes. Because of the low-flow velocities and volumes that occur in these systems, any solids introduced would accumulate, causing blockages and reduction of the internal cross-sectional area of the piping. Therefore, water closets, sinks equipped with food waste grinders, clinical sinks and fixtures that discharge solid waste are not allowed to be served by a combination waste and vent system. Ground food waste might collect in the drain, causing blockages. Combination waste and vent systems are oversized drains; therefore, the velocity of the flow will be low, thereby reducing normally available scouring action.

915.2 Installation. The only vertical pipe of a combination waste and vent system shall be the connection between the fixture drain and the horizontal combination waste and vent pipe. The vertical distance shall not exceed 8 feet (2438 mm).

❖ In a combination waste and vent system, the drain serves as the vent for the fixture. The system is intended to be a horizontal piping system, with the only vertical piping [limited to 8 feet (2438 mm) in height] being the vertical portion of a fixture drain located above the level of the combination waste and vent [see Commentary Figure 915.2(5)].

Each fixture drain connected to a combination waste and vent system is sized in accordance with Chapter 7. Depending on the method by which it is sized, the fixture drain may or may not be considered as a portion of the combination waste and vent system. Where the fixture drain is sized as a combination waste and vent in accordance with Table 915.2.2, the trap-to-vent distances listed in Table 909.1 do not apply because the drain is also a vent for its entire length.

Additional installation requirements for the combination waste and vent and the fixture branch or drain connections are contained in Sections 915.2.1 through 915.2.5 [see Commentary Figures 915.2(1) through 915.2(5)].

For SI: 1 inch = 25.4 mm.

**Figure 915.2(1)
COMBINATION WASTE AND VENT**

VENTS

For SI: 1 inch = 25.4 mm.

**Figure 915.2(2)
COMBINATION WASTE AND VENT**

For SI: 1 inch = 25.4 mm.

**Figure 915.2(3)
COMBINATION WASTE AND VENT**

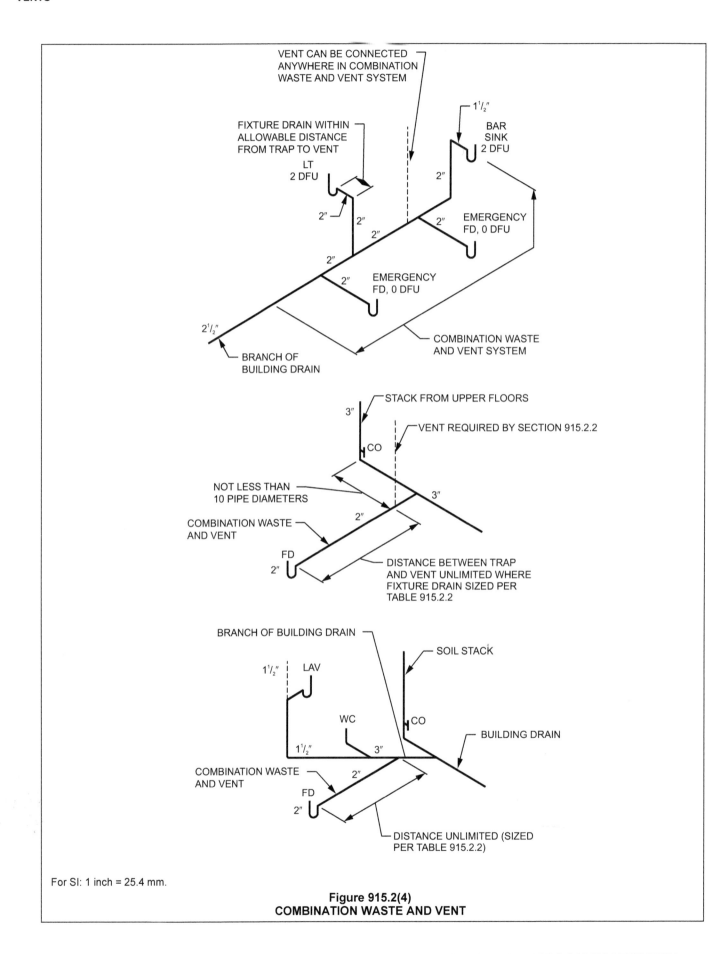

Figure 915.2(4)
COMBINATION WASTE AND VENT

VENTS

For SI: 1 foot = 304.8 mm.

Figure 915.2(5)
VERTICAL FIXTURE DRAIN COMPONENT IN COMBINATION DRAIN AND VENT SYSTEM

915.2.1 Slope. The slope of a horizontal combination waste and vent pipe shall not exceed one-half unit vertical in 12 units horizontal (4-percent slope) and shall not be less than that indicated in Table 704.1.

❖ This section specifies the maximum and minimum slopes for the combination waste and vent to regulate the flow velocity of the waste. The intent of this section is to provide a minimum slope to facilitate drainage by gravity while providing a maximum slope to limit flow velocity to a range that minimizes adverse pressure fluctuations. This section and Section 915.2 act together to make sure such systems maintain a low-flow velocity by incorporating height limitations on vertical piping.

915.2.2 Size and length. The size of a combination waste and vent pipe shall be not less than that indicated in Table 915.2.2. The horizontal length of a *combination waste and vent system* shall be unlimited.

❖ The referenced table sizes the drain serving as a combination waste and vent. Because of the degree of oversizing and the low slope for such piping, the flow velocity and depth will be low. These facts indicate that such systems cannot tolerate waste discharge containing solids, hence the reason for excluding the discharge from water closets, food waste grinders, clinical sinks and other fixtures that might discharge solids.

TABLE 915.2.2
SIZE OF COMBINATION WASTE AND VENT PIPE

DIAMETER PIPE (inches)	MAXIMUM NUMBER OF DRAINAGE FIXTURE UNITS (dfu)	
	Connecting to a horizontal branch or stack	Connecting to a building drain or building subdrain
2	3	4
2^1/$_2$	6	26
3	12	31
4	20	50
5	160	250
6	360	575

For SI: 1 inch = 25.4 mm.

❖ The system requires that the combination drain and vent pipe be oversized drainage for two reasons. First, the velocity of the drainage flowing in the pipe is greatly reduced by oversizing the pipe, which reduces the possibility of creating a siphon or blowback action. Second, the cross-sectional area of waste flow (or depth of drainage) in the pipe is reduced by increasing the drainage pipe size, which results in a greater area of the pipe above the flow of drainage to permit the movement of air for venting.

Note that when connecting to a stack or an unvented horizontal drain, a combination drain and vent system must have a dry vent connection (see commentary, Sections 915.2.3 and 915.2.4).

VENTS

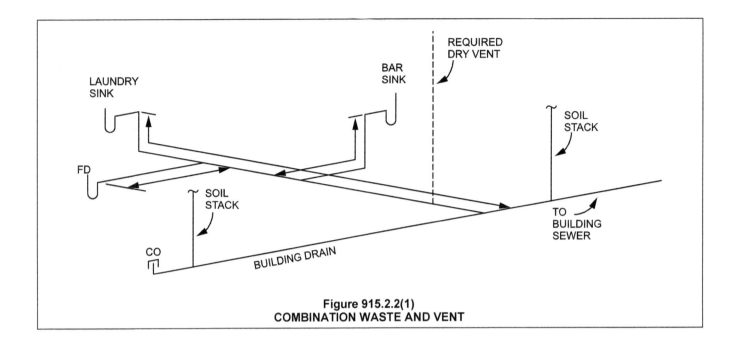

Figure 915.2.2(1)
COMBINATION WASTE AND VENT

For SI: 1 inch = 25.4 mm.

Figure 915.2.2(2)
COMBINATION WASTE AND VENT

VENTS

For SI: 1 inch = 25.4 mm.

**Figure 915.2.2(3)
COMBINATION WASTE AND VENT**

For SI: 1 inch = 25.4 mm.

**Figure 915.2.2(4)
COMBINATION WASTE AND VENT**

2015 INTERNATIONAL PLUMBING CODE® COMMENTARY 9-55

For SI: 1 inch = 25.4 mm.

**Figure 915.2.2(5)
COMBINATION WASTE AND VENT**

915.2.3 Connection. The *combination waste and vent system* shall be provided with a dry vent connected at any point within the system or the system shall connect to a horizontal drain that serves vented fixtures located on the same floor. *Combination waste and vent systems* connecting to building drains receiving only the discharge from one or more stacks shall be provided with a dry vent. The vent connection to the combination waste and vent pipe shall extend vertically to a point not less than 6 inches (152 mm) above the flood level rim of the highest fixture being vented before offsetting horizontally.

❖ A dry vent pipe must connect somewhere within the combination waste and vent system, or the system must connect to a horizontal drain that is vented. A horizontal drain that is vented is any horizontal drain that permits the free flow of air above the liquid flow level in the drain pipe. The code does not define "horizontal drain that is vented," and this has been and still is the subject of much debate. This section implies that any horizontal drain is vented if it is serving fixtures within the same story that are vented by a method in this chapter [see Commentary Figures 915.2.2(1) through 915.2.2(5)]. For example, a branch of a building drain that serves only an individually vented laundry sink would be considered a horizontal drain that is vented. Keep in mind that the sink drain would have to be oversized to allow for both the simultaneous flow of waste and air and the connection of a 2-inch (51 mm) or larger combination drain and vent pipe. Building drains and branches of building drains that receive only the discharge of stacks are not considered to be horizontal drains that are vented. Clearly, by their nature, wet-vent systems, circuit-vent systems and combination waste and vent systems are "horizontal drains that are vented." Section 704.3 prohibits the connection of horizontal branches at or near waste and soil stacks unless those connections are 10 pipe (stack) diameters downstream from the stack.

If a building drain receives only the discharge of stacks, it cannot be considered as a "horizontal drain that is vented" because of the positive pressures expected to occur in such piping. Because the combination waste and vent system is both a drain and a vent for its entire length, it is not important where the dry vent connects to the system. In fact, where these systems connect to a horizontal drain that is vented, the combination waste and vent may be considered as "end vented" because the vent source is at the termination of the combination system. If a combination waste and vent system connects to a stack, a dry vent must connect directly to the combination waste and vent system (see the column titled "Connecting to a horizontal branch or stack" in Table 915.2.2).

This section expresses the same concern as Section 905.4 regarding vertical rise of the dry vent above the flood level rim. The vertical rise of the vent to a point above the flood level rim is necessary to protect the vent against clogging (see commentary, Sections 905.4 and 905.5).

915.2.4 Vent size. The vent shall be sized for the total *drainage fixture unit* load in accordance with Section 906.2.

❖ Where a dry vent connects directly to the combination waste and vent, the vent must be sized in accordance with Section 906.2. The dry vent sizing need not be based on the installed size of the combination waste and vent pipe, but rather on the size of a drain pipe required to carry the dfu load served by the combination drain and vent system.

915.2.5 Fixture branch or drain. The *fixture branch* or *fixture drain* shall connect to the combination waste and vent within a distance specified in Table 909.1. The combination waste and vent pipe shall be considered the vent for the fixture.

❖ The oversized horizontal drain branch is considered to be the vent for any individual fixture it serves. The distance from the fixture trap to the connection to the combination waste and vent system is limited to the length specified in Table 909.1 (see commentary, Section 909.1). Note that the length of a fixture drain sized in accordance with Table 915.2.2 can be unlimited because it, too, is considered a combination waste and vent.

SECTION 916
ISLAND FIXTURE VENTING

916.1 Limitation. Island fixture venting shall not be permitted for fixtures other than sinks and lavatories. Residential kitchen sinks with a dishwasher waste connection, a food waste disposer, or both, in combination with the kitchen sink waste, shall be permitted to be vented in accordance with this section.

❖ This section contains a venting option for plumbing fixtures located where an individual vent cannot be installed without horizontal sections of piping below the flood level rim of the fixture served.

Island fixture venting is another method of venting island sinks and lavatories. The other options are a combination drain and vent regulated by Section 912 and air admittance valves in accordance with Section 918. This section also clarifies that a food waste grinder, a dishwasher or both may be connected to a kitchen sink drain that is served by an island vent where the installation occurs in a residential occupancy.

916.2 Vent connection. The island fixture vent shall connect to the *fixture drain* as required for an individual or common vent. The vent shall rise vertically to above the drainage outlet of the fixture being vented before offsetting horizontally or vertically downward. The vent or *branch vent* for multiple island fixture vents shall extend to a point not less than 6 inches (152 mm) above the highest island fixture being vented before connecting to the outside vent terminal.

❖ For an island fixture vent, the installation is similar to a conventional drainage system that is either individually or common vented. The only difference is that the vent cannot rise 6 inches (152 mm) above the flood level rim before offsetting horizontally. The vent offsets horizontally below the flood level rim so that the piping can be contained under the cabinet or other fixture support. The concern with a vent offsetting horizontally below the flood level rim of the fixture is the possibility of waste backing up into the vent. To cause a stoppage in the drain to be noticeable to the occupants, the vent must rise above the bottom of the fixture bowl before offsetting. This will result in the water rising in the fixture if there is a stoppage (see Commentary Figure 916.2).

The island vent must rise to at least 6 inches (152 mm) above the fixture flood level rim before connect-

Figure 916.2
ISLAND FIXTURE VENT

ing to any other vent or vent terminal. In the event of drain stoppage, waste may be discharged to other vents or the vent terminal if this installation precaution was not taken.

916.3 Vent installation below the fixture flood level rim. The vent located below the *flood level rim* of the fixture being vented shall be installed as required for drainage piping in accordance with Chapter 7, except for sizing. The vent shall be sized in accordance with Section 906.2. The lowest point of the island fixture vent shall connect full size to the drainage system. The connection shall be to a vertical drain pipe or to the top half of a horizontal drain pipe. Cleanouts shall be provided in the island fixture vent to permit rodding of all vent piping located below the *flood level rim* of the fixtures. Rodding in both directions shall be permitted through a cleanout.

❖ Because it is possible to have the vent fill with waste when it is located below the flood level rim, the vent must be installed as a drain. The installation will require drainage pattern fittings and a cleanout. The cleanout may be used to rod in both directions because the piping is a vent. A connection to the drainage piping is required at the lowest point of the vent to prevent it from retaining water and becoming blocked by water or waste (see commentary, Section 916.2). The vent will have to be at a higher elevation than the drain pipe to allow the vent to drain any waste water that has collected therein.

SECTION 917
SINGLE STACK VENT SYSTEM

917.1 Where permitted. A drainage *stack* shall serve as a single stack vent system where sized and installed in accordance with Sections 917.2 through 917.9. The drainage *stack* and *branch* piping shall be the vents for the drainage system. The drainage *stack* shall have a *stack vent*.

❖ The single stack venting system is a method of venting similar to the waste stack venting method, combined with the combination waste and vent method. In other words, the venting of the drainage stack and fixtures is accomplished by oversizing of the drainage stack and the vertical sections of fixture drains so that sufficient venting air is available for the fixtures located within relative close proximity of the stack. The roots of this venting method are in the Philadelphia Single Stack Method that has been successfully used in Philadelphia, Pennsylvania for over 100 years.

At a minimum, the single stack venting system has only a stack vent that must terminate to the outdoors. Where the length of fixture drains exceeds certain limits, additional venting for those drains is required to be vented in accordance with any one of the many venting methods in this chapter (see Section 917.6).

The single stack method is ideal for multistory buildings that have small-to-medium size toilet rooms that are "stacked" one above the other. Such arrangements are typical in hotels, condominiums and apartment buildings (see Commentary Figure 917.8).

917.2 Stack size. Drainage *stacks* shall be sized in accordance with Table 917.2. *Stacks* shall be uniformly sized based on the total connected *drainage fixture unit* load. The *stack vent* shall be the same size as the drainage *stack*. A 3-inch (76 mm) *stack* shall serve not more than two water closets.

❖ The size of a single stack drainage stack is based upon the height and the total number of dfu connected to the stack. Table 917.2 provides the stack dfu limitations according to stack height. The height of the stack is from the connection to the building drain to the most upper drainage connection to the stack. Drainage stacks must be the same size from the base to the termination of the stack vent to the outdoors. The requirement for the stack vent to be the same size as the drainage stack allows the stack vent to be of any length on its path to its termination point outdoors.

For a 3-inch (76 mm) stack, there is a limitation of two water closets on the stack to limit the amount of solids in this small diameter stack so that airflow is not impeded.

TABLE 917.2
SINGLE STACK SIZE

STACK SIZE (inches)	MAXIMUM CONNECTED DRAINAGE FIXTURE UNITS		
	Stacks less than 75 feet in height	Stacks 75 feet to less than 160 feet in height	Stacks 160 feet and greater in height
3	24	NP	NP
4	225	24	NP
5	480	225	24
6	1,015	480	225
8	2,320	1,015	480
10	4,500	2,320	1,015
12	8,100	4,500	2,320
15	13,600	8,100	4,500

For SI: 1 inch = 25.4 mm, 1 foot = 304.8 mm.

❖ Table 917.2 is used simply by identifying the column that includes the height of the stack and then reading down that column to identify the row that has a dfu value that meets or exceeds the total dfu load on the stack. Once the row is identified, the user will find the stack size required from the first left-hand column. For example, if the total load on the stack is 1,000 dfu and the height of the stack is 100 feet (30 480 mm), the third column from the left is chosen and the row for an 8-inch (203 mm) stack size is identified. The "NP" in the table means "not possible."

917.3 Branch size. Horizontal *branches* connecting to a single stack vent system shall be sized in accordance with Table 710.1(2). Not more than one water closet shall discharge into a 3-inch (76 mm) horizontal *branch* at a point within a developed length of 18 inches (457 mm) measured horizontally from the *stack*.

Where a water closet is within 18 inches (457 mm) measured horizontally from the *stack* and not more than one fixture with a drain size of not more than 1¹/₂ inches (38 mm) connects to a 3-inch (76 mm) horizontal *branch*, the *branch* drain connection to the *stack* shall be made with a sanitary tee.

❖ The sizing of the horizontal branches that connect to the stack is in accordance with Table 710.1(2). Only the first and second columns of this table are to be used for sizing. The remaining columns of the table covering stacks have no relationship to this section. For example, if a horizontal branch is required to carry 200 dfu, then a 5-inch (127 mm) horizontal branch is required.

A 3-inch (76 mm) horizontal branch with a water closet located within 18-inch (457 mm) horizontal

Figure 917.3(1)
WATER CLOSET WITHIN 18 INCHES OF STACK AND 3-INCH HORIZONTAL BRANCH CONNECTING TO WYE ON STACK

Figure 917.3(2)
WATER CLOSET WITHIN 18 INCHES OF STACK AND 3-INCH HORIZONTAL BRANCH CONNECTING TO SANITARY TEE ON STACK

VENTS

developed length of the stack has special limitations. If a water closet is located within 18-inch (457 mm) horizontal developed length of the stack and the connection is not made with a sanitary tee (i.e., the connection to the stack is made through a wye), then the only fixture that can be served by that horizontal branch is that water closet [see Commentary Figure 917.3(1)]. If the connection to the stack of a 3-inch (76 mm) horizontal branch serving a single water closet is made using a sanitary tee, then one fixture having a 1¹/₂-inch (38 mm) drain can also be connected to the 3-inch (76 mm) horizontal branch [see Commentary Figure 917.3(2)].

917.4 Length of horizontal branches. The length of horizontal *branches* shall conform to the requirements of Sections 917.4.1 through 917.4.3.

❖ This section refers the reader to subsections covering the length limitations for horizontal branches before additional venting is required.

917.4.1 Water closet connection. Water closet connections shall be not greater than 4 feet (1219 mm) in *developed length* measured horizontally from the *stack*.

Exception: Where the connection is made with a sanitary tee, the maximum *developed length* shall be 8 feet (2438 mm).

❖ Where a horizontal branch connects to the stack through a wye connection and the branch carries the discharge of one or more water closets, the water closets must be not more than 4 feet (1219 mm) in horizontal developed length from the stack [see Commentary Figure 917.4.1(1)]. If the connection at the stack is made with a sanitary tee, the distance limitation is extended to 8 feet (2438 mm) in horizontal developed length [see Commentary Figure 917.4.1(2)].

917.4.2 Fixture connections. Fixtures other than water closets shall be located not greater than 12 feet (3657 mm) in developed length, measured horizontally from the *stack*.

❖ See Commentary Figure 917.4.2.

For SI: 1 foot = 304.8 mm.

Figure 917.4.1(1)
WATER CLOSETS NOT MORE THAN 4 FEET FROM STACK WHEN BRANCH CONNECTS THROUGH WYE TO STACK

917.4.3 Vertical piping in branch. The length of vertical piping in a *fixture drain* connecting to a horizontal *branch* shall not be considered in computing the fixture's distance in *developed length* measured horizontally from the *stack*.

❖ This section simply confirms what is already stated by the phrase "distance in developed length measured horizontally from the stack."

917.5 Minimum vertical piping size from fixture. The vertical portion of piping in a *fixture drain* to a horizontal *branch* shall be 2 inches (51 mm). The minimum size of the vertical portion of piping for a water-supplied urinal or standpipe shall be 3 inches (76 mm). The maximum vertical drop shall be 4 feet (1219 mm). *Fixture drains* that are not increased in size, or have a vertical drop in excess of 4 feet (1219 mm), shall be individually vented.

❖ This section specifies the minimum sizes for the vertical portions of the fixture drains (other than water closets) that are connected to horizontal branches. Vertical sections of fixture drains are required for fixtures such as lavatories, wall-hung urinals and standpipes with traps above the horizontal branch. The minimum size for any vertical section of fixture drain is 2 inches (51 mm), except where standpipes and water-supplied urinals have vertical sections of fixture drains, then the vertical section must be 3 inches (76 mm) (see Commentary Figure 917.4.2). If the vertical sections of the fixture drains cannot be of these minimum sizes (i.e., they are of normal size), then the fixture drains must be vented by a method other than the single stack method.

The height of the vertical section must not be greater than 4 feet (1219 mm). If the height is greater than 4 feet (1219 mm), then the fixture requires venting by a method other than the single stack method. In most situations, fixtures will not be greater than 4 feet (1219 mm) above the horizontal branch so additional venting methods are typically not required.

For SI: 1 foot = 304.8 mm.

Figure 917.4.1(2)
WATER CLOSETS NOT MORE THAN 8 FEET FROM STACK
WHEN BRANCH CONNECTS THROUGH SANITARY TEE TO STACK

Figure 917.4.2
MAXIMUM HORIZONTAL DEVELOPED LENGTH FROM STACK FOR FIXTURES OTHER THAN WATER CLOSETS

917.6 Additional venting required. Additional venting shall be provided where more than one water closet discharges to a horizontal *branch* and where the distance from a fixture trap to the stack exceeds the limits in Section 917.4. Where additional venting is required, the fixture(s) shall be vented by individual vents, common vents, wet vents, circuit vents, or a combination waste and vent pipe. The dry vent extensions for the additional venting shall connect to a *branch vent*, vent *stack*, *stack vent*, air admittance valve, or shall terminate outdoors.

❖ In arrangements where a horizontal branch carries the discharge of two or more water closets and the horizontal developed length of other fixtures discharging to the horizontal branch exceeds the limitations of Section 917.4, then those fixtures (other than the water closets) located beyond the developed length limitation require venting by a method other than the single stack method. The section indicates all the types of venting that can be used for this purpose.

917.7 Stack offsets. Where *fixture drains* are not connected below a horizontal offset in a *stack*, a horizontal offset shall not be required to be vented. Where horizontal *branches* or *fixture drains* are connected below a horizontal offset in a *stack*, the offset shall be vented in accordance with Section 907. Fixture connections shall not be made to a *stack* within 2 feet (610 mm) above or below a horizontal offset.

❖ This section addresses horizontal offsets in a single stack vent system. Where horizontal branches or fixture drains connect below a horizontal offset of a stack, the offset requires venting in accordance with Section 907. Fixture drain connections to the stack must not be made in the area of the stack that is 2 feet (610 mm) above or below the horizontal offset. Where there are no fixtures below the horizontal offset, the offset does not require venting (see Commentary Figure 917.7).

917.8 Prohibited lower connections. *Stacks* greater than 2 *branch intervals* in height shall not receive the discharge of horizontal *branches* on the lower two floors. There shall be no connections to the *stack* between the lower two floors and a distance of not less than 10 pipe diameters downstream from the base of the single stack vented system.

❖ See Commentary Figure 917.8.

VENTS

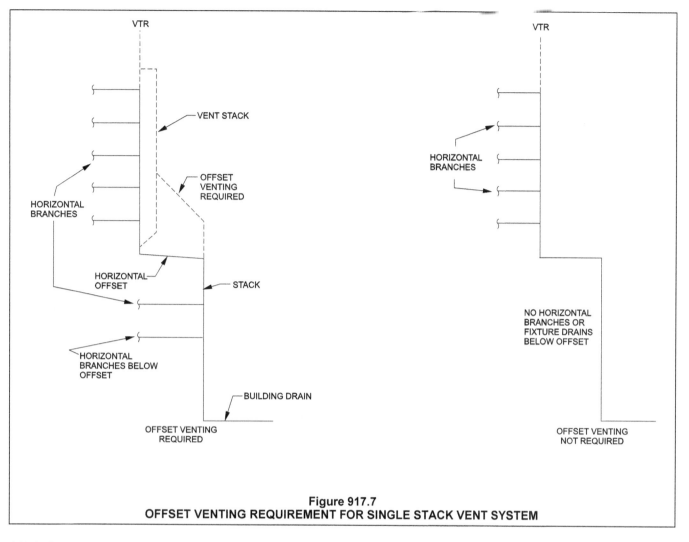

Figure 917.7
OFFSET VENTING REQUIREMENT FOR SINGLE STACK VENT SYSTEM

917.9 Sizing building drains and sewers. The *building drain* and *building sewer* receiving the discharge of a single stack vent system shall be sized in accordance with Table 710.1(1).

❖ The total number of dfu coming from the stack is used to size the building drain and building sewer in accordance with Table 710.1(1).

SECTION 918
AIR ADMITTANCE VALVES

918.1 General. Vent systems utilizing air admittance valves shall comply with this section. Stack-type air admittance valves shall conform to ASSE 1050. Individual and branch-type air admittance valves shall conform to ASSE 1051.

❖ This section contains the requirements for air admittance valves. Air admittance valves must comply with the provisions of ASSE 1050 or ASSE 1051, both of which require marking of each device with the name of the manufacturer, model number or device description and the device classification.

The ASSE 1051 standard divides the classification of air admittance valves into two types: ASSE 1051— Type A and ASSE 1051—Type B. Type A air admittance valves are permitted to serve only as a vent for an individual fixture, and Type B air admittance valves are permitted to serve one or more individual vents (see the definition of "Air admittance valve" in Section 202).

918.2 Installation. The valves shall be installed in accordance with the requirements of this section and the manufacturer's instructions. Air admittance valves shall be installed after the DWV testing required by Section 312.2 or 312.3 has been performed.

❖ The installation of air admittance valves must conform to the requirements of Sections 918.3 through 918.8 and the manufacturer's instructions. Where differences occur between the provisions of the code and the manufacturer's instructions, the most restrictive provisions must be applied.

Air admittance valves are designed to withstand limited positive pressures and because extreme pressures may damage the devices, they must not be subjected to the pressures of system testing. Drain, waste and vent testing involves pressures far in excess of the pressures of normal system operation.

VENTS

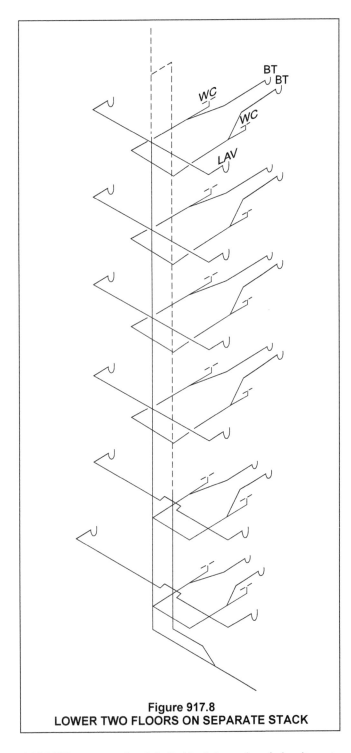

Figure 917.8
LOWER TWO FLOORS ON SEPARATE STACK

918.3 Where permitted. Individual, *branch* and circuit vents shall be permitted to terminate with a connection to an individual or branch-type air admittance valve in accordance with Section 918.3.1. *Stack vents* and vent *stacks* shall be permitted to terminate to stack-type air admittance valves in accordance with Section 918.3.2.

❖ This section directs the reader to the appropriate code sections according to the type of venting application. Section 918.3.1 is for individual and branch venting applications. Section 918.3.2 is for stack vent and vent stack applications.

918.3.1 Horizontal branches. Individual and branch-type air admittance valves shall vent only fixtures that are on the same floor level and connect to a *horizontal branch drain*. Where the horizontal *branch* is located more than four branch intervals from the top of the stack, the horizontal *branch* shall be provided with a relief vent that shall connect to a vent stack or stack vent, or extend outdoors to the open air. The relief vent shall connect to the *horizontal branch drain* between the stack and the most downstream *fixture drain* connected to the *horizontal branch drain*. The relief vent shall be sized in accordance with Section 906.2 and installed in accordance with Section 905. The relief vent shall be permitted to serve as the vent for other fixtures.

❖ As drainage flows down a stack, it tends to draw air downward with it. Air located in the drainage stack ahead of the flow also tends to be pushed downward by the flowing drainage—similar to the action of a piston in a cylinder. Measurements at various elevations of the air pressure in a stack (flowing at near full capacity) with many branch intervals will indicate that the flow creates negative pressures in the upper part of the stack and positive pressures in the lower part of the stack [see Commentary Figure 918.3.1(1)].

A horizontal branch drain served by an air admittance valve must be located within the top four branch intervals of the stack where pressure differentials would be tolerable. Otherwise, a method of relieving pressure is required because of the greater pressure differentials expected in lower portions of a stack (see commentary, Section 918.3.2). Commentary Figures 918.3.1(2) through 918.3.1(5) show various applications for using air admittance valves. Commentary Figure 918.3.1(6) shows a prohibited use of an air admittance valve.

Figure 918.3.1(1)
AIR ADMITTANCE—INDIVIDUAL VENT

918.3.2 Stack. Stack-type air admittance valves shall be prohibited from serving as the vent terminal for vent *stacks* or *stack vents* that serve drainage *stacks* having more than six *branch intervals*.

❖ When a building has a drainage stack that has more than six branch intervals, stack-type air admittance valves must not be used to vent these drains. The pressure fluctuations in stacks with more than six branch intervals can be excessive, and could cause the loss of trap water seals, thereby allowing sewer gases into the building (see Commentary Figure 918.3.2).

918.4 Location. Individual and branch-type air admittance valves shall be located a minimum of 4 inches (102 mm) above the *horizontal branch drain* or *fixture drain* being vented. Stack-type air admittance valves shall be located not less than 6 inches (152 mm) above the *flood level rim* of the highest fixture being vented. The air admittance valve shall be located within the maximum *developed length* permitted for the vent. The air admittance valve shall be installed not less than 6 inches (152 mm) above insulation materials.

❖ An air admittance valve has one moving part (a seal), which must be maintained a safe distance above the drain served. In the event of a drain stoppage, the seal may become inoperable or operate improperly if waste is permitted to rise into the air admittance valve assembly. Note that the air admittance valve need not extend above the flood level rim of the fixture served (see Commentary Figure 918.4) because in the event of a drain blockage, such device will trap air between it and the rising waste, thereby protecting the device from contamination.

The maximum developed length of a vent pipe equipped with an air admittance valve is regulated by

Figure 918.3.1(2)
COMMON VENT

Figure 918.3.1(4)
CIRCUIT VENT

Figure 918.3.1(3)
WET VENT

Figure 918.3.1(5)
BRANCH VENT

VENTS

Figure 918.3.1(6)
VENT CONNECTION NOT PERMITTED

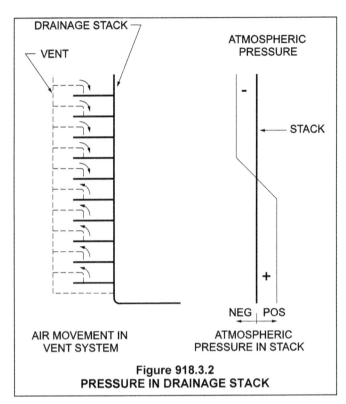

Figure 918.3.2
PRESSURE IN DRAINAGE STACK

For SI: 1 inch = 25.4 mm.

Figure 918.4
DISTANCE ABOVE DRAIN

Sections 906.2 and 906.3. Section 906.2 requires increasing the size of a vent exceeding 40 feet (12 192 mm) in developed length (measured from the air admittance valve to the point of connection to the drain pipe) by one nominal pipe size.

The distance between a trap and its protecting vent must also meet the maximum developed length requirements of Table 909.1.

An air admittance valve must be located a safe distance above insulation materials that may block air inlets or otherwise impair the operation of the device.

918.5 Access and ventilation. Access shall be provided to all air admittance valves. Such valves shall be installed in a location that allows air to enter the valve.

❖ Because an air admittance valve is a device with a moving part, it must be accessible for inspection, service, repair or replacement. The code defines "Access (to)" as being able to be reached but that first may require the removal of a panel, door or similar obstruction. Access to an air admittance valve is not achieved if it is necessary to remove any portion of the structure's permanent finish materials such as drywall, plaster, paneling, built-in furniture or cabinets or any other similar permanently affixed building component.

Where air admittance valves are located in confined spaces, ventilation openings are required. Because these devices admit air into the drainage system, an unimpeded air supply must be available at all times. Additionally, access and ventilation requirements are addressed in the manufacturer's installation instructions.

918.6 Size. The air admittance valve shall be rated in accordance with the standard for the size of the vent to which the valve is connected.

❖ The size of the vent pipe to which the air admittance valve is connected is regulated by Section 916. The air admittance valve is sized based on the required size of the vent pipe and the dfu load limitations found in the manufacturer's installation instructions. The dfu load rating is based on testing performed in accordance with ASSE 1051. The air admittance valves now available have dfu limitations that are consistent with the maximum number permitted on a horizontal branch as found in Table 710.1(2).

918.7 Vent required. Within each plumbing system, not less than one *stack vent* or vent *stack* shall extend outdoors to the open air.

❖ Conventional venting methods are intended to relieve both positive and negative pressures that occur in the drainage system. The primary function of an air admittance valve is relieving negative pressures. The intent of this section is to require at least one conventional vent pipe per plumbing system to terminate outside the building. The vent to the open air serves both to enhance the positive pressure neutralization in the drainage system and to protect the plumbing system from pressure excursions caused by the use of public sewer cleaning equipment (see Commentary Figure 918.7). This section does not limit the size of the plumbing system or the distances to the required stack vent or vent stack and states no required size for the stack vent or vent stack.

918.8 Prohibited installations. Air admittance valves shall not be installed in nonneutralized special waste systems as described in Chapter 8 except where such valves are in compliance with ASSE 1049, are constructed of materials *approved* in accordance with Section 702.5 and are tested for chemical resistance in accordance with ASTM F 1412. Air admittance valves shall not be located in spaces utilized as supply or return air plenums. Air admittance valves without an engineered design shall not be utilized to vent sumps or tanks of any type.

❖ Because of the deleterious nature of the material and vapors associated with corrosive wastes, standard construction air admittance valves are prohibited for use with such systems. Special chemically resistant air admittance valves that comply with ASSE 1049 and are tested for chemical resistance in accordance with ASTM F 1412 are the only type that can be used for venting nonneutralized chemical waste systems.

Air admittance valves are not permitted in spaces where pressure conditions adversely affect the valve's operation. They are also not allowed in concealed wall and ceiling spaces used as plenums as part of an air distribution system that is under negative or positive pressure. An air admittance valve located in this environment will be affected and may not function properly. For example, if located in a return plenum, greater negative pressure will be allowed to develop in the drain before the device will open and admit air to the drainage system. If located in a supply plenum, the device may be held open continuously because the pressure at the device inlet will be greater than the internal pressure of the drainage system; this is the case when the device functions normally to neutralize negative pressures.

Air admittance valves must not be used to vent sumps or tanks as there is positive pressure created when the tank or sump is filling. Because air admittance valves cannot relieve positive pressure within the sump or tank, the positive pressure might be relieved through the traps of the fixtures connected to the sump or tank. However, where air admittance valves are used as part of an engineered design that allows for positive pressure relief in other ways, air admittance valves can be used to allow air into the tank or sump (see Commentary Figures 918.8(1) and (2)].

Figure 918.7
VENT TO THE OPEN AIR REQUIRED

Figure 918.8(1)
PROHIBITED USE OF AIR ADMITTANCE VALVE

Figure 918.8(2)
ONE MANUFACTURER'S ENGINEERED DESIGN FOR USE OF AN AIR ADMITTANCE VALVE ON A SUMP

SECTION 919
ENGINEERED VENT SYSTEMS

919.1 General. Engineered vent systems shall comply with this section and the design, submittal, approval, inspection and testing requirements of Section 105.4.

❖ This section contains an alternative sizing method for individual vents. An engineered vent system is considered to be an alternative engineered design. As such, it must comply with both the requirements specifically for this engineered vent system as well as the general design, documentation, inspection, testing and approval of an alternative engineered design (see commentary, Section 105.4). Note that the engineered vent system must be designed, signed and sealed by a registered design professional (see commentary, Section 105.4.4).

919.2 Individual branch fixture and individual fixture header vents. The maximum *developed length* of individual fixture vents to vent branches and vent headers shall be determined in accordance with Table 919.2 for the minimum pipe diameters at the indicated vent airflow rates.

The individual vent airflow rate shall be determined in accordance with the following:

$$Q_{h,b} = N_{n,b} Q_v \quad \text{(Equation 9-1)}$$

For SI: $Q_{h,b} = N_{n,b} Q_v$ (0.4719 L/s)

where:

$N_{n,b}$ = Number of fixtures per header (or vent *branch*) ÷ total number of fixtures connected to vent *stack*.

$Q_{h,b}$ = Vent *branch* or vent header airflow rate (cfm).

Q_v = Total vent *stack* airflow rate (cfm).

Q_v (gpm) = $27.8 \, r_s^{2/3} (1 - r_s) D^{8/3}$

Q_v (cfm) = $0.134 \, Q_v$ (gpm)

where:

D = Drainage *stack* diameter (inches).

Q_w = Design discharge load (gpm).

r_s = Waste water flow area to total area.

$= \dfrac{Q_w}{27.8 \, D^{8/3}}$

Individual vent airflow rates are obtained by equally distributing $Q_{h,b}$ into one-half the total number of fixtures on the *branch* or header for more than two fixtures; for an odd number of total fixtures, decrease by one; for one fixture, apply the full value of $Q_{h,b}$.

Individual vent *developed length* shall be increased by 20 percent of the distance from the vent *stack* to the fixture vent connection on the vent *branch* or header.

❖ The individual vent pipe size could be reduced to as small as $1/2$ inch (12.7 mm) in diameter by calculating the vent airflow rate. The sizing method is based on every fixture being individually vented and connected to a vent header or vent stack. The reduced size applies only to the length of the individual vent up to the point where it connects to either the branch vent or vent stack. This distance is also the individual vent-developed length.

Each fixture or group of fixtures must be individually calculated to determine the individual vent airflow rate that is applied to Section 918.2 to determine the individual vent minimum pipe size. For example purposes, the fifth floor of a 10-story building will be

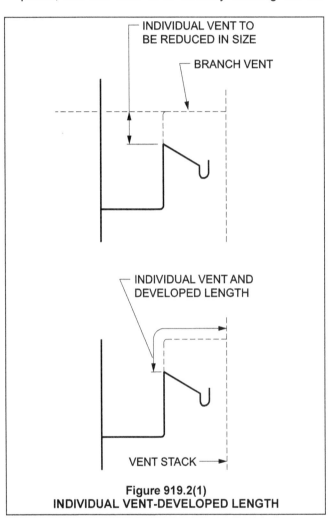

Figure 919.2(1)
INDIVIDUAL VENT-DEVELOPED LENGTH

TABLE 919.2
MINIMUM DIAMETER AND MAXIMUM LENGTH OF INDIVIDUAL BRANCH FIXTURE VENTS AND INDIVIDUAL FIXTURE HEADER VENTS FOR SMOOTH PIPES

DIAMETER OF VENT PIPE (inches)	INDIVIDUAL VENT AIRFLOW RATE (cubic feet per minute)																			
	Maximum developed length of vent (feet)																			
	1	2	3	4	5	6	7	8	9	10	11	12	13	14	15	16	17	18	19	20
$1/2$	95	25	13	8	5	4	3	2	1	1	1	1	1	1	1	1	1	1	1	1
$3/4$	100	88	47	30	20	15	10	9	7	6	5	4	3	3	3	2	2	2	2	1
1	—	—	100	94	65	48	37	29	24	20	17	14	12	11	9	8	7	7	6	6
$1 1/4$	—	—	—	—	—	—	—	100	87	73	62	53	46	40	36	32	29	26	23	21
$1 1/2$	—	—	—	—	—	—	—	—	—	—	100	96	84	75	65	60	54	49	45
2	—	—	—	—	—	—	—	—	—	—	—	—	—	—	—	—	—	—	—	100

For SI: 1 inch = 25.4 mm, 1 cubic foot per minute = 0.4719 L/s, 1 foot = 304.8 mm.

analyzed to determine the minimum individual vent pipe size [see Commentary Figures 918.2(1) and 918.2(2)].

Example:

Building statistics:

Building height	10 stories
Total dfus to drainage stack	320
Size of drainage stack [Table 710.1(2)]	4 inches
Size of vent stack with developed length exceeding 40 feet (Section 916.1)	2$\frac{1}{2}$ inches
Total number of plumbing fixtures	108

Fifth Floor

Number of fixtures (1 sink, 3 water closets, 5 lavatories, 3 showers)	12
Total dfu for horizontal branch	31
Size of horizontal branch [Table 710.1(2)]	4 inches
Size of branch vent (Section 916.2)	2 inches
Maximum developed length of individual vent	25 feet

The Q (gpm) must first be calculated using the equation:

$$Q\ (gpm) = 27.8 r_s^{2/3}(1-r_s)D^{8/3}$$

The diameter of the drainage stack (D) is 4 inches (102 mm). The r_s may be determined by calculation or by use of the values in Commentary Figure 919.2 (an excerpt from *National Bureau of Standards Monograph 31*).

Commentary Figure 919.2 lists an r_s of 0.25 for a 4-inch (102 mm) stack with a total fixture unit load of 320. Substituting the values into the equation, it reads:

$$Q\ (gpm) = 27.8\ (0.25)^{2/3}\ (1-0.25)(4)^{8/3}$$

$$Q\ (gpm) = 333.6\ gpm$$

For SI: 1 inch = 25.4 mm.

Figure 919.2(2)
EXAMPLE OF ENGINEERED REDUCED-SIZE VENT

STACK DIAMETER (inches)	EQUIVALENT LOAD (fixture units)	PROPORTION OF CROSS SECTION OCCUPIED BY FALLING WATER	WATER DISCHARGE RATE (gpm)
4	43	0.15	47.5
4	140	0.20	76.7
4	320	0.25	111.0
4	530	0.29	144.0
4	580	0.30	151.0
4	850	0.33	180.0

For SI: 1 inch = 25.4 mm, 1 gallon per minute = 3.785 L/m.

Figure 919.2
MAXIMUM LOADS ON MULTIPLE-STORY DRAINAGE STACKS

The value Q (gpm) must be converted to Q (cfm) by the following equation:

$$Q \text{ (cfm)} = 0.134\ Q \text{ (gpm)}$$

Substituting the value for Q (gpm):

$$Q \text{ (cfm)} = 0.134\ (333.6 \text{ gpm})$$

$$Q \text{ (cfm)} = 44.7 \text{ cfm}$$

The vent header airflow rate is calculated using the equation:

$$Q_{h,b} = N_{n,b}\ Q$$

The value of $N_{n,b}$ for the fifth floor is equal to the number of fixture vents on the fifth floor divided by the total number of fixtures connected to the drainage stack.

Substituting the values into the vent header equation:

$$Q_{h,b} = (0.11)(44.7 \text{ cfm})$$

$$Q_{h,b} = 4.9 \text{ cfm}$$

The individual vent airflow rate is calculated by dividing the branch airflow rate by half the number of fixtures connected to the branch. The fifth floor has 12 fixtures, half of which equals six. The individual vent airflow rate becomes:

Individual vent airflow rate = 4.9 cfm/6 = 0.82

The value used for Table 919.2 is rounded up to 1. With a maximum developed length of 25 feet (7620 mm), each individual vent on the fifth floor may be a minimum size of $^1/_2$ inch (12.7 mm).

For SI: 1 inch = 25.4 mm, 1 foot = 304.8 mm,
1 gallon per minute = 3.785 L/m,
1 cubic foot per minute = 0.4719 L/s.

Note that the example listed here would be solved in a similar manner for a design using metric values or "System International" (SI). The information shown in the code text as "For SI" includes the necessary conversion factor to change the resultant value from cubic feet per minute (cfm) to liters per second (L/s).

When the individual vent airflow rate is determined by the equation in Section 919.2, the size and developed length are determined by Table 919.2. The values in the table indicate the maximum developed length for a given pipe size and the individual vent airflow rate. Note that as a precaution against vent blockage, it may be desirable to use larger-diameter vent piping from the connection to the fixture drain to the point 6 inches (152 mm) above the flood level rim of the fixture served.

SECTION 920
COMPUTERIZED VENT DESIGN

920.1 Design of vent system. The sizing, design and layout of the vent system shall be permitted to be determined by *approved* computer program design methods.

❖ This section provides an alternative sizing method for venting systems, similar to Section 918.

This section is the logical complement to Section 714, which addresses computerized drainage system design (see commentary, Section 714).

920.2 System capacity. The vent system shall be based on the air capacity requirements of the drainage system under a peak load condition.

❖ This section says to design for the worst-case scenario. The airflow capacity of the venting system must be equal to or greater than the capacity demanded by a drainage system under its maximum anticipated load condition.

Bibliography

The following resource materials are referenced in this chapter or are relevant to the subject matter addressed in this chapter.

ASHRAE, *Fundamentals Handbook 2001.* Atlanta, GA: American Society of Heating, Refrigerating and Air-Conditioning Engineers, Inc., 2001.

ASSE 1051-02, *Performance Requirements for Air Admittance Valves for Plumbing Drainage Systems, Fixtures and Branch Devices.* Westlake, OH: American Society of Sanitary Engineers, 2002.

"Vents and Venting" (Chapter 17). *ASPE Data Book.* Westlake, CA: American Society of Plumbing Engineers, 1991.

Chapter 10:
Traps, Interceptors and Separators

General Comments

The plumbing trap is, perhaps, the single most important device in a sanitary plumbing system as it prevents exposure of a building's occupants to sewer gases and vapors contained within the drainage system. Many requirements in the code are based upon protecting the "liquid seal" created by traps located at every plumbing fixture. Many more complicated designs of trap devices have been developed, used and abandoned throughout plumbing history in favor of the simplicity and reliability of the "U"-shaped trap.

If only "domestic" waste water generated by humans for bathing, household cooking and cleaning and the flushing of waste eliminations was to be conveyed and treated, there would be no need for interceptors and separators. However, our society also generates commercial and industrial waste waters from manufacturing processes, mass food production, vehicle cleaning, chemical production and product recycling that has components that are detrimental to the piping systems and waste-water treatment processes. These materials must be prevented from entering waste-water systems by the use of interceptors and separators.

Purpose

This chapter contains design requirements and installation limitations for traps. Prohibited types of traps are specifically identified. Where fixtures do not frequently replenish the liquid in traps, a method is provided to insure that the trap seal of the trap will be replenished.

Requirements for the design and location of various types of interceptors and separators are provided. Specific venting requirements are given for separators and interceptors as those requirements are not addressed in Chapter 9.

SECTION 1001
GENERAL

1001.1 Scope. This chapter shall govern the material and installation of traps, interceptors and separators.

❖ This section contains the scope of the chapter, which is to provide and maintain a physical barrier between a building and its drainage system, protect the drainage system from prohibited materials and prevent the emission of sewer gases into the building.

SECTION 1002
TRAP REQUIREMENTS

1002.1 Fixture traps. Each plumbing fixture shall be separately trapped by a liquid-seal trap, except as otherwise permitted by this code. The vertical distance from the fixture outlet to the trap weir shall not exceed 24 inches (610 mm), and the horizontal distance shall not exceed 30 inches (610 mm) measured from the centerline of the fixture outlet to the centerline of the inlet of the trap. The height of a clothes washer standpipe above a trap shall conform to Section 802.4. A fixture shall not be double trapped.

Exceptions:

1. This section shall not apply to fixtures with integral traps.

2. A combination plumbing fixture is permitted to be installed on one trap, provided that one compartment is not more than 6 inches (152 mm) deeper than the other compartment and the waste outlets are not more than 30 inches (762 mm) apart.

3. A grease interceptor intended to serve as a fixture trap in accordance with the manufacturer's installation instructions shall be permitted to serve as the trap for a single fixture or a combination sink of not more than three compartments where the vertical distance from the fixture outlet to the inlet of the interceptor does not exceed 30 inches (762 mm) and the *developed length* of the waste pipe from the most upstream fixture outlet to the inlet of the interceptor does not exceed 60 inches (1524 mm).

4. Floor drains in multilevel parking structures that discharge to a building storm *sewer* shall not be required to be individually trapped. Where floor drains in multilevel parking structures are required to discharge to a combined *building sewer* system, the floor drains shall not be required to be individually trapped provided that they are connected to a main trap in accordance with Section 1103.1.

❖ This section provides design and installation criteria for fixture traps, which are needed to protect building occupants from the hazards associated with the contents of drainage systems. A trap is the method used to keep sewer gases from emanating out of the drainage system. The liquid seal prevents sewer gases and aerosol-borne bacteria from entering the building space. Sewer gases often contain methane gas and could cause an explosion when exposed to an ignition source.

The configuration of a trap interferes with the flow of drainage; however, the interference is minimal because of the construction of the trap and the relatively high inlet velocity of the waste flow [see Commentary Figure 1002.1(1), Exception 1]. Double trapping of a fixture is prohibited because of the additional obstruction of flow and potential for stoppages. Double trapping will cause air to be trapped between two trap seals, and the "air-bound" drain will impede the flow.

The primary purpose of the horizontal distance of 30 inches (762 mm) from a fixture outlet to the trap weir is to limit the amount of bacterial growth and resulting odor. It is desirable to locate the trap as close as possible to the fixture. Buildup on the wall of the fixture outlet pipe will breed bacteria and cause odors to develop. The vertical distance is also limited to control the velocity of the drainage flow. If the trap has a long vertical separation distance from the fixture, the velocity of flow at the trap inlet can create a self-siphoning of the trap [see Commentary Figure 1002.1(1), Exception 2].

Exception 1 addresses fixtures such as water closets and urinals that have integral traps and, therefore, do not require a field-installed trap [see Commentary Figure 1002.1(1), Exception 1].

Exception 2 allows a single trap to serve multiple compartment sinks and other combination fixtures. A difference in elevation of sink wells (compartments) in combination fixtures will affect the vertical distance between the fixture outlet and the trap. The head pressure difference caused by elevation could affect the drainage rate of the fixture or cause waste from one well of a sink to discharge into a lower well. The limitation on distance between outlets is intended to minimize the amount of waste piping on the inlet side of the trap [see Commentary Figure 1002.1(1), Exception 2].

Exception 3 in Commentary Figure 1002.1(2) intends to permit the common practice of allowing a grease interceptor to act as the fixture trap where recommended by the grease interceptor manufacturer. Because grease interceptors function like a trap, double trapping will result where the typically designed grease interceptor is preceded by a fixture trap. Grease Interceptors of the submerged inlet design will not allow the gases in the holding tank to enter the fixture drain; therefore, the source of odor would be limited to the section of piping between the fixture outlet and the interceptor inlet [see Commentary Figure 1002.1(2)].

For SI: 1 inch = 25.4 mm.

Figure 1002.1(1)
TRAP REQUIREMENTS

TRAPS, INTERCEPTORS AND SEPARATORS

Figure 1002.1(2)
TRAP REQUIREMENTS

Exception 4 exempts the requirement for providing traps for floor drains in parking garages where the floor drains connect to a storm sewer (but not a combined sewer). Traps can be damaged where the garage air temperature is below freezing even though the traps are heat traced and insulated. Traps are for preventing the emission of sanitary sewer gases from entering a building. There is no need for a trap where the trap is not connected to a sanitary sewer system. Exception 4 also exempts the requirement for providing traps for each floor drain in a parking garage where the floor drain system is required to connect to a combined sewer system. Traps for floor drains in parking garages are subject to freeze damage and typically do not receive enough flow during drier weather to keep individual traps with sufficient liquid seal to prevent sewer gases from escaping. By having one main trap to isolate the combined public sewer from the parking garage floor drains, the garage is protected from sewer gas so individual floor drain traps are not necessary.

1002.2 Design of traps. Fixture traps shall be self-scouring. Fixture traps shall not have interior partitions, except where such traps are integral with the fixture or where such traps are constructed of an *approved* material that is resistant to corrosion and degradation. Slip joints shall be made with an *approved* elastomeric gasket and shall be installed only on the trap inlet, trap outlet and within the trap seal.

❖ A trap must have a pattern allowing unobstructed flow to the drain. Interior partitions are allowed where constructed of a material resistant to corrosion and degradation. Bottle traps, having an interior partition, are commonly used in areas where the tailpiece of the fixture is very close to the wall such as is sometimes encountered with pedestal lavatory installation. However, the use of bottle traps is not limited to only close quarter installations as they are also used where a more "refined" appearance is desired for an exposed trap application.

Most traps are premanufactured; however, a trap can be field fabricated with pipe and fittings (see commentary, Section 705.16 and Commentary Figure 1002.2).

1002.3 Prohibited traps. The following types of traps are prohibited:

1. Traps that depend on moving parts to maintain the seal.
2. Bell traps.
3. Crown-vented traps.
4. Traps not integral with a fixture and that depend on interior partitions for the seal, except those traps constructed of an *approved* material that is resistant to corrosion and degradation.
5. "S" traps.
6. Drum traps.

Exception: Drum traps used as solids interceptors and drum traps serving chemical waste systems shall not be prohibited.

❖ A trap is intended to be a simple U-shaped piping arrangement that offers minimal resistance to flow. Based on their design or configuration, prohibited traps are not a simple U-shape piping design. Such traps typically impede drainage flow by moving parts,

Figure 1002.2
SLIP JOINTS IN TRAPS

TRAPS, INTERCEPTORS AND SEPARATORS

Figure 1002.3
PROHIBITED TRAPS

design (tends to clog with debris), flow pattern and ability to lose a trap seal (see Commentary Figure 1002.3).

Item 1 refers to mechanical traps that use moving parts such as floats or flappers. Such designs are not dependable because of corrosion, clogging and waste deposits that interfere with the operation of moving components and seals.

Item 2 prohibits bell traps that, because of their design, tend to clog with debris, and their trap seal has a larger exposed surface area that accelerates evaporation.

Item 3 more accurately describes a venting arrangement than a type of trap. A crown-vented trap is a factory-made or field-constructed "P" or "S" trap that has a vent opening located in the crown of the outlet weir of the trap. The flow pattern and momentum of waste flow will cause waste to enter and eventually clog the vent connection.

Item 4 refers to independent traps that are not integral with fixtures. For example, water closet traps are partition traps but, of course, are not prohibited because they are regulated as part of the fixture.

Partition traps are single-wall traps that rely on a single partition to separate the house side of the trap from the sewer side of the trap. Partition traps made of materials that are not corrosion resistant can structurally fail without any leakage that would indicate a failure of the partition; therefore, the inability to maintain a trap seal would not be indicated by a partition trap.

Item 5 prohibits factory-made and field-constructed "S" configuration traps because of their inherent ability to self-siphon, which results in a loss of trap seal.

Item 6 prohibits drum traps because they are not self-cleaning and therefore clog easily. An exception provides for the use of such traps where installed as a solids interceptor or serving chemical waste systems. Many drum traps are used to prevent solids from entering the drainage system. Hair traps, metal traps, plaster traps and single-fixture acid neutralizers are variations of the drum-trap design.

1002.4 Trap seals. Each fixture trap shall have a liquid seal of not less than 2 inches (51 mm) and not more than 4 inches (102 mm), or deeper for special designs relating to accessible fixtures.

❖ A liquid seal of 2 inches (51 mm) is standard for most traps. Some larger pipes, 3 through 6 inches (76 through 152 mm), have a greater seal of up to 4 inches (102 mm) to construct a smooth pattern of flow for the given pipe size [See Commentary Figure 1002.1(1), Exception 1].

A trap seal must be deep enough to resist the pressures that can develop in a properly vented drainage system, but not deep enough to promote the retention of solids or the growth of bacteria.

Traps that do not periodically receive waste discharge will eventually lose their seal as a result of evaporation. The rate of trap seal evaporation is somewhat dependent on the location of the trap. For example, water in fixture traps in environments with high ambient temperatures or high-volume air movement will evaporate rapidly. Fixtures, such as floor drains, typically have trap seals that are subject to loss by evaporation and, therefore, must be protected by trap seal primers (see Commentary Figure 1002.4). A deep seal will not prevent the loss of a trap seal, but will simply lengthen the time it will take for it to evaporate.

1002.4.1 Trap seal protection. Trap seals of *emergency floor drain* traps and trap seals subject to evaporation shall be protected by one of the methods in Sections 1002.4.1.1 through 1002.4.1.4.

❖ There are two methods to protect the trap seal against evaporation: 1) a trap seal primer, and 2) a trap seal protection device. Sections 1002.4.1.1 through 1002.4.1.4 cover these methods.

1002.4.1.1 Potable water-supplied trap seal primer valve. A potable water-supplied trap seal primer valve shall supply water to the trap. Water-supplied trap seal primer valves shall conform to ASSE 1018. The discharge pipe from the trap seal primer valve shall connect to the trap above the trap seal on the inlet side of the trap.

❖ The trap primer concept is simple: add a small amount of water to the trap every so often and the trap seal will remain intact. A potable water-supplied trap primer connects to the potable water distribution system and supplies a small amount of potable water to the trap when a triggering event occurs. For one type of potable water trap seal primer, the triggering event is the use of cold water at a fixture. The cold water distribution piping near the fixture experiences a slight drop of pressure when the cold water valve at the fixture is opened. The trap seal primer device senses the drop in pressure and briefly opens a valve to "waste" a small amount of potable water into a tube that is connected to the trap. Frequent use of the fixture ensures that the trap seal doesn't evaporate. The discharge tube must connect at a level higher than the trap seal so that the tube doesn't become blocked by waste debris floating in or on top of the trap seal (see Commentary Figure 1002.4.1.1).

1002.4.1.2 Reclaimed or gray water-supplied trap seal primer valve. A reclaimed or gray water-supplied trap seal primer valve shall supply water to the trap. Water-supplied trap seal primer valves shall conform to ASSE 1018. The quality of reclaimed or gray water supplied to trap seal primer valves shall be in accordance with the requirements of the manufacturer of the trap seal primer valve. The discharge pipe from the trap seal primer valve shall connect to the trap above the trap seal, on the inlet side of the trap.

❖ See the commentary for Section 1002.4.1.1, except the valve is connected to a pressurized reclaimed water source or a pressurized (and treated) graywater source. The triggering fixture could be a water closet or a urinal that is supplied with reclaimed or gray water for flushing. The discharge tube must connect at a level higher than the trap seal, so that the tube doesn't become blocked by waste debris floating in or on top of the trap seal. Note that trap seal primer

For SI: 1 inch = 25.4 mm.

**Figure 1002.4
TRAP SUBJECT TO EVAPORATION**

Figure 1002.4.1.1
TRAP PRIMER FOR FLOOR DRAINS

valve manufacturers provide the needed water quality requirements for the nonpotable water to be used in the trap seal primer valve.

1002.4.1.3 Waste water-supplied trap primer device. A waste water-supplied trap primer device shall supply water to the trap. Waste water-supplied trap primer devices shall conform to ASSE 1044. The discharge pipe from the trap seal primer device shall connect to the trap above the trap seal on the inlet side of the trap.

❖ This type of device simply diverts some raw gray water from a fixture that discharges gray water. The fixture is typically a lavatory. Usually such lavatory has a grid strainer outlet to keep debris from being discharged from the fixture. However, a grid strainer is not a code requirement.

1002.4.1.4 Barrier-type trap seal protection device. A barrier-type trap seal protection device shall protect the floor drain trap seal from evaporation. Barrier-type floor drain trap seal protection devices shall conform to ASSE 1072. The devices shall be installed in accordance with the manufacturer's instructions.

❖ This device is an elastomeric "check valve" that is inserted into the pipe opening of a floor drain. When the floor drain is accepting water, the flapper allows the water to flow into the pipe. Otherwise, the flapper returns to its closed position to prevent trap seal evaporation from exposure to the air in the room.

1002.5 Size of fixture traps. Fixture trap size shall be sufficient to drain the fixture rapidly and not less than the size indicated in Table 709.1. A trap shall not be larger than the drainage pipe into which the trap discharges.

❖ The minimum fixture trap sizes are listed in Table 709.1. The minimum trap sizes for fixtures not listed in Table 709.1 must be the size of the fixture outlet, but in no case less than $1^1/_4$ inches (32 mm). The code does not prescribe a maximum trap size; however, if a trap is oversized, it will not scour (cleanse) itself and is therefore prone to clogging. A trap that is larger than the drainage pipe into which it discharges is also subject to frequent clogging because the reduced size outlet pipe will not allow a waste flow velocity that is adequate to scour and clean the trap (see commentary, Table 709.1 and Section 704.2).

1002.6 Building traps. Building (house) traps shall be prohibited.

❖ A building trap creates a problem because it is an obstruction to flow, possibly causing a complete stoppage in the drainage system.

1002.7 Trap setting and protection. Traps shall be set level with respect to the trap seal and, where necessary, shall be protected from freezing.

❖ Traps must be set level to maintain the trap seal depth and reduce the possibility of self-siphoning (see Commentary Figure 1002.7).

1002.8 Recess for trap connection. A recess provided for connection of the underground trap, such as one serving a bathtub in slab-type construction, shall have sides and a bottom of corrosion-resistant, insect- and verminproof construction.

❖ Where a trap is installed below the building, such as in a recess below a slab on grade, requirements similar to those in the *International Building Code®* (IBC®) for

**Figure 1002.6
BUILDING TRAP**

waterproofing, ratproofing and termite protection apply (see Commentary Figure 1002.6).

1002.9 Acid-resisting traps. Where a vitrified clay or other brittleware, acid-resisting trap is installed underground, such trap shall be embedded in concrete extending 6 inches (152 mm) beyond the bottom and sides of the trap.

❖ Because of their brittle nature, earthenware-type traps, including vitrified clay, must be embedded in concrete where installed below grade (see Section 702.5).

1002.10 Plumbing in mental health centers. In mental health centers, pipes and traps shall not be exposed.

❖ In certain mental health centers, patients are considered a danger to themselves. As a result of this concern, special care is required to secure the plumbing fixtures and piping in these facilities.

SECTION 1003
INTERCEPTORS AND SEPARATORS

1003.1 Where required. Interceptors and separators shall be provided to prevent the discharge of oil, grease, sand and other substances harmful or hazardous to the public sewer, the private sewage system or the sewage treatment plant or processes.

❖ The design and installation of sanitary drain and sewer systems and the waste treatment systems for disposal of waste water are based upon the known general characteristics of "domestic" waste water. Domestic waste water is largely that which is generated by the use of water by humans for bathing, household cooking and cleaning and for the flushing of waste eliminations from humans. Waste waters generated from commercial and industrial sources such as manufacturing processes, mass food production, vehicle cleaning, chemical production and product recycling can have detrimental effects to the materials and methods used to convey the waste water as well as the mechanical, biological and chemical processes used to make the waste water safe to discharge into the environment. Therefore, these

**Figure 1002.7
INSTALLATION OF TRAP**

"commercial/industrial" waste waters must have the detrimental components removed in order for the waste water to be safe for drainage systems and less problematic for domestic waste-water treatment biological and chemical processes.

The terms "separator" and "interceptor" are used interchangeably in this code, this commentary and by many involved in the separator/interceptor industry. Both terms mean a device that causes one or more entrained materials in a waste-water flow to divert from the flow for retention within the device or follow a different flow path for capture and retention outside the device. The retained materials require periodic removal from the devices or storage vessels in order to maintain the efficiency and function of the separating device. The removal and disposal of these detrimental wastes from the devices or storage vessels is typically performed by specialized service companies that are state licensed to transport and dispose of the wastes in an environmentally safe manner. Note that separators/interceptors do not chemically alter commercial/industrial waste water but only mechanically separate out physical components from waste-water flows. Section 803 covers applications where waste-water flows require adjustment of pH, temperature reduction, noxious odor abatement or removal of hazardous chemicals.

This section identifies grease, oil and sand as specific substances of concern that require removal from waste water before discharge to public sewers, private sewage systems or public sewage treatment plants. The congealing nature of grease can cause severe clogging problems in sewers, resulting in waste-water overflows into buildings or to the environment. Oil can be detrimental to sewage treatment processes as it tends to float on water surfaces making it difficult to capture and remove in the treatment process. Sand can be abrasive to most sewer line materials and the pumps used to move waste water. Sand can also build up and cause clogging problems in manholes located where turns occur in gravity sewer lines. Sections 1003.3 and 1003.4 indicate what occupancy uses specifically require oil and grease interceptors, respectively. Sections 1003.6, 1003.7, and 1003.8 require separators for commercial laundries, bottling establishments and slaughterhouses, respectively. The code does not require sand interceptors for a specific occupancy use; however, where sand interceptors are installed, Section 1003.5 has design requirements for those interceptors. Other occupancies that might require separators include: hospitals; fowl-, fish- or meat-packaging plants; canneries; cement producers; beauty salons; refineries and machine shops. The code official is responsible for identifying any other occupancy uses where similar discharges might require a separator.

It is not the intent of this section to require interceptors to protect the drain piping system within a building or the building sewer. However, locating interceptors as close to the point where harmful substances enter the building drainage system has the benefit of reducing maintenance required for the building drainage system and the building sewer. The designer of the building or tenant space is responsible for locating interceptors with respect to the occupant's expectations for frequency and degree of building drainage system and building sewer maintenance, ease of interceptor cleaning and building floor space allocation for such equipment.

Where waste-water flows are destined for a private sewage disposal system which ultimately discharges to the environment through various ground infiltration methods, state and county ground-water environmental control departments might require additional separators, special local treatment methods, holding tanks, or monitoring stations. Those requirements are not covered by this code and can vary from state to state.

1003.2 Approval. The size, type and location of each interceptor and of each separator shall be designed and installed in accordance with the manufacturer's instructions and the requirements of this section based on the anticipated conditions of use. Wastes that do not require treatment or separation shall not be discharged into any interceptor or separator.

❖ Interceptors are devices that are designed to remove a specific substance from a waste-water stream flowing at a maximum rate of flow. Although the code has a few requirements for the design of certain types of interceptors, the interceptor designer/manufacturer is responsible for all other details of construction such as

Figure 1003.2
INTERCEPTORS AND SEPARATORS

size and materials as well as the installation alignments, elevations and access requirements for maintenance. Because the size and, therefore, the cost of interceptors are directly proportional to the flow rate, and some waste-water flows not having substances that require removal might have substances that will cause problems in separation; only the waste-water flows having the substances that need to be removed are allowed to be routed through an interceptor (see Commentary Figure 1003.2). The commentary to Section 1003.1 provides information regarding the location of interceptors.

1003.3 Grease interceptors. Grease interceptors shall comply with the requirements of Sections 1003.3.1 through 1003.3.5.

❖ The need for and the performance of grease interceptors have received increased visibility over the years due to costly sewer overflows as a result of congealed grease clogging public sewer lines. Although grease problems in public sewer systems have existed since the first commercial kitchens discharged to public sewers, the extreme attention to preventing grease discharges into public sewers is due largely to the severe fines levied by state environmental health departments for waste-water overflows, which are in violation of the Federal Clean Air and Water Act. Once hot grease enters the sewer system, it cools and congeals in the pipes creating blockages for waste-water flow. Even though sewer system operators are accustomed to routine maintenance cleaning of sewer lines, it is nearly impossible for any reasonable cleaning program to keep sewers free of grease so that overflows do not occur. Therefore, many sewer system operators are establishing severe limits for grease concentration in discharges, extraordinary inspection/reporting requirements and strong enforcement procedures for the producers of grease-laden waste waters in order to attack the problem at the source.

The code does not prescribe grease concentration limits for waste-water discharge or inspection/maintenance requirements for grease interceptors. The only code requirement for the performance of grease interceptors is provided in the applicable referenced standards in Section 1003.3.4. The performance requirements provided in the standards are not a guarantee that any grease interceptor covered by those standards will limit the grease concentration in the waste water to the levels required by local sewer system operators or waste-water discharge enforcement agencies. The designer/manufacturer is responsible for providing a device design that meets customer-specified removal rates for the concentration of grease-laden waste water as specified by the customer.

There are two types of grease interceptors that are commonly used for grease-laden waste-water applications: 1) gravity and 2) hydromechanical. The terms for these two types of grease interceptors have recently come into use because the code-defined term for one of these devices was changed in the 2003 edition. Prior to that edition, gravity-type grease interceptors were called grease interceptors and hydromechanical-type grease interceptors were called grease traps. The term "trap" was changed to interceptor because not all grease trap designs provide a "trap-like" function in so far as preventing sewer gases from backflowing through the device and escaping from untrapped fixture drains. Changing the term "trap" to "interceptor" did not recognize that and in doing so, the distinction between the two types of interceptors was blurred.

The gravity-type grease interceptor is large in size [e.g., at least 500 gallons (1893 L) in volume], and provides for extremely low flow velocity through the unit to allow the grease to collect in a compartment [see Commentary Figure 1003.3(1)]. The hydromechanical-type grease interceptor is small in size [e.g., typically less than 250 gallons (946 L) in volume], of fabricated steel construction and utilizes changes in flow direction along with air-entrainment to cause grease to be retained in the device [see Commentary Figure 1003.3(2)]. Both rely upon gravity as the means of separation. As the code requires all grease interceptors to comply with the referenced standards as stated in Section 1003.3.4, this automatically excludes gravity-type grease interceptors from complying with the code. The code does not reference any standards for gravity-type grease interceptors.

However, gravity-type grease interceptors are sometimes required by jurisdictions instead of what is required by the code simply due to local experience, familiarity and comfort with gravity-type units. In some areas, grease interceptor installations are controlled not by code officials but by departments of health, waste-water system utility operators or environmental discharge departments, none of which are bound by the code. There are few published "standards" for the design of gravity-type grease interceptors. Decades of design experience by credible engineering concerns have resulted in design philosophies and construction criteria that have proven that gravity-type grease interceptors can provide for adequate grease removal from most waste-water flows. However, due to the variety of fats, oils and greases (FOG) produced by some applications and tighter limitations by certain jurisdictions on discharges of FOG, gravity-type grease interceptors alone might not produce the desired results. Hydromechanical devices fitted to the discharge of upstream grease-producing fixtures and equipment might be necessary in addition to gravity-type interceptors in order to attain compliance with local discharge limits of FOG.

Note that an automatic grease removal device is not a third type of grease interceptor but only a variation of the hydromechanical type that offers automatic removal of collected grease to an external container.

TRAPS, INTERCEPTORS AND SEPARATORS

Figure 1003.3(1)
TYPICAL GRAVITY-TYPE GREASE INTERCEPTOR

Figure 1003.3(2)
TYPICAL HYDROMECHANICAL GREASE INTERCEPTOR

1003.3.1 Grease interceptors and automatic grease removal devices required. A grease interceptor or automatic grease removal device shall be required to receive the drainage from fixtures and equipment with grease-laden waste located in food preparation areas, such as in restaurants, hotel kitchens, hospitals, school kitchens, bars, factory cafeterias and clubs. Fixtures and equipment shall include pot sinks, prerinse sinks; soup kettles or similar devices; wok stations; floor drains or sinks into which kettles are drained; automatic hood wash units and dishwashers without prerinse sinks. Grease interceptors and automatic grease removal devices shall receive waste only from fixtures and equipment that allow fats, oils or grease to be discharged. Where lack of space or other constraints prevent the installation or replacement of a grease interceptor, one or more grease interceptors shall be permitted to be installed on or above the floor and upstream of an existing grease interceptor.

❖ This section requires grease interceptors for the waste water generated by specific fixtures, equipment and certain floor drains in food preparation areas for occupancy uses where food is commercially prepared. Although the text provides a list of occupancy uses, this list is not all-inclusive but an indication as to the types of facilities that might have a food preparation area. For example, church kitchens, catering kitchens, delis, meat markets and recreation center kitchens could be occupancy uses that could be generators of grease-laden waste waters. The code official has the responsibility to define the fixtures and equipment and their intended uses that constitute the need for grease interception. For example, a bar that only serves beverages in glasses and uses a dishwasher or three-compartment sink to clean only the glasses would not require a grease interceptor. However, a bar that also serves food (if only heated in a microwave or convection oven) on reusable plates with silverware might require a grease interceptor for the three-compartment sink or dishwasher that washes those items.

Only the waste-water flows that require grease interception are allowed to be connected to a grease interceptor. Although this requirement is generally stated in Section 1003.2, it is even a more critical requirement for grease interceptors as cooler waste water, as well as waste-water flows with high concentrations of cleaning chemicals, sugars or acids, can affect the separation efficiency of a grease interceptor (see Commentary Figure 1003.2).

The last line of this section is intended to cover situations where gravity-type grease interceptors cannot be installed or the existing gravity grease interceptor is determined to be too small, where replacement with a larger unit is not possible for a variety of reasons (i.e., not enough space, too costly, etc.). Where these situations are encountered, the code allows grease interceptors (typically hydromechanical type) to be installed upstream of the existing gravity grease interceptor (or where a gravity grease interceptor would normally be located) in order to satisfy the jurisdictional requirements for a gravity grease interceptor [see Commentary Figures 1003.3.1(1) through 1003.3.1(5)].

1003.3.2 Food waste disposers. Where food waste disposers connect to grease interceptors, a solids interceptor shall separate the discharge before connecting to the grease interceptor. Solids interceptors and grease interceptors shall be sized and rated for the discharge of the food waste disposers. Emulsifiers, chemicals, enzymes and bacteria shall not discharge into the food waste disposer.

❖ As the code only recognizes the hydromechanical type of grease interceptors, the intent of this section is to prevent large particles of ground-up food discharged by the food waste disposer from clogging or rapidly filling these types of grease interceptors. Hydromechanical-type grease interceptors have numerous baffles, sometimes with small openings for flow such that food particles could block the flow. Also, the storage capacity for solids in these relatively small units is limited. Without a solids interceptor on the discharge of the food waste disposer, a hydromechanical grease interceptor could become inefficient or blocked within hours of being cleaned. Therefore, a solids interceptor is required for the food waste disposer discharge before connection to a hydromechanical grease interceptor (see Commentary Figure 1003.3.2).

Figure 1003.3.1(1)
TYPICAL SINGLE-FIXTURE INSTALLATIONS

TRAPS, INTERCEPTORS AND SEPARATORS

**Figure 1003.3.1(2)
GREASE INTERCEPTOR FOR FIXTURES**

**Figure 1003.3.1(3)
AUTOMATIC GREASE INTERCEPTOR**

Even though a solids interceptor "filters out" large particles of ground-up food, there is still a grease and flow load that must be accommodated by the grease interceptor. The specifier of the interceptor must include this information on the request for the interceptor or provide for an interceptor somewhat larger than the size selected to accommodate all other flows to provide for more storage capacity.

TRAPS, INTERCEPTORS AND SEPARATORS

Figure 1003.3.1(4)
SENSOR-CONTROLLED GREASE RECOVERY DEVICE

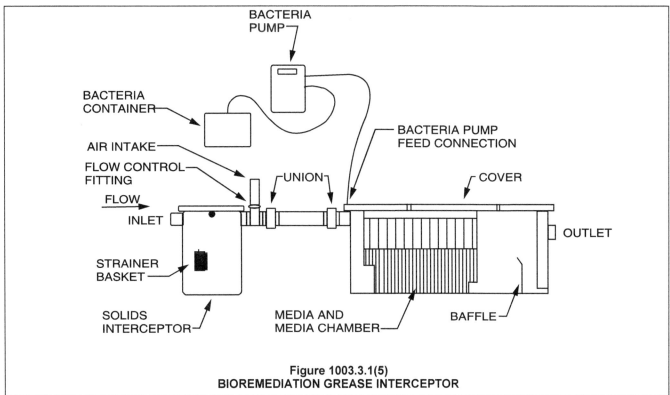

Figure 1003.3.1(5)
BIOREMEDIATION GREASE INTERCEPTOR

This section does not require that flow from a food waste disposer be routed through a grease interceptor. The code does not prohibit the discharge from a food waste disposer from connecting downstream of a grease interceptor (either of the hydromechanical type or of the gravity type). There are mixed opinions concerning where food waste disposer discharges should be connected. Because studies have indicated that dishwasher prerinse (food scrap) sinks are a significant source of grease waste water and that many food waste disposer installations receive the waste from prerinse sinks, the disposer discharge should go through the grease interceptor. However, since a food waste disposer operates best with a cold water flow and chops the food waste (and the grease contained within) into solidified particles, the particles will readily flow in the sewer system without the congealing problems associated with hot or warm grease-laden wastewater discharges. Manufacturers of food waste disposers understandably find the requirement for a solids interceptor for the disposer discharge illogical, especially where jurisdictions require them for gravity-type grease interceptors. Ultimately, the code official or the public sewer system operators will make the decision on where food waste disposer will be connected.

As reducing problems associated with grease clogging in plumbing piping systems as well as the cost of interceptor cleaning is a marketable concept, many producers of emulsifiers, chemicals, enzymes and bacteria products offer products that are marketed to aid in the reduction of grease. While the effectiveness of these products is beyond the discussion of this commentary, the manufacturers of hydromechanical grease interceptors are certain that the addition of these products to the flow stream and into the grease interceptor are detrimental to the operation of the interceptor. Thus, this code section, with the assumption that the grinder is discharging to a hydromechanical grease interceptor, prohibits the addition of these products to a food waste grinder. The intent of the prohibition has no relevance to the protection of food waste grinders nor is it intended to apply to food waste grinders discharging only to gravity-type grease interceptors.

Note that for gravity-type interceptors with large pathways for internal flow, limited number of baffles and significant storage capacity, clogging and solids buildup due to food waste disposer discharge is usually a nonissue unless maintenance cleaning (usually on a weekly or monthly basis) is ignored.

1003.3.3 Grease interceptors and automatic grease removal devices not required. A grease interceptor or an automatic grease removal device shall not be required for individual dwelling units or any private living quarters.

❖ The volume and concentration of grease in the wastewater discharge from a dwelling unit or private living quarters are usually not a significant source of grease discharged to public sewer systems. Therefore, the complications with installation and the resulting expense (initially and ongoing) of a grease interceptor are not warranted. However, this does not mean that users and owners of individual or private living quarters should not be encouraged to "can the grease (cooking oils and fats)" rather than pour such waste down the drain. Many public sewer system operators realize that even small quantities of grease from many locations can result in sewer system problems. With advance techniques in real-time grease monitoring and the ease of camera inspection of sewer lines, users who are chronic dumpers of fats, oils and greases into the public sewer system or even clandestine commercial kitchen operations might be identified for enforcement follow-up.

1003.3.4 Hydromechanical grease interceptors, fats, oils and greases disposal systems and automatic grease removal devices. Hydromechanical grease interceptors; fats, oils, and greases disposal systems and automatic grease removal devices shall be sized in accordance with ASME A112.14.3, ASME 112.14.4, ASME A112.14.6, CSA B481.3 or PDI G101. Hydromechanical grease interceptors; fats, oils, and greases disposal systems and automatic grease removal devices shall be designed and tested in accordance with ASME A112.14.3, ASME 112.14.4, CSA B481.1, PDI G101 or PDI G102. Hydromechanical grease interceptors; fats, oils,

Figure 1003.3.2
FOOD WASTE GRINDER CONNECTING TO GREASE INTERCEPTOR

and greases disposal systems and automatic grease removal devices shall be installed in accordance with the manufacturer's instructions. Where manufacturer's instructions are not provided, hydromechanical grease interceptors; fats, oils, and greases disposal systems and automatic grease removal devices shall be installed in compliance with ASME A112.14.3, ASME 112.14.4, ASME A112.14.6, CSA B481.3 or PDI G101.

❖ The referenced standards PDI G101 and ASME A112.14.3 are nearly identical and describe the testing and certification ratings (flow-through and grease retention) for a specific manufacturer's design of a (hydromechanical) grease interceptor. The CSA B481.3 standard is a similar testing standard for hydromechanical) grease interceptors. These standards do not specify any construction requirements other than the use of standard pipe threads for any external threaded connections. ASME A112.14.4 is the standard for automatic grease removal devices, an "automatic purging" variation of hydromechanical grease interceptors (see Commentary Figure 1003.3.4).

The standard ASME A112.14.6 covers FOG disposal systems. The acronym "FOG" stands for "fats, oils and greases." This is a somewhat recent plumbing system solution that processes grease-laden waste into gray water so that FOG doesn't create problems in public sewer systems, waste-water treatment facilities and private sewage disposal systems. Rather than having a grease interceptor that needs frequent periodic cleaning (and the subsequent waste disposal) to comply with a jurisdiction's limitations on grease discharge to the sewer, these systems biomediate the FOG at the source. Although these systems might have higher initial costs, in the long run, such systems could have lower operational costs while providing better effluent quality (lower grease concentrations) that some jurisdictions have mandated.

The standards do not provide any mandatory sizing methods for hydromechanical grease interceptors or automatic grease removal devices. Some of these standards provide a sizing method in a nonmandatory appendix of each standard. Sizing methods for grease interceptors have been debated for decades. Although there is most likely a point at which a grease interceptor that is too small does reduce the removal efficiency under ideal conditions (i.e., just cleaned), there is no penalty (other than initial cost) for the selection of a unit larger than required. A larger unit than "required by the numbers" might cost only slightly more, with the benefit of less frequent cleanings (cleaning being a cost as well). Note that when choosing any interceptor, coordination of the outlet connection size with the drainage piping downstream of the interceptor is critical. Larger inceptors might have a larger (standard) outlet connection than what is planned for the facility. Section 704.2 prohibits the reduction in pipe size in the direction of flow. Ultimately, the performance of any grease interceptor is directly related to adherence to an appropriate maintenance schedule for actual loading conditions. Maintenance schedules are usually developed after

Figure 1003.3.4
TYPICAL HYDROMECHANICAL GREASE INTERCEPTOR

monitoring grease collection amounts over numerous cleanings to determine an optimum cleaning frequency. However, local jurisdictions might require that cleanings occur at intervals no less than weekly or monthly (regardless of whether the cleaning is actually needed or not) based upon their experience with the type of facility operation. Although a discussion of a "best" method for sizing grease interceptors is not within the scope of this commentary, the following discussion of sizing methods is provided for informational purposes only. Grease interceptor manufacturers might also provide sizing methods for their interceptor design. The authority having jurisdiction is responsible for providing or approving the methods used for sizing grease interceptors.

Generally, there are four sizing methods that are used in the prefabricated grease interceptor industry (both gravity and hydromechanical types): 1) the PDI method; 2) the 1980 USEPA "seats" method; 3) the 1980 USEPA "meals" method and 4) the "dfu" method. Some jurisdictions use methods similar to these but have added other requirements or changed the "constants" used in the equations. Others have developed completely different methods that will not be discussed in this commentary. Grease interceptor designs can also be "engineered." There are no consensus standards for the sizing of grease interceptors. The authority having jurisdiction must provide or approve the sizing method to be used.

The PDI method

This method is published in the nonmandatory Appendix A of the Plumbing and Drainage Institute's (PDI) standard PDI G101. This method derives a gallon-per-minute (gpm) flow-through rate for selecting a hydromechanical grease interceptor.

Sample Problem 1: A delicatessen has a three-compartment sink for washing utensils, pots and pans used for preparing meats and other deli products for sale. Each compartment of the sink is 24 inches wide by 24 inches long by 18 inches deep (610 mm by 610 mm by 457 mm) (maximum water depth). The three-compartment sink will be directly connected through a flow control located at the same elevation as the inlet to the interceptor. The highest water elevation in the sink will be no greater than 6 feet (1829 mm) above the inlet to the grease interceptor. Determine the required grease interceptor size

Problem Approach

Calculate the total water holding volume of the sink in gallons, assume a drain time for the sink and choose a "certified" size from Table 1003.3.4.1. Check with the manufacturer to determine if special flow-control orifice size is required for the elevation difference between the water level and the interceptor inlet.

Solution

Total sink volume = [3 × (24 × 24 × 18)]/231 cubic inches per gallon

Total sink volume = 134 gallons

Consider the fact that the actual water capacity in the sink will be no greater than approximately 75 percent of the sink due to the volume taken up by the items being washed and the typical fill levels. Therefore,

Actual sink volume = (75/100) × 134 gallons

Actual sink volume = 101 gallons

Consider the fact that the time to drain the sink is not critical. Therefore, assume a 2-minute drain time. (To be more conservative, a 1-minute drain time could be assumed). Therefore,

Flow rate through interceptor = 101 gallons/2 minutes = 51 gallons per minute

Refer to Table 1003.3.4.1 to determine that a 75 gpm flow-through rating exceeds the required 51 gallons per minute. Although a 50-gpm unit is technically not large enough for the application, use of a 50-gpm-rated interceptor (with a 50-gpm flow control) would be suitable as the only drawback would be that the drain time for the sink would be slightly greater.

Sample Problem 2: Determine the required grease interceptor size for a hotel restaurant kitchen having the following fixtures. The grease interceptor(s) will be located in the floor of the one-level kitchen:

One three-compartment sink for washing cooking utensils, pots and pans. Each compartment of the sink is 24 inches wide by 24 inches long by 18 inches (610 mm by 610 mm by 457 mm) deep (maximum water depth).

One two-compartment sink for food preparation. Each compartment of the sink is 24 inches wide by 24 inches long by 18 inches (610 mm by 610 mm by 457 mm) deep (maximum water depth).

One one-compartment sink for food preparation. The sink compartment is 24 inches wide by 24 inches long by 12 inches (610 mm by 610 mm by 304.8 mm) deep (maximum water depth).

One steam kettle with a 60-gallon capacity.

One dishwasher with a 30-gpm rating for washing plates, bowls and silverware.

One glassware washer with a 20-gpm rating for washing glasses and cups.

One prewash station (no food waste grinder) with 1.6 gpm.

All of the above fixtures (and compartments of sinks) will be individually indirect connected to the drain system.

Three hand sinks.

One mop (service sink).

Three floor drains.

Problem Approach

Eliminate fixtures that do not produce a grease load from the calculations, calculate the total water holding volume of the sinks in gallons, assume a drain time for the sinks, add equipment discharge rates, add up all the discharge rates and choose a "certified" size from Table 1003.3.4.1.

<u>Solution—Part I</u>

The glassware washer, hand sinks, mop sink and floor drains are not significant grease contributors and therefore are eliminated from the computation. The drain piping from these fixtures must not discharge to the grease interceptor.

<u>Solution—Part 2</u>

Three-compartment sink:

Total sink volume = [3 × (24 × 24 × 18)]/231 cubic inches per gallon

Total sink volume = 134 gallons

Consider the fact that the actual water capacity in the sink will be no greater than approximately 75 percent of the sink due to the volume taken up by the items being washed. Therefore,

Actual sink volume = (75/100) × 134 gallons
= 101 gallons

Because the discharge of the sink compartments is individually indirect connected into floor sinks, the draining of the sink will be rapid. Assuming a 1-minute drain time:

Flow rate from entire sink = 101 gallons/1 minute
= 101 gallons per minute

Two-compartment sink:

Total sink volume = [2 compartments × (24 × 24 × 18)] / 231 cubic inches per gallon

Total sink volume = 90 gallons

Consider the fact that the actual water capacity in the sink will be no greater than approximately 75 percent of the sink due to the volume taken up by the items being washed and necessary "freeboard" space above the waterline. Therefore,

Actual sink volume = (75/100) × 90 gallons

Actual sink volume = 68 gallons

Because the discharge of the sink compartments is individually indirect connected into floor sinks, the draining of the sink will be rapid. Assuming a 1-minute drain time:

Flow rate through interceptor = 68 gallons/1 minute
= 68 gallons per minute

One-compartment sink:

Total sink volume = [1 compartment × (24 × 24 × 18)]/231 cubic inches per gallon
= 45 gallons

Consider the fact that the actual water capacity in the sink will be no greater than approximately 75 percent of the sink due to the volume taken up by the items being washed and necessary "freeboard" space above the waterline. Therefore,

Actual sink volume = (75/100) × 45 gallons
= 34 gallons

Because the discharge of the sink compartments is individually indirect connected into floor sinks, the draining of the sink will be rapid. Assuming a 1-minute drain time:

Flow rate through interceptor = 45 gallons/1 minute
= 45 gallons per minute

<u>Solution—Part 3</u>

The following equipment discharge rates are identified:

Dishwasher: 30 gpm

Steam kettle: 60 gallons (assume a drain time of 2 minutes) = 30 gpm

Prerinse station: 1.6 gpm

Total flow rate from all sinks and equipment = 101 + 68 + 45 + 30 + 30 + 1.6
= 276 gpm

Because the total flow rate exceeds 100 gpm (the maximum allowable flow rate for a certified grease interceptor), several grease interceptors will be required. The following example is one of many possible ways to configure the drain system for this kitchen.

Grease interceptor No. 1: 100-gpm rating with 100-gpm flow control to accommodate the three-compartment sink.

Grease interceptor No. 2: 100-gpm rating with 100-gpm flow control to accommodate the two-compartment sink, the dishwasher and the prerinse station.

Grease interceptor No. 3: 75-gpm rating with 75-gpm flow control to accommodate the one-compartment sink and the steam kettle.

The "seats" method for restaurants

In 1980, the U.S. Environmental Protection Agency (EPA) released the *Design Manual: Onsite Wastewater Treatment and Disposal Systems* (U.S. EPA document no. 625/1-80-012). The derived interceptor volume (in gallons) is intended for choosing a gravity-type grease interceptor. Note that the interceptor is not sized for a flow-through rating (gallons per minute). Section 8.2 of the manual offers the following formula for deriving the size of a gravity-type grease interceptor for a restaurant:

Grease interceptor size,
gallons[1] = $D \times GL \times ST \times (HR/2) \times LF$

where:

D = number of seats in dining area

GL = gallons of waste water per meal, normally 5 gallons

ST = storage capacity factor, minimum value = 1.7 maximum value = 2.5

HR = number of hours open
LF = loading factor,
  1.25 for interstate freeway locations
  1.0 for other freeways
  1.0 for recreational areas
  0.8 main highways
  0.8 other highways

Note 1: minimum volume is 750 gallons.

The "meals" method for hospitals, nursing homes and other types of commercial kitchens

This method was also published in the 1980 U.S. EPA's *Design Manual: Onsite Wastewater Treatment and Disposal Systems* (U.S. EPA document no. 625/1-80-012). The derived interceptor volume (in gallons) is intended for choosing a gravity-type grease interceptor. Note that the interceptor is not sized for a flow-through rating (gpm). Section 8.2 of the manual offers the following formula for deriving the size of a gravity-type grease interceptor for commercial kitchens other than a restaurant kitchen:

Grease interceptor size, gallons^1 =

$M \times GL \times ST \times 2.5 \times LF$

where:
M = number of meals served per day
GL = gallons of waste water per meal, normally 4.5 gallons
LF = loading factor,
  1.25 for kitchens with food waste grinders and dishwashing
  1.0 for kitchens without food waste grinders but with dishwashers
  0.75 for kitchens with food waste grinders but without dishwashers
  0.5 for kitchens without food waste grinders and dishwashers
ST = storage capacity factor, minimum value = 1.7
  maximum value = 2.5

Note 1: minimum volume is 750 gallons.

The "dfu" method

This method provides a table that equates the number of drainage fixture units (dfu) for the total drainage flow requiring grease interception, to the required interceptor capacity. Separate tables are often provided to size hydromechanical and gravity-type grease interceptors, respectively. Although some tables are developed by arbitrary assignments of sizes to flows, other tables are developed by converting the number of dfu (as assigned by the adopted plumbing code) to a gpm flow. The choice of a hydromechanical grease interceptor is straightforward whereas, for a gravity-type interceptor, sizing is based upon providing at least a 30-minute retention time through the grease interceptor vessel. Because dfu assignments vary between model codes, the reader is cautioned about applying one code's dfu assignments to another model code's interceptor sizing tables.

1003.3.4.1 Grease interceptor capacity. Grease interceptors shall have the grease retention capacity indicated in Table 1003.3.4.1 for the flow-through rates indicated.

❖ Table 1003.3.4.1 comes directly from the PDI G101 and ASME A112.14.3 standards. Note a is not included in the standard's table but was added to the code in the 2006 edition. The minimum grease retention capacity for a hydromechanical grease interceptor provides for a "reasonable" interceptor cleaning interval for an "average" concentration of grease within the waste water at the interceptor's rated flow. The minimum grease retention capacity was determined by extensive testing long ago and establishes a performance baseline for all manufacturers to comply with. Because the grease concentration in grease-laden waste water varies widely, the actual cleaning interval necessary for the maximum desired grease concentration at the outlet of the interceptor is determined by observation or testing.

The PDI G101 and ASME A112.14.3 standards only provide for testing and certification for flow-through ratings of up to 100 gpm. Because some interceptor manufacturers produce designs having flow-through ratings greater than 100 gpm, Note a was added to Table 1003.3.4.1 to extend the same baseline of performance to these larger designs. However, since the standards do not provide for testing of these larger units, hydromechanical grease interceptor designs over 100 gpm must be submitted for approval under Section 105.2 since they cannot meet the testing and certification requirements of Section 1003.3.4.

TABLE 1003.3.4.1
CAPACITY OF GREASE INTERCEPTORS[a]

TOTAL FLOW-THROUGH RATING (gpm)	GREASE RETENTION CAPACITY (pounds)
4	8
6	12
7	14
9	18
10	20
12	24
14	28
15	30
18	36
20	40
25	50
35	70
50	100
75	150
100	200

For SI: 1 gallon per minute = 3.785 L/m, 1 pound = 0.454 kg.

a. For total flow-through ratings greater than 100 (gpm), double the flow-through rating to determine the grease retention capacity (pounds).

❖ Table 1003.3.4.1 indicates the standard "sizes" (flow-through ratings) of hydromechanical grease interceptors to which numerous manufacturers have designed, tested and certified their units to meet the requirements of PDI G101 and ASME A112.14.3. The use of standard sizes allows for manufacturers to produce a tested and certified unit without having to test and certify the actual unit being constructed for a customer. This allows for faster delivery times and allows manufacturers to build frequently purchased sizes in advance of customer orders. While it might be possible to order a "nonstandard" size, the delivery time and cost for special testing and certification of such a unit will most likely be substantially more than if the next larger size of standard unit was ordered. This is not to say that a nonstandard unit could not be ordered, but given that most projects have significant time and financial constraints, ordering standard sizes should offer the best economy.

The grease retention capacity for each of the standard interceptor sizes is the minimum required capacity. Larger grease retention capacities offer the advantage of decreased cleaning frequency; however, if the local jurisdiction requires a cleaning interval of not less than a certain frequency, the additional grease capacity might never be used.

1003.3.4.2 Rate of flow controls. Grease interceptors shall be equipped with devices to control the rate of water flow so that the water flow does not exceed the rated flow. The flow-control device shall be vented and terminate not less than 6 inches (152 mm) above the flood rim level or be installed in accordance with the manufacturer's instructions.

❖ A hydromechanical grease interceptor is designed for a maximum flow-through rate to allow for gravity separation to occur and so that velocities within the unit do not cause excessive turbulence, which might cause "carryover" of grease to the outlet. The rate of discharge from an individual plumbing fixture varies depending on the "head" (height of water) in each receptor (sink compartment), the number of sink compartments discharged at once and the configuration of the compartment outlets and the drain piping. The probability of simultaneous discharge of multiple plumbing fixtures along with other equipment (e.g., dishwashers, kettles) also varies. The selection of a grease interceptor size might be predicated on only certain combinations of fixtures discharging simultaneously in order to optimize interceptor sizing for the most probable flow condition. Note that the sample problems for the PDI sizing method provided in the commentary for Section 1003.3.4 include assumptions for the drain times of sinks. In reality, sinks could drain faster than the assumed time and more fixtures/equipment could discharge simultaneously than was considered in the selection of the interceptor size. Therefore, the interceptor must be provided with a flow-control device to limit flow rate to the interceptor (see Commentary Figure 1003.3.4.2).

Although an interceptor might be sized to accommodate the maximum possible flow rate of all plumbing fixtures and equipment (i.e., simultaneous discharge), the flow control is still required as it also serves to entrain air into the waste flow to aid in the grease separation process. Entrained air in the waste causes the grease, oils and fats to become more

Figure 1003.3.4.2
FLOW CONTROL DEVICE

buoyant in order to more readily rise to the retention area within the interceptor.

The requirement in the second sentence in this section for the flow control to be "vented" ensures that the flow control will be installed with a means to readily obtain air for entrainment into the waste flow. The requirement for the air intake to terminate (with the same room as the fixture) no less than 6 inches (152 mm) above the flood level rim of the fixture applies only to a specific fixture/interceptor arrangement as follows: The grease interceptor must be capable of serving as a fixture trap for a single fixture or combination sink of no more than three compartments and must be within a specific proximity to the fixture (see Exception 3 of Section 1002.1). Terminating the flow control vent 6 inches (152 mm) above the fixture flood level rim prevents waste overflow through the vent should the grease interceptor become clogged. Otherwise, all fixture and equipment drains must be trapped and vented as required by Chapter 9 and the flow control air intake must be connected to the sanitary drain vent system or must terminate outdoors in accordance with the requirements for vent terminations in Chapter 9.

The manufacturer's installation instructions will indicate whether the grease interceptor can serve as a trap for a single fixture or combination sink and might require additional vent connections to the interceptor as well.

The code does not prohibit a grease interceptor from being indirectly connected to a sanitary drainage system through a waste receptor. For example, where a grease interceptor serves an adjacent untrapped single-fixture or multicompartment combination sink (as allowed by Exception 3 of Section 1002.1), the interceptor could discharge, through an air break, into a waste receptor such as a floor sink. This is especially useful where a change in kitchen configuration requires the addition of a floor-mounted interceptor for a new or relocated fixture. Note that in this arrangement, if a vent is required for the interceptor, it must vent to the room or to the outdoors. The vent must not connect to the sanitary vent system.

1003.3.5 Automatic grease removal devices. Where automatic grease removal devices are installed, such devices shall be located downstream of each fixture or multiple fixtures in accordance with the manufacturer's instructions. The automatic grease removal device shall be sized to pretreat the measured or calculated flows for all connected fixtures or equipment. Ready *access* shall be provided for inspection and maintenance.

❖ Automatic grease removal devices serve the same function as grease interceptors except that the unpleasant task of removing the grease is performed by automated internal means such as a pump or skimmer that discharges the collected grease to a container outside of the interceptor. The container, when full, is replaced with an empty one. Although ASME A112.14.4 is the standard for automatic grease removal devices, the standard requires that grease removal devices comply with the ratings and grease retention capacities required by ASME A112.14.3 for hydromechanical grease interceptors.

Note that this section requires ready access for grease removal devices so that maintenance and repairs can be properly performed.

1003.3.6 Gravity grease interceptors and gravity grease interceptors with fats, oils, and greases disposal systems. The required capacity of gravity grease interceptors and gravity grease interceptors with fats, oils, and greases disposal systems shall be determined by multiplying the peak drain flow into the interceptor in gallons per minute by a retention time of 30 minutes. Gravity grease interceptors shall be designed and tested in accordance with IAPMO/ANSI Z1001. Gravity grease interceptors with fats, oils, and greases disposal systems shall be designed and tested in accordance with ASME A112.14.6 and IAPMO/ANSI Z1001. Gravity grease interceptors and gravity grease interceptors with fats, oils, and greases disposal systems shall be installed in accordance with manufacturer's instructions. Where manufacturer's instructions are not provided, gravity grease interceptors and gravity grease interceptors with fats, oils, and greases disposal systems shall be installed in compliance with ASME A112.14.6 and IAPMO/ANSI Z1001.

❖ Gravity grease interceptors remove grease from waste-water flows by reducing the waste-water flow velocity so that the grease floats and solids settle to the bottom of the tank. Waste water between these two layers is allowed to pass into the sewer system. Gravity grease interceptors are usually located outside a building or under a parking lot or driveway so that there is ease of access for the pumping truck that must come periodically to remove the accumulated grease (see Commentary Figure 1003.3.6).

Gravity grease interceptors can be fitted with a fats, oils and greases (FOG) disposal system so that grease material can be reduced into a gray water form, thus extending the required cleaning intervals for the tank (see Section 1003.3.4).

1003.3.7 Direct connection. The discharge piping from a grease interceptor shall be directly connected to the sanitary drainage system.

❖ Although it may seem like common sense, grease interceptors must not indirectly connect to the sanitary drainage system. The discharge can be extremely unpleasant, especially where the interceptor is located inside of a building.

1003.4 Oil separators required. At repair garages where floor or trench drains are provided, car washing facilities, factories where oily and flammable liquid wastes are produced and hydraulic elevator pits, oil separators shall be installed into which oil-bearing, grease-bearing or flammable wastes shall be discharged before emptying into the building drainage system or other point of disposal.

Exception: An oil separator is not required in hydraulic elevator pits where an approved alarm system is installed.

Figure 1003.3.6
TYPICAL GRAVITY-TYPE GREASE INTERCEPTOR

For SI: 1 inch = 25.4 mm.

Such alarm systems shall not terminate the operation of pumps utilized to maintain emergency operation of the elevator by fire fighters.

❖ Readers of this section must remember the overall intent of Section 1003 (see Section 1003.1), which is to prevent oil, grease and other harmful substances (e.g., flammable liquid wastes) from being discharged to a public sewer or a private sewage disposal system. As the scope of the code covers the installation of plumbing fixtures including floor and trench drains, all of which would ultimately be connected to the building's drainage system (leading to the sewer), the concern of this section is only for situations where fixtures will exist and those fixtures are likely to "capture" the subject detrimental wastes. For example, consider an automotive repair shop with a toilet (or bath) facility but without floor drains or trench drains in the vehicle service area. Simply because work in the facility could cause oil, grease or flammable liquid wastes to be generated does not mean that an oil separator is required for the facility. There has to be a likely method for those materials to enter the drainage system. Is it likely that the toilet (or bath) facility will be capturing significant detrimental wastes? Probably not. However, if the repair area has floor or trench drains, then it is very likely that an oil or fuel spill could occur at some point (or the floor be "hosed down," flushing the detrimental wastes into floor or trench drains), such drains would need to be connected to an oil separator in the drain piping system before that drain piping is connected to the building drainage system (or building sewer).

The same is true for factories that produce oily or flammable liquid wastes and car-washing facilities. A separator is only required for those types of facilities that have a likely method, such as floor or trench drains, for the wastes to enter the drainage system. Most car-washing facilities will require floor or trench drains for the simple reason that wash/rinse water must be recycled to minimize operational costs and to comply with local and state environmental regulations (that typically would not allow wash water into storm sewers or onto the ground). Whether or not the oil or flammable liquid wastes-producing factory will have floor or trench drains is a decision made by the designer of the facility.

This section does not require that the indicated facilities have floor or trench drains installed so that an oil separator can be installed.

A question often asked is whether any plumbing fixture, other than floor or trench drains, might be likely to capture oil, grease and other harmful substances. In other words, could someone pour waste oil into a water closet? Certainly, anything might be possible, but is it likely to happen? And if there was concern about a water closet used for that purpose, how could an oil separator function without clogging because the separator would eventually plug with human wastes?

It is not the intent of this section to require that an oil separator serve all plumbing fixtures installed in the indicated facilities. Although the occupants (soiled from working) will wash their hands or even shower at these facilities in order to aid in the complete removal of oil or grease, it is understood that the majority of oily and greasy substances will have already been removed from the body by the use of shop towels and absorbent cloths. The quantity of oil and grease from showering or hand washing will be negligible such that those fixtures can safely discharge directly to the sanitary drain system (without the need to be connected to a separator).

Service sinks, especially those used for washing/rinsing floor mops and for the discharging of mop water, might seem to be a "borderline" fixture for consideration to be connected to an oil separator. However, mopping usually requires detergents/chemicals and at the point that the mopped wastes combine with those detergents/chemicals, the oil and grease becomes an emulsion. Therefore, the oil cannot be separated from the water.

Most (perhaps all) elevators have a sump (a pit) at the bottom of the elevator shaft because the elevator "code" (standard), ASME A17.1/CSA B44 as referenced by the IBC, requires that there be a method for draining liquids from the bottom of the shaft. This is to remove any rainwater that might enter through an above-roof penthouse that houses traction equipment but more so, for water that could enter through door openings during a building fire event. Also, water could enter the shaft when a water pipe breaks or a water heater tank leaks. For hydraulic elevators, hydraulic fluid (an oil-based product) could leak because of a leaking pipe connection or worn lift piston seals.

The sump must be evacuated by either draining by gravity or by an automatic pump. The end of the gravity drainage piping or the sump pump discharge pipe must indirectly connect to the building drainage system. In other words, the termination end of the piping must be at a waste receptor. Note that although Section 301.6 prohibits plumbing from being located in the elevator shaft or equipment room, it is understood that where a sump pump is located at the bottom of the shaft, there will need to be pump discharge from the pump to outside of the shaft. The expectation is that such pump piping be routed as directly as possible to the side of the shaft and not turn upward and alongside of where the elevator car travels (see commentary, Section 301.6).

For hydraulic elevators, because of the potentially oily nature of the sump discharge, an oil separator is required on the sump discharge pipe before the indirect connection to the building drainage system. Where the discharge is from a sump pump, the oil separator would need to be rated for the maximum flow rate of the pump, adjusted for the flowing pressure losses through the piping and oil separator. The ASME A17.1/CSA B44 standard requires the pump to have a capacity of not less than 50 gpm (2013 edition). However, depending on the pump chosen, the actual flow rate could be much greater. To be effective, the oil separator should be sized for actual flow rate. Properly sized oil separators are large, depending on the manufacturer and model chosen. For a gravity flow discharge arrangement, ample elevation difference must be provided to accommodate the height of the separator.

The exception allows for an oil separator to not be installed on the discharge of a sump for a hydraulic elevator where there is an approved oil alarm system installed in the sump. The code is silent concerning the details of an approved oil alarm system. As the intent of the base code section is to prevent oil discharge to the sanitary drain system, an approved alarm system should have the means to prevent such discharge.

Some sump pump manufacturers offer an "oil alarm" control for their sump pumps and those controls do have the capability of turning off the pump (and also indicating an alarm in some manner) when oil is sensed. As oil floats on top of water, it seems reasonable to assume that such an alarm system could accommodate the addition of water to the sump and turn on the pump (to pump water) until the water level decreased, oil is sensed and the control again, shuts off the pump. The control alarm, indicating that oil was sensed, is a warning to the building maintenance staff, that action needs to be taken (seek out the source of the oil leak). However, control arrangements can vary and there is the possibility that some controls might not shut off the pump and oil is pumped to the drainage (only indicating that oil was sensed at some point in time) or some controls might "lock out" the pump until the problem is resolved and the control is manually reset. Note that a gravity-drained sump could also be stopped from draining by the installation of an automatically controlled valve.

However, this does not necessarily mean that the sump should not be evacuated when oil is sensed in the sump. Shutting off the pump (or drain) could conflict with how the authority having jurisdiction for the elevator "code" (standard ASME A17.1/CSA B44) interprets that code. In addition, the last sentence of the exception does not allow stopping of a (sump) pump for a hydraulic elevator that is used for emergency operation by fire fighters.

If the interpretation of ASME A17.1/CSA B44 is that under all conditions (emergency or otherwise) the sump must be evacuated, either an oil separator would be required or some other means for preventing oil from being discharged to the drainage system would have to be provided. Those means might be an intelligent control system that pumps oil to a holding tank and only water is pumped to the drainage system. Other means would require approval in accordance with Section 105.2.

1003.4.1 Separation of liquids. A mixture of treated or untreated light and heavy liquids with various specific gravities shall be separated in an *approved* receptacle.

❖ There is a variety of designs for oil separators. Typically an oil separator is a large tank designed to retain oil by minimizing turbulent flow. Oil is removed through a draw-off pipe and grease, sludge and solids settle to the bottom or collect on the filter screen. Oil separators function by relying on the differences in specific gravity (density) of the waste water and oil.

1003.4.2 Oil separator design. Oil separators shall be listed and labeled, or designed in accordance with Sections 1003.4.2.1 and 1003.4.2.2.

❖ The design criteria for effective performance depend on the minimum cross-sectional area and the minimum ratio of depth to width. In turn, design and shape of the separator depend on the character and quality of the oils to be separated. A properly sized separator allows the separation and prevents the evacuation of solids into the drainage system. Where an oil separator is not designed in accordance with Sections 1003.4.2.1 and 1003.4.2.2, then the oil separator must be listed and labeled. Listing and labeling ensures that the design has been tested against some performance criteria and certified that the design meets or exceeds that criteria.

1003.4.2.1 General design requirements. Oil separators shall have a depth of not less than 2 feet (610 mm) below the invert of the discharge drain. The outlet opening of the separator shall have not less than an 18-inch (457 mm) water seal.

❖ The sizing specified in this section relates to the open-tank-type separator. The minimum depth is necessary to provide sludge and solids retention capacity and to obtain efficient retention of oil or other volatile liquid wastes by minimizing turbulent flow through the separator. The outlet must be designed to provide at least 18 inches (457 mm) of liquid seal to provide for the storage of oil to at least that depth.

1003.4.2.2 Garages and service stations. Where automobiles are serviced, greased, repaired or washed or where gasoline is dispensed, oil separators shall have a capacity of not less than 6 cubic feet (0.168 m^3) for the first 100 square feet (9.3 m^2) of area to be drained, plus 1 cubic foot (0.028 m^3) for each additional 100 square feet (9.3 m^2) of area to be drained into the separator. Parking garages in which servicing, repairing or washing is not conducted, and in which gasoline is not dispensed, shall not require a separator. Areas of commercial garages utilized only for storage of automobiles are not required to be drained through a separator.

❖ In addition to the general requirements of Section 1003.4.2.1, this section specifically applies to motor vehicle garages or service stations where lubrication, oil changing, fuel dispensing, repair work or hand or mechanical washing takes place. The intent of this section is to require separators in locations where flammable or combustible liquids are to be discharged into the drainage system, typically in occupancies provided with floor or trench drains.

The requirement for a separator does not, in itself, create the requirement for floor or trench drains. Where a drainage system is provided in these occupancies, provisions must be made to prevent flammable and combustible liquids from entering the drainage system, sewers and waste treatment facilities.

The sizing of a separator can be reduced to two equations as follows:

Floor Area ≤ 100 square feet $C = 6$

Floor Area > 100 square feet $C = 5 + (A/100)$

where:

C = Minimum capacity of separator (ft^3).

A = Floor area served (ft^2).

For SI: 1 square foot = 0.0929 m^2, 1 cubic foot = 0.0283 m^3.

Garages used exclusively for parking or storage of automobiles do not normally produce wastes requiring separation; therefore, separators are not required for drains in such locations.

1003.5 Sand interceptors in commercial establishments. Sand and similar interceptors for heavy solids shall be designed and located so as to be provided with ready *access* for cleaning, and shall have a water seal of not less than 6 inches (152 mm).

❖ Like grease interceptors, sand interceptors need periodic maintenance to function correctly. They must be installed to allow access for cleaning and servicing. Note that this section does not require installation of sand or heavy solids interceptors but only regulates their installation and design.

1003.6 Clothes washer discharge interceptor. Clothes washers shall discharge through an interceptor that is provided with a wire basket or similar device, removable for cleaning, that prevents passage into the drainage system of solids $^1/_2$ inch (12.7 mm) or larger in size, string, rags, buttons or other materials detrimental to the public sewage system.

Exceptions:

1. Clothes washers in individual dwelling units shall not be required to discharge through an interceptor.

2. A single clothes washer designed for use in individual dwelling units and installed in a location other than an individual dwelling unit shall not be required to discharge through an interceptor.

❖ Clothes washer discharge must pass through an interceptor that is capable of preventing string, lint and other solids from entering a drainage system that leads to a public sewage system. These materials can foul rotating equipment and other mechanisms in public sewer pumping systems and waste-water treatment plants. The filter, screen or basket will accumulate this debris, so the interceptor must allow for cleaning and removing of such materials. Although this section implies that this requirement is only for drainage systems that lead to a public sewage system, discharge of these materials would be just as detrimental to private sewage disposal systems.

Exception 1 is an allowance for not providing an interceptor for clothes washers in an individual dwelling unit such as an apartment or condominium. There could be large single-family homes (required to be covered by the IBC and not the IRC) where either multiple "household-use" clothes washers or clothes washers that are similar to larger washers (that can handle quilts and rugs) found in a self-serve laundries. These are not required to discharge to an interceptor.

Exception 2 is an allowance for not providing an interceptor for two "household-use"-type clothes washers that are in a location other than an individual dwelling unit. For example, a pair of washers in a common laundry room for an entire small apartment building, or a pair of washers on each floor of an apartment building.

1003.7 Bottling establishments. Bottling plants shall discharge process wastes into an interceptor that will provide for the separation of broken glass or other solids before discharging waste into the drainage system.

❖ Drains from areas in both bottling and food processing plants where glass bottles and jars are used as containers present a greater problem from broken glass than generally recognized. Separators should be located to intercept and treat this waste separately from the remainder of the processing area.

1003.8 Slaughterhouses. Slaughtering room and dressing room drains shall be equipped with *approved* separators. The separator shall prevent the discharge into the drainage system of feathers, entrails and other materials that cause clogging.

❖ The type of interceptors necessary to protect slaughtering room drains depends to a great extent on the particular operation. For example, in a poultry processing plant, separators have various-size screens designed to prevent feathers from the killing and defeathering area from entering the drainage system. Additionally, the drainage system will need protection against entrails and other heavy solids that could gain entry from the cutting room. In such operations, a screen or basket-type interceptor is necessary to prevent the passage of feathers, entrails and similar animal matter.

1003.9 Venting of interceptors and separators. Interceptors and separators shall be designed so as not to become air bound. Interceptors and separators shall be vented in accordance with one of the methods in Chapter 9.

❖ Most interceptors and separators require venting of the containment tank and the outlet trap if so equipped. Typically, a vent is required to allow the escape or admittance of air to compensate for the variable fluid level in the tank. Interceptors and separators with integral traps must be vented to prevent the tank from emptying its contents by siphon action. Oil and gasoline separators require vents independent of the sanitary drain venting system to help control and dissipate vapor buildup that may occur in the holding tank. (see Commentary Figure 1003.9).

1003.10 Access and maintenance of interceptors and separators. *Access* shall be provided to each interceptor and separator for service and maintenance. Interceptors and separators shall be maintained by periodic removal of accumulated grease, scum, oil, or other floating substances and solids deposited in the interceptor or separator.

❖ For an interceptor or separator to function properly, it must be cleaned periodically. The solids and liquids, such as grease, oil, large solids, feathers or glass, must be removed before they exceed the capacity of the interceptor or separator. An interceptor or separator that has not been cleaned will cause stoppage in the drainage system.

When a grease interceptor or oil separator reaches its maximum retention capacity, it will cease to function, which will allow oil and grease to pass through to the sewage system and possibly clog the drainage pipes downstream. An interceptor or separator that is no longer functioning properly will not provide any indication of malfunction; therefore, such devices must be periodically monitored to determine the volume of substances being retained. Oil separators are typically equipped with test cocks to determine the need for oil removal from the device.

Grease interceptors that have not been properly maintained have caused some jurisdictions and health authorities to prohibit the location of these devices in kitchens. Without proper maintenance, interceptors would become full of grease and eventually restrict or plug the interceptors and drain lines downstream, causing greasy waste to leak from the cover of the interceptor. The resulting mess is considered a health hazard.

Some manufacturers have addressed the problems associated with the lack of maintenance by providing various methods of removing or drawing off the grease from the grease interceptor semiautomatically or automatically. Additionally, the use of bacteria (bioremediation) traps and digests greases in the interceptor to convert them into the byproducts of digestion; however, these devices still require periodic maintenance and cleaning

**Figure 1003.9
VENTING OF INTERCEPTORS AND SEPARATORS**

SECTION 1004
MATERIALS, JOINTS AND CONNECTIONS

1004.1 General. The materials and methods utilized for the construction and installation of traps, interceptors and separators shall comply with this chapter and the applicable provisions of Chapters 4 and 7. The fittings shall not have ledges, shoulders or reductions capable of retarding or obstructing flow of the piping.

❖ The materials and methods used to install traps, interceptors and separators are regulated by different sections in Chapters 4 and 7. Section 405.8 requires that slip joint connections be made only on the trap inlet, trap outlet and within the trap seal. Additionally, fixtures with such connections must have an access opening at least 12 inches (305 mm) in diameter. Section 405.7.1 requires that the overflow from any fixture discharge into the drainage system on the inlet or fixture side of the trap (except water closets and urinals). Section 702 requires the material used to complete the sanitary system to conform to the standards listed in Table 702.1, 702.2, 702.3 or 702.4. Section 705 regulates the method of joining pipe and fittings. Section 707 states the joining methods that are prohibited. Section 708 governs the location, installation, maintenance, size and clearances of drainage pipe cleanouts. Section 709 regulates drain pipe sizing.

Bibliography

The following resource materials are referenced in this chapter or are relevant to the subject matter addressed in this chapter.

PDI G101 (2003), *Testing and Rating Procedure for Grease Interceptors with Appendix of Sizing and Installation Data.* South Easton, MA: Plumbing and Drainage Institute, 2003

Chapter 11:
Storm Drainage

General Comments

Storm drainage systems are generally assumed to be systems for the collection and transport of rainwater and ground water away from a building. The final disposal of this water normally occurs at some remote location, typically a stream, river or lake. This discharged water is normally left untreated.

The code prohibits the connection of floor drains to the storm drainage system (see commentary, Section 1101.3). If the purpose of this prohibition is to prevent the connection of floor drains located within the building to the storm drainage system, it raises the question of whether floor drains that are part of an open parking garage are included in this limitation. When a parking garage does not contain repair and service operations or other similar activity, it can be considered equivalent to a parking lot that is commonly drained to the storm drainage system. The code is silent on this specific issue and, therefore, the code will be interpreted and applied by the code official, based on the facts related to the individual parking garage. If the parking garage will function as a parking lot, with no increased possibility that repair and service operations will occur, floor drains in the garage could be connected to the storm drainage system the same as for a parking lot.

This chapter also covers the collection and discharge of subsoil water at the exterior of the building. This subsoil water is to be prevented from entering the building and causing possible damage to the structure and its contents. This subsoil water collection and discharge requirement, while not specifically stated, needs to be coordinated with the *International Building Code*® (IBC®) and its requirement for foundation wall dampproofing. These two elements, when properly designed and installed, will work together to keep subsoil water out of the building.

Purpose

Chapter 11 regulates the removal of storm water typically associated with rainfall. The proper installation of a storm drainage system reduces the possibility of structural collapse on a flat roof, prevents the leakage of water through the roof, prevents damage to the footings and foundation of the building and prevents flooding of the lower levels of the building.

SECTION 1101
GENERAL

1101.1 Scope. The provisions of this chapter shall govern the materials, design, construction and installation of storm drainage.

❖ Chapter 11 is one of the few chapters in the code that contains requirements that are applicable beyond the building itself. The main concern is protection of the building from the effects of rainwater; therefore, the regulations apply to roofs and paved areas around the building. Once driveways and walkways are paved, the ground does not have the ability to absorb the rainwater. A storm water system is necessary to control the runoff from roofs and paved areas.

1101.2 Disposal. Rainwater from roofs and storm water from paved areas, yards, courts and courtyards shall drain to an *approved* place of disposal. For one- and two-family dwellings, and where *approved*, storm water is permitted to discharge onto flat areas, such as streets or lawns, provided that the storm water flows away from the building.

❖ This section requires the safe control and disposal of rainwater that has fallen on roofs, paved areas, yards, courts and courtyards. Note that once rainwater has fallen on any surface other than the roof, the water is called storm water. The disposal location of rainwater and storm water is of upmost importance, as these flows can be significant at times. The presence of a new structure along with its associated driveways, parking lots and walkways creates a much greater flow rate during a rainfall event as compared to the original undisturbed natural landscape. The water cannot soak into the earth, be absorbed by plants and cannot be slowed by grasses and natural contours on its way to eventually being collected by streams and rivers. The increase in flow rate is capable of causing severe erosion of the land resulting in significant volumes of earth (with its contaminants) being moved into natural waterways. These events upset the natural ecosystem of the region and create many problems in the water ways that are further downstream. Many jurisdictions have stringent requirements for controlling the flow of storm water into streams, rivers or ponds. Storm water detention basins are often required to slow the storm flow discharge so that storm water is released to the environment in a controlled manner that doesn't impact the surrounding ecosystems.

Next to any building, it is paramount to control storm water and rainwater so that the building foundation is not compromised. This includes keeping water away from ground surfaces adjacent to foundation walls that are basement walls so that they are not subjected to undue water pressure that might cause wall leaks, resulting in a "wet basement."

Where public storm sewers are available, storm water detention basins are also used to prevent the sewers from being overloaded during peak periods of rainfall.

The majority of buildings will require a storm water drainage system to manage the rainwater and storm water at the building site. However, some buildings might be approved for draining storm water onto flat areas adjacent to the building, provided that the water flows away from the building. One- and two-family dwellings do not require such approval, provided that the water is directed away from the building.

1101.3 Prohibited drainage. Storm water shall not be drained into *sewers* intended for sewage only.

❖ A sanitary sewer system will become overloaded during a rainstorm if storm water is discharged into it. Typically, sewage treatment plants are not designed to take the surcharge imposed on them from additional storm water discharge. Discharging storm water into the sanitary system could result in either flooding and backups or the polluting of nearby waterways that receive inadequately treated sanitary discharge. The intent is to discharge into sanitary sewers only that which requires treatment.

The only exception to the prohibition stated in this section occurs where combined sanitary and storm sewers exist, as they do in many older cities (see commentary, Section 1108), and the code official approves such discharge connection.

1101.4 Tests. The conductors and the building *storm drain* shall be tested in accordance with Section 312.

❖ This section references Section 312, which establishes testing and inspection requirements for the completed, installed storm drainage system. The required testing is the same as would be required for the sanitary drainage system. When the storm system handles the discharge from a large roof area, and the storm discharge piping is contained on the interior of the building, the potential for vast building damage exists if the storm drainage system fails. The volume of water, pressures and dynamic forces that are possible in these systems oftentimes greatly exceed that for sanitary drainage systems. Unlike sanitary systems, storm drainage systems can be designed for full-bore flow, meaning that the storm system would have to function under more severe service conditions.

1101.5 Change in size. The size of a drainage pipe shall not be reduced in the direction of flow.

❖ One of the fundamental requirements of any form of drainage system is that reduction of the piping size is not permitted in the direction of drainage flow. This is also the case with storm drainage systems. A size reduction creates an obstruction to the flow of discharge, possibly resulting in a backup, an interruption of service in the drainage system or a stoppage occurring in the piping.

1101.6 Fittings and connections. All connections and changes in direction of the storm drainage system shall be made with *approved* drainage-type fittings in accordance with Table 706.3. The fittings shall not obstruct or retard flow in the system.

❖ This section requires the use of approved drainage-type fittings in the storm drainage system to make necessary directional changes. Although the flow through the storm drainage system is significantly different from that of the sanitary drainage system, a smooth transition of flow is still necessary (see commentary, Section 706.3).

1101.7 Roof design. Roofs shall be designed for the maximum possible depth of water that will pond thereon as determined by the relative levels of roof deck and overflow weirs, scuppers, edges or serviceable drains in combination with the deflected structural elements. In determining the maximum possible depth of water, all primary roof drainage means shall be assumed to be blocked. The maximum possible depth of water on the roof shall include the height of the water required above the inlet of the secondary roof drainage means to achieve the required flow rate of the secondary drainage means to accommodate the design rainfall rate as required by Section 1106.

❖ The roof structure must be capable of supporting the maximum ponding of water that will occur when the primary roof drainage means are blocked. Section 1108, requires the installation of a secondary (emergency) roof drainage system to limit the amount of rainwater that could be retained on top of the roof. This also limits the potential increased load to be structurally supported by the roof. Refer to Section 1108 to determine how elements such as scuppers, overflow weirs, serviceable drains and other similar openings, along with the roof edge itself, are addressed in the code.

Although not totally defined, the primary storm drainage system is generally interpreted to include roof drains, leaders, conductors and horizontal storm drains within the structure. Although not immune to failure, the storm sewer system is usually not considered to be part of the building's primary roof drainage means. On large roofs with multiple, independently drained areas or bays, the primary drainage means for each such area could be considered separately.

The maximum depth of the ponding water should be measured to the highest flow depth allowed by the design of the secondary (emergency) roof drainage system. The volume of flow into a scupper or roof drain is a function of water depth at the scupper or roof drain. A scupper or roof drain, for example, will typically not reach its maximum flow capacity until the water has reached a certain depth above the weir of the drain. This means that water may necessarily pond on a roof in order for the drain to reach design flow conditions for the drainage system. Commentary Figure 1101.7.1 shows several roof drains and the depth of water (hydraulic heads) above the drain inlet.

Consideration must be given to roof structure deflection, which will increase the volume of water held by ponding. An improperly designed roof could collapse under the weight of the water held on it [see Commentary Figure 1101.7(1)]. The roof structure in Commentary Figure 1101.7(2) must be designed to support the weight of water at the maximum possible depth. The maximum depth is determined assuming that the primary drainage system is blocked and the water must rise above the scuppers to a height sufficient to cause flow. Because of the many variables, each roof drainage system design must be evaluated individually, taking into consideration such factors as

DRAINAGE SYSTEM	FLOW RATE (gpm)									
	Depth of water above drain inlet (hydraulic head) (inches)									
	1	2	2.5	3	3.5	4	4.5	5	7	8
4-inch-diameter drain	80	170	180							
6-inch-diameter drain	100	190	270	380	540					
8-inch-diameter drain	125	230	340	560	850	1,100	1,170			
6-inch-wide, open-top scupper	18	50	*	90	*	140	*	194	321	393
24-inch-wide, open-top scupper	72	200	*	360	*	560	*	776	1,284	1,572
6-inch-wide, 4-inch-high, closed top scupper	18	50	*	90	*	140	*	177	231	253
24-inch-wide, 4-inch-high, closed-top scupper	72	200	*	360	*	560	*	708	924	1,012
6-inch-wide, 6-inch-high, closed-top scupper	18	50	*	90	*	140	*	194	303	343
24-inch-wide, 6-inch-high, closed-top scupper	72	200	*	360	*	560	*	776	1,212	1,372

For SI: 1 inch = 25.4 mm, 1 gallon per minute = 3.785 L/m.
Source: Factory Mutual Engineering Corp. Loss Prevention Data 1-54.

Figure 1101.7(1)
FLOW RATE, IN GALLONS PER MINUTE, OF VARIOUS SCUPPERS AT VARIOUS WATER DEPTHS

a: THE LEVEL MUST RISE ABOVE THE WEIR TO INDUCE FLOW. THE LEVEL WILL CONTINUE TO RISE UNTIL THE DISCHARGE AND RAINFALL RATE REACH EQUILIBRIUM. THE MAXIMUM DEPTH MUST BE CONSIDERED IN CALCULATING ROOF LOADS.

Figure 1101.7(2)
ROOF DRAIN WITH OVERFLOW

roof drain capacity at ponding depth, roof type, roof configurations, mains and sewer sizes, drain capacities, locations of trees, climate and probability of system failures.

The secondary (emergency) roof drainage system must be completely independent of the primary roof drainage system so that a failure in the primary system does not influence the proper functioning of the secondary system. The secondary (emergency) roof drainage system discharge must occur in a location above grade that will be noticed by either the users or service personnel of the building. The discharge point becomes a visual indication that the primary system has failed or become blocked.

1101.8 Cleanouts required. Cleanouts shall be installed in the storm drainage system and shall comply with the provisions of this code for sanitary drainage pipe cleanouts.

Exception: Subsurface drainage system.

❖ The cleanout requirements apply to drainage systems without making a distinction between sanitary and storm with the exception of subsurface drainage systems. These systems serve to drain water away from foundations and use perforated pipe. A storm drain is susceptible to stoppage in the same way as a sanitary system; hence, cleanouts are necessary as required by Section 708.

1101.9 Backwater valves. Storm drainage systems shall be provided with backwater valves as required for sanitary drainage systems in accordance with Section 715.

❖ Although backwater valves are typically associated with the installation of the sanitary sewer and the connection between the building and the sanitary sewer, they are also installed on storm systems connected to combination storm and sanitary sewers. There is often a danger of backflow of storm water into a building using combination storm and sanitary drainage systems where the public sewer becomes overloaded, clogged or surcharged. To prevent damage from backflow, a backwater valve should be installed in accordance with Section 715 [see Commentary Figure 1111.1(1)].

SECTION 1102
MATERIALS

1102.1 General. The materials and methods utilized for the construction and installation of storm drainage systems shall comply with this section and the applicable provisions of Chapter 7.

❖ This section requires the use of approved sanitary drainage-type piping, fittings and any other components in the storm drainage system. Although the flow through the storm drainage system is significantly different from that of the sanitary drainage system (see commentary, Section 1101.6), the need to provide a system with a smooth flow of discharge is still a valid concern; therefore, this section requires compliance with the requirements in Chapter 7.

1102.2 Inside storm drainage conductors. Inside storm drainage conductors installed above ground shall conform to one of the standards listed in Table 702.1.

❖ Any of the materials acceptable for sanitary drainage are allowable for storm drainage; hence, the reference to Table 702.1, which is the table for above-ground sanitary drainage and vent pipe.

1102.3 Underground building storm drain pipe. Underground building *storm drain* pipe shall conform to one of the standards listed in Table 702.2.

❖ Similar to above-ground piping, as discussed in Section 1102.2, any of the piping materials listed in Table 702.2 for underground building drainage and vent pipe are acceptable for use as building storm drain piping underground.

1102.4 Building storm sewer pipe. Building storm *sewer* pipe shall conform to one of the standards listed in Table 1102.4.

❖ The referenced table specifies acceptable piping materials for storm sewers (see commentary, Table 1102.4).

**TABLE 1102.4
BUILDING STORM SEWER PIPE**

MATERIAL	STANDARD
Acrylonitrile butadiene styrene (ABS) plastic pipe	ASTM D 2661; ASTM D 2751; ASTM F 628; CSA B181.1; CSA B182.1
Cast-iron pipe	ASTM A 74; ASTM A 888; CISPI 301
Concrete pipe	ASTM C 14; ASTM C 76; CSA A257.1M; CSA A257.2M
Copper or copper-alloy tubing (Type K, L, M or DWV)	ASTM B 75; ASTM B 88; ASTM B 251; ASTM B 306
Polyethylene (PE) plastic pipe	ASTM F 2306/F 2306M
Polyvinyl chloride (PVC) plastic pipe (Type DWV, SDR26, SDR35, SDR41, PS50 or PS100)	ASTM D 2665; ASTM D 3034; ASTM F 891; CSA B182.4; CSA B181.2; CSA B182.2
Vitrified clay pipe	ASTM C 4; ASTM C 700
Stainless steel drainage systems, Type 316L	ASME A112.3.1

❖ The table identifies the materials that are permitted and the standards with which these materials must comply.

ASTM D 2751 is the referenced standard for typical acrylonitrile butadiene styrene (ABS) sewer pipe and fittings. The pipe may be produced from ABS material with a cell classification of 1-3-3, 3-2-2 or 2-2-3. This pipe is produced in standard dimension ratios (SDRs) of 23.5, 35 and 42. SDR is the ratio between the average outside diameter of the pipe and the minimum wall thickness. The other standards referenced are for Schedule 40 plastic pipe fittings and Schedule 40 drain, waste and vent (DWV) pipe with a cellular core.

Cast-iron pipe is available in two pipe styles: hubbed and hubless. Hubbed cast-iron pipe is the oldest type. It uses a hub at joints and connections; hubless pipe is joined using mechanical couplings. Hubbed cast iron is produced in two weights: service and extra heavy. Extra heavy is characterized by a thicker pipe wall.

Previously, other grades of cast-iron pipe were produced: medium weight and "victory pipe." Medium-weight pipe was a class between extra-heavy and service-weight pipe. A pipe referred to as "victory pipe" was manufactured during World War II—it was lighter and had a thinner pipe wall than service-weight pipe.

The wall thickness of hubless cast-iron pipe is approximately equal to service-weight hub pipe. Hubless pipe is produced both with and without a spigot bead. Cast-iron pipe is commonly referred to as "soil pipe" or "cast-iron soil pipe." The term "soil pipe" was used when cast iron was essentially the only material used for soil drainage pipes. The term is still in use today when differentiating between cast-iron pipe used for drainage and cast-iron pipe used for water. Cast iron is typically produced in 5-foot (1524 mm) and 10-foot (3048 mm) lengths. The pipe is marked with the manufacturer's name and designated as hubless, extra heavy (XH) or service (SV).

Concrete pipe is made from portland cement, blast-furnace slag cement or portland-pozzolan cement and aggregate mixture. The pipe is produced both reinforced and nonreinforced.

ASTM C 14 and CAN/CSA A257.1M are the referenced standards for unreinforced concrete pipe. ASTM C 14 concrete pipe is available in three classes: 1, 2 or 3. The pipe classification relates to the concrete strength based on diameter. Class 1 has the lowest strength requirements and Class 3 has the highest. The pipe is manufactured in diameters of 4 through 36 inches (102 through 914 mm).

ASTM C 76 and CAN/CSA A257.2M are the referenced standards for reinforced concrete pipe, which is strengthened with either steel wire, hot rod or welded wire fabric. ASTM C 76 pipe is produced in five classes: 1, 2, 3, 4 or 5. The class determines the strength of the concrete, pipe and reinforcement geometry. The pipe is manufactured in sizes 12 inches (305 mm) and larger.

Copper or copper-alloy tubing (Type K, L, M or DWV) is manufactured in two different tempers: drawn (called hard copper with the designation "H") and annealed (called soft copper with the designation "O"). Drawn copper tubing typically comes in straight lengths, where annealed copper tubing typically comes in both straight lengths and coils. Annealed and tempered drawn copper can be formed by bending. Bending-tempered copper tubing is identified on the pipe with the designation "BT."

The four types of copper tubing are identified with a colored marking system for ease of identification, as follows:

- Green—Type K copper tubing
- Blue—Type L copper tubing
- Red—Type M copper tubing
- Yellow—Type DWV copper tubing

Corrugated double wall polyethylene (PE) pipe (ASTM F 2306) for storm water conveyance has been in widespread use by utilities and road building contractors for over 20 years. Single wall corrugated PE pipe has been approved for use in subsoil drain systems for many years but for reasons unexplained, the double wall version was never proposed for inclusion in the list of approved materials for building storm drains. The double wall version of PE pipe covered by the referenced ASTM International (ASTM) number has a smooth, uniform bore with a corrugated exterior that imparts strength to resist crushing.

Polyvinyl chloride (PVC) plastic pipe (Type DWV, SDR 26, SDR 35, SDR 41, PS 50 or PS 100) is regulated by ASTM F 891, which specifies that sewer grade pipe be designated by a "PS" in addition to a Schedule 40 designation (Type DWV). PS indicates pipe stiffness. The grades of PVC plastic pipes used in sewers are PS 50 and 100. The numerical designation indicates the pipe stiffness in pounds per square inch (psi). The higher value indicates plastic pipe with a thicker pipe wall.

ASTM D 2665 and ASTM D 3034 are the standards regulating sewer-grade PVC pipe and fittings. The materials may be used interchangeably. Both pipes are produced in SDRs of 35 and 41, with a hub end. The PVC material used to produce the pipe is classified as 12454-B, 12454-C or 13364-B. The pipe is produced in diameters of 4 through 15 inches (102 mm through 381 mm). CSA B182.2 is the referenced standard for SDR 28, 35 and 41 PVC sewer pipe and fittings, and CSA B182.4 is the referenced standard for PVC ribbed sewer pipe and fittings.

Vitrified clay pipe is surface clay, fire clay, shale or a combination of the materials, and is classified as standard strength, extra strength, extra quality or heavy duty. The classification determines the strength of the pipe. It is produced in diameters of 3 inches (76 mm) and larger. Each length of pipe is marked with the manufacturer's name and the classification of the pipe.

Type 316L stainless steel provides excellent corrosion resistance when installed either above or below ground in a storm sewer system. Stainless steel pipe and fittings have been used for years worldwide to convey sanitary wastes because of the strength, durability and corrosion resistance of the material. The additional cost of installing stainless steel piping can often be offset by the piping's longer expected life.

STORM DRAINAGE

1102.5 Subsoil drain pipe. Subsoil drains shall be open-jointed, horizontally split or perforated pipe conforming to one of the standards listed in Table 1102.5.

❖ The materials not described herein are described in detail in the commentary to Table 1102.4. PE plastic pipe: ASTM F 405 is the referenced standard for corrugated pipe and PE tubing. The pipe is available with perforations and is classified as either standard or heavy-duty tubing.

PVC plastic pipe (type sewer pipe, PS 25, PS 50 or PS 100): ASTM D 2729 regulates thin-wall drain pipe, which is available as both perforated and solid wall pipe.

TABLE 1102.5
SUBSOIL DRAIN PIPE

MATERIAL	STANDARD
Cast-iron pipe	ASTM A 74; ASTM A 888; CISPI 301
Polyethylene (PE) plastic pipe	ASTM F 405; CSA B182.1; CSA B182.6; CSA B182.8
Polyvinyl chloride (PVC) Plastic pipe (type sewer pipe, SDR35, PS25, PS50 or PS100)	ASTM D 2729; ASTM D 3034; ASTM F 891; CSA B182.2; CSA B182.4
Stainless steel drainage systems, Type 316L	ASME A 112.3.1
Vitrified clay pipe	ASTM C 4; ASTM C 700

1102.6 Roof drains. Roof drains shall conform to ASME A112.6.4 or ASME A112.3.1.

❖ ASME A112.6.4 regulates the design and construction of roof drains. Roof drains must have either a dome or grate covering the opening. This helps restrict leaves and debris from entering the storm drainage piping and creating a stoppage. Because roof drains are exposed to extreme temperatures, it is common practice to install an expansion joint between the roof drain and the storm drainage pipe.

ASME A112.3.1 provides details for Types 304 and 316L stainless steel roof drains.

1102.7 Fittings. Pipe fittings shall be *approved* for installation with the piping material installed, and shall conform to the respective pipe standards or one of the standards listed in Table 1102.7. The fittings shall not have ledges, shoulders or reductions capable of retarding or obstructing flow in the piping. Threaded drainage pipe fittings shall be of the recessed drainage type.

❖ Each fitting is designed to be installed in a particular system with a given material or combination of materials. Drainage pattern fittings must be installed in drainage systems. Vent fittings are limited to the venting system. Fittings are also available that could be used in any type of plumbing system.

Fittings must be the same material as, or compatible with, the pipe. This prevents any chemical or corrosive action between dissimilar materials.

Many pipe standards also include provisions regulating fittings. There are, however, a number of standards that strictly regulate pipe fittings.

TABLE 1102.7
PIPE FITTING

MATERIAL	STANDARD
Acrylonitrile butadiene styrene (ABS) plastic	ASTM D 2661; ASTM D 3311; CSA B181.1
Cast-iron	ASME B16.4; ASME B16.12; ASTM A 888; CISPI 301; ASTM A 74
Coextruded composite ABS and drain DR-PS in PS35, PS50, PS100, PS140, PS200	ASTM D 2751
Coextruded composite ABS DWV Schedule 40 IPS pipe (solid or cellular core)	ASTM D 2661; ASTM D 3311; ASTM F 628
Coextruded composite PVC DWV Schedule 40 IPS-DR, PS140, PS200 (solid or cellular core)	ASTM D 2665; ASTM D 3311; ASTM F 891
Coextruded composite PVC sewer and drain DR-PS in PS35, PS50, PS100, PS140, PS200	ASTM D 3034
Copper or copper alloy	ASME B16.15; ASME B16.18; ASME B16.22; ASME B16.23; ASME B16.26; ASME B16.29
Gray iron and ductile iron	AWWA C110/A21.10
Malleable iron	ASME B16.3
Plastic, general	ASTM F 409
Polyethylene (PE) plastic pipe	ASTM F 2306/F 2306M
Polyvinyl chloride (PVC) plastic	ASTM D 2665; ASTM D 3311; ASTM F 1866
Steel	ASME B16.9; ASME B16.11; ASME B16.28
Stainless steel drainage systems, Type 316L	ASME A112.3.1

SECTION 1103
TRAPS

1103.1 Main trap. Leaders and *storm drains* connected to a combined sewer shall be trapped. Individual storm water traps shall be installed on the storm water drain *branch* serving each conductor, or a single trap shall be installed in the main *storm drain* just before its connection with the combined *building sewer* or the *public sewer*. Leaders and storm drains connected to a building storm sewer shall not be required to be trapped.

❖ This section regulates the installation of traps on combination sanitary and storm sewer systems (see Commentary Figure 1103.1).

The trap on the storm drain is intended to prevent sewer gas from entering the system and is required only where the storm drain discharges into the sanitary sewer system. Unlike a sanitary system, a storm drainage system is not designed for any precautions against sewer gas. Any storm drain connection to a combination sanitary and storm sewer must be trapped to prevent the escape of sewer gases. For example, an air intake opening for an air-handling unit may be located next to a roof drain. A properly located trap prevents sewer gas from escaping through a roof drain and entering the outdoor air intake.

Figure 1103.1
TRAPS ON STORM DRAINS THAT CONNECT TO COMBINATION SANITARY AND STORM SEWER SYSTEMS

1103.2 Material. Storm water traps shall be of the same material as the piping system to which they are attached.

❖ This section simply requires that the material for the trap on the storm sewer line connecting to the combination sanitary and storm sewer be the same material as the combination sanitary and storm sewer.

1103.3 Size. Traps for individual conductors shall be the same size as the horizontal drain to which they are connected.

❖ This section requires that the trap be the same size as the horizontal drain pipe to which it discharges to prevent any obstruction of flow in the piping.

1103.4 Cleanout. An accessible cleanout shall be installed on the building side of the trap.

❖ This section requires the installation of an accessible cleanout and also states where it should be located.

SECTION 1104
CONDUCTORS AND CONNECTIONS

1104.1 Prohibited use. Conductor pipes shall not be used as soil, waste or vent pipes, and soil, waste or vent pipes shall not be used as conductors.

❖ This section regulates the installation of conductors, leaders and connections.

A sanitary drainage stack and storm drainage conductors are designed differently and serve different purposes. They cannot be interconnected and the components of the two different systems cannot be interchanged (see commentary, Sections 1101.3 and 1104.3).

1104.2 Floor drains. Floor drains shall not be connected to a *storm drain*.

❖ To prevent contamination of the storm water discharge with chemicals, waste, sewage, etc., the code prohibits the connection of floor drains to a storm drain. A floor drain connected to a storm drainage system invites the discharge of sanitary waste into the storm system, which will create an environmental hazard, considering that storm drainage is discharged to the environment without treatment. Another concern is that under flow conditions, a floor drain connection will act as a relief opening, thereby allowing water flow into the building.

Areaway drains serving subsurface spaces outside of the building, such as window wells or basement or cellar entrance wells, are typically not considered to be floor drains when applying this section. Additionally, where approved by the code official, if a parking garage will function only as a parking lot, with no increased possibility that repair and service operations will occur, floor drains in the garage could be connected to the storm drainage system the same as for a parking lot.

SECTION 1105
ROOF DRAINS

1105.1 General. Roof drains shall be installed in accordance with the manufacturer's instructions. The inside opening for the roof drain shall not be obstructed by the roofing membrane material.

❖ Because roof drains penetrate the roof of the building, their proper installation is necessary to prevent water leaking into the structure. The manufacturer's installation instructions provide the necessary procedures for making sure that the roof drain is properly mounted and sealed to the roof assembly. The last line of this section is a reminder for inspectors to make sure that roofing membrane materials are properly trimmed to allow full flow to pass through the roof drain.

1105.2 Roof drain flow rate. The published roof drain flow rate, based on the head of water above the roof drain, shall be used to size the storm drainage system in accordance with Section 1106. The flow rate used for sizing the storm drainage piping shall be based on the maximum anticipated ponding at the roof drain.

❖ A research project by the American Society of Plumbing Engineers (ASPE) Research Foundation revealed that roof drains are the limiting factor in the design of storm drain systems. The previous design philosophy (for over 40 years) was that the diameter of the piping system was the limiting factor. Any roof drain sized for the required pipe size was assumed to be capable of accepting the design flow rate for the system. Actual testing of many different sizes and configurations of roof drains from a variety of manufacturers indicated that the roof drain assembly is the limiting factor.

SECTION 1106
SIZE OF CONDUCTORS, LEADERS AND STORM DRAINS

1106.1 General. The size of the vertical conductors and leaders, building *storm drains*, building storm *sewers* and any horizontal branches of such drains or *sewers* shall be based on the 100-year hourly rainfall rate indicated in Figure 1106.1 or on other rainfall rates determined from *approved* local weather data.

❖ This section regulates the size of storm drainage systems as a function of a 60-minute duration, 100-year rainfall rate on the roof or paved area to be drained. Selection of the correct rainfall rate is critical for the proper sizing of the storm drain system.

Figure 1106.1 spans five consecutive pages of the code and is broken into major regions of the country. This figure indicates the rainfall rates for a storm of 1-hour duration and a 100-year return period. Rainfall rates indicate the maximum rate of rainfall within a given period of time occurring every so many years. For example, Figure 1106.1 of the code indicates a rainfall rate of 2.1 inches (53 mm) per hour for Burlington, Vermont. Thus, it is predicted that it will rain 2.1 inches (53 mm) within 1 hour once every 100 years.

The rainfall rates are calculated by a statistical analysis of weather records. Because the statistics are based on previous or historical weather conditions, it is conceivable to have two 100-year storms in one week's time. The probability of this occurring, however, is very low.

1106.2 Size of storm drain piping. Vertical and horizontal *storm drain* piping shall be sized based on the flow rate through the roof drain. The flow rate in *storm drain* piping shall not exceed that specified in Table 1106.2.

Example:

Building location: Charleston, South Carolina

Horizontal projected roof area: 50,000 square feet

Number of conductors: 5

Rainfall rate: 4.3 inches per hour (from Figure 1106.1 or Appendix B of the code)

For SI: 1 inch = 25.4 mm, 1 square foot = 0.0929 m².

1106.3 Vertical leader sizing. Vertical leaders shall be sized based on the flow rate from horizontal gutters or the maximum flow rate through roof drains. The flow rate through vertical leaders shall not exceed that specified in Table 1106.3.

❖ Unlike a sanitary drainage system, a horizontal storm drain or sewer is sized for a full-flow condition. The pipe may be filled to capacity under a worst-case condition, and is still sized for gravity flow. Under "normal" operation, the storm sewer is not full and functions more like a sanitary sewer.

1106.4 Vertical walls. In sizing roof drains and storm drainage piping, one-half of the area of any vertical wall that diverts rainwater to the roof shall be added to the projected roof area for inclusion in calculating the required size of vertical conductors, leaders and horizontal storm drainage piping.

❖ This section includes the requirement for the mandatory inclusion of half of the area of vertical walls, including parapet walls, that are adjacent to and above a roof area, into the calculation for projected horizontal roof area when sizing storm drainage components. This acknowledges that vertical walls can catch and divert rainwater onto the roof. This additional water must be accounted for in the sizing of the components that will catch and discharge this water (see Commentary Figure 1106.4). This additional water must also be accounted for in the design of a controlled flow system (see commentary, Section 1110).

The example in Commentary Figure 1106.4 illustrates how to compute the wall area that must be added to the lower projected roof area for sizing the storm drains.

1106.5 Parapet wall scupper location. Parapet wall roof drainage scupper and overflow scupper location shall comply with the requirements of Section 1503.4 of the *International Building Code*.

❖ This section requires that the parapet wall scuppers be installed to satisfy the requirements of the IBC. This is a statement of mandatory coordination between two parts of the code; in this case, plumbing and building. There are other sections that require this coordination, but are not specifically referenced. Section 1101.7, for example, requires that a roof be designed structurally to support the maximum depth of rainwater that can pond on it, but does not require designing a roof in accordance with the requirements of the IBC, even though it would apply as well.

1106.6 Size of roof gutters. Horizontal gutters shall be sized based on the flow rate from the roof surface. The flow rate in horizontal gutters shall not exceed that specified in Table 1106.6.

❖ Roof gutters are designed for full-flow drainage, assuming the volume of rainwater associated with a 60-minute duration, 100-year rain. The sizing table is based on full capacity of the gutters. The size and number of conductors or leader connections to a roof gutter system must be adequate to prevent overflow of the gutters.

STORM DRAINAGE

For SI: 1 inch = 25.4 mm.
Source: National Weather Service, National Oceanic and Atmospheric Administration, Washington D.C.

**FIGURE 1106.1
100-YEAR, 1-HOUR RAINFALL (INCHES) EASTERN UNITED STATES**

STORM DRAINAGE

For SI: 1 inch = 25.4 mm.
Source: National Weather Service, National Oceanic and Atmospheric Administration, Washington D.C.

**FIGURE 1106.1—continued
100-YEAR, 1-HOUR RAINFALL (INCHES) CENTRAL UNITED STATES**

STORM DRAINAGE

For SI: 1 inch = 25.4 mm.
Source: National Weather Service, National Oceanic and Atmospheric Administration, Washington D.C.

FIGURE 1106.1—continued
100-YEAR, 1-HOUR RAINFALL (INCHES) WESTERN UNITED STATES

STORM DRAINAGE

For SI: 1 inch = 25.4 mm.
Source: National Weather Service, National Oceanic and Atmospheric Administration, Washington D.C.

FIGURE 1106.1—continued
100-YEAR, 1-HOUR RAINFALL (INCHES) ALASKA

STORM DRAINAGE

For SI: 1 inch = 25.4 mm.
Source: National Weather Service, National Oceanic and Atmospheric Administration, Washington, D.C.

FIGURE 1106.1—continued
100-YEAR, 1-HOUR RAINFALL (INCHES) HAWAII

STORM DRAINAGE

**TABLE 1106.2
STORM DRAIN PIPE SIZING**

PIPE SIZE (inches)	CAPACITY (gpm)				
	VERTICAL DRAIN	SLOPE OF HORIZONTAL DRAIN			
		$^1/_{16}$ inch per foot	$^1/_8$ inch per foot	$^1/_4$ inch per foot	$^1/_2$ inch per foot
2	34	15	22	31	44
3	87	39	55	79	111
4	180	81	115	163	231
5	311	117	165	234	331
6	538	243	344	487	689
8	1,117	505	714	1,010	1,429
10	2,050	927	1,311	1,855	2,623
12	3,272	1,480	2,093	2,960	4,187
15	5,543	2,508	3,546	5,016	7,093

For SI: 1 inch = 25.4 mm, 1 foot = 304.8 mm, 1 gallon per minute = 3.785 L/m.

**TABLE 1106.3
VERTICAL LEADER SIZING**

SIZE OF LEADER (inches)	CAPACITY (gpm)
2	30
2 × 2	30
$1^1/_2$ × $2^1/_2$	30
$2^1/_2$	54
$2^1/_2$ × $2^1/_2$	54
3	92
2 × 4	92
$2^1/_2$ × 3	92
4	192
3 × $4^1/_4$	192
$3^1/_2$ × 4	192
5	360
4 × 5	360
$4^1/_2$ × $4^1/_2$	360
6	563
5 × 6	563
$5^1/_2$ × $5^1/_2$	563
8	1208
6 × 8	1208

For SI: 1 inch = 25.4 mm, 1 gallon per minute = 3.785 L/m.

Figure 1106.4
VERTICAL WALLS

SECTION 1107
SIPHONIC ROOF DRAINAGE SYSTEMS

1107.1 General. Siphonic roof drains and drainage systems shall be designed in accordance with ASME A112.6.9 and ASPE 45.

❖ Siphonic roof drainage systems are vastly different than gravity roof drainage systems. Special roof drain designs are necessary to make siphonic roof drainage systems work efficiently. The piping for siphonic roof drainage systems is smaller in diameter than is required for gravity roof drainage systems because the piping is designed to flow completely full. Full flow is developed by the action of storm water falling down through vertical piping so that a negative pressure is developed above the falling water. The negative pressure "draws" additional flow through the special roof drain until all of the piping in the siphonic roof drainage system is flowing full and is creating the maximum "suction" possible at the roof drain.

SECTION 1108
SECONDARY (EMERGENCY) ROOF DRAINS

1108.1 Secondary (emergency overflow) drains or scuppers. Where roof drains are required, secondary (emergency overflow) roof drains or scuppers shall be provided where the roof perimeter construction extends above the roof in such a manner that water will be entrapped if the primary drains allow buildup for any reason. Where primary and secondary roof drains are manufactured as a single assembly, the inlet and outlet for each drain shall be independent.

❖ This section requires all buildings to have some method for preventing the accumulation of unplanned excessive rainwater. A secondary drainage system is required where the building has parapet walls or other construction on the perimeter of the building that would cause ponding. See Section 1101.7 regarding roof loads resulting from ponding on roofs.

The intent of this section is to limit the amount of ponding water that will be placed on the roof by rainfall. If the building is designed so water cannot pond

TABLE 1106.6
HORIZONTAL GUTTER SIZING

GUTTER DIMENSIONS[a] (inches)	SLOPE (inch per foot)	CAPACITY (gpm)
$1^1/_2 \times 2^1/_2$	$1/_4$	26
$1^1/_2 \times 2^1/_2$	$1/_2$	40
4	$1/_8$	39
$2^1/_4 \times 3$	$1/_4$	55
$2^1/_4 \times 3$	$1/_2$	87
5	$1/_8$	74
$4 \times 2^1/_2$	$1/_4$	106
$3 \times 3^1/_2$	$1/_2$	156
6	$1/_8$	110
3×5	$1/_4$	157
3×5	$1/_2$	225
8	$1/_{16}$	172
8	$1/_8$	247
$4^1/_2 \times 6$	$1/_4$	348
$4^1/_2 \times 6$	$1/_2$	494
10	$1/_{16}$	331
10	$1/_8$	472
5×8	$1/_4$	651
4×10	$1/_2$	1055

For SI: 1 inch = 25.4 mm, 1 foot = 304.8 mm, 1 gallon per minute = 3.785 L/m, 1 inch per foot = 83.3 mm/m.

a. Dimensions are width by depth for rectangular shapes. Single dimensions are diameters of a semicircle.

on the roof, such as roofs sloped toward the edge of the building, secondary drainage is not required (see Commentary Figure 1108.1). This simple concept should be carried to all portions of the roof, so that if any portion is designed so water can pond, secondary drains would be required.

Bear in mind that a roof could be intentionally designed to allow ponding to some design depth such as for a rainwater harvesting system.

1108.2 Separate systems required. Secondary roof drain systems shall have the end point of discharge separate from the primary system. Discharge shall be above grade, in a location that would normally be observed by the building occupants or maintenance personnel.

❖ This section requires the installation of a completely separate system for secondary roof drains. This eliminates past confusion over how far the secondary roof drain piping had to run before it could be connected to the primary drainage system. This section requires that the two systems be completely independent and that they have separate points of discharge. This requirement is consistent with the IBC as well as ASCE 7, which is referenced by the IBC.

This system is designed to function in an emergency situation; therefore, it is required to discharge above grade. By requiring this discharge above grade, the discharge will be a means of signaling that there is a problem with the primary drain that requires immediate attention. Additionally, because this discharge is not a constant means of relieving the roof rainwater, it would not normally constitute a runoff problem. Discharge into an open ditch is considered "above grade." Even though this section has been interpreted to prohibit open ditch discharge, that is incorrect. The intent is to prohibit hidden discharge points. Commentary Figure 1108.2(1) illustrates typical examples of separate secondary drains. Commentary Figure 1108.2(2) illustrates the secondary drain improperly connecting to the primary roof drain.

Figure 1108.1
SECONDARY ROOF DRAINAGE NOT REQUIRED

Figure 1108.2(1)
SEPARATE PRIMARY AND SECONDARY ROOF DRAINS

STORM DRAINAGE

Figure 1108.2(2)
PROHIBITED SECONDARY ROOF DRAIN CONNECTION

HEAD (H) (inches)	CAPACITY OF SCUPPER (gallons per minute)									
	Length (L) of scupper (inches)									
	4	6	8	10	12	18	24	30	36	48
1	10.7	17.4	23.4	29.3	35.4	53.4	71.5	89.5	107.5	143.7
2	30.5	47.5	64.4	81.4	98.5	149.4	200.3	251.1	302.1	404.0
3	52.9	84.1	115.2	146.3	177.8	271.4	364.9	458.5	552.0	739.0
4	76.7	124.6	172.6	220.5	269.0	413.3	557.5	701.8	846.0	1135.0
6	123.3	211.4	299.5	387.5	476.5	741.1	1005.8	1270.4	1535.0	2067.5

For SI: 1 inch = 25.4 mm, 1 foot = 304.8 mm, 1 gallon per minute = 3.785 L/m.

Based on the Francis formula:

$Q = 3.33 (L - 0.2H) H^{1.5}$

Where:

Q = Flow rate (cubic feet per second).
L = Length of scupper opening (feet).
H = Head on scupper [feet (measured 6 feet back from opening)].

Figure 1108.3
SIZE OF SCUPPERS

1108.3 Sizing of secondary drains. Secondary (emergency) roof drain systems shall be sized in accordance with Section 1106 based on the rainfall rate for which the primary system is sized. Scuppers shall be sized to prevent the depth of ponding water from exceeding that for which the roof was designed as determined by Section 1101.7. Scuppers shall have an opening dimension of not less than 4 inches (102 mm). The flow through the primary system shall not be considered when sizing the secondary roof drain system.

❖ This section gives the method for sizing the secondary (emergency) roof drainage system and its associated components. Sizing the components of a secondary drainage system is identical to sizing the primary system. The same tables and rainfall rates are used.

Instead of installing a secondary system of roof drains and pipes, many designers install scuppers to allow rainwater to overflow the roof. The height and length of a scupper is critical in determining the maximum depth of ponding on a roof. Although the code states that the minimum scupper dimension is 4 inches (102 mm), it would be necessary to increase the length dimension to keep the ponding depth at or below the level for which the roof was designed. Commentary Figure 1108.3 contains a table and equation that can be used to design the scupper.

SECTION 1109
COMBINED SANITARY AND STORM PUBLIC SEWER

1109.1 General. Where the *public sewer* is a combined system for both sanitary and storm water, the *storm sewer* shall be connected independently to the *public sewer*.

❖ Eventually, most municipalities having combined public sewers will need to separate the combined sewer into separate sanitary sewers and storm sewers to reduce sewage overflows during rainfall events. New construction must prepare for this by making the building storm sewer separate, even though currently the building sewer connects to the combined public sewer system.

SECTION 1110
CONTROLLED FLOW ROOF DRAIN SYSTEMS

1110.1 General. The roof of a structure shall be designed for the storage of water where the storm drainage system is engineered for controlled flow. The controlled flow roof drain system shall be an engineered system in accordance with this section and the design, submittal, approval, inspection and testing requirements of Section 316.1. The controlled flow system shall be designed based on the required rainfall rate in accordance with Section 1106.1.

❖ This section contains requirements for a storm drainage system that is designed as a controlled-flow system. Often such a system is used to maintain a small level of water on the roof to help with cooling. Section 1107 also contains a requirement for emergency systems.

Controlled-flow roof drainage systems have smaller diameter storm drainage pipes relying on the roof to serve as a temporary storage or retention basin. Rainwater collects on the roof while the vertical conductors discharge at an assumed rate.

The principles of design for a controlled-flow roof drainage system are the same as for a retention basin or a dam for flood control. The water discharges at a gradual rate and continues after the storm has subsided.

Many engineers will size such a system for a maximum period of storage, usually 24 hours. The storm used for sizing is typically a 25-year frequency storm. (When listed as a 25-year storm without a period of intensity, it includes the total amount of rainfall for the duration of the storm.) The engineer may further design for a maximum depth of water of 3 inches (76 mm).

A controlled-flow system is an alternative engineering design practice and is not required by the code. In accordance with Section 1101.7, the structural engineer must design for the roof loads imposed by any depth of ponding. Such a system replaces the primary system of Section 1105 and does not replace the emergency system of Section 1107.

This section also requires the testing, inspecting and approval requirements contained in Sections 105.3 and 105.4 (see commentary, Sections 105.3 and 105.4).

1110.2 Control devices. The control devices shall be installed so that the rate of discharge of water per minute shall not exceed the values for continuous flow as indicated in Section 1110.1.

❖ A controlled-flow system relies on a gradual discharge to the drainage system. A controlling device, usually a properly sized roof drain, is necessary to control the discharge to the system.

1110.3 Installation. Runoff control shall be by control devices. Control devices shall be protected by strainers.

❖ Control devices are usually roof drains that are properly sized (see commentary, Section 1110.2). This section requires that whatever control device is used must be protected by a strainer against blockage.

1110.4 Minimum number of roof drains. Not less than two roof drains shall be installed in roof areas 10,000 square feet (929 m^2) or less and not less than four roof drains shall be installed in roofs over 10,000 square feet (929 m^2) in area.

❖ For controlled-flow systems, a system redundancy safety precaution is built into the design by the requirement for a minimum of two drains for roof areas of 10,000 square feet (929 m^2) or less. The objective is that if one drain becomes clogged, the other would allow water to drain. For larger buildings in excess of 10,000 square feet (929 m^2) of roof area, a minimum of four roof drains is required (see Commentary Figure 1110.4). Although no specific requirements exist in this section covering design of the controlled-flow system, the potential for partial clogging of part of the sys-

STORM DRAINAGE

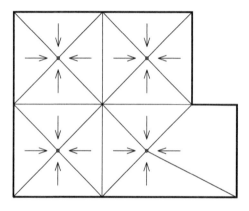

For SI: 1 square foot = 0.0929 m².

Figure 1110.4
MINIMUM NUMBER OF ROOF DRAINS FOR CONTROLLED FLOW

tem must be evaluated so that the contained water will drain off the roof without any structural or other problems.

SECTION 1111
SUBSOIL DRAINS

1111.1 Subsoil drains. Subsoil drains shall be open-jointed, horizontally split or perforated pipe conforming to one of the standards listed in Table 1102.5. Such drains shall not be less than 4 inches (102 mm) in diameter. Where the building is subject to backwater, the subsoil drain shall be protected by an accessibly located backwater valve. Subsoil drains shall discharge to a trapped area drain, sump, dry well or *approved* location above ground. The subsoil sump shall not be required to have either a gas-tight cover or a vent. The sump and pumping system shall comply with Section 1113.1.

❖ This section requires that subsoil drain material conform to the requirements for subsoil drain pipe contained in Table 1102.5. The collection of water into these subsoil drain pipes should be either open jointed, horizontally split or perforated. The minimum allowable size for a subsoil drain pipe is 4 inches (102 mm).

As with other below-grade piping, subsoil drain piping must have an accessible backwater valve if the area being drained is subject to backwater conditions [see Commentary Figure 1111.1(1)].

The discharge of the subsoil drain system is regulated in this section. Discharge to either another storm water discharge area drain, a storm water sump, a storm water dry well or an approved above-ground discharge point in the adjacent grading of the property is allowed. When a subsoil sump is used, a gas-tight cover is not required on top of the sump or to vent the sump because it is not connected to a sanitary sewer; thus, there are no sewer gases to be concerned about [see Commentary Figure 1111.1(2)]. If the discharge is to a sump, the sump and pumping systems must be installed in accordance with Section 1113.1.

STORM DRAINAGE

Figure 1111.1(1)
SUBSOIL DRAIN SUBJECT TO BACKFLOW

Figure 1111.1(2)
SUBSOIL DRAIN

STORM DRAINAGE

SECTION 1112
BUILDING SUBDRAINS

1112.1 Building subdrains. *Building subdrains* located below the *public sewer* level shall discharge into a sump or receiving tank, the contents of which shall be automatically lifted and discharged into the drainage system as required for building sumps. The sump and pumping equipment shall comply with Section 1113.1.

❖ This section regulates the installation of building subdrains located below the public sewer.

The building subdrains identified include subsoil drainage around the perimeter of the building (foundation drains) and other clear-water storm drains, such as basement window well and outdoor stairwell drains.

Storm rainwater systems that discharge into sumps and sump pump systems need not conform to Sections 712 and 906.5 for sanitary sump drainage systems. The sump and pumping systems must comply with Section 1113.1.

This section requires that the collected storm water that is fed into the sump automatically be discharged from the sump. This requirement is intended to protect against flooding and potential damage to the building should someone fail to turn on the pump when necessary. Additionally, this will help minimize damage to the pump caused by unnecessary extended operation.

The discharge from the sump can be to grade in the surrounding area of the building if the discharge does not create a nuisance to the surrounding area and is approved by the code official. The sump pump system receiving subsoil drainage can also discharge to the storm drainage system.

Subsoil drainage should not be discharged into the sanitary drainage system because it places a substantial burden on the sanitary treatment facility (see commentary, Section 1108). This is also a concern in areas that still use combined sanitary and storm sewer systems.

SECTION 1113
SUMPS AND PUMPING SYSTEMS

1113.1 Pumping system. The sump pump, pit and discharge piping shall conform to Sections 1113.1.1 through 1113.1.4.

❖ This section regulates the sumps and pumping systems associated with subsoil drains and building subdrains.

1113.1.1 Pump capacity and head. The sump pump shall be of a capacity and head appropriate to anticipated use requirements.

❖ This section requires that the pump be sized properly to function under the anticipated load. The sump pump system must be designed to remove the subsoil drainage at the anticipated inflow rate under a worst-case condition. The sizing is determined by evaluating the local ground conditions, the length of piping and the weather in the area. A worst-case condition for subsoil drainage usually happens in the spring when heavy rains occur and any snow accumulation melts. The selection of a sump pump should include consideration for the on-off cycle time and the duration of operation of the pump in the "on" phase.

1113.1.2 Sump pit. The sump pit shall be not less than 18 inches (457 mm) in diameter and not less than 24 inches (610 mm) in depth, unless otherwise *approved*. The pit shall be accessible and located such that all drainage flows into the pit by gravity. The sump pit shall be constructed of tile, steel, plastic, cast iron, concrete or other *approved* material, with a removable cover adequate to support anticipated loads in the area of use. The pit floor shall be solid and provide permanent support for the pump.

❖ This section requires that the minimum dimensions of the sump pit be 18 inches (457 mm) in diameter and 24 inches (610 mm) deep. The pit size is very important in the design of the system because it determines the amount of water that is stored in the system prior to discharge. It also determines the operation time of the pump for each "on" cycle (see commentary, Section 712.3.2).

The pit must be constructed of tile, steel, plastic, cast iron, concrete or any material that performs like these and is approved by the code official. The pit must have a removable cover to allow access to the pump and other related items, such as the check valve. The cover is to be structurally capable of holding anticipated loads. The pit must also have a floor that is capable of permanently supporting the pump, including piping that bears on the pump. This requirement basically eliminates the use of gravel as the bottom of the sump because gravel would not allow for the permanent support of the pump.

1113.1.3 Electrical. Electrical service outlets, when required, shall meet the requirements of NFPA 70.

❖ Where a pump is power by electricity, the electrical receptacles are required to be installed in accordance with NFPA 70. The commentary user should consult that document for specific requirements.

1113.1.4 Piping. Discharge piping shall meet the requirements of Section 1102.2, 1102.3 or 1102.4 and shall include a gate valve and a full flow check valve. Pipe and fittings shall be the same size as, or larger than, the pump discharge tapping.

Exception: In one- and two-family dwellings, only a check valve shall be required, located on the discharge piping from the pump or ejector.

❖ The piping material used for the sump discharge must conform to the requirements contained in Sections 1102.2, 1102.3 and 1102.4, which refer to Tables 702.1, 702.2 and 1102.4, respectively. The piping in the sump system discharge line must have a check valve and a full-open valve installed to permit maintenance on the other components of the system. The requirement for a full-open valve does not apply to one- and two-family dwellings. This check valve should be placed as close as practical to the sump pump and is required to prevent discharge water from the pump from running back into the pit after the pump stops, thus preventing the constant cycling of the sump pump.

Access must be provided to such valves and located above the sump cover. Where the discharge pipe is below grade, the valve must be located outside the sump below grade in an access pit with a removable cover.

Pipe and fittings of the discharge piping are not to be smaller than the size of the discharge outlet of the pump itself. Reduced-size discharge piping will affect pumping capacity, thereby causing the system not to function as designed.

Bibliography

The following resource materials are referenced in this chapter or are relevant to the subject matter addressed in this chapter.

Storm Drainage System Research Project Flow Rate Through Roof Drains. Rosemont, IL: American Society of Plumbing Engineers, 2012.

Chapter 12:
Special Piping and Storage Systems

General Comments

Even though the gases used in medical and inhalation anesthetic systems are nonflammable, they are frequently oxidizing gases and, as such, create an environment that supports combustion and where ignition of combustible materials might occur. Therefore, the proper design, installation and performance of these systems is essential to protect against the physical hazard they could pose. The portion of these systems not regulated by this chapter and the *International Fire Code*® (IFC®) is the mechanical exhaust of vacuum systems, which must be designed and installed in accordance with the requirements of the *International Mechanical Code*® (IMC®).

The nonmedical oxygen systems regulated by this chapter are bulk oxygen storage systems and oxygen-fuel gas systems used for welding and cutting. Bulk oxygen storage systems contain oxygen in either a liquid or gaseous state, which though nonflammable, possesses properties that support combustion. In a pure oxygen environment, many materials that do not burn in air can burn under pressure. The proper selection of materials used as components of bulk oxygen systems as well as the location of bulk oxygen storage containers are essential to minimizing the potential fire hazards posed by an oxygen-enriched environment.

Welding and cutting operations using fuel gas and oxygen pose an inherent risk of fire or explosion because of the combined presence of flammable gases, flames and oxygen. Additionally, the generation of acetylene gas, the most common flammable gas used with oxygen for welding and cutting operations, and the storage and use of calcium carbide to produce acetylene gas and other hazardous processes, are regulated by this chapter. Calcium carbide, when exposed to water, generates acetylene gas, which poses an explosion hazard where not properly stored and handled.

Purpose

This chapter sets the requirements for the design, installation, storage, handling and use of nonflammable medical gas systems, including inhalation anesthetic and vacuum piping systems, bulk oxygen storage systems and oxygen-fuel gas systems used for welding and cutting operations. The intent of these requirements is to minimize the potential fire and explosion hazards associated with the gases used in these systems.

SECTION 1201
GENERAL

1201.1 Scope. The provisions of this chapter shall govern the design and installation of piping and storage systems for nonflammable medical gas systems and nonmedical oxygen systems. All maintenance and operations of such systems shall be in accordance with the *International Fire Code*.

❖ This chapter contains the requirements and referenced standards that regulate nonflammable medical gas, inhalation anesthetic, vacuum piping and nonmedical oxygen systems. The referenced standards contain the design, installation, system components, materials, testing, operation, maintenance and safety requirements for these systems. The IFC regulates the maintenance and operation of these systems.

SECTION 1202
MEDICAL GASES

[F] 1202.1 Nonflammable medical gases. Nonflammable medical gas systems, inhalation anesthetic systems and vacuum piping systems shall be designed and installed in accordance with NFPA 99.

Exceptions:

1. This section shall not apply to portable systems or cylinder storage.

2. Vacuum system exhaust terminations shall comply with the *International Mechanical Code*.

❖ This section contains the requirements and referenced standards that apply to nonflammable medical gas, inhalation anesthetic and vacuum piping systems.

NFPA 99 applies to nonflammable medical gas, inhalation anesthetic and permanently installed vacuum piping systems, with the exception of portable systems and the storage of gas cylinders. The mechanical exhaust of vacuum piping systems is also exempt from NFPA 99 requirements and must comply with the requirements of the IMC.

Gases such as oxygen, nitrous oxide, compressed air, carbon dioxide, helium, nitrogen and various mixtures of these gases are typically used in medical gas-piping systems.

Oxygen and nitrous oxide are two of the most common gases used in medical gas and inhalation anesthetic systems. Even though these two gases are nonflammable, they are strong oxidizing gases, and their presence creates an environment where the potential for fire, chemical, mechanical and electrical hazards exists. Being a strong oxidizer, these gases, either by themselves or in combination, provide an environment that lowers the ignition energies of combustibles.

Because of the hazards that medical gases and inhalation anesthetics pose, NFPA 99 contains specific criteria for connection of the gas source to the distribution piping system, system identification and system alarm requirements. These requirements help reduce the potential for mistaken interconnection of a gas system to a vacuum system, which could present a fire and explosion hazard.

NFPA 99 requires that pipe and fittings used in a medical gas system must be of Type K or L (ASTM B 819) copper tubing, precleaned and capped. Additionally, pipe must be identified with one of the following markings: OXY, MED, OXY/MED, ACR/OXY or ACR/MED. Valves, fittings and other piping components must be cleaned for "oxygen service" by the manufacturer or an approved supplier, with the result a sealed, oil-free fitting enclosed in a dry nitrogen-purged container.

The piping system must be continuously purged prior to, during and after installation with oil-free dry nitrogen to prevent the formation of scale within the tubing. Openings must be plugged or capped to prevent debris or contamination on the inside of the tube.

The soldered or brazed joint is to be visually checked after it has been washed with hot water and a stainless steel wire brush has been used to remove scale. The following conditions are considered unacceptable:

1. Flux or flux residue present.
2. Excessive oxidation of the joint.
3. Cracks in the tube or component.
4. Cracks in the braze filler metal.
5. Presence of unmelted filler metal.
6. Presence of filler metal clearly visible around the entire joint and at the interface between the socket and tube.
7. Failure of the joint to hold the test pressure.

Vacuum system piping must be corrosion-resistant metal such as seamless copper water tube (ASTM B 88, Types K, L and M), copper ACR tube (ASTM B 280), copper medical gas tube (ASTM B 819), stainless steel tube or galvanized steel pipe (ASTM A 53). Joints in copper must be soldered or brazed. Joints in stainless steel tube must be brazed or welded. Joints in galvanized piping must be flanged or threaded. The vacuum piping must be marked or labeled prior to installation with one of the following markings: Medical-Surgical Vacuum or MED/SURG VAC. They must be marked not more than 20 feet (6096 mm) apart. Vacuum systems must never be converted for use as gas systems.

Where flux is used for soldered or brazed joints, it must be used sparingly to avoid excess flux inside the joint. If flux has been used, it must be washed with hot water.

NFPA 99 contains requirements regarding flammable inhalation anesthetic gases; however, use of flammable inhalation gases is not permitted in a piped distribution system. According to B.R. Klein, in 1950 the NFPA Committee on Gases prohibited piped distribution of flammable anesthetic gases to eliminate one possible source of explosion and fire in health care facilities. The IFC also contains requirements applicable to the storage, use and handling of compressed gases used in connection with nonflammable medical gases and inhalation anesthetic systems. The mechanical exhaust discharge and discharge termination of vacuum piping systems must comply with the IMC.

SECTION 1203
OXYGEN SYSTEMS

[F] 1203.1 Design and installation. Nonmedical oxygen systems shall be designed and installed in accordance with NFPA 55 and NFPA 51.

❖ This section contains the general requirements and referenced standards that apply to nonmedical oxygen systems.

NFPA 55 applies to bulk oxygen systems that are located at a consumer site and have a storage capacity exceeding 20,000 cubic feet (566 m^3) of liquid or gaseous oxygen, including unconnected reserves. The bulk oxygen originates off site, and the supply is delivered to the premises by mobile delivery equipment. This standard does not apply to oxygen manufacturing plants, oxygen suppliers or systems having capacities less than 20,000 cubic feet (566 m^3).

Oxygen is an odorless, tasteless, colorless, nontoxic nonflammable gas. Although oxygen gas is nonflammable, it is an oxidizer, and as such creates an environment where the ignition of combustibles could occur.

The location of bulk storage systems is critical to protect them from becoming involved in fires that originate from sources adjacent to the system. NFPA 55 contains the minimum clearance requirements between bulk oxygen storage containers and adjacent structures, combustible materials stored on site, places of public assembly, public sidewalks, parked vehicles, adjoining property lines and other storage containers containing flammable and combustible liquids or gases.

The standard requires that bulk oxygen storage containers comply with Section VIII of the ASME *Boiler and Pressure Vessel Code* and be corrosion resistant, or designed, constructed, tested and maintained in accordance with U.S. Department of Transportation (DOT) specifications and regulations for 4L containers. The containers must also be equipped with safety relief valves. Piping, tubing and fittings used for bulk oxygen systems must be suitable for oxygen service and must conform to the requirements of ASME B 31.3 for chemical plant and petroleum refinery piping joints in piping, and tubing must be either welded or brazed, flanged, threaded, socket, slip or compression fittings. The use of gaskets and threaded sealants must be suitable for oxygen service. Brazing materials must have a melting point above 1,000°F (538°C). Field-assembled piping must be tested prior to being placed in service, and the bulk oxygen storage location must be permanently identified with signage stating "Oxygen—No Smoking—No Open Flames" or an equivalent warning.

The IFC also contains requirements applicable to the storage, handling and transportation of liquid oxygen.

NFPA 51 applies to oxygen-fuel gas systems for welding, cutting and allied processes, the use and storage of fuels, acetylene generation and calcium carbide storage. The standard does not apply to systems consisting of regulators, hoses, a torch and a single tank each of oxygen and fuel gas. Additionally, in systems where oxygen is not used with fuel gases, the manufacture of gases and filling of cylinders, the storage of empty cylinders and compressed air-fuel systems are not governed by this standard.

Oxygen-fuel gas systems for welding and cutting consist of flammable fuel gases such as acetylene, hydrogen, LP-gas, natural gas or stabilized methylacetylene-propadiene used with oxygen to create a high-temperature flame. Although NFPA 51 applies to fuel gases used with oxygen, very specific regulations governing the generation of acetylene and the storage of calcium carbide are stated because acetylene is the most commonly used gas for welding and cutting.

Acetylene is a colorless, flammable gas produced by the direct combination of hydrogen and carbon. The generation of acetylene gas is limited by NFPA 51 to the carbide-to-water process, in which pieces of calcium carbide are placed in water. The calcium carbide, a gray, stone-like material produced by the smelting of lime and coke, reacts with water to produce gas bubbles that are the acetylene gas.

Acetylene generators must be listed and are not permitted to generate acetylene at pressures exceeding 15 pounds per square inch (103 kPa) gauge. Acetylene generated at higher pressures is unstable and will, under certain circumstances, decompose into hydrogen and carbon with explosive violence.

Acetylene generators must be installed on a level foundation, protected from freezing and have sufficient area around them for maintenance, operation, adjustment and charging. Rooms or buildings in which acetylene generators are located must have explosion relief vents at not less than 1 square foot per 50 cubic feet (0.0929 m^2/1.4 m^3) of room volume.

Cylinders used for the storage of fuel gases or oxygen are required to conform to U.S. DOT or Canada Transport Commission regulations.

Pipe and fittings must comply with ANSI B31.3. Pipe must be Schedule 40 steel, brass, copper or stainless steel, and fittings must be standard weight in sizes up to and including 6 inches (152 mm) in diameter. Copper tubing must be Type K or L in accordance with ASTM B 88. Piping for acetylene and methylacetylene-propadiene systems must be steel. Joints in steel piping must be welded, threaded or flanged. Joints in brass or copper pipe must be welded, brazed, threaded or flanged. Where brazed, joints must be made with a silver-brazing alloy or similar high-melting-point filler metal.

Piping must be visually inspected for contaminants prior to assembly and the assembled must be system blown out with compressed air. Prior to being placed in service, the entire system must be pressure tested to a minimum 1.5 times the maximum operating pressure.

Finally, piping systems must be provided with a protective equipment device designed to prevent not only backflow of oxygen into the fuel gas supply but also a flashback of flame into the gas supply. The device must also prevent the development of an oxygen-fuel gas mixture at pressures that if ignited will achieve combustion pressure that will cause the device to fail and thus not prevent both of the previous items from occurring. Pressure relief devices are also required, and listed shutoff valves and backflow check valves must be installed at every fuel gas and oxygen outlet station.

The IFC contains requirements applicable to welding and cutting operations and equipment, calcium carbide and acetylene generators and related piping manifolds, hose systems and other system components for fuel gas and oxygen systems. It also contains provisions for the storage, use and handling of both compressed gases used in cutting and welding operations and calcium carbide.

Bibliography

The following resource materials are referenced in this chapter or are relevant to the subject matter addressed in this chapter.

ASME B31.3-02, *Process Piping*. New York: American Society of Mechanical Engineers, 2002.

NFPA 50-01, *Bulk Oxygen Systems at Consumer Sites*. Quincy, MA: National Fire Protection Association, 2001.

NFPA 51-02, *Design and Installation of Oxygen-Fuel Gas Systems for Welding, Cutting, and Allied Processes*. Quincy, MA: National Fire Protection Association, 2002.

NFPA 99-15, *Health Care Facilities Code*, Quincy, MA: National Fire Protection Association, 2015.

Chapter 13: Nonpotable Water Systems

General Comments

Inadequate quantities of fresh water supplies in many areas of the world have created a need to reduce the demand on potable water supply systems. One way to reduce the potable water demand is to use nonpotable water instead of potable water for various applications. Because flows of raw nonpotable water are intermittent, collection systems are necessary to store the water until needed. Such storage systems must be designed to accommodate a reasonable quantity of these waters. Depending on the source of the nonpotable water and the intended end use, various water treatment methods may be required.

Purpose

This chapter specifies the collection systems, storage tank sizing and the treatment requirements for reusing nonpotable water.

SECTION 1301
GENERAL

1301.1 Scope. The provisions of Chapter 13 shall govern the materials, design, construction and installation of systems for the collection, storage, treatment and distribution of nonpotable water. The use and application of nonpotable water shall comply with laws, rules and ordinances applicable in the jurisdiction.

❖ This chapter covers the required details about collecting, storing, treating and distributing nonpotable water. The end use for these waters is controlled by the rules established in each jurisdiction. For example, collecting and using rainwater might be allowed in Atlanta, Georgia, whereas that practice might not be allowed in Denver, Colorado.

1301.2 Water quality. Nonpotable water for each end use application shall meet the minimum water quality requirements as established for the intended application by the laws, rules and ordinances applicable in the jurisdiction. Where nonpotable water from different sources is combined in a system, the system shall comply with the most stringent of the requirements of this code that are applicable to such sources.

❖ The quality required for the nonpotable water to be used in a particular application must be established by each jurisdiction. It is hoped that, eventually, quality requirements for nonpotable water used in specific applications will become somewhat standardized, similar to the standardization of the quality for potable drinking water. However, unlike potable water where the ultimate use is human ingestion, the use of nonpotable water can be quite varied such that setting an ultimate quality standard for any one use will most likely be impractical. Another compounding issue is the extremely wide (quality) variations of raw sources of nonpotable water. For example, raw water sources for potable water are typically lakes, rivers and underground aquifers that provide "nature cleansed" water for starting the potable water treatment process. Treatment is fairly straightforward because the quality and consistency of the raw water source are known. For nonpotable water, the raw water source can vary greatly depending on what nonpotable waters are collected, the combinations of different nonpotable waters and the volumes of each, all of which can vary over any period of time. Of course, the goal is not to "overtreat" the nonpotable water for an end use application, as this can be costly in terms of initial equipment installation and long-term operation. Conversely, there is the inherent obligation for ensuring that the use of the nonpotable water does not create hazards for living organisms that come into contact with it. For example, some raw nonpotable waters might have a pH characteristic that can be accommodated by flush valves for water closets and urinals, but are unsuitable for drip irrigation for landscaping. A treatment process to correct the pH so that the nonpotable water can be used for both the landscaping and the flushing applications might be appropriate. However, why use pH-adjusting treatment for that volume of nonpotable water that is used only for flushing (when the flush valves can accommodate the non pH-adjusted water)? Treating all of the (combined) water for the highest quality required for any one use might make sense, depending on what treatments are necessary. However, the code does not prohibit combining the raw sources, treating that water to a certain quality level and then sending part of that water to another location for additional treatment for a specific end use application that requires a "higher" quality level. This provides a better understanding of the last sentence of the section: Only where the raw water sources are combined, and treated, stored and used as a combined treated source do the most stringent code requirements apply for treatment.

1301.2.1 Residual disinfectants. Where chlorine is used for disinfection, the nonpotable water shall contain not more than 4 ppm (4mg/L) of chloramines or free chlorine when tested in accordance with ASTM D1253. Where ozone is used for disinfection, the nonpotable water shall not contain gas bubbles having elevated levels of ozone at the point of use.

> **Exception:** Reclaimed water sources shall not be required to comply with these requirements.

❖ Chlorinating and ozonizing are two of the most common methods for disinfection of water. Chlorine is inexpensive to obtain and ozone is easy to make on site. They are effective, very controllable and the results (bacteria kill) are well established for the quantities and detention times needed for volumes of water. Some "overdisinfection" is necessary to ensure that a minimum level of disinfected product is maintained. But too much disinfected product in the water is costly in the long run and does not yield any better results (the bacteria is dead—more disinfection is unnecessary). The excess disinfection product off-gases out of the water and produces undesirable effects in the short term. For chlorine, that effect is an unpleasant odor, similar to the air around most indoor public swimming pools. For ozone, that effect is an effervescence bubbling up and out of the water along with a sweet, pungent smell that is similar to how the air smells on a calm summer day just prior to the beginning of thunderstorm. These effects in an outdoor environment are typically of no concern. However, in an indoor environment, the effects can be a nuisance to the building occupants and, depending on the concentrations and length of exposure, could impact the health of the occupants. Therefore, this section requires control in the use of these products.

The use of ultraviolet light for disinfection instead of chlorine or ozone will require approval by the code official in accordance with Section 105.2.

The exception for reclaimed water recognizes that such water is disinfected at the source by the water utility that provides the water. See the definition for "Reclaimed water" in Chapter 2.

1301.2.2 Filtration required. Nonpotable water utilized for water closet and urinal flushing applications shall be filtered by a 100-micron or finer filter.

> **Exception:** Reclaimed water sources shall not be required to comply with these requirements.

❖ Water closet fill valves (and flush valves) and urinal flush valves have small-diameter waterways that could become clogged with particulate matter in nonpotable water. The 100-micron filter requirement ensures that valves will function as the manufacturer intended.

The exception for reclaimed water recognizes that such water is filtered at the source by the water utility that provides the water. See the definition for "Reclaimed water" in Chapter 2.

1301.3 Signage required. Nonpotable water outlets such as hose connections, open ended pipes and faucets shall be identified at the point of use for each outlet with signage that reads as follows: "Nonpotable water is utilized for [application name]. CAUTION: NONPOTABLE WATER – DO NOT DRINK." The words shall be legibly and indelibly printed on a tag or sign constructed of corrosion-resistant waterproof material or shall be indelibly printed on the fixture. The letters of the words shall be not less than 0.5 inch (12.7 mm) in height and in colors in contrast to the background on which they are applied. In addition to the required wordage, the pictograph shown in Figure 1301.3 shall appear on the signage required by this section.

❖ This section is identical to Section 608.8.1. See the commentary for that section.

**FIGURE 1301.3
PICTOGRAPH—DO NOT DRINK**

1301.4 Permits. Permits shall be required for the construction, installation, alteration and repair of nonpotable water systems. Construction documents, engineering calculations, diagrams and other such data pertaining to the nonpotable water system shall be submitted with each permit application.

❖ Although Section 106 requires permits for the work covered by the code, this section reinforces that requirement, as not everyone considers all nonpotable water system work as needing a permit. For example, rainwater collection and distribution systems have been installed for many decades without a thought towards compliance with a code. There have been guidelines developed by numerous jurisdictions where water shortages are a significant problem. Some guidelines have been adopted and made part of the laws so that code officials could enforce them. In other areas, the guidelines are not law. This section makes it clear that this chapter is enforceable when the code is adopted.

1301.5 Potable water connections. Where a potable system is connected to a nonpotable water system, the potable water supply shall be protected against backflow in accordance with Section 608.

❖ This section simply refers back to the backflow prevention section for where nonpotable water might possibly contaminate a potable water supply system when a backflow event (in the potable water system) occurs.

1301.6 Approved components and materials. Piping, plumbing components and materials used in collection and conveyance systems shall be manufactured of material approved for the intended application and compatible with any disinfection and treatment systems used.

❖ The use of the term "approved" may or may not mean approved by the code official, as the code does not provide a basis for which the materials can be evaluated for some applications. For example, a gray water collection system inside of (and below) the building will be constructed of piping materials that comply with Tables 702.1 and 702.1 and 702.4. However, should the continuation of that piping outside of the building be allowed to comply with Table 702.3 (building sewer piping)? The code official might agree that it does comply with the code but manufacturers of systems that "process" the nonpotable waters might not agree.

1301.7 Insect and vermin control. The system shall be protected to prevent the entrance of insects and vermin into storage tanks and piping systems. Screen materials shall be compatible with contacting system components and shall not accelerate the corrosion of system components.

❖ Large floating "mats" of dead insects and vermin in storage tanks will decompose and create odors through the tank vent. Although it is usually not possible to eliminate all insects from entry, those that do enter hopefully will become waterlogged and sink to the bottom of the tank. Sections 1302.5 and 1303.8 provide additional requirements for on-site nonpotable water and rainwater, respectively. Filtering to remove insects and vermin from reclaimed water is performed at the water utility that distributes the reclaimed water.

Screens for storage tank vents must not be of materials that are incompatible with the material that they are in contact with so that the screen does not corrode and fall off the opening.

It is uncertain whether this section requires screens at each vent pipe termination (of the vent piping required for equalizing pressures in drain piping system) to prevent the entry of insects and vermin into the piping system. Any insect and vermin would be captured by filters installed before the storage tank. The code official will need to make this decision.

1301.8 Freeze protection. Where sustained freezing temperatures occur, provisions shall be made to keep storage tanks and the related piping from freezing.

❖ Because storage tanks have a large volume of water, it would take an "amount of time" of exposure to freezing temperatures for the water in the tank to begin freezing. The amount of time it will take for the water to freeze will depend on many factors, such as sunlight (see Section 1301.9.2), wind, movement of water in the tank and the lowest temperature. For these reasons, many nonpotable water tanks are buried underground. However, some climates could have only "mildly freezing" temperatures that only last for a few hours a day. In regions that do have freezing temperatures, storage tanks could be located outdoors under open shelters or enclosed in shelter buildings that have very little or no heat. Tank contents that are kept moving by a circulation system could endure "mildly freezing" overnight conditions.

The exposure of piping, pumps and other equipment to freezing conditions is a more serious issue as in most cases, the water is not moving. See the commentary for Section 305.4 concerning protection of piping against freezing. Note that while the piping, pumps and other equipment might need protection (insulation, heat or both) against freezing conditions, that is not an indication that the storage tank needs similar protection.

1301.9 Nonpotable water storage tanks. Nonpotable water storage tanks shall comply with Sections 1301.9.1 through 1301.9.11.

❖ A storage tank can be of any shape and be of a multitude of materials. The following sections provide the requirements for storage tanks.

1301.9.1 Sizing. The holding capacity of the storage tank shall be sized in accordance with the anticipated demand.

❖ Ideally, all nonpotable water collected and stored will (eventually) be reused for another purpose. However, that is not the case with raw gray water, as Section 1302.6 prohibits untreated gray water from being retained for more than 24 hours. Another factor is that there is not enough demand (use) for the collected nonpotable water. For example, a system collects large volumes of rainwater in a continuously "rainy" region; however, the rainwater is only used to flush water closets (the installed nonwater urinals not needing water). Does it make sense to install a large storage tank to retain all of the rainwater? Perhaps not. After careful analysis of the rainwater flows (historical weather data) and the demands (anticipated use of the fixtures that will use the water), the designer of the system will make the decision on the size of the storage tank.

1301.9.2 Location. Storage tanks shall be installed above or below grade. Above-grade storage tanks shall be protected from direct sunlight and shall be constructed using opaque, UV-resistant materials such as, but not limited to, heavily tinted plastic, fiberglass, lined metal, concrete, wood, or painted to prevent algae growth, or shall have specially constructed sun barriers including, but not limited to, installation in garages, crawl spaces or sheds. Storage tanks and their manholes shall not be located directly under soil piping, waste piping or any source of contamination.

❖ Storage tanks can be located anywhere, such as inside of buildings, on roofs or on grade (with proper shelter) or underground (inside or outside of the building footprint). The designer chooses the location that best suits the needs of the system that he or she is designing.

Tanks above grade need to be protected so that the material of the tank's outer wall is not degraded by exposure to sunlight, as sunlight includes wavelengths in the ultraviolet (UV) range. Most untinted thermoplastic materials such as polypropylene and polyethylene can start cracking because the sun's UV radiation causes degradation of the connection points within the molecules of plastic materials. The cracking continues until the material loses enough strength to fail. Although "above grade" usually means outdoors and above grade, the situation would be the same for a plastic storage tank located inside of a standard glass greenhouse (plain glass being able to pass some UV radiation). Does this mean that storage tanks located in a room inside of a building where a wall window allows natural light into the room must be heavily tinted? Perhaps not, as the exposure to sunlight will probably be incidental or only be directly shining on the tank for a very limited amount of time (as the sun moves though the sky). However, if the room has many clear skylights (not tinted—such as might be found in a warehouse), that thought might have to be re-evaluated. Note that this section is only concerned about exposure of the tank to sunlight. However, the sun is not the only generator of UV radiation. A shop that performs open welding could cause significant UV radiation towards a plastic tank if the tank was not shielded in some manner.

Plastic tanks can be protected against UV radiation by the addition of pigments that absorb the UV rays. As UV absorbers (UVAs) will degrade over long periods of time, significant UVA pigments must be added to compensate for the UVA degradation. Thus, the code requires "heavily tinted" plastic tank materials.

Another reason for keeping direct sunlight off of a translucent tank is to prevent algae growth in the water. Algae need sunlight to propagate. Unchecked algae growth causes significant equipment problems and can create difficult-to-solve water quality issues.

Lined metal tanks are one type of metal tank. This specifically refers to carbon steel metal that will rust and produce scale (corrode). Typical linings are epoxy coatings that are applied after the carbon steel metal is sandblasted clean. Presumably, stainless steel tanks would not require a lining because stainless steel (if the proper series of stainless steel is chosen) will not corrode.

The requirement for storage tanks (and their manholes) to not be located under soil or waste piping, or any "source of contamination," is to make sure that the quality of the water in the tank does not become something other than expected.

1301.9.3 Materials. Where collected on site, water shall be collected in an *approved* tank constructed of durable, nonabsorbent and corrosion-resistant materials. The storage tank shall be constructed of materials compatible with any disinfection systems used to treat water upstream of the tank and with any systems used to maintain water quality in the tank. Wooden storage tanks that are not equipped with a makeup water source shall be provided with a flexible liner.

❖ Storage tanks must be approved by the code official. There are many types of arrangements of "tanks" that are not of the durability, nonabsorbency and corrosion resistance needed for long-term use as a permanent system for the intended use. For example, a field-built brick (or block) structure with a rubber lining that was adhered to the walls might hold water for the initial test, but is that an appropriate long-term solution? The code official must make the decision.

The tank materials must be compatible with the disinfection and treatment methods for the water that will be stored in the tank. For example, if alternate approval was obtained for the use of UV lamps in the tank for disinfection of the water, then a tank made of untinted plastic might not be compatible with the disinfection method (see commentary, Section 1301.9.2).

The watertightness of wooden storage tanks depends upon the wood planking "swelling" and closing off the gaps between the planks. If the water level in a wood tank is allowed to decrease for long periods of time such as a day, the planks dry out, shrink and the gaps open up to cause leaks. Makeup water added to the tank keeps the planks wet. The makeup water would not be needed at any time that the water level in the tank becomes low, but only when the tank level remains low for a period of time where the planks would start drying out. If adding makeup water does not make sense in the scheme of collecting as much nonpotable water as possible (and reducing the use of potable water), then a wood tank would have to be lined with a flexible liner so that the water level could remain low for long periods of time. For example, rainwater might be used from rainwater storage at a very slow rate while rainwater is collected in "surges" (rainfall events). Filing the wood tank to the overflow point to keep the wood from drying out would be senseless.

1301.9.4 Foundation and supports. Storage tanks shall be supported on a firm base capable of withstanding the weight of the storage tank when filled to capacity. Storage tanks shall be supported in accordance with the *International Building Code*.

❖ Water weighs 62.4 pounds per cubic feet or 8.34 pounds per gallon. A 100-gallon tank will weigh 834 pounds (378 kg) plus the weight of the tank itself. The weight of a filled water tank is always significant, but especially so when supported by a building structure. Chapter 16 of the *International Building Code*[a] (IBC[a]) covers the structural design of buildings, including wind, ice build-up and earthquake loads. For underground tanks, Chapter 18 of the IBC covers soils and foundations.

1301.9.4.1 Ballast. Where the soil can become saturated, an underground storage tank shall be ballasted, or otherwise secured, to prevent the tank from floating out of the ground when empty. The combined weight of the tank and hold down ballast shall meet or exceed the buoyancy force of the tank. Where the installation requires a foundation, the foundation shall be flat and shall be designed to support the weight of the storage tank when full, consistent with the bearing capability of adjacent soil.

❖ Uplift forces of an empty underground tank vary significantly when the ground around the tank becomes saturated with water. The effect is no different than placing a ping pong ball under water and releasing it. It will return to the surface. The effect was scientifically discovered by Archimedes of Syracuse (Greece) around 250 B.C. His principle states: "Any object, wholly or partially immersed in a fluid, is buoyed up by a force equal to the weight of the fluid displaced by the object." In other words, objects have "buoyancy"; however, when the weight of the object is greater than the weight of the fluid displaced, the object is not buoyant (and is not uplifted towards the surface of the water).

Consider a tank of 5,000-gallon (18927 L) capacity. The uplift force of that tank (empty and neglecting the empty weight of the tank) in fully saturated soil is 5,000 gallons x 8.34 pounds per gallon or 41700 pounds (18914 kg). That amount of force will cause a $1^1/_8$-inch nominal diameter reinforcing bar (40 ksi yield strength) to begin to pull apart. The design of the straps and their connections to the tank and to the concrete foundation must be carefully considered to prevent underground tanks from "popping up and out of the ground."

The (concrete) foundation that the tank rests upon must also be properly designed to accept the loads from the tank (filled to capacity) without failure and to not overload the soil that the foundation rests upon.

1301.9.4.2 Structural support. Where installed below grade, storage tank installations shall be designed to withstand earth and surface structural loads without damage and with minimal deformation when empty or filled with water.

❖ A underground tank must be able to support the earth loads against the sides and top of the tank. If the tank is under a road, driveway or parking lot, vehicle loads add to the loads that the tank must withstand. Where the tank is adjacent to the road, driveway or parking lot, some of those loads could distribute to areas beyond those directly under the tank. Water-saturated soils also impart significant loads to the sides of an empty tank.

1301.9.5 Makeup water. Where an uninterrupted supply is required for the intended application, potable or reclaimed water shall be provided as a source of makeup water for the storage tank. The makeup water supply shall be protected against backflow in accordance with Section 608. A full-open valve located on the makeup water supply line to the storage tank shall be provided. Inlets to the storage tank shall be controlled by fill valves or other automatic supply valves installed to prevent the tank from overflowing and to prevent the water level from dropping below a predetermined point. Where makeup water is provided, the water level shall not be permitted to drop below the source water inlet or the intake of any attached pump.

❖ A makeup water supply for the storage tank is necessary where a constant supply of nonpotable water is depended upon, for example, where rainwater or treated gray water is used to flush water closets and urinals. If the volume of raw nonpotable water supply is not sufficient to keep up with the demand, then additional water must be added to the tank to keep the nonpotable water supply system going. Makeup water does not necessarily need to be potable water, but often it is. The connection of makeup water to a storage tank must comply with Section 608 to protect the higher quality makeup water from contamination in a backflow event.

The intent of this section is that the makeup water be supplied to the tank by automatic means. The term "fill valve" implies the type of mechanical "float-type" fill valve that is found in a flush tank water closet or at the top of a vat. The valve also could be the type that is electrically controlled (such as a solenoid valve) through float switches or other water level detection devices. A full open valve is required to be on the makeup water line so that the "fill valve" or other automatic control valve can be removed, repaired or replaced.

Obviously, the makeup water fill valve must shut before water in the tank begins to empty out through the overflow of the tank. However, the shutoff point does not need to be at the overflow level in order to leave room for the addition of nonpotable water from the original sources. The valve must open before the water in the tank reaches a certain "low" level as determined by the designer of the system. In most nonpotable water storage tanks, especially those storing water with some particulate matter, the incoming water (both the original water and makeup) is introduced into the tank at a point below the water surface and somewhat above the bottom of the tank. This is to reduce "stirring" of the floating debris on the surface of water in the tank. Entry of water into the tank should be as calm as possible. Allowing the level of water to drop below the inlet level would only introduce turbulence on the surface of the water resulting in the particulate becoming entrained throughout the tank. This could later negatively impact the operation of pumps, valves and filters in the distribution system.

1301.9.6 Overflow. The storage tank shall be equipped with an overflow pipe having a diameter not less than that shown in Table 606.5.4. The overflow pipe shall be protected from insects or vermin and shall discharge in a manner consistent with storm water runoff requirements of the jurisdiction. The overflow pipe shall discharge at a sufficient distance from the tank to avoid damaging the tank foundation or the adjacent property. Drainage from overflow pipes shall be directed to prevent freezing on roof walkways. The overflow drain shall not be equipped with a shutoff valve. A cleanout shall be provided on each overflow pipe in accordance with Section 708.

❖ Storage tanks can become filled to capacity and additional water added to the tank has to go somewhere or the collection system will back up and cause problems. The overflow is another location where insects and vermin can enter the storage tank so a screen is necessary (see commentary, Section 1301.7). In some situations, such as rainwater storage tanks, the overflow could discharge to a storm sewer or a natural waterway. Because it is rainwater, it would seem logical to allow it to flow into the storm sewer or natural waterway. However, that depends on the storm water runoff requirements of the jurisdiction. If the rainwater has been treated (other than screening for particulate) or potable (or other nonpotable) water is used for the makeup water supply, there could be restrictions against discharging to a storm sewer or natural waterway. If the local jurisdiction does not know, state environmental agencies should be consulted to determine the answer. In the United States, the Environmental Protection Agency (EPA) considers storm water discharges as point sources of pollution because rainwater could contain harmful pollutants that could wash into streams, rivers, lakes and coastal water. Most states administer the National Pollutant Discharge Elimination System (NPDES) stormwater program. However, in some states and territories and all tribal lands, the EPA directly administers the program.

Although this section implies that any nonpotable water (other than rainwater) could be allowed to overflow to a storm sewer or waterway, it is doubtful that this would be allowed in any jurisdiction, pursuant to the overriding requirements of the EPA. Additionally, even though reclaimed water is of an equal quality (or perhaps of higher quality) than what exits the tertiary stage of the municipal wastewater treatment facility that distributes the reclaimed water, reclaimed water may not necessarily be allowed to flow into storm sewers or waterways.

If the flow of water from the overflow is allowed to drain onto the ground around the outside of a storage tank, it must be routed to drain away from the tank area and not toward a building or an adjacent lot. Foundations for the tank and a building can be quickly undermined when excess water surcharges the soil at these locations. Adjacent lot owners would not appreciate such overflows.

Where a water tank is on a roof (as many water tanks have been in New York City), water coming out of an overflow during freezing water creates hazardous conditions on the roof walkways. Prior planning to put the overflow and walkways in different locations is necessary.

Overflows could become clogged, especially where the overflow has an outlet screen. A cleanout is necessary to clear blockages.

1301.9.7 Access. Not less than one access opening shall be provided to allow inspection and cleaning of the tank interior. Access openings shall have an *approved* locking device or other approved method of securing access. Below-grade storage tanks, located outside of the building, shall be provided with a manhole either not less than 24 inches (610 mm) square or with an inside diameter not less than 24 inches (610 mm). Manholes shall extend not less than 4 inches (102 mm) above ground or shall be designed to prevent water infiltration. Finished grade shall be sloped away from the manhole to divert surface water. Manhole covers shall be secured to prevent unauthorized access. Service ports in manhole covers shall be not less than 8 inches (203 mm) in diameter and shall be not less than 4 inches (102 mm) above the finished grade level. The service port shall be secured to prevent unauthorized access.

Exception: Storage tanks less than 800 gallons (3028 L) in volume and installed below grade shall not be required to be equipped with a manhole, but shall have a service port not less than 8 inches (203 mm) in diameter.

❖ Storage tanks, big and small, need to be accessed for inspection. Above-ground tanks are not required to have a manhole access, only an inspection access. Below-grade tanks must have manhole access except where the tank has a capacity of less than 800 gallons (3028 L). All openings must be secured to prevent unauthorized access.

1301.9.8 Venting. Storage tanks shall be provided with a vent sized in accordance with Chapter 9 and based on the aggregate diameter of all tank influent pipes. The reservoir vent shall not be connected to sanitary drainage system vents. Vents shall be protected from contamination by means of an *approved* cap or U-bend installed with the opening directed downward. Vent outlets shall extend not less than 4 inches (102 mm) above grade or as necessary to prevent surface water from entering the storage tank. Vent openings shall be protected against the entrance of vermin and insects in accordance with the requirements of Section 1301.7.

❖ As the water level in the tank rises and drops, air needs to escape and enter the tank to prevent pressure (especially negative pressure) from building up.

1301.9.9 Draining of tanks. Where tanks require draining for service or cleaning, tanks shall be drained by using a pump or by a drain located at the lowest point in the tank. The tank drain pipe shall discharge as required for overflow pipes and shall not be smaller in size than specified in Table 606.5.7. Not less than one cleanout shall be provided on each drain pipe in accordance with Section 708.

❖ A tank can be drained by a pump or by a drain near the bottom of the tank. Most of the time, draining is required to clean the tank of accumulated debris. In-tank pumps and level controls are usually accessible by lifting them out of the tank so draining is not required for those repairs. Where there is a tank drain, this section requires a "tank drain pipe." In other words, there needs to be a pipe connected near the bottom of the tank, presumably with a valve, so that drain flow can be controlled. A "drain plug" alone would not meet the intent of this section as once the plug was removed, there would not be any way to control the discharge. This is reinforced by Item 2 of Section 1301.9.11.

1301.9.10 Marking and signage. Each nonpotable water storage tank shall be labeled with its rated capacity. The contents of storage tanks shall be identified with the words "CAUTION: NONPOTABLE WATER – DO NOT DRINK." Where an opening is provided that could allow the entry of personnel, the opening shall be marked with the words, "DANGER – CONFINED SPACE." Markings shall be indelibly printed on the tank or on a tag or sign constructed of corrosion-resistant waterproof material that is mounted on the tank. The letters of the words shall be not less than 0.5 inch (12.7 mm) in height and shall be of a color in contrast with the background on which they are applied.

❖ The requirements of this section are straightforward. A capacity label, especially for underground tanks, is useful when the contents have to be pumped out and hauled to a disposal point.

1301.9.11 Storage tank tests. Storage tanks shall be tested in accordance with the following:

Storage tanks shall be filled with water to the overflow line prior to and during inspection. All seams and joints shall be left exposed and the tank shall remain water tight without leakage for a period of 24 hours.

1. After 24 hours, supplemental water shall be introduced for a period of 15 minutes to verify proper drainage of the overflow system and that there are no leaks.
2. The tank drain shall be observed for proper operation.
3. The makeup water system shall be observed for proper operation and successful automatic shutoff of the system at the refill threshold shall be verified.

❖ The requirements of this section are straightforward. Although 24 hours seems like a long test period, it might take that amount of time for a small leak to develop enough of a puddle to be observed. For a lined carbon steel tank, even a small leak could become a much larger leak over time (because of corrosion), so it is useful to wait a day to find leaks. For wood tanks, the tank will have to filled well ahead of the test period so that the wood can swell and close up any gaps.

1301.10 System abandonment. If the owner of an on-site nonpotable water reuse system or rainwater collection and conveyance system elects to cease use of, or fails to properly maintain such system, the system shall be abandoned and shall comply with the following:

1. All system piping connecting to a utility-provided water system shall be removed or disabled.

2. The distribution piping system shall be replaced with an *approved* potable water supply piping system. Where an existing potable pipe system is already in place, the fixtures shall be connected to the existing system.

3. The storage tank shall be secured from accidental access by sealing or locking tank inlets and access points, or filling with sand or equivalent.

❖ Nonpotable water reuse systems do require maintenance and must be kept in good repair to provide water for the intended fixtures or other uses. However, there will be times that abandonment of some systems will be necessary.

1301.11 Trenching requirements for nonpotable water piping. Nonpotable water collection and distribution piping and reclaimed water piping shall be separated from the *building sewer* and potable water piping underground by 5 feet (1524 mm) of undisturbed or compacted earth. Nonpotable water collection and distribution piping shall not be located in, under or above cesspools, septic tanks, septic tank drainage fields or seepage pits. Buried nonpotable water piping shall comply with the requirements of Section 306.

Exceptions:

1. The required separation distance shall not apply where the bottom of the nonpotable water pipe within 5 feet (1524 mm) of the *sewer* is not less than 12 inches (305 mm) above the top of the highest point of the *sewer* and the pipe materials conform to Table 702.3.

2. The required separation distance shall not apply where the bottom of the potable water service pipe within 5 feet (1524 mm) of the nonpotable water pipe is a minimum of 12 inches (305 mm) above the top of the highest point of the nonpotable water pipe and the pipe materials comply with the requirements of Table 605.4.

3. Nonpotable water pipe is permitted to be located in the same trench with a *building sewer*, provided that such *sewer* is constructed of materials that comply with the requirements of Table 702.2.

4. The required separation distance shall not apply where a nonpotable water pipe crosses a *sewer* pipe, provided that the pipe is sleeved to at least 5 feet (1524 mm) horizontally from the *sewer* pipe centerline on both sides of such crossing, with pipe materials that comply with Table 702.2.

5. The required separation distance shall not apply where a potable water service pipe crosses a nonpotable water pipe, provided that the potable water service pipe is sleeved for a distance of at least 5 feet (1524 mm) horizontally from the centerline of the nonpotable pipe on both sides of such crossing, with pipe materials that comply with Table 702.2.

6. Irrigation piping located outside of a building and downstream of the backflow preventer is not required to meet the trenching requirements where nonpotable water is used for outdoor applications.

❖ These are the same requirements (excluding Exception 5) as for the installation of a potable water supply system. Nonpotable water irrigation piping does not need separation from the building sewer because this would be too difficult to accomplish and would be of limited benefit.

1301.12 Outdoor outlet access. Sillcocks, hose bibbs, wall hydrants, yard hydrants and other outdoor outlets supplied by nonpotable water shall be located in a locked vault or shall be operable only by means of a removable key.

❖ Rainwater (or possibly reclaimed water) could have a variety of uses outside of a building. Rainwater could be used to water gardens and vegetation, or to wash vehicles. Reclaimed water might be able to be used for outdoor heavy equipment washdowns. The outlets to connect to these water sources need to be controlled so that unauthorized use does not occur.

SECTION 1302
ON-SITE NONPOTABLE WATER REUSE SYSTEMS

1302.1 General. The provisions of Section 1302 shall govern the construction, installation, alteration and repair of on-site nonpotable water reuse systems for the collection, storage, treatment and distribution of on-site sources of nonpotable water as permitted by the jurisdiction.

❖ This section is about the specific details for on-site nonpotable water reuse systems. Refer to the definition of "On-site nonpotable water reuse system" in Chapter 2.

1302.2 Sources. On-site nonpotable water reuse systems shall collect waste discharge from only the following sources: bathtubs, showers, lavatories, clothes washers and laundry trays. Water from other *approved* nonpotable sources including swimming pool backwash operations, air conditioner condensate, rainwater, cooling tower blow-down water, foundation drain water, steam system condensate, fluid cooler discharge water, food steamer discharge water, combination oven discharge water, industrial process water and fire pump test water shall also be permitted to be collected for reuse by on-site nonpotable water reuse systems, as *approved* by the code official and as appropriate for the intended application.

❖ Essentially, an on-site nonpotable water reuse system only collects gray water. However, there are many other kinds of nonpotable water that could be available at the building site. Some of these might be generated in quantities that overshadow the flow of gray water into the system. The "qualities" of gray water are well understood. Other nonpotable waters are quite different from gray water and any one could have a significant impact on the suitably of the water for an end use application. For example, cooling tower blow-down water could have such a high level of total dissolved solids that for an end use such as water closet flushing, the bowls of the water closets become scaled and hard to clean. Problems with water closet fill and flush valves, and clogging of the internal waterways of the water closet could occur. Therefore, the code official needs to approve these other nonpotable water sources for inclusion.

1302.2.1 Prohibited sources. Waste water containing urine or fecal matter shall not be diverted to on-site nonpotable water reuse systems and shall discharge to the sanitary drainage system of the building or premises in accordance with Chapter 7. Reverse osmosis system reject water, water softener discharge water, kitchen sink waste water, dishwasher waste water and waste water discharged from wet-hood scrubbers shall not be collected for reuse in an on-site nonpotable water reuse system.

❖ There are certain nonpotable waters already known to be detrimental to on-site nonpotable water reuse systems. They are prohibited to be included.

1302.3 Traps. Traps serving fixtures and devices discharging waste water to on-site nonpotable water reuse systems shall comply with Section 1002.4.

❖ Plumbing fixtures (and clothes washer waste receptors) are already required to have traps that comply with Section 1002.4. Where other nonpotable water is collected by the on-site nonpotable water reuse systems, a trap is required between the connection of the piping from the nonpotable water source and the collection piping leading to the on-site nonpotable water reuse systems. The trap is required to also to comply with Section 1002.4. In most cases, these additional nonpotable waters will be connected indirectly to the gray water piping system. Indirect connections and waste receptors are covered in Chapter 8.

1302.4 Collection pipe. On-site nonpotable water reuse systems shall utilize drainage piping *approved* for use in plumbing drainage systems to collect and convey untreated water for reuse. Vent piping *approved* for use in plumbing venting systems shall be utilized for vents in the gray water system. Collection and vent piping materials shall comply with Section 702.

❖ The piping used for collecting all on-site nonpotable water must comply with Section 702. Vent piping must also be in accordance with Section 702.

1302.4.1 Installation. Collection piping conveying untreated water for reuse shall be installed in accordance with Section 704.

❖ Installation of the collection piping must comply with Section 704.

1302.4.2 Joints. Collection piping conveying untreated water for reuse shall utilize joints *approved* for use with the distribution piping and appropriate for the intended applications as specified in Section 705.

❖ Joints in collection piping must comply with Section 705.

1302.4.3 Size. Collection piping conveying untreated water for reuse shall be sized in accordance with drainage sizing requirements specified in Section 710.

❖ Although this section indicates that the drain piping is to be sized in accordance with Section 710, the flows from nonpotable water sources (other than the allowed plumbing fixtures) are not measured in drainage fixture units (dfu). In most cases, a maximum flow rate in gallons per minute can be computed for these flows. Section 709.3 can be used to determine an approximate number of dfu for a gallon-per-minute flow rate so the tables in Section 710 can be used.

1302.4.4 Labeling and marking. Additional marking of collection piping conveying untreated water for reuse shall not be required beyond that required for sanitary drainage, waste and vent piping by Chapter 7.

❖ As the code does not require labeling/marking of sanitary drainage piping, including piping that conveys only gray water, the code does not require labeling/marking for nonpotable water collection piping.

1302.5 Filtration. Untreated water collected for reuse shall be filtered as required for the intended end use. Filters shall be accessible for inspection and maintenance. Filters shall utilize a pressure gauge or other *approved* method to provide indication when a filter requires servicing or replacement. Filters shall be installed with shutoff valves immediately upstream and downstream to allow for isolation during maintenance.

❖ Filters are typically installed to filter the flow before entering the storage tank so that debris is kept out of the tank. Some type of alert system is required to indicate when the filter needs to be cleaned or changed.

1302.6 Disinfection and treatment. Where the intended application for nonpotable water collected on site for reuse requires disinfection or other treatment or both, it shall be disinfected as needed to ensure that the required water quality is delivered at the point of use. Nonpotable water collected on site containing untreated gray water shall be retained in collection reservoirs for a maximum of 24 hours.

❖ Some nonpotable water, such as gray water, will need disinfection before being used for certain applications. Other nonpotable waters, such as rainwater, will not require disinfection for certain applications. The designer decides what waters need disinfection in order to meet the jurisdiction's water quality requirements for the end use application. Note that untreated gray water cannot be stored for longer than 24 hours because of the significant bacteria present in the water.

1302.6.1 Gray water used for fixture flushing. Gray water used for flushing water closets and urinals shall be disinfected and treated by an on-site water reuse treatment system complying with NSF 350.

❖ Untreated gray water, even though disinfected, is too much of a health risk to be used inside of a building for the purposes of flushing water closets and urinals. Gray water must be processed through treatment equipment that complies with NSF 350.

1302.7 Storage tanks. Storage tanks utilized in on-site nonpotable water reuse systems shall comply with Sections 1301.9 and 1302.7.1 through 1302.7.3.

❖ In addition to Section 1301.9, there are requirements for nonpotable water tanks in other sections of the code.

1302.7.1 Location. Storage tanks shall be located with a minimum horizontal distance between various elements as indicated in Table 1302.7.1.

❖ Table 1302.7.1 indicates the minimum required clearance between nonpotable water storage tanks and sources of possible contamination, water services that could be contaminated, adjacent lot lines and the critical root zone of trees that must be protected.

1302.7.2. Design and construction. Storage tanks shall be designed and constructed in accordance with Chapters 16 through 22 of the *International Building Code* and in accordance with the following standards, as appropriate for the material of the storage tank: AWWA D100, AWWA D115, AWWA D120, UL 58, UL 1746, UL 1316, UL 142, API 12F or API 12D.

❖ The reference to Chapters 16 through 22 of the IBC is threefold, and intended to make sure that the reader understands the implication of the loads that the fully filled tank will have on the 1) building, 2) foundation (either the building or the tank foundation) and 3) adjacent structures, such as retaining walls. The tanks are required to be designed and constructed to standards.

TABLE 1302.7.1
LOCATION OF NONPOTABLE WATER REUSE STORAGE TANKS

ELEMENT	MINIMUM HORIZONTAL DISTANCE FROM STORAGE TANK (feet)
Critical root zone (CRZ) of protected trees	2
Lot line adjoining private lots	5
Seepage pits	5
Septic tanks	5
Water wells	50
Streams and lakes	50
Water service	5
Public water main	10

For SI: 1 foot = 304.8 mm.

The construction of some tanks might need to have special inspection at the point of manufacture (see Chapter 17 of the IBC).

1302.7.3 Outlets. Outlets shall be located not less than 4 inches (102 mm) above the bottom of the storage tank and shall not skim water from the surface.

❖ The "outlet" referred to is not the drain outlet of the tank. The purpose of the section is to allow some volume for sediment to settle in the bottom of the tank and to not have pump intakes or intake piping (or floating intake mechanisms) pulling water (and sediment) from this area. This is the same reason for prohibiting intake at the surface of the water: to avoid pulling in floating debris from the water surface.

1302.8 Valves. Valves shall be supplied on on-site nonpotable water reuse systems in accordance with Sections 1302.8.1 and 1302.8.2.

❖ There are two valves that must be provided on nonpotable water reuse systems: a bypass valve and a backwater valve for the overflow and drain of the storage tank.

1302.8.1 Bypass valve. One three-way diverter valve listed and labeled to NSF 50 or other approved device shall be installed on collection piping upstream of each storage tank, or drainfield, as applicable, to divert untreated on-site reuse sources to the sanitary *sewer* to allow servicing and inspection of the system. Bypass valves shall be installed downstream of fixture traps and vent connections. Bypass valves shall be marked to indicate the direction of flow, connection and storage tank or drainfield connection. Bypass valves shall be installed in accessible locations. Two shutoff valves shall not be installed to serve as a bypass valve.

❖ There will be situations where flow to the nonpotable water storage tank must be diverted. One of those times will be during cleaning of the tank. The flow must be positively diverted so that it is known that the flow is not backing up in the building but is being diverted to an appropriate location. The only way to positively know is by installing a three-way diverter valve where the flow can only be diverted from one outlet to the other outlet (without the incoming flow being blocked). Note the last sentence of the section that prohibits two independent valves from being used to divert the flow. Opening and closing two independent valves could result in both valves being inadvertently closed. The requirement for listing and labeling of the three-way valve to NSF 50 ensures that the valve is of quality construction and meets durability requirements, as this valve might be installed outdoors.

1302.8.2 Backwater valve. One or more backwater valves shall be installed on each overflow and tank drain pipe. Backwater valves shall be in accordance with Section 715.

❖ Contamination of a nonpotable water storage tank by any type of external water must be avoided. A backwater valve is a check valve for nonpotable water in gravity flow piping. Although the reference is to Section 715, Section 715.1 does not apply because this section always requires backwater valves. Sections 715.2 through 715.5 apply.

1302.9 Pumping and control system. Mechanical equipment including pumps, valves and filters shall be easily accessible and removable in order to perform repair, maintenance and cleaning. The minimum flow rate and flow pressure delivered by the pumping system shall be appropriate for the application and in accordance with Section 604.

❖ Equipment must be maintained, repaired or replaced so access will be needed to perform that work. How easy that access will be is up to the designer of the system and the area that he or she has to fit the system into. Every effort should be made to think about how components can be taken apart while still connected to the system. Where components need to be replaced, the designer needs to think about the clearances necessary for removal of the component. For example, many underground storage tanks have submersible pumps attached to discharge piping such that the entire vertical section of discharge piping has to be lifted out to bring the pump up to, and out of, the tank. Will there be sufficient overhead clearance at the completion of the entire project? Sometimes, at the "rough-in" installation of a large tank, it does appear that there is sufficient clearance. However, afterwards a large duct or a fire water line is routed across the area counted on for removal of the pump.

The pumps must have the capacity to provide the necessary flow and pressure for proper operation of the plumbing fixtures (water closets and urinals).

1302.10 Water pressure-reducing valve or regulator. Where the water pressure supplied by the pumping system exceeds 80 psi (552 kPa) static, a pressure-reducing valve shall be installed to reduce the pressure in the nonpotable water distribution system piping to 80 psi (552 kPa) static or less. Pressure-reducing valves shall be specified and installed in accordance with Section 604.8.

❖ Most pumps have a variation in pressure output that changes with the flow rate. Typically, the highest output pressure is when the flow rate is zero (or nearly zero). As outlets in the distribution system open, the flow rate from the pump increases to match the demand and the pressure that the pump generates is somewhat less. As more outlets open, the flow becomes greater and the outlet pressure of the pump becomes even less. Because of this effect, a pump model is chosen to supply the minimum required pressure when the demand (flow rate) of the system is at the greatest. Thus, at low demands (flow rates), the pressure in the system can easily exceed 80 psi (552 kPa) and a pressure regulator is required to keep the system pressure from exceeding 80 psi (552 kPa). Note that in distribution systems of multistory buildings, the pressure in the riser(s) might need to be significantly greater than 80 psi (552 kPa) to counter the effect of pressure loss caused by elevation change. Pressure reducers are installed for the branches for the fixtures on lower floor levels to maintain the pressure in those branches at or below 80 psi (552 kPa) for those floors.

Variable speed pumps for water distribution can resolve the problem of needing to choose a pump that has a much higher pressure at low/no-flow conditions so that there is adequate pressure at peak flow conditions. A variable speed pump changes the rotational speed pump impeller to better match the pump's outlet pressure in response to the flow rate, However, for multistory buildings, there will still be the need for reducing pressures for the branches serving the lower floors.

1302.11 Distribution pipe. Distribution piping utilized in on-site nonpotable water reuse systems shall comply with Sections 1302.11.1 through 1302.11.3.

Exception: Irrigation piping located outside of the building and downstream of a backflow preventer.

❖ Distribution piping begins at the treated water storage tank and ends at the termination of the fixture supply pipe such as at a water closet or urinal. Where the end use of the nonpotable water is not a plumbing fixture, then the piping ends at a control valve for the system that utilizes the nonpotable water.

The exception for irrigation systems allows for piping not complying with this section to begin after the backflow preventer for the irrigation system.

1302.11.1 Materials, joints and connections. Distribution piping shall conform to the standards and requirements specified in Section 605.

❖ In order to maintain consistency in pipe sizes, materials and connections of water distribution piping, both potable and nonpotable, the material must comply with the requirements of Section 605. Although there is very little difference between Tables 605.3 and 605.4, one notable difference is that PVC piping is not indicated in Table 605.4. For a potable water supply system, PVC is not allowed to be used for water distribution piping (piping that is downstream of the main water shutoff valve for the building) primarily because the material is not rated for hot water service. Where piping potable water systems inside of a building, there is the concern that PVC piping could be inadvertently used for a hot water line and that the piping could fail under that service condition. Because nonpotable water would not be heated prior to use in a building, it is questionable as to whether this (potable water system) limitation for use of PVC piping for nonpotable water distribution (in a building) is necessary or appropriate. The code official will have to make this decision.

1302.11.2 Design. On-site nonpotable water reuse distribution piping systems shall be designed and sized in accordance with Section 604 for the intended application.

❖ Section 604 requires that water distribution systems be designed in accordance with an accepted engineering practice and provides basic design criteria. A nonpotable water distribution system serving fixtures (water closets and urinals) needs to provide the same pressures and flow rates to the fixtures as if those fixtures were supplied by a potable water system.

1302.11.3 Marking. On-site nonpotable water distribution piping labeling and marking shall comply with Section 608.8.

❖ The term "labeling" in this section is not to be confused with requirements for third-party inspection and certification of products (e.g., the "listing and labeling" to a standard) (see commentary, Section 608.8).

1302.12 Tests and inspections. Tests and inspections shall be performed in accordance with Sections 1302.12.1 through 1302.12.6.

❖ Piping must be tested after it is installed to ensure the integrity of the installation. Note that the tests of piping are not necessarily attempting to replicate an actual condition of service. This section is also concerned with water quality at the point of use.

1302.12.1 Collection pipe and vent test. Drain, waste and vent piping used for on-site water reuse systems shall be tested in accordance with Section 312.

❖ Sections 312.2 and 312.3 cover testing of drainage and vent piping inside a building (including under the building). Note that Section 312.3 (covering air testing of piping) prohibits air testing of plastic piping. Where drainage piping is outside of a building, either above or below ground, Sections 312.6 and 312.7 apply even though the piping is not sewer piping (carrying sewage with fecal matter). See the commentary for Section 1302.12.2 concerning storage tank vent piping testing.

1302.12.2 Storage tank test. Storage tanks shall be tested in accordance with Section 1301.9.11.

❖ This section simply addresses the storage tank tests in Section 1301.9.11. In some situations, vent piping for storage tanks will be completely outdoors. The storage tank test does not include testing of the vent piping because the vent piping is completely outdoors, including the vent terminal. However, in other situations, the storage tank vent piping will be indoors (primarily because the storage tank is inside of, or under, the building). This vent piping should most likely be tested, however, not as a part of testing the storage tank. Although it would be possible to temporarily block the tank overflow and fill the storage tank completely until the water spilled out of the top of the vent pipe, storage tanks might not be designed for more than the pressure of just the tank being filled to capacity (tank overflow level). Adding the head pressure of a vertical run of vent piping, perhaps several stories or more, could overstress the storage tank walls.

This does not necessarily mean that storage tank (indoor) vent piping should not be tested. Testing of the tank vent piping might require that a valve be installed in the vent piping just above the tank, the valve closed and the vent piping filled with water for the leak check. After the test, the valve could be opened, the water in the piping drained to the tank, and the valve removed so that final vent-to-tank connections can be completed.

1302.12.3 Water supply system test. The testing of makeup water supply piping and distribution piping shall be conducted in accordance with Section 312.5.

❖ Makeup water supply piping could be another nonpotable water supply or, more often, a potable water supply. In either situation, the piping must be leak free. Note that air testing of plastic piping is prohibited. Additionally, where piping other than plastic is tested, test air pressure is only required to be 50 psi (345 kPa). Where water testing is performed, the test pressure must be the working pressure of the system.

However, in service, the piping is only required to be leak free at the normal water pressures encountered. This does not mean that all water piping to the storage tank will experience the same pressure. For example, in a large nonpotable water reuse system, the makeup water control valve might be a 3-inch (76 mm) motorized ball valve located some distance away from the storage tank. The valve might even be below the top of the storage tank. When the valve opens, the piping downstream of the valve will only be subjected to the pressure of the flow resistance in the downstream piping and any difference in the elevation between the valve and the outlet of the piping. That pressure will most likely be much less than the static pressure upstream of the makeup water valve. The point is that for the piping from the makeup water valve to the storage tank, the "test" of the piping should simply be a visual test. There is no need to "temporarily install a test valve" right at the tank for testing this piping at full line pressure. In this example, this would be a 3-inch (76 mm) valve, a needless expenditure.

Testing of distribution piping is not any different than for potable water.

1302.12.4 Inspection and testing of backflow prevention assemblies. The testing of backflow preventers and backwater valves shall be conducted in accordance with Section 312.10.

❖ Section 312.10 covers the inspection and testing of backflow prevention assemblies and hose connection backflow preventers. The section also covers the inspection of air gaps. A backflow prevention assembly is a type of backflow preventer that can be field tested using sensitive, calibrated gauges to determine if the assembly is operational. Backflow prevention assemblies have "test cocks" for attachment of the gauge's test hoses.

A hose connection backflow preventer is a very simple device to test. All that is necessary is to hold the open outlet of a connected hose above the elevation of the hose connection backflow preventer (doing this all with the valve before the backflow preventer, shut off). The contents of the hose will flow back to the backflow preventer and out its ports.

An air gap can be visually inspected to verify that the required open air distance is provided (see Section 608.15.1).

All other types of backflow preventers cannot be field tested nor should any attempts be made to perform a field test, as this could damage the device such that it might not perform the intended function under service conditions.

1302.12.5 Inspection of vermin and insect protection. Inlets and vents to the system shall be inspected to verify that each is protected to prevent the entrance of insects and vermin into the storage tank and piping systems in accordance with Section 1301.7.

❖ This section simply refers back to Section 1301.7 so that verification can be made that protection has been provided.

1302.12.6 Water quality test. The quality of the water for the intended application shall be verified at the point of use in accordance with the requirements of the jurisdiction.

❖ The jurisdiction allowed the installation of the nonpotable water reuse system based upon the premise that a certain quality level would be achieved by the treatment and disinfection of the system. The installer has to perform tests to show that the system provides that quality.

1302.13 Operation and maintenance manuals. Operation and maintenance materials shall be supplied with nonpotable on-site water reuse systems in accordance with Sections 1302.13.1 through 1302.13.4.

❖ The following sections indicate the minimum requirements for operation and maintenance manuals supplied with on-site nonpotable water reuse systems. As these systems can be very diverse and some are very customized, exceeding these minimum requirements is encouraged. For example, photographs of components, photograph sequences of specialized repair procedures, dimensional drawings including orientation references, common component manufacturer's part numbers and electronic files of the entire O&M package would be well received by the system owner. These systems can be complex, and because numerous repair and maintenance firms will be required to work on them, perhaps for many generations, information must be complete and clear. It is likely with the current diversity and number of firms providing these systems, there will be future mergers and acquisitions that will make future identification of particular systems difficult with dwindling numbers of people remaining who will remember those "older" systems, let alone have access to the original records.

1302.13.1 Manual. A detailed operations and maintenance manual shall be supplied in hardcopy form with all systems.

❖ A hard copy (a booklet or binder) of the manual must be provided for system users.

1302.13.2 Schematics. The manual shall include a detailed system schematic, and the locations and a list of all system components, including manufacturer and model number.

❖ A schematic can be in many forms, but in terms of these types of systems, the schematic would most likely be in the form of primarily showing tanks, pumps and treatment units in their approximate "physical forms" with the piping shown as lines connecting those components. Valves, meters and controls are typically shown in a common "legend" format for quick comprehension of their functionality. A numbered list similar to a "bill of material" identifies every item on the schematic.

1302.13.3 Maintenance procedures. The manual shall provide a schedule and procedures for all system components requiring periodic maintenance. Consumable parts, including filters, shall be noted along with part numbers.

❖ An on-site nonpotable water reuse system is much more complex than a potable water system for a building. Scheduled maintenance is paramount to making sure that the nonpotable water delivered for an end use does not become a health hazard for the occupants. The other reason for performing required maintenance of these systems is that where the end use for the water is for flushing water closets and urinals, the building owner can ill afford to have these fixtures not operational. If those fixtures do not work, the building will be condemned and evacuated until such time that the system is repaired and the fixtures are operational. This is a totally different concept for most building owners, unaware of the implications of poor or no maintenance of these systems. If the public water supply to a building is shut off because of a utility piping problem (e.g., water line break in the street, fire hydrant broken), the building owner is not to blame and the occupants will "make do" with the inconvenience. However, with an on-site nonpotable water reuse system, the building owner is fully responsible for providing the nonpotable water to the water closets and urinals. The health department and code enforcement department might not look as favorably at a building owner's lack of nonpotable water.

1302.13.4 Operations procedures. The manual shall include system startup and shutdown procedures. The manual shall include detailed operating procedures for the system.

❖ This requirement about indicating what the system components do over a period of time is so that the operator knows when the system is not properly operating. That period of time also includes startup and shutdown of the system where the operator is usually needed to override certain automatic functions.

SECTION 1303
NONPOTABLE RAINWATER COLLECTION AND DISTRIBUTION SYSTEMS

1303.1 General. The provisions of Section 1303 shall govern the construction, installation, alteration and repair of rainwater collection and conveyance systems for the collection, storage, treatment and distribution of rainwater for nonpotable applications, as permitted by the jurisdiction.

❖ Rainwater is very plentiful in many areas of the United States, and so there is a heavy reliance on this water source. However, other areas do not have much water fall so there is no advantage to capturing what little does fall. Additionally, there are areas of the country that have legal restrictions about capturing and reusing rainwater. These areas are concerned about yearly replenishment of lakes and underground aquifers that provide the raw water source for making potable water. The more rainwater that is captured and used in a building, the less of that water finds its way back into the natural waterways that feed the lakes and rivers. It is a very delicate problem for some areas. Thus, the jurisdiction decides whether rainwater collection and distribution systems can be installed depending on the end use application. For example, a rainwater collection system for outdoor irrigation might be approved because the water will drain to the natural waterways. However, that same rainwater collection system might not be approved where the system is used to supply water closets and urinals.

1303.2 Collection surface. Rainwater shall be collected only from above-ground impervious roofing surfaces constructed from *approved* materials. Collection of water from vehicular parking or pedestrian surfaces shall be prohibited except where the water is used exclusively for landscape irrigation. Overflow and bleed-off pipes from roof-mounted appliances including, but not limited to, evaporative coolers, water heaters, and solar water heaters shall not discharge onto rainwater collection surfaces.

❖ Before it comes in contact with a surface, rainwater is certainly of much higher "quality" than gray water. One can drink collected rainwater out of a clean drinking cup that was put outside during a rainfall event and be fairly confident that they would not experience ill effects. People have been doing this for thousands of years and even today, some people still capture rainwater for drinking and cooking. However, rainwater could become nonpotable before reaching a collection surface because of airborne pollutants that the rainwater picks up as it falls from the sky. Imagine the condition of rainwater in an area around a large coal-fired boiler such as found in a power plant. Furthermore, the collection of rainwater (contacting a roof surface) adds more pollutants to the water, as collection surfaces (roofs) have bird droppings, airborne particulates, dirt with pesticides and a host of many other things that would make the rainwater unsafe for drinking.

The requirement for roofing surfaces (used for the collection of rainwater) to be of materials approved by the code official is to make sure that the roof material itself does not add any more contamination to the rainwater that the end use application cannot have or that the treatment processes for the rainwater cannot remove before end use. For example, a roof made of copper with lead-soldered joints might not be appropriate for the end use of the rainwater to be drinking water because of the potential for copper and lead contamination that might not be able to be removed by the proposed treatment process.

Any other potential sources of contamination, such as bleedoffs and overflows from mechanical systems, and relief valves from water heaters, must not be allowed to discharge to the roof surface also to make sure that the roof material doesn't add contaminants to the rainwater.

1303.3 Debris excluders. Downspouts and leaders shall be connected to a roof washer and shall be equipped with a debris excluder or equivalent device to prevent the contamination of collected rainwater with leaves, sticks, pine needles and similar material. Debris excluders and equivalent devices shall be self-cleaning.

❖ The debris excluder can be an integral part of the roof washer or it can be separate. The debris excluder screens out sticks, leaves and other large particles from the rainwater flow. As debris captured in the debris excluder can build up quickly, these devices are required to be self-cleaning.

1303.4 Roof washer. A sufficient amount of rainwater shall be diverted at the beginning of each rain event, and not allowed to enter the storage tank, to wash accumulated debris from the collection surface. The amount of rainfall to be diverted shall be field adjustable as necessary to minimize storage tank water contamination. The roof washer shall not rely on manually operated valves or devices, and shall operate automatically. Diverted rainwater shall not be drained to the roof surface, and shall be discharged in a manner consistent with the storm water runoff requirements of the jurisdiction. Roof washers shall be accessible for maintenance and service.

❖ The term "roof washer" can be misleading. The roof washer is actually a device that wastes the initial flow of rainwater coming from the roof so that the highly contaminated rainwater is diverted from the storage tank. This device must be adjustable so that the amount of water wasted can be varied to meet the actual conditions of service. The basis of adjustment is usually water testing to determine the level of contaminants in the water going to the storage tank. Once set, the roof washer may not need adjustment unless conditions of the captured rainwater change.

1303.5 Roof gutters and downspouts. Gutters and downspouts shall be constructed of materials that are compatible with the collection surface and the rainwater quality for the desired end use. Joints shall be water tight.

❖ The construction materials for gutters and downspouts must be "compatible" with the raw rainwater quality that is needed for the end uses. The intent is the same as for roof materials that collect rainwater (see commentary, Section 1303.3). As gutters are often in direct contact with roofing materials (the collection surface), material compatibility between the two surfaces must be considered. For example, an aluminum collection surface (roofing) might not be compatible with galvanized steel gutters (see commentary, Section 1304.6).

1303.5.1 Slope. Roof gutters, leaders and rainwater collection piping shall slope continuously toward collection inlets. Gutters and downspouts shall have a slope of not less than $^1/_8$ inch per foot (10.4 mm/m) along their entire length, and shall not permit the collection or pooling of water at any point.

Exception: Siphonic drainage systems installed in accordance with the manufacturer's instructions shall not be required to have a slope.

❖ Once rainwater is collected by gutters, the rainwater must be kept moving towards the storage tank so that particulate matter and debris do not drop out of the flow. Pockets and pools of nonmoving water will begin to create bacteria "nests" that will contaminate the stored rainwater after the roof washer has wasted the highly contaminated initial flow from the roof.

Siphonic drain systems are not required to have sloped piping as they operate, when under full siphon flow, as full flow piping.

1303.5.2 Size. Gutters and downspouts shall be installed and sized in accordance with Section 1106.6 and local rainfall rates.

❖ The code specifies the requirements for gutter sizes in Section 1106.6. For sizing for flow, downspouts are considered to be vertical and horizontal storm drain piping, which is covered by Section 1106.2. Downspouts not of a round shape, such as corrugated or rectangular, can be approximated as round shape. A corrugated-material round downspout could be approximated as the pipe size having the same or smaller internal diameter as the smallest inside diameter of the corrugated shape. A rectangular size could be approximated as a round shape using Equation 11-1 in Table 1106.2(2) of the 2012 IPC.

1303.5.3 Cleanouts. Cleanouts shall be provided in the water conveyance system to allow *access* to all filters, flushes, pipes and downspouts.

❖ There needs to be access to the collection and conveyance systems and filters for conveying rainwater, so that debris can be periodically cleared from those systems.

1303.6 Drainage. Water drained from the roof washer or debris excluder shall not be drained to the sanitary sewer. Such water shall be diverted from the storage tank and discharge in a location that will not cause erosion or damage to property in accordance with the *International Building Code*. Roof washers and debris excluders shall be provided with an automatic means of self-draining between rain events, and shall not drain onto roof surfaces.

❖ The first sentence reinforces Section 1101.3 to prohibit storm water (rainwater) from entering a drain system for sewage only. Note that the code no longer allows combined (both sewer and storm) building sewers, as Section 1109.1 requires separate building storm sewers and building sewers, even though the public sewer may be a combined sewer (see commentary, Section 1109.1). Control of rainwater runoff from a roof washer or debris excluder is required to avoid erosion and other possible negative effects (see commentary, Section 1301.9.1).

Roof washers and debris excluders must be self-draining so that there is not captive water that could harbor mosquitos. Also, captive water during freezing temperatures could freeze and damage these devices.

The water and debris that is prevented from going to the storage tank should not be drained to the roof surface, as this will only reintroduce the contaminants to the collection system after the roof washer has cycled for the initial rainwater flow.

1303.7 Collection pipe. Rainwater collection and conveyance systems shall utilize drainage piping *approved* for use within plumbing drainage systems to collect and convey captured rainwater. Vent piping *approved* for use within plumbing venting systems shall be utilized for vents within the rainwater system. Collection and vent piping materials shall comply with Section 702.

❖ Where piping is used to convey rainwater, or to vent rainwater systems, the piping must comply with materials that are in accordance with Section 702. See the commentary for Section 1303.4 for gutters and downspouts.

1303.7.1 Installation. Collection piping conveying captured rainwater shall be installed in accordance with Section 704.

❖ See the commentary for Section 1302.4.1.

1303.7.2 Joints. Collection piping conveying captured rainwater shall utilize joints *approved* for use with the distribution piping and appropriate for the intended applications as specified in Section 705.

❖ See the commentary for Section 1302.4.2.

1303.7.3 Size. Collection piping conveying captured rainwater shall be sized in accordance with drainage sizing requirements specified in Section 710.

❖ See the commentary for Section 1302.4.3.

1303.7.4 Marking. Additional marking of collection piping conveying captured rainwater for reuse shall not be required beyond that required for sanitary drainage, waste and vent piping by Chapter 7.

❖ See the commentary for Section 1302.4.4.

1303.8 Filtration. Collected rainwater shall be filtered as required for the intended end use. Filters shall be accessible for inspection and maintenance. Filters shall utilize a pressure gauge or other *approved* method to provide indication when a filter requires servicing or replacement. Filters shall be installed with shutoff valves installed immediately upstream and downstream to allow for isolation during maintenance.

❖ See the commentary for Section 1302.5.

1303.9 Disinfection. Where the intended application for rainwater requires disinfection or other treatment or both, it shall be disinfected as needed to ensure that the required water quality is delivered at the point of use. Where chlorine is used for disinfection or treatment, water shall be tested for residual chlorine in accordance with ASTM D1253. The levels of residual chlorine shall not exceed that allowed for the intended use in accordance with the requirements of the jurisdiction.

❖ Collected rainwater will not require disinfection for many end use applications. The designer decides whether there is need for disinfection in order to meet the jurisdiction's water quality requirements for the end use application.

1303.10 Storage tanks. Storage tanks utilized in nonpotable rainwater collection and conveyance systems shall comply with Sections 1301.9 and 1303.10.1 through 1303.10.3.

❖ In addition to Section 1301.9, there are requirements for nonpotable water tanks in other sections of the code.

1303.10.1 Location. Storage tanks shall be located with a minimum horizontal distance between various elements as indicated in Table 1303.10.1.

❖ This section refers to Table 1303.10.1 which is similar except there is not a concern for the rainwater contaminating other water sources. The distance to the critical root zone is reduced as compared to Table 1302.7.1 because rainwater doesn't offer the roots as much "incentive" to travel towards the storage tank where the tank is underground.

1303.10.2 Inlets. Storage tank inlets shall be designed to introduce collected rainwater into the tank with minimum tur-

**TABLE 1303.10.1
LOCATION OF RAINWATER STORAGE TANKS**

ELEMENT	MINIMUM HORIZONTAL DISTANCE FROM STORAGE TANK (feet)
Critical root zone (CRZ) of protected trees	2
Lot line adjoining private lots	5
Seepage pits	5
Septic tanks	5

For SI: 1 foot = 304.8 mm.

bulence, and shall be located and designed to avoid agitating the contents of the storage tank.

❖ The introduction of rainwater into a storage tank (already having some water in the tank) must be in a manner that limits agitation of the water. Typically, the water is introduced below a certain water "low point" level but above the bottom of the tank. Although there is usually not too much floating debris on the surface of the water, a biofilm does develop on the surface of the water and the bottom of the tank. The biofilms are helpful in maintaining the water in good condition. Therefore, avoiding unnecessary turbulence contributes to the health of these biofilms.

1303.10.3 Outlets. Outlets shall be located at least 4 inches (102 mm) above the bottom of the storage tank and shall not skim water from the surface.

❖ The purpose of the section is to allow some volume for sediment to settle in the bottom of the tank and to not have pump intakes or intake piping (or floating intake mechanisms) pulling water (and sediment) from this area.

1303.11 Valves. Valves shall be supplied on rainwater collection and conveyance systems in accordance with Section 1303.11.1.

❖ The code only requires a valve (a backwater valve) on the overflow outlet of a rainwater storage tank. This is to prevent contamination of the water contents of the tank should the tank overflow be submerged (usually by floodwaters).

The code does not require a valve for the inlet of the tank; however, valves might be necessary so that maintenance can be performed on the storage tank. Such inlet valves would not necessarily need to be a three-way valve (see Section 1302.8.1). One valve could be used for the tank inlet and the inlet piping arranged so that the inlet piping (before the valve) "overflows" to a storm sewer or alternate discharge point while the tank is out of service. However, such an overflow would have to be protected by a backwater valve.

1303.11.1 Backwater valve. Backwater valves shall be installed on each overflow and tank drain pipe. Backwater valves shall be in accordance with Section 715.

❖ See the commentary for Section 1302.8.2.

1303.12 Pumping and control system. Mechanical equipment including pumps, valves and filters shall be easily accessible and removable in order to perform repair, maintenance and cleaning. The minimum flow rate and flow pressure delivered by the pumping system shall be appropriate for the application and in accordance with Section 604.

❖ See the commentary for Section 1302.9.

1303.13 Water pressure-reducing valve or regulator. Where the water pressure supplied by the pumping system exceeds 80 psi (552 kPa) static, a pressure-reducing valve shall be installed to reduce the pressure in the rainwater distribution system piping to 80 psi (552 kPa) static or less. Pressure-reducing valves shall be specified and installed in accordance with Section 604.8.

❖ See the commentary for Section 1302.10.

1303.14 Distribution pipe. Distribution piping utilized in rainwater collection and conveyance systems shall comply with Sections 1303.14.1 through 1303.14.3.

Exception: Irrigation piping located outside of the building and downstream of a backflow preventer.

❖ See the commentary for Section 1302.11.

1303.14.1 Materials, joints and connections. Distribution piping shall conform to the standards and requirements specified in Section 605 for nonpotable water.

❖ See the commentary for Section 1302.11.1.

1303.14.2 Design. Distribution piping systems shall be designed and sized in accordance with Section 604 for the intended application.

❖ See the commentary for Section 1302.11.2.

1303.14.3 Marking. Nonpotable rainwater distribution piping labeling and marking shall comply with Section 608.8.

❖ See the commentary for Section 1302.11.3

1303.15 Tests and inspections. Tests and inspections shall be performed in accordance with Sections 1303.15.1 through 1303.15.8.

❖ Piping must be tested after it is installed to ensure the integrity of the installation. Note that the tests of piping are not necessarily attempting to replicate an actual condition of service. Gutters need to be inspected to make sure they have proper slope. This section is also concerned with water quality at the point of use.

1303.15.1 Roof gutter inspection and test. Roof gutters shall be inspected to verify that the installation and slope is in accordance with Section 1303.5.1. Gutters shall be tested by pouring not less than 1 gallon (3.8 L) of water into the end of the gutter opposite the collection point. The gutter being tested shall not leak and shall not retain standing water.

❖ Water cannot be collected from roof surfaces unless the water can make it to the storage tank. Improperly sloped gutters (usually backwardly sloped or without slope) will cause overflow during peak rainfall events resulting in loss of rainwater. Standing water in gutters will evaporate (another loss) or create pools where mosquitos can proliferate. Leaky gutters also waste rainwater. Pouring water into the most remote end of a gutter system is the simplest way determine proper slope and to find leaks. Note that the minimum quantity of 1 gallon (3.8 L) is to prevent someone from pouring a smaller amount of water into the gutter and, because the water did not leak out of the gutter (because it did not arrive at all points in the gutter system), the gutter is considered "leak free." The intent of this section is to introduce sufficient water to the gutter to test the gutter system for leaks, flow direction and standing water.

1303.15.2 Roofwasher test. Roofwashers shall be tested by introducing water into the gutters. Proper diversion of the first quantity of water in accordance with the requirements of Section 1303.4 shall be verified.

❖ At the same time that the gutters are being tested, the roof washer operation can be tested. The initial amount of rainwater diversion is chosen by the designer. Knowing that the roof washer is adjustable, verification of the amount for this test is not necessary. The only verification necessary is that the device wastes the initial flow and returns to a position to allow the flow to be sent to the storage tank.

1303.15.3 Collection pipe and vent test. Drain, waste and vent piping used for rainwater collection and conveyance systems shall be tested in accordance with Section 312.

❖ See the commentary for Section 1302.12.1.

1303.15.4 Storage tank test. Storage tanks shall be tested in accordance with Section 1301.9.11.

❖ See the commentary for Section 1302.12.2.

1303.15.5 Water supply system test. The testing of makeup water supply piping and distribution piping shall be conducted in accordance with Section 312.5.

❖ Makeup water supply piping will be a potable water supply. The piping must be leak free. Note that air testing of plastic piping is prohibited. Where piping other than plastic is tested, test air pressure is only required to be 50 psi (345 kPa). Where water testing is performed, the test pressure must be the working pressure of the system.

However, in service, the piping is only required to be leak free at the normal water pressures encountered. This does not mean that all water piping to the storage tank will experience the same pressure. For example, in a large nonpotable water reuse system, the makeup water control valve might be a 3-inch (76 mm) motorized ball valve located some distance away from the storage tank. The valve might even be below the top of the storage tank. When the valve opens, the piping downstream of the valve will only be subjected to the pressure of the flow resistance in the downstream piping and any difference in the elevation between the valve and the outlet of the piping. That pressure will most likely be much less than the static pressure upstream of the makeup water valve. The point is that for the piping from the makeup water valve to the storage tank, the "test" of the piping should simply be a visual test. There is no need to "temporarily install a test valve" right at the tank for testing this piping at full line pressure. In this example, this would be a 3-inch (76 mm) valve, a needless expenditure.

Testing of distribution piping is not any different than for potable water.

1303.15.6 Inspection and testing of backflow prevention assemblies. The testing of backflow preventers and backwater valves shall be conducted in accordance with Section 312.10.

❖ See the commentary for Section 1302.12.4.

1303.15.7 Inspection of vermin and insect protection. Inlets and vents to the system shall be inspected to verify that each is protected to prevent the entrance of insects and vermin into the storage tank and piping systems in accordance with Section 1301.7.

❖ This section simply refers back to Section 1301.7 so that verification can be made that protection has been provided.

1303.15.8 Water quality test. The quality of the water for the intended application shall be verified at the point of use in accordance with the requirements of the jurisdiction. Except where site conditions as specified in ASTM E2727 affect the rainwater, collected rainwater shall be considered to have the parameters indicated in Table 1303.15.8.

❖ The jurisdiction allowed the installation of the nonpotable rainwater reuse system based upon the premise that a certain quality level, at the end use point, would be achieved by the treatment and disinfection systems. The installer has to perform tests to show that the system provides that quality at the end use point. This does not mean the collected rainwater in the storage tank must meet these requirements because the treatment of rainwater occurs after the storage tank.

The designers of nonpotable rainwater collection and distribution systems must rely upon a standard for raw rainwater quality in order to design downstream treatment systems to provide the required quality (required by the jurisdiction) at the end use point. The raw rainwater quality is specified in ASTM E2727 with the parameters in that standard duplicated in Table 1303.15.8. Where the raw rainwater quality (that which comes out of the storage tank) is not within these specifications, then the downstream treatment systems will be unable to make the necessary corrections to provide the desired end use water quality.

TABLE 1303.15.8
RAINWATER QUALITY

PARAMETER	VALUE
pH	6.0-7.0
BOD	Not greater than 10 mg/L
NTU	Not greater than 2
Fecal coliform	No detectable fecal coli in 100 mL
Sodium	No detectable sodium in 100 mL
Chlorine	No detectable chlorine in 100 mL
Enteroviruses	No detectable enteroviruses in 100 mL

1303.16 Operation and maintenance manuals. Operation and maintenance manuals shall be supplied with rainwater collection and conveyance systems in accordance with Sections 1303.16.1 through 1303.16.4.

❖ See the commentary for Section 1302.13.

1303.16.1 Manual. A detailed operations and maintenance manual shall be supplied in hardcopy form with all systems.

❖ See the commentary for Section 1302.13.1.

1303.16.2 Schematics. The manual shall include a detailed system schematic, and locations and a list of all system components, including manufacturer and model number.

❖ See the commentary for Section 1302.13.2.

1303.16.3 Maintenance procedures. The manual shall provide a maintenance schedule and procedures for all system components requiring periodic maintenance. Consumable parts, including filters, shall be noted along with part numbers.

❖ See the commentary for Section 1302.13.3.

1303.16.4 Operations procedures. The manual shall include system startup and shutdown procedures, as well as detailed operating procedures.

❖ See the commentary for Section 1302.16.4.

SECTION 1304
RECLAIMED WATER SYSTEMS

1304.1 General. The provisions of this section shall govern the construction, installation, alteration and repair of systems supplying nonpotable reclaimed water.

❖ In most situations, reclaimed water is supplied to a building site by a public water supply utility (see the definition for "Reclaimed water" in Chapter 2). The systems that provide reclaimed water to the building site are similar to how a public utility supplies potable water to a building site. The public water supply utility goes to great lengths to make sure that these nonpotable water supplies are well marked/identified in order to avoid inadvertent connection to those systems. In most situations, piping, valves and backflow preventers are colored purple.

However, the code does not limit the use of reclaimed water to only that which is supplied by the public water supply utility. A building site could have a waste water facility (licensed by authorities having jurisdiction over such facilities) where the effluent is of a quality that meets the jurisdiction's requirements for reclaimed water.

1304.2 Water pressure-reducing valve or regulator. Where the reclaimed water pressure supplied to the building exceeds 80 psi (552 kPa) static, a pressure-reducing valve shall be installed to reduce the pressure in the reclaimed water distribution system piping to 80 psi (552 kPa) static or less. Pressure-reducing valves shall be specified and installed in accordance with Section 604.8.

❖ See the commentary for Section 1302.10.

1304.3 Reclaimed water systems. The design of the reclaimed water systems shall conform to ASTM E2635 and *accepted engineering practice*.

❖ ASTM E2635 covers the portions of a system to make reclaimed water from available nonpotable water on site. As each of these systems will be custom engineered for the specific building site, accepted engineering practices must be used.

1304.3.1 Distribution pipe. Distribution piping shall comply with Sections 1304.3.1.1 through 1304.3.1.3.

Exception: Irrigation piping located outside of the building and downstream of a backflow preventer.

❖ See the commentary for Section 1302.11.

1304.3.1.1 Materials, joints and connections. Distribution piping conveying reclaimed water shall conform to standards and requirements specified in Section 605 for nonpotable water.

❖ See the commentary for Section 1302.11.1.

1304.3.1.2 Design. Distribution piping systems shall be designed and sized in accordance with Section 604 for the intended application.

❖ See the commentary for Section 1302.11.2.

1304.3.1.3 Labeling and marking. Nonpotable rainwater distribution piping labeling and marking shall comply with Section 608.8.

❖ See the commentary for Section 1302.11.3.

1304.4 Tests and inspections. Tests and inspections shall be performed in accordance with Sections 1304.4.1 and 1304.4.2.

❖ Piping must be tested after it is installed to ensure the integrity of the installation and that the piping does not leak.

1304.4.1 Water supply system test. The testing of makeup water supply piping and reclaimed water distribution piping shall be conducted in accordance with Section 312.5.

❖ See the commentary for Section 1302.12.3.

1304.4.2 Inspection and testing of backflow prevention assemblies. The testing of backflow preventers shall be conducted in accordance with Section 312.10.

❖ See the commentary for Section 1302.12.4.

Chapter 14:
Subsurface Landscape Irrigation Systems

General Comments

Section 1302.6 prohibits the storage of untreated gray water for more than 24 hours. Although untreated gray water could be discharged to the public sewer system (or a private sewage disposal system), another possible building site use for this water would be subsurface irrigation. Reusing gray water on site lessens the load on public sewer systems and septic tank systems, while at the same time adding beneficial nutrients to the soil and eventual recharging of local aquifers.

Purpose

This chapter specifies the design and construction of systems that can accept gray water without compromising the environment.

SECTION 1401
GENERAL

1401.1 Scope. The provisions of Chapter 14 shall govern the materials, design, construction and installation of subsurface landscape irrigation systems connected to nonpotable water from on-site water reuse systems.

❖ The chapter provides specific details for designing and constructing subsurface irrigation systems.

1401.2 Materials. Above-ground drain, waste and vent piping for subsurface landscape irrigation systems shall conform to one of the standards listed in Table 702.1. Subsurface landscape irrigation, underground building drainage and vent pipe shall conform to one of the standards listed in Table 702.2.

❖ The piping for above-ground landscape systems must comply with the standards in Tables 702.1 and 702.2. The appropriate pipe fittings from Table 702.4 must be used to make connections in the pipe

1401.3 Tests. Drain, waste and vent piping for subsurface landscape irrigation systems shall be tested in accordance with Section 312.

❖ Sections 312.2 and 312.3 cover testing of drainage and vent piping inside a building (including under the building). Note that Section 312.3 covering air testing of piping prohibits air testing of plastic piping. Where drainage piping is outside of a building, either above or below ground, Sections 312.6 and 312.7 apply even though the piping is not sewer piping (carrying sewage with fecal matter).

1401.4 Inspections. Subsurface landscape irrigation systems shall be inspected in accordance with Section 107.

❖ Systems need to be inspected prior to concealment to make sure that the installation meets the requirements of the code.

1401.5 Disinfection. Disinfection shall not be required for on-site nonpotable water reuse for subsurface landscape irrigation systems.

❖ Water sent to a subsurface irrigation system is not required to be disinfected as it will not come in contact with humans.

1401.6 Coloring. On-site nonpotable water reuse for subsurface landscape irrigation systems shall not be required to be dyed.

❖ Coloring of the water is not required because the water is not visible to humans.

SECTION 1402
SYSTEM DESIGN AND SIZING

1402.1 Sizing. The system shall be sized in accordance with the sum of the output of all water sources connected to the subsurface irrigation system. Where gray water collection piping is connected to subsurface landscape irrigation systems, gray water output shall be calculated according to the gallons-per-day-per-occupant number based on the type of fixtures connected. The gray water discharge shall be calculated by the following equation:

$C = A \times B$ (Equation 14-1)

where:

A = Number of occupants:

Residential—Number of occupants shall be determined by the actual number of occupants, but not less than two occupants for one bedroom and one occupant for each additional bedroom.

Commercial—Number of occupants shall be determined by the *International Building Code*.

B = Estimated flow demands for each occupant:

Residential—25 gallons per day (94.6 lpd) per occupant for showers, bathtubs and lavatories and 15 gallons per day (56.7 lpd) per occupant for clothes washers or laundry trays.

Commercial—Based on type of fixture or water use records minus the discharge of fixtures other than those discharging gray water.

SUBSURFACE LANDSCAPE IRRIGATION SYSTEMS

C = Estimated gray water discharge based on the total number of occupants.

❖ The volume of gray-water discharge from residential buildings is the estimated number of occupants living in the building times a fixed amount of 25 gallons per day (94.6 L/day). An additional 15 gallons per day (56.7 L/day) per occupant is added if clothes washers or laundry trays exist. The number of occupants for the building must be not less than two for the first bedroom and one for each additional bedroom.

The volume of gray-water discharge from commercial buildings can be determined by two methods. The first method requires the designer to determine the number of times that a gray-water-producing fixture is used by the occupants of the building and estimate the volume produced by such fixtures. The second method involves observing water use records of a similar existing building and subtracting out the estimated volume used for nongray-water-producing fixtures. Chapter 7 of the *International Green Construction Code®* (IGCC®) offers generally accepted numbers for daily uses per occupant for common plumbing fixtures, including water closets and urinals. By knowing the flush volume of each fixture and the maximum building occupant load, the anticipated daily volume of gray water needed can be determined.

1402.2 Percolation tests. The permeability of the soil in the proposed absorption system shall be determined by percolation tests or permeability evaluation.

❖ In order to determine how large the subsurface irrigation system (the absorption system) must be to accommodate the maximum daily flow of gray water, tests must be performed to determine the subsurface soil's capacity for accepting the flow. Soil testing can be by the percolation test method or permeability evaluation method.

1402.2.1 Percolation tests and procedures. At least three percolation tests in each system area shall be conducted. The holes shall be spaced uniformly in relation to the bottom depth of the proposed absorption system. More percolation tests shall be made where necessary, depending on system design.

❖ A percolation test involves digging at least three holes to a depth where the proposed absorption system is to be installed. The undisturbed soil at the bottom of the hole will be representative of the soil that must absorb the gray-water flow.

1402.2.1.1 Percolation test hole. The test hole shall be dug or bored. The test hole shall have vertical sides and a horizontal dimension of 4 inches to 8 inches (102 mm to 203 mm). The bottom and sides of the hole shall be scratched with a sharp-pointed instrument to expose the natural soil. All loose material shall be removed from the hole and the bottom shall be covered with 2 inches (51 mm) of gravel or coarse sand.

❖ The percolation test holes must be of the same size and shape so that the test results from multiple holes can be compared. The surfaces of holes must be scratched to break up any soil smearing that could block the flow of test water. Test holes are made purposely small in horizontal dimension so a limited amount of water is necessary to perform the test. The placement of gravel or coarse sand in the bottom of the hole prevents scouring of the bottom of the test hole as test water is poured into it.

1402.2.1.2 Test procedure, sandy soils. The hole shall be filled with clear water to a minimum of 12 inches (305 mm) above the bottom of the hole for tests in sandy soils. The time for this amount of water to seep away shall be determined, and this procedure shall be repeated if the water from the second filling of the hole seeps away in 10 minutes or less. The test shall proceed as follows: Water shall be added to a point not more than 6 inches (152 mm) above the gravel or coarse sand. Thereupon, from a fixed reference point, water levels shall be measured at 10-minute intervals for a period of 1 hour. Where 6 inches (152 mm) of water seeps away in less than 10 minutes, a shorter interval between measurements shall be used, but in no case shall the water depth exceed 6 inches (152 mm). Where 6 inches (152 mm) of water seeps away in less than 2 minutes, the test shall be stopped and a rate of less than 3 minutes per inch (7.2 s/mm) shall be reported. The final water level drop shall be used to calculate the percolation rate. Soils not meeting the above requirements shall be tested in accordance with Section 1303.7.1.3.

❖ To prepare sandy soil for the percolation test, 12 inches (305 mm) of water is poured into the test hole and allowed to drain out. Another 12 inches (305 mm) of water is added to the hole and, if the time that it takes to drain out is 10 minutes or less, then the hole is repeatedly filled with 12 inches (305 mm) of water until the time that it takes for the water to drain away is greater than 10 minutes.

For the sandy soil percolation test, the hole is filled with 6 inches (152 mm) of water and the level measured from a fixed reference point is taken every 10 minutes. If the 6 inches (152 mm) of water completely drains from the hole in less than 10 minutes, then the measurement interval must be shortened so that some water remains in the hole at the end of the measurement interval. If the 6 inches (152 mm) of water completely drains out in 2 minutes or less, then a percolation rate of 3 minutes per inch (7.2 s/mm) is assigned. If the 6 inches (152 mm) of water drains out slower than 2 minutes, then the percolation rate is determined from the final water level drop in the test hole. For example, assume that the measurement interval is 5 minutes and the last water level drop (before the hole becomes completely empty) was $^3/_4$ inch (19.1 mm). The percolation rate would be 5 minutes/0.75 inch (19.1 mm) = 6.7 minutes per inch.

1402.2.1.3 Test procedure, other soils. The hole shall be filled with clear water, and a minimum water depth of 12 inches (305 mm) shall be maintained above the bottom of the hole for a 4-hour period by refilling whenever necessary or by use of an automatic siphon. Water remaining in the hole after 4 hours shall not be removed. Thereafter, the soil shall be allowed to swell not less than 16 hours or more than 30 hours. Immediately after the soil swelling period, the measurements for determining the percolation rate shall be made as follows: any soil sloughed into the hole shall be removed and the water level shall be adjusted to 6 inches (152 mm) above the gravel or coarse sand. Thereupon, from a fixed reference point, the water level shall be measured at 30-minute intervals for a period of 4 hours, unless two successive water level drops do not vary by more than $^1/_{16}$ inch (1.59 mm). At least three water level drops shall be observed and recorded. The hole shall be filled with clear water to a point not more than 6 inches (152 mm) above the gravel or coarse sand whenever it becomes nearly empty. Adjustments of the water level shall not be made during the three measurement periods except to the limits of the last measured water level drop. When the first 6 inches (152 mm) of water seeps away in less than 30 minutes, the time interval between measurements shall be 10 minutes and the test run for 1 hour. The water depth shall not exceed 5 inches (127 mm) at any time during the measurement period. The drop that occurs during the final measurement period shall be used in calculating the percolation rate.

❖ For other than sandy soils, the soil in the test hole has to be prepared by soaking with not less than 12 inches (305 mm) of water for 4 hours. After the soaking period, the water is left in the hole and the soil allowed to swell for not less than 16 hours or more than 30 hours. Immediately after the swelling period, the water in the hole must be adjusted to a 6-inch (152 mm) depth above the bottom of the hole and measurements of the water level drop taken at 30-minute intervals for 4 hours. If two consecutive water level drops do not vary more than $^1/_{16}$ inch (1.6 mm), then the test can be stopped and the percolation rate can be determined from the last water level drop.

If the water in the hole drains out in less than 30 minutes, then water level drop measurements must be taken at 10-minute intervals for a total of 1 hour. After each measurement, the hole must be refilled to at least 5 inches (127 mm) above the bottom of the hole.

1402.2.1.4 Mechanical test equipment. Mechanical percolation test equipment shall be of an *approved* type.

❖ Another way to determine soil percolation rates is by using mechanical percolation test equipment such as a permeameter. Such equipment must be approved by the code official.

1402.2.2 Permeability evaluation. Soil shall be evaluated for estimated percolation based on structure and texture in accordance with accepted soil evaluation practices. Borings shall be made in accordance with Section 1402.2.1.1 for evaluating the soil.

❖ The soil from the bored percolation test hole must be evaluated for the type of soil structure and texture to determine if the soil is suitable for absorbing water. Evidence of soil mottling could indicate that a seasonal high water table exists, even though the test hole soil is dry.

1402.3 Subsurface landscape irrigation site location. The surface grade of all soil absorption systems shall be located at a point lower than the surface grade of any water well or reservoir on the same or adjoining lot. Where this is not possible, the site shall be located so surface water drainage from the site is not directed toward a well or reservoir. The soil absorption system shall be located with a minimum horizontal distance between various elements as indicated in Table 1402.3. Private sewage disposal systems in compacted areas, such as parking lots and driveways, are prohibited. Surface water shall be diverted away from any soil absorption site on the same or neighboring lots.

❖ Because gray water contains bacteria, it is important to make sure that the soil absorption system be located away and preferably downhill from water reservoirs and water wells. Table 1402.3 indicates the minimum distances that tanks and absorption systems must be located from certain features on the lot.

TABLE 1402.3
LOCATION OF SUBSURFACE IRRIGATION SYSTEM

ELEMENT	MINIMUM HORIZONTAL DISTANCE	
	HOLDING TANK (feet)	IRRIGATION DISPOSAL FIELD (feet)
Buildings	5	2
Lot line adjoining private property	5	5
Water wells	50	100
Streams and lakes	50	50
Seepage pits	5	5
Septic tanks	0	5
Water service	5	5
Public water main	10	10

For SI: 1 foot = 304.8 mm.

SECTION 1403
INSTALLATION

1403.1 Installation. Absorption systems shall be installed in accordance with Sections 1403.1.1 through 1403.1.5 to provide landscape irrigation without surfacing of water.

❖ This section directs the reader to the subsections that regulate the installation of absorption systems.

1403.1.1 Absorption area. The total absorption area required shall be computed from the estimated daily gray water discharge and the design-loading rate based on the percolation rate for the site. The required absorption area equals the estimated gray water discharge divided by the design-loading rate from Table 1403.1.1.

❖ The area of the absorption system is determined by dividing the estimated number of gray-water gallons per day by the design loading rate from Table 1403.1.1. The design loading rate is a function of the percolation test results.

**TABLE 1403.1.1
DESIGN LOADING RATE**

PERCOLATION RATE (minutes per inch)	DESIGN LOADING FACTOR (gallons per square foot per day)
0 to less than 10	1.2
10 to less than 30	0.8
30 to less than 45	0.72
45 to 60	0.4

For SI: 1 minute per inch = min/25.4 mm, 1 gallon per square foot = 40.7 L/m^2.

1403.1.2 Seepage trench excavations. Seepage trench excavations shall be not less than 1 foot (304 mm) in width and not greater than 5 feet (1524 mm) in width. Trench excavations shall be spaced not less than 2 feet (610 mm) apart. The soil absorption area of a seepage trench shall be computed by using the bottom of the trench area (width) multiplied by the length of pipe. Individual seepage trenches shall be not greater than 100 feet (30 480 mm) in *developed length*.

❖ One method for making a subsurface irrigation system is to dig a trench 1 to 5 feet (304 to 1524 mm) in width and place a distribution pipe along the center of the width of the trench. Trenches are dug to a depth of where the percolation tests were performed. Where multiple trenches are required, they must be not less than 2 feet (610 mm) apart. Trenches are limited to 100 feet (30 480 mm) in length as this is a practical limit for keeping the entire trench bottom level. The absorption area of the trench is the width of the trench multiplied by the length of the pipe to be placed in the trench.

1403.1.3 Seepage bed excavations. Seepage bed excavations shall be not less than 5 feet (1524 mm) in width and have more than one distribution pipe. The absorption area of a seepage bed shall be computed by using the bottom of the trench area. Distribution piping in a seepage bed shall be uniformly spaced not greater than 5 feet (1524 mm) and not less than 3 feet (914 mm) apart, and greater than 3 feet (914 mm) and not less than 1 foot (305 mm) from the sidewall or headwall.

❖ Another method for making a subsurface irrigation system is to excavate an area that is not less than 5 feet (1524 mm) wide and place multiple distribution pipes, equally spaced from 3 to 5 feet (914 to 1524 mm) apart in the excavated area. The distribution pipes must be not closer than 1 foot (305 mm) and not more than 3 feet (914 mm) from the headwall and sidewalls. The absorption area of the trench is the width of the excavated bed area multiplied by the length of the bed area.

1403.1.4 Excavation and construction. The bottom of a trench or bed excavation shall be level. Seepage trenches or beds shall not be excavated where the soil is so wet that such material rolled between the hands forms a soil wire. All smeared or compacted soil surfaces in the sidewalls or bottom of seepage trench or bed excavations shall be scarified to the depth of smearing or compaction and the loose material removed. Where rain falls on an open excavation, the soil shall be left until sufficiently dry so a soil wire will not form when soil from the excavation bottom is rolled between the hands. The bottom area shall then be scarified and loose material removed.

❖ Care must be taken to not excavate or work in trenches or beds when the soil is wet so that a soil wire can be formed between the hands. Smeared and compacted soil must be scarified (scratched) to break up the smearing and compaction. Loose soil must be removed from the trench or bed area. The bottom of bed and trench areas must be level.

1403.1.5 Aggregate and backfill. Not less than 6 inches in depth of aggregate, ranging in size from $^1/_2$ to $2^1/_2$ inches (12.7 mm to 64 mm), shall be laid into the trench below the distribution piping elevation. The aggregate shall be evenly distributed not less than 2 inches (51 mm) in depth over the top of the distribution pipe. The aggregate shall be covered with *approved* synthetic materials or 9 inches (229 mm) of uncompacted marsh hay or straw. Building paper shall not be used to cover the aggregate. Not less than 9 inches (229 mm) of soil backfill shall be provided above the covering.

❖ The bottom of the bed or trench area must have aggregate placed to a depth of 6 inches (152 mm). Then the distribution pipe is laid on top of the aggregate and covered with 2 inches (51 mm) of aggregate. The installer has the choice of covering the aggregate with synthetic materials or 9 inches (229 mm) of marsh hay or straw. The trench or bed is then covered by not less than 9 inches (229 mm) of soil backfill.

1403.2 Distribution piping. Distribution piping shall be not less than 3 inches (76 mm) in diameter. Materials shall comply with Table 1403.2. The top of the distribution pipe shall be not less than 8 inches (203 mm) below the original surface. The slope of the distribution pipes shall be not less than 2 inches (51 mm) and not greater than 4 inches (102 mm) per 100 feet (30 480 mm).

❖ Distribution piping must be at least 3 inches (76 mm) in diameter and be of one of the materials listed in Table 1403.2. The elevation of the top of the pipe in the trench or bed must be not less than 8 inches (203 mm) below the original grade elevation. The distribution pipe must be sloped at least 2 inches (51 mm) per 100 feet (30 480 mm) but not more than 4 inches (102 mm) per 100 feet (30 480 mm).

**TABLE 1403.2
DISTRIBUTION PIPE**

MATERIAL	STANDARD
Polyethylene (PE) plastic pipe	ASTM F405
Polyvinyl chloride (PVC) plastic pipe	ASTM D729
Polyvinyl chloride (PVC) plastic pipe with a 3.5-inch O.D. and solid cellular core or composite wall	ASTM F1488

For SI: 1 inch = 25.4 mm.

1403.2.1 Joints. Joints in distribution pipe shall be made in accordance with Section 705 of this code.

❖ Distribution piping must be joined by one of the methods in Section 705, as applicable to the type of pipe being used.

Bibliography

The following resource materials are referenced in this chapter or are relevant to the subject matter addressed in this chapter.

Art Ludwig. *Builder's Greywater Guide*. Oasis Design, 1995-2006.

Graywater Guide. Department of Water Resources, State of California, January 1995.

Chapter 15: Referenced Standards

General Comments

This chapter lists the standards that are referenced in various sections of this document. The standards are listed herein by the promulgating agency of the standard, the standard identification, the date and title, and the section or sections of this document that reference the standard. The application of the referenced standards is as specified in Section 102.8.

It is important to understand that not every document related to building design and construction is qualified to be a "referenced standard." The International Code Council® (ICC®) has adopted a criterion that standards referenced in the *International Codes®* and standards intended for adoption into the *International Codes* must meet to qualify as a referenced standard. The policy is summarized as follows:

- Code references: The scope and application of the standard must be clearly identified in the code text.
- Standard content: The standard must be written in mandatory language and be appropriate for the subject covered. The standard shall not have the effect of requiring proprietary materials or prescribing a proprietary testing agency.
- Standard promulgation: The standard must be readily available and developed and maintained in a consensus process such as ASTM or ANSI.

The ICC Code Development Procedures, of which the standards policy is a part, are updated periodically. A copy of the latest version can be obtained from the ICC offices.

Once a standard is incorporated into the code through the code development process, it becomes an enforceable part of the code. When the code is adopted by a jurisdiction, the standard also is part of that jurisdiction's adopted code. It is for this reason that the criteria were developed. Compliance with this policy means that documents incorporated into the code are developed through the use of a consensus process, written in mandatory language, and do not mandate the use of proprietary materials or agencies. The requirement for a standard to be developed through a consensus process is vital because it means that the standard will be representative of the most current body of available knowledge on the subject as determined by a broad spectrum of interested or affected parties without dominance by any single interest group. A true consensus process has many attributes, including but not limited to:

- An open process that has formal (published) procedures that allow for the consideration of all viewpoints,
- A definitive review period that allows for the standard to be updated or revised,
- A process of notification to all interested parties and
- An appeals process.

Many available documents related to design, installation and construction, though useful, are not "standards" and are not appropriate for reference in the code. Often, these documents are developed or written with the intention of being used for regulatory purposes and are unsuitable for use as a regulation because of extensive use of recommendations, advisory comments and non-mandatory terms. Typical examples of such documents include installation instructions, guidelines and practices.

The objective of ICC's standards policy is to provide regulations that are clear, concise and enforceable—thus the requirement for standards to be written in mandatory language. This requirement is not intended to mean that a standard cannot contain informational or explanatory material that will aid the user of the standard in its application. When it is the desire of the standard's promulgating agency for such material to be included, however, the information must appear in a nonmandatory location, such as an annex or appendix, and be clearly identified as not being part of the standard.

Overall, standards referenced by the code must be authoritative, relevant, up to date and, most important, reasonable and enforceable. Standards that comply with the ICC standards policy fulfill these expectations.

Purpose

As a performance-oriented code, the code contains numerous references to documents that are used to regulate materials and methods of construction. The references to these documents within the code text consist of the promulgating agency's acronym and its publication designation (e.g., ASME A112.1.3) and a further indication that the document being referenced is the one that is listed in Chapter 15. Chapter 15 contains all of the information that is necessary to identify the specific referenced document. Included is the following information on a document's promulgating agency (see Figure 15):

- The promulgating agency (i.e., the agency's title),
- The promulgating agency's acronym and
- The promulgating agency's address. For example, a reference to an ASME standard within the code indicates that the document is promulgated by the American Society of Mechanical Engineers

- (ASME), which is located in New York City. Chapter 15 lists the standards agencies alphabetically for ease of identification.

Chapter 15 also includes the following information on the referenced document itself (see Figure 15):
- The document's publication designation,
- The document's edition year,
- The document's title,
- Any addenda or revisions to the document that are applicable and
- Every section of the code in which the document is referenced.

For example, a reference to ASME A112.1.3 indicates that this document can be found in Chapter 15 under the heading ASME. The specific standards designation is A112.1.3. For convenience, these designations are listed in alphanumeric order. Chapter 15 identifies that ASME A112.1.3 is titled *Air Gap Fittings for Use with Plumbing Fixtures, Appliances and Appurtenances*; the applicable edition (i.e., its year of publication) is 2000; and it is referenced in two sections of the code.

The key aspect of the manner in which standards are referenced by the code is that a specific edition of a specific standard is clearly identified. In this manner, the requirements necessary for compliance can be readily determined. The basis for code compliance is, therefore, established and available on an equal basis to the code official, the contractor, the designer and the owner.

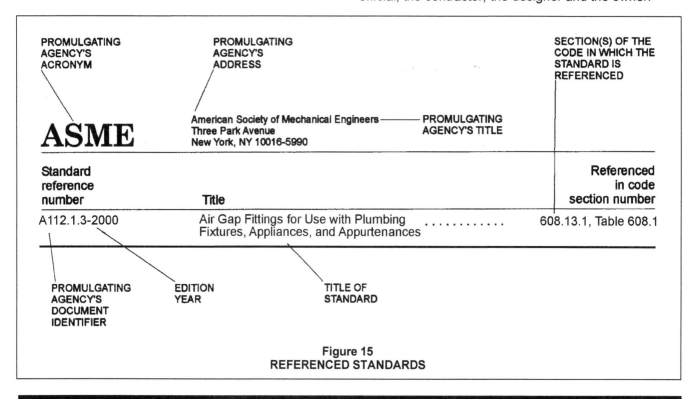

Figure 15
REFERENCED STANDARDS

This chapter lists the standards that are referenced in various sections of this document. The standards are listed herein by the promulgating agency of the standard, the standard identification, the effective date and title, and the section or sections of this document that reference the standard. The application of the referenced standards shall be as specified in Section 102.8.

AHRI

Air-Conditioning, Heating, & Refrigeration Institute
4100 North Fairfax Drive, Suite 200
Arlington, VA 22203

Standard reference number	Title	Referenced in code section number
1010—02	Self-contained, Mechanically Refrigerated Drinking-Water Coolers	410.1

REFERENCED STANDARDS

ANSI

American National Standards Institute
25 West 43rd Street, Fourth Floor
New York, NY 10036

Standard reference number	Title	Referenced in code section number
A118.10—99	Specifications for Load Bearing, Bonded, Waterproof Membranes for Thin Set Ceramic Tile and Dimension Stone Installation	417.5.2.5, 417.5.2.6
Z4.3—95	Minimum Requirements for Nonsewered Waste-disposal Systems	311.1
Z21.22—99 (R2003)	Relief Valves for Hot Water Supply Systems with Addenda Z21.22a—2000 (R2003) and Z21.22b—2001 (R2003)	504.2, 504.4, 504.4.1
CSA B45.5—11/ IAPMO Z124—11	Plastic Plumbing Fixtures	407.1, 415.1, 416.1, 416.2, 417.1, 418.1, 419.1, 420.1

API

American Petroleum Institute
1220 L Street NW
Washington, DC 20005-4070

Standard reference number	Title	Referenced in code section number
12D—2008	Specification for Field Welded Tanks for Storage of Production Liquids, effective April 1, 2009	1302.7.2
12F—2008	Specification for Shop Welded Tanks for Storage of Production Liquids, effective April 1 2009	1302.7.2

ASME

American Society of Mechanical Engineers
Three Park Avenue
New York, NY 10016-5990

Standard reference number	Title	Referenced in code section number
A112.1.2—2004	Air Gaps in Plumbing Systems	406.1, 409.2, Table 608.1, 608.13.1
A112.1.3—2000 (R2011)	Air Gap Fittings for Use with Plumbing Fixtures, Appliances and Appurtenances	406.1, 409.2, Table 608.1, 608.13.1, 1102.6
A112.3.1—2007	Stainless Steel Drainage Systems for Sanitary, DWV, Storm and Vacuum Applications Above and Below Ground	412.1, Table 702.1, Table 702.2, Table 702.3, Table 702.4, Table 1102.4, Table 1102.5, 1102.6, Table 1102.7
ASME A112.3.4—2013/ CSA B45.9—2013	Macerating Toilet Systems and Related Components	712.4.1
A112.4.1—2009	Water Heater Relief Valve Drain Tubes	504.6
A112.4.2—2009	Water Closet Personal Hygiene Devices	424.9
A112.4.3—1999 (R2010)	Plastic Fittings for Connecting Water Closets to the Sanitary Drainage System	405.4
A112.4.14—2004 (R2010)	Manually Operated, Quarter-Turn Shutoff Valves for Use in Plumbing Systems	Table 605.7
A112.6.1M—1997 (R2008)	Floor-affixed Supports for Off-the-floor Plumbing Fixtures for Public Use	405.4.3
A112.6.2—2000 (R2010)	Framing-affixed Supports for Off-the-floor Water Closets with Concealed Tanks	405.4.3
A112.6.3—2001 (R2007)	Floor and Trench Drains	412.1
A112.6.4—2003 (R2008)	Roof, Deck, and Balcony Drains	1102.6
A112.6.7—2010	Enameled and Epoxy-coated Cast-iron and PVC Plastic Sanitary Floor Sinks	427.1
A112.6.9—2005 (R2010)	Siphonic Roof Drains	1107.1
A112.14.1—2003	Backwater Valves	715.2
A112.14.3—2000	Grease Interceptors	1003.3.4
A112.14.4—2001 (Reaffirmed 2007)	Grease Removal Devices	1003.3.4
A112.14-14.6—2010	FOG (Fats, Oils and Greases) Disposal Systems	1003.3.4, 1003.3.6
A112.18.1—2012/ CSA B125.1—2012	Plumbing Supply Fittings	424.1, 424.2, 424.3, 424.4, 424.6, 424.8, 605.7, 607.4, 608.2
A112.18.2—2011/ CSA B125.2—2011	Plumbing Waste Fittings	424.1.2
A112.18.3—2002 (Reaffirmed 2008)	Performance Requirements for Backflow Protection Devices and Systems in Plumbing Fixture Fittings	424.2, 424.6

REFERENCED STANDARDS

ASME—continued

Standard	Title	Section
A112.18.6/ CSA B125.6—2009	Flexible Water Connectors	605.6
A112.18.7—1999 (Reaffirmed 2004)	Deck mounted Bath/Shower Transfer Valves with Integral Backflow Protection	424.8
A112.18.9—2011	Protectors/Insulators for Exposed Waste and Supplies on Accessible Fixtures	404.3
A112.19.1/—2013 CSA B45.2—2013	Enameled Cast Iron and Enameled Steel Plumbing Fixtures	407.1, 410.1, 415.1, 416.1, 418.1
A112.19.2—2013/ CSA B45.1—13	Ceramic Plumbing Fixtures	401.2, 405.9, 407.1, 408.1, 410.1, 415.1, 416.1, 417.1, 418.1, 419.1, 420.1
A112.19.3—2008/ CSA B45.4—08(R2013)	Stainless Steel Plumbing Fixtures	405.9, 407.1, 415.1, 416.1, 418.1, 420.1
A112.19.5/—2011 CSA B45.15—2011	Flush Valves and Spuds for Water-closets, Urinals, and Tanks	425.4
A112.19.7M—2012/ CSA B45.10—2012	Hydromassage Bathtub Systems	421.1, 421.4
A112.19.12—2006	Wall Mounted and Pedestal Mounted, Adjustable, Elevating, Tilting and Pivoting Lavatory, Sink and Shampoo Bowl Carrier Systems and Drain Systems	416.4, 418.3
A112.19.14-2006(R2011)	Six-Liter Water Closets Equipped with a Dual Flushing Device	420.1
A112.19.15—2005	Bathtub/Whirlpool Bathtubs with Pressure Sealed Doors	407.4, 421.6
A112.19.19—2006	Vitreous China Nonwater Urinals	419.1, 421.6
A112.21.3-1985(R2007)	Hydrants for Utility and Maintenance Use	Table 608.1, 608.13.6
A112.36.2M—1991 (R2008)	Cleanouts	708.1.10.2
ASSE 1016/ ASME A112.1016/ CSA B125.16-2011	Performance Requirements for Individual Thermostatic, Pressure Balancing and Combination Control Valves for Individual Fixture Fittings	424.3, 424.4, 607.4
B1.20.1—1983(R2006)	Pipe Threads, General Purpose (inch)	605.10.3, 605.11.3, 605.13.4, 605.15.3, 605.18.1, 605.22.4, 705.2.3, 705.3.3, 705.6.4, 705.19.1, 705.11.3
B 16.3—2011	Malleable Iron Threaded Fittings Classes 150 and 300	Table 702.4, Table 1102.7
B 16.4—2011	Gray Iron Threaded Fittings Classes 125 and 250	Table 605.5, Table 702.4, Table 1102.7
B 16.9—2007	Factory-made Wrought Steel Buttwelding Fittings	Table 605.5, Table 702.4, Table 1102.7
B 16.11—2011	Forged Fittings, Socket-welding and Threaded	Table 605.5, Table 702.4, Table 1102.7
B 16.12—2009	Cast-iron Threaded Drainage Fittings	Table 702.4, Table 1102.7
B 16.15—2011	Cast Bronze Threaded Fittings	Table 605.5, Table 702.4, Table 1102.7
B 16.18—2012	Cast Copper Alloy Solder Joint Pressure Fittings	Table 605.5, Table 702.4, Table 1102.7
B 16.22—2001(R2010)	Wrought Copper and Copper Alloy Solder Joint Pressure Fittings	Table 605.5, Table 702.4, Table 1102.7
B 16.23—2011	Cast Copper Alloy Solder Joint Drainage Fittings DWV	Table 702.4, Table 1102.7
B 16.26—2011	Cast Copper Alloy Fittings for Flared Copper Tubes	Table 605.5, Table 702.4, Table 1102.7
B 16.28—1994	Wrought Steel Buttwelding Short Radius Elbows and Returns	Table 605.5, Table 702.4, Table 1102.7
B 16.29—2012	Wrought Copper and Wrought Copper Alloy Solder Joint Drainage Fittings (DWV)	Table 702.4, Table 1102.7
B16.34—2009	Valves Flanged, Threaded and Welding End	Table 605.7
B16.51—2011	Copper and Copper Alloy Press-Connect Pressure Fittings	Table 605.5

ASPE

American Society of Plumbing Engineers
8614 Catalpa Avenue, Suite 1007
Chicago, IL 60656-1116

Standard reference number	Title	Referenced in code section number
45—2013	Siphonic Roof Drainage Systems	1107.1

REFERENCED STANDARDS

ASSE

American Society of Sanitary Engineering
901 Canterbury Road, Suite A
Westlake, OH 44145

Standard reference number	Title	Referenced in code section number
1001—08	Performance Requirements for Atmospheric Type Vacuum Breakers	425.2, Table 608.1, 608.13.6, 608.16.4.1
1002—08	Performance Requirements for Antisiphon Fill Valves (Ballcocks) for Gravity Water Closet Flush Tanks	425.3.1, Table 608.1
1003—09	Performance Requirements for Water Pressure Reducing Valves	604.8
1004—08	Performance Requirements for Backflow Prevention Requirements for Commercial Dishwashing Machines	409.1
1005—99	Performance Requirements for Water Heater Drain Valves	501.3
1008—06	Performance Requirements for Plumbing Aspects of Food Waste Disposer Units	413.1
1010—04	Performance Requirements for Water Hammer Arresters	604.9
1011—04	Performance Requirements for Hose Connection Vacuum Breakers	Table 608.1, 608.13.6
1012—09	Performance Requirements for Backflow Preventers with Intermediate Atmospheric Vent	Table 608.1, 608.13.3, 608.16.2
1013—09	Performance Requirements for Reduced Pressure Principle Backflow Preventers and Reduced Pressure Principle Fire Protection Backflow Preventers	Table 608.1, 608.13.2, 608.16.2
1015—09	Performance Requirements for Double Check Backflow Prevention Assemblies and Double Check Fire Protection Backflow Prevention Assemblies	Table 608.1, 608.13.7
ASSE 1016/ ASME A112.1016/ CSA B125.16—2011	Performance Requirements for Individual Thermostatic, Pressure Balancing and Combination Control Valves for Individual Fixture Fittings	424.3, 424.4, 607.4
1017—2010	Performance Requirements for Temperature Actuated Mixing Valves for Hot Water Distribution Systems	501.2, 613.1
1018—2010	Performance Requirements for Trap Seal Primer Valves; Potable Water Supplied	1002.4.1.2
1019—2010	Performance Requirements for Vacuum Breaker Wall Hydrants, Freeze Resistant, Automatic Draining Type	Table 608.1, 608.13.6, 608.13.10
1020—04	Performance Requirements for Pressure Vacuum Breaker Assembly	Table 608.1, 608.13.5
1022—03	Performance Requirements for Backflow Preventer for Beverage Dispensing Equipment	Table 608.1, 608.16.1, 608.16.10
1024—04	Performance Requirements for Dual Check Valve Type Backflow Preventers (for Residential Supply Service or Individual Outlets)	605.3.1, Table 608.1, 608.13.10
1035—08	Performance Requirements for Laboratory Faucet Backflow Preventers	Table 608.1, 608.13.6
1037—90	Performance Requirements for Pressurized Flushing Devices for Plumbing Fixtures	425.2
1044—2010	Performance Requirements for Trap Seal Primer Devices Drainage Types and Electronic Design Types	1002.4.1.3
1047—2009	Performance Requirements for Reduced Pressure Detector Fire Protection Backflow Prevention Assemblies	Table 608.1, 608.13.2
1048—2009	Performance Requirements for Double Check Detector Fire Protection Backflow Prevention Assemblies	Table 608.1, 608.13.7
1049—2009	Performance Requirements for Individual and Branch Type Air Admittance Valves for Chemical Waste Systems	901.3, 918.8
1050—2009	Performance Requirements for Stack Air Admittance Valves for Sanitary Drainage Systems	918.1
1051—2009	Performance Requirements for Individual and Branch Type Air Admittance Valves for Sanitary Drainage Systems-fixture and Branch Devices	918.1
1052—04	Performance Requirements for Hose Connection Backflow Preventers	Table 608.1, 608.13.6
1055—2009	Performance Requirements for Chemical Dispensing Systems	608.13.9
1056—2010	Performance Requirements for Spill Resistant Vacuum Breaker	Table 608.1, 608.13.5, 608.13.8
1060—2006	Performance Requirements for Outdoor Enclosures for Fluid Conveying Components	608.14.1
1061—2010	Performance Requirements for Removable and Nonremovable Push Fit Fittings	Table 605.5
1062—2006	Performance Requirements for Temperature Actuated, Flow Reduction Valves to Individual Supply Fittings	424.7
1066—2009	Performance Requirements for Individual Pressure Balancing In-line Valves for Individual Fixture Fittings	604.11
1069—05	Performance Requirements for Automatic Temperature Control Mixing Valves	424.4
1070—04	Performance Requirements for Water-temperature Limiting Devices	408.3, 416.5, 423.3, 424.5, 607.1.2
1072—07	Performance Requirements for Barrier Type Floor Drain Tap Seal Protection Devices	1002.4.14
1079—2005	Performance Requirements for Dielectric Pipe Unions	605.24.1, 605.24.3
5013—2009	Performance Requirements for Testing Reduced Pressure Principle Backflow Prevention Assembly (RPA) and Reduced Pressure Fire Protection Backflow Preventers (RFP)	312.10.2

REFERENCED STANDARDS

ASSE—continued

5015—2009	Performance Requirements for Testing Double Check Valve Backflow Prevention Assemblies (DC) and Double Check Fire Protection Backflow Prevention Assemblies (DCF)	312.10.2
5020—2009	Performance Requirements for Testing Pressure Vacuum Breaker Assemblies (PVBA)	312.10.2
5047—98	Performance Requirements for Testing Reduced Pressure Detector Fire Protection Backflow Prevention Assemblies (RPDA)	312.10.2
5048—2009	Performance Requirements for Testing Double Check Valve Detector Assembly (DCDA)	312.10.2
5052—98	Performance Requirements for Testing Hose Connection Backflow Preventers	312.10.2
5056—98	Performance Requirements for Testing Spill Resistant Vacuum Breaker (SRVB)	312.10.2

ASTM

ASTM International
100 Barr Harbor Drive
West Conshohocken, PA 19428-2959

Standard reference number	Title	Referenced in code section number
A53/A 53M—12	Specification for Pipe, Steel, Black and Hot-dipped, Zinc-coated Welded and Seamless	Table 605.3, Table 605.4, Table 702.1
A74—13a	Specification for Cast-iron Soil Pipe and Fittings	Table 702.1, Table 702.2, Table 702.3, Table 702.4, 708.2, 708.7, Table 1102.4, Table 1102.5, Table 1102.7
A312/A 312M—13a	Specification for Seamless, Welded, And Heavily Cold Worked Austenitic Stainless Steel Pipes	Table 605.3, Table 605.4, Table 605.5, 605.23.2
A733—2003(2009)e1	Specification for Welded and Seamless Carbon Steel and Austenitic Stainless Steel Pipe Nipples	Table 605.8
A778—01(2009)e1	Specification for Welded Unannealed Austenitic Stainless Steel Tubular Products	Table 605.3, Table 605.4, Table 605.5
A888—13a	Specification for Hubless Cast-iron Soil Pipe and Fittings for Sanitary and Storm Drain, Waste, and Vent Piping Application	Table 702.1, Table 702.2, Table 702.3, Table 702.4, 708.7, Table 1102.4, Table 1102.5, Table 1102.7
B32—08	Specification for Solder Metal	605.13.3, 605.14.6, 705.6.3, 705.17.3
B42—10	Specification for Seamless Copper Pipe, Standard Sizes	Table 605.3, Table 605.4, Table 702.1
B43—09	Specification for Seamless Red Brass Pipe, Standard Sizes	Table 605.3, Table 605.4, Table 702.1
B75—11	Specification for Seamless Copper Tube	Table 605.3, Table 605.4, Table 702.1, Table 702.2, Table 702.3, Table 1102.4
B88—09	Specification for Seamless Copper Water Tube	Table 605.3, Table 605.4, Table 702.1, Table 702.2, Table 702.3, Table 1102.4
B152/B 152M—13	Specification for Copper Sheet, Strip Plate and Rolled Bar	402.3, 417.5.2.4, 425.3.3, 902.2
B251—10	Specification for General Requirements for Wrought Seamless Copper and Copper-alloy Tube	Table 605.3, Table 605.4, Table 702.1, Table 702.2, Table 702.3, Table 1102.4
B302—12	Specification for Threadless Copper Pipe, Standard Sizes	Table 605.3, Table 605.4, Table 702.1
B306—09	Specification for Copper Drainage Tube (DWV)	Table 702.1, Table 702.2, Table 1102.4
B447—12a	Specification for Welded Copper Tube	Table 605.3, Table 605.4
B687—99(2011)	Specification for Brass, Copper and Chromium-plated Pipe Nipples	Table 605.8
B813—10	Specification for Liquid and Paste Fluxes for Soldering of Copper and Copper Alloy Tube	605.13.3, 605.14.6, 705.6.3, 705.7.3
B828—02(2010)	Practice for Making Capillary Joints by Soldering of Copper and Copper Alloy Tube and Fittings	605.13.3, 605.14.6, 705.6.3, 705.17.3
C4—04(2009)	Specification for Clay Drain Tile and Perforated Clay Drain Tile	Table 702.3, Table 1102.4, Table 1102.5
C14—11	Specification for Nonreinforced Concrete Sewer, Storm Drain and Culvert Pipe	Table 702.3, Table 1102.4
C76—13a	Specification for Reinforced Concrete Culvert, Storm Drain and Sewer Pipe	Table 702.3, Table 1102.4
C425—04(2009)	Specification for Compression Joints for Vitrified Clay Pipe and Fittings	705.12, 705.16
C443—12	Specification for Joints for Concrete Pipe and Manholes, Using Rubber Gaskets	705.5, 705.16
C564—12	Specification for Rubber Gaskets for Cast-iron Soil Pipe and Fittings	705.4.2, 705.4.3, 705.16
C700—13	Specification for Vitrified Clay Pipe, Extra Strength, Standard Strength, and Perforated	Table 702.3, Table 702.4, Table 1102.4, Table 1102.5
C1053—00(2010)	Specification for Borosilicate Glass Pipe and Fittings for Drain, Waste, and Vent (DWV) Applications	Table 702.1, Table 702.4
C1173—10e1	Specification for Flexible Transition Couplings for Underground Piping System	705.2.1, 705.5, 705.11.1, 705.12, 705.13.2, 705.16

ASTM—continued

C1277—12	Specification for Shielded Coupling Joining Hubless Cast-iron Soil Pipe and Fittings	705.4.3
C1440—08	Specification for Thermoplastic Elastomeric (TPE) Gasket Materials for Drain, Waste, and Vent (DWV), Sewer, Sanitary and Storm Plumbing Systems	705.16
C1460—08	Specification for Shielded Transition Couplings for Use with Dissimilar DWV Pipe and Fittings Above Ground	705.16
C1461—08	Specification for Mechanical Couplings Using Thermoplastic Elastomeric (TPE) Gaskets for Joining Drain, Waste and Vent (DWV) Sewer, Sanitary and Storm Plumbing Systems for Above and Below Ground Use	705.19
C1540—11	Specification for Heavy Duty Shielded Couplings Joining Hubless Cast-iron Soil Pipe and Fittings	705.4.3
C1563—13	Standard Test Method for Gaskets for Use in Connection with Hub and Spigot Cast Iron Soil Pipe and Fittings for Sanitary Drain, Waste, Vent and Storm Piping Applications	705.4.2
D1253—08	Standard Test Method For Residual Chlorine in Water	1301.2.1, 1303.9
D1527—99(2005)	Specification for Acrylonitrile-Butadiene-Styrene (ABS) Plastic Pipe, Schedules 40 and 80	Table 605.3
D1785—12	Specification for Poly (Vinyl Chloride) (PVC) Plastic Pipe, Schedules 40, 80 and 120	Table 605.3
D2235—04(2011)	Specification for Solvent Cement for Acrylonitrile-Butadiene-Styrene (ABS) Plastic Pipe and Fittings	605.10.2, 705.2.2, 705.2.2, 705.7.2
D2239—12a	Specification for Polyethylene (PE) Plastic Pipe (SIDR-PR) Based on Controlled Inside Diameter	Table 605.3
D2241—09	Specification for Poly (Vinyl Chloride) (PVC) Pressure-rated Pipe (SDR-Series)	Table 605.3
D2282—99(2005)	Specification for Acrylonitrile-Butadiene-Styrene (ABS) Plastic Pipe (SDR-PR)	Table 605.3
D2464—06	Specification for Threaded Poly (Vinyl Chloride) (PVC) Plastic Pipe Fittings, Schedule 80	Table 605.5
D2466—06	Specification for Poly (Vinyl Chloride) (PVC) Plastic Pipe Fittings, Schedule 40	Table 605.5
D2467—06	Specification for Poly (Vinyl Chloride) (PVC) Plastic Pipe Fittings, Schedule 80	Table 605.5
D2468—96a	Specification for Acrylonitrile-Butadiene-Styrene (ABS) Plastic Pipe Fittings, Schedule 40	Table 605.5
D2564—12	Specification for Solvent Cements for Poly (Vinyl Chloride) (PVC) Plastic Piping Systems	605.22.3, 705.11.2, 705.14.2
D2609—02(2008)	Specification for Plastic Insert Fittings for Polyethylene (PE) Plastic Pipe	Table 605.5
D2657—07	Practice for Heat Fusion-joining of Polyolefin Pipe and Fitting Waste, and Vent Pipe and Fittings	605.19.2, 705.13.1
D2661—11	Specification for Acrylonitrile-Butadiene-Styrene (ABS) Schedule 40 Plastic Drain, Waste, and Vent Pipe and Fittings	Table 702.1, Table 702.2, Table 702.3, Table 702.4, 705.2.2, 705.7.2, Table 1102.4, Table 1102.7
D2665—12	Specification for Poly (Vinyl Chloride) (PVC) Plastic Drain, Waste, and Vent Pipe and Fittings	Table 702.1, Table 702.2, Table 702.3, Table 702.4, Table 1102.4, Table 1102.7
D2672—96a(2009)	Specification for Joints for IPS PVC Pipe Using Solvent Cement	Table 605.3
D2683—10e1	Standard Specification for Socket-type Polyethylene fittings for Outside Diameter-controlled Polyethylene Pipe and Tubing	Table 605.5, 717.5
D2729—11	Specification for Poly (Vinyl Chloride) (PVC) Sewer Pipe and Fittings	Table 1102.5, Table 1403.2
D2737—2012a	Specification for Polyethylene (PE) Plastic Tubing	Table 605.3
D2751—05	Specification for Acrylonitrile-Butadiene-Styrene (ABS) Sewer Pipe and Fittings	Table 702.3, Table 702.4, Table 1102.4, Table 1102.7
D2846/D 2846M—09be1	Specification for Chlorinated Poly (Vinyl Chloride) (CPVC) Plastic Hot and Cold Water Distribution Systems	Table 605.3, Table 605.4, Table 605.5, 605.15.2, 605.16.2
D2855—96(2010)	Standard Practice for Making Solvent-cemented Joints with Poly (Vinyl Chloride) (PVC) Pipe and Fittings	605.22.3, 705.11.2
D2949—10	Specification for 3.25-in Outside Diameter Poly (Vinyl Chloride) (PVC) Plastic Drain, Waste, and Vent Pipe and Fittings	Table 702.1, Table 702.2, Table 702.3, Table 702.4, Table 1102.7
D3034—08	Specification for Type PSM Poly (Vinyl Chloride) (PVC) Sewer Pipe and Fittings	Table 702.3, Table 702.4, Table 1102.4, Table 1102.5, Table 1102.7
D3035—2012e1	Standard Specification for Polyethylene (PE) Plastic Pipe (DR-PR) Based on Controlled Outside Diameter	Table 605.3
D3139—98(2011)	Specification for Joints for Plastic Pressure Pipes Using Flexible Elastomeric Seals	605.10.1, 605.22.1
D3212—07	Specification for Joints for Drain and Sewer Plastic Pipes Using Flexible Elastomeric Seals	705.2.1, 705.11.1, 705.13.2

REFERENCED STANDARDS

ASTM—continued

D3261—12	Specification for Butt Heat Fusion Polyethylene (PE) Plastic fittings for Polyethylene (PE) Plastic Pipe and Tubing.	Table 605.5
D3311—11	Specification for Drain, Waste and Vent (DWV) Plastic Fittings Patterns	Table 1102.7
D4068—09	Specification for Chlorinated Polyethlene (CPE) Sheeting for Concealed Water-containment Membrane	417.5.2.2
D4551—12	Specification for Poly (Vinyl Chloride) (PVC) Plastic Flexible Concealed Water-containment Membrane	417.5.2.1
E2635—08	Standard Practice for Water Conservation in Buildings Through In-Situ Water Reclamation	1304.3
E2727—10	Standard Practice for the Assessment of Rainwater Quality	1302.8.1
F405—05	Specification for Corrugated Polyethylene (PE) Pipe and Fittings.	Table 1102.5, Table 1403.2
F409—12	Specification for Thermoplastic Accessible and Replaceable Plastic Tube and Tubular Fittings	424.1.2, Table 1102.7
F437—09	Specification for Threaded Chlorinated Poly (Vinyl Chloride) (CPVC) Plastic Pipe Fittings, Schedule 80.	Table 605.5
F438—09	Specification for Socket-type Chlorinated Poly (Vinyl Chloride) (CPVC) Plastic Pipe Fittings, Schedule 40.	Table 605.5
F439—12	Standard Specification for Chlorinated Poly (Vinyl Chloride) (CPVC) Plastic Pipe Fittings, Schedule 80.	Table 605.5
F441/F 441M—13	Specification for Chlorinated Poly (Vinyl Chloride) (CPVC) Plastic Pipe, Schedules 40 and 80	Table 605.3, Table 605.4
F442/F 442M—13	Specification for Chlorinated Poly (Vinyl Chloride) (CPVC) Plastic Pipe (SDR-PR)	Table 605.3, Table 605.4
F477—10	Specification for Elastomeric Seals (Gaskets) for Joining Plastic Pipe	605.24, 705.16
F493—10	Specification for Solvent Cements for Chlorinated Poly (Vinyl Chloride) (CPVC) Plastic Pipe and Fittings	605.15.2, 605.16.2
F628—08	Specification for Acrylonitrile-Butadiene-Styrene (ABS) Schedule 40 Plastic Drain, Waste, and Vent Pipe with a Cellular Core	Table 702.1, Table 702.2, Table 702.3, Table 702.4, 705.2.2, Table 1102.4, Table 1102.7
F656—10	Specification for Primers for Use in Solvent Cement Joints of Poly (Vinyl Chloride) (PVC) Plastic Pipe and Fittings	605.22.3, 705.11.2, 705.14.2
F714—2013	Specification for Polyethylene (PE) Plastic Pipe (SDR-PR) Based on Outside Diameter	Table 702.3, 717.4
F876—2013	Specification for Cross-linked Polyethylene (PEX) Tubing.	Table 605.3, Table 605.4
F877—2011a	Specification for Cross-linked Polyethylene (PEX) Plastic Hot and Cold Water Distribution Systems	Table 605.3, Table 605.4, Table 605.5
F891—10	Specification for Coextruded Poly (Vinyl Chloride) (PVC) Plastic Pipe with a Cellular Core	Table 702.1 Table 702.2, Table 702.3, Table 1102.4, Table 1102.5, Table 1102.7
F1055—13	Standard Specification for Electrofusion Type Polyethylene Fittings for Outside Diameter Controlled Polyethylene and Crosslinked Polyethylene Pipe and Tubing	Table 605.5
F1281—11	Specification for Cross-linked Polyethylene/Aluminum/ Cross-linked Polyethylene (PEX-AL-PEX) Pressure Pipe	Table 605.3, Table 605.4, Table 605.5, 605.21.1
F1282—10	Specification for Polyethylene/Aluminum/Polyethylene (PE-AL-PE) Composite Pressure Pipe.	Table 605.3, Table 605.4, Table 605.5, 605.21.1
F1412—09	Specification for Polyolefin Pipe and Fittings for Corrosive Waste Drainage.	Table 702.1, Table 702.2, Table 702.4, 705.14.1, 901.3
F1476—07	Standard Specification for Performance of Gasketed Mechanical Couplings for Use in Piping Applications	Table 605.5, 605.14.3, 605.18.3, 605.22.2, 605.23.3
F1488—09e1	Specification for Coextruded Composite Pipe	Table 702.1, Table 702.2, Table 702.3, Table 1403.2
F1548—01(2006)	Standard Specification for the Performance of Fittings for Use with Gasketed Mechanical Couplings Used in Piping Applications	Table 605.5
F1673—10	Polyvinylidene Fluoride (PVDF) Corrosive Waste Drainage Systems	Table 702.1, Table 702.2, Table 702.3, Table 702.4, 705.15.1
F1807—13	Specification for Metal Insert Fittings Utilizing a Copper Crimp Ring for SDR9 Cross-linked Polyethylene (PEX) Tubing and SDR9 Polyethylene of Raised Temperature (PE-RT) Tubing.	Table 605.5
F1866—07	Specification for Poly (Vinyl Chloride) (PVC) Plastic Schedule 40 Drainage and DWV Fabricated Fittings	Table 702.4, Table 1102.7
F1960—12	Specification for Cold Expansion Fittings with PEX Reinforcing Rings for use with Cross-linked Polyethylene (PEX) Tubing.	Table 605.5
F1970—12	Special Engineered Fittings, Appurtenances or Valves for use in Poly (Vinyl Chloride) (PVC) OR Chlorinated Poly (Vinyl Chloride) (CPVC) Systems.	Table 605.7
F1974—09	Specification for Metal Insert Fittings for Polyethylene/Aluminum/Polyethylene and Cross-linked Polyethylene/Aluminum/Cross-linked Polyethylene Composite Pressure Pipe.	Table 605.5, 605.21.1
F1986—01(2011)	Specification for Multilayer Pipe, Type 2, Compression Fittings and Compression Joints for Hot and Cold Drinking Water Systems	Table 605.3, Table 605.4, Table 605.5

REFERENCED STANDARDS

ASTM—continued

Standard	Title	Referenced in code section number
F2080—12	Specifications for Cold-expansion Fittings with Metal Compression-sleeves for Cross-linked Polyethylene (PEX) Pipe	Table 605.5
F2098—08	Standard specification for Stainless Steel Clamps for Securing SDR9 Cross-linked Polyethylene (PEX) Tubing to Metal and Plastic Insert Fittings	Table 605.5
F2159—11	Specification for Plastic Insert Fittings Utilizing a Copper Crimp Ring for SDR9 Cross-linked Polyethylene (PEX) Tubing and SDR9 Polyethylene of Raised Temperature (PE-RT) Tubing	Table 605.5
F2262—09	Specification for Cross-linked Polyethylene/Aluminum/ Cross-linked Polyethylene Tubing OD Controlled SDR9	Table 605.3, Table 605.4
F2306/F 2306M—13	12" to 60" Annular Corrugated Profile-wall Polyethylene (PE) Pipe and Fittings for Gravity Flow Storm Sewer and Subsurface Drainage Applications	Table 1102.4, Table 1102.7
F2389—10	Specification for Pressure-rated Polypropylene (PP) Piping Systems	Table 605.3, Table 605.4, Table 605.5, Table 605.7, 605.20.1
F2434—09	Standard Specification for Plastic Insert Fittings Utilizing a Copper Crimp Ring for SDR9 Cross-linked Polyethylene (PEX) Tubing and SDR9 Cross-linked Polyethylene/Aluminum/ Cross-linked Polyethylene (PEX AL-PEX) Tubing	Table 605.5
F2735—09	Standard Specification for Plastic Insert Fittings for SDR9 Cross-linked Polyethylene (PEX) and Polyethylene of Raised Temperature (PE-RT) Tubing	Table 605.5
F2769—10	Polyethylene or Raised Temperature (PE-RT) Plastic Hot and Cold Water Tubing and Distribution Systems	Table 605.3, Table 605.4, Table 605.5
F2831—11	Standard Practice for Internal Non Structural Epoxy Barrier Coating Material Used In Rehabilitation of Metallic Pressurized Piping Systems	601.5
F2855—11	Specification Poly (Vinyl Chloride)/Aluminum/Poly (Vinyl Chloride) (CPVC/AL/CPVC) Composite Pressure Tubing	Table 605.3, Table 605.4

AWS

American Welding Society
8669 NW 36 Street, #130
Doral, FL 33166

Standard reference number	Title	Referenced in code section number
A5.8M/A5.8—2011	Specifications for Filler Metals for Brazing and Braze Welding	605.11.1, 605.13.1, 605.14.1, 705.3.1, 705.6.1, 705.7.1

AWWA

American Water Works Association
6666 West Quincy Avenue
Denver, CO 80235

Standard reference number	Title	Referenced in code section number
C104/A21.4-08	Cement-mortar Lining for Ductile-iron Pipe and Fittings for Water	605.3, 605.5
C110/A21.10—12	Ductile-iron and Gray-iron Fittings	Table 605.5, Table 702.4, Table 1102.7
C111/A21.11-12	Rubber-gasket Joints for Ductile-iron Pressure Pipe and Fittings	605.12
C115/A21.15—11	Flanged Ductile-iron Pipe with Ductile-iron or Gray-iron Threaded Flanges	Table 605.3, Table 605.4
C151/A21.51—09	Ductile-iron Pipe, Centrifugally Cast for Water	Table 605.3, Table 605.4
C153—00/A21.53—11	Ductile-iron Compact Fittings for Water Service	Table 605.5
C500—09	Standard for Metal-Seated Gate Valves for Water Supply Service	Table 605.7
C504—10	Standard for Rubber-Seated Butterfly Valves	Table 605.7
C507—11	Standard for Ball Valves, 6 In. Through 60 In.	Table 605.7
C510—07	Double Check Valve Backflow Prevention Assembly	Table 608.1, 608.13.7
C511—07	Reduced-pressure Principle Backflow Prevention Assembly	Table 608.1, 608.13.2, 608.16.2
C651—05	Disinfecting Water Mains	610.1
C652—11	Disinfection of Water-storage Facilities	610.1
C901—08	Polyethylene (PE) Pressure Pipe and Tubing $^1/_2$ inch (13 mm) Through 3 inch (76 mm) for Water Service	Table 605.3
C904—08	Cross-linked Polyethylene (PEX) Pressure Pipe $^1/_2$ inch (13 mm) Through 3 inch (76 mm) for Water Service	Table 605.3
D100—05	Standard for Welded Carbon Steel Tanks for Water Storage	1302.7.2
D115—06	Standard for Tendon Prestressed-Concrete Water Tanks	1302.7.2
D120—09	Standard for Thermosetting Fiberglass-Reinforced Plastic Tanks	1302.7.2

REFERENCED STANDARDS

CISPI

Cast Iron Soil Pipe Institute
5959 Shallowford Road, Suite 419
Chattanooga, TN 37421

Standard reference number	Title	Referenced in code section number
301—04a	Specification for Hubless Cast-iron Soil Pipe and Fittings for Sanitary and Storm Drain, Waste and Vent Piping Applications	Table 702.1, Table 702.2, Table 702.3, Table 702.4, 708.7, Table 1102.4, Table 1102.5, Table 1102.7
310—04	Specification for Coupling for Use in Connection with Hubless Cast-iron Soil Pipe and Fittings for Sanitary and Storm Drain, Waste and Vent Piping Applications	705.4.3

CSA

CSA Group
8501 East Pleasant Valley
Cleveland, OH 44131-5516

Standard reference number	Title	Referenced in code section number
A257.1M—2009	Circular Concrete Culvert, Storm Drain, Sewer Pipe and Fittings	Table 702.3, Table 1102.4
A257.2M—2009	Reinforced Circular Concrete Culvert, Storm Drain, Sewer Pipe and Fittings	Table 702.3, Table 1102.4
A257.3M—2009	Joints for Circular Concrete Sewer and Culvert Pipe, Manhole Sections and Fittings Using Rubber Gaskets	705.5, 705.16
ASME A112.18.1—2012/ CSA B125.1—2012	Plumbing Supply Fittings	424.1, 424.2, 424.3, 424.4, 424.6, 424.8, 605.7, 607.4, 608.2
ASME A112.18.2—2011/ CSA B125.2—2011	Plumbing Waste Fittings	424.1.2
ASME A112.19.2—2013/ B45.1—2013	Ceramic Plumbing Fixtures	401.2, 405.9, 407.1, 408.1, 410.1, 415.1, 416.1, 417.1, 418.1, 419.1, 420.1
ASME A112.19.1—2013/ CSA B45.2—2013	Enameled Cast-iron and Enameled Steel Plumbing Fixtures	407.1, 410.1, 415.1, 416.1, 418.1
ASME A112.19.3—2008/ CSA B45.4—08(R2013)	Stainless-steel Plumbing Fixtures	405.9, 407.1, 415.1, 416.1, 418.1, 420.1
ASME A112.19.5—2011/ CSA B45.15—2011	Flush Valves and Spuds for Water Closets, Urinals and Tanks	425.4
ASME A112.19.7—2012/ CSA B45.10—2012	Hydromassage Bathtub Systems	421.1, 421.4
CSA B45.5—11/ IAPMO Z124-2011	Plastic Plumbing Fixtures	407.1, 415.1, 416.1, 416.2, 417.1, 419.1, 420.1
ASME A112.3.4—2013/ CSA B45.9—2013	Macerating Systems and Related Components	712.4.1
ASSE 1016/ ASME A112.1016/ CSA B125.16—2011	Performance Requirements for Individual Thermostatic, Pressure Balancing and Combination Control Valves for Individual Fixture Fittings	424.3, 424.4, 607.4
B64.1.1—11	Vacuum Breakers, Atmospheric Type (AVB)	425.2, Table 608.1, 608.13.6, 608.16.4.1
B64.1.2—11	Pressure Vacuum Breakers, (PVB)	Table 608.1, 608.13.5
B64.1.3—11	Spill Resistant Pressure Vacuum Breaks (SRPVB)	608.13.8
B64.2—11	Vacuum Breakers, Hose Connection Type (HCVB)	Table 608.1, 608.13.6
B64.2.1—11	Vacuum Breakers, Hose Connection (HCVB) with Manual Draining Feature	Table 608.1, 608.13.6
B64.2.1.1—11	Hose Connection Dual Check Vacuum Breakers, (HCDVB)	Table 608.1, 608.13.6
B64.2.2—11	Vacuum Breakers, Hose Connection Type (HCVB) with Automatic Draining Feature	Table 608.1, 608.13.6
B64.3—11	Backflow Preventers, Dual Check Valve Type with Atmospheric Port (DCAP)	Table 608.1, 608.13.3, 608.16.2
B64.4—11	Backflow Preventers, Reduced Pressure Principle Type (RP)	Table 608.1, 608.13.2, 608.16.2
B64.4.1—11	Reduced Pressure Principle for Fire Sprinklers (RPF)	Table 608.1, 608.13.2
B64.5—11	Double Check Backflow Preventers (DCVA)	Table 608.1, 608.13.7
B64.5.1—11	Double Check Valve Backflow Preventer for Fire Systems (DCVAF)	Table 608.1 608.13.7
B64.6—11	Dual Check Backflow PreventersValve (DuC)	605.3.1, Table 608.1, 608.13.10
B64.7—11	Laboratory Faucet Vacuum Breakers (LFVB)	Table 608.1, 608.13.6
B64.10—11	Manual for the Selection and Installation of Backflow Prevention Devices	312.10.2

REFERENCED STANDARDS

CSA—continued

Standard	Title	Referenced in code section number
B64.10.1—11	Maintenance and Field Testing of Backflow Preventers	312.10.2
B79—08(R2013)	Commercial and Residential Drains, and Cleanouts	412.1
B125.3—2012	Plumbing Fittings	408.3, 416.5, 423.3, 424.4, 424.5, 425.2, 425.3.1, Table 605.7, Table 608.1
B137.1—13	Polyethylene (PE) Pipe, Tubing and Fittings for Cold Water Pressure Services	Table 605.3, Table 605.5
B137.2—13	Polyvinylchloride PVC Injection-moulded Gasketed Fittings for Pressure Applications	Table 605.5
B137.3—13	Rigid Poly (Vinyl Chloride) (PVC) Pipe for Pressure Applications	Table 605.3, Table 605.5, 605.22.3, 705.11.2, 705.14.2
B137.5—13	Cross-linked Polyethylene (PEX) Tubing Systems for Pressure Applications	Table 605.3, Table 605.4, Table 605.5
B137.6—13	CPVC Pipe, Tubing and Fittings for Hot and Cold Water Distribution Systems	Table 605.3, Table 605.4, Table 605.5
B137.9—13	Polyethylene/Aluminum/Polyethylene Composite Pressure Pipe Systems	Table 605.3, Table 605.5, 605.21.1
B137.10—13	Cross-linked Polyethylene/Aluminum/Polyethylene Composite Pressure Pipe Systems	Table 605.3, Table 605.4, Table 605.5, 605.21.1
B137.11—13	Polypropylene (PP-R) Pipe and Fittings for Pressure Applications	Table 605.3, Table 605.4, Table 605.5
B181.1—11	Acrylonitrile-butadiene-styrene ABS Drain, Waste and Vent Pipe and Pipe Fittings	Table 702.1, Table 702.2, Table 702.3, Table 702.4, 705.2.2, 715.2, Table 1102.4, Table 1102.7
B 181.2—11	Polyvinylchloride PVC and chlorinated polyvinylchloride (CPVC) Drain, Waste, and Vent Pipe and Pipe Fittings	Table 702.1 Table 702.2, 705.11.2, 705.14.2, 715.2
B 181.3—11	Polyolefin and Polyvinylidene Fluoride (PVDF) Laboratory Drainage Systems	Table 702.1, Table 702.2, Table 702.3, Table 702.4, 705.14.1
B182.1—11	Plastic Drain and Sewer Pipe and Pipe Fittings	705.11.2, 705.14.2, Table 1102.4, Table 1102.5
B182.2—11	PSM Type Polyvinylchloride PVC Sewer Pipe and Fittings	Table 702.3, Table 1102.4, Table 1102.5
B182.4—11	Profile Polyvinylchloride PVC Sewer Pipe and Fittings	Table 702.3, Table 1102.4, Table 1102.5
B182.6—11	Profile Polyethylene (PE) Sewer Pipe and Fittings for Leak-proof Sewer Applications	Table 1102.5
B182.8—11	Profile Polyethylene (PE) Storm Sewer and Drainage Pipe and Fittings	Table 1102.5
B356—10	Water Pressure Reducing Valves for Domestic Water Systems	604.8
B483.1—14	Drinking Water Treatment Units	611.1, 611.2
B481.1—12	Testing and Rating of Grease interceptors Using Lard	1003.3.4
B481.3—12	Sizing, Selection, Location and Installation of Grease Interceptors	1003.3.4
B602—10	Mechanical Couplings for Drain, Waste and Vent Pipe and Sewer Pipe	705.2.1, 705.4.3, 705.5, 705.11.1, 705.12, 705.13.2, 705.16

IAPMO

IAPMO Group
4755 E. Philadelphia
Ontario, CA 91761

Standard reference number	Title	Referenced in code section number
Z1001-2007	Prefabricated Gravity Grease Interceptors	1003.3.6
CSA B45.5—11/ IAPMO Z124—2011	Plastic Plumbing Fixtures	407.1, 415.1, 416.1, 416.2, 417.1, 419.1, 420.1

ICC

International Code Council, Inc.
500 New Jersey Ave, NW
6th Floor
Washington, DC 20001

Standard reference number	Title	Referenced in code section number
A117.1—2009	Accessible and Usable Buildings and Facilities	404.2
IBC—15	International Building Code®	201.3, 307.1, 307.2, 307.3, 308.2, 309.1, 309.2, 310.1, 310.3, 315.1, 403.1, Table 403.1, 403.4, 404.1, 407.3, 417.6, 502.4, 606.5.2, 1106.5, 1301.9.4, 1307.7.2, 1303.6, 1402.1
IECC—15	International Energy Conservation Code®	313.1, 607.5
IFC—15	International Fire Code®	201.3, 1201.1

REFERENCED STANDARDS

ICC—continued

IFGC—15	International Fuel Gas Code®	101.2, 201.3, 502.1, 502.1.1
IMC—15	International Mechanical Code®	201.3, 307.7, 310.1, 422.9, 502.1, 502.1.1, 612.1, 1202.1
IPSDC—15	International Private Sewage Disposal Code®	701.2
IRC—15	International Residential Code®	101.2

ISEA
International Safety Equipment Association
1901 N. Moore Street, Suite 808
Arlington, VA 22209

Standard reference number	Title	Referenced in code section number
ANSI/ISEA Z358.1—2009	Emergency Eyewash and Shower Equipment	411.1

MSS
Manufacturers Standardization Society of the Valve and Fittings Industry, Inc.
127 Park Street, NE
Vienna, VA 22180-4602

Standard reference number	Title	Referenced in code section number
SP-67—2011	Butterfly Valves	Table 605.7
SP-70—2011	Gray Iron Gate Valves, Flanged and Threaded Ends	Table 605.7
SP-71—2011	Gray Iron Swing Check Valves, Flanged and Threaded Ends	Table 605.7
SP-72—2010	Ball Valves with Flanged or Butt-Welding Ends for General Service	Table 605.7
SP-78—2011	Cast Iron Plug Valves, Flanged and Threaded Ends	Table 605.7
SP-80—2008	Bronze Gate, Globe, Angle and Check Valves	Table 605.7
SP-110—2010	Ball Valves, Threaded, Socket Welded, Solder Joint, Grooved and Flared Ends	Table 605.7

NFPA
National Fire Protection Association
1 Batterymarch Park
Quincy, MA 02169-7471

Standard reference number	Title	Referenced in code section number
51—13	Design and Installation of Oxygen-fuel Gas Systems for Welding, Cutting and Allied Processes	1203.1
55—13	Compressed Gases and Cryogenic Fluids Code	1203.1
70—14	National Electric Code	502.1, 504.3, 1113.1.3
99—15	Health Care Facilities Code	1202.1

NSF
NSF International
789 Dixboro Road, P.O. Box 130140
Ann Arbor, MI 48105

Standard reference number	Title	Referenced in code section number
3—2010	Commercial Warewashing Equipment	409.1
14—2011	Plastic Piping System Components and Related Materials	303.3, 611.3
18—2012	Manual Food and Beverage Dispensing Equipment	426.1
42—2011	Drinking Water Treatment Units-Aesthetic Effects	611.1, 611.3
44—2012	Residential Cation Exchange Water Softeners	611.1, 611.3
50—2012	Equipment for Swimming Pools, Spas, Hot Tubs and other Recreational Facilities	1302.8.1
53—2011a	Drinking Water Treatment Units—Health Effects	611.1, 611.3
58—2012	Reverse Osmosis Drinking Water Treatment Systems	611.2, 611.3

REFERENCED STANDARDS

NSF—continued

Reference number	Title	Referenced in code section number
61—2012	Drinking Water System Components—Health Effects	410.1, 424.1, 605.3, 605.4, 605.5, 605.7, 611.3
62—2012	Drinking Water Distillation Systems	611.1
350—2011	Onsite Residential and Commercial Water Reuse Treatment Systems	1302.6.1
359—2011	Valves for Crosslinked Polyethylene (PEX) Water Distribution Tubing Systems	Table 605.7
372—2010	Drinking Water Systems Components—Lead Content	605.2.1

PDI

Plumbing and Drainage Institute
800 Turnpike Street, Suite 300
North Andover, MA 01845

Standard reference number	Title	Referenced in code section number
PDI G101 (2012)	Testing and Rating Procedure for Grease Interceptors with Appendix of Sizing and Installation Data	1003.3.4
PDI G102 (2009)	Testing and Certification for Grease Interceptors with Fog Sensing and Alarm Devices	1003.3.4

UL

UL LLC
333 Pfingsten Road
Northbrook, IL 60062-2096

Standard reference number	Title	Referenced in code section number
58—1996	Steel Underground Tanks for Flammable and Combustible Liquids—with revisions through July 27, 1998	1302.7.2
142—2006	Steel Aboveground Tanks for Flammable and Combustible Liquids—with revisions through February 12, 2010	1302.7.2
399—2008	Drinking-Water Coolers—with revisions through January 14, 2011	410.1
430—2009	Waste Disposers—with revisions through March 23, 2011	413.1
508—99	Industrial Control Equipment—with revisions through March 2013	314.2.3
1316—1994	Glass-Fiber Reinforced Plastic Underground Storage Tanks for Petroleum Products, Alcohols and Alcohol Gasoline Mixtures—with revisions through May 12, 2006	1302.7.2
1746—2007	External Corrosion Protection Systems for Steel Underground Storage Tanks	1302.7.2
1795—2009	Hydromassage Bathtubs including revisions through August 23, 2011	421.1

Appendix A: Plumbing Permit Fee Schedule

General Comments

Appendix A is a sample, guide or prototype of a permit fee schedule. Once established, the schedule should be made available to all plumbing contractors to aid them in preparing bids and applying for permits. This sample fee schedule will more than likely have to be modified to account for local contractor licensing laws, local practices and the typical work seen in the particular jurisdiction. The ultimate goal is to develop a fee schedule that is fair to all permit applicants. The permit fee schedule needs to balance the revenues necessary for department funding without becoming a financial burden to the contractors and building owners purchasing the plumbing permits.

Purpose

Section 106.6 requires that fees be collected for the issuance of plumbing work permits. These fees support the department of plumbing inspection. Many department costs must be funded, including inspector salaries, clerical worker salaries, legal assistance costs, office space costs, equipment costs, insurance costs, education costs and transportation costs. The majority of plumbing inspection departments are either totally or partially funded by permit fees.

Section 106.6.2 requires the creation of a permit fee schedule for permit issuance in the jurisdiction. This schedule sets the fees for permits based on the total cost of the proposed work or the total number of fixtures, devices, water and drainage outlets and systems or some combination of the above. Some jurisdictions simply charge a fee for the number of each type of fixture; others charge a fee based on a percentage of the total plumbing job cost. For example, a $10 fee per $1,000 of job cost would be a 1-percent fee. In all cases, the fees must be determined at the local level based on the needs of the plumbing department. The fees will depend on the amount of funding, if any, that the plumbing department receives from sources other than permit fees.

The provisions contained in this appendix are not mandatory unless specifically referenced in the adopting ordinance.

Permit Issuance

1. For issuing each permit... $ _____

2. For issuing each supplemental permit .... _____

Unit Fee Schedule

1. For each plumbing fixture or trap or set of fixtures on one trap (including water, drainage piping and backflow protection thereof)... _____

2. For each building sewer and each trailer park sewer .... _____

3. Rainwater systems—per drain (inside building) .... _____

4. For each cesspool (where permitted) .... _____

5. For each private sewage disposal system .... _____

6. For each water heater and/or vent... _____

7. For each industrial waste pretreatment interceptor including its trap and vent, excepting kitchen-type grease interceptors functioning as fixture traps .... _____

8. For installation, alteration or repair of water-piping and/or water-treating equipment, each........ _____

9. For repair or alteration of drainage or vent piping, each fixture.............................. _____

10. For each lawn sprinkler system on any one meter including backflow protection devices therefor .... _____

(continued)

Unit Fee Schedule—continued

11. For atmospheric-type vacuum breakers not included in Item 2:

    1 to 5 .... _____

    over 5, each .... _____

12. For each backflow protective device other than atmospheric-type vacuum breakers:

    2 inches (51 mm) and smaller .... _____

    Over 2 inches (51 mm) .... _____

Other Inspections and Fees

1. Inspections outside of normal business hours (minimum charge two hours) .... _____ per hour
2. Reinspection fee assessed under provisions of Section 107.4.3 .... _____ each
3. Inspections for which no fee is specifically indicated (minimum charge one-half hour) .... _____ per hour
4. Additional plan review required by changes, additions or revisions to approved plans (minimum charge one-half hour) .... _____ per hour

Appendix B:
Rates of Rainfall for Various Cities

General Comments

Gutters, leaders, conductors and downspouts are all components of the storm drainage system and are sized based on the rainfall rate. The rainfall rates, in inches per hour, are based on a storm of 1-hour duration and a 100-year return period. The rainfall rates shown in this appendix are derived from Figure 1106.1. Rainfall rates indicate the maximum rate of rainfall within a given period of time. The rainfall rates are calculated from a statistical analysis of weather records. The roof area listed must be matched to the local rainfall rate.

Purpose

Appendix B provides the hourly rainfall rates indicated in Figure 1106.1.

This appendix is informative and is not part of the code.

Rainfall rates, in inches per hour, are based on a storm of 1-hour duration and a 100-year return period. The rainfall rates shown in the appendix are derived from Figure 1106.1.

Alabama:
- Birmingham 3.8
- Huntsville 3.6
- Mobile 4.6
- Montgomery 4.2

Alaska:
- Fairbanks 1.0
- Juneau 0.6

Arizona:
- Flagstaff 2.4
- Nogales 3.1
- Phoenix 2.5
- Yuma 1.6

Arkansas:
- Fort Smith 3.6
- Little Rock 3.7
- Texarkana 3.8

California:
- Barstow 1.4
- Crescent City 1.5
- Fresno 1.1
- Los Angeles 2.1
- Needles 1.6
- Placerville 1.5
- San Fernando 2.3
- San Francisco 1.5
- Yreka 1.4

Colorado:
- Craig 1.5
- Denver 2.4
- Durango 1.8
- Grand Junction 1.7
- Lamar 3.0
- Pueblo 2.5

Connecticut:
- Hartford 2.7
- New Haven 2.8
- Putnam 2.6

Delaware:
- Georgetown 3.0
- Wilmington 3.1

District of Columbia:
- Washington 3.2

Florida:
- Jacksonville 4.3
- Key West 4.3
- Miami 4.7
- Pensacola 4.6
- Tampa 4.5

Georgia:
- Atlanta 3.7
- Dalton 3.4
- Macon 3.9
- Savannah 4.3
- Thomasville 4.3

Hawaii:
- Hilo 6.2
- Honolulu 3.0
- Wailuku 3.0

Idaho:
- Boise 0.9
- Lewiston 1.1
- Pocatello 1.2

Illinois:
- Cairo 3.3
- Chicago 3.0

- Peoria 3.3
- Rockford 3.2
- Springfield 3.3

Indiana:
- Evansville 3.2
- Fort Wayne 2.9
- Indianapolis 3.1

Iowa:
- Davenport 3.3
- Des Moines 3.4
- Dubuque 3.3
- Sioux City 3.6

Kansas:
- Atwood 3.3
- Dodge City 3.3
- Topeka 3.7
- Wichita 3.7

Kentucky:
- Ashland 3.0
- Lexington 3.1
- Louisville 3.2
- Middlesboro 3.2
- Paducah 3.3

Louisiana:
- Alexandria 4.2
- Lake Providence 4.0
- New Orleans 4.8
- Shreveport 3.9

Maine:
- Bangor 2.2
- Houlton 2.1
- Portland 2.4

Maryland:
- Baltimore 3.2
- Hagerstown 2.8
- Oakland 2.7
- Salisbury 3.1

Massachusetts:
- Boston 2.5
- Pittsfield 2.8
- Worcester 2.7

Michigan:
- Alpena 2.5
- Detroit 2.7
- Grand Rapids 2.6
- Lansing 2.8
- Marquette 2.4
- Sault Ste. Marie 2.2

Minnesota:
- Duluth 2.8
- Grand Marais 2.3
- Minneapolis 3.1
- Moorhead 3.2
- Worthington 3.5

Mississippi:
- Biloxi 4.7
- Columbus 3.9
- Corinth 3.6
- Natchez 4.4
- Vicksburg 4.1

Missouri:
- Columbia 3.2
- Kansas City 3.6
- Springfield 3.4
- St. Louis 3.2

Montana:
- Ekalaka 2.5
- Havre 1.6
- Helena 1.5
- Kalispell 1.2
- Missoula 1.3

Nebraska:
- North Platte 3.3
- Omaha 3.8
- Scottsbluff 3.1
- Valentine 3.2

Nevada:
- Elko 1.0
- Ely 1.1
- Las Vegas 1.4
- Reno 1.1

New Hampshire:
- Berlin 2.5
- Concord 2.5
- Keene 2.4

New Jersey:
- Atlantic City 2.9
- Newark 3.1
- Trenton 3.1

New Mexico:
- Albuquerque 2.0
- Hobbs 3.0
- Raton 2.5
- Roswell 2.6
- Silver City 1.9

New York:
- Albany 2.5
- Binghamton 2.3
- Buffalo 2.3
- Kingston 2.7
- New York 3.0
- Rochester 2.2

North Carolina:
- Asheville 4.1
- Charlotte 3.7
- Greensboro 3.4
- Wilmington 4.2

North Dakota:
- Bismarck 2.8
- Devils Lake 2.9
- Fargo 3.1
- Williston 2.6

Ohio:
- Cincinnati 2.9
- Cleveland 2.6
- Columbus 2.8
- Toledo 2.8

Oklahoma:
- Altus 3.7
- Boise City 3.3
- Durant 3.8
- Oklahoma City 3.8

Oregon:
- Baker 0.9
- Coos Bay 1.5
- Eugene 1.3
- Portland 1.2

Pennsylvania:
- Erie 2.6
- Harrisburg 2.8
- Philadelphia 3.1
- Pittsburgh 2.6
- Scranton 2.7

Rhode Island:
- Block Island 2.75
- Providence 2.6

South Carolina:
- Charleston 4.3
- Columbia 4.0
- Greenville 4.1

South Dakota:
- Buffalo 2.8
- Huron 3.3
- Pierre 3.1
- Rapid City 2.9
- Yankton 3.6

Tennessee:
- Chattanooga 3.5
- Knoxville 3.2
- Memphis 3.7
- Nashville 3.3

Texas:
- Abilene 3.6
- Amarillo 3.5
- Brownsville 4.5
- Dallas 4.0
- Del Rio 4.0
- El Paso 2.3
- Houston 4.6
- Lubbock 3.3
- Odessa 3.2
- Pecos 3.0
- San Antonio 4.2

Utah:
- Brigham City 1.2
- Roosevelt 1.3
- Salt Lake City 1.3
- St. George 1.7

Vermont:
- Barre 2.3
- Bratteboro 2.7
- Burlington 2.1
- Rutland 2.5

Virginia:
- Bristol 2.7
- Charlottesville 2.8
- Lynchburg 3.2
- Norfolk 3.4
- Richmond 3.3

Washington:
- Omak 1.1
- Port Angeles 1.1
- Seattle 1.4
- Spokane 1.0
- Yakima 1.1

West Virginia:
- Charleston 2.8
- Morgantown 2.7

Wisconsin:
- Ashland 2.5
- Eau Claire 2.9
- Green Bay 2.6
- La Crosse 3.1
- Madison 3.0
- Milwaukee 3.0

Wyoming:
- Cheyenne 2.2
- Fort Bridger 1.3
- Lander 1.5
- New Castle 2.5
- Sheridan 1.7
- Yellowstone Park 1.4

Appendix C: Structural Safety

General Comments

Although the title of Section C101 is "Cutting, Notching and Boring in Wood Members," it also includes some regulations pertaining to structural steel and cold-formed steel structural members.

Purpose

This appendix contains criteria limiting the cutting away of a structural member during the installation of a plumbing system. These provisions are needed for the safety of the joists, studs, beams, columns or other structural members that support the building.

The provisions contained in this appendix are not mandatory unless specifically referenced in the adopting ordinance.

SECTION C101
CUTTING, NOTCHING AND BORING IN WOOD MEMBERS

[BS] C101.1 Joist notching. Notches on the ends of joists shall not exceed one-fourth the joist depth. Holes bored in joists shall not be within 2 inches (51 mm) of the top or bottom of the joist, and the diameter of any such hole shall not exceed one-third the depth of the joist. Notches in the top or bottom of joists shall not exceed one-sixth the depth and shall not be located in the middle third of the span.

❖ Wood is comprised of relatively small elongated cells oriented parallel to each other and bound together physically and chemically by lignin. The cells are continuous from end to end of a piece of lumber. When a beam or joist is notched or cut, the cell strands are interrupted. In notched beams, the effective depth is reduced equal to the depth of the notch (see Commentary Figure C101.1).

Some designs and installation practices require that limited notching and cutting occur. Notching should be avoided when possible. Holes bored in beams and joists create the same problems as notches. When necessary, the holes should be located in areas with the least stress concentration, generally along the neutral axis of the joist. Beams subject to high horizontal shear stress (short span/heavy load) should never be cut. Wood members having a thickness greater than 4 inches (102 mm) should never be notched on the tension side of the member, except at the ends of the members.

[BS] C101.2 Stud cutting and notching. In exterior walls and bearing partitions, any wood stud is permitted to be cut or notched to a depth not exceeding 25 percent of its width. Cutting or notching of studs to a depth not greater than 40 percent of the width of the stud is permitted in nonbearing partitions supporting no loads other than the weight of the partition.

❖ To maintain the cross-sectional bearing area, studs should not be cut, notched or bored. When the alteration exceeds that specified in this section, the studs should be doubled or otherwise reinforced to provide the required strength (see Commentary Figure C101.2).

For SI: 1 inch = 25.4 mm.

Figure C101.1
CUTTING, NOTCHING AND BORING JOISTS

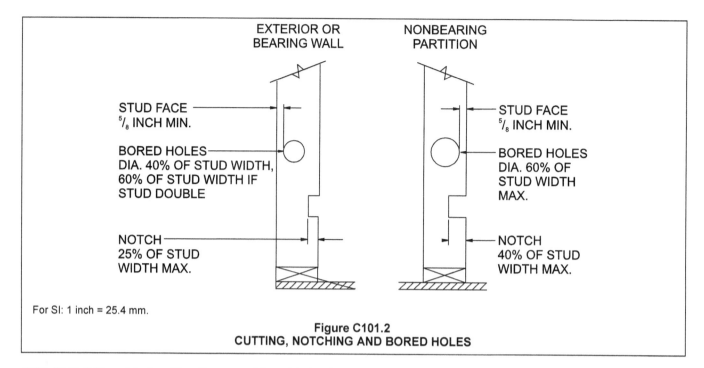

For SI: 1 inch = 25.4 mm.

Figure C101.2
CUTTING, NOTCHING AND BORED HOLES

[BS] C101.3 Bored holes. The diameter of bored holes in wood studs shall not exceed 40 percent of the stud depth. The diameter of bored holes in wood studs shall not exceed 60 percent of the stud depth in nonbearing partitions. The diameter of bored holes in wood studs shall not exceed 60 percent of the stud depth in any wall where each stud is doubled, provided that not more than two such successive doubled studs are so bored. The edge of the bored hole shall not be closer than $^5/_8$ inch (15.9 mm) to the edge of the stud. Bored holes shall be not located at the same section of stud as a cut or notch.

❖ Because of the redundancy of wood construction, limited hole boring is encouraged. This section describes the limits for bored holes in load-bearing and nonload-bearing wood studs. Commentary Figure C101.2 illustrates the requirements of this section.

[BS] C101.4 Cutting, notching and boring holes in structural steel framing. The cutting, notching and boring of holes in structural steel framing members shall be as prescribed by the registered design professional.

❖ This section does not allow the cutting, notching or boring of structural steel girders, beams, joists, columns or other structural steel members unless it is approved first by the registered design professional responsible for the structural design of the building. The code does not provide any prescriptive limits as it does for wood framing. Since there are so many conditions that could be encountered, it is best for the design professional to consider them on a case-by-case basis.

[BS] C101.5 Cutting, notching and boring holes in cold-formed steel framing. Flanges and lips of load-bearing cold-formed steel framing members shall not be cut or notched. Holes in webs of load-bearing cold-formed steel framing members shall be permitted along the centerline of the web of the framing member and shall not exceed the dimensional limitations, penetration spacing or minimum hole edge distance as prescribed by the registered design professional. Cutting, notching and boring holes of steel floor/roof decking shall be as prescribed by the registered design professional.

❖ No cutting, notching or boring is allowed at all in the flanges of the cold-formed lips of load-bearing studs, beams, columns or other load-carrying members, as that would result in a significant reduction of the load-carrying capacity.

This section requires the approval of the registered design professional prior to the boring of any holes in webs of any cold-formed steel member that is load bearing. The code does not provide any prescriptive limits as it does for wood framing.

[BS] C101.6 Cutting, notching and boring holes in nonstructural cold-formed steel wall framing. Flanges and lips of nonstructural cold-formed steel wall studs shall not be cut or notched. Holes in webs of nonstructural cold-formed steel wall studs shall be permitted along the centerline of the web of the framing member, shall not exceed $1^1/_2$ inches (38 mm) in width or 4 inches (102 mm) in length, and the holes shall not be spaced less than 24 inches (610 mm) center to center from another hole or less than 10 inches (254 mm) from the bearing end.

❖ This section applies to nonload-bearing studs in partition walls. The holes in the studs are limited to a series of holes in the web, as described in this section. Some cold-formed studs are manufactured with these web openings already cut out for the installation of piping and electrical wiring.

Appendix D:
Degree Day and Design Temperatures

This appendix is informative and is not part of the code.

General Comments

Appendix D specifies outdoor design conditions. Although the data were extracted from the 1985 ASHRAE *Handbook of Fundamentals*, updated values for design conditions can be obtained from the 1987, 1993, 1997 and 2001 editions of the ASHRAE handbook. The exterior design conditions are described below in three groups. The groups are established according to how the data are used.

Purpose

Appendix D is reprinted for informational purposes and provides climate conditions for 223 locations throughout the United States. These summaries include sources for heating degree days and values of dry-bulb and wet-bulb temperatures to be used for evaluating the protection of vent terminals or the thickness of insulation required for recirculating hot water distribution systems as required by the code. The appendix also assigns the outdoor design conditions required for the calculations called for in the *International Property Maintenance Code®* (IPMC®) for minimum space heating requirements.

Load Calculations

Outdoor design temperatures are used in the load calculations required by the IPMC for determining minimum heating facilities for residential structures (see Sections 602.2 and 602.3 of the IPMC).

Winter Design Dry-Bulb ($97^1/_2$ percent). This is the outdoor dry-bulb temperature used to calculate heating loads. The value means that, on average, the outdoor temperature will be above this temperature for $97^1/_2$ percent of the hours during the heating season. For the northern hemisphere, the heating season in the United States is the three-month winter period from December through February.

Summer Design Dry-Bulb ($2^1/_2$ percent). This is the outdoor dry-bulb temperature used to calculate cooling loads. The value means that, on average, the outdoor temperature will be above this temperature for $2^1/_2$ percent of the hours during the cooling season. For the northern hemisphere, the cooling season in the United States is the four-month summer period from May through August.

Mean Coincident Wet-Bulb ($2^1/_2$ percent). This is the outdoor wet-bulb temperature used to calculate cooling loads. The value means that, on average, the outdoor temperature will be above this temperature for $2^1/_2$ percent of the hours during the cooling season.

Piping Insulation

Distribution losses impact building energy use both in the energy required to make up for the lost heat and in the additional load that can be placed on the space cooling system if the heat is released to the air-conditioned space. In circulating systems, hot water is exposed to loss throughout the entire distribution system as long as the water is circulating. These losses can be limited by insulating the entire hot water supply and return piping system to the requirements of the *International Energy Conservation Code®* (IECC®).

Protection of Vent Terminals

The possibility of frost closure of vent terminals predominantly occurs in areas of the country having cold climates, as determined where the winter design dry-bulb ($97^1/_2$ percent) temperature is less than 0°F (-18°C) (see Sections 903.2 and 903.7).

APPENDIX D

TABLE D101
DEGREE DAY AND DESIGN TEMPERATURES[a] FOR CITIES IN THE UNITED STATES

STATE	STATION[b]	HEATING DEGREE DAYS (yearly total)	DESIGN TEMPERATURES			DEGREES NORTH LATITUDE[c]
			Winter	Summer		
			97$\frac{1}{2}$%	Dry bulb 2$\frac{1}{2}$%	Wet bulb 2$\frac{1}{2}$%	
AL	Birmingham	2,551	21	94	77	33°30'
	Huntsville	3,070	16	96	77	34°40'
	Mobile	1,560	29	93	79	30°40'
	Montgomery	2,291	25	95	79	32°20'
AK	Anchorage	10,864	-18	68	59	61°10'
	Fairbanks	14,279	-47	78	62	64°50'
	Juneau	9,075	1	70	59	58°20'
	Nome	14,171	-27	62	56	64°30'
AZ	Flagstaff	7,152	4	82	60	35°10'
	Phoenix	1,765	34	107	75	33°30'
	Tuscon	1,800	32	102	71	33°10'
	Yuma	974	39	109	78	32°40'
AR	Fort Smith	3,292	17	98	79	35°20'
	Little Rock	3,219	20	96	79	34°40'
	Texarkana	2,533	23	96	79	33°30'
CA	Fresno	2,611	30	100	71	36°50'
	Long Beach	1,803	43	80	69	33°50'
	Los Angeles	2,061	43	80	69	34°00'
	Los Angeles[d]	1,349	40	89	71	34°00'
	Oakland	2,870	36	80	64	37°40'
	Sacramento	2,502	32	98	71	38°30'
	San Diego	1,458	44	80	70	32°40'
	San Francisco	3,015	38	77	64	37°40'
	San Francisco[d]	3,001	40	71	62	37°50'
CO	Alamosa	8,529	-16	82	61	37°30'
	Colorado Springs	6,423	2	88	62	38°50'
	Denver	6,283	1	91	63	39°50'
	Grand Junction	5,641	7	94	63	39°10'
	Pueblo	5,462	0	95	66	38°20'
CT	Bridgeport	5,617	9	84	74	41°10'
	Hartford	6,235	7	88	75	41°50'
	New Haven	5,897	7	84	75	41°20'
DE	Wilmington	4,930	14	89	76	39°40'
DC	Washington	4,224	17	91	77	38°50'
FL	Daytona	879	35	90	79	29°10'
	Fort Myers	442	44	92	79	26°40'
	Jacksonville	1,239	32	94	79	30°30'
	Key West	108	57	90	79	24°30'
	Miami	214	47	90	79	25°50'
	Orlando	766	38	93	78	28°30'
	Pensacola	1,463	29	93	79	30°30'
	Tallahassee	1,485	30	92	78	30°20'
	Tampa	683	40	91	79	28°00'
	West Palm Beach	253	45	91	79	26°40'

(continued)

TABLE D101—continued
DEGREE DAY AND DESIGN TEMPERATURES[a] FOR CITIES IN THE UNITED STATES

STATE	STATION[b]	HEATING DEGREE DAYS (yearly total)	DESIGN TEMPERATURES			DEGREES NORTH LATITUDE[c]
			Winter	Summer		
			97$\frac{1}{2}$%	Dry bulb 2$\frac{1}{2}$%	Wet bulb 2$\frac{1}{2}$%	
GA	Athens	2,929	22	92	77	34°00'
	Atlanta	2,961	22	92	76	33°40'
	Augusta	2,397	23	95	79	33°20'
	Columbus	2,383	24	93	78	32°30'
	Macon	2,136	25	93	78	32°40'
	Rome	3,326	22	93	78	34°20'
	Savannah	1,819	27	93	79	32°10'
HI	Hilo	0	62	83	74	19°40'
	Honolulu	0	63	86	75	21°20'
ID	Boise	5,809	10	94	66	43°30'
	Lewiston	5,542	6	93	66	46°20'
	Pocatello	7,033	-1	91	63	43°00'
IL	Chicago (Midway)	6,155	0	91	75	41°50'
	Chicago (O'Hare)	6,639	-4	89	76	42°00'
	Chicago[d]	5,882	2	91	77	41°50'
	Moline	6,408	-4	91	77	41°30'
	Peoria	6,025	-4	89	76	40°40'
	Rockford	6,830	-4	89	76	42°10'
	Springfield	5,429	2	92	77	39°50'
IN	Evansville	4,435	9	93	78	38°00'
	Fort Wayne	6,205	1	89	75	41°00'
	Indianapolis	5,699	2	90	76	39°40'
	South Bend	6,439	1	89	75	41°40'
IA	Burlington	6,114	-3	91	77	40°50'
	Des Moines	6,588	-5	91	77	41°30'
	Dubuque	7,376	-7	88	75	42°20'
	Sioux City	6,951	-7	92	77	42°20'
	Waterloo	7,320	-10	89	77	42°30'
KS	Dodge City	4,986	5	97	73	37°50'
	Goodland	6,141	0	96	70	39°20'
	Topeka	5,182	4	96	78	39°00'
	Wichita	4,620	7	98	76	37°40'
KY	Covington	5,265	6	90	75	39°00'
	Lexington	4,683	8	91	76	38°00'
	Louisville	4,660	10	93	77	38°10'
LA	Alexandria	1,921	27	94	79	31°20'
	Baton Rouge	1,560	29	93	80	30°30'
	Lake Charles	1,459	31	93	79	30°10'
	New Orleans	1,385	33	92	80	30°00'
	Shreveport	2,184	25	96	79	32°30'
ME	Caribou	9,767	-13	81	69	46°50'
	Portland	7,511	-1	84	72	43°40'
MD	Baltimore	4,654	13	91	77	39°10'
	Baltimore[d]	4,111	17	89	78	39°20'
	Frederick	5,087	12	91	77	39°20'

(continued)

TABLE D101—continued
DEGREE DAY AND DESIGN TEMPERATURES[a] FOR CITIES IN THE UNITED STATES

STATE	STATION[b]	HEATING DEGREE DAYS (yearly total)	DESIGN TEMPERATURES			DEGREES NORTH LATITUDE[c]
			Winter	Summer		
			97 1/2 %	Dry bulb 2 1/2 %	Wet bulb 2 1/2 %	
MA	Boston	5,634	9	88	74	42°20'
	Pittsfield	7,578	-3	84	72	42°30'
	Worcester	6,969	4	84	72	42°20'
MI	Alpena	8,506	-6	85	72	45°00'
	Detroit (City)	6,232	6	88	74	42°20'
	Escanaba[d]	8,481	-7	83	71	45°40'
	Flint	7,377	1	87	74	43°00'
	Grand Rapids	6,894	5	88	74	42°50'
	Lansing	6,909	1	87	74	42°50'
	Marquette[d]	8,393	-8	81	70	46°30'
	Muskegon	6,696	6	84	73	43°10'
	Sault Ste. Marie	9,048	-8	81	70	46°30'
MN	Duluth	10,000	-16	82	70	46°50'
	Minneapolis	8,382	-12	89	5	44°50'
	Rochester	8,295	-12	87	75	44°00'
MS	Jackson	2,239	25	95	78	32°20'
	Meridian	2,289	23	95	79	32°20'
	Vicksburg[d]	2,041	26	95	80	32°20'
MO	Columbia	5,046	4	94	77	39°00'
	Kansas City	4,711	6	96	77	39°10'
	St. Joseph	5,484	2	93	79	39°50'
	St. Louis	4,900	6	94	77	38°50'
	St. Louis[d]	4,484	8	94	77	38°40'
	Springfield	4,900	9	93	77	37°10'
MT	Billings	7,049	-10	91	66	45°50'
	Great Falls	7,750	-15	88	62	47°30'
	Helena	8,129	-16	88	62	46°40'
	Missoula	8,125	-6	88	63	46°50'
NE	Grand Island	6,530	-3	94	74	41°00'
	Lincoln[d]	5,864	-2	95	77	40°50'
	Norfolk	6,979	-4	93	77	42°00'
	North Platte	6,684	-4	94	72	41°10'
	Omaha	6,612	-3	91	77	41°20'
	Scottsbluff	6,673	-3	92	68	41°50'
NV	Elko	7,433	-2	92	62	40°50'
	Ely	7,733	-4	87	59	39°10'
	Las Vegas	2,709	28	106	70	36°10'
	Reno	6,332	10	92	62	39°30'
	Winnemucca	6,761	3	94	62	40°50'
NH	Concord	7,383	-3	87	73	43°10'
NJ	Atlantic City	4,812	13	89	77	39°30'
	Newark	4,589	14	91	76	40°40'
	Trenton[d]	4,980	14	88	76	40°10'
NM	Albuquerque	4,348	16	94	65	35°00'
	Raton	6,228	1	89	64	36°50'
	Roswell	3,793	18	98	70	33°20'
	Silver City	3,705	10	94	64	32°40'

(continued)

TABLE D101—continued
DEGREE DAY AND DESIGN TEMPERATURES[a] FOR CITIES IN THE UNITED STATES

STATE	STATION[b]	HEATING DEGREE DAYS (yearly total)	DESIGN TEMPERATURES			DEGREES NORTH LATITUDE[c]
			Winter	Summer		
			97$^1/_2$%	Dry bulb 2$^1/_2$%	Wet bulb 2$^1/_2$%	
NY	Albany	6,875	-1	88	74	42°50′
	Albany[d]	6,201	1	88	74	42°50′
	Binghamton	7,286	1	83	72	42°10′
	Buffalo	7,062	6	85	73	43°00′
	NY (Central Park)[d]	4,871	15	89	75	40°50′
	NY (Kennedy)	5,219	15	87	75	40°40′
	NY (LaGuardia)	4,811	15	89	75	40°50′
	Rochester	6,748	5	88	73	43°10′
	Schenectady[d]	6,650	1	87	74	42°50′
	Syracuse	6,756	2	87	73	43°10′
NC	Charlotte	3,181	22	93	76	35°10′
	Greensboro	3,805	18	91	76	36°10′
	Raleigh	3,393	20	92	77	35°50′
	Winston-Salem	3,595	20	91	75	36°10′
ND	Bismarck	8,851	-19	91	71	46°50′
	Devils Lake[d]	9,901	-21	88	71	48°10′
	Fargo	9,226	-18	89	74	46°50′
	Williston	9,243	-21	88	70	48°10′
OH	Akron-Canton	6,037	6	86	73	41°00′
	Cincinnati[d]	4,410	6	90	75	39°10′
	Cleveland	6,351	5	88	74	41°20′
	Columbus	5,660	5	90	75	40°00′
	Dayton	5,622	4	89	75	39°50′
	Mansfield	6,403	5	87	74	40°50′
	Sandusky[d]	5,796	6	91	74	41°30′
	Toledo	6,494	1	88	75	41°40′
	Youngstown	6,417	4	86	73	41°20′
OK	Oklahoma City	3,725	13	97	77	35°20′
	Tulsa	3,860	13	98	78	36°10′
OR	Eugene	4,726	22	89	67	44°10′
	Medford	5,008	23	94	68	42°20′
	Portland	4,635	23	85	67	45°40′
	Portland[d]	4,109	24	86	67	45°30′
	Salem	4,754	23	88	68	45°00′
PA	Allentown	5,810	9	88	75	40°40′
	Erie	6,451	9	85	74	42°10′
	Harrisburg	5,251	11	91	76	40°10′
	Philadelphia	5,144	14	90	76	39°50′
	Pittsburgh	5,987	5	86	73	40°30′
	Pittsburgh[d]	5,053	7	88	73	40°30′
	Reading[d]	4,945	13	89	75	40°20′
	Scranton	6,254	5	87	73	41°20′
	Williamsport	5,934	7	89	74	41°10′
RI	Providence	5,954	9	86	74	41°40′
SC	Charleston	2,033	27	91	80	32°50′
	Charleston[d]	1,794	28	92	80	32°50′
	Columbia	2,484	24	95	78	34°00′

(continued)

APPENDIX D

TABLE D101—continued
DEGREE DAY AND DESIGN TEMPERATURES[a] FOR CITIES IN THE UNITED STATES

STATE	STATION[b]	HEATING DEGREE DAYS (yearly total)	Winter 97 1/2 %	Summer Dry bulb 2 1/2 %	Summer Wet bulb 2 1/2 %	DEGREES NORTH LATITUDE[c]
SD	Huron	8,223	-14	93	75	44°30′
	Rapid City	7,345	-7	92	69	44°00′
	Sioux Falls	7,839	-11	91	75	43°40′
TN	Bristol	4,143	14	89	75	36°30′
	Chattanooga	3,254	18	93	77	35°00′
	Knoxville	3,494	19	92	76	35°50′
	Memphis	3,232	18	95	79	35°00′
	Nashville	3,578	14	94	77	36°10′
TX	Abilene	2,624	20	99	74	32°30′
	Austin	1,711	28	98	77	30°20′
	Dallas	2,363	22	100	78	32°50′
	El Paso	2,700	24	98	68	31°50′
	Houston	1,396	32	94	79	29°40′
	Midland	2,591	21	98	72	32°00′
	San Angelo	2,255	22	99	74	31°20′
	San Antonio	1,546	30	97	76	29°30′
	Waco	2,030	26	99	78	31°40′
	Wichita Falls	2,832	18	101	76	34°00′
UT	Salt Lake City	6,052	8	95	65	40°50′
VT	Burlington	8,269	-7	85	72	44°30′
VA	Lynchburg	4,166	16	90	76	37°20′
	Norfolk	3,421	22	91	78	36°50′
	Richmond	3,865	17	92	78	37°30′
	Roanoke	4,150	16	91	74	37°20′
WA	Olympia	5,236	22	83	66	47°00′
	Seattle-Tacoma	5,145	26	80	64	47°30′
	Seattle[d]	4,424	27	82	67	47°40′
	Spokane	6,655	2	90	64	47°40′
WV	Charleston	4,476	11	90	75	38°20′
	Elkins	5,675	6	84	72	38°50′
	Huntington	4,446	10	91	77	38°20′
	Parkersburg[d]	4,754	11	90	76	39°20′
WI	Green Bay	8,029	-9	85	74	44°30′
	La Crosse	7,589	-9	88	75	43°50′
	Madison	7,863	-7	88	75	43°10′
	Milwaukee	7,635	-4	87	74	43°00′
WY	Casper	7,410	-5	90	61	42°50′
	Cheyenne	7,381	-1	86	62	41°10′
	Lander	7,870	-11	88	63	42°50′
	Sheridan	7,680	-8	91	65	44°50′

a. All data were extracted from the 1985 ASHRAE Handbook, Fundamentals Volume.
b. Design data developed from airport temperature observations unless noted.
c. Latitude is given to the nearest 10 minutes. For example, the latitude for Miami, Florida, is given as 25°50′, or 25 degrees 50 minutes.
d. Design data developed from office locations within an urban area, not from airport temperature observations.

Bibliography

The following resource materials are referenced in this chapter or are relevant to the subject matter addressed in this chapter.

ASHRAE, *Handbook of Fundamentals*. Atlanta, GA: American Society of Heating, Refrigerating and Air-Conditioning Engineers, Inc., 2001.

IECC-15, *International Energy Conservation Code*. Washington, DC: International Code Council, 2014.

IPMC-15, *International Property Maintenance Code*. Washington, DC: International Code Council, 2014.

Owenby, J.R., D.S. Ezell, and R.R. Heim, Jr. *Annual Degree Days to Selected Bases Derived from the 1961 to 1990 Normals: Climatography of the United States No. 81—Supplement No. 2*. Asheville, N.C.: U.S. Department of Commerce, National Oceanic and Atmospheric Administration, National Climatic Data Center, 1992.

U.S. Department of Energy (DOE) 90.1, *Code Compliance Manual*. Prepared by Eley Associates under contract to the Building Standards and Guidelines Program. Richland, WA: Pacific Northwest National Laboratory, 1995.

Appendix E:
Sizing of Water Piping System

General Comments

The design of a water supply system ensures adequate water supply to all fixtures at all times and achieves economic sizing of the piping. To do this, the maximum probable rate of flow or demand for which provisions should be made to every portion of the system is used. The demand cannot be determined exactly because most fixtures are used intermittently, and the probability of simultaneous use of such fixtures cannot be established. To be acceptable for estimating the demand of the water supply system, a method must produce the estimated demand for all fixtures on the system during periods of peak demand, avoid oversizing of the piping and be adaptable for estimating the demand of like fixtures as well as different types of fixtures and building occupancy classifications. This procedure takes into consideration the pressure, velocity limitations, materials and characteristics of the water source and demand.

Purpose

Appendix E describes two methods of sizing a water supply system. These methods provide plumbing fixtures and equipment with water in sufficient volume and adequate pressures to enable them to function properly without excessive noise under normal conditions of use. There are many other sizing or design methods conforming to good engineering practices.

The provisions contained in this appendix are not mandatory unless specifically referenced in the adopting ordinance.

SECTION E101
GENERAL

E101.1 Scope.

E101.1.1 This appendix outlines two procedures for sizing a water piping system (see Sections E103.3 and E201.1). The design procedures are based on the minimum static pressure available from the supply source, the head changes in the system caused by friction and elevation, and the rates of flow necessary for operation of various fixtures.

E101.1.2 Because of the variable conditions encountered in hydraulic design, it is impractical to specify definite and detailed rules for sizing of the water piping system. Accordingly, other sizing or design methods conforming to good engineering practice standards are acceptable alternatives to those presented herein.

SECTION E102
INFORMATION REQUIRED

E102.1 Preliminary. Obtain the necessary information regarding the minimum daily static service pressure in the area where the building is to be located. If the building supply is to be metered, obtain information regarding friction loss relative to the rate of flow for meters in the range of sizes likely to be used. Friction loss data can be obtained from most manufacturers of water meters.

E102.2 Demand load.

E102.2.1 Estimate the supply demand of the building main and the principal branches and risers of the system by totaling the corresponding demand from the applicable part of Table E103.3(3).

E102.2.2 Estimate continuous supply demands in gallons per minute (L/m) for lawn sprinklers, air conditioners, etc., and add the sum to the total demand for fixtures. The result is the estimated supply demand for the building supply.

❖ The determination of water demand by the occupants of a building can be very difficult. Studies by Roy B. Hunter starting in 1923 resulted in a method of weighting plumbing fixtures with their water supply load producing effects. Based on experience of using this method for the plumbing design in large federal buildings, Hunter revised the method slightly and then published it in 1940. In 1968, Elmer Jones of the Agricultural Research Service further revised the method using data from small housing developments. This method of estimating demand has been widely recognized as the generally accepted method to produce reliable water demand for all types of occupancies.

Determining occupant water demand involves the probability of use, the rate of use and the length of use. Not all fixtures will be used simultaneously (although halftime at spectator-viewed sporting events may come close), the fixtures are not necessarily operated in the same manner, and the time spent using a fixture varies from person to person. Although sizing a system for continuous operation of all fixtures would be the easiest solution, the resulting pipe sizes might be unrealistically large and thus costly to install. However, on the same note, undersize piping could result in user complaints or even dangerous conditions caused by low water pressure. Special occupancies such as health care, schools, restaurants and stadiums usually require significant engineering expertise to properly design water systems for these applications.

SECTION E103
SELECTION OF PIPE SIZE

E103.1 General. Decide from Table 604.3 what is the desirable minimum residual pressure that should be maintained at the highest fixture in the supply system. If the highest group of fixtures contains flushometer valves, the pressure for the group should be not less than 15 pounds per square inch (psi) (103.4 kPa) flowing. For flush tank supplies, the available pressure should be not less than 8 psi (55.2 kPa) flowing, except blowout action fixtures must be not less than 25 psi (172.4 kPa) flowing.

E103.2 Pipe sizing.

E103.2.1 Pipe sizes can be selected according to the following procedure or by other design methods conforming to acceptable engineering practice and *approved* by the administrative authority. The sizes selected must not be less than the minimum required by this code.

❖ In general, plumbing fixtures require at least 8 psig (55 kPa) of flowing pressure to perform adequately. There are special fixtures that require higher pressures under flowing conditions such as flush valve-type water closets and temperature-controlling (e.g., balanced-pressure/thermostatic control valves) shower controls. The fixture that has the highest flowing pressure demand in the system controls the limit for pressure loss used for system sizing.

E103.2.2 Water pipe sizing procedures are based on a system of pressure requirements and losses, the sum of which must not exceed the minimum pressure available at the supply source. These pressures are as follows:

1. Pressure required at fixture to produce required flow. See Sections 604.3 and 604.5.

2. Static pressure loss or gain (due to head) is computed at 0.433 psi per foot (9.8 kPa/m) of elevation change.

   Example: Assume that the highest fixture supply outlet is 20 feet (6096 mm) above or below the supply source. This produces a static pressure differential of 20 feet by 0.433 psi/foot (2096 mm by 9.8 kPa/m) and an 8.66 psi (59.8 kPa) loss.

3. Loss through water meter. The friction or pressure loss can be obtained from meter manufacturers.

4. Loss through taps in water main.

5. Losses through special devices such as filters, softeners, backflow prevention devices and pressure regulators. These values must be obtained from the manufacturers.

6. Loss through valves and fittings. Losses for these items are calculated by converting to equivalent length of piping and adding to the total pipe length.

7. Loss due to pipe friction can be calculated when the pipe size, the pipe length and the flow through the pipe are known. With these three items, the friction loss can be determined. For piping flow charts not included, use manufacturers' tables and velocity recommendations.

Note: For the purposes of all examples, the following metric conversions are applicable:

1 cubic foot per minute = 0.4719 L/s

1 square foot = 0.0929 m^2

1 degree = 0.0175 rad

1 pound per square inch = 6.895 kPa

1 inch = 25.4 mm

1 foot = 304.8 mm

1 gallon per minute = 3.785 L/m

❖ Frictional losses resulting from the flow of fluids in piping, fittings and standard components are well established from equations of fluid mechanics and testing. Textbooks, handbooks and other reputable publications as well as specific manufacturer's data are resources for frictional loss values. The most important aspect of the information gathering process is to make sure that no component is missed or considered to be "negligible" so that it is not reflected on the computation sheet. As projects come to life, actual component data sheet information should be compared to that used for calculation purposes to assure that all losses are accounted for.

E103.3 Segmented loss method. The size of water service mains, *branch* mains and risers by the segmented loss method, must be determined according to water supply demand [gpm (L/m)], available water pressure [psi (kPa)] and friction loss caused by the water meter and *developed length* of pipe [feet (m)], including equivalent length of fittings. This design procedure is based on the following parameters:

- Calculates the friction loss through each length of the pipe.
- Based on a system of pressure losses, the sum of which must not exceed the minimum pressure available at the street main or other source of supply.
- Pipe sizing based on estimated peak demand, total pressure losses caused by difference in elevation, equipment, *developed length* and pressure required at most remote fixture, loss through taps in water main, losses through fittings, filters, backflow prevention devices, valves and pipe friction.

Because of the variable conditions encountered in hydraulic design, it is impractical to specify definite and detailed rules for sizing of the water piping system. Current sizing methods do not address the differences in the probability of use and flow characteristics of fixtures between types of occupancies. Creating an exact model of predicting the demand for a building is impossible and final studies assessing the impact of water conservation on demand are not yet complete. The following steps are necessary for the segmented loss method.

1. **Preliminary.** Obtain the necessary information regarding the minimum daily static service pressure in the area where the building is to be located. If the building supply is to be metered, obtain information regarding

friction loss relative to the rate of flow for meters in the range of sizes to be used. Friction loss data can be obtained from manufacturers of water meters. It is essential that enough pressure be available to overcome all system losses caused by friction and elevation so that plumbing fixtures operate properly. Section 604.6 requires the water distribution system to be designed for the minimum pressure available taking into consideration pressure fluctuations. The lowest pressure must be selected to guarantee a continuous, adequate supply of water. The lowest pressure in the public main usually occurs in the summer because of lawn sprinkling and supplying water for air-conditioning cooling towers. Future demands placed on the public main as a result of large growth or expansion should also be considered. The available pressure will decrease as additional loads are placed on the public system.

2. **Demand load.** Estimate the supply demand of the building main and the principal branches and risers of the system by totaling the corresponding demand from the applicable part of Table E103.3(3). When estimating peak demand sizing methods typically use water supply fixture units (w.s.f.u.) [see Table E103.3(2)]. This numerical factor measures the load-producing effect of a single plumbing fixture of a given kind. The use of such fixture units can be applied to a single basic probability curve (or table), found in the various sizing methods [Table E103.3(3)]. The fixture units are then converted into gallons per minute (L/m) flow rate for estimating demand.

   2.1. Estimate continuous supply demand in gallons per minute (L/m) for lawn sprinklers, air conditioners, etc., and add the sum to the total demand for fixtures. The result is the estimated supply demand for the building supply. Fixture units cannot be applied to constant use fixtures such as hose bibbs, lawn sprinklers and air conditioners. These types of fixtures must be assigned the gallon per minute (L/m) value.

❖ The code is silent on how to select and apply the values from Table E103.3(2) for an application. One approach to selection would be to use the individual cold and hot water supply fixture unit values for each fixture with no consideration given to the column headed "Total." This would be a conservative method that yields higher overall water demand from the system and may result in pipe sizes larger than if values from the "Total" column were used with appropriate guidance. Note that the "Total" column is not the sum of the fixture hot and cold values, but is a weighted total representing the "load producing effect" of that fixture's combined hot and cold water demand on the water distribution system. A fixture "total" water demand value could be used when the water flow through the pipe section is the result of both hot and cold water being used at the fixture. But if the pipe is carrying water flow as a result of only hot water flow, individual hot and cold fixture water supply fixture units need to be used for the demand loading.

3. **Selection of pipe size.** This water pipe sizing procedure is based on a system of pressure requirements and losses, the sum of which must not exceed the minimum pressure available at the supply source. These pressures are as follows:

   3.1. Pressure required at the fixture to produce required flow. See Section 604.3 and Section 604.5.

   3.2. Static pressure loss or gain (because of head) is computed at 0.433 psi per foot (9.8 kPa/m) of elevation change.

   3.3. Loss through a water meter. The friction or pressure loss can be obtained from the manufacturer.

   3.4. Loss through taps in water main [see Table E103.3(4)].

   3.5. Losses through special devices such as filters, softeners, backflow prevention devices and pressure regulators. These values must be obtained from the manufacturers.

   3.6. Loss through valves and fittings [see Tables E103.3(5) and E103.3(6)]. Losses for these items are calculated by converting to equivalent length of piping and adding to the total pipe length.

   3.7. Loss caused by pipe friction can be calculated when the pipe size, the pipe length and the flow through the pipe are known. With these three items, the friction loss can be determined using Figures E103.3(2) through E103.3(7). When using charts, use pipe inside diameters. For piping flow charts not included, use manufacturers' tables and velocity recommendations. Before attempting to size any water supply system, it is necessary to gather preliminary information which includes available pressure, piping material, select design velocity, elevation differences and *developed length* to most remote fixture. The water supply system is divided into sections at major changes in elevation or where branches lead to fixture groups. The peak demand must be determined in each part of the hot and cold water supply system which includes the corresponding water supply fixture unit and conversion to gallons per minute (L/m) flow rate to be expected through each section. Sizing methods require the determination of the "most hydraulically remote" fixture to compute the pressure loss caused by pipe and fittings. The hydraulically remote fixture represents the most downstream fixture along the circuit of piping requiring the most available pressure to operate properly. Consideration must be given to all pressure demands and losses, such as friction caused by pipe, fittings and equipment, elevation and the residual pressure required by Table 604.3. The two most common and frequent complaints about the water supply system operation are lack of adequate pressure and noise.

❖ The "Segmented Loss Method" is also known as the "Velocity-Pressure Drop" method. Once flow through a pipe is known in gallons per minute (L/m) for any particular pipe section under consideration, a pipe size can be selected based on a recommended maximum velocity of the water through a specific pipe material.

Experience and extensive research on all types of pipe materials have produced general guidelines for recommended maximum velocities in domestic water piping. For cold water piping, maximum velocities in the range of 5 to 8 feet per second (1.5 to 2.4 m/s) are typical and for hot water piping, especially for temperatures above 150°F (66°C), maximum velocities of 4 feet per second (1.2 m/s) are typical. However, the type of pipe material has a significant impact on the maximum recommended velocities. For example, in general, manufacturers of cross-linked polyethylene tubing (PEX) allow velocities to approach 12 feet per second (3.7 m/s) in both hot [maximum 140°F (60°F)] and cold piping, whereas, copper piping manufacturers recommend using the lower velocity general guidelines as mentioned above.

Control of velocity in piping is necessary for a variety of reasons. High velocities create flowing "noise" and increased possibility for erosion of the pipe material. High velocities also contribute to water hammer problems because rapidly flowing fluid has very high energy. When high-energy fluid is stopped abruptly, that energy dissipates in a shock wave moving upstream from the valve that closed, causing noise and instantaneous high pressures (and possibly damaging forces) within the pipe. But most importantly, high fluid velocities cause a loss in flowing pressure. The inside wall of new building water distribution pipe, while visually appearing smooth, has a roughness that fluid moving though the pipe tends to "drag" on. The higher the velocity is in the pipe, the greater the drag (friction) forces are on the fluid. For any fluid flowing full in a pipe, given an initial flowing pressure and velocity at the beginning of a section of pipe, at any downstream point in the pipe, there will be a decrease in the flowing pressure in the pipe. The friction takes energy away from the fluid. This loss of energy (i.e., pressure) caused by fluid flow dragging on the inside walls of a pipe is called pressure drop or pressure loss.

Friction loss (or pressure drop) graphs are available for a variety of pipe materials. Figures E103.3(2) through (4) are for new copper water piping (Types K, L and M, respectively) with recessed, streamline soldered joints (see graph Note a). Note that these graphs show velocity lines only up to 10 feet per second (3 m/s). The code does not provide specific friction loss graphs for any other types of water pipe as listed in Tables 605.3 and 605.4, but provides only the graphs in Figures E103.3(5) through (7). Figure E103.3(5), although not specifically labeled as such, is the friction loss graph for Schedule 40 new steel pipe with inside walls of "fairly smooth" condition. New galvanized steel piping, rarely used in modern construction for water distribution, is considered to have a "fairly smooth" inside wall condition when new. However, once galvanized steel pipe has been in service for an extended length of time, the inside wall condition usually degrades to what is considered to be a "fairly rough or rough" condition as a result of corrosion and deposit build up. The pipe material and wall thickness used as the basis for Figures E103.3(6) and (7) is not specified, only the condition of the pipe interior. Figure E103.3(7) also has an apparent discrepancy in specifying the inside wall condition at the top of the graph as "rough." The caption of the graph says "fairly rough" and Note a says "very rough."

Perhaps of most importance is not what the friction losses are when piping is new but what the friction losses will be after extended service. Friction loss graphs for a range of inside wall surface conditions are available from many sources to aid the system designer in determining accurate values for expected pressure drop.

Problem: What size Type L copper water pipe, service and distribution will be required to serve a two-story factory building having on each floor, back-to-back, two toilet rooms each equipped with hot and cold water? The highest fixture is 21 feet (6401 mm) above the street main, which is tapped with a 2-inch (51 mm) corporation cock at which point the minimum pressure is 55 psi (379.2 kPa). In the building basement, a 2-inch (51 mm) meter with a maximum pressure drop of 11 psi (75.8 kPa) and 3-inch (76 mm) reduced pressure principle backflow preventer with a maximum pressure drop of 9 psi (621 kPa) are to be installed. The system is shown by Figure E103.3(1). To be determined are the pipe sizes for the service main and the cold and hot water distribution pipes.

Solution: A tabular arrangement such as shown in Table E103.3(1) should first be constructed. The steps to be followed are indicated by the tabular arrangement itself as they are in sequence, columns 1 through 10 and lines A through L.

Step 1

Columns 1 and 2: Divide the system into sections breaking at major changes in elevation or where branches lead to fixture groups. After point B [see Figure E103.3(1)], separate consideration will be given to the hot and cold water piping. Enter the sections to be considered in the service and cold water piping in Column 1 of the tabular arrangement. Column 1 of Table E103.3(1) provides a line-by-line recommended tabular arrangement for use in solving pipe sizing.

The objective in designing the water supply system is to ensure an adequate water supply and pressure to all fixtures and equipment. Column 2 provides the pounds per square inch (psi) to be considered separately from the minimum pressure available at the main. Losses to take into consideration are the following: the differences in elevations between the water supply source and the highest water supply outlet, meter pressure losses, the tap in main loss, special fixture devices such as water softeners and backflow prevention devices and the pressure required at the most remote fixture outlet. The difference in elevation can result in an increase or decrease in available pressure at the main. Where the water supply outlet is located above the source, this results in a loss in the available pressure and is subtracted from the pressure at the water source. Where the highest water supply outlet is located below the water supply source, there will be an increase in pressure that is added to the available pressure of the water source.

Column 3: According to Table E103.3(3), determine the gpm (L/m) of flow to be expected in each section of the system. These flows range from 28.6 to 108 gpm. Load values for fixtures must be determined as water supply fixture units and then converted to a gallon-per-minute (gpm) rating to determine peak demand. When calculating peak demands, the water supply fixture units are added and then converted to the gallon-per-minute rating. For continuous flow fixtures such as hose bibbs and lawn sprinkler systems, add the gallon-per-minute demand to the intermittent demand of fixtures. For example, a total of 120 water supply fixture units is converted to a demand of 48 gallons per minute. Two hose bibbs × 5 gpm demand = 10 gpm. Total gpm rating = 48.0 gpm + 10 gpm = 58.0 gpm demand.

Step 2

Line A: Enter the minimum pressure available at the main source of supply in Column 2. This is 55 psi (379.2 kPa). The local water authorities generally keep records of pressures at different times of day and year. The available pressure can also be checked from nearby buildings or from fire department hydrant checks.

Line B: Determine from Table 604.3 the highest pressure required for the fixtures on the system, which is 15 psi (103.4 kPa), to operate a flushometer valve. The most remote fixture outlet is necessary to compute the pressure loss caused by pipe and fittings, and represents the most downstream fixture along the circuit of piping requiring the available pressure to operate properly as indicated by Table 604.3.

Line C: Determine the pressure loss for the meter size given or assumed. The total water flow from the main through the service as determined in Step 1 will serve to aid in the meter selected. There are three common types of water meters; the pressure losses are determined by the American Water Works Association Standards for displacement type, compound type and turbine type. The maximum pressure loss of such devices takes into consideration the meter size, safe operating capacity (gpm) and maximum rates for continuous operations (gpm). Typically, equipment imparts greater pressure losses than piping.

Line D: Select from Table E103.3(4) and enter the pressure loss for the tap size given or assumed. The loss of pressure through taps and tees in pounds per square inch (psi) are based on the total gallon-per-minute flow rate and size of the tap.

Line E: Determine the difference in elevation between the main and source of supply and the highest fixture on the system. Multiply this figure, expressed in feet, by 0.43 psi (2.9 kPa). Enter the resulting psi loss on Line E. The difference in elevation between the water supply source and the highest water supply outlet has a significant impact on the sizing of the water supply system. The difference in elevation usually results in a loss in the available pressure because the water supply outlet is generally located above the water supply source. The loss is caused by the pressure required to lift the water to the outlet. The pressure loss is subtracted from the pressure at the water source. Where the highest water supply outlet is located below the water source, there will be an increase in pressure which is added to the available pressure of the water source.

Lines F, G and H: The pressure losses through filters, backflow prevention devices or other special fixtures must be obtained from the manufacturer or estimated and entered on these lines. Equipment such as backflow prevention devices, check valves, water softeners, instantaneous or tankless water heaters, filters and strainers can impart a much greater pressure loss than the piping. The pressure losses can range from 8 psi to 30 psi.

Step 3

Line I: The sum of the pressure requirements and losses that affect the overall system (Lines B through H) is entered on this line. Summarizing the steps, all of the system losses are subtracted from the minimum water pressure. The remainder is the pressure available for friction, defined as the energy available to push the water through the pipes to each fixture. This force can be used as an average pressure loss, as long as the pressure available for friction is not exceeded. Saving a certain amount for available water supply pressures as an area incurs growth, or because of aging of the pipe or equipment added to the system is recommended.

Step 4

Line J: Subtract Line I from Line A. This gives the pressure that remains available from overcoming friction losses in the system. This figure is a guide to the pipe size that is chosen for each section, incorporating the total friction losses to the most remote outlet (measured length is called developed length).

> **Exception:** When the main is above the highest fixture, the resulting psi must be considered a pressure gain (static head gain) and omitted from the sums of Lines B through H and added to Line J.

The maximum friction head loss that can be tolerated in the system during peak demand is the difference between the static pressure at the highest and most remote outlet at no-flow conditions and the minimum flow pressure required at that outlet. If the losses are within the required limits, then every run of pipe will also be within the required friction head loss. Static pressure loss is the most remote outlet in feet × 0.433 = loss in psi caused by elevation differences.

Step 5

Column 4: Enter the length of each section from the main to the most remote outlet (at Point E). Divide the water supply system into sections breaking at major changes in elevation or where branches lead to fixture groups.

Step 6

E103.3.3. Selection of pipe size, Step 6 Column 5: When selecting a trial pipe size, the length from the water service or meter to the most remote fixture outlet must be measured to determine the developed length. However, in systems having a flushometer valve or temperature controlled shower at the top most floors the developed length would be from the water meter to the most remote flushometer valve on the system. A rule of thumb is that size will become progressively smaller as the system extends farther from the main source of supply. Trial pipe size may be arrived at by the following formula:

Line J: (Pressure available to overcome pipe friction) × 100/equivalent length of run total *developed length* to most remote fixture × percentage factor of 1.5 (note: a percentage factor is used only as an estimate for friction losses imposed for fittings for initial trial pipe size) = psi (average pressure drops per 100 feet of pipe).

For trial pipe size see Figure E 103.3(3) (Type L copper) based on 2.77 psi and a 108 gpm = $2^1/_2$ inches. To determine the equivalent length of run to the most remote outlet, the *developed length* is determined and added to the friction losses for fittings and valves. The developed lengths of the designated pipe sections are as follows:

A - B	54 ft
B - C	8 ft
C - D	13 ft
D - E	150 ft
Total developed length = 225 ft	

The equivalent length of the friction loss in fittings and valves must be added to the *developed length* (most remote outlet). Where the size of fittings and valves is not known, the added friction loss should be approximated. A general rule that has been used is to add 50 percent of the *developed length* to allow for fittings and valves. For example, the equivalent length of run equals the *developed length* of run (225 ft × 1.5 = 338 ft). The total equivalent length of run for determining a trial pipe size is 338 feet.

Example: 9.36 (pressure available to overcome pipe friction) × 100/ 338 (equivalent length of run = 225 × 1.5) = 2.77 psi (average pressure drop per 100 feet of pipe).

Step 7

Column 6: Select from Table E103.3(6) the equivalent lengths for the trial pipe size of fittings and valves on each pipe section. Enter the sum for each section in Column 6. (The number of fittings to be used in this example must be an estimate.) The equivalent length of piping is the developed length plus the equivalent lengths of pipe corresponding to friction head losses for fittings and valves. Where the size of fittings and valves is not known, the added friction head losses must be approximated. An estimate for this example is found in Table E.1.

Step 8

Column 7: Add the figures from Column 4 and Column 6, and enter in Column 7. Express the sum in hundreds of feet.

Step 9

Column 8: Select from Figure E103.3(3) the friction loss per 100 feet (30 480 mm) of pipe for the gallon-per-minute flow in a section (Column 3) and trial pipe size (Column 5). Maximum friction head loss per 100 feet is determined on the basis of total pressure available for friction head loss and the longest equivalent length of run. The selection is based on the gallon-per-minute demand, the uniform friction head loss, and the maximum design velocity. Where the size indicated by hydraulic table indicates a velocity in excess of the selected velocity, a size must be selected which produces the required velocity.

Step 10

Column 9: Multiply the figures in Columns 7 and 8 for each section and enter in Column 9.

Total friction loss is determined by multiplying the friction loss per 100 feet (30 480 mm) for each pipe section in the total *developed length* by the pressure loss in fittings expressed as equivalent length in feet. Note: Section C-F should be considered in the total pipe friction losses only if greater loss occurs in Section C-F than in pipe section D-E. Section C-F is not considered in the total *developed length*. Total friction loss in equivalent length is determined in Table E.2.

Step 11

Line K: Enter the sum of the values in Column 9. The value is the total friction loss in equivalent length for each designated pipe section.

Step 12

Line L: Subtract Line J from Line K and enter in Column 10.

The result should always be a positive or plus figure. If it is not, repeat the operation using Columns 5, 6, 8 and 9 until a balance or near balance is obtained. If the difference between Lines J and K is a high positive number, it is an indication that the pipe sizes are too large and should be reduced, thus saving materials. In such a case, the operations using Columns 5, 6, 8 and 9 should again be repeated.

The total friction losses are determined and subtracted from the pressure available to overcome pipe friction for trial pipe size. This number is critical as it provides a guide to whether the pipe size selected is too large and the process should be repeated to obtain an economically designed system.

Answer: The final figures entered in Column 5 become the design pipe size for the respective sections. Repeating this operation a second time using the same sketch but considering the demand for hot water, it is possible to size the hot water distribution piping. This has been worked up as a part of the overall problem in the tabular arrangement used for sizing the service and water distribution piping. Note that consideration must be given to the pressure losses from the street main to the water heater (Section A-B) in determining the hot water pipe sizes.

TABLE E.1

COLD WATER PIPE SECTION	FITTINGS/VALVES	PRESSURE LOSS EXPRESSED AS EQUIVALENT LENGTH OF TUBE (feet)	HOT WATER PIPE SECTION	FITTINGS/ VALVES	PRESSURE LOSS EXPRESSED AS EQUIVALENT OF TUBE (feet)
A-B	3-2$\frac{1}{2}$" Gate valves	3	A-B	3-2$\frac{1}{2}$" Gate valves	3
	1-2$\frac{1}{2}$" Side branch tee	12		1-2$\frac{1}{2}$" Side branch tee	12
B-C	1-2$\frac{1}{2}$" Straight run tee	0.5	B-C	1-2" Straight run tee	7
				1-2" 90-degree ell	0.5
C-F	1-2$\frac{1}{2}$" Side branch tee	12	C-F	1-1$\frac{1}{2}$" Side branch tee	7
C-D	1-2$\frac{1}{2}$" 90-degree ell	7	C-D	1-1$\frac{1}{2}$" 90-degree ell	4
D-E	1-2$\frac{1}{2}$" Side branch tee	12	D-E	1-1$\frac{1}{2}$" Side branch tee	7

For SI: 1 foot = 304.8 mm, 1 inch = 25.4 mm.

TABLE E.2

PIPE SECTIONS	FRICTION LOSS EQUIVALENT LENGTH (feet)	
	Cold Water	Hot Water
A-B	0.69 × 3.2 = 2.21	0.69 × 3.2 = 2.21
B-C	0.085 × 3.1 = 0.26	0.16 × 1.4 = 0.22
C-D	0.20 × 1.9 = 0.38	0.17 × 3.2 = 0.54
D-E	1.62 × 1.9 = 3.08	1.57 × 3.2 = 5.02
Total pipe friction losses (Line K)	5.93	7.99

For SI: 1 foot = 304.8 mm, 1 gpm = 3.785 L/m.

APPENDIX E

For SI: 1 foot = 304. mm, 1 gpm = 3.785 L/m.

FIGURE E103.3(1)
EXAMPLE—SIZING

APPENDIX E

TABLE E103.3(1)
RECOMMENDED TABULAR ARRANGEMENT FOR USE IN SOLVING PIPE SIZING PROBLEMS

COLUMN	1		2	3	4	5	6	7	8	9	10
Line	Description		Lb per square inch (psi)	Gal. per min through section	Length of section (feet)	Trial pipe size (inches)	Equivalent length of fittings and valves (feet)	Total equivalent length col. 4 and col. 6 (100 feet)	Friction loss per 100 feet of trial size pipe (psi)	Friction loss in equivalent length col. 8 × col. 7 (psi)	Excess pressure over friction losses (psi)
A	Service and cold water distribution piping[a]	Minimum pressure available at main	55.00								
B		Highest pressure required at a fixture (Table 604.3)	15.00								
C		Meter loss 2″ meter	11.00								
D		Tap in main loss 2″ tap (Table E103A)	1.61								
E		Static head loss 21 × 43 psi	9.03								
F		Special fixture loss backflow preventer	9.00								
G		Special fixture loss—Filter	0.00								
H		Special fixture loss—Other	0.00								
I		Total overall losses and requirements (Sum of Lines B through H)	45.64								
J		Pressure available to overcome pipe friction (Line A minus Lines B to H)	9.36								
	DESIGNATION Pipe section (from diagram) Cold water Distribution piping	FU	264								
		AB	288	108.0	54	2½	15.00	0.69	3.2	2.21	—
		BC	264	104.5	8	2½	0.5	0.85	3.1	0.26	—
		CD	132	77.0	13	2½	7.00	0.20	1.9	0.38	—
		CF[b]	132	77.0	150	2½	12.00	1.62	1.9	3.08	—
		DE[b]	132	77.0	150	2½	12.00	1.62	1.9	3.08	—
K	Total pipe friction losses (cold)		—	—	—	—	—	—	—	5.93	—
L	Difference (Line J minus Line K)		—	—	—	—	—	—	—	—	3.43
	Pipe section (from diagram) Diagram Hot water Distribution Piping	A′B′	288	108.0	54	2½	12.00	0.69	3.3	2.21	—
		B′C′	24	38.0	8	2	7.5	0.16	1.4	0.22	—
		C′D′	12	28.6	13	1½	4.0	0.17	3.2	0.54	—
		C′F′[b]	12	28.6	150	1½	7.00	1.57	3.2	5.02	—
		D′E′[b]	12	28.6	150	1½	7.00	1.57	3.2	5.02	—
K	Total pipe friction losses (hot)		—	—	—	—	—	—	—	7.99	—
L	Difference (Line J minus Line K)		—	—	—	—	—	—	—	—	1.37

For SI: 1 inch = 25.4 mm, 1 foot = 304.8 mm, 1 psi = 6.895 kPa, 1 gpm = 3.785 L/m.
a. To be considered as pressure gain for fixtures below main (to consider separately, omit from "I" and add to "J").
b. To consider separately, in K use C-F only if greater loss than above.

TABLE E103.3(2)
LOAD VALUES ASSIGNED TO FIXTURES[a]

FIXTURE	OCCUPANCY	TYPE OF SUPPLY CONTROL	LOAD VALUES, IN WATER SUPPLY FIXTURE UNITS (wsfu)		
			Cold	Hot	Total
Bathroom group	Private	Flush tank	2.7	1.5	3.6
Bathroom group	Private	Flushometer valve	6.0	3.0	8.0
Bathtub	Private	Faucet	1.0	1.0	1.4
Bathtub	Public	Faucet	3.0	3.0	4.0
Bidet	Private	Faucet	1.5	1.5	2.0
Combination fixture	Private	Faucet	2.25	2.25	3.0
Dishwashing machine	Private	Automatic	—	1.4	1.4
Drinking fountain	Offices, etc.	$^3/_8$" valve	0.25	—	0.25
Kitchen sink	Private	Faucet	1.0	1.0	1.4
Kitchen sink	Hotel, restaurant	Faucet	3.0	3.0	4.0
Laundry trays (1 to 3)	Private	Faucet	1.0	1.0	1.4
Lavatory	Private	Faucet	0.5	0.5	0.7
Lavatory	Public	Faucet	1.5	1.5	2.0
Service sink	Offices, etc.	Faucet	2.25	2.25	3.0
Shower head	Public	Mixing valve	3.0	3.0	4.0
Shower head	Private	Mixing valve	1.0	1.0	1.4
Urinal	Public	1" flushometer valve	10.0	—	10.0
Urinal	Public	$^3/_4$" flushometer valve	5.0	—	5.0
Urinal	Public	Flush tank	3.0	—	3.0
Washing machine (8 lb)	Private	Automatic	1.0	1.0	1.4
Washing machine (8 lb)	Public	Automatic	2.25	2.25	3.0
Washing machine (15 lb)	Public	Automatic	3.0	3.0	4.0
Water closet	Private	Flushometer valve	6.0	—	6.0
Water closet	Private	Flush tank	2.2	—	2.2
Water closet	Public	Flushometer valve	10.0	—	10.0
Water closet	Public	Flush tank	5.0	—	5.0
Water closet	Public or private	Flushometer tank	2.0	—	2.0

For SI: 1 inch = 25.4 mm, 1 pound = 0.454 kg.

a. For fixtures not listed loads should be assumed by comparing the fixture to one listed using water in similar quantities and at similar rates. The assigned loads for fixtures with both hot and cold water supplies are given for separate hot and cold water loads and for total load. The separate hot and cold water loads being three-fourths of the total load for the fixture in each case.

APPENDIX E

TABLE E103.3(3)
TABLE FOR ESTIMATING DEMAND

SUPPLY SYSTEMS PREDOMINANTLY FOR FLUSH TANKS			SUPPLY SYSTEMS PREDOMINANTLY FOR FLUSHOMETER VALVES		
Load	Demand		Load	Demand	
(Water supply fixture units)	(Gallons per minute)	(Cubic feet per minute)	(Water supply fixture units)	(Gallons per minute)	(Cubic feet per minute)
1	3.0	0.04104	—	—	—
2	5.0	0.0684	—	—	—
3	6.5	0.86892	—	—	—
4	8.0	1.06944	—	—	—
5	9.4	1.256592	5	15.0	2.0052
6	10.7	1.430376	6	17.4	2.326032
7	11.8	1.577424	7	19.8	2.646364
8	12.8	1.711104	8	22.2	2.967696
9	13.7	1.831416	9	24.6	3.288528
10	14.6	1.951728	10	27.0	3.60936
11	15.4	2.058672	11	27.8	3.716304
12	16.0	2.13888	12	28.6	3.823248
13	16.5	2.20572	13	29.4	3.930192
14	17.0	2.27256	14	30.2	4.037136
15	17.5	2.3394	15	31.0	4.14408
16	18.0	2.90624	16	31.8	4.241024
17	18.4	2.459712	17	32.6	4.357968
18	18.8	2.513184	18	33.4	4.464912
19	19.2	2.566656	19	34.2	4.571856
20	19.6	2.620128	20	35.0	4.6788
25	21.5	2.87412	25	38.0	5.07984
30	23.3	3.114744	30	42.0	5.61356
35	24.9	3.328632	35	44.0	5.88192
40	26.3	3.515784	40	46.0	6.14928
45	27.7	3.702936	45	48.0	6.41664
50	29.1	3.890088	50	50.0	6.684
60	32.0	4.27776	60	54.0	7.21872
70	35.0	4.6788	70	58.0	7.75344
80	38.0	5.07984	80	61.2	8.181216
90	41.0	5.48088	90	64.3	8.595624
100	43.5	5.81508	100	67.5	9.0234
120	48.0	6.41664	120	73.0	9.75864
140	52.5	7.0182	140	77.0	10.29336
160	57.0	7.61976	160	81.0	10.82808
180	61.0	8.15448	180	85.5	11.42964
200	65.0	8.6892	200	90.0	12.0312
225	70.0	9.3576	225	95.5	12.76644
250	75.0	10.026	250	101.0	13.50168

(continued)

APPENDIX E

TABLE E103.3(3)-continued
TABLE FOR ESTIMATING DEMAND

SUPPLY SYSTEMS PREDOMINANTLY FOR FLUSH TANKS			SUPPLY SYSTEMS PREDOMINANTLY FOR FLUSHOMETER VALVES		
Load	Demand		Load	Demand	
(Water supply fixture units)	(Gallons per minute)	(Cubic feet per minute)	(Water supply fixture units)	(Gallons per minute)	(Cubic feet per minute)
275	80.0	10.6944	275	104.5	13.96956
300	85.0	11.3628	300	108.0	14.43744
400	105.0	14.0364	400	127.0	16.97736
500	124.0	16.57632	500	143.0	19.11624
750	170.0	22.7256	750	177.0	23.66136
1,000	208.0	27.80544	1,000	208.0	27.80544
1,250	239.0	31.94952	1,250	239.0	31.94952
1,500	269.0	35.95992	1,500	269.0	35.95992
1,750	297.0	39.70296	1,750	297.0	39.70296
2,000	325.0	43.446	2,000	325.0	43.446
2,500	380.0	50.7984	2,500	380.0	50.7984
3,000	433.0	57.88344	3,000	433.0	57.88344
4,000	525.0	70.182	4,000	525.0	70.182
5,000	593.0	79.27224	5,000	593.0	79.27224

For SI: 1 inch = 25.4 mm, 1 gallon per minute = 3.785 L/m, 1 cubic foot per minute = 0.28 m^3 per minute.

TABLE E103.3(4)
LOSS OF PRESSURE THROUGH TAPS AND TEES IN POUNDS PER SQUARE INCH (psi)

GALLONS PER MINUTE	SIZE OF TAP OR TEE (inches)						
	$^5/_8$	$^3/_4$	1	$1^1/_4$	$1^1/_2$	2	3
10	1.35	0.64	0.18	0.08	—	—	—
20	5.38	2.54	0.77	0.31	0.14	—	—
30	12.10	5.72	1.62	0.69	0.33	0.10	—
40	—	10.20	3.07	1.23	0.58	0.18	—
50	—	15.90	4.49	1.92	0.91	0.28	—
60	—	—	6.46	2.76	1.31	0.40	—
70	—	—	8.79	3.76	1.78	0.55	0.10
80	—	—	11.50	4.90	2.32	0.72	0.13
90	—	—	14.50	6.21	2.94	0.91	0.16
100	—	—	17.94	7.67	3.63	1.12	0.21
120	—	—	25.80	11.00	5.23	1.61	0.30
140	—	—	35.20	15.00	7.12	2.20	0.41
150	—	—	—	17.20	8.16	2.52	0.47
160	—	—	—	19.60	9.30	2.92	0.54
180	—	—	—	24.80	11.80	3.62	0.68
200	—	—	—	30.70	14.50	4.48	0.84
225	—	—	—	38.80	18.40	5.60	1.06
250	—	—	—	47.90	22.70	7.00	1.31
275	—	—	—	—	27.40	7.70	1.59
300	—	—	—	—	32.60	10.10	1.88

For SI: 1 inch = 25.4 mm, 1 pound per square inch = 6.895 kpa, 1 gallon per minute = 3.785 L/m.

TABLE E103.3(5)
ALLOWANCE IN EQUIVALENT LENGTHS OF PIPE FOR FRICTION LOSS IN VALVES AND THREADED FITTINGS (feet)

FITTING OR VALVE	PIPE SIZE (inches)							
	1/2	3/4	1	1 1/4	1 1/2	2	2 1/2	3
45-degree elbow	1.2	1.5	1.8	2.4	3.0	4.0	5.0	6.0
90-degree elbow	2.0	2.5	3.0	4.0	5.0	7.0	8.0	10.0
Tee, run	0.6	0.8	0.9	1.2	1.5	2.0	2.5	3.0
Tee, branch	3.0	4.0	5.0	6.0	7.0	10.0	12.0	15.0
Gate valve	0.4	0.5	0.6	0.8	1.0	1.3	1.6	2.0
Balancing valve	0.8	1.1	1.5	1.9	2.2	3.0	3.7	4.5
Plug-type cock	0.8	1.1	1.5	1.9	2.2	3.0	3.7	4.5
Check valve, swing	5.6	8.4	11.2	14.0	16.8	22.4	28.0	33.6
Globe valve	15.0	20.0	25.0	35.0	45.0	55.0	65.0	80.0
Angle valve	8.0	12.0	15.0	18.0	22.0	28.0	34.0	40.0

For SI: 1 inch = 25.4 mm, 1 foot = 304.8 mm, 1 degree = 0.0175 rad.

TABLE E103.3(6)
PRESSURE LOSS IN FITTINGS AND VALVES EXPRESSED AS EQUIVALENT LENGTH OF TUBE[a] (feet)

NOMINAL OR STANDARD SIZE (inches)	FITTINGS					VALVES			
	Standard Ell		90-Degree Tee		Coupling	Ball	Gate	Butterfly	Check
	90 Degree	45 Degree	Side Branch	Straight Run					
3/8	0.5	—	1.5	—	—	—	—	—	1.5
1/2	1	0.5	2	—	—	—	—	—	2
5/8	1.5	0.5	2	—	—	—	—	—	2.5
3/4	2	0.5	3	—	—	—	—	—	3
1	2.5	1	4.5	—	—	0.5	—	—	4.5
1 1/4	3	1	5.5	0.5	0.5	0.5	—	—	5.5
1 1/2	4	1.5	7	0.5	0.5	0.5	—	—	6.5
2	5.5	2	9	0.5	0.5	0.5	0.5	7.5	9
2 1/2	7	2.5	12	0.5	0.5	—	1	10	11.5
3	9	3.5	15	1	1	—	1.5	15.5	14.5
3 1/2	9	3.5	14	1	1	—	2	—	12.5
4	12.5	5	21	1	1	—	2	16	18.5
5	16	6	27	1.5	1.5	—	3	11.5	23.5
6	19	7	34	2	2	—	3.5	13.5	26.5
8	29	11	50	3	3	—	5	12.5	39

For SI: 1 inch = 25.4 mm, 1 foot = 304.8 mm, 1 degree = 0.01745 rad.

a. Allowances are for streamlined soldered fittings and recessed threaded fittings. For threaded fittings, double the allowances shown in the table. The equivalent lengths presented above are based on a C factor of 150 in the Hazen-Williams friction loss formula. The lengths shown are rounded to the nearest half-foot.

APPENDIX E

PRESSURE DROP PER 100 FEET OF TUBE, POUNDS PER SQUARE INCH

Note: Fluid velocities in excess of 5 to 8 feet/second are not usually recommended.

For SI: 1 inch = 25.4 mm, 1 foot = 304.8 mm, 1 gpm = 3.785 L/m, 1 psi = 6.895 kPa, 1 foot per second = 0.305 m/s.
a. This chart applies to smooth new copper tubing with recessed (streamline) soldered joints and to the actual sizes of types indicated on the diagram.

**FIGURE E103.3(2)
FRICTION LOSS IN SMOOTH PIPE[a] (TYPE K, ASTM B88 COPPER TUBING)**

APPENDIX E

PRESSURE DROP PER 100 FEET OF TUBE, POUNDS PER SQUARE INCH

Note: Fluid velocities in excess of 5 to 8 feet/second are not usually recommended.

For SI: 1 inch = 25.4 mm, 1 foot = 304.8 mm, 1 gpm = 3.785 L/m, 1 psi = 6.895 kPa, 1 foot per second = 0.305 m/s.

a. This chart applies to smooth new copper tubing with recessed (streamline) soldered joints and to the actual sizes of types indicated on the diagram.

FIGURE E103.3(3)
FRICTION LOSS IN SMOOTH PIPE[a] (TYPE L, ASTM B88 COPPER TUBING)

APPENDIX E

PRESSURE DROP PER 100 FEET OF TUBE, POUNDS PER SQUARE INCH

Note: Fluid velocities in excess of 5 to 8 feet/second are not usually recommended.

For SI: 1 inch = 25.4 mm, 1 foot = 304.8 mm, 1 gpm = 3.785 L/m, 1 psi = 6.895 kPa, 1 foot per second = 0.305 m/s.
a. This chart applies to smooth new copper tubing with recessed (streamline) soldered joints and to the actual sizes of types indicated on the diagram.

FIGURE E103.3(4)
FRICTION LOSS IN SMOOTH PIPE[a] (TYPE M, ASTM B88 COPPER TUBING)

APPENDIX E

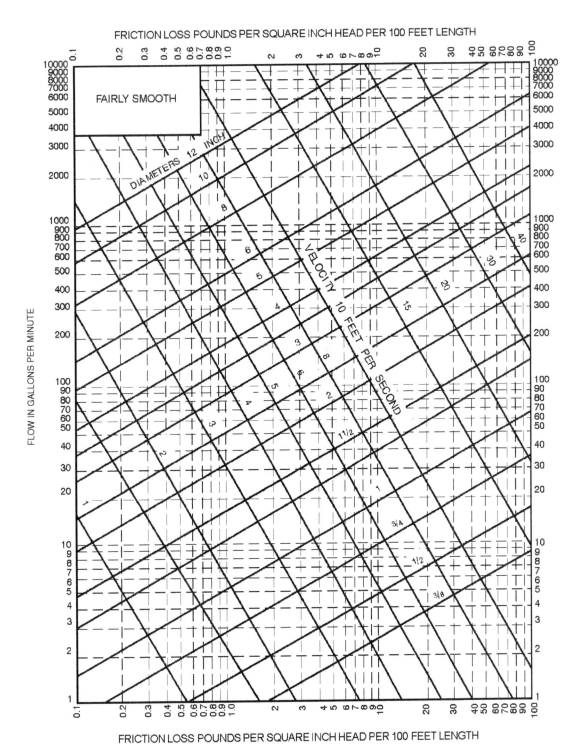

For SI: 1 inch = 25.4 mm, 1 foot = 304.8 mm, 1 gpm = 3.785 L/m, 1 psi = 6.895 kPa, 1 foot per second = 0.305 m/s.
a. This chart applies to smooth new steel (fairly smooth) pipe and to actual diameters of standard-weight pipe.

**FIGURE E103.3(5)
FRICTION LOSS IN FAIRLY SMOOTH PIPE[a]**

APPENDIX E

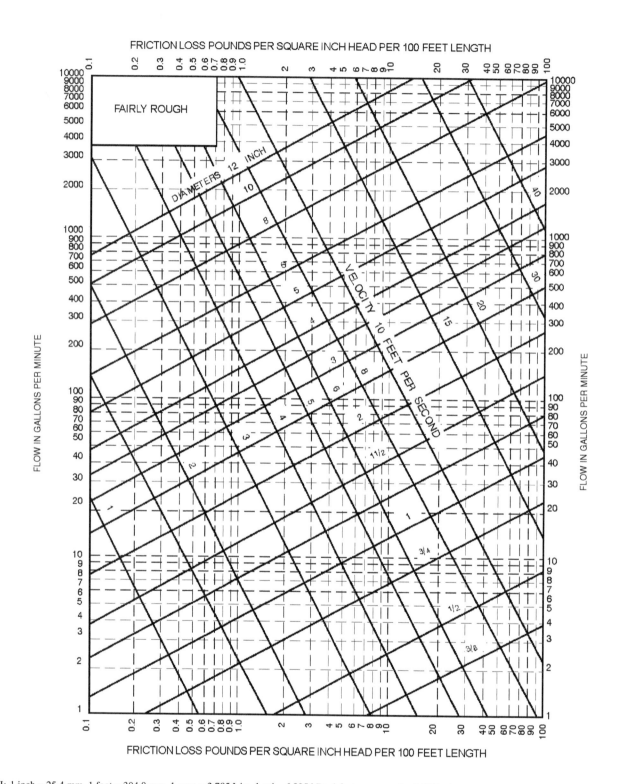

For SI: 1 inch = 25.4 mm, 1 foot = 304.8 mm, 1 gpm = 3.785 L/m, 1 psi = 6.895 kPa, 1 foot per second = 0.305 m/s.
a. This chart applies to fairly rough pipe and to actual diameters which in general will be less than the actual diameters of the new pipe of the same kind.

**FIGURE E103.3(6)
FRICTION LOSS IN FAIRLY ROUGH PIPE[a]**

APPENDIX E

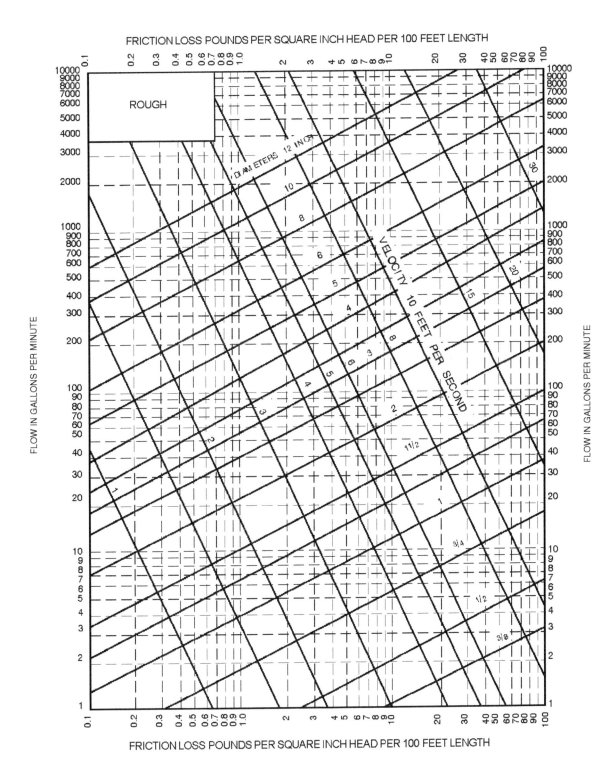

For SI: 1 inch = 25.4 mm, 1 foot = 304.8 mm, 1 gpm = 3.785 L/m, 1 psi = 6.895 kPa, 1 foot per second = 0.305 m/s.
a. This chart applies to very rough pipe and existing pipe and to their actual diameters.

**FIGURE E103.3(7)
FRICTION LOSS IN ROUGH PIPE[a]**

SECTION E201
SELECTION OF PIPE SIZE

E201.1 Size of water-service mains, branch mains and risers. The minimum size water service pipe shall be $^3/_4$ inch (19.1 mm). The size of water service mains, *branch* mains and risers shall be determined according to water supply demand [gpm (L/m)], available water pressure [psi (kPa)] and friction loss due to the water meter and *developed length* of pipe [feet (m)], including equivalent length of fittings. The size of each water distribution system shall be determined according to the procedure outlined in this section or by other design methods conforming to acceptable engineering practice and *approved* by the code official:

1. Supply load in the building water-distribution system shall be determined by total load on the pipe being sized, in terms of water-supply fixture units (w.s.f.u.), as shown in Table E103.3(2). For fixtures not listed, choose a w.s.f.u. value of a fixture with similar flow characteristics.

2. Obtain the minimum daily static service pressure [psi (kPa)] available (as determined by the local water authority) at the water meter or other source of supply at the installation location. Adjust this minimum daily static pressure [psi (kPa)] for the following conditions:

    2.1. Determine the difference in elevation between the source of supply and the highest water supply outlet. Where the highest water supply outlet is located above the source of supply, deduct 0.5 psi (3.4 kPa) for each foot (0.3 m) of difference in elevation. Where the highest water supply outlet is located below the source of supply, add 0.5 psi (3.4 kPa) for each foot (0.3 m) of difference in elevation.

    2.2. Where a water pressure reducing valve is installed in the water distribution system, the minimum daily static water pressure available is 80 percent of the minimum daily static water pressure at the source of supply or the set pressure downstream of the pressure reducing valve, whichever is smaller.

    2.3. Deduct all pressure losses due to special equipment such as a backflow preventer, water filter and water softener. Pressure loss data for each piece of equipment shall be obtained through the manufacturer of such devices.

    2.4. Deduct the pressure in excess of 8 psi (55 kPa) due to installation of the special plumbing fixture, such as temperature controlled shower and flushometer tank water closet. Using the resulting minimum available pressure, find the corresponding pressure range in Table E201.1.

3. The maximum *developed length* for water piping is the actual length of pipe between the source of supply and the most remote fixture, including either hot (through the water heater) or cold water branches multiplied by a factor of 1.2 to compensate for pressure loss through fittings. Select the appropriate column in Table E201.1 equal to or greater than the calculated maximum *developed length*.

4. To determine the size of water service pipe, meter and main distribution pipe to the building using the appropriate table, follow down the selected "maximum *developed length*" column to a fixture unit equal to, or greater than the total installation demand calculated by using the "combined" water supply fixture unit column of Table E103.3(2). Read the water service pipe and meter sizes in the first left-hand column and the main distribution pipe to the building in the second left-hand column on the same row.

5. To determine the size of each water distribution pipe, start at the most remote outlet on each *branch* (either hot or cold *branch*) and, working back toward the main distribution pipe to the building, add up the water supply fixture unit demand passing through each segment of the distribution system using the related hot or cold column of Table E103.3(2). Knowing demand, the size of each segment shall be read from the second left-hand column of the same table and maximum *developed length* column selected in Steps 1 and 2, under the same or next smaller size meter row. In no case does the size of any *branch* or main need to be larger that the size of the main distribution pipe to the building established in Step 4.

SECTION E202
DETERMINATION OF PIPE VOLUMES

E202.1 Determining volume of piping systems. Where required for engineering design purposes, Table E202.1 shall be used to determine the approximate internal volume of water distribution piping.

TABLE E201.1
MINIMUM SIZE OF WATER METERS, MAINS AND DISTRIBUTION PIPING BASED ON WATER SUPPLY FIXTURE UNIT VALUES (w.s.f.u.)

METER AND SERVICE PIPE (inches)	DISTRIBUTION PIPE (inches)	MAXIMUM DEVELOPMENT LENGTH (feet)									
Pressure Range 30 to 39 psi		40	60	80	100	150	200	250	300	400	500
$3/4$	$1/2$ [a]	2.5	2	1.5	1.5	1	1	0.5	0.5	0	0
$3/4$	$3/4$	9.5	7.5	6	5.5	4	3.5	3	2.5	2	1.5
$3/4$	1	32	25	20	16.5	11	9	7.8	6.5	5.5	4.5
1	1	32	32	27	21	13.5	10	8	7	5.5	5
$3/4$	$1 1/4$	32	32	32	32	30	24	20	17	13	10.5
1	$1 1/4$	80	80	70	61	45	34	27	22	16	12
$1 1/2$	$1 1/4$	80	80	80	75	54	40	31	25	17.5	13
1	$1 1/2$	87	87	87	87	84	73	64	56	45	36
$1 1/2$	$1 1/2$	151	151	151	151	117	92	79	69	54	43
2	$1 1/2$	151	151	151	151	128	99	83	72	56	45
1	2	87	87	87	87	87	87	87	87	87	86
$1 1/2$	2	275	275	275	275	258	223	196	174	144	122
2	2	365	365	365	365	318	266	229	201	160	134
2	$2 1/2$	533	533	533	533	533	495	448	409	353	311

METER AND SERVICE PIPE (inches)	DISTRIBUTION PIPE (inches)	MAXIMUM DEVELOPMENT LENGTH (feet)									
Pressure Range 40 to 49 psi		40	60	80	100	150	200	250	300	400	500
$3/4$	$1/2$ [a]	3	2.5	2	1.5	1.5	1	1	0.5	0.5	0.5
$3/4$	$3/4$	9.5	9.5	8.5	7	5.5	4.5	3.5	3	2.5	2
$3/4$	1	32	32	32	26	18	13.5	10.5	9	7.5	6
1	1	32	32	32	32	21	15	11.5	9.5	7.5	6.5
$3/4$	$1 1/4$	32	32	32	32	32	32	32	27	21	16.5
1	$1 1/4$	80	80	80	80	65	52	42	35	26	20
$1 1/2$	$1 1/4$	80	80	80	80	75	59	48	39	28	21
1	$1 1/2$	87	87	87	87	87	87	87	78	65	55
$1 1/2$	$1 1/2$	151	151	151	151	151	130	109	93	75	63
2	$1 1/2$	151	151	151	151	151	139	115	98	77	64
1	2	87	87	87	87	87	87	87	87	87	87
$1 1/2$	2	275	275	275	275	275	275	264	238	198	169
2	2	365	365	365	365	365	349	304	270	220	185
2	$2 1/2$	533	533	533	533	533	533	533	528	456	403

(continued)

TABLE E201.1—continued
MINIMUM SIZE OF WATER METERS, MAINS AND DISTRIBUTION PIPING BASED ON WATER SUPPLY FIXTURE UNIT VALUES (w.s.f.u.)

METER AND SERVICE PIPE (inches)	DISTRIBUTION PIPE (inches)	MAXIMUM DEVELOPMENT LENGTH (feet)									
Pressure Range 50 to 60 psi		40	60	80	100	150	200	250	300	400	500
$3/4$	$1/2$ [a]	3	3	2.5	2	1.5	1	1	1	0.5	0.5
$3/4$	$3/4$	9.5	9.5	9.5	8.5	6.5	5	4.5	4	3	2.5
$3/4$	1	32	32	32	32	25	18.5	14.5	12	9.5	8
1	1	32	32	32	32	30	22	16.5	13	10	8
$3/4$	$1 1/4$	32	32	32	32	32	32	32	32	29	24
1	$1 1/4$	80	80	80	80	80	68	57	48	35	28
$1 1/2$	$1 1/4$	80	80	80	80	80	75	63	53	39	29
1	$1 1/2$	87	87	87	87	87	87	87	87	82	70
$1 1/2$	$1 1/2$	151	151	151	151	151	151	139	120	94	79
2	$1 1/2$	151	151	151	151	151	151	146	126	97	81
1	2	87	87	87	87	87	87	87	87	87	87
$1 1/2$	2	275	275	275	275	275	275	275	275	247	213
2	2	365	365	365	365	365	365	365	329	272	232
2	$2 1/2$	533	533	533	533	533	533	533	533	533	486

METER AND SERVICE PIPE (inches)	DISTRIBUTION PIPE (inches)	MAXIMUM DEVELOPMENT LENGTH (feet)									
Pressure Range Over 60		40	60	80	100	150	200	250	300	400	500
$3/4$	$1/2$ [a]	3	3	3	2.5	2	1.5	1.5	1	1	0.5
$3/4$	$3/4$	9.5	9.5	9.5	9.5	7.5	6	5	4.5	3.5	3
$3/4$	1	32	32	32	32	32	24	19.5	15.5	11.5	9.5
1	1	32	32	32	32	32	28	28	17	12	9.5
$3/4$	$1 1/4$	32	32	32	32	32	32	32	32	32	30
1	$1 1/4$	80	80	80	80	80	80	69	60	46	36
$1 1/2$	$1 1/4$	80	80	80	80	80	80	76	65	50	38
1	$1 1/2$	87	87	87	87	87	87	87	87	87	84
$1 1/2$	$1 1/2$	151	151	151	151	151	151	151	144	114	94
2	$1 1/2$	151	151	151	151	151	151	151	151	118	97
1	2	87	87	87	87	87	87	87	87	87	87
$1 1/2$	2	275	275	275	275	275	275	275	275	275	252
2	2	365	368	368	368	368	368	368	368	318	273
2	$2 1/2$	533	533	533	533	533	533	533	533	533	533

For SI: 1 inch = 25.4, 1 foot = 304.8 mm.

a. Minimum size for building supply is $3/4$-inch pipe.

APPENDIX E

TABLE E202.1
INTERNAL VOLUME OF VARIOUS WATER DISTRIBUTION TUBING

Size Nominal, Inch	OUNCES OF WATER PER FOOT OF TUBE						
	Copper Type M	Copper Type L	Copper Type K	CPVC CTS SDR 11	CPVC SCH 40	Composite ASTM F1281	PEX CTS SDR 9
$^3/_8$	1.06	0.97	0.84	N/A	1.17	0.63	0.64
$^1/_2$	1.69	1.55	1.45	1.25	1.89	1.31	1.18
$^3/_4$	3.43	3.22	2.90	2.67	3.38	3.39	2.35
1	5.81	5.49	5.17	4.43	5.53	5.56	3.91
$1^1/_4$	8.70	8.36	8.09	6.61	9.66	8.49	5.81
$1^1/_2$	12.18	11.83	11.45	9.22	13.20	13.88	8.09
2	21.08	20.58	20.04	15.79	21.88	21.48	13.86

For SI: 1 ounce = 0.030 liter.

INDEX

A

ABS PIPE
  Approved standards . . . . Table 605.3,
               Table 605.5, Table 702.1, Table 702.2,
                     Table 702.3, Table 702.4,
                     Table 1102.4, Table 1102.7
  Support . . . . Table 308.5

ACCESS TO
  Air admittance valve 918.5
  Appliances 502.5
  Backflow preventers 608.14
  Backwater valve 715.5
  Cleanouts . . . 708.1.10, 708.4, 708.9, 1103.4, 1301.8
  Filters 1302.5, 1302.9, 1303.8, 1303.12
  Fixtures 405.2
  Flexible water connectors 605.6
  Floor drain 412.2
  Flush tank components 425.3.4
  Grease removal devices 1003.3.5
  Interceptors and separators 1003.5, 1003.10
  Manifolds 604.10.3
  Piping 422.5, 422.9.1
  Plumbing fixtures 405.2
  Pumps 602.3.5, 1302.9, 1302.12
  Reservoir or interior 1301.9.7
  Roof washers 1303.4
  Shower compartment 417.4.2
  Slip joints 405.8
  Standpipes 802.3.3
  Sump 712.3.2
  Vacuum breakers 425.2
  Vacuum station 713.4
  Valves 503.1, 604.11, 606.3, 712.2,
         715.5, 918.5, 1302.8.1, 1302.9, 1303.12
  Waste receptors 802.3
  Water heaters 502.3, 502.5
  Whirlpool pump 421.5

ACCESSIBLE PLUMBING FACILITIES
  Clearances 404.2
  Protection required 404.3
  Route 403.5
  Signs 403.4
  Where required 404.1

ADMINISTRATION
  Applicability 102
  Approval 105, 106.5.7, 107.5
  Inspections 107

Intent 101.3, 109.1
Maintenance 102.3
Means of appeal 109
Scope 101.2
Validity 106.5.2
Violations 108

AIR ADMITTANCE VALVES
  Chemical waste 901.3
  Definition 202
  Where permitted 903.4, 905.1, 917.6, 918

AIR BREAK
  Definition 202
  Method of providing 802.1.5, 802.1.6,
                               802.1.7, 802.1.8, 802.2
  Required 406.2, 802.1.2, 802.1.5, 802.3.3

AIR GAP
  Annual inspection of 312.10.1
  Application of . . . . Table 608.1
  Bidet 408.2
  Clothes washer 406.1
  Definition 202
  Dishwasher 409.2
  Method of providing 608.15.1, 802.2.1
  Required 413.4, 414.1, 506.4, 606.5.6, 608.3.1,
              608.9, 608.13.1, 608.13.3, 608.13.9,
            608.14.2.1, 608.15.1, 608.16.1, 608.16.2,
            608.16.4.1, 608.16.10, 609.7, 611.2, 713.3,
            802.1, 802.1.1, 802.1.2, 802.1.3, 802.1.4,
     802.1.5, 802.1.6, 802.1.7, 802.1.8, 802.2, 802.2.1

AIR TEST 312.3, 312.5, 312.8

ALTERNATIVE ENGINEERED DESIGN
  Definition 202
  Requirements for 316
  Special inspections of 107.3

ALTERNATE ON-SITE NONPOTABLE WATER
  Definition 202

ALTERATIONS AND REPAIRS . . . 102.4, 307.4, 612.1

**ALTERNATIVE MATERIALS AND
EQUIPMENT** 105.2

APPROVED
  Definition 202

ATMOSPHERIC VACUUM BREAKER 424.8,
                   Table 608.1, 608.13.6, 608.15.4,
                   608.15.4.2, 608.16.4.1, 608.16.5

AUTOMATIC CLOTHES WASHER 301.3,
                     Table 403.1, 406, 412.4,
     608.15.4.2, Table 709.1, 1002.4, 1003.6, 1302.2

INDEX

B

BACKFILLING 306.3
BACKFLOW PREVENTER
   Definition 202
   Required 608.1
   Standards . . . . Table 608.1
   Testing 312.10.2
BACKFLOW PROTECTION 608.1
BACKWATER VALVES
   Where required 715
BALL COCK (see FILL VALVE)
BAROMETRIC LOOP 608.13.4
BATHTUBS 407
BEDPAN STEAMERS 202
BEDPAN WASHERS 202
BIDETS 408
BOOSTER SYSTEM 606.5
BOTTLING ESTABLISHMENT 1003.7
BRASS PIPE
   Approved standards . . . . . Table 605.3, Table 605.4,
                      Table 605.8, Table 702.1
BUILDING DRAIN
   Branches of . . . . Table 710.1(2)
   Definition 202
   Material . . . . Table 702.1, Table 702.2
   Sizing . . . . Table 710.1(1)
BUILDING DRAINAGE SYSTEMS
   Below sewer level 712.1
BUILDING SEWER
   Definition 202
   Material . . . . Table 702.3
   Sizing . . . . Table 710.1(1)
BUILDING SYSTEMS
   Connection to public sewer 701.2
BUILDING TRAPS 1002.6

C

CAST-IRON PIPE
   Approved standards . . . . . Table 702.1, Table 702.2,
                      Table 702.3, Table 702.4
CAULKING FERRULES 705.18
CHANGE IN DIRECTION OF DRAINAGE PIPING . . . . Table 706.3
CHEMICAL WASTE
   Exclusion from the sewer 803.2
   Neutralizing 803.1
   Requirements 702.6
   System venting 901.3
CIRCUIT VENT 914

CIRCULATING HOT WATER SYSTEM 202
CLAY PIPE . . . . Table 702.3
CLEANOUTS
   Building drain 708.3.1.1
   Change of direction 708.1.4
   Clearances 708.1.9
   Condensate drain 314.2.5
   Definition 202
   Direction of flow 708.1.8
   Floor accessed 708.1.10.2
   Horizontal drains 708.1.1
   Location of 708.1.10
   Manholes 708.1.1, 708.1.2, 708.1.7
   On stacks 708.1.5
   Plugs, materials for 708.1.6
   Prohibited use of 708.1.11
   Sewers 708.1.2
   Size 708.1.5
   Trim covers for plugs 708.1.10.1
   Where required 708.1
CLEARANCES 502.5
CLEAR WATER WASTE 802.1.5
CODE OFFICIAL
   Application for permit 106.3
   Appointment 103.2
   Definition 202
   Department records 104.7
   Duties and powers 104
   General 104.1
   Identification 104.5
   Inspections 107
COLLECTION PIPE 202
COMBINATION WASTE AND VENT SYSTEM . . . 915
COMBINED BUILDING DRAIN 202
COMMON VENT 202, 911
COMPARTMENTS 405.3.1, 405.3.4
CONCRETE PIPE
   Where permitted . . . . Table 702.3
CONDENSATE DRAINS
   Cleanouts 314.2.5
   Piping material 314.2.2
   Mini-split systems 314.2.4.1
   Traps 314.2.4
CONDUCTOR 202
CONFLICTS 301.7
CONNECTION TO PLUMBING SYSTEM REQUIRED 701.2
CONNECTIONS TO SUMPS AND EJECTORS
   Below sewer level 712.1

CONSERVATION, WATER AND ENERGY
   Fixtures 604.4
   Flow rates . . . . Table 604.3
   Insulation 505, 607.5
CONSTRUCTION DOCUMENTS 106.3.1
CONTROLLED FLOW STORM DRAINAGE 1110
COPPER OR COPPER-ALLOY
PIPE OR TUBING
   Approved standards . . . . . . Table 605.3, Table 605.4
CPVC PIPE OR TUBING
   Approved standards . . . . . . Table 605.3, Table 605.4
CROSS CONNECTIONS 608
CUTTING OR NOTCHING,
STRUCTURAL MEMBERS 307.2, 307.4,
                                                          Appendix C

D

DEMAND RECIRCULATION WATER SYSTEM
   Definition 202
   Controls 607.2.1.2
DESIGN PROFESSIONAL 105.4.2, 105.4.3,
                 105.4.4, 105.4.5, 107.2.2, 109.2.1
DETRIMENTAL WASTES 302
DISHWASHING MACHINE 409
DISINFECTION OF POTABLE
WATER SYSTEM 610
DISTANCE OF TRAP FROM VENT . . . . . . Table 909.1
DISTRIBUTION SYSTEM (See WATER)
DRAIN
   Roof 1105
   Storm 1106
DRAINAGE FIXTURE UNITS
   Definition 202
   Values for continuous flow 709.3
   Values for fixtures . . . . Table 709.1
DRAINAGE SYSTEM
   Connection to sewer or private
      disposal system 701.2
   Determining size 710.1
   Fixture units . . . . Table 709.1
   Indirect waste 802
   Joints 705
   Materials 702
   Material detrimental to 302
   Obstructions 706.2
   Offset sizing 711
   Provisions for future fixtures 710.2
   Sizing . . . . Table 710.1(1), Table 710.1(2)
   Slope of piping . . . . Table 704.1
   Sumps and ejectors 712

   Testing 312
DRINKING FOUNTAINS
   Accessibility requirements 410.3
   Approvals 410.1
   Definition 202
   Location 403.5
   Number required . . . . Table 403.1
   Prohibited location 410.5
   Small occupancies 410.2
   Substitution for 410.4

E

EJECTORS 712
EJECTOR CONNECTION 712.3.5
ELEVATOR SHAFT 301.6
EMPLOYEE FACILITIES 403.3
ENGINEERED DESIGN, ALTERNATIVE 316
EXCAVATION 306
EXISTING BUILDING
PLUMBING SYSTEMS 102.2

F

FACILITIES, TOILET 403
FAMILY AND ASSISTED USE 403.1.2, 403.2.1
FAUCETS 424
FEES 106.6
FERRULES (See CAULKING FERRULES)
FILL VALVE 425.3.1
FITTINGS
   Drainage system 706
   Approved standards . . . . . . Table 605.5, Table 702.4
FIXTURE CALCULATIONS 403.1.1
FIXTURE FITTINGS 424
FIXTURE LOCATION 405.3.1
FIXTURE TRAPS 1002
   Acid-resisting 1002.9
   Building 1002.6
   Design of 1002.2
   For each fixture 1002.1
   Prohibited 1002.3
   Seals 1002.4
   Seal protection 1002.4.1
   Setting and protection 1002.7
   Size 1002.5
FIXTURE UNITS, DRAINAGE
   Definition 202
   Values for continuous flow 709.3
   Values for fixtures . . . . 709.1, Table 709.1
FIXTURES (See PLUMBING FIXTURES)

INDEX

FLOOD LEVEL RIM 202
FLOOD-RESISTANT CONSTRUCTION
   Design flood elevation 202
   Flood hazard area 202
   Flood hazard resistance 309
FLOOR DRAINS . . . . 412, Table 709.1
FLOOR DRAINS, EMERGENCY . . . . 202, Table 709.1
FLOOR FLANGES 405.4.1, 405.4.2
FLOW RATES . . . . 604.3, Table 604.3,
                            Table 604.4, 604.10
FLUSHING DEVICES
   Dual 420.1
   Flush tanks 425.3, 604.3, 604.5
   Flushometer tanks 425.2, 604.3, 604.5
   Flushometer valves 425.2, 604.3, 604.5
   Required 425
FOOD HANDLING DRAINAGE 802.1.1, 802.1.8
FOOD WASTE DISPOSER 413
FOOTBATHS 423.3
FOOTINGS, PROTECTION OF 307.5
FREEZING, PROTECTION OF
   BACKFLOW DEVICES 608.14.2
FREEZING, PROTECTION OF PIPES 305.4
FROST CLOSURE 903.2
FUTURE FIXTURES 704.4

G

GALVANIZED STEEL PIPE . . . . Table 605.3,
                             Table 605.4, 605.18
GARBAGE CAN WASHERS 414
GARBAGE DISPOSALS
   (See FOOD WASTE DISPOSER)
GENERAL REGULATIONS 301
   Conflicts 301.7
   Connection to plumbing system 301.3
   Connection to public water and sewer . . 301.3, 301.4
   Elevator machinery rooms 301.6
   Health and safety . . . 108.7, 108.7.1, 108.7.2, 108.7.3
   Materials detrimental 302, 302.1, 302.2
   Piping sizes indicated 301.5
   Protection of pipes 305
   Rodentproofing 304, 304.1, 304.2, 304.3, 304.4
   Sleeves 305.3
   Strains and stresses in pipe 305.2
   Toilet facilities for workers 311
   Trenching, excavation and backfill 306
   Washroom requirements 310
GRAY WATER (See also ALTERNATE
   ON-SITE NONPOTABLE WATER)
   Definition 202
   Color of distribution piping 608.8.2.1

   Flushing 1302.6.1
   Subsurface irrigation 1402.1
   Trap priming 1002.4.1.2
GREASE INTERCEPTORS 202, 1003
   Approved standards 1003.3.4, 1003.3.6
   Capacity . . . . Table 1003.3.4.1,
                         1003.4.2.1, 1003.4.2.2
   Discharge 1003.3.7
   Gravity-type 1003.3.6
   Hydromechanical-type 1003.3.4
   Not required 1003.3.3
   Required 1003.1
GUTTERS . . . . 1106.6, Table 1106.6

H

HANGERS AND SUPPORTS
   Attachment to buildings 308.4
   Definitions 202
   Material 308.3
   Seismic 308.2
   Spacing . . . . Table 308.5
HEALTH AND SAFETY 101.3
HEALTH CARE PLUMBING 422, 609, 713
HEAT EXCHANGERS 608.16.3
HORIZONTAL PIPE
   Definition 202
   Slope for drainage 704.1
HOSPITAL PLUMBING FIXTURES 422
   Clinical sinks 422.6
   Height of special controls 422.10
   Sterilizer equipment 422.9
HOT WATER
   Circulating system 202, 607.2.1
   Definition 202
   Demand recirculation water system 607.2.1.2
   Flow of hot water to fixtures 607.4
   Heaters and tanks . . . . Chapter 5
   Pipe insulation 607.5
   Pumps for water heaters 607.2.1
   Recirculating systems with
      thermostatic mixing valves 607.2.2
   Supply system 607
   Temperature maintenance (Heat trace) 607.2.1
HOUSE TRAP (See BUILDING TRAPS)

I

INDIRECT WASTE 802
   Air gap or break 802.2.1, 802.2.2
   Food handling establishment 802.1.1
   Receptors 802.3

Special wastes 803
Waste water temperature 702.5
Where required 802.1
INDIVIDUAL VENT 910
INSPECTION 107.1
   Final 107.1
   Reinspection 107.4.3
   Required 107.2
   Rough-in 107.2
   Scheduling of 107.2
   Testing 107.4
INSPECTOR (See CODE OFFICIAL)
INSULATION 505, 607.5
INTERCEPTORS AND SEPARATORS 1003
   Access to 1003.10
   Approved standards 1003.3.4, 1003.3.6
   Bottling establishments 1003.7
   Capacity of grease interceptors .... Table 1003.3.4.1
   Clothes washers 1003.6
   Connection of discharge 1003.3.7
   Definition 202
   Fats, oils and greases systems 1003.3.6
   Not required 1003.3.3
   Oil and flammable liquids separators 1003.4
   Rate of flow control for
      grease interceptors 1003.3.4.2
   Required 1003.1
   Slaughterhouses 1003.8
   Venting 905.4, 1003.9
IRRIGATION, LAWN 608.16.5
ISLAND FIXTURE VENT 916

J

JOINTS AND CONNECTIONS 605, 705
   ABS plastic pipe 605.10, 705.2
   Between different materials 605.24, 705.16
   Brass pipe 605.11, 705.3
   Brazed joints 605.11.1, 605.13.1,
      605.14.1, 705.4.1, 705.6.1. 705.7.1
   Cast-iron pipe 705.4
   Caulked joint 705.4.1, 705.8.1
   Concrete pipe 705.5
   Copper or copper-alloy pipe 605.13, 705.6
   Copper or copper-alloy tubing 605.14, 705.7
   CPVC plastic pipe 605.15, 605.16
   Expansion joints 308.8
   Galvanized steel pipe 605.18
   Grooved and shouldered 605.14.3, 605.18.3,
      605.22.2, 605.23.3
   Mechanical joints 605, 705

Polyethylene plastic pipe or tubing (PE) 605.19
Polyethylene of
   raised temperature (PE-RT) 605.25
Polypropylene plastic pipe or tubing (PP) 605.20
Prohibited 605.9, 707
PVC plastic pipe 605.22, 705.11
Slip joints 405.8, 1002.2
Soldered joints 605.13.3, 605.14.6,
      705.6.3, 705.7.3
Solvent cementing 605, 705
Stainless steel 605.23
Threaded 605.10.3, 605.11.3, 605.13.4,
      605.15.3, 605.18.1, 605.22.4, 705.2.3,
      705.3.3, 705.6.4, 705.9.1, 705.11.3
Vitrified clay pipe 705.12
Welded 605.11.4, 605.13.5,
      605.23.2, 705.3.4, 705.6.5

K

KITCHEN 418

L

LABELING, BUNDLED WATER PIPING 606.7
LAUNDRIES (See CLOTHES WASHERS)
LAUNDRY TRAY 415
LAUNDRY TUB (See LAUNDRY TRAY)
LAVATORIES 416
LEADERS 1106
LIGHT AND VENTILATION REQUIREMENT ... 310.1
LOADING, DRAINAGE
   FIXTURE UNITS .... Table 709.1
LOCATION
   Anchorage for drain piping 308.7
   Drinking fountain 403.5
   Fixtures 405.3
   Fixtures, obstruction 405.3.3
   Potable water supply tanks 606.5.8
   Prohibited, drinking fountain 410.5
   Secondary roof drain discharge 1108.2
   Storage tanks 1302.7.1, 1303.10.1
   Subsoil irrigation 1402.3
   Toilet facilities, malls 403.3.4
   Toilet facilities, other than in malls 403.3.3
   Toilet room and kitchen 403.3.2
   Trench 307.5
   Valves 606.1, 606.2
   Vent terminal 903.5
   Water heater 501.4
   Wells 608.7.1

M

MANHOLES 708.1.1, 708.1.7
MANIFOLDS 604.10
MATERIAL
   Above-ground drainage and
      vent pipe . . . . Table 702.1
   Alternative 105.2
   Approval 107.2.3
   Building sewer pipe 702.3
   Building storm sewer pipe 1102.4
   Chemical waste system 702.6
   Fittings . . . . Table 605.5, Table 702.4,
                                        Table 1102.7
   Identification 303.1
   Joints 605, 705
   Roof drains 1105
   Sewer pipe 702.3
   Standards . . . . .Chapter 15
   Storm drainage pipe . . . . Table 1102.4
   Subsoil drain pipe 1102.5
   Underground building
      drainage and vent pipe 702.2
   Vent pipe 702.1, 702.2
   Water distribution pipe . . . . .Table 605.4
   Water service pipe . . . . .Table 605.3
MATERIAL, FIXTURES
   Quality 402.1
   Special use 402.2
MATERIAL, SPECIAL
   Caulking ferrules 705.18
   Cleanout plugs 708.1.6
   Sheet copper 402.3
   Sheet lead 402.4
   Soldering bushings 705.19
MECHANICAL JOINTS 605, 705
MEDICAL GAS, NONFLAMMABLE 1202.1

N

**NONPOTABLE WATER
REUSE SYSTEMS** 1302
NONPOTABLE WATER SYSTEMS
   Identification 608.8
   Disinfection 1301.2.1
   Filtration 1301.2.2
   Piping color 608.2.1
   Protection of potable water from 608.1, 1301.5
   Requirements for . . . . .Chapter 13
   Signage 1301.3
   Tanks 1301.9

O

OFFSETS
   Definitions 202
   Drainage 711
   Venting 907
OPENINGS
   Through walls or roofs 304.5, 305.5, 315
OVERFLOW **407.2**

P

PARKING GARAGE 403.3, 1002.1
PAN, WATER HEATER 504.7
PARTITIONS 405.3.4, 405.3.5
PEDICURE BATHS 423.3
PERMIT
   Application for 106.3
   Conditions of 106.5
   Fees 106.6
   Suspension of 106.5.5
PENETRATIONS 315
PIPE BURSTING 717
PIPING
   Construction documents 105.4.4
   Drainage, horizontal slope . . . . .Table 704.1
   Drainage piping installation 704
   Drainage piping offset, size 710, 711
   Joints 605, 705
   Rehabilitation of the inside of 601.5
PIPING PROTECTION
   Backfilling 306.3
   Corrosion 305.1
   Expansion and contraction 305.2
   Exposed locations 305.6
   Foundation wall 305.3
   Freezing 305.6
   Penetration by fasteners 305.6
   Structural settlement 305.2
   Tunneling 306.4
PLUMBING FACILITIES 403
PLUMBING FIXTURES
   Accessible 404
   Automatic clothes washers 406
   Bidet 408
   Clearances 404.2, 405.3
   Definition 202
   Dishwashing machine 409
   Drainage fixture unit values . . . . .Table 709.1
   Drinking fountain 202, 410
   Emergency showers 411

Eyewash stations 411
Floor sinks 427
Floor and trench drains. 412
Food waste disposer 413
Future fixtures. 704.4
Garbage can washer 414
Hospital. 422
Installation 405
Joints at wall or floor. .405.5
Kitchen sink 418
Laundry tray 415
LAUNDRY TUB (See LAUNDRY TRAY)
Lavatories. 416
Minimum number of .403.1
Materials. 402.1
Ornamental pools 423.1
Quality 402.1
Separate facilities 403.2
Setting 405.3
Showers 417
Sinks. 418
Slop sinks 418
Urinals 419
Water closets 420
Water coolers 202, 410
Water supply protection 608
PLUMBING INSPECTOR (See CODE OFFICIAL)
PNEUMATIC EJECTORS 712
POLYETHYLENE PIPE OR TUBING
PE-AL-PE approved standards . . . . .Table 605.3, Table 605.4
PEX approved standards . . . . Table 605.3, Table 605.4
PEX-AL-PEX approved standards . . . . . Table 605.3, Table 605.4
(PE-RT) approved standards . . . . Table 605.3, Table 605.4
POLYPROPYLENE (PP) PIPE OR TUBING
Approved standards . . . . . . Table 605.3, Table 605.4
POTABLE WATER, PROTECTION OF 608
PRESSURE OF WATER DISTRIBUTION. 604.6
PROHIBITED
Joints and connections. 605.9, 707
Traps 1002.3
PROTECTION OF POTABLE WATER 603.2, 608
PUMPING EQUIPMENT 712
PVC PIPE . . . . Table 605.3, Table 605.4, Table 702.1, Table 702.2, Table 702.3

Q
QUALITY OF WATER
General. 608.1
QUICK CLOSING VALVES (See WATER, HAMMER)

R
**RAINWATER COLLECTION AND
DISTRIBUTION SYSTEMS** 1303
RAINWATER DRAINAGE . . . . .Chapter 11
RECLAIMED WATER SYSTEMS 1304
**REDUCED PRESSURE PRINCIPLE
BACKFLOW PREVENTER**
Definition. 202
Periodic inspections 312.10
Standard . . . . Table 608.1
Where required 608.15.2, 608.16.2, 608.16.4, 608.16.4.1, 608.16.5, 608.16.6
REFERENCED STANDARDS. . . . . Chapter 15
RELIEF VALVES 504.4, 504.5, 504.6
Definition. 202
RELIEF VENT 906.2, 908, 914.4
Definition. 202
ROOF DRAINS
Definition. 202
Rating for water height above. 1101.7, 1105.2
ROUGH-IN INSPECTION 107.2

S
SANITARY SEWER (See BUILDING SEWER)
SANITARY TEES. 706.3
**SCREWED JOINTS
(See JOINTS AND CONNECTIONS)**
SEPARATE FACILITIES 403.2
SERVICE SINKS . . . . .Table 403.1, 418
SEWER (See BUILDING SEWER)
SHEET COPPER. 402.3
SHEET LEAD 402.4
SHAMPOO SINKS. 423.3
SHOWER LINER
Material. 312.9
Testing 417.5.2
SHOWERS
Approvals for prefabricated types. 417.1
Compartment size. 417.4
Required number 403.1
Wall protection for. .417.4.1
Water supply riser for 417.2
SILL COCK 424, 608.15.4.2

INDEX

SINGLE STACK VENT SYSTEM 917
SINKS 418
SIPHONIC ROOF DRAINAGE SYSTEMS 1107
SIZING
    Drainage system 710
    Fixture drain. 709
    Fixture water supply. 604.5
    Vent system. 906
    Water distribution system 604
    Water service. 603.1
SLAUGHTERHOUSES 1003.8
SLEEVES 305.3
SLIP JOINTS (See JOINTS AND CONNECTIONS)
SOLDERING BUSHINGS 705.19
SOLVENT CEMENTING
    (See JOINTS AND CONNECTIONS)
SPECIAL PLUMBING FIXTURES 423
SPECIAL WASTES 803
STAINLESS STEEL PIPE
    Approved standards for
       drainage systems . . . . . . Table 702.1, Table 702.2,
       Table 702.3, Table 702.4,
       Table 1102.4, Table 1102.5, Table 1102.7
    Approved standards for
       water systems . . . . Table 605.3, Table 605.4
STANDARDS (See REFERENCED STANDARDS)
STANDPIPE DRAIN 802.3.3
STEEL PIPE (See also MATERIAL)
    Galvanized, approved standards . . . . Table 605.4
    Stainless, approved standards . . . . . . Table 702.1,
       Table 702.2, Table 702.3,
       Table 702.4, Table 1102.4,
       Table 1102.5, Table 1102.7
STERILIZER PIPE 422.9
STORM DRAINS
    Building size. 1106.1
    Building subdrains 1112.1
    Building subsoil drains 1111.1
    Conductors and connections 1104
    Definition 202
    General 1101
    Prohibited drainage 1101.3
    Roof drains 1102.6, 1105
    Secondary roof drains 1108
    Sizing of conductors, leaders and storm drains. 1106
    Sizing of roof gutters 1106.6
    Sizing of vertical conductors and leaders 1106.2
    Traps 1103
    Where required 1101.2
STRAPS (See HANGERS AND SUPPORTS)
STRUCTURAL SAFETY . . . . 307, Appendix C
SUBDRAIN BUILDING 1112.1

SUBSOIL DRAIN PIPE 1111.1
SUBSOIL LANDSCAPE
    IRRIGATION SYSTEMS . . . . Chapter 14
SUMP VENT 906.5
SUMP PUMP DISCHARGE PIPE 712.3.3
SUMPS 1113
SUPPORTS (See also PIPING PROTECTION) . . . 308
SWIMMING POOL
    Definition 202
SWIMMING POOL DRAINAGE 802.1.4

T

TEMPERATURE AND
    PRESSURE RELIEF VALVE 504
TEMPERATURE CONTROL
    Mixing valves . . . 424.3, 424.4, 424.5, 607.2.2, 613.1
TEMPERATURE OF WASTE WATER 702.5
TEMPERED WATER 202, 607.1.1, 607.1.2
TEST
    Drainage and vent air test 312.3
    Drainage and vent final test 312.4
    Drainage and vent water test 312.2
    Forced sewer test 312.7
    Gravity sewer test 312.6
    Percolation test 1402.2
    Required tests 312.1
    Shower liner 312.19
    Test gauges 312.1.1
    Test of backflow prevention devices 312.10
    Test of conductors 312.8
THERMAL EXPANSION CONTROL 607.3
THREADED JOINTS
    (See JOINTS AND CONNECTIONS)
TOILET FACILITIES
    Definition 202
    Required 403
    Signs 403.4
    Travel distance 403.3.3, 403.3.4
    Workers' 311
TOILET ROOM DOOR LOCKING 403.3.6
TOILETS (See WATER CLOSETS)
TRAP SEAL PROTECTION 1002.4.1
TRAPS
    Acid-resisting 1002.9
    Building 1002.6
    Design 1002.2
    Prohibited types 1002.3
    Seal 1002.4
    Separate for each fixture 1002.1
    Size . . . . Table 709.1, Table 709.2
TRENCH DRAINS 412

INDEX

TRENCHING, EXCAVATION AND BACKFILL 306
TUNNELING 306.4

U

**UNDERGROUND DRAINAGE
AND VENT PIPE** 702.2
URINAL PARTITIONS................... 405.3.5
URINALS................................. 419

V

VACUUM BREAKERS .... Table 608.1, 608.13.5,
608.13.6, 608.13.8, 608.15.4
VACUUM DRAINAGE SYSTEMS............ 716
VACUUM SYSTEM, MEDICAL............ 1202.1
VALVES
   Air admittance........................ 202, 918
   Approved standards .... Table 605.7
   Backwater............................... 715
   Definition................................ 202
   Pressure reducing 604.8
   Relief 202, 504.4
   Stop and waste......................... 608.7
   Temperature-actuated flow reduction valves ...424.6
   Transfer valves......................... 424.8
VENT
   Individual 910
   Sizing 906
VENT RELIEF SIZE......................... 908
VENT STACK
   Definition......................... 202, 904.2
VENTS AND VENTING
   Branch vents......................... 906.4
   Circuit 914
   Combination waste 915
   Common 911
   Definition.............................. 202
   Distance from trap .... Table 909.1
   Engineered systems.................... 919
   Island fixture.......................... 916
   Required vent........................ 904.1
   Single stack 917
   Vent stack........................... 904.2
   Vent terminal 903
   Waste stack 913
   Wet vent 912

W

WALL-HUNG WATER CLOSETS 420

WASTE
   Indirect 802
   Special 803
WASTE RECEPTORS
   Definition.............................. 202
WASTE STACK 913
WASTE WATER TEMPERATURE............ 702.5
WATER
   Cross connection 608.6
   Distribution system design 604
   Distribution piping material .... Table 605.4
   Excessive pressure..................... 604.8
   Hammer 604.9
   Pressure booster...................... 606.5
   Rain.................................. 202
   Reclaimed............................. 202
   Service 603
   Service piping material .... Table 605.3
   Storm 202
WATER CLOSETS.......................... 420
WATER CLOSET COMPARTMENTS 405.3.4
**WATER CLOSET PERSONAL
HYGIENE DEVICE**....................... 424.9
WATER CONSERVATION
   Fixtures.............................. 604.4
   Flow rates .... Table 604.4
WATER COOLER
   Approvals 410.1
   Definition.............................. 202
   Prohibited location 410.5
WATER DISPENSER
   Definition.............................. 202
   Prohibited location 410.5
   Substitution for drinking fountain.......... 410.4
WATER HEATER PAN..................... 504.7
WATER HEATERS
   Definition.............................. 202
   Requirements for .... Chapter 5
WATER TREATMENT
   Drinking water systems................. 611
   Softeners............................ 608.3
WET VENT............................... 912
WIPED JOINTS 705.10.2
WORKERS' TOILET FACILITIES 311

EDITORIAL CHANGES – SECOND PRINTING

Page 6-73, Figure 607.3 now reads as shown.

Page 8-2, Section 802.1.1: commentary now reads . . .All food must be protected from possible contamination caused by the sanitary drainage system. The requirement for indirect waste connections extends to all storage, cooking and preparation equipment, including vegetable sinks; food-washing sinks; refrigerated cases and cabinets; ice boxes; ice-making machines; steam kettles; steam tables; potato peelers; egg boilers; coffee urns and brewers; drink dispensers and similar types of equipment and fixtures [see Commentary Figures 802.1.1(1), 802.1.1(2) and 802.1.1(3)]. Food stored in walk-in coolers and freezers is specifically addressed in Section 802.1.2. Note that sinks used for washing, rinsing and sanitizing utensils, dishes, pots, pans or service ware are also considered as if they were food preparation fixtures because of the connection requirements of Section 802.1.8. Domestic and commercial dishwashing machines must discharge indirectly through an air gap or air break into a standpipe or waste receptor or as allowed by Section 802.1.6 (see Commentary Figure 802.1.6 or 802.1.7).

Occasionally, commercial kitchen designers will desire to mount a food waste grinder to a food preparation sink for ease of disposal of food prep wastes such as leaves, skins, spoiled produce, bones and meat trimmings. Unfortunately, this creates several code conflicts. A food waste grinder is not a plumbing "fixture" that is required to be indirect connected; therefore, it must be direct connected to the drain system. However, because the discharge of the food waste grinder is also the outlet for the food-prep sink, this section requires the drain to be indirectly connected. The concern of the code for indirectly connecting drains of food prep sinks is to avoid contamination of food from a backup of the drainage system. Therefore, the code's intent is to protect food from bacterial contamination. Without question, the grinding chamber of a food waste grinder will always contain significant bacteria regardless of any cleaning or sanitation practices. Although, not directly stated in the code, the requirement for food waste grinders to be directly connected and the code's intent to protect food from contamination suggests that food waste grinders must not be installed on food-prep sinks.

People Helping People Build a Safer World®

Valuable Guides to Changes in the 2015 I-Codes®

FULL COLOR! HUNDREDS OF PHOTOS AND ILLUSTRATIONS!

SIGNIFICANT CHANGES TO THE 2015 INTERNATIONAL CODES®

Practical resources that offer a comprehensive analysis of the critical changes made between the 2012 and 2015 editions of the codes. Authored by ICC code experts, these useful tools are "must-have" guides to the many important changes in the 2015 International Codes.

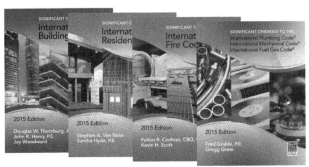

SIGNIFICANT CHANGES TO THE IBC, 2015 EDITION
#7024S15

SIGNIFICANT CHANGES TO THE IRC, 2015 EDITION
#7101S15

SIGNIFICANT CHANGES TO THE IFC, 2015 EDITION
#7404S15

SIGNIFICANT CHANGES TO THE IPC/IMC/IFGC, 2015 EDITION
#7202S15

Changes are identified then followed by in-depth discussion of how the change affects real world application. Photos, tables and illustrations further clarify application.

ORDER YOUR HELPFUL GUIDES TODAY!
1-800-786-4452 | www.iccsafe.org/books

HIRE ICC TO TEACH

Want your group to learn the Significant Changes to the I-Codes from an ICC expert instructor? Schedule a seminar today!
email: **ICCTraining@iccsafe.org** | phone: **1-888-422-7233 ext. 33818**

14-09379

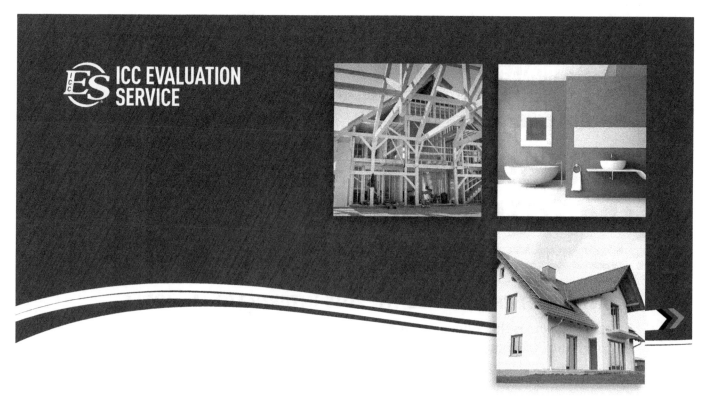

APPROVE WITH
CONFIDENCE

- ICC-ES® Evaluation Reports are the most widely accepted and trusted in the nation.
- ICC-ES is dedicated to providing evaluation reports, PMG and Building Product Listings to the highest quality and technical excellence.
- ICC-ES is a subsidiary of ICC®, the publisher of the IBC®, IRC®, IPC® and other I-Codes®.

We do thorough evaluations. You approve with confidence.

Look for the ICC-ES marks of conformity before approving for installation

 Code Officials — Contact Us With Your Questions About ICC-ES Reports at:
800-423-6587 ext. ES help (374357) or email EShelp@icc-es.org
www.icc-es.org

Subsidiary of **INTERNATIONAL CODE COUNCIL®**

The International Code Council's Training and Education Department

Industry professionals look to ICC to provide the critical knowledge and experiences necessary to excel in today's challenging world. ICC's educational programs cover a broad spectrum of code and construction related topics, offering some of the highest quality instruction direct from the source.

INSTITUTES
Acquire Skills, Share Experiences, Earn CEUs
ICC's training institutes offer a comprehensive education experience and a great way to earn much needed CEUs. Learn best practices from the leading experts in the field, build your network of professional contacts and share experiences. Institutes are held across the country throughout the year. To see the full schedule go to www.iccsafe.org/training.

HIRE ICC TO TEACH
Bring ICC On-Site and Earn CEUs
Give your group the confidence they need to meet tough challenges head-on so they can reach their full potential. ICC's course catalog contains a wide variety of educational topics that are available for contract at a location of your choice. Customized training is also available to fit your needs. For our full course catalogue go to www.iccsafe.org/hireicc.

ICC ONLINE CAMPUS
Earn CEUs at Your Own Pace
Online courses offer access to a comprehensive training and education portal, providing you with an effective and convenient tool to enhance your professional skills. Courses are available anytime for 99 days following registration. For a quick and easy way to earn CEUs towards your certification, try our Online Certifications Renewal section. Go to www.iccsafe.org/onlinecampus for our full offerings.

For more information about ICC's Training please contact us at
888-ICC-SAFE (422-7233) ext. 33818 or email us at **icctraining@iccsafe.org.**

GET IMMEDIATE DOWNLOADS OF THE STANDARDS YOU NEED

Browse hundreds of industry standards adopted by reference. Available to you 24/7!

Count on ICC for standards from a variety of publishers, including:

ACI	**CPSC**	**GYPSUM**
AISC	**CSA**	**HUD**
ANSI	**DOC**	**ICC**
APA	**DOJ**	**ISO**
APSP	**DOL**	**NSF**
ASHRAE	**DOTn**	**SMACNA**
ASTM	**FEMA**	**USC**
AWC	**GBI**	

DOWNLOAD YOUR STANDARDS TODAY!
WWW.ICCSAFE.ORG/BOOKS